ULTRA-WIDEBAND RADIO TECHNOLOGIES FOR COMMUNICATIONS, LOCALIZATION AND SENSOR APPLICATIONS

Edited by **Reiner Thomä, Reinhard H. Knöchel, Jürgen Sachs, Ingolf Willms** and **Thomas Zwick**

Ultra-Wideband Radio Technologies for Communications, Localization and Sensor Applications
http://dx.doi.org/10.5772/2648
Edited by Reiner Thomä, Reinhard H. Knöchel, Jürgen Sachs, Ingolf Willms and Thomas Zwick

Contributors

Rainer Moorfeld, Adolf Finger, Hanns-Ulrich Dehner, Holger Jäkel, Martin Braun, Friedrich K. Jondral, María Dolores Pérez Guirao, Josef A. Nossek, Peter Russer, Tobias Noll, Amine Mezghani, Michel T. Ivrlač, Matthias Korb, Farooq Mukhtar, Hristomir Yordanov, Johannes A. Russer , Nuan Song, Mike Wolf, Martin Haardt, Andreas Schenk, Robert F.H. Fischer, Mohamed El-Hadidy, Mohammed El-Absi, Yoke Leen Sit, Markus Kock, Thomas Zwick, Holger Blume, Thomas Kaiser, Markus Grimm, Dirk Manteuffel, Gholamreza Alirezaei, Rudolf Mathar, Daniel Bielefeld, Rudolf Zetik, Honghui Yan, Elke Malz, Snezhana Jovanoska, Guowei Shen, Reiner S. Thomä, Rahmi Salman, Thorsten Schultze, Robert Tobera, Hans-Ingolf Willms, Lars Reichardt, Malgorzata Janson, Thomas Zwick, Werner Wiesbeck, Tobias Deißler, Jörn Thielecke, Steffen Heuel, Hermann Rohling, Ingrid Hilger, Katja Dahlke, Gabriella Rimkus, Christiane Geyer, Frank Seifert, Olaf Kosch, Florian Thiel, Matthias Hein, Francesco Scotto di Clemente, Ulrich Schwarz, Marko Helbig, Jürgen Sachs, Henning Mextorf, Frank Daschner, Mike Kent, Reinhard Knöchel, Alexander Esswein, Robert Weigel, Christian Carlowitz, Martin Vossiek, Stefan Heinen, Ralf Wunderlich, Markus Robens, Martin Kmec, Robert Weigel, Thomas Ußmüller, Benjamin Sewiolo, Mohamed Hamouda, Rolf Kraemer, Johann-Christoph Scheytt, Yevgen Borokhovych, Dayang Lin, Michael Mirbach, Thanawat Thiasiriphet, Jürgen Lindner, Wolfgang Menzel, Hermann Schumacher, Mario Leib, Bernd Schleicher

Published by InTech

Janeza Trdine 9, 51000 Rijeka, Croatia

Copyright © 2013 InTech

Notice

Publishing Process Manager Viktorija Zgela
Technical Editor InTech DTP team
Cover InTech Design team

Additional hard copies can be obtained from orders@intechopen.com

Ultra-Wideband Radio Technologies for Communications, Localization and Sensor Applications, Edited by Reiner Thomä, Reinhard H. Knöchel, Jürgen Sachs, Ingolf Willms and Thomas Zwick
p. cm.
ISBN 978-953-51-0936-5

For EU product safety concerns:
IN TECH d.o.o., Prolaz Marije Krucifikse Kozulić 3, 51000 Rijeka, Croatia,
info@intechopen.com or visit our website at intechopen.com.

Contents

Reiner Thomä received the Dipl.-Ing. (M.S.E.E.), Dr.-Ing. (Ph.D.E.E.), and the Dr.-Ing. habil. degrees in electrical engineering and information technology from Technische Hochschule Ilmenau, Germany, in 1975, 1983, and 1989, respectively.

From 1975 to 1988, he was a Research Associate in the fields of electronic circuits, measurement engineering, and digital signal processing at the same university. From 1988 to 1990, he was a Research Engineer at the Akademie der Wissenschaften der DDR (Zentrum für Wissenschaftlichen Gerätebau). During this period he was working in the field of radio surveillance. In 1991, he spent a three-month sabbatical leave at the University of Erlangen-Nürnberg (Lehrstuhl für Nachrichtentechnik). Since 1992, he has been a Professor of electrical engineering (electronic measurement) at TU Ilmenau where he was the Director of the Institute of Communications and Measurement Engineering from 1999 until 2005. With his group, he has contributed to many European and German research projects and clusters such as WINNER, PULSERS, EUWB, NEWCOM, COST 273, 2100, IC 1004, EASY-A, EASY-C. Currently he is the speaker of the German nation-wide DFG priority funding project UKoLOS, Ultra-Wideband Radio Technologies for Communications, Localization and Sensor Applications (SPP 1202).

His research interests include measurement and digital signal processing methods (correlation and spectral analysis, system identification, sensor arrays, compressive sensing, time-frequency and cyclostationary signal analysis), their application in mobile radio and radar systems (multidimensional channel sounding, propagation measurement and parameter estimation, MIMO-, mm-wave-, and ultra-wideband radar), measurement-based performance evaluation of MIMO transmission systems including over-the-air testing in virtual electromagnetic environments, and UWB sensor systems for object detection, tracking and imaging.

Since 1999 he has been serving as chair of the IEEE-IM TC-13 on Measurement in Wireless and Telecommunications. In 2007 he was awarded IEEE Fellow Member and received the Thuringian State Research Award for Applied Research both for contributions to high-resolution multidimensional channel sounding.

Reinhard H. Knöchel received the Dipl.-Ing. in Electrical Engineering in 1975, and the Dr.-Ing. in 1980 from the Technical University of Braunschweig, Germany. From 1980 to 1986 he was a principal scientist at the Philips Research Laboratory, Hamburg, Germany. In 1986 he joined the Technical University Hamburg-Harburg, where he was a Full Professor in Microwave Electronics until November 1993. Since December 1993 he holds the Chair in Microwave Engineering with the University of Kiel, Kiel, Germany. From July 2010-July 2012 he was Dean of the Department. His research interests include active and passive microwave components, ultra-wideband technology, microwave and field measurement techniques, industrial microwave sensors, radar and magnetic field sensors. Dr. Knöchel is a Fellow of the IEEE "for contributions to microwave systems and sensors for industrial process control".

Jürgen Sachs is a Senior Lecturer at Ilmenau University of Technology, Germany. He teaches "Basics of Electrical Measurement Technology", "Measurements in Communications" and "Ultra-Wideband Radar Sensing". He is a head of several research projects, and inter alia coordinator of European projects for humanitarian demining and disaster relief. His research areas cover RF-signal analysis and RF-system identification; Surface Penetrating Radar for non-destructive testing and medical engineering, ultra wideband methods and their application in high resolution radar and impedance spectroscopy, digital processing of ultra wideband signals, array processing; and design and implementation of new RF device approaches.

Ingolf Willms received the diploma degree in electrical engineering from the RWTH Aachen University in 1977 and the Ph.D. with honours from the former Gerhard-Mercator University Duisburg in 1983. He was awarded for both degrees. He then worked for Dräger in Lübeck for 6 years before he returned to University Duisburg-Essen in 1990 taking up the position of Professor in Information Technology. His research interests include automatic fire alarm systems, especially video detectors and detector test systems, and ultra-wideband radar systems for fire and security.
He is co-recipient of the Best Paper Award presented at 2011 IEEE International Conference on Ultra-Wideband. Since 1996 he is member of the Executive Committee of the European Society for Automatic Alarm Systems (EUSAS) and is secretary of this society since 2006.

Thomas Zwick received the Dipl.-Ing. (M.S.E.E.) and the Dr.-Ing. (Ph.D.E.E.) degrees from the Universität Karlsruhe (TH), Germany in 1994 and 1999, respectively. From 1994 to 2001 he was a Research Assistant at the Institut für Hochfrequenztechnik und Elektronik (IHE) at the Universität Karlsruhe (TH), Germany. February 2001 he joined IBM as research staff member at the IBM T. J. Watson Research Center in Yorktown Heights, NY, USA. From October 2004 to September 2007 T. Zwick was with Siemens AG, Lindau, Germany. During this period he managed the RF development team for automotive radars. In October 2007 he became appointed as a Full Professor at the Karlsruhe Institute of Technology (KIT), Germany. T. Zwick is the Director of the Institut für Hochfrequenztechnik und Elektronik (IHE) at the KIT.
His research topics include wave propagation, stochastic channel modeling, channel measurement techniques, material measurements, microwave techniques, millimeter wave antenna and system design, wireless communication and radar system design. He participated as an expert in the European COST231 Evolution of Land Mobile Radio (Including Personal) Communications and COST259 Wireless Flexible Personalized Communications. For the Carl Cranz Series for Scientific Education he served as a lecturer for Wave Propagation. He received the best paper award on the Intern. Symp. on Spread Spectrum Techn. and Appl. ISSSTA 1998. In 2005 he received the Lewis award for outstanding paper at the IEEE International Solid State Circuits Conference. Since 2008 he is president of the Institute for Microwaves and Antennas (IMA). T. Zwick became selected as a distinguished microwave lecturer for the 2013 – 2015 period. He is author or co-author of over 200 technical papers and over 20 patents.

Preface

Sometimes history seems to repeat. Even in the so-called 'mature' technological fields. When the radio pioneers such as Heinrich Hertz, Guglielmo Marconi, and Alexander Stepanovich Popov made their first experiments of wireless transmission more than a hundred years ago using spark-gap transmitters with simple coherer-detectors they did not care which 'frequency band' they were using, nor did they worry about their signals being 'spectrally efficient' or 'band limited'. The world of radio frequency regulation was very simple then since regulations have not yet existed. Over the years this has dramatically changed. The frequency band was subdivided into small 'boxes' of different sizes, regulated and supervised. The rules governing these 'bands' are strict and vary with the respective region, time and demand. Sometimes even frequency bands of a few tens or hundreds MHz are sold by auction for millions or even billions of dollars. Hence 'Spectral Dividend' became a key word in the media. Scientists worldwide have begun an intensive search for a more efficient usage of the available frequency spectrum. One of the ideas, which came more and more into the center of attention, was to use signals with very low spectral density yet huge instantaneous bandwidth. This 'underlaying' technique allows the reuse of the spectrum, which is already occupied by other narrowband users. The proactive release of a 'new' frequency band of several GHz (3.1 GHz – 10.6 GHz) in February 2002 by the Federal Communications Commission (FCC) hastened research in this field immensely. With such a technique the ultra-wide frequency band can be used without any further spectral slicing even though there are already a large number of established users and services within it! Thus, contrary to the mainstream of contemporary wireless technology, bandwidth efficiency becomes of minor concern again for interference mitigation as in the early days of Hertz and Marconi. However, severe limitations in terms of power spectral density emission are placed on the emitted signals as the first measure of interference mitigation and to avoid a slipshod use of our limited spectral resources as in the early days of Marconi.

So, what is it that makes ultra-wideband (UWB) so interesting for research and emerging applications? What are the paradigm shifts and challenges for circuit and system design? What does it hold for new and pioneering applications? In order to answer these questions the German Research Foundation (Deutsche Forschungsgemeinschaft, DFG) has funded a nation-wide priority-funding program called 'Ultra-Wideband Radio Technologies for Communications, Localization and

Sensor Applications (UKoLoS)'. UKoLoS started in 2006 and ended in 2012. Altogether 14 research partners participated in the research program, which are mainly universities. Most of the projects are conducted as a cooperative research between two to four partners. The initial aim of UKoLoS was the joint UWB research in the areas of communications, localization, and sensors. Remarkable synergy effects and technological advancement and development are expected. This book gives an overview of the scope and results of the UKoLoS program.

In contrast to the conventional frequency multiplexed radio approach, ultra-wideband radio systems earmark a completely new technological philosophy. Since UWB frequencies are already occupied by other radio services, coexistence with the existing systems is a serious concern. Hence it must be made clear that the aim of the UWB technologies is not to replace the current existing systems but to simply coexist with them. Therefore, the transmit power for UWB systems is strictly limited and intelligent interference mitigation methods and cognitive access schemes are being investigated and developed.

For short-range communication and sensor networks, the UWB technology offers a very interesting alternative to the current conventional systems since very high data rates at low power radio interfaces can be achieved. The current research in UWB communications addresses optimal energy-efficient modulation-, access-, coding-, and detection schemes. New results from information theory are needed to determine the basic capabilities of UWB systems under real network- and propagation conditions as well as to unveil optimal system concepts. Further research into variable data rates in sensor networks, dedicated short-range access, secure communication, cooperative detection and integrated communication, sensor and location functions for sensor networks, etc., are also being done.

For localization and sensing applications, the huge UWB bandwidth allows the unprecedented time delay resolution. Precise range estimates in strong multipath environments become possible making UWB the key technology for indoor radio localization, be it infrastructure-based localization, relative inter-terminal localization in ad-hoc networks, or passive localization (e.g. radar imaging). Interaction of the UWB wave field with materials and objects delivers vast information about object's shape, position, motion dynamics, structural time variance, material composition, etc. Since the extremely large bandwidth of UWB is provided at a comparably low frequency (for sensor applications the lower frequency limit may be as low as several hundreds of MHz), UWB can also penetrate materials and obstacles. Information about the inner structure of objects can be made available and the investigation of objects that are hidden by obstacles becomes feasible. Such capabilities open up many applications for use in the industry, e.g. in civil engineering, surveillance, security and safety operations, and even medicine. However basic research is still required to investigate the theory behind the interaction between UWB radio signals and objects, material, environments, technical, or biomedical processes, etc. This will then lead to many different interdisciplinary questions, since non-electrical properties and their relation to the ultra-wideband electro-magnetic field will become essential knowledge.

As mentioned before, UWB utilizes an extremely large bandwidth of potentially several GHz at a comparably low frequency of a few GHz. Associated with the extremely large bandwidth is the potential for super high data rates, yet limited by low power constraints. Therefore, and especially due to the extremely high relative bandwidth, the UWB technology not only promises new and outstanding performance features but also adds new and highly challenging design demands on the envisioned UWB circuits and systems. UWB radio interfaces require innovative integrated hardware architecture, design and implementation. This includes UWB front-ends, antennas and data processing units as a whole. Efficient small antennas with optimal time-domain behavior in real propagation environments are just but one example. Known space-time signal processing algorithms must be tested against the properties of UWB signals and subsequently improved and new processing schemes have to be developed as well. Generally speaking, UWB requires a change of paradigm from narrowband to wideband principles in both algorithms and hardware. Linearity, impulse response, stability, robustness, and power consumption are very important properties, which lead to new requirements in design strategies for UWB circuits. Modern microwave semiconductor technologies such as Silicon-Germanium (SiGe) together with new manufacturing processes and packaging technologies play a key role in the implementation of complex UWB systems on a single integrated circuit.

When the FCC published their report and order that authorizes the unlicensed use of the ultra-wideband (UWB) of 3.1–10.6 GHz a great storm of research, publications, standard proposals, etc. emerged. The usage of the UWB band seemed extremely promising since the very large bandwidth could support high data rates in wireless communications or a good range resolution in sensors. However the hype is now over. On one hand, there are indeed some UWB radio access devices on the market. UWB has become the foundation of Wireless USB and WiMedia access. Yet the anticipated big economic success of UWB still remains to be seen. The last ten years of UWB research has however brought us many insights to a completely new and alternative radio access philosophy. Whereas the first hype was driven by the expectation of a big economic success in the electronic mass market, now the motivation is clearly driven by the physical advantage of such a huge bandwidth at low frequencies. New and innovative applications are generated, which are not yet mainstream. So the initial idea of high data rate wireless UWB systems may have taken a backseat in favor of UWB based systems for medical diagnostics, localization, sensing etc. But the coming years will show if these new ideas will launch UWB into a broad commercial success or at least as indispensable technology in the niche markets. This book at hand is meant to provide some important basics for that goal.

Acknowledgement

All the authors and their corresponding researchers from the various institutions supported by this program would like to express their utmost gratitude to DFG for funding these research projects over 6 years and to enable many revolutionary discoveries to be made in the field of UWB. We would also like to thank the panel of reviewers for their time and effort in reviewing all the submitted project proposals. Last but not least much appreciation goes to Dr. Klaus Wefelmeier and his successor Dr. Damian Dudek for their great support and their personal commitment. We hope that this collective research effort will propel the technology to greater heights and to inspire new innovations.

Reiner Thomä, Reinhard Knöchel, Jürgen Sachs, Ingolf Willms, Thomas Zwick

Chapter 1

MIRA – Physical Layer Optimisation for the Multiband Impulse Radio UWB Architecture

Rainer Moorfeld, Adolf Finger, Hanns-Ulrich Dehner, Holger Jäkel,
Martin Braun and Friedrich K. Jondral

Additional information is available at the end of the chapter

http://dx.doi.org/10.5772/55076

1. Introduction

Future wireless communication systems have to be realised in a simple and energy efficient manner while guaranteeing sufficient performance. Furthermore, the available frequency resources have to be used flexibly and efficiently. In this context two different approaches have been considered in recent years: On one hand OFDM-based overlay systems in which a primary user dynamically allocates unused frequencies to one or more secondary users [57] and on the other hand unlicensed, easy-to-realise and low-cost ultra-wideband (UWB) systems. This underlying technology operates with an extremely low transmission power over a wide frequency range and does not interfere with existing licensed systems [15].

In order to establish UWB on the consumer market it has to get along with some challenges. Such challenges are, e.g., the realisation of practical, low-complex and energy-efficient transceiver architectures, the investigation of methods for accurate synchronization and channel estimation or the handling of high sample rates. To meet these requirements this chapter considers a non-coherent multiband impulse radio UWB (MIR-UWB) system [11, 45, 46]. The MIR-UWB system focuses on short-range high data rate communication applications. The MIR-UWB system is an alternative to the architectures Multiband OFDM UWB [2] and Direct Sequence UWB [16] which have been proposed within the IEEE 802.15.3a standardization process.

The chapter is organised as follows: Section 2 gives a short introduction into the physical layer architecture of the non-coherent MIR-UWB system. In the following section 3 the performance of the energy detection receiver is analysed with respect to different aspects. In contrast section 4 deals with interference investigations for the non-coherent MIR-UWB system aiming at an efficient and intelligent interference handling. The chapter concludes with section 5 in which a summary is given.

2. Multiband impulse radio

The idea of the MIR-UWB architecture is based on [45, 46]. The architecture proposed there comprises a transmitter using multiple bands and impulse radio within the bands to transmit data and a receiver, which detects only the energy of the transmitted impulses. The combination of energy detection receiver and multiband enables a flexible high data rate system with low power consumption.

2.1. Transmitter

The MIR-UWB transmitter is based on a multiband pulse generation followed by a modulator. The multiband pulse generator generates a pulse with a specified bandwidth for every subband. Subbands can be activated or deactivated using the bandplan. Different possibilities to generate theses pulses are shown in [30]. Each subband pulse will be modulated with different data, all subband pulses are summed up to a multiband pulse, amplified and transmitted. Figure 1 shows a transmitter based on an *oscillator bank* pulse generator.

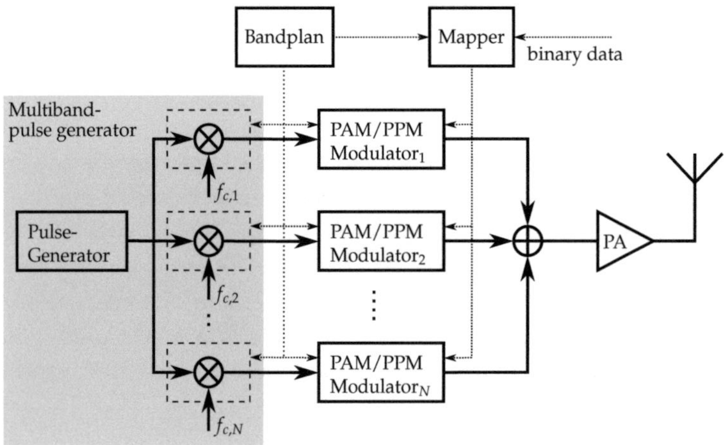

Figure 1. MIR-UWB Transmitter based on *oscillator bank* pulse generation

2.2. Receiver

The MIR-UWB receiver is based on N parallel energy detection receivers. A filter bank separates the individual subband pulses and an energy detector measures the energy in every subband. Based on the measured energy, the demodulator makes his decision. For pulse amplitude modulation (PAM) and its special case of on-off-keying (OOK) the demodulation process needs to know the SNR in each subband. This can be estimated using a preamble [46]. The channel state information can be used for Detect and Avoid (DAA) algorithms [34] and to increase the performance of the multiband system [28]. Pulse position modulation (PPM) and transmit reference (TR) do not need any channel state information.

3. Energy detection

The MIR-UWB architecture is based on energy detection. Thus the receiver detects *only* the energy of the received signal in a specified window. The disadvantage in performance is

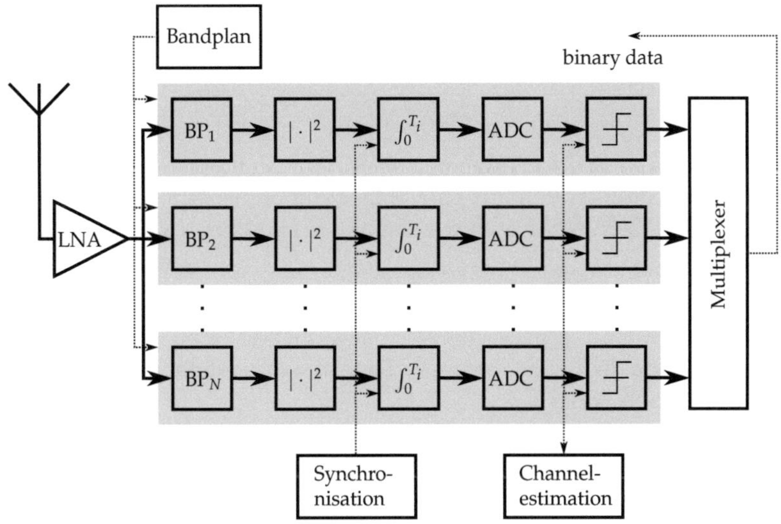

Figure 2. MIR-UWB receiver

accompanied by a very simple receiver design [36, 38, 54]. The performance measure is based on the average symbol error probability (SEP) or bit error probability (BEP) and will be derived in the following section.

3.1. Demodulation

In the additive white gaussian noise (AWGN) channel the received signal R is the sum of the transmitted signal s and white Gaussian noise W with the power spectral density of $N_0/2$:

$$R(t) = s(t) + W(t) = a_m p(t) + W(t). \tag{1}$$

The transmitted signal s is a weighted pulse p. The amplitude of the pulse is a_m. Thus, the energy of a transmitted signal is:

$$\begin{aligned} E_S &= \int_{t_0}^{t_0+T_i} s^2(t)\,\mathrm{d}t = a_m^2 \int_{t_0}^{t_0+T_i} p^2(t)\,\mathrm{d}t \\ &= \sum_{i=1}^{2D} s_i = a_m^2 \sum_{i=1}^{2D} p_i, \end{aligned}$$

where $s_i := s(i/(2B))$ and $p_i := p(i/(2B))$ and B denoting the bandwidth. The received energy can be approximated by a finite sum of $2D = 2T_iB$ samples [56]. The event $\{A = a_m\}$ with the range $\mathcal{A} = \{a_0, \ldots, a_{M-1}\}$ describes a transmitted symbol with the amplitude a_m. Without loss of generality the integration starts at $t_0 = 0$. In order to measure the energy, the detector squares the received signal $R(t)$ and integrates the result over the time interval T_i. The received energy, normalized by the power spectral density $N_0/2$, is:

$$Y = \frac{2}{N_0} \int_0^{T_i} R^2(t)\,\mathrm{d}t. \tag{2}$$

A time discrete representation of the received energy Y is:

$$Y = \frac{1}{N_0 B} \sum_{i=1}^{2D} (s_i + W_i)^2 , \tag{3}$$

where $W_i := W(i/(2B))$. If the symbol energy is $E_S = 0$, the received energy Y will be χ^2 distributed with the degree of freedom of $2D$. If the symbol energy $E_S > 0$, the received energy will be noncentral χ^2 distributed with the degree of freedom of $2D$ and the noncentrality parameter μ:

$$\mu = \sum_{i=1}^{2D} s_i^2 = \frac{2}{N_0} \int_0^T s^2(t)\, \mathrm{d}t =: \frac{2E_s}{N_0} = 2\gamma, \tag{4}$$

where $\gamma = E_S/N_0$ is the SNR at the receiver. Thus, the distribution of the received energy Y depends on the symbol energy E_S:

$$Y \sim \begin{cases} \chi^2_{2D} & \text{for } E_S = 0 \\ \chi^2_{2D}(2\gamma) & \text{for } E_S > 0 . \end{cases}$$

The conditional probability density function $f_{Y|A}(\cdot|a_0)$ with $a_0 = 0$ and $E_S = 0$ of the received energy Y is:

$$f_{Y|A}(y|a_0) = \frac{1}{2^D \Gamma(D)} y^{D-1} \exp\left(-\frac{y}{2}\right). \tag{5}$$

The conditional probability density function $f_{Y|A}(\cdot|a_m)$ with $a_m > 0$ and $E_S > 0$ of the received energy Y is:

$$f_{Y|A}(y|a_m) = \frac{1}{2} \left(\frac{y}{2\gamma}\right)^{\frac{D-1}{2}} \exp\left(-\frac{2\gamma + y}{2}\right) \mathrm{I}_{D-1}\left(\sqrt{2\gamma y}\right), \tag{6}$$

where Γ is the gamma function [18, eq. 8.310.1] and I_n is the modified Bessel function of the first kind of order n [1, eq. 9.6.3].

3.2. AWGN channel

First we calculate the bit error probability of the energy detection receiver in the AWGN channel (1). This receiver detects only the energy of the received signal (2), (3).

3.2.1. Pulse amplitude modulation

The M-PAM modulated signal is:

$$s(t) = \sum_{k=-\infty}^{\infty} a_{m,k} p(t - kT_r)$$

and transmits $\log_2(M)$ bit per symbol. The energy of the mth symbol is:

$$E_{S_m} = \underbrace{\int_0^{T_i} s^2(t)\, \mathrm{d}t}_{=\sum_{i=1}^{2D} s_i^2} = a_m^2 \underbrace{\int_0^{T_i} p^2(t)\, \mathrm{d}t}_{=\sum_{i=1}^{2D} p_i^2} = a_m^2 E_p,$$

where E_p is the energy of an unmodulated pulse p. The demodulator has to decide, which symbol m with the energy E_{S_m} and the amplitude a_m has been transmitted, based on the observation of the random variable Y. The optimal receiver, i. e. the receiver with the lowest probability to make a wrong decision, makes the decision for the symbol that has been sent most likely, given a certain energy y at the receiver. Thus, the receiver makes the decision for the symbol m with the amplitude a_m, when [25]:

$$\mathbb{P}\{A = a_m | Y = y\} \geq \mathbb{P}\{A = a_k | Y = y\}, \quad \forall\, m \neq k. \tag{7}$$

This is the maximum a posteriori probability (MAP) decision rule. If all transmitted symbols are equal probable, it can be reduced to the maximum-likelihood (ML) decision rule:

$$f_{Y|A}(y|a_m) \geq f_{Y|A}(y|a_k), \quad \forall\, m \neq k,$$

using the Bayes theorem:

$$\mathbb{P}(A = a_m | Y = y) = \frac{f_{Y|A}(y|a_m)\mathbb{P}(A = a_m)}{f_Y(y)} = \frac{f_{Y|A}(y|a_m)1/M}{f_Y(y)},$$

because $\mathbb{P}(A = a_m) = 1/M$ for all $m \in \{0, 1, \ldots, M-1\}$ and $f_Y(y)$ are independent of m. For the M-PAM modulated signal, we use the ML receiver with multiple hypothesis testing:

$$m = \underset{k \in [0, M-1]}{\arg\max}\, f_{Y|A}(y|a_k), \tag{8}$$

with the conditional probability density function $f_{Y|A}$ based on (5) and (6).

The SEP P_e for the energy detection receiver in the AWGN channel with M-PAM signals can be calculated as:

$$\begin{aligned} P_e(\overline{\gamma}, \mathbf{a}, \boldsymbol{\rho}, D) &= 1 - P_c(\overline{\gamma}, \mathbf{a}, \boldsymbol{\rho}, D) \\ &= 1 - \sum_{m=0}^{M-1} \mathbb{P}(\rho_m \leq Y < \rho_{m+1} | A = a_m)\mathbb{P}(A = a_m), \end{aligned} \tag{9}$$

where P_c is the probability of a correct decision and $\mathbb{P}(\rho_m < Y \leq \rho_{m+1} | A = a_m)$ is the conditional probability, that the received energy Y is in the interval $[\rho_m, \rho_{m+1})$, with the optimal interval thresholds ρ. Thus, the decision has been made using the ML decision rule (8). $\mathbb{P}(A = a_m)$ is the a priori probability, that the symbol m has been sent and $\mathbb{P}(A = a_m) = 1/M$ for all $m \in \{0, 1, \ldots, M-1\}$. The conditional probability $\mathbb{P}(\rho_m \leq Y < \rho_{m+1} | A = a_m)$ is:

$$\begin{aligned} \mathbb{P}(\rho_m < y \leq \rho_{m+1} | A = a_m) &= \int_{\rho_m}^{\rho_{m+1}} f_{Y|A}(y|a_m)\, \mathrm{d}y \\ &= F_{Y|A}(\rho_{m+1}|a_m) - F_{Y|A}(\rho_m|a_m). \end{aligned} \tag{10}$$

The related distribution function $F_{Y|A}(\cdot|0)$ can be calculated in closed form:

$$F_{Y|A}(y|0) = \int_0^y \frac{1}{2^D \Gamma(D)} u^{D-1} \exp\left(-\frac{u}{2}\right) \mathrm{d}u = \frac{\Gamma\left(D, \frac{y}{2}\right)}{\Gamma(D)}, \tag{11}$$

where $\Gamma(\cdot)$ is the Gamma function [18, eq. 8.310.1] and $\Gamma(\cdot,\cdot)$ is the incomplete Gamma function [18, eq. 8.350.1]. The distribution function (11) can be also displayed with the help of the Marcum-$\mathcal{Q}$ function:

$$\mathcal{Q}_m(a,b) = \int_b^{\infty} x\left(\frac{x}{a}\right)^{m-1} \exp\left(-\frac{x^2+a^2}{2}\right) \mathrm{I}_{m-1}(ax)\, \mathrm{d}x, \tag{12}$$

where I_n is the modified Bessel function of the first kind of order n [1, eq. 9.6.3]. Thus, an alternative representation for the distribution function (11) is for $D \in \mathbb{Z}^+$ with [49, eq. 2.1.124] and [52, eq. 4.71]:

$$F_{Y|A}(y|0) = 1 - \mathcal{Q}_D(0, \sqrt{y}). \tag{13}$$

If a symbol with an amplitude $a_m > 0$ has been sent, the conditional probability density function $f_{Y|A}(\cdot|a_m)$ has the form (6). There does not exist a closed form for the distribution function in general. But for $D \in \mathbb{Z}^+$ it can also be solved in closed form with the help of the Marcum-$\mathcal{Q}$ function (12):

$$\begin{aligned} F_{Y|A}(y|a_m) &= \int_0^y \frac{1}{2}\left(\frac{u}{2a_m^2\gamma}\right)^{\frac{D-1}{2}} \exp\left(-\frac{2a_m^2\gamma+u}{2}\right) \mathrm{I}_{D-1}\left(\sqrt{2a_m^2\gamma u}\right) du \\ &= 1 - \mathcal{Q}_D\left(a_m\sqrt{2\gamma}, \sqrt{y}\right). \end{aligned} \tag{14}$$

Combining (13) and (14) with (9), the SEP P_e for an energy detection receiver with M-PAM for an SNR of $\gamma = E_p/N_0$ is [35]:

$$P_e(\gamma, \mathbf{a}, \boldsymbol{\rho}, M, D) = 1 - \frac{1}{M}\left[\sum_{m=0}^{M-1} \mathcal{Q}_D\left(a_m\sqrt{2\gamma}, \sqrt{\rho_{m+1}}\right) - \mathcal{Q}_D\left(a_m\sqrt{2\gamma}, \sqrt{\rho_m}\right)\right] \tag{15}$$

with the M symbol amplitudes $\boldsymbol{a} = (a_0, a_1, \dots, a_{M-2}, a_{M-1})$, $M+1$ interval thresholds $\boldsymbol{\rho} = (\rho_0, \rho_1, \dots, \rho_{M-1}, \rho_M)$ and the degree of freedom $2D$. Applying the interval thresholds $\rho_0 = 0$ und $\rho_M \to \infty$, we get $\mathcal{Q}_D\left(a_m\sqrt{2\gamma}, \sqrt{\rho_0}\right) = 0$ and $\mathcal{Q}_D\left(a_m\sqrt{2\gamma}, \sqrt{\rho_M}\right) = 1$. Combining this with (15), it reduces to:

$$P_e(\gamma, \mathbf{a}, \boldsymbol{\rho}, M, D) = \frac{1}{M}\left[M - 1 + \sum_{m=1}^{M-1} \mathcal{Q}_D\left(a_m\sqrt{2\gamma}, \sqrt{\rho_m}\right) - \sum_{m=0}^{M-2} \mathcal{Q}_D\left(a_m\sqrt{2\gamma}, \sqrt{\rho_{m+1}}\right)\right]. \tag{16}$$

Figure 3 shows the influence of different degrees of freedom on the BEP of an energy detection receiver with OOK and 2-PPM. For large degrees of freedom, a higher SNR is necessary to achieve the same BEP. This is due to an increasing amount of noise at the detector. OOK shows a slightly better performance than 2-PPM. Figure 4 shows the influence of higher order modulation on the BEP.

3.2.1.1. Optimal interval thresholds

The optimal interval thresholds to minimise the SEP have to fulfil the following optimisation problem:

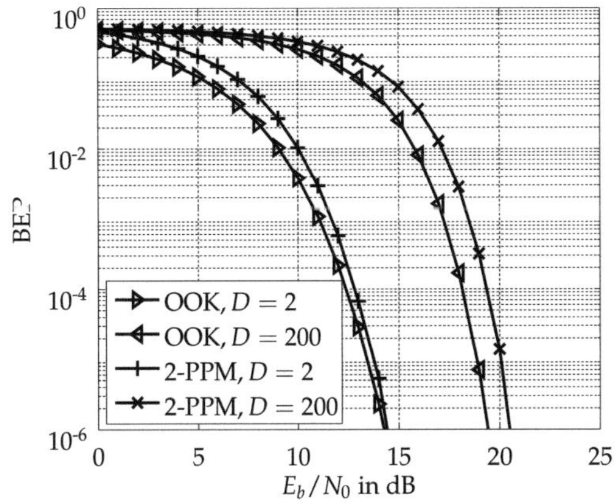

Figure 3. BEP for OOK and 2-PPM with different degrees of freedom

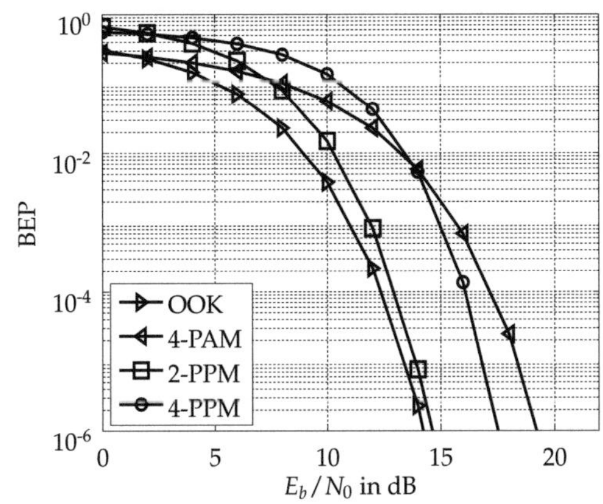

Figure 4. BEP for multilevel M-PAM and M-PPM for $D = 2$

$$\underset{\boldsymbol{\rho}}{\text{minimise}} \quad P_e(\gamma, \boldsymbol{a}, \boldsymbol{\rho}, M, D)$$

$$\text{subject to} \quad \frac{1}{M}\sum_{m=0}^{M-1} a_m^2 = 1,$$

Thus, the optimal interval threshold ρ_{opt} between the two symbol amplitudes a_m and a_{m+1} has to fulfil the following equation:

$$f_{Y|A}(\rho_{\text{opt}}|a_m) = f_{Y|A}(\rho_{\text{opt}}|a_{m+1}), \tag{17}$$

where $f_{Y|A}$ are the conditional probability density functions based on (5) and (6). With these optimal interval thresholds, the symbol decision is based on the ML-criteria (8). Unfortunately, there is no closed form solution for determining the optimal interval thresholds. Thus, they have to be calculated numerically. Figure 5 shows the conditional probability density functions with equidistant symbol amplitudes a_m and optimal interval thresholds ρ_1, ρ_2 and ρ_3 with an SNR of $\gamma = 10\,\text{dB}$.

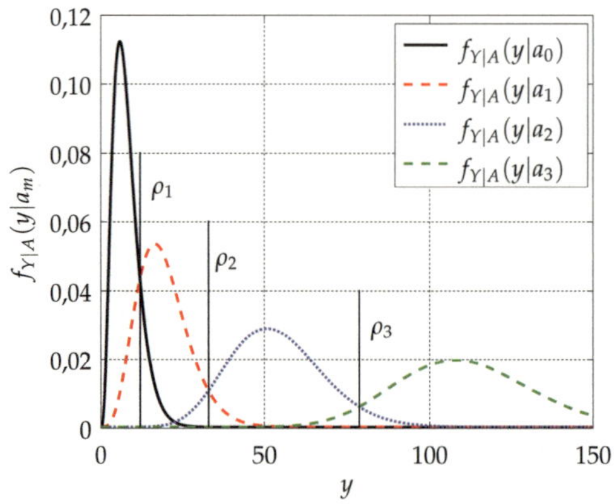

Figure 5. Optimal interval thresholds for 4-PAM ($\gamma = 10\,\text{dB}$)

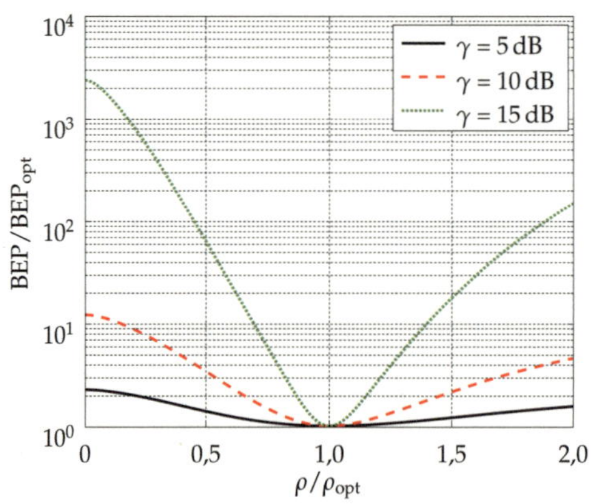

Figure 6. Sensitivity of the BEP related to the interval threshold ρ_1 for OOK

Figure 6 shows the influence of a non optimal interval threshold ρ_1 on the SEP for OOK ($M = 2$). In such a case, the SEP gets more sensitive for high SNR.

3.2.1.2. Optimal amplitudes

Now we try to minimise the SEP by optimising the symbol amplitudes $\boldsymbol{a}$. The optimisation problem is now [33, 37]:

$$\begin{aligned} \underset{\boldsymbol{a}}{\text{minimise}} \quad & P_e(\gamma, \boldsymbol{a}, \boldsymbol{\rho}, M, D) \\ \text{subject to} \quad & \frac{1}{M} \sum_{m=0}^{M-1} a_m^2 = 1, \end{aligned}$$

This optimisation problem can only be solved numerically, because the optimal interval thresholds $\boldsymbol{\rho}$ are based on the amplitudes $\boldsymbol{a}$ and there exists no closed form solution for the optimal interval thresholds $\boldsymbol{\rho}$ (17). For OOK ($M = 2$) the optimal amplitudes are $\boldsymbol{a}_{\text{opt}} = (0, 2)$. In this case they are independent of the SNR γ. For $M > 2$ it is possible to calculate a set of optimal amplitudes $\boldsymbol{a}_{\text{opt}}$ for every SNR γ. Figure 7 shows the SEP for 4-PAM for different symbol amplitudes a_1 and a_2 for a SNR of 16 dB. For figure 7 the amplitudes $a_0 = 0$ and

$a_3 = 1$ are set. The minimal SEP has been reached for $\boldsymbol{a} = (0, 0.35, 0.67, 1)$. Figure 8 shows

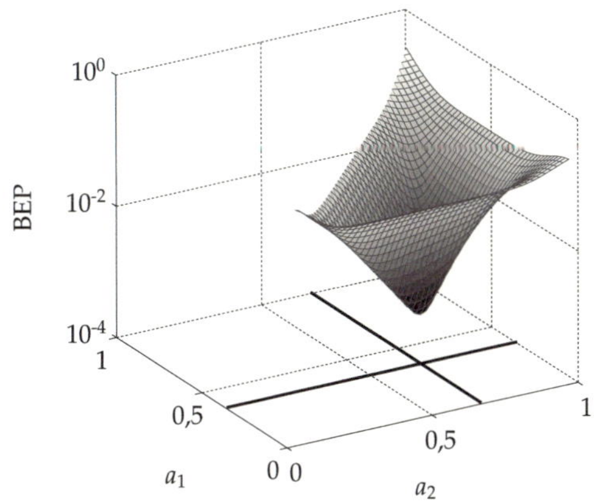

Figure 7. SEP for 4-PAM with different interval thresholds

the gain for 4-PAM with optimal amplitudes for different degrees of freedom. The results show impressive gains for large degrees of freedom. Figure 9 shows the optimal amplitudes for different degrees of freedom. For $D = 2$ the amplitudes are almost equidistant but for $D = 200$ the amplitudes are adjusted and not equidistant any more.

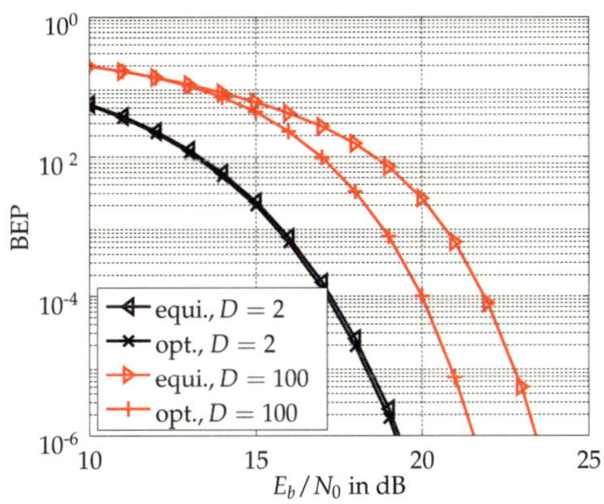

Figure 8. BEP for 4-PAM with equidistant and optimal amplitudes

3.3. Flat fading channel

To analyse the performance of an energy detection receiver we need a channel model that enables a good approximation of the energy at the receiver. Investigations of the IEEE channel model (802.15.3a) show that the energy at the receiver can be approximated by a random variable which is constant for one symbol.

Figure 10 compares the channel's magnitude (denoted as CIR) to a moving average of width 100 MHz and 1 GHz of the energy in the IEEE channel model. Figure 11 shows the magnitude at the receiver for a detector with 100 MHz and 1 GHz bandwidth. Thus we can use the

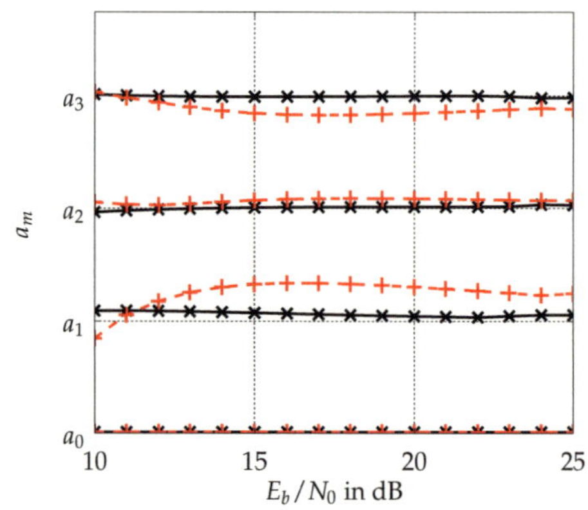

Figure 9. Optimal amplitudes for different degrees of freedom ($D = 2$, $D = 200$)

flat fading channel model to model the energy at the receiver in a frequency selective fading channel.

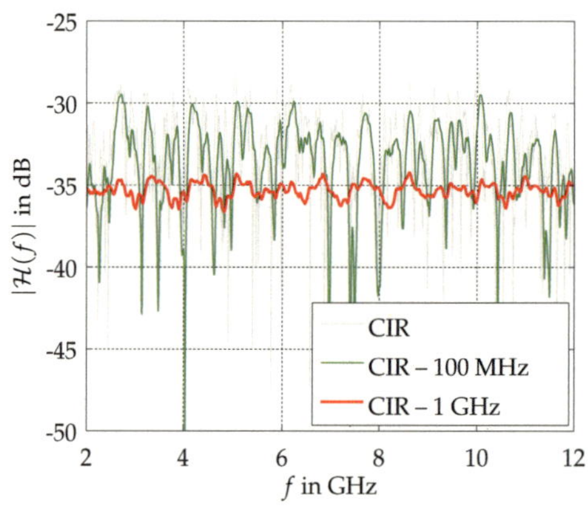

Figure 10. Moving average of the energy at the receiver (IEEE 802.15.3a, CM1)

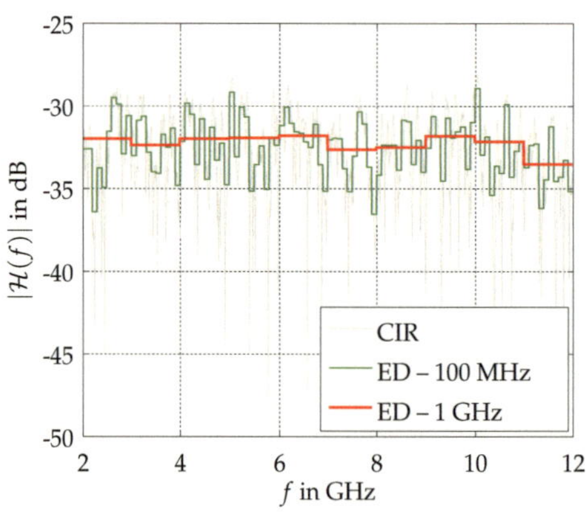

Figure 11. Energy at the receiver for different bandwidth (IEEE 802.15.3a, CM1)

In a flat fading channel, the random path attenuation H is assumed to be constant for the duration of a symbol. Thus, the received signal $R(t)$ in the interval $0 \leq t \leq T_S$ is:

$$R(t) = Hs(t) + W(t),$$

where $W(t)$ denotes the additive white Gaussian noise (AWGN) with a spectral power density of $N_0/2$.

The instantaneous SNR γ in the flat fading channel is:

$$\gamma = h^2 \frac{E_S}{N_0},$$

where h denotes the instantaneous path attenuation and E_S denotes the symbol energy. Since the SNR γ is random, γ is a realization of a a random variable Γ. The average SNR $\overline{\gamma}$ can be calculated using the expectation of the random variable H^2:

$$\overline{\gamma} = \Omega \frac{\overline{E_S}}{N_0} = \int_0^\infty \gamma f_\Gamma(\gamma)\,\mathrm{d}\gamma$$

with $\Omega = \mathbb{E}(H^2)$. Introducing a change of variable, the probability density function of the random variable Γ is [52, eq. 2.3]:

$$f_\Gamma(\gamma) = \frac{f_H(\sqrt{\frac{\Omega\gamma}{\overline{\gamma}}})}{2\sqrt{\frac{\gamma\overline{\gamma}}{\Omega}}}. \tag{18}$$

To calculate the average SEP in a flat fading channel, we have to solve the following integral [52, eq. 1.8]:

$$\overline{P}_e(\overline{\gamma}) = \int_0^\infty P_{e,AWGN}(\gamma) f_\Gamma(\gamma)\,\mathrm{d}\gamma. \tag{19}$$

3.3.1. Pulse amplitude modulation

Using the SEP in the AWGN channel (16) and the probability density function of the random SNR Γ in the flat fading channel, the average SEP for M-PAM is:

$$\begin{aligned}\overline{P}_e(\overline{\gamma}, \boldsymbol{a}, \boldsymbol{\rho}, M, D) = \frac{1}{M} \Bigg[& M - 1 - \mathcal{Q}_D\left(0, \sqrt{\rho_1}\right) \\ & + \sum_{m=1}^{M-1} \int_0^\infty \mathcal{Q}_D\left(a_m\sqrt{2\gamma}, \sqrt{\rho_m}\right) f_\Gamma(\gamma)\,\mathrm{d}\gamma \\ & - \sum_{m=1}^{M-2} \int_0^\infty \mathcal{Q}_D\left(a_m\sqrt{2\gamma}, \sqrt{\rho_{m+1}}\right) f_\Gamma(\gamma)\,\mathrm{d}\gamma \Bigg],\end{aligned} \tag{20}$$

with the symbol amplitudes $\boldsymbol{a} = (a_0, a_1, \ldots, a_{M-2}, a_{M-1})$ and the interval thresholds $\boldsymbol{\rho} = (\rho_0, \rho_1, \ldots, \rho_{M-1}, \rho_M)$. (20) is a general solution for the average SEP of an energy detection receiver in a flat fading channel with M-PAM.

3.3.2. Rayleigh fading

Rayleigh distributed path gains are used to model fading channels with no line-of-sight (NLOS) [49, 50, 55]. Thus, the random variable H is *Rayleigh* distributed:

$$f_H(h) = \frac{2h}{\Omega} \exp\left(-\frac{h^2}{\Omega}\right), \quad h \geq 0. \tag{21}$$

In UWB channels with a large bandwidth and a corresponding high temporal resolution, it is questionable, if the central limit theorem is applicable [4, 32, 59]. Nevertheless, some UWB channel measurements show a good fit to the *Rayleigh* distribution [17, 24, 51]. Using (18), the probability density function of the random SNR is:

$$f_\Gamma(\gamma) = \frac{1}{\overline{\gamma}} \exp\left(-\frac{\gamma}{\overline{\gamma}}\right), \quad \gamma \geq 0. \tag{22}$$

Combining (20) and (22) we get an integral of the form:

$$\Upsilon_D(\overline{\gamma}, a_m, \rho_m) = \frac{1}{\overline{\gamma}} \int_0^\infty \mathcal{Q}_D\left(a_m\sqrt{2\gamma}, \sqrt{\rho_m}\right) \exp\left(-\frac{\gamma}{\overline{\gamma}}\right) \mathrm{d}\gamma. \tag{23}$$

The integral in (23) can be solved using [40, eq. 12], given that the interval thresholds do not depend on the instantaneous SNR γ but on the average SNR $\overline{\gamma}$. This is the case for an energy detection receiver with limited channel state information (only knowledge of the average SNR). The closed form solution for Υ_D is then:

$$\Upsilon_D(\overline{\gamma}, a_m, \rho_m) = \exp\left(-\frac{\rho_m}{2}\right) \left\{ \left(\frac{1/\overline{\gamma} + a_m^2}{a_m^2}\right)^{D-1} \right.$$
$$\left. \cdot \left[\exp\left(\frac{\rho_m}{2} \frac{a_m^2}{a_m^2 + 1/\overline{\gamma}}\right) - \sum_{d=0}^{D-2} \frac{1}{d!} \left(\frac{\rho_m}{2} \frac{a_m^2}{a_m^2 + 1/\overline{\gamma}}\right)^d\right] + \sum_{d=0}^{D-2} \frac{1}{d!} \left(\frac{\rho_m}{2}\right)^d \right\}. \tag{24}$$

Combining (20) and (24) yields to the closed form solution for the energy detection receiver in a *Rayleigh* fading channel with M-PAM:

$$\overline{P}_{e,ray}(\overline{\gamma}, \mathbf{a}, \boldsymbol{\rho}, M, D) = \frac{1}{M} \left[M - 1 - \mathcal{Q}_D\left(0, \sqrt{\rho_1}\right) \right.$$
$$\left. + \sum_{m=1}^{M-1} \Upsilon_D\left(\overline{\gamma}, a_m, \rho_m\right) - \sum_{m=1}^{M-2} \Upsilon_D\left(\overline{\gamma}, a_m, \rho_{m+1}\right) \right]. \tag{25}$$

3.3.3. Rician fading

Rice distributed path gains are used to model line-of-sight (LOS) fading channels [49, 50, 55]. Thus, the random variable H is *Rice* distributed:

$$f_H(h) = \frac{2h(k+1)}{\Omega} \exp\left(-k - \frac{(k+1)\,h^2}{\Omega}\right) \mathrm{I}_0\left(2h\sqrt{\frac{k\,(k+1)}{\Omega}}\right), \quad h \geq 0, \tag{26}$$

where I_0 denotes the modified Bessel function of the first kind of order zero. The *Rician*-k-factor is the ratio between the power in the direct path and the power in the scattered paths. For $k = 0$ the *Rice* distribution is equal to the *Rayleigh* distribution. For $k \to \infty$ the *Rician* fading channel converges to the AWGN channel. Different UWB measurement campaigns show a good fit with the distribution of the path gains with a *Rice* distribution [20, 26, 43]. Using (18), the probability density function of the random SNR is:

$$f_\Gamma(\gamma) = \frac{k+1}{\overline{\gamma}} \exp\left(-k - \frac{(k+1)\,\gamma}{\overline{\gamma}}\right) \mathrm{I}_0\left(2\sqrt{\frac{k\,(k+1)\,\gamma}{\overline{\gamma}}}\right), \quad \gamma \geq 0. \tag{27}$$

Combining (20) and (27) we get an integral of the form:

$$\Phi_D(\overline{\gamma}, a_m, \rho_m, k) = \frac{k+1}{\overline{\gamma}} \int_0^\infty \mathcal{Q}_D\left(a_m u, \sqrt{\rho_m}\right) \exp\left(-K - \frac{(K+1)u^2}{\overline{\gamma}}\right) \cdot \mathrm{I}_0\left(2\sqrt{\frac{K\,(K+1)\,u^2}{\overline{\gamma}}}\right) u\,du. \tag{28}$$

There exists only a closed form solution for $D = 1$ [39, eq. 45]. In this case, we get:

$$\Phi_1(\overline{\gamma}, a_m, \rho_m, k) = \mathcal{Q}_1\left(\sqrt{\frac{2ka_m^2\gamma}{(k+1)+a_m^2\gamma}}, \sqrt{\frac{(k+1)\rho_m}{(k+1)+a_m^2\gamma}}\right). \tag{29}$$

Thus, the average SEP in a *Rician* fading channel is with a degree of freedom of $2D = 2$:

$$\overline{P}_{e,ric}(\overline{\gamma}, \mathbf{a}, \boldsymbol{\rho}, k, M) = \frac{1}{M}\Bigg[M - 1 - \mathcal{Q}_1\left(0, \sqrt{\rho_1}\right) + \sum_{m=1}^{M-1} \Phi_1\left(\overline{\gamma}, a_m, \rho_m, k\right) - \sum_{m=1}^{M-2} \Phi_1\left(\overline{\gamma}, a_m, \rho_{m+1}, k\right)\Bigg]. \tag{30}$$

A closed form solution for the integral in (28) is not known for $D > 1$. In this case, the integral has to be calculated numerically.

3.3.4. *Nakagami-m fading*

The probability density function of the *Nakagami-m* distribution of the random path gains is related to the χ^2 distribution:

$$f_H(h) = \frac{2h^{2m-1}}{\Gamma(m)} \left(\frac{m}{\Omega}\right)^m \exp\left(-\frac{m}{\Omega}h^2\right), \quad h \geq 0, \tag{31}$$

where m denotes the *Nakagami-m* fading parameter with $m \in [1/2, \infty)$ and Γ denotes the Gamma function. The *Nakagami-m* distribution includes as special cases the one-sided normal distribution ($m = 1/2$) and the *Rayleigh*-distribution ($m = 1$). For $m \to \infty$ the *Nakagami-m* fading channel converges to the AWGN channel. Different UWB measurement campaigns show a good fit to the *Nakagami-m* distribution [4, 19]. *Nakagami-m* distribution is also used in the IEEE channel model 802.15.4a to model the path gains [31]. The probability density function of the random SNR Γ is with (18) and (31):

$$f_\Gamma(\gamma) = \frac{\gamma^{m-1}}{\Gamma(m)} \left(\frac{m}{\overline{\gamma}}\right)^m \exp\left(-\frac{m}{\overline{\gamma}}\gamma\right), \quad \gamma \geq 0. \tag{32}$$

In this case we have to solve the following integral by substituting $\gamma = u^2/2$:

$$\int_0^\infty u^{2m-1} \exp\left(-\frac{mu^2}{a\overline{\gamma}}\right) \mathcal{Q}_D\left(au, \sqrt{\rho_m}\right) du. \tag{33}$$

(33) can be solved recursively [13]. The average symbol error rate in a fading channel with *Nakagami-m* distributed fading gains is:

$$\begin{aligned} \overline{P}_{e,nak}(\overline{\gamma}, m, \mathbf{a}, \boldsymbol{\rho}, M, D) = \frac{1}{M} \Bigg[& M - 1 - \mathcal{Q}_D\left(0, \sqrt{\rho_1}\right) \\ & + \sum_{v=1}^{M-1} \left(A\left(\overline{\gamma}, m, a_v, \rho_v\right) + \beta_v^m B\left(\overline{\gamma}, m, a_v, \rho_v\right)\right) \\ & - \sum_{v=1}^{M-2} \left(A\left(\overline{\gamma}, m, a_v, \rho_{v-1}\right) + \beta_v^m B\left(\overline{\gamma}, m, a_v, \rho_{v+1}\right)\right) \Bigg], \end{aligned} \tag{34}$$

where

$$A\left(\overline{\gamma}, m, a, \rho\right) = \exp\left(-\frac{\beta\rho}{2}\right) \left[\beta^{m-1} L_{m-1}\left(-\frac{(1-\beta)\rho}{2}\right) + (1-\beta)\sum_{i=0}^{m-2} \beta^i L_i\left(-\frac{(1-\beta)\rho}{2}\right)\right]$$

and

$$B\left(\overline{\gamma}, m, a, \rho\right) = \exp\left(-\frac{\rho}{2}\right) \sum_{n=1}^{D-1} \frac{\rho_w^n}{2^n n!} {}_1F_1\left(m; n+1; \frac{(1-\beta)\rho}{2}\right) \text{ with } \beta_v = \frac{2m}{2m + a_v\overline{\gamma}}.$$

L_i is the *Laguerre* polynomial of degree i [18, eq. 8.970] and ${}_1F_1$ is the confluent hypergeometric function [18, eq. 9.210.1].

Figure 12 shows the bit error probability in a flat fading channel with *Rayleigh* and *Rice* distributed channel gains. Figure 13 shows the bit error probability in a flat fading channel with *Nakagami-m* distributed channel gains.

3.4. Diversity reception

Now we analyse the SEP of an energy detection receiver with diversity reception. The goal is to increase the SNR to improve its performance. Because of the architecture of the receiver, detecting only the energy of the received signal, the possibilities to improve its performance are limited and many combining techniques like maximum ratio combining (MRC) or equal gain combining (EGC) are not feasible. Thus we concentrate on square law combining (SLC) and square law selection (SLS) [29].

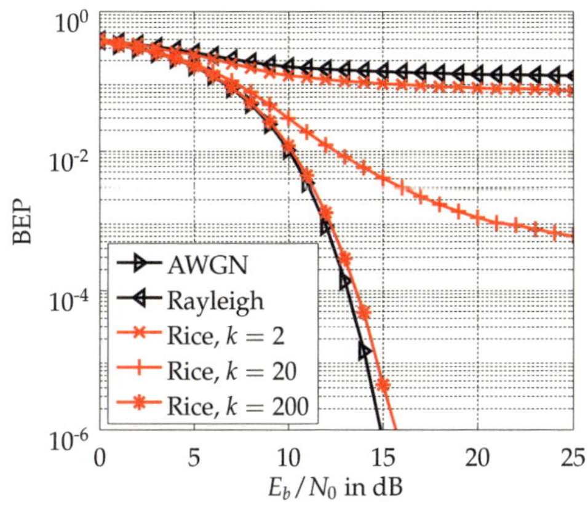

Figure 12. BEP in a flat fading channel with *Rayleigh* and *Rice* distributed channel gains (OOK, $D = 2$)

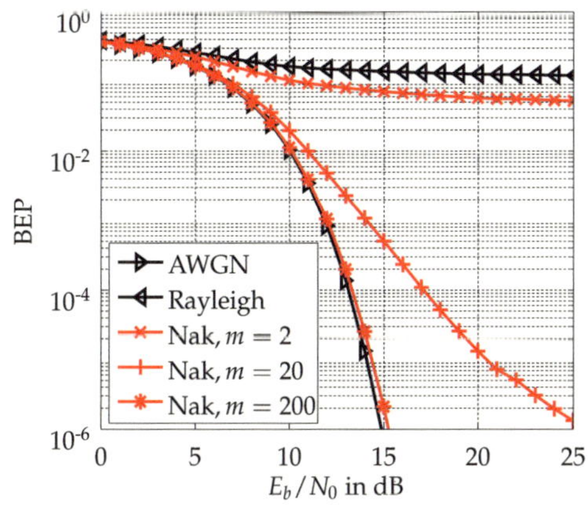

Figure 13. BEP in a flat fading channel with *Nakagami-m* distributed channel gains (OOK, $D = 2$)

The channel model used here is based on flat fading with independent and correlated fading gains H_l for all l diversity paths. The instantaneous SNR at the energy detector l is:

$$\gamma_l = h_l^2 E_S / N_{0,l}$$

and the average SNR at the lth detector is:

$$\overline{\gamma}_l = \Omega_l E_S / N_{0,l}$$

with $\Omega_l = \mathbb{E}(H_l^2)$. Figure 14 shows the model of multichannel receiver.

3.4.1. *Square law combining*

At the SLC receiver we have a new SNR Y_{SLC} at the receiver output based on the sum of the SNR Y_l at the l detectors:

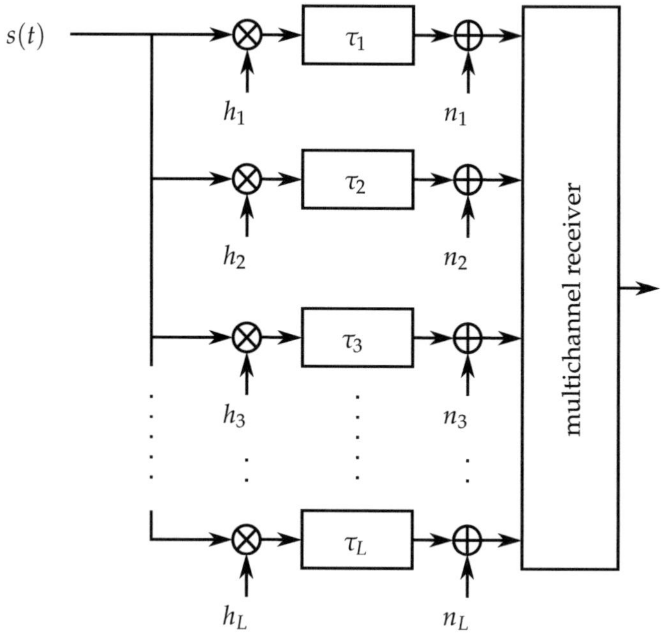

Figure 14. Channel model for multichannel receiver

$$Y_{\text{SLC}} = \sum_{l=1}^{L} Y_l.$$

The new random variable Y_{SLC} has a central χ^2 distribution for $a_0 = 0$ and a noncentral χ^2 distribution for $a_m > 0$ with the degree of freedom of $2LD$. The conditional distribution function for SLC is:

$$F_{Y|A}(y|a_m) = \mathcal{Q}_{LD}(a_m\sqrt{\gamma_{\text{SLC}}}, \sqrt{y}) \tag{35}$$

with the non centrality parameter:

$$\mu_{\text{SLC}} = 2\gamma_{\text{SLC}} = 2\sum_{l=1}^{L} \gamma_l.$$

Thus the SEP in L parallel AWGN channels with different SNR γ_l using SLC is:

$$P_e(\gamma_{\text{SLC}}, \mathbf{a}, \tilde{\boldsymbol{\rho}}, M, D, L) = \frac{1}{M}\Bigg[M - 1 + \sum_{m=1}^{M-1} \mathcal{Q}_{LD}\left(\sqrt{a_m^2\gamma_{\text{SLC}}}, \sqrt{\tilde{\rho}_m}\right) - \sum_{m=0}^{M-2} \mathcal{Q}_{LD}\left(\sqrt{a_m^2\gamma_{\text{SLC}}}, \sqrt{\tilde{\rho}_{m+1}}\right)\Bigg],$$

where $\tilde{\rho}$ are the optimal interval thresholds (section 3.2.1.1), based on the new random variable Y_{SLC}.

3.4.1.1. Independent and identically distributed Rayleigh distributed channel gains

Now we derive the SEP for SLC with independent and identical distributed (i.i.d.) Rayleigh distributed channel gains H_l. The SNR Γ_l is exponential distributed for all l. The probability density function $f_{\Gamma_{\text{SLC}}}$ of $\Gamma_{\text{SLC}} = \sum_{l=1}^{L} \Gamma_l$ with $\overline{\gamma} = \mathbb{E}(\Gamma_l)$ is [27, eq. 10.61]:

$$f_{\Gamma_{\text{SLC}}}(\gamma_{\text{SLC}}) = \frac{\gamma_{\text{SLC}}^{L-1}}{(L-1)!\overline{\gamma}^{L}} \exp\left(\frac{\gamma_{\text{SLC}}}{\overline{\gamma}}\right). \tag{36}$$

Comparing (36) with the probability density function of the SNR with *Nakagami-m* distributed channel gains we get the average SEP in I.I.D. *Rayleigh* fading channels:

$$\overline{P}_{e,SLC}(\overline{\gamma}, \mathbf{a}, \tilde{\boldsymbol{\rho}}, M, D, L) = \overline{P}_{e,nak}(L\overline{\gamma}, L, \mathbf{a}, \tilde{\boldsymbol{\rho}}, M, LD),$$

substituting $\overline{\gamma}$ by $L\overline{\gamma}$, m by L, $\boldsymbol{\rho}$ by $\tilde{\boldsymbol{\rho}}$ and $2D$ by $2LD$. The intervall thresholds $\tilde{\boldsymbol{\rho}}$ are based on $L\overline{\gamma}$.

3.4.1.2. Correlated Rayleigh distributed channel gains

In the next step, we assume correlated *Rayleigh* distributed channel gains H_l. In this case the probability density function of the SNR Γ is a sum of weighted exponential distributions [27, eq. 10.60]:

$$f_{\Gamma_{\text{SLC}}}(\gamma_{\text{SLC}}) = c_1 \sum_{l=1}^{L} c_{2,l} \exp\left(\frac{\gamma_{\text{SLC}}}{\overline{\gamma}}\right) \tag{37}$$

with

$$c_1 = \frac{1}{\prod_{l=1}^{L} \lambda_l}, \quad c_{2,l} = \frac{1}{\prod_{l \neq k}^{L} (1/\lambda_k - 1/\lambda_l)},$$

where λ_l are the eigenvalues of the $L \times L$ covariance matrix $\boldsymbol{\Sigma}$ of the normalized received signal. Using the function Υ (24) the SEP in a fading channel with correlated *Rayleigh* distributed fading gains can be written as:

$$\begin{aligned}\overline{P}_{e,\text{SLC}}(\overline{\gamma}, \mathbf{a}, \tilde{\boldsymbol{\rho}}, M, D, L) = \frac{1}{M} \Bigg[& M - 1 - Q_{LD}\left(0, \sqrt{\rho_1}\right) \\ & + c_1 \sum_{m=1}^{M-1} \sum_{l=1}^{L} \lambda_l c_{2,l} \Upsilon_{LD}\left(\lambda_l, a_m, \tilde{\rho}_m\right) \\ & - c_1 \sum_{m=1}^{M-2} \sum_{l=1}^{L} \lambda_l c_{2,l} \Upsilon_{LD}\left(\lambda_l, a_m, \tilde{\rho}_{m+1}\right) \Bigg].\end{aligned}$$

3.4.2. *Square law selection*

The receiver based on SLS chooses the detector with the highest received energy Y_{max}:

$$Y_{\text{max}} = \max(Y_1, Y_2, \ldots, Y_L). \tag{38}$$

Thus, this receiver collects only a fraction of the total received energy. Using (9) and (38), the SEP with SLS in the AWGN channel is:

$$P_e(\gamma, \mathbf{a}, \boldsymbol{\rho}, D, L) = 1 - P_c(\gamma, \mathbf{a}, \boldsymbol{\rho}, D)$$

$$= 1 - \sum_{m=0}^{M-1} \mathbb{P}(\rho_m < Y_{\max} \leq \rho_{m+1} | A = a_m) \mathbb{P}(A = a_m). \quad (39)$$

In order to calculate (39), we calculate the probability that $Y_{\max}$ is in the interval $(\rho_m, \rho_{m+1}]$:

$$\mathbb{P}(\rho_m < Y_{\max} \leq \rho_{m+1} | A = a_m)$$

$$= \mathbb{P}\left(\bigcap_{l=1}^{L} \{Y_l < \rho_{m+1} | A = a_m\}\right) - \mathbb{P}\left(\bigcap_{l=1}^{L} \{Y_l < \rho_m | A = a_m\}\right). \quad (40)$$

If all received energies Y_l are independent, (40) reduces to:

$$\mathbb{P}\left(\bigcap_{l=1}^{L} \{Y_l < \rho_m | A = a_m\}\right) = \prod_{l=1}^{L} \mathbb{P}\left(Y_l < \rho_m | \, A = a_m\right). \quad (41)$$

Using the distribution function (14), the probability can be written as:

$$\mathbb{P}\left(Y_l < \rho_m | \, A = a_m\right) = 1 - Q_D\left(a_m \sqrt{\gamma}, \sqrt{\rho_m}\right). \quad (42)$$

Combining (40), (41) and (42) and assume $\mathbb{P}(A = a_m) = 1/M$, (39) can be written as:

$$P_e(\gamma, \mathbf{a}, \boldsymbol{\rho}, D, L) = 1 - \frac{1}{M} \sum_{m=0}^{M-1} \left[\prod_{l=1}^{L} \left(1 - Q_D\left(a_m \sqrt{\gamma_l}, \sqrt{\rho_{m+1,l}}\right)\right) \right.$$

$$\left. - \prod_{l=1}^{L} \left(1 - Q_D\left(a_m \sqrt{\gamma_l}, \sqrt{\rho_{m,l}}\right)\right) \right], \quad (43)$$

where $\rho_{m,l}$ denotes the optimal interval thresholds which are based on the SNR γ_l. If all received energies Y_l are independent and identical distributed, (41) reduces to:

$$\prod_{l=1}^{L} \mathbb{P}\left(Y_l < \rho_m | A = a_m\right) = \left(1 - Q_D\left(a_m \sqrt{\gamma}, \sqrt{\rho_m}\right)\right)^L. \quad (44)$$

In this case, (43) reduces to:

$$P_e(\gamma, \mathbf{a}, \boldsymbol{\rho}, D, L) = 1 - \frac{1}{M} \sum_{m=0}^{M-1} \left[\left(1 - Q_D\left(a_m \sqrt{\gamma}, \sqrt{\rho_{m+1}}\right)\right)^L \right.$$

$$\left. - \left(1 - Q_D\left(a_m \sqrt{\gamma}, \sqrt{\rho_m}\right)\right)^L \right].$$

If the SNR is independent and identically *Rayleigh* distributed with $\overline{\gamma} = \mathbb{E}(\Gamma_l)$ for all l, the SEP in a fading channel with i.i.d. *Rayleigh* distributed channel gains can be written using the function Y (24):

$$P_e(\overline{\gamma}, \mathbf{a}, \boldsymbol{\rho}, D, L) = \frac{1}{M}\Bigg[M + 1 + \left(1 - Q_D\left(a_0\sqrt{\overline{\gamma}}, \sqrt{\rho_1}\right)\right)^L + \sum_{m=1}^{M-2}\left(1 - \Upsilon_D(\overline{\gamma}, a_m, \rho_{m+1})\right)^L - \sum_{m=1}^{M-1}\left(1 - \Upsilon_D(\overline{\gamma}, a_m, \rho_m)\right)^L\Bigg].$$

Figure 15 shows the gain that can be achieved using SLC. Figure 16 shows the gain that can be achieved using SLS. The SLC based receiver can collect more energy, but also more noise than the SLS based receiver. The SLS based receiver chooses only the dominant path.

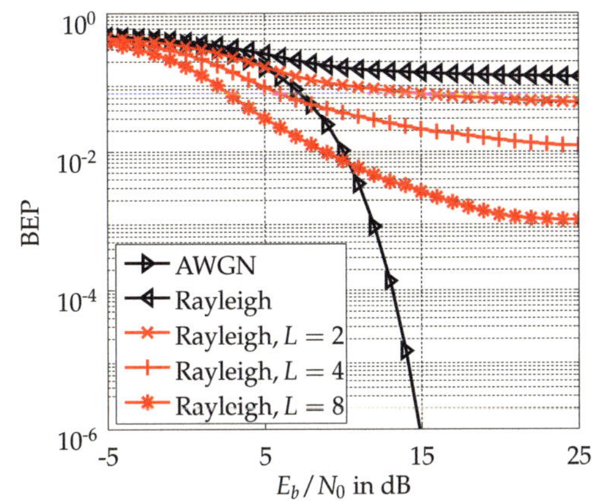

Figure 15. BEP in the *Rayleigh* channel with SLC

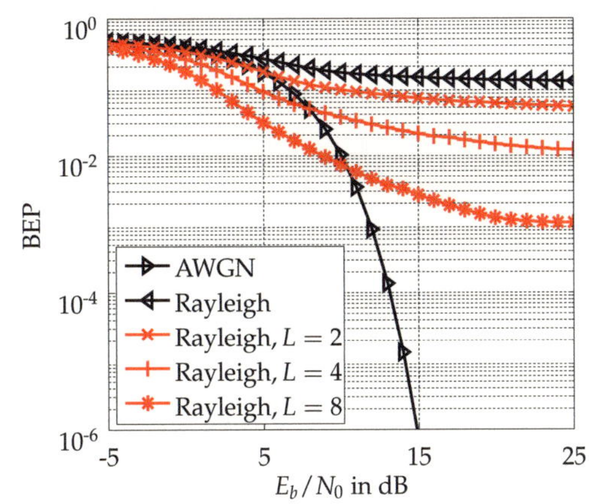

Figure 16. BEP in the *Rayleigh* channel with SLS

3.5. Frequency selective fading channel

Now we focus on the frequency selectivity of a channel and the effect on the SEP for an energy detection receiver. The received signal $\boldsymbol{R}$ is in the baseband equivalent model:

$$\boldsymbol{R} = \underbrace{\boldsymbol{x}\boldsymbol{H}}_{\boldsymbol{S}} + \boldsymbol{W},$$

where $\boldsymbol{H} = (H_1, \ldots, H_L)^T$ denotes the random channel impulse response, circular symmetric complex white Gaussian noise $\boldsymbol{W} \sim \mathcal{CN}_L\left(\mathbf{0}, \sigma_n^2 \mathbf{I}\right)$ and $\boldsymbol{x}$ denotes the convolution matrix, containing shifted versions of the transmitted signal:

$$\mathbf{x} = \begin{pmatrix} s_1 & 0 & \ldots & 0 \\ s_2 & s_1 & \ldots & 0 \\ \vdots & \vdots & \ddots & \vdots \\ s_L & s_{L-1} & \ldots & s_1 \end{pmatrix}.$$

The effect of intersymbol interference has not been taken into account by choosing a gap between symbols larger than the channel delay spread. Using the quadratic form of $\boldsymbol{S}$ the SNR Γ at the receiver can be written as:

$$\Gamma = \frac{E_R}{N_0} = \frac{\boldsymbol{S}^H \boldsymbol{S}}{4 N_0 B}.$$

3.5.1. *NLOS fading channel*

In an environment with no line of sight between transmitter and receiver, the expectation of the random samples of the channel impulse response $\boldsymbol{H}$ is zero:

$$\boldsymbol{H} \sim \mathcal{CN}_D\left(\mathbf{0}, \boldsymbol{\Sigma}\right) \quad \text{with } \boldsymbol{\Sigma}(i,j) = h_i h_j e^{-\frac{|i-j|}{\beta}},$$

where $\beta > 0$ describes the correlation of the elements in the covariance matrix $\boldsymbol{\Sigma}$. If the samples of the channel impulse response are uncorrelated, $\boldsymbol{\Sigma}$ is a diagonal matrix. The expectation the noise free received signal is also zero:

$$\boldsymbol{S} = \boldsymbol{x}\boldsymbol{H} \sim \mathcal{CN}_D\left(\mathbf{0}, \boldsymbol{\Sigma}_s\right) \quad \text{with } \boldsymbol{\Sigma}_s = \boldsymbol{x}\boldsymbol{\Sigma}\boldsymbol{x}^H.$$

Let $\boldsymbol{\Sigma}_s^{\frac{1}{2}}$ denote a matrix that fulfils $\boldsymbol{\Sigma}_s^{\frac{1}{2}} \boldsymbol{\Sigma}_s^{\frac{1}{2}} = \boldsymbol{\Sigma}_s$, then a whitened vector $\mathbf{S}' \sim \mathcal{CN}_L\left(\mathbf{0}, \mathbf{I}\right)$ is defined as

$$\boldsymbol{\Sigma}_s^{-\frac{1}{2}} \mathbf{S} = \mathbf{S}'. \tag{45}$$

Note that the existence of $\boldsymbol{\Sigma}_s^{\frac{1}{2}}$ is guaranteed for any positive definite matrix $\boldsymbol{\Sigma}_s$. To calculate the distribution of instantaneous SNR Γ, we need to analyse the distribution of $\mathbf{S}^H\mathbf{S}$ first. Using (45) yields

$$\mathbf{S}^H \mathbf{S} = \mathbf{S}'^H \boldsymbol{\Sigma}_s \mathbf{S}'. \tag{46}$$

Using eigenvalue decomposition and special properties of the central χ^2-distribution, a closed form expression for the PDF of $\mathbf{r}^H\mathbf{r}$ can be found [21]. Performing an eigenvalue decomposition of $\mathbf{\Sigma}_S$ in (46) leads to

$$\mathbf{S}^H\mathbf{S} = \mathbf{S}'^H\mathbf{U}^H\mathbf{\Lambda}\mathbf{U}\mathbf{S}'$$

with $\mathbf{\Lambda}$ being a diagonal matrix containing the eigenvalues $\lambda_1, \ldots, \lambda_L$ of $\mathbf{\Sigma}_s$ and the rows of unitary matrix $\mathbf{U}$ being the corresponding eigenvectors. Substituting $\mathbf{G} = \mathbf{U}\mathbf{S}'$, the random variable V is:

$$V = \mathbf{S}^H\mathbf{S} = \mathbf{G}^H\mathbf{\Lambda}\mathbf{G} = \sum_{l=1}^{L} \lambda_l G_l G_l^* \tag{47}$$

with G_l representing the elements of $\mathbf{G}$ and G_l^* denoting the complex conjugate of G_l. Note that $\mathbf{G}$ and $\mathbf{H}$ are equally distributed because the rows of $\mathbf{U}$ are orthonormal among each other. Therefore

$$\mathbf{G} \sim \mathcal{CN}_L\left(\mathbf{0}, \mathbf{\Sigma}_s\right).$$

The random variable $G_l' = G_l G_l^* = |G_l|^2$ is central χ^2 distributed with two degrees of freedom. That is a special case of the central χ^2 distribution which is equivalent to an exponential distribution. As a linear combination of independent and identically exponential distributed variates, V is general *Gamma* or general *Erlang* distributed with the PDF given as [21, eq. 19.147]:

$$f_V(v) = \sum_{j=1}^{D} \left(\prod_{k \neq j} \left(\lambda_j - \lambda_k\right)^{-1} \right) \lambda_j^{D-2} e^{-\frac{v}{\lambda_j}}. \tag{48}$$

The average SNR $\overline{\gamma}$ is based on the expectation of the random variable V:

$$\overline{\gamma} = \frac{\mathbb{E}(V)}{4N_0B}. \tag{49}$$

Because all random variables G_i are independent $\mathbb{E}(V)$ can be written as:

$$\mathbb{E}(V) = \sum_{i=1}^{D} \lambda_i \mathbb{E}(|G_i|^2) = 2\sum_{i=1}^{D} \lambda_i \quad \text{with } \mathbb{E}(|G_i|^2) = 2.$$

Introducing a change of variable in (48) yields:

$$f_\Gamma(\gamma) = 2N_0B\, f_V\left(2N_0B\gamma\right). \tag{50}$$

Using the relative SNR $\gamma' = \frac{\gamma}{\overline{\gamma}}$, (50) can be expressed as:

$$f_{\Gamma'}\left(\gamma'\right) = 2N_0B\overline{\gamma}\, f_V\left(2N_0B\overline{\gamma}\gamma'\right). \tag{51}$$

The SEP in a fading channel can be calculated using (20). Combining (20) and (51) yields to the following integral:

$$2N_0B \sum_{j=1}^{D} \left(\prod_{k \neq j} \left(\lambda_j - \lambda_k\right)^{-1} \right) \lambda_j^{D-2} \int_0^\infty \mathcal{Q}_D\left(\sqrt{a_m^2\overline{\gamma}\gamma'}, \sqrt{\rho_m}\right) e^{-\frac{2N_0B}{b\lambda_j}\overline{\gamma}\gamma'}\, \mathrm{d}\gamma'.$$

In the case of partial CSI, the integral can be solved using [40, eq. 12] with adequate substitutions:

$$\Theta_D(\overline{\gamma}, N_0, \boldsymbol{\lambda}, B, a_m, \rho_m) = \frac{1}{\overline{\gamma}} \sum_{j=1}^{D} \left(\prod_{k \neq j} \left(\lambda_j - \lambda_k \right)^{-1} \right) \lambda_j^{D-2} \mathrm{e}^{-\frac{\rho_m}{2}} \left\{ \left(1 + \frac{2N_0B}{a_m^2 \lambda_j} \right)^{D-1} \right.$$

$$\left. \cdot \left[\exp\left(\frac{\rho_m}{2} \frac{1}{1+\frac{2N_0B}{a_m^2\lambda_j}} \right) - \sum_{n=0}^{D-2} \frac{1}{n!} \left(\frac{\rho_m}{2} \frac{1}{1+\frac{2N_0B}{a_m^2\lambda_j}} \right)^n \right] + \sum_{n=0}^{D-2} \frac{1}{n!} \left(\frac{\rho_m}{2} \right)^n \right\}. \quad (52)$$

Using the function Θ_D (52), the average SEP in a frequency selective fading channel can be expressed in closed form [3]:

$$\overline{P}_e(\overline{\gamma}, N_0, B, \boldsymbol{a}, \boldsymbol{\rho}, \boldsymbol{\lambda}, M, D) = \frac{1}{M} \left[M - 1 + \mathcal{Q}_D\left(0, \sqrt{\rho_1}\right) + \sum_{m=1}^{M-1} \Theta_D(\overline{\gamma}, N_0, \boldsymbol{\lambda}, B, a_m, \rho_m) - \sum_{m=1}^{M-2} \Theta_D(\overline{\gamma}, N_0, \boldsymbol{\lambda}, B, a_m, \rho_{m+1}) \right].$$

This equation is only valid if two constraints are met:

1. The receiver may only use $\overline{\rho}$ to determine the decision threshold (partial CSI) and
2. all eigenvalues λ_j are pairwise disjunct.

If there exist two or more identical eigenvalues, one might rearrange (47) and decrease D used in the subsequent calculations to meet the second constraint. Figure 17 shows the BEP in the frequency selective fading channel with i.i.d. and correlated channel gains.

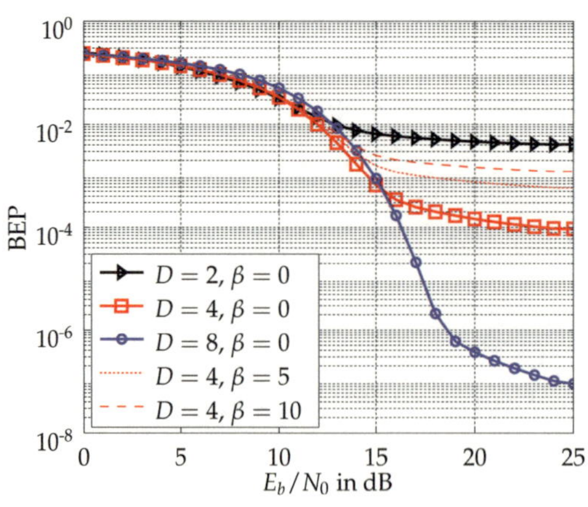

Figure 17. BEP for OOK in the frequency selective fading channel

3.5.2. LOS fading channel

In a channel model with LOS, $\mathbf{H}$ is expected to have a non-zero mean vector $\mathbf{M}$. Therefore V is equal in distribution to a linear combination of noncentral χ^2 distributed variates. Unfortunately, there is no such convenient correspondence for that distribution as the general *Erlang* distribution for V in section 3.5.1.

Even though the PDF of a linear combination of noncentral χ^2 distributed variates is given in [48], there is no known closed-form expression for the integral of the product of this PDF and the MARCUM-Q-function in (20).

The average SEP may still be computed using a discrete approximation of the PDF of V (histogram). This histogram can be generated with arbitrary precision by sampling the PDF given in [48] or by performing a Monte-Carlo-simulation based on the distribution of $\mathbf{H}$. For the Monte-Carlo-simulations the samples of the LOS channel impulse response $\mathbf{H}$ follow a normal distribution

$$\mathbf{H} \sim \mathcal{CN}_L\left(\mathbf{M}, \boldsymbol{\Sigma}_{\mathbf{h}}\right)$$

with the mean $\mathbf{M} = (M_1, \ldots, M_D)$ with $M_i = \sqrt{\frac{\kappa}{\kappa+1}}\ \forall\, i$ and the covariance matrix $\boldsymbol{\Sigma}_{\mathbf{h}} = \frac{1}{\kappa+1}\mathbf{I}$.

4. Interference investigations for non-coherent multiband UWB

As the unlicensed MIR-UWB system has no exclusive frequency range there is an increased interference potential from present and from future radio systems operating in the same frequency range. Hence, the performance of the MIR-UWB system can be reduced and a reliable communication cannot be guaranteed at any time. For this reason it is imperative to realise an efficient and low complex interference mitigation.

Section 4 is structured as follows: In section 4.1 an analysis of the interference robustness of an energy detection receiver is presented. This allows the identification of suitable MIR-UWB system parameters which can be configured preferably robust against interferences before initial operation of the MIR-UWB system. The following section 4.2 deals with coexistence-based approaches which are focused on an efficient and adaptive interference mitigation with low complexity. As the mitigation of narrowband interference (NBI) is a crucial issue of the MIR-UWB section 4.3 analyses the non-linear Teager-Kaiser (TK) operation. Thereby, the potential to mitigate NBI and to integrate the TK operation into the existing MIR UWB system is shown.

4.1. Interference robustness of energy detection

A basic issue of the MIR-UWB's energy detection receiver is its high sensitivity with respect to interferences passing the analogue front-end. A significant reduction of the instantaneous Signal-to-Interference-and Noise Ratio (SINR) can occur so that a reliable communication is not guaranteed.

For this reason it is required to investigate the interference robustness of an OOK and BPPM specific energy detection [7, 9]. The analysis bases on an analytical investigation of the interference robustness of an energy detector within an arbitrary but fixed subband. Thereby, dependencies between system- and interference specific parameters can be identified which promise an increase of interference robustness. The analysis bases on one hand on [53]. Therein, the performance of a BPPM specific energy detector is analysed in presence of out-of-band interference. On the other hand it relies on [60] in which the performance of a BPPM specific correlation receiver is investigated in presence of interference.

4.1.1. Signal model

Binary data transmission within an MIR-UWB subband of bandwidth B is considered. Thereby, based on OOK/BPPM the rectangular pulse

$$p(t) = \begin{cases} \sqrt{\frac{2}{T_\mathrm{p}}} \cos(2\pi f_\mathrm{c} t) & , 0 < t < T_\mathrm{p} \\ 0 & , \text{else} \end{cases} \tag{53}$$

with carrier frequency f_c and pulse duration T_p is emitted with energy 1 ($f_\mathrm{c} \gg 1/T_\mathrm{p}$). The resulting signal to be transmitted conducts to

$$s_\mathrm{O}(t) = \sqrt{E_\mathrm{p}^\mathrm{O}} \sum_{n=-\infty}^{\infty} b_n p_i (t - nT_\mathrm{b}) \tag{54}$$

for OOK and

$$s_\mathrm{P}(t) = \sqrt{E_\mathrm{p}^\mathrm{P}} \sum_{n=-\infty}^{\infty} p_i \left(t - nT_\mathrm{b} - b_n \frac{T_\mathrm{b}}{2} \right) \tag{55}$$

for BPPM. The uniformly distributed data bit $b_n \in \{0, 1\}$ is characterised by bit energy E_b as well as by bit duration $T_\mathrm{b} = \frac{T_\mathrm{p}}{d_\mathrm{s}}, d_\mathrm{s} > 0$ with duty cycle $d_\mathrm{s} \leq \frac{1}{2}$. Finally, $E_\mathrm{p}^i, i \in \{\mathrm{O}, \mathrm{P}\}$ stands for the modulation specific pulse energy which equals $E_\mathrm{p}^\mathrm{O} = 2E_\mathrm{p}^\mathrm{P} = 2E_\mathrm{b}$.

Assuming perfect synchronisation between the transmitter and the receiver, the signal $s_i(t), i \in \{\mathrm{O}, \mathrm{P}\}$ is superposed with zero mean white Gaussian noise $n(t)$ of two-sided spectral density $\frac{N_0}{2}$ and interference $j(t)$ leading to (no fading)

$$y_i(t) = s_i(t) + n(t) + j(t). \tag{56}$$

Interference is described as band-limited wide-sense stationary, time-continuous zero mean Gaussian process $J(t)$ characterised by the autocorrelation function ($\tau = t_1 - t_2$)

$$R_\mathrm{J}(\tau) = P_\mathrm{J} \frac{\sin(\pi B_\mathrm{J} \tau)}{\pi B_\mathrm{J} \tau} \cos(2\pi f_\mathrm{J} \tau). \tag{57}$$

It depends on the mean interference power P_J determined by the ratio of the interferer's bit energy $E_\mathrm{b,J}$ and bit duration $T_\mathrm{b,J} = qT_\mathrm{b}, q > 0$. Further parameters are the interference center frequency f_J as well as its bandwidth B_J.[1] The resulting interferer's signal duration $T_\mathrm{p,J} \approx \frac{1}{B_\mathrm{J}} \leq T_\mathrm{b,J}$ leads to an interference duty cycle of $d_\mathrm{J} = \frac{T_\mathrm{p,J}}{T_\mathrm{b,J}} = \frac{d_\mathrm{s} T_\mathrm{p,J}}{q T_\mathrm{p}}$.

[1] Eq. (57) holds if the interference source is completely inside the MIR-UWB subband. In case B_J overlaps completely or only partially with the subband f_J, B_J and P_J have to be properly modified. However, as it can be ascribed to (57) the following investigations focus solely on an interference source being completely inside the subband.

The received signal of (56) is first bandpass filtered and afterwards subjected to non-coherent energy detection with integration time T_P. At its input stage the available SINR is given by

$$\mathrm{SINR_E} = 10\log_{10}\frac{E_b}{T_b\left(P_J + P_N\right)}, \tag{58}$$

which is identical for OOK/BPPM. In (58) P_N stands for the mean noise power of the passband noise signal, which is modelled as band-limited wide-sense stationary, time-continuous zero mean Gaussian process $N(t)$: $(\tau = t_1 - t_2)$

$$R_N(\tau) = P_N \frac{\sin(\pi B\tau)}{\pi B\tau}\cos(2\pi f_c \tau). \tag{59}$$

At the output of energy detection the decision variable differs for OOK/BPPM.

Energy detection of OOK
For OOK, the asymmetric decision variable

$$x^O = \int_0^{T_P} y^2(t)\,dt = x_s^O + \Delta x^O \tag{60}$$

occurs. The resulting energy value x^O consists of a deterministic signal-only part

$$x_s^O = \begin{cases} 0 & , b_n = 0 \\ 2E_b & , b_n = 1 \end{cases} \tag{61}$$

of mean E_b and second order moment $2E_b^2$. The component $\Delta x^O = x_{sjn}^O + x_{jn}^O$ contains the mixed signal-noise and signal-interference term

$$x_{sjn}^O = \begin{cases} 0 & , b_n = 0 \\ 2\sqrt{\frac{2E_P^O}{T_P}} \int_0^{T_P} \cos(2\pi f_c t)\left(n(t) + j(t)\right) dt dt & , b_n = 1 \end{cases} \tag{62}$$

as well as the contribution

$$x_{jn}^O = \int_0^{T_P} \left(n(t) + j(t)\right)^2 dt \quad , b_n = 0, 1 \tag{63}$$

due to noise and interference-only.

Energy detection of BPPM
In contrast to OOK the BPPM specific decision variable at the output of energy detection is symmetric:

$$x^P = \int_0^{T_P} y_P^2(t)\,dt - \int_{\frac{T_b}{2}}^{\frac{T_b}{2}+T_P} y_P^2(t)\,dt = x_s^P + \Delta x^P. \tag{64}$$

The decision variable x^{P} compares energy values within two observation intervals of duration T_{p}.[2] It is composed of a signal-only contribution

$$x_s^{\mathrm{P}} = \begin{cases} E_{\mathrm{b}} & , b_n = 0 \\ -E_{\mathrm{b}} & , b_n = 1, \end{cases} \tag{65}$$

which is characterised by mean zero and second order moment E_{b}^2. The additional term $\Delta x^{\mathrm{P}} = x_{\mathrm{sjn}}^{\mathrm{P}} + x_{\mathrm{jn}}^{\mathrm{P}}$ is on one hand composed of a mixed signal-noise and signal-interference component

$$x_{\mathrm{sjn}}^{\mathrm{P}} = \begin{cases} a \int\limits_0^{T_{\mathrm{p}}} \cos\left(2\pi f_{\mathrm{c}} t\right)\left(n\left(t\right) + j\left(t\right)\right) \mathrm{d}t & , b_n = 0 \\ -a \int\limits_{\frac{T_{\mathrm{b}}}{2}}^{\frac{T_{\mathrm{b}}}{2}+T_{\mathrm{p}}} \cos\left(2\pi f_{\mathrm{c}}\left(t - \frac{T_{\mathrm{b}}}{2}\right)\right)\left(n\left(t\right) + j\left(t\right)\right) \mathrm{d}t & , b_n = 1 \end{cases} \tag{66}$$

with $a = 2\sqrt{2E_{\mathrm{p}}^{\mathrm{P}}/T_{\mathrm{p}}}$. On the other hand it consists of the noise and interference-only part

$$x_{\mathrm{jn}}^{\mathrm{P}} = \int\limits_0^{T_{\mathrm{p}}} \left(n\left(t\right) + j\left(t\right)\right)^2 \mathrm{d}t - \int\limits_{\frac{T_{\mathrm{b}}}{2}}^{\frac{T_{\mathrm{b}}}{2}+T_{\mathrm{p}}} \left(n\left(t\right) + j\left(t\right)\right)^2 \mathrm{d}t. \tag{67}$$

4.1.2. Statistical analysis of interference robustness

To make statements on the interference robustness of an OOK and BPPM specific energy detection a proper quality criterion has to be introduced. A possible measure is the processing gain (PG) of the energy detector. It refers the available SINR at its output to the $\mathrm{SINR_E}$ at its input. For OOK this can be described as

$$\mathrm{PG^O} = 10\log_{10}\left(\frac{2E_{\mathrm{b}}^2}{0{,}5Q_1^{\mathrm{O}} + Q_2^{\mathrm{O}}}\right) - 10\log_{10}\left(\mathrm{SINR_E}\right), \tag{68}$$

which differs from the PG of the BPPM based energy detection receiver expressed as

$$\mathrm{PG^P} = 10\log_{10}\left(\frac{E_{\mathrm{b}}^2}{Q_1^{\mathrm{P}} + Q_2^{\mathrm{P}}}\right) - 10\log_{10}\left(\mathrm{SINR_E}\right). \tag{69}$$

In (68) and (69) $Q_1^i, i \in \{\mathrm{O},\mathrm{P}\}$ stands for the second order moment of the mixed signal-noise and signal-interference component $x_{\mathrm{sjn}}^i, i \in \{\mathrm{O},\mathrm{P}\}$. In contrast, $Q_2^i, i \in \{\mathrm{O},\mathrm{P}\}$ describes the second order moment of the noise and interference-only part $x_{\mathrm{jn}}^i, i \in \{\mathrm{O},\mathrm{P}\}$.

Based on PG, separate statements on the detection performance can be made for each modulation scheme, i.e., a low modulation specific PG indicates an increased error probability

[2] It is assumed that the pulse of duration $T_{\mathrm{p}} \leq \frac{T_{\mathrm{b}}}{2}$ occurs at the beginning of an interval of duration $\frac{T_{\mathrm{b}}}{2}$. Hence, the position of a pulse within the interval is perfectly known.

and vice versa. Hence, the smaller Q_1^i and $Q_2^i, i \in \{\mathrm{O}, \mathrm{P}\}$ the lower the modulation related error detection probability. In the following, Q_1^i and $Q_2^i, i \in \{\mathrm{O}, \mathrm{P}\}$ are determined for both modulation schemes.

OOK: For OOK the second order moment of the signal-noise and signal-interference part $x_{\mathrm{sjn}}^{\mathrm{O}}$ can be formulated as ($\tau = t_1 - t_2$)

$$Q_1^{\mathrm{O}} = \int_{-t_1}^{T_\mathrm{P}-t_1} \int_0^{T_\mathrm{P}} \left(R_\mathrm{J}(\tau) + R_\mathrm{N}(\tau)\right) \cdot p(t_1) \cdot p(t_1+\tau)\,\mathrm{d}t_1 \mathrm{d}\tau. \tag{70}$$

A general solution of Q_1^{O} can be found using Parseval's theorem under the assumptions $2|f_\mathrm{c} + f_\mathrm{J}| \gg B_\mathrm{J}$ and $4f_\mathrm{c} \gg B$. This leads to the closed-form expression

$$\begin{aligned} Q_1^{\mathrm{O}} = {} & E_\mathrm{P}^{\mathrm{O}} P_\mathrm{J} \sum_{n=0}^{\infty} (-1)^n \left(2\pi T_\mathrm{P}\right)^{2n} \left(\frac{r_{2n+1} \sum_{l=0}^{2n+1} u_{n,l}}{2\pi f_\mathrm{c}(2n+1)} \right. \\ & \left. - \frac{2\pi T_\mathrm{P} r_{2n+2} \sum_{l=0}^{2n+2} v_{n,l}}{2\pi f_\mathrm{c}(2n+2)} + \frac{4T_\mathrm{P} r_{2n+1}}{(2n+2)!\,(2n+1)} \right) \\ & + \frac{E_\mathrm{P}^{\mathrm{O}} P_\mathrm{N}}{2\pi f_\mathrm{c}} \sum_{n=0}^{\infty} \frac{(-1)^n \left(\pi T_\mathrm{P} B_\mathrm{T}\right)^{2n}}{(2n+1)} \left(\frac{8\pi T_\mathrm{P} f_\mathrm{c}}{(2n+2)!} + (1+ \right. \\ & \left. \cos\left(4\pi f_\mathrm{c} T_\mathrm{P}\right)) \sum_{l=0}^{2n+1} w_{n,l} - \sin\left(4\pi f_\mathrm{c} T_\mathrm{P}\right) \sum_{l=0}^{2n+1} z_{n,l} \right), \end{aligned} \tag{71}$$

whereas, with $\Delta_{f_{\mathrm{c,J}}} = f_\mathrm{c} - f_\mathrm{J}$, the following notations are used:

$$\begin{aligned} r_\nu &= \frac{1}{B_\mathrm{J}} \left(\left(\frac{B_\mathrm{J}}{2} + \Delta_{f_{\mathrm{c,J}}} \right)^\nu - \left(-\frac{B_\mathrm{J}}{2} + \Delta_{f_{\mathrm{c,J}}} \right)^\nu \right), \\ w_{n,l} &= \frac{\sin\left(4\pi f_\mathrm{c} T_\mathrm{P} + \frac{1}{2} l\pi\right)}{(2n+1-l)!\,\left(4\pi f_\mathrm{c} T_\mathrm{P}\right)^l}, \\ u_{n,l} &= w_{n,l} + \frac{(-1)^l}{(2n+1-l)!} \sum_{k=0}^{l} \frac{\sin\left(4\pi f_\mathrm{c} T_\mathrm{P} + \frac{1}{2} k\pi\right)}{(l-k)!\,\left(4\pi f_\mathrm{c} T_\mathrm{P}\right)^k}, \\ z_{n,l} &= \frac{\cos\left(4\pi f_\mathrm{c} T_\mathrm{P} + \frac{1}{2} l\pi\right)}{(2n+1-l)!\,\left(4\pi f_\mathrm{c} T_\mathrm{P}\right)^l}, \\ v_{n,l} &= \frac{z_{n,l}}{(2n+2-l)} - \frac{(-1)^l}{(2n+2-l)!} \sum_{k=0}^{l} \frac{\cos\left(4\pi f_\mathrm{c} T_\mathrm{P} + \frac{1}{2} k\pi\right)}{(l-k)!\,\left(4\pi f_\mathrm{c} T_\mathrm{P}\right)^k}. \end{aligned}$$

Eq. (71) shows that Q_1^{O} is depending from the system parameters $E_\mathrm{P}^{\mathrm{O}}$, T_P, f_c, B as well as from the interference parameters P_J, B_J, f_J. In addition, concerning the special case $B_\mathrm{J} \to 0$, e.g., a cosine tone, r_ν has to be replaced with $r_\nu^m = \lim_{B_\mathrm{J} \to 0} r_\nu = \nu \Delta_{f_{\mathrm{c,J}}}^{\nu-1}$. Note that this result is consistent to [9] if $P_N = 0$. In this case (71) simplifies to

$$Q_1^O = \frac{E_p^O P_J}{T_p \pi^2 \left(f_c^2 - f_J^2\right)^2} \left[f_c^2 + 3f_J^2 + \left(f_J^2 - f_c^2\right) \cos\left(4\pi f_c T_p\right) - 2f_J \left(f_J + f_c\right) \cos\left(2\pi \left(f_c - f_J\right) T_p\right) - 2f_J \left(f_J - f_c\right) \cos\left(2\pi \left(f_c + f_J\right) T_p\right)\right]. \tag{72}$$

The second order moment of the noise and interference-only part x_{jn}^O can be generally described as

$$Q_2^O = \int_{-t_1}^{T_p - t_1} \int_0^{T_p} \left[\left(P_N + P_J\right)^2 + 2\left(R_N(\tau) + R_J(\tau)\right)^2\right] dt_1 d\tau. \tag{73}$$

Thereby, using the theorem of Price [44], (73) can be written in terms of the noise and interference related autocorrelation functions. With Parseval and the assumptions $2f_c \gg B$, $2f_J \gg B_J$ and $|f_c + f_J| \gg (B_J \text{ or } (B - B_J))$ (73) results in

$$Q_2^O = 2T_p^2 \left[P_J^2 + P_J P_N + P_N^2 + \sum_{k=1}^{\infty} \frac{(-1)^k (2\pi T_p)^{2k} \left(P_J^2 B_J^{2k} + P_N^2 B_T^{2k}\right)}{(2k+1)!\,(2k+1)(k+1)} + \sum_{k=2}^{\infty} \frac{(-1)^k (2\pi T_p)^{2k-2} \left(P_J^2 B_J^{2k-2} + P_N^2 B_T^{2k-2}\right)}{(k)\,(2k)!}\right] + \frac{2P_J P_N}{\pi B \left(f_p - f_m\right)} \sum_{k=0}^{\infty} \frac{(-1)^k (2\pi)^{2k+1} T_p^{2k+2}}{(2k+2)!} \cdot \left(\frac{f_1(k)}{2k+2} + \frac{f_2(k)}{2k+1}\right). \tag{74}$$

Thereby, with $f_p = \frac{B}{2} + \frac{B_J}{2}$ and $f_m = \frac{B}{2} - \frac{B_J}{2}$, $f_1(k)$ and $f_2(k)$ are defined as:

$$f_1(k) = \left(-f_m - \Delta_{f_{c,J}}\right)^{2k+2} - \left(-f_p - \Delta_{f_{c,J}}\right)^{2k+2} + \left(-f_m + \Delta_{f_{c,J}}\right)^{2k+2} - \left(-f_p + \Delta_{f_{c,J}}\right)^{2k+2}$$

and

$$f_2(k) = \left(f_p + \Delta_{f_{c,J}}\right)\left[\left(-f_m - \Delta_{f_{c,J}}\right)^{2k+1} - \left(-f_p - \Delta_{f_{c,J}}\right)^{2k+1}\right] + \left(f_p - \Delta_{f_{c,J}}\right)\left[\left(-f_m + \Delta_{f_{c,J}}\right)^{2k+1} - \left(-f_p + \Delta_{f_{c,J}}\right)^{2k+1}\right] + \left(f_p - f_m\right)\left[\left(f_m - \Delta_{f_{c,J}}\right)^{2k+1} - \left(-f_m - \Delta_{f_{c,J}}\right)^{2k+1}\right].$$

Q_2^O is influenced by the system parameters T_p, f_c, B as well as by the interference parameters P_J, B_J, f_J. However, in contrast to the second order moment Q_1^O it cannot be reduced via E_p^O. Eq. (74) simplifies for $B_J \rightarrow 0$ due to $P_J^2 B_J^{2k} = P_J^2 B_J^{2k-2} = 0$, $\frac{f_1(k)}{f_p - f_m} = (2k+2)\left[\left(-\frac{B}{2} - \Delta f_{c,J}\right)^{2k+1} + \left(-\frac{B}{2} + \Delta f_{c,J}\right)^{2k+1}\right]$ as well as $\frac{f_2(k)}{f_p - f_m} = (2k+1)\left(\frac{B}{2} + \Delta f_{c,J}\right)\left(-\frac{B}{2} - \Delta f_{c,J}\right)^{2k}$. Assuming $P_N = 0$ (74) equals the result of [9]:

$$Q_2^O = 2T_p^2 P_J^2 + \frac{P_J^2}{8\pi^2 f_J^2}\left[1 - \cos\left(4\pi f_J T_p\right)\right]. \tag{75}$$

BPPM: Considering BPPM the second order moment of the signal-noise and signal-interference part x_{sjn}^P is: $Q_1^P = \frac{1}{2} Q_1^O$. Q_1^P differs from Q_1^O solely in a factor of two which can be ascribed to the reduced modulation specific pulse energy. In contrast to Q_1^P there is a significant difference concerning the second order moment of the noise and interference-only part x_{jn}^P. With the theorem of Price this can be generally described in terms of the noise and interference specific autocorrelation functions: ($\tau = t_1 - t_2$)

$$\begin{aligned} Q_2^P = {} & 4 \int_{-t_1}^{T_p - t_1} \int_{0}^{T_p} \left[R_J^2(\tau) + R_N^2(\tau) + R_J(\tau) R_N(\tau)\right] \mathrm{d}t_1 \mathrm{d}\tau \\ & - 4 \int_{-t_1}^{T_p - t_1} \int_{\frac{T_b}{2}}^{T_p + \frac{T_b}{2}} \left[R_J^2(\tau) + R_N^2(\tau) + R_J(\tau) R_N(\tau)\right] \mathrm{d}t_1 \mathrm{d}\tau. \end{aligned} \tag{76}$$

Therefore, using the theorem of Parseval for $2f_J \gg B_J$, the closed-form result

$$\begin{aligned} Q_2^P = {} & 2\sum_{k=1}^{\infty} \frac{(-1)^k (2\pi)^{2k} \left(P_J^2 B_J^{2k} + P_N^2 B_T^{2k}\right) g_{2k+2}}{(2k+1)!\,(2k+1)\,(k+1)} \\ & + \sum_{k=2}^{\infty} \frac{(-1)^k 2^{2k} (\pi)^{2k-2} \left(P_J^2 B_J^{2k-2} + P_N^2 B_T^{2k-2}\right) g_{2k}}{(2k)!\,(2k)} \\ & + \frac{2P_J P_N}{\pi B_T (f_p - f_m)} \sum_{k=0}^{\infty} \frac{(-1)^k (2\pi)^{2k+1} g_{2k+2}}{(2k+2)!} \cdot \left(\frac{f_1(k)}{2k+2} + \frac{f_2(k)}{2k+1}\right) \end{aligned} \tag{77}$$

with

$$g_\nu = 2T_p^\nu - \left(T_p - \frac{T_b}{2}\right)^\nu + 2\left(\frac{T_b}{2}\right)^\nu - \left(T_p + \frac{T_b}{2}\right)^\nu$$

can be found. Q_2^P is influenced by the system parameters T_p, T_b, f_c, B as well as by the interference parameters P_J, B_J, f_J. Similar to Q_2^O it cannot be reduced via E_p^P. Eq. (77) reveals that for low data rates ($T_b \rightarrow \infty$) $g_\nu \approx 2T_p^\nu$ resulting in a negligible influence of Q_2^P. In contrast, the larger the data rate the higher its impact, e.g., for $T_b = 2T_p$ g_ν conducts to $g_\nu = 4T_p^\nu - (2T_p)^\nu$. In addition, for $B_J \rightarrow 0$ and $P_N = 0$, (77) allows the same simplifications as for

Q_2^{O}. In this case (77) conducts to

$$Q_2^{\mathrm{P}} = \frac{P_{\mathrm{J}}^2}{8\pi^2 f_{\mathrm{J}}^2} \left[2 - 2\cos\left(2\pi f_{\mathrm{J}} T_{\mathrm{b}}\right) + \cos\left(2\pi f_{\mathrm{J}}\left(T_{\mathrm{b}} - 2T_{\mathrm{p}}\right)\right) - 2\cos\left(4\pi f_{\mathrm{J}} T_{\mathrm{p}}\right) + \cos\left(2\pi f_{\mathrm{J}}\left(T_{\mathrm{b}} + 2T_{\mathrm{p}}\right)\right)\right]. \tag{78}$$

4.1.3. Results

Based on the previous analysis the interference robustness of an OOK/BPPM based energy detection receiver can be identified. Thereby, assuming regulation of ECC [14] an MIR-UWB system with four subbands of equal bandwidth $B = 625\,\mathrm{MHz}$ is taken into account. Without loss of generality, the analysis focuses solely on the first subband located at $f_{\mathrm{c}} = 6.3125\,\mathrm{GHz}$. However, an extension to other subbands or other MIR-UWB system configurations, which are possibly based on other frequency masks is easily possible. Further common system parameters used in the following are the pulse duration $T_{\mathrm{p}} = 3.2\,\mathrm{ns}$, a mean transmit power normalized to one, the modulation specific pulse energy $E_{\mathrm{p}}^i, i \in \{\mathrm{O}, \mathrm{P}\}$ as well as a constant $\mathrm{SNR_E} = 10\,\mathrm{dB}$ at the input of the energy detector. Fixed interference parameter is the interference specific bit duration $T_{\mathrm{b,J}} = 16T_{\mathrm{b}} = 102.4\,\mathrm{ns}$.

In Fig. 18 (a) the PG is plotted vs. the $\mathrm{SINR_E}$ for the duty cycle $d_{\mathrm{s}} = \frac{1}{2}$. An interference with the two bandwidths $B_{\mathrm{J,1}} = 20\,\mathrm{MHz}$ or $B_{\mathrm{J,2}} = 400\,\mathrm{MHz}$ located at $f_{\mathrm{J}} = f_{\mathrm{c}} + 50\,\mathrm{MHz}$ is considered leading to the fixed duty cycles $d_{\mathrm{J,1}} = \frac{1}{B_{\mathrm{J,1}} T_{\mathrm{b,J}}} = 0.4883$ or $d_{\mathrm{J,2}} = \frac{1}{B_{\mathrm{J,2}} T_{\mathrm{b,J}}} = 0.0244$. For OOK/BPPM, the PG increases with higher $\mathrm{SINR_E}$ up to the interference-free PG at $\mathrm{SNR_E} = 10\,\mathrm{dB}$. Furthermore, it can be observed that the OOK/BPPM based PG varies with the interference bandwidth. For OOK, the PG increases with a larger interference bandwidth because of the minor impact of the mixed signal-interference as well as the interference-only component involved in the energy detection. A PG of energy detection can be achieved from an $\mathrm{SINR_E} = -3.5\,\mathrm{dB}$ ($B_{\mathrm{J,1}} = 20\,\mathrm{MHz}$) and from an $\mathrm{SINR_E} = -5.5\,\mathrm{dB}$ ($B_{\mathrm{J,2}} = 400\,\mathrm{MHz}$), respectively. For strong interference no PG results as the energy detector's decision variable is significantly corrupted. In contrast, for BPPM a PG can be achieved for small interference bandwidths, e.g., $B_{\mathrm{J,1}} = 20\,\mathrm{MHz}$, over nearly the complete $\mathrm{SINR_E}$ range. For $B_{\mathrm{J,2}} = 400\,\mathrm{MHz}$ a PG occurs from $\mathrm{SINR_E} = -2\,\mathrm{dB}$. The reason for this behaviour lies in a different amount of energy resulting from the mixed signal-interference and interference-only term within the two observation periods of duration T_{p}. Finally, the consideration of OOK/BPPM with respect to their relative PG shows that for strong NBI BPPM is more robust whereas OOK is more robust for mean and low interference.

The detector efficiency in terms of PG can be increased by increasing T_{b}, e.g., with multiples of T_{p}. This is illustrated in Fig. 18 (b) for $d_{\mathrm{s}} = \frac{1}{4}$. The enlargement of T_{b} via d_{s} results in an increase of E_{b} for fixed signal power. Thereby, the interference related second order moments Q_1^i and $Q_2^i, i \in \{\mathrm{O}, \mathrm{P}\}$ will be reduced as the interferer's energy is only collected during integration time T_{p} within T_{b}. Larger T_{b} can be implemented into an MIR-UWB transmitter with minor complexity. However, the trade-off to increase the detector's interference robustness is a reduction of data rate.

Fig. 19 (a) shows the PG of an OOK and BPPM specific energy detection vs. f_{J} for $d_{\mathrm{s}} = \frac{1}{2}$, which varies from $f_{\mathrm{c}} - \frac{B}{2}$ to $f_{\mathrm{c}} + \frac{B}{2}$ for fixed $\mathrm{SINR_E} = 0\,\mathrm{dB}$, $\mathrm{SNR_E} = 10\,\mathrm{dB}$, $T_{\mathrm{b,J}} = 102.4\,\mathrm{ns}$ as

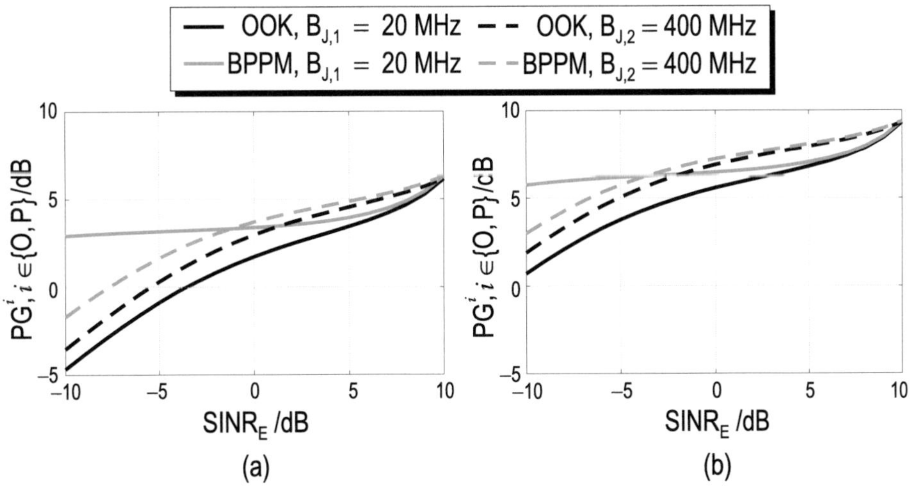

Figure 18. Processing gain of an OOK and BPPM specific energy detection vs. SINR$_E$ with $B_{J,1} = 20\,\text{MHz}$ and $B_{J,2} = 400\,\text{MHz}$, $f_J = f_c + 50\,\text{MHz}$, $\text{SNR}_E = 10\,\text{dB}$, $T_{b,J} = 102.4\,\text{ns}$, $d_J\,(B_{J,1}) = 0.4883$ and $d_J\,(B_{J,2}) = 0.0244$ for $T_b = 2T_p$ (a) and $T_b = 4T_p$ (b).

well as for $B_{J,1} = 20\,\text{MHz}$ and $B_{J,2} = 400\,\text{MHz}$. As long as the interference is completely inside the subband $T_{p,J}$, d_J and hence P_J are fix. In particular, the modulation specific PG at $f_J = f_c + 50\,\text{MHz}$ coincides with the one of Fig. 18. In addition, both modulation schemes show an increase of PG the more the interference is located at the subband's boundary ($f_J = f_c \pm \frac{1}{T_p}$). This can be on one hand ascribed to the subband pulse's sinc spectrum which is zero at the subband's boundary. On the other hand, the more f_J is located at the subband's boundary the minor the interference bandwidth falling into the subband. In case interference overlaps with the subband's boundary, the effective interference parameters B_J, f_J and d_J changes resulting in a reduction of the actual interference power P_J. Fig. 19 (b) shows again that a significant increase of the modulation specific PG can be achieved for $d_s = \frac{1}{4}$. However, without proper pulse shaping the increase depends strongly on the interferer's position inside the subband.

4.2. Coexistence-based approaches

The MIR-UWB system has no exclusive frequency range within the available transmission bandwidth. For this reason there is an increased interference potential from a possible large number of radio systems operating in the same frequency domain. As shown in [12] the impact of interference can result in a significant decrease of the bit error rate (BER) performance of the MIR-UWB system. To maintain system performance, the following section gives a short overview of various coexistence-based approaches with respect to an efficient and easy-to-realise interference mitigation.

Coexistence-based approaches aim at the reliable on-line mitigation of interferences which occur in the environment of the MIR-UWB system. Thus, a best possible trade-off between a maximum data rate and a minimum BER can be obtained for arbitrary interference situations. In this context, an essential requirement is the integration of coexistence-based approaches into the existing MIR-UWB system with only minor complexity increase. Thus, the MIR-UWB system configuration should not be changed in presence of interference on one hand; on

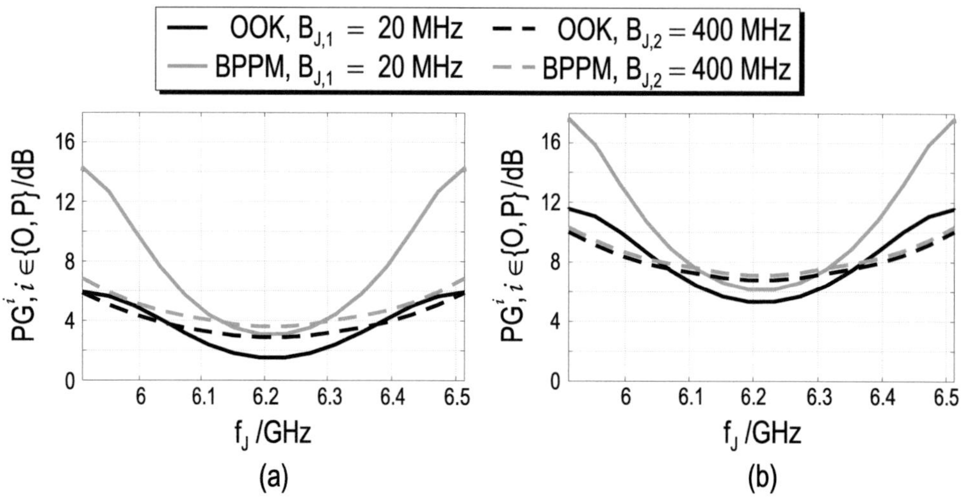

Figure 19. Processing gain of an OOK and BPPM specific energy detection vs. f_J for $\mathrm{SINR_E} = 0\,\mathrm{dB}$, $\mathrm{SNR} = 10\,\mathrm{dB}$, $B_{J,1} = 20\,\mathrm{MHz}$, $B_{J,2} = 400\,\mathrm{MHz}$ and $T_{b,J} = 102.4\,\mathrm{ns}$ for $T_b = 2T_p$ (a) as well as $T_b = 4T_p$ (b).

the other hand it should be possible to realise coexistence without complex estimations of interference specific parameters such as the interference power, the interference bandwidth, the interference carrier frequency or the number of instantaneous available interferences.

4.2.1. Static coexistence approach

From a complexity point of view a static coexistence approach should be used. Thereby interfered subbands are deactivated by means of the system specific band plan before any data transmission occurs. Best trade-off between system performance and effort for interference handling will be achieved. However, this approach does not consider dynamic interference situations and hence does not contribute to efficient spectrum usage. Therefrom, the necessity for further efficient and low complex, but more flexible alternatives arises.

4.2.2. Detect and Avoid (DAA)

In [6, 34] an easy-to-realise DAA approach is presented allowing a reliable detection of temporary NBI after system initialisation or within data transmission.

For this purpose, the regularly transmitted preamble is adjusted to simultaneous subband specific signal and noise energy estimation used by the DAA approach. Thereby, a static interference-free working point (WP) is defined, ensuring a determinated BER in each subband. After initialisation, the estimated SNRs are compared with the WP, leading to subband deactivation, if the SNR is lower than the WP. Otherwise, the subband is (re)enabled for data transmission.

During data transmission, the initially estimated signal and noise energy values of enabled subbands are recursively updated in dependence of the actual subband specific bit decision to adapt the initial decision threshold. In addition, this process allows the possibility of fast and reliable detection of suddenly weak or strong interfered subbands. For this reason, a characteristic weighting factor k is used to manipulate the instantaneous energy value's influence on the recursive estimation. Simulation results [6] show a robust interference detection with only a marginal BER performance loss for $k = 10$.

4.2.3. Image-based thresholding

In [10], a simple cluster-based coexistence approach (coexistence approach 1) is analysed resulting in a decision threshold ϵ_{th} with respect to interfered or not interfered noise energy values. Based on the image-based thresholding method of [41] occupied time-frequency slots can be automatically detected and recorded within an extended time-frequency band plan used for initialisation or data transmission.

Assuming knowledge of the interferers' periodicity and perfect synchronisation, m_0 binary zeros are sent over each subband $i \in \{1, \ldots, N_{\text{sub}}\}$. The measured energy values are written as an energy matrix $\underline{X} = \left[x_{j,i}\right], j = 1, \ldots, m_0, i = 1, \ldots, N_{\text{sub}}$ and assigned to their nearest quantized energy levels $E_{\min} = \epsilon_1 < \epsilon_2 < \ldots < \epsilon_U = E_{\max}$. Thereby, $E_{\min}$ and $E_{\max}$ stands for the minimum and the maximum occurring energy value in $\underline{X}$. The allocation leads to an energy distribution, which is described as

$$p_i = \frac{n_i}{m_0 N_{\text{sub}}}, i \in 1, \ldots, U. \tag{79}$$

To obtain a separation between interfered and not interfered energy values, two energy classes $C_0 = \{\epsilon_1, \ldots, \epsilon_u\}$ and $C_1 = \{\epsilon_{u+1}, \ldots, \epsilon_U\}$ can be defined, which include interfered and not interfered energy values (Fig. 20 (a))

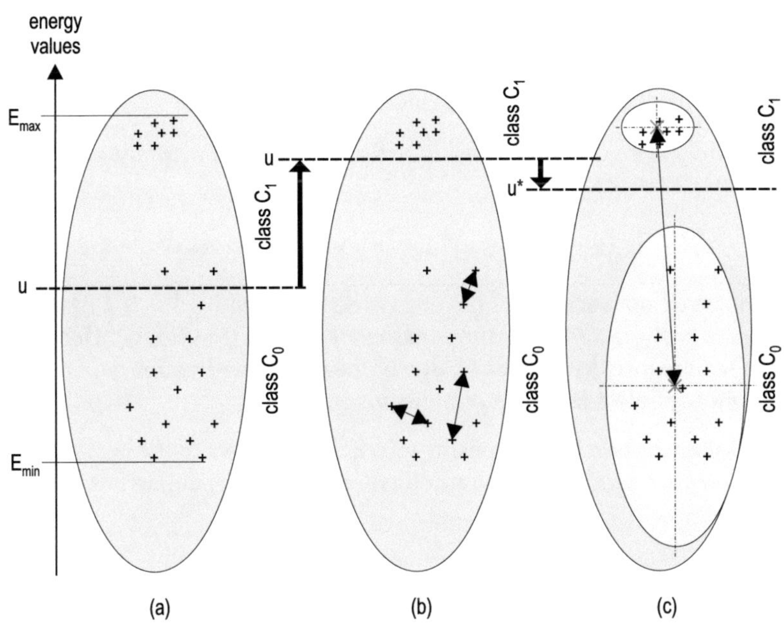

Figure 20. Energy classes, threshold determination. Sub-figure (a) shows the average of all values, sub-figure (b) optimizes according to eq. (81) and sub-figure (c) according to eq. (84).

In order to classify energy values, index u has to be determined using two optimisation criteria.

The first optimisation criterion consists of the minimisation of the combined empirical energy class variance

$$s_{\mathrm{w}}^2(u) = s_{\mathrm{C}_0}^2(u)\, P_{\mathrm{C}_0}(u) + s_{\mathrm{C}_1}^2(u)\, P_{\mathrm{C}_1}(u)\,. \tag{80}$$

This step aims at an adjustment of the initially set index u, whereas a correct allocation of energy outliers to the corresponding other class is achieved (Fig. 20 (b)). It is based on a weighted sum of the classes' occurrence probabilities

$$P_{\mathrm{C}_0}(u) = \sum_{i=1}^{u} p_i, \qquad P_{\mathrm{C}_1}(u) = 1 - P_{\mathrm{C}_0}(u)\,, \tag{81}$$

as well as on the empirical energy classes' variances

$$s_{\mathrm{C}_l}^2(u) = \begin{cases} \sum\limits_{i=1}^{u} \left(\epsilon_i - \bar{x}_{\mathrm{C}_l}(u)\right)^2 \frac{p_i}{P_{\mathrm{C}_l}(u)}, & l = 0 \\ \sum\limits_{i=u+1}^{U} \left(\epsilon_i - \bar{x}_{\mathrm{C}_l}(u)\right)^2 \frac{p_i}{P_{\mathrm{C}_l}(u)}, & l = 1 \end{cases} \tag{82}$$

with

$$\bar{x}_{\mathrm{C}_l}(u) = \begin{cases} \sum\limits_{i=1}^{u} \frac{\epsilon_i p_i}{P_{\mathrm{C}_l}(u)}, & l = 0 \\ \sum\limits_{i=u+1}^{U} \frac{\epsilon_i p_i}{P_{\mathrm{C}_l}(u)}, & l = 1 \end{cases} \tag{83}$$

describing the empirical energy class mean levels.

The second optimisation criterion considers the maximisation of the empirical variance between both energy classes

$$s_{\mathrm{b}}^2(u) = \left(\bar{x}_{\mathrm{C}_0}(u) - \bar{x}_{tot}\right)^2 P_{\mathrm{C}_0}(u) + \left(\bar{x}_{\mathrm{C}_1}(u) - \bar{x}_{tot}\right)^2 P_{\mathrm{C}_1}(u) \tag{84}$$

standing for the weighted variance of the energy class means $\bar{x}_{\mathrm{C}_i}, i = 0, 1$ themselves around the total mean $\bar{x}_{\mathrm{tot}} = \bar{x}_{\mathrm{C}_0} + \bar{x}_{\mathrm{C}_1}$ of the time-frequency pattern (Fig. 20 (c)). Hence, a separation of both classes with respect to the mean value of the total time-frequency pattern is obtained, leading to a more accurate adaptation of index u.

As both optimisation criteria have opposing effects with respect to the best index $u^\star$, they are combined into one characteristic optimisation criterion which is defined as [41]:

$$u^\star = \arg \max_{u=1,\ldots,U} \frac{s_{\mathrm{b}}^2(u)}{s_{\mathrm{w}}^2(u)}. \tag{85}$$

This leads to an adjustment of the initially arbitrary index u. Thereby, a correct allocation of energy outliers to the corresponding energy classes is achieved. Evaluated index $u^\star$ leads to interference detection threshold $\epsilon_{\mathrm{th}} = \epsilon_{u^\star}$. Afterwards a binary decision has to be done for all received energy values $x_{j,i}, j = 1, \ldots, m_0, i = 1, \ldots, N_{\text{sub}}$, which are logged in band plan $\underline{L}$. In this context, we define an energy cell as interfered if $x_{j,i} \geq \epsilon_{\text{th}}$.

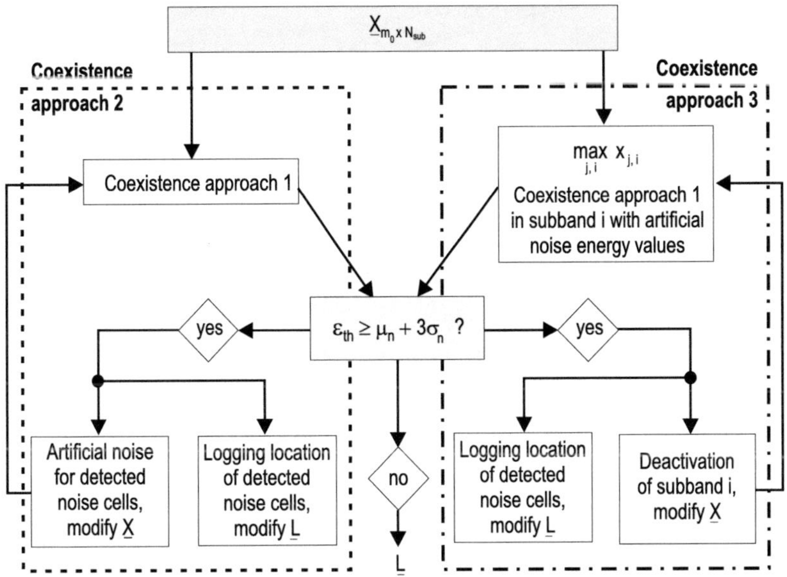

Figure 21. Schematic procedure of global iterative method (coexistence approach 2) and local hierarchical method (coexistence approach 3).

Since the main drawback of the global coexistence approach is its reduced efficiency in case of simultaneously operating broad- or narrowband interference, it should be improved by using iterative coexistence methods [12]. In a global iterative method (coexistence approach 2, left side of Fig. 21) the plausibility of the resulting interference detection threshold is verified via the $3\sigma_{\text{n}}$ standard deviation termination criterion

$$\epsilon_{\text{th}} \stackrel{?}{\geq} \mu_{\text{n}} + 3\sigma_{\text{n}} \tag{86}$$

for which Gaussian distribution as well as an interference-free noise source $n(t)$ with mean μ_{n} and standard deviation σ_{n} is assumed at the receiver side. If ϵ_{th} exceeds the confidence interval, deactivation flags are logged within the band plan.

Simultaneously, all labelled possibly corrupted noise energy values are replaced with artificially generated noise energy values resulting from the noise source mentioned above. This procedure is iteratively repeated until the termination criterion is achieved. In this case, the final iterative band plan is delivered and used for initialisation and data transmission.

In contrast, a local hierarchical iterative method (coexistence approach 3, right side of Fig. 21) handles every subband individually to achieve a local based interference detection threshold. The subband with maximum occurring energy value is identified and delivered to receiver side interference detection in conjunction with a sufficient number of additional artificial noise energy values of the above mentioned noise source. For the detection of the coexistence approach's termination, the $3\sigma_n$ standard deviation is used again. If the interference threshold lies above the confidence interval, detected noise cells of the considered subband are labelled

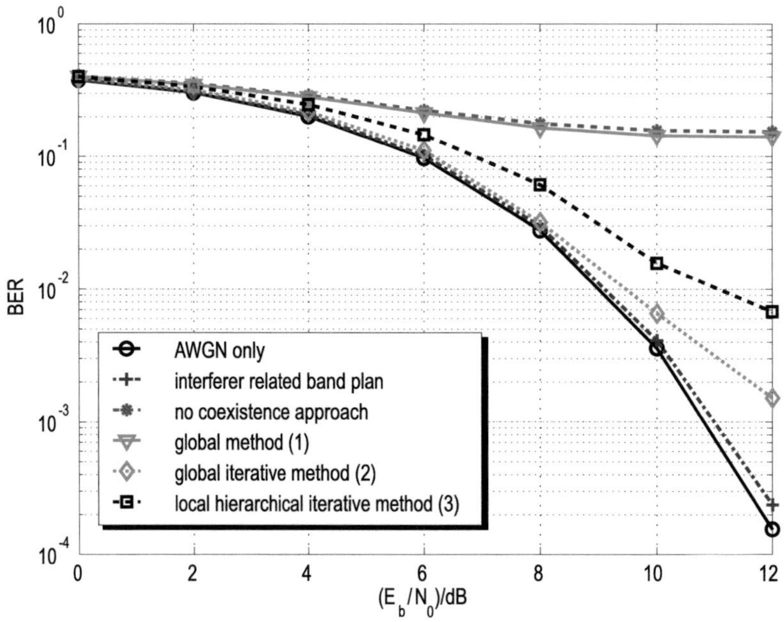

Figure 22. BER vs. E_b/N_0 for AWGN and fixed SIRs of -10 dB (IEEE 802.11a WLAN [47]), 0 dB and 5 dB (IEEE 802.15.3a MB OFDM UWB 1 and 2 [2]) regarding an interferer related band plan, no coexistence approach and the coexistence approaches 1 to 3 [12].

within the band plan. In parallel, the corresponding subband is deactivated and the procedure is iteratively repeated. If the coexistence approach terminates, the binary band plan is allocated to the transmitter for initialisation and data transmission.

Performance analysis [12] demonstrates the high capability of both iterative coexistence methods in presence of multiple interferers having the same or different interference powers (Fig. 22).

4.3. Narrowband interference mitigation

The MIR-UWB system is characterised by a particular high vulnerability to NBI as all interferences inside the passband of the analogue front-end are considered by the energy detection receiver. Its performance can decrease significantly resulting in an increase of the error probability. Hence, a crucial issue concerns the efficient mitigation of NBI [58].

This section analyses the potential of the non-linear Teager-Kaiser (TK) operation [22] to mitigate NBI without the knowledge of the interference related carrier frequency. Based on the definition of the TK operation and some of its most important properties the mitigation potential of a TK based energy detector is analysed. In this context a modified TK operation is introduced and compared with the traditional TK operation [8]. The analysis of the traditional and the modified TK operation considers first one narrowband signal in the baseband. Finally, the analysis is extended to the bandpass domain for one and multiple NBI [5]. Thereby, the potential to integrate the TK operation into the existing MIR-UWB system is discussed for one NBI. Based on the proposal of [42] it is shown that the integration can be realised with only minor complexity increase.

4.3.1. *Teager-Kaiser operation*

The continuous TK operation is a non-linear differential operator of order two defined as

$$\Psi\left(x\left(t\right)\right) = \dot{x}^2\left(t\right) - x\left(t\right)\ddot{x}\left(t\right). \tag{87}$$

To illustrate the effectiveness of the TK operation the harmonic oscillation $x\left(t\right) = A\cos\left(\omega_0 t + \phi_0\right)$ is considered. Using (87) the signal at the output of the TK operation can be described as

$$\Psi\left(x\left(t\right)\right) = A^2\omega_0^2\left(\sin^2\left(\omega_0 t + \phi_0\right) + \cos^2\left(\omega_0 t + \phi_0\right)\right) = A^2\omega_0^2.$$

Hence, for the special case of a simple harmonic oscillation the TK operation leads to a frequency shift to DC. Fig. 23 highlights the spectrum of a harmonic oscillation (a) as well as the resulting spectrum at the output of TK operation (b).

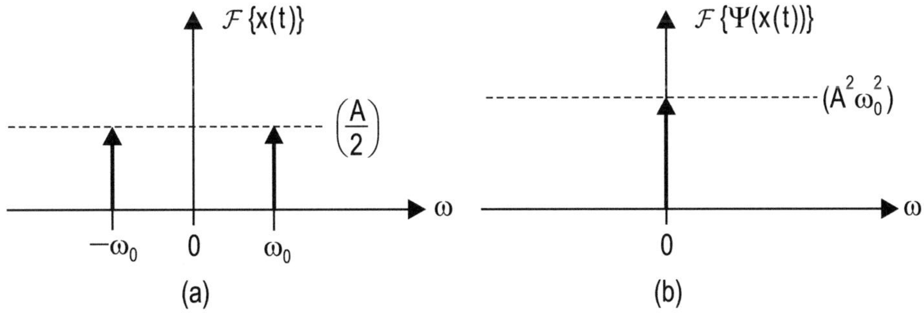

Figure 23. Effectiveness of TK operation for the harmonic oscillation $x\left(t\right) = A\cos\left(\omega_0 t + \phi_0\right)$.

An interpretation of this behaviour is given in [22]. In case of a harmonic oscillation the output of the TK operation describes the required energy to generate the oscillation. Considering (88) the energy depends not only by the amplitude A but also by the oscillation frequency ω_0. Thus, for a constant A the required energy to generate an exemplary 10 Hz signal is lower than the one of an 1000 Hz signal.

When using the continuous TK operation of (87) basic operator specific rules have to be generally considered. A detailed description can be found in [23]. In context with interference mitigation the most important property can be the behaviour of the TK operation in presence of several overlapping signals. In case of K_S overlapping signals $x_1\left(t\right)$, $x_2\left(t\right)$, ..., $x_{K_S}\left(t\right)$ the resulting signal at the output of TK operation conducts to

$$\Psi\left(\sum_{i=1}^{K_S} x_i\left(t\right)\right) = \sum_{i=1}^{K_S}\Psi\left(x_i\left(t\right)\right) + \sum_{j=1}^{K_S-1}\sum_{i=j+1}^{K_S}\Psi_c\left(x_j\left(t\right), x_i\left(t\right)\right). \tag{88}$$

It consists on one hand of K_S summands describing the TK operation of the signals $x_1\left(t\right)$, $x_2\left(t\right)$, ..., $x_{K_S}\left(t\right)$. Furthermore, due to the non-linearity of the TK operation the additional cross component

$$\begin{aligned}\Psi_c\left(x_j\left(t\right), x_i\left(t\right)\right) &= \Psi_{c_j}\left(x_j\left(t\right), x_i\left(t\right)\right) + \Psi_{c_i}\left(x_i\left(t\right), x_j\left(t\right)\right) \\ &= 2\dot{x}_j\left(t\right)\dot{x}_i\left(t\right) - x_j\left(t\right)\ddot{x}_i\left(t\right) - x_i\left(t\right)\ddot{x}_j\left(t\right)\end{aligned} \tag{89}$$

occurs. This component considers the mutual influence of the two signals $x_j(t)$ and $x_i(t)$. It is composed of the generally non-symmetric signal parts $\Psi_{c_1}(x_1(t), x_2(t)) = \dot{x}_1(t)\dot{x}_2(t) - x_1(t)\ddot{x}_2(t)$ and $\Psi_{c_2}(x_2(t), x_1(t)) = \dot{x}_2(t)\dot{x}_1(t) - x_2(t)\ddot{x}_1(t)$. From (89) two special cases can be immediately concluded. Firstly, the cross component $\Psi_c(x(t), x(t)) = 2\Psi(x(t))$ if $x(t)$ is overlapped with itself. Secondly, $\Psi_c(a, x(t)) = -a\ddot{x}(t)$. The cross component of a constant a and a signal $x(t)$ can be expressed as the product of a with the second derivative of $x(t)$. In particular, the cross component disappears completely, if $a = 0$.

The TK operation of (87) can be modified with the weighting factor $k \neq 0$ as follows [8]:

$$\Psi_k(x(t)) = k\dot{x}(t)^2 - x(t)\ddot{x}(t). \tag{90}$$

This definition contains the traditional TK operation if $k = 1$. Thereby, the Fourier transform of $\Psi_k(x(t))$ of a signal $x(t)$ is generally given as

$$\mathcal{F}\{\Psi_k(x(t))\} = 4\pi^2 \left(X(f) * f^2 X(f) - kfX(f) * fX(f)\right). \tag{91}$$

Considering again the special case of a harmonic oscillation $x(t) = A\cos(\omega_0 t)$ the Fourier transform conducts to

$$\mathcal{F}\{\Psi_k(x(t))\} = \frac{1}{2}A^2(2\pi f_0)^2 \left((k+1)\delta(f) + \frac{1}{2}(1-k)(\delta(f+2f_0) + \delta(f-2f_0))\right). \tag{92}$$

The effectiveness of the modified TK operation is highlighted in Fig. 24.

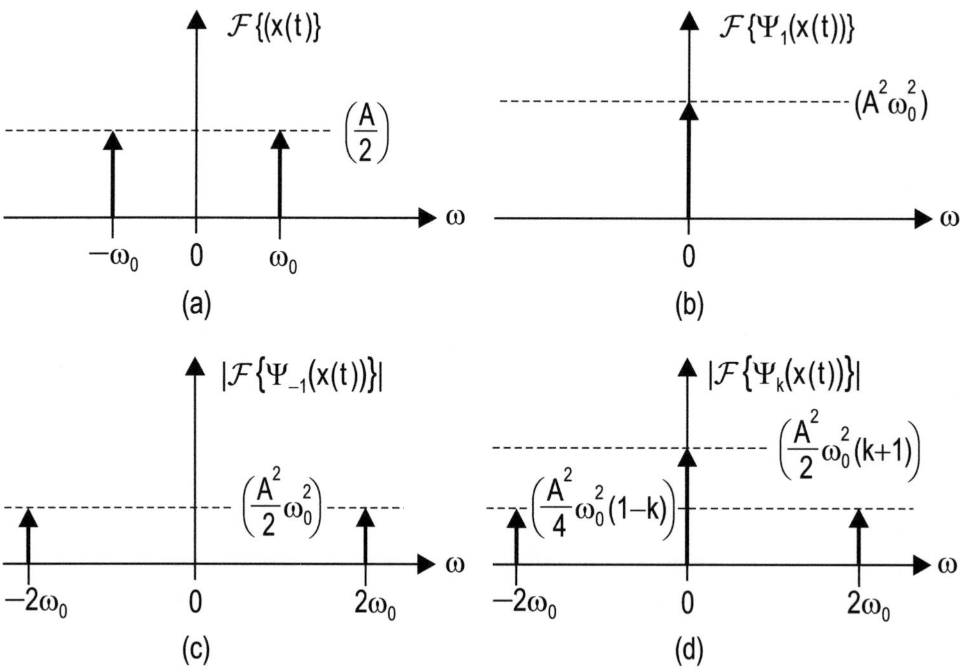

Figure 24. Effectiveness of the traditional and the modified TK operation for $x(t) = A\cos(\omega_0 t)$.

In contrast to the traditional TK operation (Fig. 24 (b)) additional spectral parts at $2\omega_0$ can be observed for $k \neq 1$ (Fig. 24 (d)). For the special case $k = -1$ the complete energy is even shifted to this frequency (Fig. 24 (c)). Hence, it can be concluded that a modified TK operation operates not only as DC frequency shifter.

Due to the modification of the TK operation the property of overlapping signals changes. If two signals $x_1(t)$ and $x_2(t)$ occur at the input of the TK operation its output conducts to [8]:

$$\Psi_k(x_1(t) + x_2(t)) = \Psi_k(x_1(t)) + \Psi_k(x_2(t)) + \Psi_k^c(x_1(t), x_2(t)), \tag{93}$$

where

$$\Psi_k^c(x_1(t), x_2(t)) = 2k\dot{x}_1(t)\dot{x}_2(t) - x_1(t)\ddot{x}_2(t) - x_2(t)\ddot{x}_1(t) \tag{94}$$

stands for the modified cross component of $x_1(t)$ and $x_2(t)$.

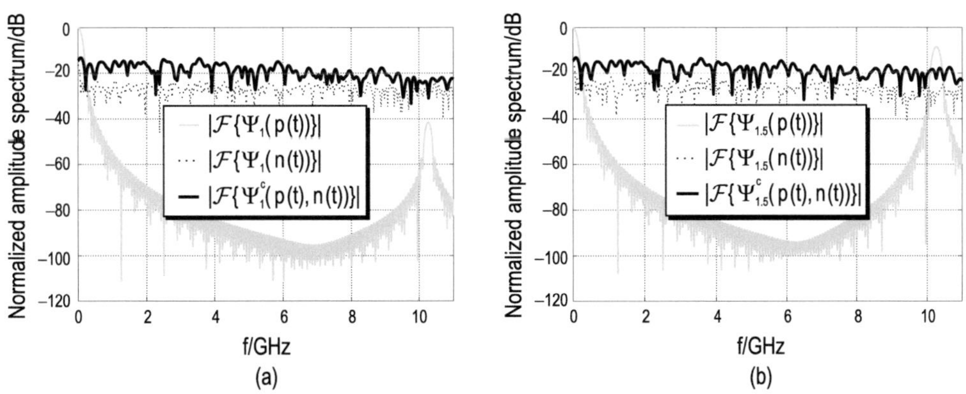

Figure 25. Normalized amplitude spectrum at the output of the traditional TK operation for $k = 1$ (a) and $k = 1.5$ (b), $E_b/N_0 = 12$ dB.

Fig. 25 shows the difference of the traditional (a) and the modified TK operation (b) for OOK in case of a binary one. Thereby, Fig. 25 (a) highlights the normalized amplitude spectrum of the occurring components for the interference-free AWGN case for $k = 1$ and $E_b/N_0 = 12$ dB. The considered pulse is a cosine-shaped pulse with pulse duration $T = 12.8$ ns and carrier frequency $f_c = 5.13$ GHz [8]. Obviously, the TK operation mixes a large signal part of the pulse to around DC. However, a minor contribution around $2f_c$ can also be observed. As this contribution is much lower than the spectral contributions of $\Psi_1(n(t))$ and $\Psi_1^c(p(t), n(t))$ a possible detection may be limited to the frequency range around DC.

In contrast, Fig. 25 (b) illustrates that the modification of the TK operation with, e.g. $k = 1.5$, has the potential to improve the detection for AWGN. Also in this case large pulse related spectral contributions around DC occur. Furthermore, a significant increase of ≈ 30 dB of the pulse's amplitude spectrum $\mathcal{F}\{\Psi_{1,5}(p(t))\}$ occurs around $2f_c$. The resulting amplitude spectrum of the pulse is both around DC and around $2f_c$ larger than the contributions of $\Psi_{1.5}(n(t))$ and $\Psi_{1.5}^c(p(t), n(t))$. Hence, an improved detection performance can be expected for AWGN if frequencies around DC as well as around $2f_c$ are considered.

The reason for this behaviour can be ascribed to the modified TK operation of the pulse. Assuming negligible influences of the first and the second derivative of $p(t)$ at $|t| = \frac{T}{2}$ the output of the TK operation on $p(t)$ can be described as [8]:

$$\begin{aligned}\Psi_k(p(t)) = \frac{2}{3T}\Big[&0.5(k+1)\left(\omega_\mathrm{T}^2+3\omega_\mathrm{c}^2\right)\\ &-0.25(k-1)(\omega_\mathrm{T}-\omega_\mathrm{c})^2\cos\left(2(\omega_\mathrm{c}-\omega_\mathrm{T})t\right)\\ &+0.5\left((k+1)\omega_\mathrm{T}^2-3(k-1)\omega_\mathrm{c}^2\right)\cos(2\omega_\mathrm{c}t)\\ &+\left(0.5\omega_\mathrm{T}^2+(k-1)\left(\omega_\mathrm{c}\omega_\mathrm{T}-\omega_\mathrm{c}^2\right)\right)\cos\left((2\omega_\mathrm{c}-\omega_\mathrm{T})t\right)\\ &+\left(0.5\omega_\mathrm{T}^2-(k-1)\left(\omega_\mathrm{c}\omega_\mathrm{T}+\omega_\mathrm{c}^2\right)\right)\cos\left((2\omega_\mathrm{c}+\omega_\mathrm{T})t\right)\\ &-0.25(k-1)(\omega_\mathrm{T}+\omega_\mathrm{c})^2\cos\left(2(\omega_\mathrm{c}+\omega_\mathrm{T})t\right)\\ &+0.5\left(\omega_\mathrm{c}^2(k+1)-\omega_\mathrm{T}^2(k-1)\right)\cos(2\omega_\mathrm{T}t)\\ &+\left(2\omega_\mathrm{c}^2(k+1)+\omega_\mathrm{T}^2\right)\cos(\omega_\mathrm{T}t)\Big] \qquad (95)\end{aligned}$$

with $\omega_\mathrm{T} = \frac{2\pi}{\mathrm{T}}$ and $\omega_\mathrm{c} = 2\pi f_\mathrm{c}$. Obviously, a modification of the TK operation leads to additional spectral parts at its output. Hence, it is possible to influence the distribution of pulse energy via a simple weighting of the TK operation. It should be noted that the traditional as well as the modified TK operation acts not only as a frequency-to-DC shifter due to additional spectral parts around $2f_\mathrm{c}$. For the traditional TK operation ($k = 1$) the two spectral components at $2(\omega_\mathrm{c}-\omega_\mathrm{T})$ and $2(\omega_\mathrm{c}+\omega_\mathrm{T})$ disappear completely resulting in a lower energy concentration around $2f_\mathrm{c}$.

4.3.2. Mitigation potential of Teager-Kaiser operation

Assuming negligible noise this section analyses the potential of the traditional and the modified TK operation to mitigate NBI. For this the TK operation's effectiveness is first described in the baseband. Afterwards it is extended to the bandpass domain.

Baseband Domain

To make statements on the effectiveness of the TK operation for a narrowband signal we first consider a narrowband baseband signal of bandwidth $2\omega_N$ which can be modelled as

$$j(t) = \sum_{i=1}^{N} A_i \sin(\omega_i t + \phi_i). \qquad (96)$$

It consists of N overlapping sinusoids of amplitudes $A_i, i = 1,\ldots,N$, of phases $\phi_i, i = 1,\ldots,N$ as well as of frequencies $\omega_i = 2\pi f_i, i = 1,\ldots,N$. Given to the modified TK operation its output can be described for $\phi_i = 0, i = 1,\ldots,N$ as[3]

$$\Psi_k(j(t)) = C_k(t) + \sum_{i=1}^{N}\sum_{j>i}^{N} z_i^j \cos\left(\left(\omega_i-\omega_j\right)t\right) + z_j^i \cos\left(\left(\omega_i+\omega_j\right)t\right), \qquad (97)$$

[3] The TK operation is generally characterised by a strong phase dependency. To simplify the descriptions we only consider the special case $\phi_i = 0, i = 1,\ldots,N$ in the following.

with

$$C_k(t) = \frac{1}{2}\sum_{i=1}^{N} A_i^2\omega_i^2\left[(k+1)+(k-1)\cos\left(2\omega_i t\right)\right],$$

$$z_i^j = kA_iA_j\omega_i\omega_j + \frac{1}{2}A_iA_j\omega_i^2 + \frac{1}{2}A_iA_j\omega_j^2$$

and

$$z_j^i = kA_iA_j\omega_i\omega_j - \frac{1}{2}A_iA_j\omega_i^2 - \frac{1}{2}A_iA_j\omega_j^2.$$

For the traditional TK operation ($k=1$) the resulting signal is composed of the DC component $C_1 = \sum_{i=1}^{N} A_i^2\omega_i^2$. In addition, further components around $\omega_i - \omega_j$ and $\omega_i + \omega_j$ occur. Its quantity depends on the two factors $z_i^j = 0,5A_iA_j\left(\omega_i+\omega_j\right)^2$ and $z_j^i = -0,5A_iA_j\left(\omega_i-\omega_j\right)^2$. Thereby, a higher energy concentration can be observed for lower frequencies. In contrast, the modified TK operation ($k \neq 1$) shows spectral contributions around DC, around $\omega_i+\omega_j$ and $\omega_i-\omega_j$ as well as around $2\omega_i, i = 1,\ldots,N$. In particular, the contributions at $\omega_i+\omega_j$ and $\omega_i-\omega_j$ are influenced with larger weighting factors z_i^j and z_j^i. Hence, using the modified TK operation a high energy concentration still occurs for low frequencies. However, due to the additional spectral components at $2\omega_i, i = 1,\ldots,N$ the relative difference of energy concentration reduces.

Bandpass Domain

NBI influencing the MIR-UWB system operates in the bandpass domain. Based on the insight for a baseband signal an analytical description of the effectiveness of the traditional and the modified TK operation is done for one or more narrowband bandpass signals. Due to analytical tractability it is assumed that the lth bandpass signal can be described as

$$\begin{aligned} j_l(t) &= 2A_l\sum_{i=1}^{N_l}\sin\left(\omega_{l,i}t\right)\cos\left(\omega_{c_l}t\right) \\ &= \underbrace{A_l\sum_{i=1}^{N_l}\sin\left(\left(\omega_{c_l}+\omega_{l,i}\right)t\right)}_{\alpha_l(t)} - \underbrace{A_l\sum_{i=1}^{N_l}\sin\left(\left(\omega_{c_l}-\omega_{l,i}\right)t\right)}_{\beta_l(t)}. \end{aligned} \tag{98}$$

The bandpass signal $j_l(t)$ of amplitude $2A_l$ consists of N_l sinusoids of frequencies $\omega_{l,1} < \omega_{l,2} < \ldots < \omega_{l,N_l}, \omega_{l,i} = 2\pi f_{l,i}, i = 1,\ldots,N_l$ located around the carrier frequency $\omega_{c_l} = 2\pi f_{c_l}$ with the bandwidth $B_l = 2\omega_{N_l}$.

If K_S bandpass signals $j_l(t), l = 1,\ldots,K_S$ are present at the input of TK operation its output can be described as

$$\Psi_k\left(\sum_{l=1}^{K_S} j_l(t)\right) = \sum_{l=1}^{K_S}\Psi_k\left(j_l(t)\right) + \Psi_{k,m}^{c}(t). \tag{99}$$

It consists of the components $\Psi_k\left(j_l(t)\right) = \Psi_k\left(\alpha_l(t)\right) + \Psi_k\left(\beta_l(t)\right) + \Psi_k^{c}\left(\alpha_l(t), \beta_l(t)\right), l = 1,\ldots,K_S$ which always occur in presence of one bandpass signal. Assuming $\omega_{c_l} \gg B_l$ the lth bandpass signal $j_l(t), l = 1, \ldots, K_S$ results in [8]

$$\begin{aligned}
&\Psi_k(j_l(t)) \approx \\
&\quad A_l^2 \omega_{c_l}^2 \sum_{u=1}^{N_l} \left[(k+1) + 0.5(k-1) \cdot \left[\cos\left(2\left(\omega_{c_l} - \omega_{l,u}\right)t\right) + \cos\left(2\left(\omega_{c_l} + \omega_{l,u}\right)t\right)\right]\right] \\
&\quad + A_l^2 \omega_{c_l}^2 \sum_{u=1}^{N_l} \sum_{v>u}^{N_l} \left[2(k+1)\cos\left(\left(\omega_{l,u} - \omega_{l,v}\right)t\right) + (k-1)\cos\left(\left(2\omega_{c_l} + \omega_{l,u} + \omega_{l,v}\right)t\right)\right. \\
&\qquad \left. + (k-1)\cos\left(\left(2\omega_{c_l} - \omega_{l,u} - \omega_{l,v}\right)t\right)\right] \\
&\quad - A_l^2 \omega_{c_l}^2 \sum_{u=1}^{N_l} \sum_{v=1}^{N_l} \left[(k+1)\cos\left(\left(\omega_{l,u} + \omega_{l,v}\right)t\right) - (k-1)\cos\left(\left(2\omega_{c_l} + \omega_{l,u} - \omega_{l,v}\right)t\right)\right]
\end{aligned} \quad (100)$$

This result reveals that the traditional TK operation ($k = 1$) acts as a frequency-to-DC shifter for each bandpass signal $j_l(t), l = 1, \ldots, K_S$. In this case the corresponding spectral range goes from DC to the largest occurring bandwidth $B_l, l = 1, \ldots, K_S$ of the K_S bandpass signals. For the modified TK operation additional spectral components around $2\omega_{c,l}, l = 1, \ldots, K_S$ occur. For this reason a mitigation of NBI by the modified TK operation is critical. Finally, for the special case $k = -1$ the complete energy is shifted to $2\omega_{c,l}, l = 1, \ldots, K_S$. This confirms the statement that energy parts can be shifted between frequency ranges with a modified TK operation.

If the output of the traditional or the modified TK operation would only consist of components from $\Psi_k(j_l(t)), l = 1, \ldots, K_S$ a mitigation of the K_S bandpass signal could be possible as $B_T \gg B_l, l = 1, \ldots, K_S$. However, as can be seen in (99), the additional signal component

$$\Psi_{k,m}^{c}(t) = \sum_{r=1}^{K_S-1} \sum_{l=r+1}^{K_S} \left[\Psi_k^{c}\left(\alpha_r(t), \alpha_l(t)\right) + \Psi_k^{c}\left(\beta_r(t), \beta_l(t)\right)\right] + \sum_{r=1}^{K_S} \sum_{l \neq r} \Psi_k^{c}\left(\alpha_r(t), \beta_l(t)\right) \quad (101)$$

occurs in case of at least two bandpass signals. The component $\Psi_{k,m}^{c}(t)$ describes the cross components between different bandpass signals $j_l(t)$ and $j_r(t), l \neq r$. Thereby, assuming $\omega_{c_l} \gg B_l$ and $\omega_{c_r} \gg B_r$ the two cross components can be described as

$$\begin{aligned}
&\Psi_k^{c}\left(\alpha_r(t), \alpha_l(t)\right) + \Psi_k^{c}\left(\beta_r(t), \beta_l(t)\right) \approx \\
&\sum_{u=1}^{N_r} \sum_{v=1}^{N_l} Z_{1,k}\left[\cos\left(\left(\omega_{c_r} - \omega_{c_l} + \omega_{r,u} - \omega_{l,v}\right)t\right) + \cos\left(\left(\omega_{c_r} - \omega_{c_l} - \omega_{r,u} + \omega_{l,v}\right)t\right)\right] \\
&\qquad + Z_{2,k}\left[\cos\left(\left(\omega_{c_r} + \omega_{c_l} + \omega_{r,u} + \omega_{l,v}\right)t\right) + \cos\left(\left(\omega_{c_r} + \omega_{c_l} - \omega_{r,u} - \omega_{l,v}\right)t\right)\right]
\end{aligned} \quad (102)$$

with the amplitudes

$$Z_{1,k} = \frac{A_r A_l}{2}\left(\omega_{c_r}^2 + 2k\omega_{c_r}\omega_{c_l} + \omega_{c_l}^2\right)$$

and

$$Z_{2,k} = -\frac{A_r A_l}{2}\left(\omega_{c_r}^2 - 2k\omega_{c_r}\omega_{c_l} + \omega_{c_l}^2\right).$$

Finally, the third cross component $\Psi_k^c(\alpha_r(t), \beta_l(t))$ is given for $r \neq l$ as

$$\Psi_k^c(\alpha_r(t), \beta_l(t)) = \Psi_{k,r>l}^c(\alpha_r(t), \beta_l(t)) + \Psi_{k,l>r}^c(\alpha_r(t), \beta_l(t))$$

$$\approx \sum_{u-1}^{N_r} \sum_{v-1}^{N_l} -Z_{1,k}\left[\cos\left(\left(\omega_{c_r} - \omega_{c_l} - \omega_{r,u} - \omega_{l,v}\right)t\right) + \cos\left(\left(\omega_{c_r} - \omega_{c_l} + \omega_{r,u} + \omega_{l,v}\right)t\right)\right]$$

$$-Z_{2,k}\left[\cos\left(\left(\omega_{c_r} + \omega_{c_l} - \omega_{r,u} + \omega_{l,v}\right)t\right) + \cos\left(\left(\omega_{c_r} + \omega_{c_l} + \omega_{r,u} - \omega_{l,v}\right)t\right)\right] \quad (103)$$

Hence, in presence of more than one bandpass signal additional spectral components occur arround $|\omega_{c_r} - \omega_{c_l}|$ and $\omega_{c_r} + \omega_{c_l}$. The frequency parts depend on the carrier frequencies of the bandpass signals. The spectral components are influenced by the weighting factor k. E.g., for the traditional TK operation ($k = 1$) the spectral component around $|\omega_{c_r} - \omega_{c_l}|$ dominates. In contrast for the modified TK operation ($k \neq 1$) additional relevant spectral components can be identified around $\omega_{c_r} + \omega_{c_l}$ making the usage of the frequency at twice the subband's center frequency $2\omega_c$ critical.

To verify the results a subband of bandwidth 625 MHz and carrier frequency 5.2 GHz is considered for $k = 1$. $K_S = 3$ bandpass signals of amplitudes $A_1 = 1$, $A_2 = 1/3$ and $A_3 = 2/3$, of bandwidths $B_1 = 5$ MHz, $B_2 = 10$ MHz and $B_3 = 1$ MHz as well as of carrier frequencies $f_{c_1} = 4.98$ GHz, $f_{c_2} = 5.04$ GHz and $f_{c_3} = 5.28$ GHz are assumed at the input of TK operation. Fig. 26 shows the positive frequency range for the resulting to one

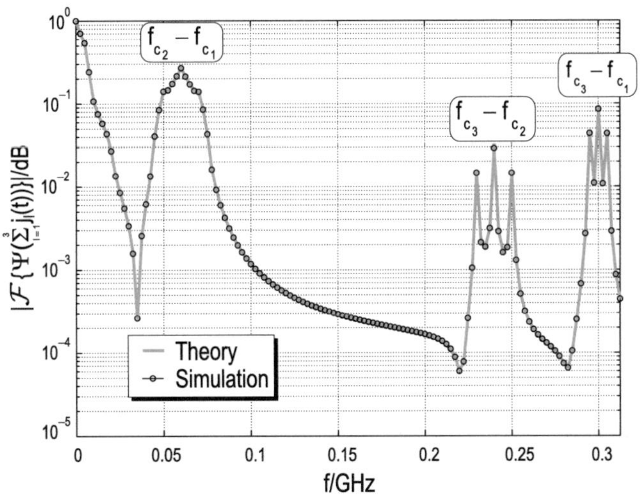

Figure 26. Normalized amplitude spectrum at the output of the traditional TK operation ($k = 1$) in presence of three narrowband bandpass signals.

normalized baseband spectrum at the output of the traditional TK operation. Simulation as well as analytical results show the spectral contributions of the three signals occurring at $f_{c_2} - f_{c_1} = 60$ MHz, $f_{c_3} - f_{c_2} = 240$ MHz and $f_{c_3} - f_{c_1} = 300$ MHz with the bandwidths 30 MHz, 22 MHz and 12 MHz. The spectral components are distributed over the complete bandwidth of the subband. For this reason the consideration of the TK operation with an additional filtering operation is critical. In particular, the mitigation scheme proposed in [42] becomes inefficient to efficiently mitigate all occurring interferences.

4.3.3. Integration of Teager-Kaiser operation

As illustrated in the previous section an integration of TK operation into the MIR-UWB system is possible with only minor complexity increase if at most one NBI occurs in each subband. In this case the approach of [42] can be used. It bases on the interplay of the TK operation with a highpass filtering. As illustrated in Fig. 27 only two additional analogue components have to be integrated into each subband of the existing non-coherent MIR-UWB receiver. Thereby, received subband signals are given to TK operation which acts as a frequency-to-DC shifter. The resulting low-frequency signal is afterwards highpass filtered to mitigate interfered signal components without any a priori information of the interference specific carrier frequency. As the bandwidth of the subband signal is larger than the interference bandwidth energy detection might be possible.

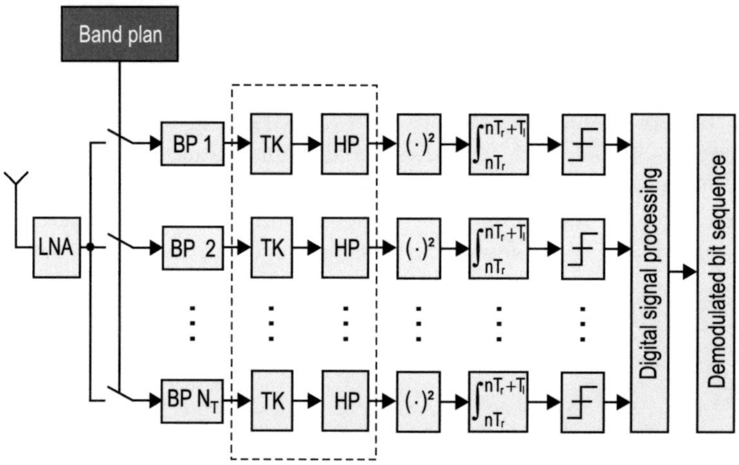

Figure 27. Integration of TK operation into the existing non-coherent MIR-UWB receiver.

In the following the potential of interference mitigation with the TK operation is shown for OOK in case of a binary one. An MIR-UWB subband of carrier frequency 5.13 GHz and effective bandwidth 162.5 MHz is considered for SNR $=$ 11 dB. It is assumed that an IEEE 802.11a WLAN signal [47] of bandwidth 20 MHz and of carrier frequency 5.14 GHz interferes the MIR-UWB subband with an SIR of -5 dB.

Fig. 28 (a) shows the to one normalized amplitude spectrum of all occurring signal components at the output of TK operation. Thereby, the UWB signal spectrum ranges from DC to 162.5 MHz whereas the lower frequency regions have a higher energy concentration. A similar behaviour occurs for the narrowband WLAN signal. Its corresponding amplitude spectrum ranges from DC to 20 MHz whereas energy is strongly distributed around DC. Furthermore, additional spectral cross components between signal, noise and interference occur which can be ascribed to the non-linearity of the TK operation. To mitigate the WLAN signal highpass filtering is done after the TK operation.

Fig. 28 (b) illustrates the to one normalized amplitude spectrum after highpass filtering. The used highpass filter is characterised by the order six, a passband ripple of 0.1 dB, a 50 dB stopband attenuation as well as a 50 MHz wide stopband. Obviously, the narrowband WLAN signal is mitigated after highpass filtering. In contrast the subband signal has an

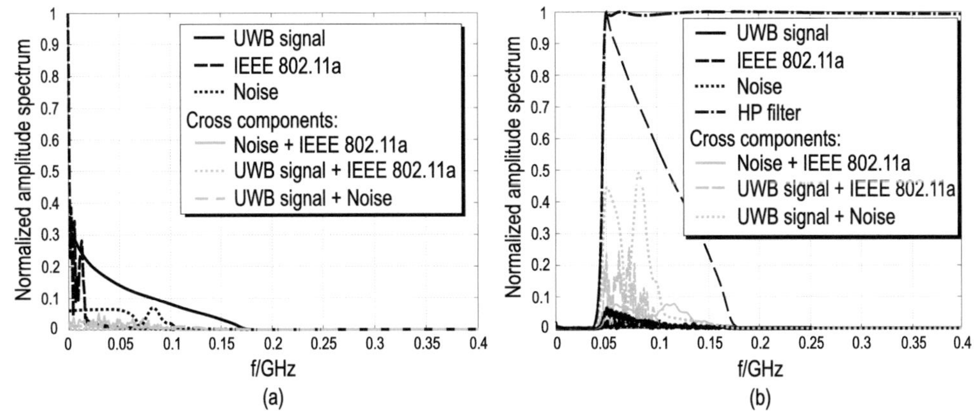

Figure 28. Normalized amplitude spectrum at the output of TK operation (a) and after highpass filtering (b), SNR = 11 dB.

amplitude spectrum which ranges from 50 MHz to 162.5 MHz. As this spectrum dominates the amplitude spectra of the occurring cross components energy detection is possible.

5. Summary

This chapter deals with an easy-to-realise non-coherent MIR-UWB system which is a promising approach for high data rate and energy efficient communication over short distances. Due to its low complexity the MIR-UWB system is an alternative to already existing UWB systems for high data rate applications such as Multiband OFDM UWB.

The MIR-UWB system is based on an energy detection receiver. Thus the first part of this chapter deals with the performance of this component. To understand the energy detection receiver we look at the bit and symbol error probability in different wireless channels.

First we introduce a closed form expression of the SEP for an energy detection receiver with M-PAM in the AWGN channel. Based on this result, we optimise the interval thresholds to minimise the SEP. Optimal interval thresholds guarantee a minimal SEP for M-PAM. In the next step we look into the optimal amplitudes for M-PAM using an energy detection receiver. This approach enables to reduce the SEP for M-PAM with medium to large degrees of freedom.

To understand the characteristics of the energy detection receiver in fading channels we look into different approaches to model the energy at the receiver. It has been shown, that the flat fading channel model can be used to model the energy at a receiver for a receiver bandwidth $B > 100$ MHz. Based on this assumption we introduce closed form expressions for the SEP of the energy detection receiver with M-PAM for different fading statistics such as *Rayleigh*, *Rice* and *Nakagami-m*. We also analyse the SEP of an multichannel receiver using different combining techniques. Square Law Combing and Square Law Selection are possible combining schemes for an energy detection receiver. A closed form solution for SLC and SLS is introduced for the AWGN and for the *Rayleigh* fading channel including i.i.d. and correlated fading gains.

The first part ends with the analysis of the SEP in a frequency selective fading channel. Based on *Rayleigh* distributed fading gains, representing a non line-of-sight channel (NLOS), we introduce a closed form expression for the energy detection receiver with M-PAM. The result also contains the possibility to analyse the effect of correlated fading gains. This is the case in typical UWB wireless channels. If the fading gains are not *Rayleigh* distributed, we present a numerical solution for any fading distributions. The results of the first part enable a precise prediction of an energy detection receiver with M-PAM in many different scenarios.

Since the MIR-UWB system is highly sensitive to interference, the second part of the chapter considers three different aspects regarding an efficient interference mitigation.

The first aspect deals with the analysis of the interference robustness of an OOK and BPPM specific energy detection receiver being the essential component of the non-coherent MIR-UWB receiver. Thereby, taking into account thermal noise a general frame work is presented which can be used to give statements on the detector's interference robustness for an interference with arbitrary bandwidth. Furthermore, possible system parameters can be identified to increase the detector's interference robustness.

The second aspect considers the coexistence capability of the MIR-UWB system. Thereby, various easy-to-realise adaptive coexistence-based approaches are presented. Starting with a static coexistence approach a DAA coexistence approach for temporary NBI is presented being integrated into the system specific initialisation and data phase. The proposed method allows a reliable adaptive mitigation of temporary NBI. A further adaptive coexistence approach bases on image-based thresholding which can be integrated into the initialisation phase of the MIR-UWB system. Based on an exemplary interference scenario the potential to efficiently mitigate multiple interferences of different interference power is shown.

Lastly, the third aspect focuses on the analytical investigation of the potential to mitigate NBI inside an energy detection receiver. Hereby, the TK operation as well as a modified TK operation is analysed. It is shown that for a narrowband baseband signal the output of the TK operation is characterised by a larger energy concentration in the lower frequency range. In contrast, for the modified TK operation further spectral components occur for higher frequencies. A subsequent analysis of one NBI in the bandpass domain shows that the TK operation acts like a frequency-to-DC shifter. This reveals the potential to mitigate a single NBI without the knowledge of the NBI's carrier frequency. In contrast, for the modified TK operation additional spectral components at twice the NBI's carrier frequency occur making interference mitigation critical. In case of multiple NBI further spectral components occur at the TK operation's output which can be ascribed to the mutual interference influences. Due to a possible distribution of the spectral components within the total MIR-UWB subband interference mitigation depends on the interference position inside the MIR-UWB subband.

Acknowledgements

This work was supported within the priority program No. 1202 (UKoLoS) by the German Research Foundation (DFG).

Author details

Rainer Moorfeld and Adolf Finger
Communications Laboratory, Dresden University of Technology, Germany

Hanns-Ulrich Dehner, Holger Jäkel, Martin Braun and Friedrich K. Jondral
Communications Engineering Lab, Karlsruhe Institute of Technology (KIT), Germany

6. References

[1] Abramowitz, M. & Stegun, I. [1964]. *Handbook of mathematical functions with formulas, graphs, and mathematical tables*, Applied mathematics series, Dover Publications.

[2] Batra, A. [2008]. *Multiband OFDM Physical Layer Proposal for IEEE 802.15 Task Group 3a*, IEEE P802.15-03/268r1.

[3] Bober, M., Moorfeld, R. & Jorswieck, E. [2011]. Performance of Energy Detection in NLOS frequency-selective Fading Channels, *Proc. of IEEE International Symposium on Personal, Indoor and Mobile Communications (PIMRC)*.

[4] Cassioli, D., Win, M. Z. & Molisch, A. F. [2002]. The ultra-wide bandwidth indoor channel: from statistical model to simulations, *IEEE Journal on Selected Areas in Communications* 20(6): 1247–1257.

[5] Dehner, H., Jäkel, H., Burgkhardt, D. & Jondral, F. K. [2010]. The Teager-Kaiser Energy Operator in Presence of Multiple Narrowband Interference, *IEEE Communications Letters* Vol. 14(No, 8): 716 – 718.

[6] Dehner, H., Jäkel, H., Burgkhardt, D., Jondral, F. K., Moorfeld, R. & Finger, A. [2010]. Treatment of temporary narrowband interference in noncoherent multiband impulse radio UWB, *Proc. of IEEE Mediterranean Electrotechnical Conference*, pp. 1335 – 1339.

[7] Dehner, H., Jäkel, H. & Jondral, F. K. [2011a]. Narrow- and broadband Interference Robustness for OOK/BPPM based Energy Detection, *Proc. of IEEE International Conference on Communications*.

[8] Dehner, H., Jäkel, H. & Jondral, F. K. [2011b]. On the modified Teager-Kaiser energy operator regarding narrowband interference, *Proc. of IEEE Wireless Telecommunications Symposium*.

[9] Dehner, H., Koch, Y., Jäkel, H., Burgkhardt, D., Jondral, F. K., Moorfeld, R. & Finger, A. [2010]. Narrow-band Interference Robustness for Energy Detection in OOK/PPM, *Proc. of IEEE International Conference on Communications*.

[10] Dehner, H., Linde, M., Moorfeld, R., Jäkel, H., Burgkhardt, D., Jondral, F. K. & Finger, A. [2009]. A low complex and efficient coexistence approach for non-coherent multiband impulse radio UWB, *Proc. of IEEE Sarnoff Symposium*.

[11] Dehner, H., Moorfeld, R., Jäkel, H., Burgkhardt, D., Finger, A. & Jondral, F. K. [2009]. Multi-band Impulse Radio – An Alternative Physical Layer for High Data Rate UWB Communication, *Frequenz, Journal of RF-Engineering and Telecommunications* Vol. 63(No, 9-10): 200–204.

[12] Dehner, H., Romero, A., Jäkel, H., Burgkhardt, D., Moorfeld, R., Jondral, F. K. & Finger, A. [2009]. Iterative coexistence approaches for noncoherent multi-band impulse radio UWB, *Proc. of IEEE International Conference on Ultra-Wideband*, pp. 734 – 738.

[13] Digham, F. F., Alouini, M.-S. & Simon, M. K. [2007]. On the Energy Detection of Unknown Signals Over Fading Channels, *IEEE Transactions on Communications* 55(1): 21–24.

[14] ECC [2006]. ECC decision of 24 march 2006 on the harmonised conditions for devices using ultra-wideband (UWB) technology in bands below 10.6 GHz, *Technical Report* .

[15] Eisenacher, M. [2006]. Optimierung von Ultra-Wideband-Signalen (UWB), *Dissertation, Forschungsberichte aus dem Institut für Nachrichtentechnik der Universität Karlsruhe (TH), Band 16, 2006* .

[16] Fisher, R., Kohno, R., Laughlin, M. & Welborn, M. [2005]. *DS-UWB Physical Layer Submission to 802.15 Task Group 3a*, IEEE P802.15-04/0137r1.

[17] Ghassemzadeh, S. S., Jana, R., Rice, C. W., Turin, W. & Tarokh, V. [2004]. Measurement and modeling of an ultra-wide bandwidth indoor channel, *IEEE Transactions on Communications* 52(10): 1786–1796.

[18] Gradshteyn, I. S. & Ryzhik, I. M. [2007]. *Table of Integrals, Series, and Products*, 7 edn, Academic Press.

[19] Hentilä, L., Taparungssanagorn, A., Viittala, H. & Hämäläinen, M. [2005]. Measurement and modelling of an UWB channel at hospital, *Proc. of IEEE International Conference on Ultra-Wideband (ICU)*.

[20] Hovinen, V., Hämäläinen, M. & Pätsi, T. [2002]. Ultra wideband indoor radio channel models: preliminary results, *IEEE Digest of Papers Ultra Wideband Systems and Technologies*.

[21] Johnson, N. L., Kotz, S. & Balakrishnan, N. [1994]. *Continuous Univariate Distributions*, Vol. 1, 2 edn, Wiley.

[22] Kaiser, J. [1990]. On a simple algorithm to calculate the energy of a signal, *Proc. of IEEE International Conference on Acoustics, Speech, and Signal Processing*, pp. 381 – 384.

[23] Kaiser, J. [1993]. Some Useful Properties of Teager's Energy Operators, *Proc. of IEEE International Conference on Acoustics, Speech, and Signal Processing*, pp. 149 – 152.

[24] Karedal, J., Wyne, S., Almers, P., Tufvesson, F. & Molisch, A. F. [2004]. Statistical analysis of the UWB channel in an industrial environment, *Proc. of IEEE Vehicular Technology Conference (VTC-Fall)*.

[25] Kay, S. M. [1998]. *Fundamentals of Statistical Signal Processing: Detection Theory*, Vol. 2, Prentice Hall.

[26] Kunisch, J. & Pamp, J. [2002]. Measurement results and modeling aspects for the UWB radio channel, *IEEE Digest of Papers Ultra Wideband Systems and Technologies*.

[27] Lee, W. C. [1997]. *Mobile Communications Engineering: Theory and Applications*, McGraw-Hill, Inc.

[28] Li, L., Moorfeld, R. & Finger, A. [2011a]. Bit and power loading for the multiband impulse radio UWB architecture, *Proc. of 8th Workshop Positioning Navigation and Communication (WPNC)*.

[29] Li, L., Moorfeld, R. & Finger, A. [2011b]. Closed form expressions for the symbol error rate for non-coherent UWB impulse radio systems with energy combining, *Proc. of 8th Workshop Positioning Navigation and Communication (WPNC)*.

[30] Mittelbach, M., Moorfeld, R. & Finger, A. [2006]. Performance of a Multiband Impulse Radio UWB Architecture, *Proc. of 3rd International conference on mobile technology, applications and systems*.

[31] Molisch, A. F., Cassioli, D., Chong, C.-C., Emami, S., Fort, A., Kannan, B., Karedal, J., Kunisch, J., Schantz, H. G., Siwiak, K. & Win, M. Z. [2006]. A comprehensive standardized model for ultrawideband propagation channels, *IEEE Transactions on Antennas and Propagation* 54(11): 3151–3166.

[32] Molisch, A. F., Foerster, J. R. & Pendergrass, M. [2003]. Channel models for ultrawideband personal area networks, *IEEE Wireless Communications Magazine* 10(6): 14–21.

[33] Moorfeld, R. & Finger, A. [2009]. Multilevel PAM with optimal amplitudes for non-coherent energy detection, *Proc. of Wireless Communications and Signal Processing (WCSP)*.

[34] Moorfeld, R., Finger, A., Dehner, H., Jäkel, H. & Jondral, F. K. [2008]. A simple and fast detect and avoid algorithm for non-coherent multiband impulse radio UWB, *Proc. of IEEE International Symposium on Spread Spectrum Techniques and Applications*, pp. 587 – 591.

[35] Moorfeld, R., Finger, A., Dehner, H.-U., Jäkel, H. & Jondral, F. K. [2009]. Performance of a high flexible non-coherent multiband impulse radio UWB system, *Proc. of 9th IASTED International Conference on Wireless and Optical Communications (WOC)*.

[36] Moorfeld, R., Finger, A. & Zeisberg, S. [2004]. High data rate UWB performance with reduced implementation complexity, *Proc. of IEEE International Symposium on Spread Spectrum Techniques and Applications (ISSSTA)*.

[37] Moorfeld, R., Lu, Y. & Finger, A. [2012]. Energy detection with optimal symbol constellation for M-PAM in UWB fading channels, *Proc. of IEEE International Conference on Ultra-Wideband (ICUWB)*.

[38] Moorfeld, R., Zeisberg, S., Pezzin, M., Rinaldi, N. & Finger, A. [2004]. Ultra-Wideband Impulse Radio (UWB-IR) Algorithm Implementation Complexity and Performance, *Proc. of IST Mobile and Wireless Communications Summit*.

[39] Nuttall, A. H. [1972]. Some integrals involving the Q function, *Technical report*, Naval Underwater Systems Center (NUSC).

[40] Nuttall, A. H. [1975]. Some integrals involving the Q_M function, *IEEE Transactions on Information Theory* 21(1): 95–96.

[41] Otsu, N. [1979]. A threshold selection method from gray-level histograms, *IEEE Transactions on Systems, Man, and Cybernetics* Vol. 20: 62 – 66.

[42] Ozdemir, O., Sahinoglu, Z. & Zhang, J. [2008]. Narrowband Interference Resilient Receiver Design for Unknown UWB Signal Detection, *Proc. of IEEE International Conference on Communications*, pp. 785 – 789.

[43] Pagani, P. & Pajusco, P. [2006]. Characterization and Modeling of Temporal Variations on an Ultrawideband Radio Link, *IEEE Transactions on Antennas and Propagation* 54(11): 3198–3206.

[44] Papoulis, A. [2002]. *Probability, Random Variables and Stochastic Processes*, 4. edn, McGraw-Hill.

[45] Paquelet, S. & Aubert, L. M. [2004]. An energy adaptive demodulation for high data rates with impulse radio, *Proc. of IEEE Radio and Wireless Conference*, pp. 323–326.

[46] Paquelet, S., Aubert, L. M. & Uguen, B. [2004]. An impulse radio asynchronous transceiver for high data rates, *Proc. of Joint Conference on Ultrawideband Systems and Technologies Ultra Wideband Systems (UWBST & IWUWBS)*.

[47] *Part 11: Wireless LAN Medium Access Control (MAC) and Physical Layer (PHY) Specifications: High-speed Phyiscal Layer in the 5 GHz Band* [1999]. IEEE Std. 802.11a-1999.

[48] Press, S. J. [1966]. Linear combinations of non-central chi-square variates, *The Annals of Mathematical Statistics* 37(2): 480–487.

[49] Proakis, J. G. [2001]. *Digital Communications*, 4th edn, McGraw-Hill.

[50] Rappaport, T. S. [2002]. *Wireless Communications: Principles and Practice*, Prentice Hall.

[51] Saleh, A. A. M. & Valenzuela, R. [1987]. A statistical model for indoor multipath propagation, *IEEE Journal on Selected Areas in Communications* 5(2): 128–137.

[52] Simon, M. K. & Alouini, M.-S. [2006]. *Digital Communication over Fading Channels: A Unified Approach to Performance Analysis*, Wiley.

[53] Steiner, C. & Wittneben, A. [2007]. On the interference robustness of ultra-wideband energy detection receivers, *Proc. of IEEE International Conference on Ultra-Wideband*.

[54] Stoica, L. [2008]. *Non-coherent energy detection transceivers for Ultra Wideband Impulse radio systems*, PhD thesis, Centre for Wireless Communications, Oulu, Finland.
[55] Tse, D. & Viswanath, P. [2005]. *Fundamentals of wireless communication*, Cambridge University Press.
[56] Urkowitz, H. [1967]. Energy detection of unknown deterministic signals, *Proc. of the IEEE* 55(4): 523–531.
[57] Weiß, T. [2004]. OFDM-basiertes Spectrum Pooling, *Dissertation, Forschungsberichte aus dem Institut für Nachrichtentechnik der Universität Karlsruhe (TH), Band 13* .
[58] Witrisal, K., Leus, G., Janssen, G., Pausini, M., Troesch, F., Zasowski, T. & Romme, J. [2009]. Noncoherent ultra-wideband systems, *IEEE Signal Processing Magazine* Vol. 26(No, 4): 48 – 66.
[59] Zhang, H., Udagawa, T., Arita, T. & Nakagawa, M. [2002]. A statistical model for the small-scale multipath fading characteristics of ultra wideband indoor channel, *Proc. of IEEE Digest of Papers Ultra Wideband Systems and Technologies*.
[60] Zhao, L. & Haimovich, A.-M. [2002]. Performance of ultra-wideband communications in the presence of interference, *IEEE Journal on Selected Areas in Communications* Vol. 20(No, 9): 1684 – 1691.

Chapter 2

Pulse Rate Control for Low Power and Low Data Rate Ultra Wideband Networks

María Dolores Pérez Guirao

Additional information is available at the end of the chapter

http://dx.doi.org/10.5772/52497

1. Introduction

The growing request for low-to-medium data rate low cost networks is raising the interest in wireless sensor networking. For instance in the field of industrial and logistic applications, in order to improve the processes' efficiency the tight monitoring of goods, tools, and machinery -as to their state and position- is required. In response to this interest, IEEE approved in 2003 the IEEE 802.15.4 standard, this being the first one for low data rate, low complexity, and low power wireless networks.

The market success of wireless sensor networks (WSN) requires inexpensive devices with low power consumption. In order to satisfy this requirement, transmission technology, protocol as well as hardware design must give a common answer. UWB radio, particularly with impulse radio transmission (IR), is especially suitable for the development of WSN. IR-UWB is expected to allow low power, low complexity and low cost implementation as well as centimeter accuracy in ranging. The low complexity and low cost characteristics arise from the essentially baseband nature of the signal transmission. The high ranging accuracy results from the large absolute bandwidth, which must be at least 500 MHz. Indeed, the introduction of ranging functionality in low data rate networks was one of the main reasons for the IEEE 802.15.4a (2007) amendment, which added an IR-UWB physical layer to the original standard.

IEEE 802.15.4a allows for the use of non-coherent receivers, and defines ALOHA[1] as the mandatory medium access control (MAC) protocol. The use of a non-coherent receiver, such as an energy detector, helps to minimise power consumption and reduces implementation complexity. The choice of ALOHA is justified by the potential robustness of IR-UWB to multi-user interference (MUI) and by the low data rate nature of the applications envisioned.

In fact, the design of the MAC layer plays a very important role in order to materialize the benefits of IR-UWB in sensor networks. From a networking perspective, one potential

[1] Random medium access scheme that does not check whether the shared medium is already busy before transmission.

benefit of IR-UWB over narrowband radio technologies is the possibility to allow concurrent transmissions by using different pseudo-random, time hopping codes (THCs) as a multiple access (MA) method. However, TH codes are not perfectly orthogonal, and even if, Multi User Interference (MUI) is still a challenge due to the presence of multipath fading and the asynchronicity between sources. Beyond it, non-coherent receivers are less robust to MUI than coherent receivers; particularly, interference coming from close-by interferers can be very harmful. Thus, and specially if non-coherent receivers are used, additional interference mitigation features at the MAC layer are required.

This work is motivated by the fact that interference management at the MAC layer has not been extensively explored in the context of IR-UWB autonomous networks yet. The chapter is organized as follows. A general introduction into the field of IR-UWB radio technology and its relevant technical fundamentals is given in section 2. Additionally, a short overview into current research activities and basic principles of MAC protocol design for low to medium data rate IR-UWB networks is given. It follows a discussion about the use of game theory as a tool to model and analyse distributed MAC algorithms in wireless networks. The section ends with the description of the investigated scenario and the simulation model.

Section 3 introduces distributed Pulse Rate Control (PRC) as a novel approach for interference mitigation in autonomous IR-UWB networks. PRC enables concurrent transmissions at full power, allowing each source to independently adapt its pulse rate - measured in pulses per second (Pps)- to control the impact of pulse collisions at nearby receivers. This section shows that it is possible to incite autonomous users to decrease their impulsive emissions and thus, prevent network resource break-down. Finally, section 4 summarizes the achievements of the work presented and gives directions for future research.

2. Theoretical background

For the understanding of this chapter it is essential to have a good foundation in IR-UWB technology, as well as a general background knowledge of wireless autonomous networks[2] (AN) and game theory. The purpose of this section is to provide a short overview on these three topics. Furthermore, this section describes the scenario and the simulation model used in the investigations.

2.1. Impulse Radio Ultra Wideband (IR-UWB)

IR-UWB is a form of UWB transmission in which data is transmitted using sequences of extremely short pulses with a duration of less than 1ns and a large pulse repetition period (PRP). Due to the extreme short duration of the pulses, IR-UWB is capable of delivering high data rates in the order of several hundred Mbps, but at the expense of reduced transmission range due to the power restrictions.

Inherent to IR-UWB signaling is a high temporal resolution that enables accurate multipath resolution and ranging capabilities. Other interesting features related to the pulsed nature of IR-UWB are its robustness against fading, as well as its low power, low complexity and low cost implementation possibilities. In this respect, IR-UWB is a key technology for providing wireless networks with joint communication and ranging capabilities.

[2] Wireless autonomous networks can be considered as a special subclass of wireless ad hoc and sensor networks with reinforced self-organizing character

This chapter assumes a generic Time-Hopping IR-UWB (TH-UWB) physical layer as described in [21]. Time is divided into frames of length T_f and each user transmits one pulse of length T_p per frame. Furthermore, by dividing the frames into non-overlapping chips of length T_c, multi-access capability is provided. Each user transmits its pulse in a randomly chosen chip, according to a pseudo-random TH sequence (THS). Data modulation follows a pulse position modulation (PPM) scheme. Thus, the signal emitted by the k-th TH-PPM transmitter consists of a sequence of pulse waveforms shifted to different times.

$$s^{(k)}(t) = \sum_{j=-\infty}^{\infty} w_{tr}(t - jT_f{}^{(k)} - c^{(k)}[j]T_c - \eta b^{(k)}_{\lfloor j|N_s \rfloor} - \tau^{(k)}), \tag{1}$$

A typical expression is given in 1, where $w_{tr}(t)$ represents the transmitted pulse waveform and T_f is the average frame time, which is also denoted as the mean pulse repetition period. The inverse of the mean pulse repetition period is referred to as the mean pulse repetition frequency, $T_f = \frac{1}{prf}$. The expression $b^{(k)}_{\lfloor j|N_s \rfloor}$ represents that each data symbol $b^{(k)}$ can be transmitted by N_s identical pulses to enhance the quality of reception. The symbol duration equals then $T_s = N_s T_f$. The TH code value for pulse j is given by $c^{(k)}[j]$. The constant term η represents the time shift step introduced by the PPM modulator. Usually, this shift is much smaller than the one due to the TH code (T_c). The time shift $\tau^{(k)}$ represents the relative delay time between the instants at which user k and a reference user i start their transmission; it can be considered as a realization of a random process determined by the actions of the users.

Figure 1 illustrates some of the mentioned parameters; in the example: $N_s = 1, c^{(k)} = (2,3,4)$, $b^{(k)} = (1,1,0)$.

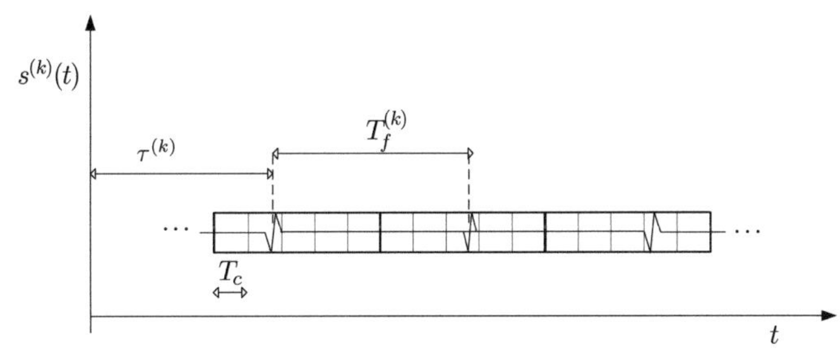

Figure 1. TH-PPM signal structure.

For more detailed information about UWB technology, the author recommends the book [3], which offers an easy-to-read, but complete introduction to the field. Concerning IR-UWB, see [21].

2.2. MAC layer design for low power low data rate IR-UWB networks

The design space for the MAC layer is large; it embraces several dimensions such as multiple access (MA), interference management, resource allocation and power saving. This section summarises the most relevant research findings concerning the design of the MAC layer for low power, low data rate (LDR) IR-UWB networks.

The unique characteristics of the IR-UWB physical layer provide both challenges and opportunities for the MAC layer design. A concrete challenge is the impossibility of carrier sensing, as practised in narrowband systems, since an IR-UWB signal has no carrier. Great opportunities such as robustness against MUI and multipath fading derive from IR-UWB's high temporal resolution. This makes uncoordinated access to the spectrum possible, provided that the local offered load is low compared with the available system bandwidth. For instance, at moderate pulse rates, the "dead time" between pulses allows several uncoordinated, concurrent transmissions to be time interleaved. As a result, ALOHA emerges as the most straightforward MA approach for low data rate IR-UWB networks [3, 9]. The inherent resilience of IR-UWB to MUI can be further increased if different links employ different pseudo-random THC in order to determine the temporal position of the transmitted pulses. The combination of ALOHA with TH coding leads to Time-Hopping Multiple-Access (THMA) [21], which is the most representative MA scheme for low data rate IR-UWB networks.

However TH codes are not perfectly orthogonal, and even if, due to asynchronicity between sources and multipath fading, THMA is still sensitive to impulsive interference. As for CDMA systems, interference coming from a near-by interferer, the so called *near-far* effect, can be very harmful and must be managed in order to avoid performance degradation. For instance, at the physical (PHY) layer multi-user receivers[3] can efficiently address the *near-far* effect at the cost of moderate to high additional hardware [19]. At the link layer, based on the estimation of the wireless link quality, it is possible to adapt the transmission rate to the level of interference experienced at the receiver; the goal can be the improvement of the data rate while satisfying a minimum BER requirement [12].

Within the MAC layer the adaptation of transmission parameters corresponds to a resource allocation task. In the technical literature some approaches on THMA resource allocation can be found, for instance in [18]. A broadly accepted result concerns the optimal power allocation in terms of *proportional fairness*[4]. Note that in autonomous networks with random topology, *proportional fairness* satisfyingly combines fairness and efficiency and outperforms other known performance metrics such as *max-min fairness* or *max total capacity*. This result indicates that the only necessary power control in a IR-UWB network, whose physical layer is in the linear regime[5], is the scheduling function $0 - P_{max}$. In other words, each node should either transmit with full power or not transmit at all [18]. Another common finding is the existence of an *exclusion region* around every destination such that the reference source and nodes within this area cannot transmit simultaneously. This is similar to the IEEE 802.11 CSMA/CA strategy. Certainly, the *exclusion region* vanishes granted that the system has infinite bandwidth ($W \to \infty$). However, in a practical IR-UWB network the bandwidth is always finite and the size of the *exclusion region* may become a critical issue forcing the MAC designer to concentrate on the scheduling task.

It can be concluded that for IR-UWB networks, and when maximizing rates is the design objective, near-far effects should be tackled by combining scheduling and rate adaptation. Particularly, for low power networks the optimal MAC layer design follows an "all at once" scheduling while it adapts the transmission rates to interference [18].

---

[3] Those which can receive on M channels concurrently.

[4] The power allocation that maximises the sum of the log of source rates, which is a concave objective function.

[5] When the rate of a link can be approximated by a linear function of the signal-to-interference-and-noise ratio.

2.3. Basics of game theory

Game theory has been applied in the recent past to model complex interactions among radio devices that have possibly conflicting interests. For the designer of wireless communication systems game theory is a powerful tool to analyse and predict the behaviour of distributed algorithms and protocols. Respected reference books are [7]. A short overview focusing on the application of game theory in the field of wireless communications can be found in [5].

A resource allocation problem can be naturally modeled as a game, in which the players are the radio devices willing to transmit or receive data. In general, there is an interest conflict since the players have to cope with a limited transmission resource such as power, bandwitdh or pulse load. In order to resolve this conflict they can make certain decisions (or take certain actions) such as changing their transmission parameters. The most familiar game form is the *strategic* form, which models a single-shot, simultaneous interaction among players. It is worth to mention that "simultaneous" is not used here in its strict temporal meaning. It does not imply that players have to choose their actions at the same point of time, but much more that no player is aware of the choice of any other player prior to making his own decision.

A *strategic* form game $\Gamma\langle \mathbf{N}, \mathbf{A}, \mathbf{u}\rangle$ is defined by the following elements[6]:

- A set of players, $\mathbf{N} = \{1, 2, ..., |N|\}$ with cardinality $|N|$.
- The possible actions that the players can choose. Assume that player i can choose among $|A_i|$ possible actions (strategies). Then, the set of possible actions of player i is $\mathbf{A_i} = \{a_{i,1}, a_{i,2}, ..., a_{i,|A_i|}\}$, and the game's global action space $\mathbf{A}$ is given by the cartesian product of the action set of each player $\mathbf{A} = A_1 \times A_2 \times ... \times A_{|N|}$. If player i chooses strategy $a_i \in \mathbf{A_i}$ in a game move, the action profile chosen by all players is denoted as the vector $\mathbf{a} = (a_1, a_2, ..., a_{|N|}) = (a_i, a_{-i})$, with $\mathbf{a_{-i}} = (a_1, a_2, ..., a_{i-1}, a_{i+1}, a_{|N|})$ representing the strategies of all players except of player i.
- The utility function, which describes the satisfaction level or payoff of player i given a certain action profile $\mathbf{a}$. The vector of utilities is denoted as $\mathbf{u}(\mathbf{a}) = \{u_1(\mathbf{a}), u_2(\mathbf{a}), ..., u_{|N|}(\mathbf{a})\}$.

To model asynchronous, continuous interactions as those which characterise resource allocation problems in AN the *strategic* form game model is however insufficient. The author agrees with the opinion advanced in [14] and considers *asynchronous, myopic, repeated* games as the most appropriate game model to analyse those problems. At this point it has to be stressed that the work in [14] provides a key result for the application of game theory in the analysis and modeling of AN, since it formalises the relation between the AN's *steady states* and the game behaviour characterised by the Nash equilibria (NE). Theorem 4.1 in [14] proofs that the *steady states* of an AN modeled by an *asynchronous, myopic, repeated* game with stage game $\Gamma\langle \mathbf{N}, \mathbf{A}, \mathbf{u}\rangle$, where all players are rational and act autonomously, coincide with the Nash equilibria of the stage game $\Gamma\langle \mathbf{N}, \mathbf{A}, \mathbf{u}\rangle$.

Notice that an NE is a strategy combination $\mathbf{a}^*$, where no player can improve his utility by individually deviating from its strategy. The existence of an NE for a game significantly

---

[6] We use bold capital letters to denote sets and bold small letters for vectors; for the sake of simplicity we have omitted the superscript T for transposed vectors

depends on the choice of the utility function; specifically on its mathematical properties. The most common existence result for a game's NE is given by the Glicksberg-Fan-Debreu fixed point theorem [7].

2.4. Investigated scenario

This chapter assumes an autonomous sensor, positioning, and identification network (SPIN) as the example system in all investigations. A SPIN is a system characterized by a medium to high node density (up to 2-3 nodes per m^2) in industrial factories or warehouses. Nodes transmit low to medium rate data (up to 1Mbps) combined with position information (position accuracy under 1m) over medium to long distances (typically less than 30 m) to a common receiver.

Concretly, we consider a cluster of up to a hundred IR-UWB sensor nodes that transmit data packets to one common receiver, called the cluster head (CH). The network operation model is based on the beacon-enabled mode of the IEEE 802.15.4a standard [8]. Each sensor node (SN) is considered a source; a link is formed by a transmitting node (source) and the cluster head (CH). Users are asynchronous among themselves. We investigate two scenarios:

- **Scenario 1- Continuous transmission**: Up to 10 UWB sensor nodes are equidistantly situated to the CH, but not necessarily to themselves, along a circle of radius 10 m. They continuously send packets to the CH.
- **Scenario 2- Factory hall**: This scenario accounts for a square simulation field with dimensions 30m×30m. In this field, 100 IR-UWB sensor nodes are considered: 5 are collocated at fixed positions, 75 move along a production line at 5m/s while the rest randomly moves at the same speed within the simulation field. It is assumed that each sensor node generates packets following an exponential distributed packet inter-arrival process and that the exponential processes of the individual sensor nodes are statistically independent. The maximum information data rate per sensor node is 50 Kbps.

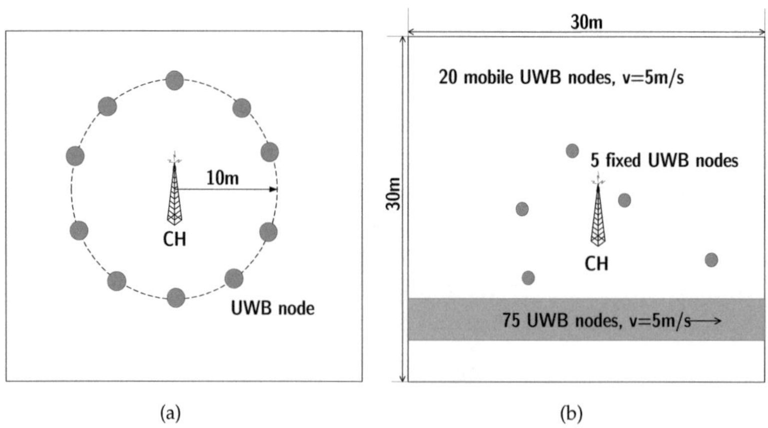

Figure 2. Investigated scenarios: (a) Continuous transmission vs. (b) Factory hall.

2.5. Simulation model

The dynamic simulation model has been developed with the discrete event simulation system OMNeT++ [15]. The CH and each SN comprise a PHY layer, a DLC layer and an application layer instance; network and transport layer operation is transparent. For the air interface the superframe structure described in Section 5.4.1 of the IEEE 802.15.4a standard [8] has been selected.

At the application layer each SN generates data packets according to an exponential distributed packet inter-arrival process. Packets are addressed to the CH; its size has been chosen to be $L_p = 400$ bits. The exponential processes are statistically independent from each other, and a maximum information data rate of 1 Mbps is considered. The DLC layer implementation corresponds with the basic "data transfer model to a coordinator" in the standard IEEE 802.15.4a [see 8, Section 5.5.2.1]. For each DLC packet a packet error rate (PER) is calculated as a function of the received power, interference from concurrent transmissions and thermal noise. At the MAC layer a link adaptation function has been implemented which aims at optimising link/system capacity under several channel and interference conditions. The development and analysis of this function is the main achievement of the work presented in this chapter and is covered in section 3.

It is assumed that the network has a fixed chip duration, T_c, so that all changes at the PHY layer transmission parameters are induced by instructions coming from the MAC/DLC layer. The selected set of PHY layer parameters remains constant for one MAC packet transmission, but can be changed from packet to packet according to the time variant channel and interference conditions.

3. MUI Mitigation by distributed pulse rate control

Interference mitigation is a fundamental problem in wireless networks. In CDMA systems, a well-known technique for this is to control the nodes' transmit powers [22]. The work in [17] has shown that for wireless networks in the *linear regime*, and that allow fine-grained rate adaptation, the optimal power allocation is to let nodes either transmit at full power or do not transmit at all. IR-UWB conforms to both attributes, and thus, according to [17], the MAC layer should concentrate on alternative interference mitigation techniques such as scheduling and rate adaptation.

The term rate adaptation embraces all technical means in a system to adapt the transmission speed (rate) to the current quality of the radio link. In IR-UWB networks, rate control can be achieved by adapting the channel coding rate, the modulation order or the processing gain. In order to adapt these parameters, the link's transmitter must have an estimate of the level of interference at its intended receiver. In autonomous networks, most approaches make use of feedback information from the receiver to the transmitter, for example within ACK packets. This information can take various forms; conventionally, it is a function of the signal-to-noise-plus-interference-ratio (SNIR). However, measuring the SNIR is difficult in practice due to the very low transmit power of UWB signals. Therefore, recent approaches [9] rely on information provided by the channel decoder, namely on BER estimations. This work follows theses approaches and also considers BER instead of SNIR feedbacks.

The processing gain of IR-UWB is twofold: the number of pulses per symbol (N_s) and the average frame duration (T_f). The adaptation of IR-UWB processing gain impacts both rate and average emitted power. Adapting the processing gain in IR-UWB systems was first suggested by Lovelace et. al. in [10]. Lovelace proposed a technique for adjusting the number of pulses per symbol in a single-hop link affected by uncoordinated near pulse interferers. This technique requires a particular type of receiver, capable of selectively and passively blank large interfering pulses from symbol decisions. Also, the approach assumes a system with a large number of pulses per symbols, so that blanking a few of them has only a minimal impact on the resulting BER. The joint adaptation of both types of processing gain was first studied in the context of IEEE 802.15.3a WPANs to reduce the mutual interference among uncoordinated, collocated WPANs [23], and was lately extended to cluster based wireless sensor networks in [20]. The basic observation in [23] was that the larger the number of collocated WPANs the larger the average frame time, T_f, has to be in order to reduce the amount of impulsive interference.

The first, general, theoretical considerations about the performance tradeoff between the two types of IR-UWB processing gain were published by Fishler and Poor in [6]. Fishler and Poor examined this tradeoff as a function of the BER in a system with fixed processing gain over flat-fading and frequency-selective channels. The study concludes that in a coded system transmitting over a flat-fading channel, the BER is independent of the ratio between the two types of processing gain. In contrast, in an uncoded system over a flat-fading channel and in frequency-selective channels there is a trade-off. This trade-off favours systems with a low number of pulses per symbol, as the system BER considerably degrades as the number of pulses per symbol increases. Thus, regarding processing gain adaptation, and assuming that the energy per transmitted pulse is the same for all users, it is preferable to extend the signal's average frame time (reduce the signal's duty cycle) than to increase the number of pulses per symbol. Moreover, using large frame times help to reduce the system complexity since a lower sampling rate can be used.

With the exception of the work in [10], which requires a particular receiver technique, the distributed adaptation of IR-UWB processing gain in autonomous networks has not been addressed in the literature before. The remaining approaches referenced in this section rely on the presence of a coordinator node that implements the adaptation algorithm and instructs other nodes on how to scale their parameters. This work focuses on autonomous networks; although hierarchical structures are not ruled out here, they are not required and therefore adaptation schemes cannot rely on the presence of coordination entities, but must be distributed.

The author claims that in autonomous IR-UWB networks, due to its self-organising and asynchronous character, and due to the monotonically increasing throughput for increasing pulse rates, the system's local pulse load can become so high that bit errors may not be longer tractable with error coding schemes. Based on this assumption, and following previous theoretical considerations on processing gain adaptation, this work develops a novel mechanism -distributed Pulse Rate Control (PRC)- to coordinate the links' pulse rate levels to optimise the overall network performance, measured in terms of total network logarithmic utility (*proportional fairness*), while satisfying a minimum per link BER requirement.

3.1. Distributed pulse rate control

Pulse rate control (PRC) can be realised in form of a link adaptation function whose goal is the improvement of the network throughput while satisfying a minimum per link BER requirement.

While adaptive modulation and channel coding are local decisions to a sender-receiver pair, PRC involves a cooperation among different links since the average[7] probability of pulse collision at the i-th receiver ($P_{coll,i}$) does not depend on its own link's pulse rate, but on the pulse rates of transmitters in its vicinity [20]. Since the collision probability is an indirect measurement of the BER, it can be further assumed that the BER at the i-th receiver, $P_{e,i}$, does not directly depend on its link's pulse rate, but on the pulse rate of the neighbouring links.

In the following we model, analyse and evaluate the distributed PRC approach. We first formulate PRC as a network logarithmic utility maximisation problem with quality of service (QoS) constraints. In order to solve the problem in a distributed manner, PRC is reformulated to a non-cooperative game with pricing. A distributed asynchronous algorithm is proposed which converges to the globally optimal solution of the original problem.

3.1.1. Problem formulation

The objective of the PRC approach is to determine the maximum pulse rate allocation, such that the QoS demands - in terms of BER- of all network links are fulfilled. This goal can be expressed by equation 2.

$$\begin{aligned} &\max_{\mathbf{prf}} \sum_{i=1}^{|\mathbf{N}|} log\left(r_i(\mathbf{prf})\right) \\ &\text{subject to:} \\ &P_{e,i}(\mathbf{prf}_{-i}) \leq \beta_i, \forall\, i \in \mathbf{N}, \\ &prf_i \in \mathbf{PRF}_i = [prf_i^{min}, prf_i^{max}], \forall\, i \in \mathbf{N} \end{aligned} \tag{2}$$

PRC assumes that each IR-UWB node can autonomously adapt its average pulse repetition frequency, prf - that is the inverse of the average frame duration ($prf = \frac{1}{T_f}$). Aditionally, it assumes that the energy per transmitted pulse, E_p^{tx}, is fixed and equal for all users, independent of the modulation scheme, and is chosen so that the IR-UWB node with the highest data rate completely exploits the FCC requirements in terms of EIRP and peak power. Controlling the source's prf is equivalent to controlling it's data rate (r) in terms of pulses per second; with fixed E_p^{tx}, it is also equivalent to controlling the source's average transmitted power.

QoS constraints in equation 2 limits the set of feasible pulse rate allocations. Since a higher pulse rate level from one transmitter increases the collision probability -and in turn the BER

[7] The average is taken over the links' asynchronism. In fact, the collision probability between two transmitters, for instance transmitter j and the reference transmitter i, depends on their pulse rates and on the relative delay time between the instants at which both transmitters start their transmissions. This relative delay is a random variable determined by the action of the users; it is represented by the time shift τ_j.

(P_e)- at neighbouring receivers, there may not be any feasible pulse rate allocation to satisfy the requirements of all users.

The logarithmic utility function in equation 2 captures the link's desire for higher data transmission rate. In an IR-UWB network, the raw data rate per link can be controlled by adapting the channel coding rate (R_i), the modulation order (m_i) or the processing gain, that is the number of pulses per symbol (N_s^i) and/or the average frame duration (T_f^i). Equation 3 depicts the dependency of the raw data rate on these parameters.

$$r_i^{raw} = \frac{1}{N_s^i \cdot T_f^i} \cdot R_i \cdot log_2(m_i) \text{ [bit/s]} \tag{3}$$

Following [6], this work focuses on systems with low number of pulses per symbol. For the sake of simplicity and without loss of generality we consider hereafter $N_s^i = 1, \forall i \in \mathbf{N}$. The useful (net) data rate per link is given in equation 4.

$$r_i(\mathbf{prf}) = r_i^{raw} \cdot \left(1 - P_{e,i}(\mathbf{prf}_{-i})\right) \text{ [bit/s]} \tag{4}$$

Since the bit error rate, P_e, is a nonlinear, and neither convex nor concave function of the links' pulse rates, the pulse rate optimisation problem is in general a nonlinear optimisation problem. The classical optimisation theory has no effective method for solving the general nonlinear optimisation problem, but several different approaches such as geometric programming (GP). Each of these approaches involves some compromise [1]; for instance, GP is limited to algorithms with a central single point of computation [2]. Game theory represents an alternative to GP and it is used in the next section to model the PRC problem (in equation 2) in a distributed manner.

3.1.2. *Pulse rate control game*

From the author's point of view, distributed PRC can be interpreted as a resource allocation mechanism that regulates the link's number of transmitted pulses per second. Hence, the framework of non-cooperative game theory can be applied to model and analyse the problem that searches for the network's maximum pulse rate allocation that satisfies a certain set of per-link BER constraints. Next we show how a game theoretical formulation helps to provide the UWB devices with incentives to minimise impulsive emissions when the cumulative system pulse load excesses certain limits, which are determined by some QoS constraints.

In PRC an increase in a link's average *prf* directly and negatively affects the probability of pulse collision, and consequently the BER, of neighbouring links [20]. A game theorist would refer to this fact by saying that there are *negative externalities* in the system. In order to deal with QoS constraints in the presence of negative externalities cooperation among the autonomous users must be enforced. Pricing is one of the most commonly used incentives to regulate selfish user behaviour and establish cooperation. Keeping this in mind, the original logarithmic utility function $u_i(\mathbf{prf}) = log(r_i(\mathbf{prf}))$ in equation 2 is modified by adding a linear pricing function of the link's *prf*. The new utility function is given in equation 5.

$$v_i(\mathbf{prf}) = log(r_i(\mathbf{prf})) - \pi_i(\mathbf{prf}) \cdot prf_i \tag{5}$$

The original logarithmic utility function reflects the level of a user's satisfaction from consuming the resource *prf* (directly related to the transmission data rate). The pricing factor, $\pi_i(\mathbf{prf})$, reflects the cost per unit of resource charged to user i. Hence, the new utility function can be interpreted as if each user i maximises the difference between its old net utility and a payment to other users in the network due to the interference (pulse collisions) it generates.

With the new utility function $v_i(\mathbf{prf})$ a non-cooperative PRC game with pricing, denoted by $\Gamma_{\text{PRC}} = \langle \mathbf{N}, \mathbf{PRF}, \mathbf{v} \rangle$, is developed. For Γ_{PRC} the set of players $\mathbf{N} = \{1, 2, ..., |N|\}$ corresponds with the set of active links (users) in the network, so that the terms "player" and "user" are used as synonym. The vector of utilities corresponds to $\mathbf{v}(\mathbf{prf}) = \{v_1(\mathbf{prf}), v_2(\mathbf{prf}), ..., v_{|N|}(\mathbf{prf})\}$, and the set of actions that players can choose, $\mathbf{PRF} = PRF_1 \times PRF_2 \times ... \times PRF_{|N|}$, is compact and convex.

As described in 2.4 this work considers a network model with a single centralised receiver (CH) and several uncoordinated sources. With a common pricing factor provided by the CH, $\pi_i(\mathbf{prf}) = \pi_j(\mathbf{prf}) = \pi_{CH}(\mathbf{prf})$, $\forall i, j \in \mathbf{N}$, each user in the network can be guided by the altruistic goal of maximising the cumulative network throughput at the CH, while keeping their average bit error rate, $P_{e,i}$, as close as possible to a target $\beta_{CH} = \beta_i, \forall i \in \mathbf{N}$. In this setting, the pricing term acts as a control parameter employed by the CH to discourage the overuse of the wireless resource *prf* and to keep the interference sustainable. It is expected that the choice of a common pricing factor for all links degenerates the *proportional-fairness* character of the original problem formulation into a *max-min fairness* solution.

Generally, an average cumulative $P_{e,CH} >> \beta_{CH}$ suggests a congestion situation caused by an overload in the local pulse density. In contrast, $P_{e,CH} << \beta_{CH}$ suggests an underload situation in which the local pulse density is below the sustainable load for the given QoS criteria. Accordingly, the CH measures the cumulative bit error rate at each superframe and continuously tracks its deviation from the target value in the variable $\triangle P^s_{e,CH}$

$$\triangle P^s_{e,CH} = \sum_{k \in \mathbf{AL}^s} P_{e,k} - \beta_{CH}, \tag{6}$$

where s represents the superframe index and $\mathbf{AL}^s$ is the set of active links during superframe s. Based on $\triangle P^s_{e,CH}$, the CH computes a congestion cost for superframe $s+1$ and feedbacks it to the UWB nodes in the next beacon frame. The computation rule for the congestion cost is very simple. If there is congestion in the current superframe s, the congestion cost for the next superframe $s+1$ must be increased, in contrast, if there is underload in the current superframe s, the congestion cost for the next superframe $s+1$ can be decreased. The congestion cost represents a common price factor for all players, $\pi_{CH} = \pi_i, \forall i \in \mathbf{N}$. The UWB nodes regulate their *prf*, and therewith their data rate, in response to the congestion cost feedback from the CH.

Specifically, the following computation rule is proposed

$$\pi^{s+1}_{CH} = \begin{cases} (1.0 - \delta)\pi^s_{CH} + \mu\pi^s_{CH}\delta, & \triangle P^s_{e,CH} > 0 \\ (1.0 - \delta)\pi^s_{CH} - \frac{\pi^s_{CH}\delta}{\mu}, & \triangle P^s_{e,CH} < (\beta_{CH}\omega|AL|^s) \end{cases}, \tag{7}$$

where μ is a weight factor and δ is the smoothing factor of a weighted exponential-moving-average (EMA) algorithm [4]. In order to improve game convergence a tolerance region for the cluster congestion level has been defined in which no adaptation is done. The tolerance range is defined by $|AL|^s$, the number of active links in superframe s, and a constant $\omega \leq 0$.

3.1.3. Pulse rate control algorithm

In a realistic distributed environment, at the start of a game it is not possible that a player $i \in \mathbf{N}$ has the complete price information (adjacent channel gains, link qualities) that is necessary to discover an NE inmediately. However, the player can make a guess, denoted by its selected action prf_i, regarding its equilibrium average prf denoted by prf_i^*. Then, assuming that the actions of the other players, $\mathbf{prf}_{-i}$, and its own price factor, π_i, remain constant while it makes a decision, player i improves its guess by selecting a new action which maximises its utility function. This new guess results in a new approximation to prf_i^*. When the deviations in all players' actions become negligibly small, the game can be assumed to have converged to an NE.

The PRC algorithm implements the adaptive behaviour described above and distributively solves the pulse rate allocation maximisation problem in equation 2.

PRC Algorithm

- **Step 1: Initialisation**

  For each user $i \in \mathbf{N}$ choose some $prf_i(0) \in PRF_i$ and a price factor $\pi_i(0) \geq 0$.
- **Step 2: Price update**

  At each iteration $t \in T_{i,\pi}$, user i updates its price according to equation 7.
- **Step 3: Pulse rate update**

  At each iteration $t \in T_{i,prf}$, user i updates prf_i according to a best response decision rule:

$$prf_i^*(t+1) = \mathrm{BR}_i\left(\mathbf{prf}_{-i}(t)\right) = \left[\frac{1}{\pi_i(t)}\right]_{prf^{min}}^{prf^{max}} \tag{8}$$

  where the notation $[x]_a^b$ means $\max\{a, \min\{b, x\}\}$ and BR_i is the set of best responses of player i to the strategy profile $\mathbf{prf}_{-i}(t)$.

Notice that the price and prf adaptations do not need to happen at the same time. The price update instants are determined by the superframe beacon raster, while the prf update instants depend on the beacon raster as well as on the source traffic model and can therefore be asynchronous across users.

In the practical implementation of the algorithm, a discretisation of the action space ($\mathbf{PRF}_i$) is unavoidable. The granularity of this discretisation process represents a trade-off for the convergence properties of the algorithm. An infinitely small granularity equals an infinite action space and guarantees that the NE action profile (if there is one) is considered in the search process; however, it increases the computation cost of the algorithm. With an increasing granularity the action space becomes finite, so that the search space for the algorithm shrinks and the computational cost is reduced. A brief discussion about the existence of Nash equilibria in discrete games can be found in [7]. In general it holds that, if the game with continuous action space has a stable equilibrium, the discrete pendant also has an equilibrium.

3.2. Game analysis

In this section, the utility function of Γ_{PRC} is analysed in terms of existence and uniqueness of Nash equilibria. The fact that Γ_{PRC} fits the framework of potential games ([13, 14]) significantly facilitates the analysis.

A potential game is characterised by the existence of a function, denoted as the potential function, $\Phi : A \rightarrow \Re$, such that the change in the utility function of a player when it unilateraly deviates in its strategy ($\triangle u_i$) is reflected in a change in value of the potential function ($\triangle\Phi$). If for all unilateral deviations, $\triangle\Phi = \triangle u_i$ the game is referred to as an *exact* potential game (EPG). If the relationship between the potential function and the utility functions is relaxed so that only sign changes are preserved, $sgn(\triangle\Phi) = sgn(\triangle u_i)$, the game is called an *ordinal* potential game (OPG).

3.2.1. Existence of an equilibrium

In order to prove the existence of an NE for game Γ_{PRC} a definition and a powerful result concerning the identification of ordinal potential games are leveraged. Definition 1, theorem 1 and 2 and their respective proofs can be found in [14].

Definition 1. *Better-Response Equivalence*

A game $\Gamma = \langle \mathbf{N}, \mathbf{A}, \mathbf{v} \rangle$ *is said to be better response equivalent to game* $\tilde{\Gamma} = \langle \mathbf{N}, \mathbf{A}, \tilde{\mathbf{v}} \rangle$, *if* $\forall i \in \mathbf{N}$, $a \in \mathbf{A}$, $v_i(a_i, \mathbf{a}_{-i}) > v_i(b_i, \mathbf{a}_{-i}) \Leftrightarrow \tilde{v}_i(a_i, \mathbf{a}_{-i}) > \tilde{v}_i(b_i, \mathbf{a}_{-i})$.

Theorem 1. *Identification of Ordinal Potential Games*

A game $\Gamma = \langle \mathbf{N}, \mathbf{A}, \mathbf{v} \rangle$ *is an ordinal potential game if and only if it is better response equivalent to an exact potential game.*

Theorem 2. *NE of Better-Response Equivalent Games*

If a game $\Gamma = \langle \mathbf{N}, \mathbf{A}, \mathbf{v} \rangle$ *is better response equivalent to game* $\tilde{\Gamma} = \langle \mathbf{N}, \mathbf{A}, \tilde{\mathbf{v}} \rangle$, *then the Nash Equilibria of* Γ, *if any exist, coincide with the Nash equilibria of* $\tilde{\Gamma}$.

With these results in mind, game $\tilde{\Gamma}_{PRC} = \langle \mathbf{N}, \mathbf{A}, \tilde{\mathbf{v}} \rangle$ with utility function

$$\tilde{v}_i(\mathbf{prf}) = log(prf_i) - \pi_{CH} prf_i, \tag{9}$$

is introduced. Since $log(xy) = log(x) + log(y)$ and through definition 1, there is a trivial better response equivalence relationship between Γ_{PRC} and $\tilde{\Gamma}_{PRC}$. Hence, from theorem 2 we know that by analysing the set of Nash equilibria of game $\tilde{\Gamma}_{PRC}$ we are at the same time solving for the Nash equilibria of Γ_{PRC}. Next, we prove that $\tilde{\Gamma}_{\text{PRC}}$ is an EPG and, consequently thanks to theorem 1, Γ_{PRC} is an OPG. Further, in [14] we find a powerful result concerning all potential games with a compact action space and a continuous potential function: The existence of at least one NE.

From [14], a sufficient condition for the existence of a potential function in game $\Gamma = \langle \mathbf{N}, \mathbf{A}, \mathbf{u} \rangle$ is

$$\frac{\partial^2 u_i(\mathbf{a})}{\partial a_i \partial a_k} = \frac{\partial^2 u_k(\mathbf{a})}{\partial a_k \partial a_i}, \forall i, k \in \mathbf{N}, \forall \mathbf{a} \in \mathbf{A} \tag{10}$$

In $\tilde{\Gamma}_{PRC}$ the utility functions are given by:

$$\tilde{v}_i(prf_i, \mathbf{prf}_{-i}) = log(prf_i) - \pi_{CH} prf_i \tag{11}$$

and

$$\tilde{v}_k(prf_k, \mathbf{prf}_{-k}) = log(prf_k) - \pi_{CH} prf_k. \tag{12}$$

In this network settings, users ignore any influence they may have on the price calculated at the CH. Hence, $\frac{\partial \pi_{CH}}{\partial prf_i} = 0 \; \forall \, i \in \mathbf{N}$, and it is easy to prove that

$$\frac{\partial^2 \tilde{v}_i(prf_i, \mathbf{prf}_{-i})}{\partial prf_i \partial prf_k} = \frac{\partial^2 \tilde{v}_k(prf_k, \mathbf{prf}_{-k})}{\partial prf_k \partial prf_i} = 0 \tag{13}$$

Characteristic for an EPG is the existence of a potential function Φ that exactly reflects any unilateral change in the utility of any player, that is $\triangle \Phi(a) = \triangle u_i(a)$. Hence, starting from any arbitrary strategy vector $\mathbf{a}$ any unilaterally player's adaptation that increases its utility $u_i(a)$ identically translates in an increase of the potential function $\Phi(a)$.

The potential function of game $\tilde{\Gamma}_{PRC}$ is given in equation 14.

$$\tilde{\Phi}_{PRC}(\mathbf{prf}) = \sum_{i \in N} log(prf_i) - \pi_{CH} \sum_{i \in N} prf_i \tag{14}$$

Based on the potential game definition given in [13], the proof that $\Gamma = \langle \mathbf{N}, \mathbf{A}, \mathbf{u} \rangle$ is an EPG requires that:

$$u_i(a_i, a_{-i}) - v_i(b_i, \mathbf{prf}_{-i}) = \Phi(a_i, a_{-i}) - \Phi(b_i, a_{-i}), \forall \, i \in \mathbf{N}, \forall \, \mathbf{a} \in \mathbf{A} \tag{15}$$

Equation 16 proves condition 15 for $\tilde{\Gamma}_{PRC}$. Note that all sum terms in the potential function that are independent of user i's strategy are constant and can be grouped in an extra term denoted as c. By the substraction of the potential functions, the term c dissapears leaving the difference of the potential functions identical to the difference of the utility functions.

$$\begin{gathered}
\tilde{v}_i(prf_i, \mathbf{prf}_{-i}) - \tilde{v}_i(prf_i^{'}, \mathbf{prf}_{-i}) = \tilde{\Phi}_{PRC}(prf_i, \mathbf{prf}_{-i}) - \tilde{\Phi}_{PRC}(prf_i^{'}, \mathbf{prf}_{-i}) \\
log(prf_i) - \pi_{CH} prf_i - log(prf_i^{'}) + \pi_{CH} prf_i^{'} = \\
log(prf_i) - \pi_{CH} prf_i \\
\underbrace{+ \sum_{j \in N \wedge j \neq i} log(prf_j) - \pi_{CH} \sum_{j \in N \wedge j \neq i} prf_j}_{=c} \\
-log(prf_i^{'}) + \pi_{CH} prf_i^{'} \\
\underbrace{- \sum_{j \in N \wedge j \neq i} log(prf_j) + \pi_{CH} \sum_{j \in N \wedge j \neq i} prf_j}_{=-c} \\
log(prf_i) - \pi_{CH} prf_i - log(prf_i^{'}) + \pi_{CH} prf_i^{'} = \\
log(prf_i) - \pi_{CH} prf_i - log(prf_i^{'}) + \pi_{CH} prf_i^{'} \\
\forall \, i \in \mathbf{N}, \forall \, prf \in \mathbf{PRF}
\end{gathered} \tag{16}$$

Equation 14 is continuous since it is the sum of continuous functions, furthermore the action space of $\tilde{\Gamma}_{\text{PRC}}$ is compact per definition. Hence, equation 16 verifies that (as any other EPG) $\tilde{\Gamma}_{\text{PRC}}$ has at least one NE. Finally, applying theorem 2 it is proven that Γ_{PRC} has at least one NE - in fact the same as $\tilde{\Gamma}_{\text{PRC}}$).

3.2.2. *Uniqueness of the equilibrium*

From [14], it is known that an EPG following an asynchronous, myopic, best response decision rule converges to a pure strategy NE that maximises the potential function. Furthermore, if the potential function is strictly concave, it has a unique global maximum which is then the unique NE of the EPG. Based on these results the following proposition can be stated.

> **Proposition 1.** *If $\tilde{\Phi}_{PRC}$ in equation 14 is strictly concave, the proposed PRC algorithm will always converge to the unique NE of game $\tilde{\Gamma}_{PRC}$, which in turn, is the unique global maximum of $\tilde{\Phi}_{PRC}$.*

Proof. The PRC algorithm can be interpreted as the players employing asynchronous myopic best response (MBR) updates. Thus, to demonstrate proposition 1 it suffices to prove the strict concavity of $\tilde{\Phi}_{\text{PRC}}$ in equation 14. As explained in [1, Section 3.1.4], this can be verified with the Hessian matrix and the second-order conditions.

For the Hessian matrix in equation 18 the second derivatives of equation 14 are required.

$$H(\tilde{\Phi}) = \begin{pmatrix} \frac{\partial^2 \tilde{\Phi}_{\text{PRC}}}{\partial^2 prf_1} & \cdots & \frac{\partial^2 \tilde{\Phi}_{\text{PRC}}}{\partial prf_1 \partial prf_j} & \cdots & \frac{\partial^2 \tilde{\Phi}_{\text{PRC}}}{\partial prf_1 \partial prf_N} \\ \vdots & \ddots & \vdots & \ddots & \vdots \\ \frac{\partial^2 \tilde{\Phi}_{\text{PRC}}}{\partial prf_N \partial prf_1} & \cdots & \frac{\partial^2 \tilde{\Phi}_{\text{PRC}}}{\partial prf_N \partial prf_j} & \cdots & \frac{\partial^2 \tilde{\Phi}_{\text{PRC}}}{\partial^2 prf_N} \end{pmatrix} \tag{17}$$

$$\frac{\partial^2 \tilde{\Phi}_{\text{PRC}}}{\partial prf_i \partial prf_j} = \begin{cases} -\frac{1}{prf_i^2} & \text{if } i = j \\ 0 & \text{if } i \neq j \end{cases} \tag{18}$$

The matrix is negative definite, $x^T H(\tilde{\Phi}) x < 0$, since all diagonal elements are negative. Hence, the proof is complete. □

Finally, proposition 1 can be reformulated to proposition 2.

> **Proposition 2.** *If $\tilde{\Phi}_{PRC}$ in equation 14 is strictly concave, the proposed PRC algorithm will always converge to the unique NE of game $\tilde{\Gamma}_{PRC}$, which, in turn, is the unique NE of game Γ_{PRC}.*

Proof. The proof of Proposition 2 results from combining the proof of Proposition 1 with Theorem 2. □

3.2.3. *Optimality of the equilibrium*

So far, there is no general result about the optimality of Nash equilibria in potential games or in any other more general class of games. However by designing the potential game in a way that its potential function complies with the network design function, a quite elegant way to demonstrate NE optimality is possible. In that case, any strategy that maximises the potential function (any NE) maximises as well the network design function.

In this sense, note that the utility function of game Γ_{PRC} in equation 5, combines the network utility function of the original resource allocation problem (cf. to equation 2) with a linear price function. By exploiting the linear space properties of EPGs, the potential function in equation 14 preserves the properties (such as concavity and uniqueness of the global maximisers) of the original network objective function $\sum_1^{|N|} log(r_i(\mathbf{prf}))$. The addition of the linear price term aims at adjusting the unique NE of $\tilde{\Gamma}_{PRC}$ so that the QoS constraint in the original problem is respected. Hence the NE is optimal from a network design perspective.

3.3. Performance evaluation

Results highlighted in this section have been computed with the simulation model presented in Section 2.5 in the scenarios depicted in Fig. 2. The aim of this section is to demonstrate the regulative character and the interference compensation functionality of the PRC approach. Therefore, its performance under increasing offered system load is compared to the one obtained with the ALOHA MAC protocol without feedback. Note that the IEEE 802.15.4a proposes ALOHA with optional feedback as the standard MAC protocol for IR-UWB physical layers.

Parameter	Value	Parameter	Value
E_p^{tx}	$2 \cdot 10^{-11}$ [Ws]	BW	1.5 [GHz]
$T_{\int rcx}$	$5 \cdot 10^{-9}$ [s]	*prf* granularity	1 [kHz]
prf_i^{max}	1 [MHz]	prf_i^{min}	10 [kHz]
m-PPM	2	β_{CH}	$5 \cdot 10^{-4}$
L_p	400 bit	μ	2
δ	$1 \cdot 10^{-2}$	ω	$2.5 \cdot 10^{-3}$

Table 1. List of main simulation parameters.

3.3.1. *Existence of an equilibrium*

First a simplified setting with a constant number of players, as usually assumed in game theory, is considered. Then, a more realistic setting where the number of players is a random variable controlled by the traffic distribution is explored. For both settings simulation results[8] confirm the existence of an NE regardless of the algorithm's initialisation parameters.

Constant number of players This setting considers the continuous transmission scenario described in Section 2.4. Recall that in this scenario up to 10 nodes are collocated along a circumference of 10m radius with the CH in the centre. It is assumed that all nodes have always packets to send; this is assured by configuring the source traffic generator to an information data rate of 1Mbps.

Figure 3 confirms the existence of a stable equilibrium in game Γ_{PRC}, and that the PRC algorithm converges to it. Figure 3(a) depicts a symmetric *prf* allocation, which agrees with the analytically predicted NE; the equilibrium *prf* per link is approximately 220 kHz. In fact, a symmetric equilibrium was expected since the congestion cost is the same for each link and

[8] These results were obtained in [11].

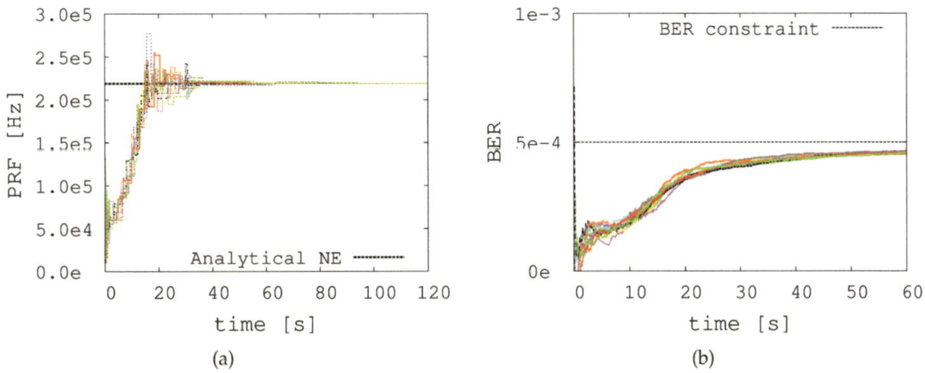

Figure 3. Existence of an equilibrium in Scenario 1 - Continuous transmission. Each link is represented with a different color: (a) Convergence to the unique equilibrium *prf* allocation; (b) Upper-bounded character of the link's BER.

the node topology is symmetric. Figure 3(b) confirms that the QoS constraint on the BER, $P_{e,i} \leq \beta_{CH}$, is fulfilled for all links.

Time-varying number of players This setting considers the factory hall scenario described in Section 2.4. There are a total of 100 heterogeneous nodes, some of them are moving along a production line, others are fixed in known positions and the rest is moving around the CH following a random waypoint movement model. The information data rate has been chosen low enough to guarantee that nodes do not always have packets to send; this results in a time-varying number of players[9].

Figure 4 presents results for two values of the information data rate which are 10 kbit/s and 15 kbit/s, respectively. With 10 kbit/s information data rate, the CH is exposed to a low system pulse load measured in pulses per second (Pps). Figure 4(a) suggests the existence of a stable equilibrium in game Γ_{PRC}, and that the PRC algorithm converges to it. The *prf* equilibrium allocation is again symmetric, with all links converging to the maximum possible *prf* of 1 MHz. Figure 4(b) confirms that the QoS constraint on the BER is satisfied for all links. Note that the BER curves of all links except one follow almost an identical course. The outlier curve corresponds to the sensor node which is located closest to the CH (at 3m) and therefore exhibits the best channel gain and the lowest BER.

By increasing the user data rate to 15 kbit/s, the system pulse load achieves a level which is not compliant with the problem's QoS constraint. The PRC algorithm should then reduce the links' pulse rates to a sustainable level, which is identified by the congestion cost factor π_{CH}. Figure 4(c) illustrates the regulative effect of the PRC algorithm. In spite of the heterogeneous network character, the *prf* equilibrium corresponds to a *max-min* socially fair allocation induced by the choice of a common pricing factor for all links. Besides, it can be observed that in Figure 4(d) some of the links violate the QoS constraint. Still, if we consider the average BER over all links (see Figure 8(b)) it is below the upper bound β_{CH}. This is

[9] Note that the number of players in the game coincide with the number of active links in each superframe, and this is a random variable controlled by the exponential traffic distribution.

due to the global altruistic behaviour of the congestion cost factor, π_{CH}, which works with an estimation of the average[10] BER at each stage of the game.

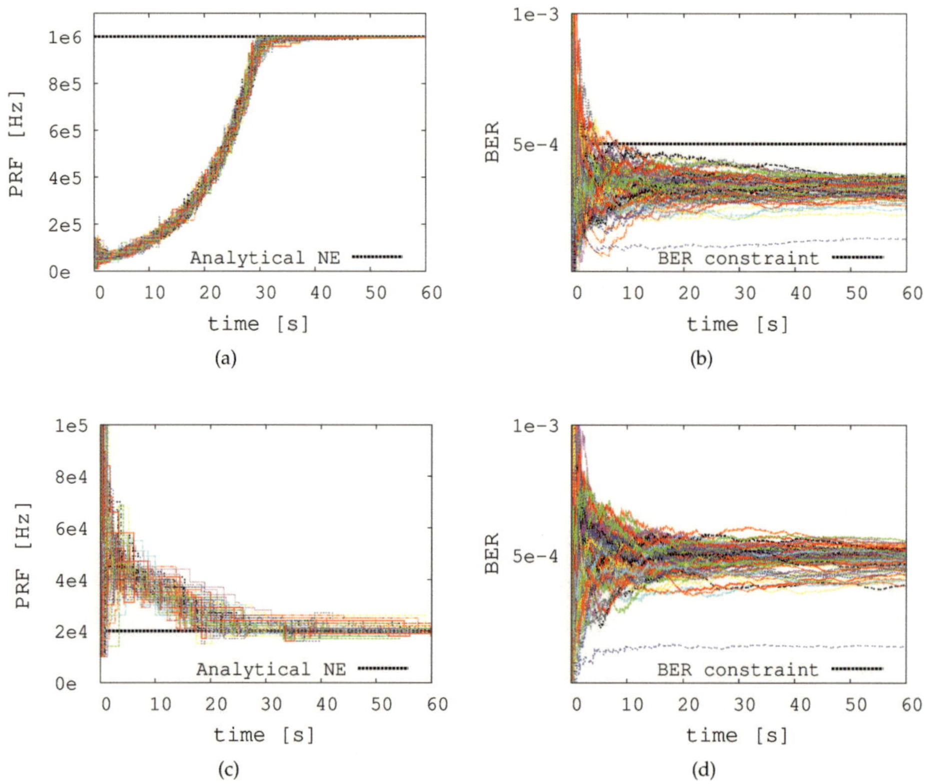

Figure 4. Existence of an equilibrium in Scenario 2 - Factory hall. Each link is represented by a different color: Graphics (a) and (c) show the convergence to the unique equilibrium *prf* allocation when the information data rate per user is 10 kbit/s and 15 kbit/s, respectively; Graphics (b) and (d) show the temporal evolution of the BER per link when the information data rate per user is 10 kbit/s and 15 kbit/s, respectively.

3.3.2. Variation of the offered load

The results in this clause show the effect of increasing offered system load on the aggregate network throughput and the average BER, both measured at the CH. Like in the previous clause, the game performance with a constant number of players is differentiated from the that with a time-varying number of players.

Constant number of players The information data rate is fixed and equal to 1 Mbps to ensure continuous packet transmission. In order to progressively increase the offered system pulse load, the number of sensor nodes collocated along the circumference of 10 m radius has been stepped up from 2 to 10.

[10] Over all links.

In general, and due to the increasing MUI, the aggregated network throughput is expected to drop as the number of nodes in the network grows. Figure 5(a) illustrates the aggregated network throughput. It can be observed that with four to ten source nodes the aggregated network throughput remains almost constant. These results suggest an interference compensation effect of the PRC algorithm; Figure 6 confirms this effect. Additionally, Figure 5(b) shows that the cumulative BER scratches the QoS upper bound in all cases except in the case of having only two nodes. In this special case, even with nodes sending with the maximum *prf* the offered system load is low enough to guarantee an average BER far below the QoS upper bound.

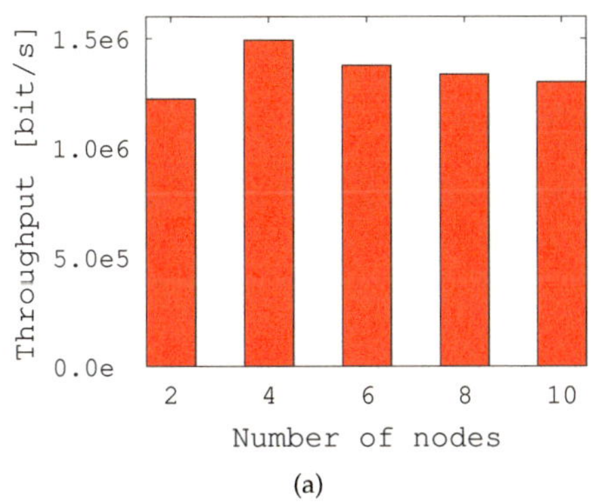

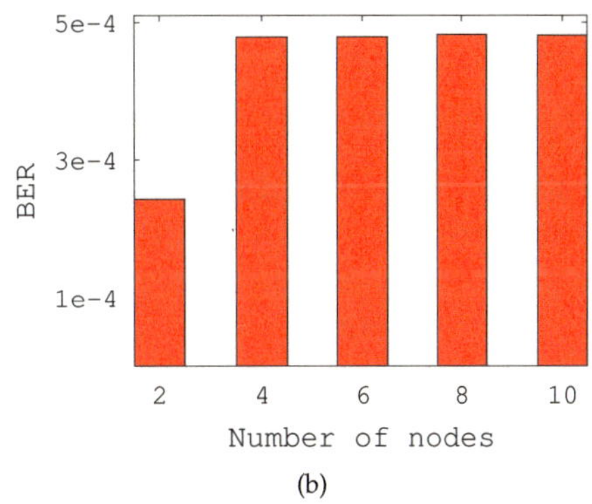

Figure 5. System performance in Scenario 1 - Continuous transmission: (a) Aggregated network throughput measured at the CH's application layer; (b) Average BER per link measured at the CH.

Figure 6 shows the mean[11] *prf* per link; where each link has been represented with a different color. It can be observed that the PRC algorithm relaxes the mean *prf* per link as the number of nodes in the network grows, since consequently the pulse density in the system increases. Recall that a higher pulse density raises the probability of pulse collisions and this, in turn, the probability of bit errors. The larger the average BER at the CH, the larger the congestion factor, π_{CH}, is. Finally, larger congestion factors lead to lower *prf* levels.

Time-varying number of players Since in Scenario 2 - Factory hall the number of sensor nodes is constant, in order to increase the offered system pulse load we progressively raise the information data rate per sensor node.

In Figure 7 two different phases can be identified. As long as the information data rate is kept below 10 kbit/s, the mean *prf* per link does not drop, but remains close to the maximum allowed *prf* value (1 MHz). When the information data rate reaches 15 kbit/s, the mean *prf* per link drops down to approximately 20 kHz and remains there despite growing information data rate. These results are consistent with the PRC algorithm's behaviour illustrated in Figure 4(a) and Figure 4(c).

In a similar way Figure 8 illustrates the regulative behaviour of the PRC algorithm. At information data rates up to 10 kbit/s the PRC algorithm converges to the maximum possible

[11] Over the whole simulation time.

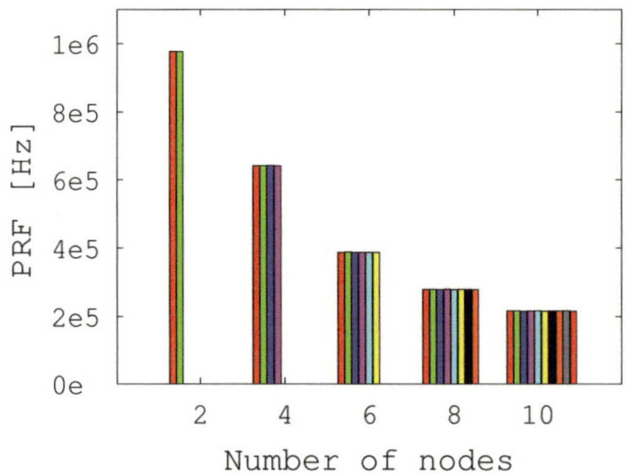

Figure 6. Average *prf* per link in Scenario 1 - Continuous transmission.

prf per link, since the offered system pulse load is low enough to guarantee the QoS constraint on the BER (see Figure 8(b)). With higher information data rates, the PRC algorithm has to relax the effective system pulse load per user to avoid network congestion, and to be able to satisfy the QoS constraint.

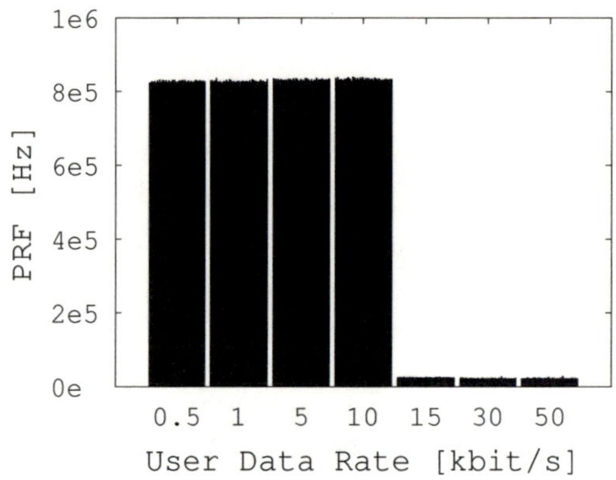

Figure 7. Average *prf* per link in Scenario 2 - Factory hall.

3.3.3. Performance comparison with IEEE 802.14.5a MAC

Finally, this clause is dedicated to compare the performance of the ALOHA MAC protocol, as described in IEEE 802.15.4a, with and without distributed PRC. For the simulations without PRC we have set the fixed *prf* per link to 1 MHz. IEEE 802.15.4a recommends ALOHA with optional feedback channel as the MAC protocol for low data rate sensor networks with IR-UWB physical layers. This work focuses on random access without feedback channel to keep the power consumption and the receiver complexity at the sensor nodes low. In exchange, the drawback has to be accepted that successful reception of data packets cannot be guaranteed since retransmission requests are not possible. Note that in the application field considered in this chapter, the relatively small throughput offered by the random access method without feedback channel is still satisfying.

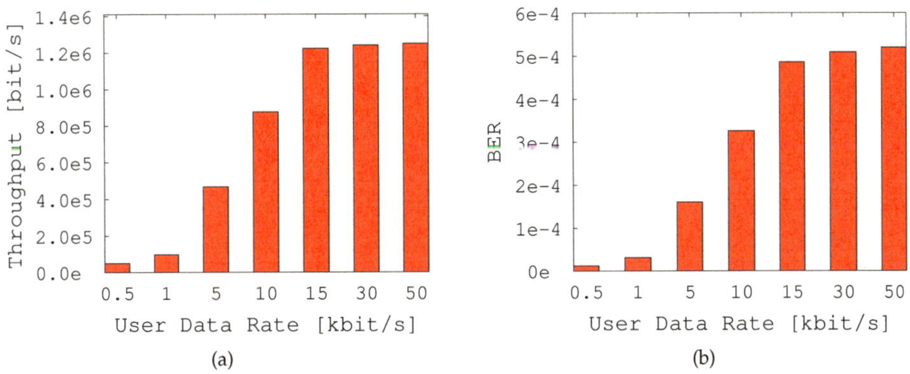

Figure 8. System performance in Scenario 2 - Factory hall: (a) Aggregated network throughput measured at the CH's application layer; (b) Average BER per link measured at the CH.

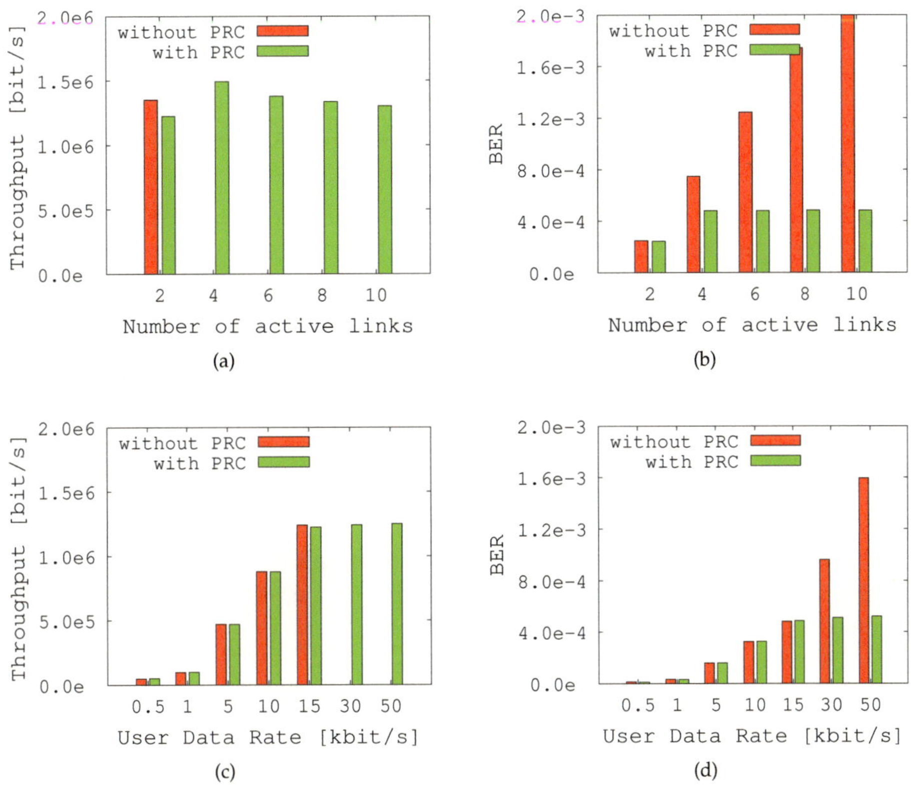

Figure 9. Performance comparison of ALOHA with and without PRC: (a) Aggregated network throughput in Scenario 1; (b) Average BER per link in Scenario 1; (c) Aggregated network throughput in Scenario 2; (d) Average BER per link in Scenario 2.

Figures 9(a) and 9(b) depict a performance comparison between ALOHA with and without PRC in the continuous transmission scenario with high offered pulse load and constant number of players. The network throughput obtained without PRC rapidly collapses[12] as the number of source nodes grows. This is comprehensible, since pulse collisions (and therewith bit errors, see Figure 9(b)) augment as the offered pulse load increases. In contrast, the interference compensation effect of the PRC approach under increasing offered load can be well observed. Notice how the PRC algorithm is able to limit the system pulse load to a level that ensures the QoS constraint.

Figures 9(c) and 9(d) depict results in the factory hall scenario with low to moderate pulse load and time-varying number of players. Notice that the regulative effect of the PRC approach (see Figure 7) limits the maximum possible cumulative network throughput, while ALOHA with fixed *prf* cannot guarantee the design QoS constraint and breaks down as the information data rate per link increases to 30 kbit/s.

4. Summary and conclusions

This chapter introduces a novel concept for impulsive interference management in low power, autonomous, IR-UWB networks. The concept enables concurrent transmissions at full power, while allowing each source to independently adapt its pulse rate (measured in pulses per second) to reduce the impact of pulse collisions at nearby receivers. The design is independent of a particular modulation scheme and can be applied to any IR-UWB physical layer. Beyond, it does not rely on any particular receiver technique and can work with a simple, low cost, non-coherent receiver.

The chapter formulates and evaluates the pulse rate allocation problem as a social rate optimisation problem with QoS constraints. It introduces a distributed algorithm implementation and analyses its performance via simulation. It has been analytically proven that the game Γ_{PRC} fits the framework of ordinal potential games, and that the NE is unique. In all considered scenarios, simulation results have confirmed the existence of an equilibrium for the game and that the distributed PRC algorithm converges to it, provided that the pricing parameters have been appropriately chosen.

We can conclude that distributed Pulse Rate Control is an appropriate means for impulsive interference management and network throughput optimisation with QoS constraints in highly loaded IR-UWB networks with a common central receiver. In [16] the author extends the work presented here and investigates the applicability of distributed PRC as well as its combination with adaptive channel coding in peer-to-peer networks, i.e. with multiple uncoordinated receivers and without any hierarchical infrastructure. Moreover she investigates a low-complexity, heuristic, alternative algorithm to the one proposed in this chapter that is more suitable for embedded hardware implementations, as preferred in the design of sensor networks.

Acknowledgements

The author thanks Prof. K. Jobmann, Prof. M. G. di Benedetto and Ralf Luebben for their valuable support to complete the work presented in this chapter.

[12] The design QoS constraint can only be guaranteed when the number of active links is below 4.

Author details

Pérez Guirao María Dolores
Institute of Communications Technology (IKT), Leibniz Universitaet Hannover (LUH), Hanover, Germany

5. References

[1] Boyd, S. & Vandenberghe, L. [2006]. *Convex Optimization*, Cambrigde University Press.

[2] Chiang, M., Tan, C. W., Palomar, D., O'Neill, D. & Julian, D. [July 2007]. Power Control by Geometric Programming, *Wireless Communications, IEEE Transactions on* 6(7): 2640–2651.

[3] di Benedetto, M. G. & Giancola, G. [2004]. *Understanding Ultra Wide Band Radio Fundamentals*, Prentice Hall PTR.

[4] Everette, S. & Gardner, J. [October-December 2006]. Exponential Smoothing: The State of the Art–Part II, *International Journal of Forecasting* 22, Issue 4: Pages 637–666.

[5] Felegyhazi, M. & Hubaux, J.-P. [2006]. Game Theory in Wireless Networks: A Tutorial, *Technical report*.

[6] Fishler, E. & Poor, V. [2005]. On the Tradeoff between two Types of Processing Gains, *IEEE Transactions on Communications* 53(9): 1744–1753.

[7] Fudenberg, D. & Tirole, J. [2001]. *Game Theory*, MIT Press.

[8] IEEE *Standard for Information Technology - Telecommunications and information exchange between systems - Local and metropolitan area networks - specific requirement Part 15.4: Wireless Medium Access Control (*MAC*) and Physical Layer (*PHY*) Specifications for Low-Rate Wireless Personal Area Networks (*WPANs*)* [2007]. *IEEE Std 802.15.4a-2007 (Amendment to IEEE Std 802.15.4-2006)* pp. 1–203.

[9] Le Boudec, J.-Y., Merz, R., Radunovic, B. & Widmer, J. [2004]. DCC-MAC: a Decentralized MAC Protocol for 802.15.4a-like UWB Mobile Ad-Hoc Networks based on Dynamic Channel Coding, *Broadband Networks, 2004. BroadNets 2004. Proceedings. First International Conference on*, pp. 396–405. TY - CONF.

[10] Lovelace, W. & Townsend, J. [16-19 Nov. 2003]. Adaptive Rate Control with Chip Discrimination in UWB Networks, *Ultra Wideband Systems and Technologies, 2003 IEEE Conference on* pp. 195–199.

[11] Luebben, R. [October 2007]. *Spieltheoretische Optimierung des Durchsatzes eines IR-UWB Sensornetzes*, Master's thesis, Institut für Kommunikationstechnik, Leibniz Universität Hannover, Germany.

[12] Merz, R., Widmer, J., Le Boudec, J.-Y. & Radunovic, B. [2005]. A joint PHY-MAC Architecture for Low-Radiated Power TH-UWB Wireless Ad Hoc Networks, *Wireless Communications and Mobile Computing Journal, Special Issue on UWB Communications, John Wiley & Sons Ltd* vol. 5: pp. 567–580.

[13] Monderer, D. & Shapley, S. [1996]. Potential Games, *Games and Economics Behaviour* 14: 124–143.

[14] Neel, J. [2006]. *Analysis and Design of Cognitive Radio Networks and Distributed Radio Resource Management Algorithms*, PhD thesis, Faculty of the Virginia Polytechnic Institute and State University.

[15] OMNeT++, *Discrete Event Simulation System*. [n.d.].
URL: *www.omnetpp.org*

[16] Perez-Guirao, M. [2009]. *Cross-Layer, Cognitive, Cooperative Pulse Rate Control for Autonomous, Low Power, IR-UWB Networks*, Shaker Verlag.

[17] Radunovic, B. & Le Boudec, J.-Y. [13-16 June 2005]. Power Control is Not Required for Wireless Networks in the Linear Regime, *World of Wireless Mobile and Multimedia Networks, 2005. WoWMoM 2005. Sixth IEEE International Symposium on a* pp. 417–427.

[18] Radunovic, B. & Le Boudec, J.-Y. [2004]. Optimal Power Control, Scheduling, and Routing in UWB Networks, *Selected Areas in Communications, IEEE Journal on* 22(7): 1252–1270.

[19] Tong, L., Naware, V. & Venkitasubramaniam, P. [Sept. 2004]. Signal Processing in Random Access, *Signal Processing Magazine, IEEE* 21(5): 29–39.

[20] Weisenhorn, M. & Hirt, W. [Sept. 2005]. Uncoordinated Rate-Division Multiple-Access Scheme for Pulsed UWB Signals, *Vehicular Technology, IEEE Transactions on* 54(5): 1646–1662.

[21] Win, M. & Scholtz, R. [2000]. Ultra-Wide Bandwidth Time-Hopping Spread-Spectrum Impulse Radio for Wireless Multiple-Access Communications, *Communications, IEEE Transactions on* 48(4): 679–689.

[22] Yates, R. [Sep 1995]. A Framework for Uplink Power Control in Cellular Radio Systems, *Selected Areas in Communications, IEEE Journal on* 13(7): 1341–1347.

[23] Yomo, H., Popovski, P., Wijting, C., Kovács, I. Z., Deblauwe, N., Baena, A. F. & Prasad, R. [Sep. 2003]. Medium Access Techniques in Ultra-Wideband Ad Hoc Networks, *in Proc. the 6th Nat. Conf. Society for Electronic, Telecommunication, Automatics, and Informatics (ETAI) of the Republic of Macedonia* .

Chapter 3

Chip-to-Chip and On-Chip Communications

Josef A. Nossek, Peter Russer, Tobias Noll, Amine Mezghani,
Michel T. Ivrlač, Matthias Korb, Farooq Mukhtar,
Hristomir Yordanov, Johannes A. Russer

Additional information is available at the end of the chapter

http://dx.doi.org/10.5772/52981

1. Introduction

In high-performance integrated circuits manufactured in CMOS deep sub-micron technology, the speed of global information exchange on the chip has developed into a bottleneck, that limits the effective information processing speed. This is caused by standard on-chip communication based on multi-conductor interconnects, e.g., implemented as parallel interconnect buses. The supported clock frequency of such wired interconnects - at best - remains constant under scaling, but - for global interconnects - reduces by a factor of four, as the structure size is reduced by half. Such multi-conductor interconnects also exhibit some undesirable properties when used for chip-to-chip communication. The much larger distances that have to be bridged, force the clock frequencies for the chip-to-chip interconnects to much lower values than those for on-chip circuitry. In widening up this bottleneck by increasing the number of parallel wires, the separation between the wires has to decrease. This causes increased mutual coupling between neighboring wires, which reduces the supported clock frequency and counters the effect of having more wires in the first place.

The high clock frequencies used in on-chip interconnects and the huge information rate of chip-to-chip communication lets possible solutions belong to the domain of ultra-wideband (UWB) technology. Pursuing suitable solutions, we explore firstly the improvement of the multi-conductor interconnect by signal processing and coding. From information theory, it is known that information can be transmitted through a noisy channel with arbitrary low probability of error as long as the rate is lower than the channel capacity given by the Shannon theorem. Achieving this capacity requires, however, sophisticated digital signal processing and coding. In particular, the DAC (Digital-to-analog converters) and the ADC (analog-to-digital converter) components which are formed by the output or the input of a logic CMOS inverter, respectively, turns to be a limiting factor. In fact, the ADC and DAC components, perform a single-bit conversion between the analog and the digital domain. With such coarse quantization, all state of the art techniques for signal processing fail. We provide information theoretic bounds on the improvements possible by coding the transmission, and propose methods to design suitable codes which allow decoding with low latency.

Thereby, an analytical field-theoretical modeling of multi-conductor interconnects is needed. Moreover, modifications to standard signal processing techniques which make them suitable for medium-low resolution quantization are developed and analyzed, and their performance is studied.

As a promising alternative solution, wireless Ultrawideband (UWB) enables high speed communication at short distances. In fact, it is anticipated that even higher performance is achievable in chip-to-chip and on-chip communication, when multi-conductor interconnects are replaced by wireless ultra-wideband multi-antenna interconnects. Hereby, the signal pulses do not necessarily increasingly disperse as they travel along their way to the receiving end of the interconnect. The propagating nature of the wireless interconnect, the extreme high available bandwidth and the very short distances can offer a much more attractive channel for chip-to-chip and on-chip communications. In addition, applying multiple antennas at the transmitter side as well as the receiver side can drastically improve the data rate and the reliability of UWB systems at the cost of certain computational complexity. This chapter provides theoretical and empirical foundations for the application of ultra-wideband multi-antenna wireless interconnects for chip-to-chip communication. Appropriate structures for integrated ultra-wideband antennas shall be developed, their properties theoretically analyzed and verified against measurements performed on manufactured prototypes. Qualified coding and signal processing techniques, which aim at efficient use of available resources of bandwidth, power, and chip area shall be developed. Since Analog-to-Digital Converters (ADCs) are considered critical components for the UWB, main focus is hereby given to low resolution signal quantization and processing. Therefore, the analysis and the design of UWB systems with low resolution signal quantization (less than 4 bits) is a vital part of this chapter, where optimized receive and transmit strategies are obtained.

On the other hand, detailed cost-models for the digital hardware architecture, which are based on signal flow charts and VLSI implementations of dedicated functional blocks shall be developed, which allow for an informative analysis of elementary trade-offs between computational speed, required chip area, and power consumption. In fact, quantitative optimization in terms of silicon area (manufacturing costs) and even more important in terms of energy dissipation (usage costs) is mandatory already in the standardization and conception phase of digital systems to be highly integrated as System-on-Chips (SoC). This is especially true for digital communication systems where e.g. in the optimization of channel coding traditionally only the transmission power has been considered. In general this leads to highly complex and energy intensive receivers. Actually a proper optimization of such systems requires a joint optimization of the transmitter and receiver cost features, e.g. the minimization of the total energy per transmitted bit. For such a quantitative optimization quite accurate cost models for the components of the transmitter and receiver are required. Instead, if any, only oversimplified cost models are applied today. While quite accurate cost models are available for many communication system components there is a lack of such models for channel decoders like Viterbi, Turbo, and Low-Density-Parity-Check (LDPC) decoders. Out of these, especially the derivation of sufficiently accurate cost models for LDPC decoders is challenging: The realization of the extensive internal exchange of messages between the so-called bit and check nodes in such a decoder results in non-linear dependencies between decoder features and code parameters. For example in high-throughput decoders the data exchange is performed via a complex dedicated interconnect structure. Its realization frequently requires an artificial expansion of silicon area. In the past various decoder architectures have been proposed to reduce the interconnect impact and trading throughput for silicon area and energy. All that together makes the derivation of LDPC decoder cost models a challenging task.

The Chapter is organized as follows. Radio frequency engineering aspects involved in wired and wireless interconnects are investigated first. There is a multitude of requirements for chip-to-chip communication, which an integrated antenna has to fulfill, like large bandwidth, small geometrical profile, and so on. Therefore, a detailed study of the possible solutions for an integrated on-chip antenna is performed. Novel solutions, which make use of the digital circuit's ground plane as a radiating element, are investigated. In the third section, the signal processing and coding aspects involved are carried out based on the obtained channel models, where both multiconductor interconnects and wireless multiantenna interconnects are interpreted as discrete-time, multi-input-multi-output (MIMO) systems. In the last section of this chapter appropriate silicon area, timing, and energy cost models for high-throughput LDPC decoders, which reproduce accurately the non-linear dependencies and being applicable to bit-parallel as well as to bit-serial decoder architectures are presented. These models allow for a quantitative comparison of different decoder architectures revealing the most area and energy efficient architecture for a given code and throughput specification. Additionally, a new highly area and energy efficient architecture based on a bit-serial interconnect is derived. This architecture is the result of a systematic architecture search and proper optimization based on the cost models.

2. Multi-conductor interconnects and on-chip antennas

2.1. Multi-conductor interconnects

With the increase of the on-chip data transfer rate to several 10 Gbit/s the spatio-temporal intersymbol interference (auto-interference) within the multiwired bus systems becomes a limiting factor for the circuit performance. Due to the limited available space for the bus systems shielding between the wires of the bus should be omitted. This allows for larger wire cross sections and thereby to reduce the signal distortion. An appropriate signal coding and signal processing will compensate for the effects of the coupling between the wires.

The wiring inside high speed MOS circuits exhibits sub-micron cross-sectional dimensions and conductor width and conductor thickness are of similar size. Within the signal frequency band the cross sectional dimensions are in the order of the skin effect penetration depth. The signal transmission properties of the bus system is detrained by the capacitances per unit of length and the resistance per unit of length.

The TEM modes of a lossless multiconductor transmission line with equidistant conductors of equal cross section and filled with homogenous isotropic dielectric material used for bus have been discussed in [18]. Figure 1 a shows the cross-sectional drawing of the bus. The quasi-electrostatic parameters of the bus embedded in the substrate between two ground planes have been computed. The bus capacitance per unit of length – matrix, describing capacitances with respect to ground and mutual capacitances, has been derived from the conductor geometry using an analytical technique based on even-odd mode analysis [10, 18]. The analytical technique is based on the inversion of the Schwarz-Christoffel conformal mapping [5, pp. 191–201]. The advantages of the proposed method are its accuracy, the lack of geometrical limitations and the algorithm efficiency. The results for the ground and coupling

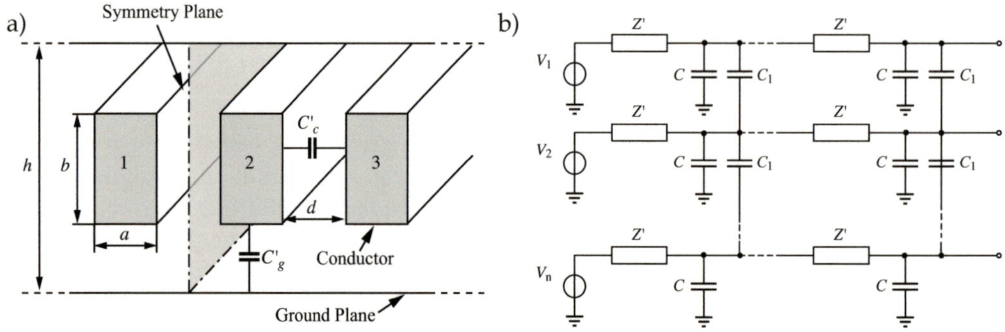

Figure 1. a) A cross section of three-wire digital bus with a coupling and a ground capacitance [10] and b) equivalent lumped element circuit [8].

capacitances per unit length for the multi conductor transmission line, filled in with silicon, are presented in Fig. 2. Since the capacitance depends only on the ratio of the line dimensions, all geometrical data are normalized to the distance h between the ground planes.

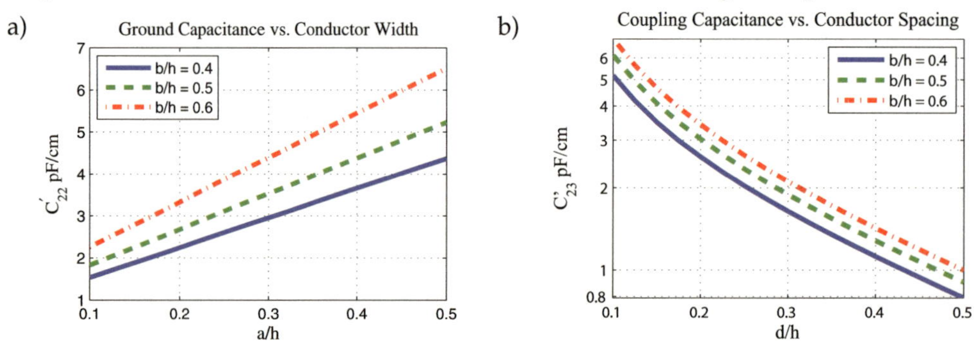

Figure 2. a) Ground capacitance vs. geometry for digital transmission line, filled in with Si, $d/h = 0.125$, b) Coupling capacitance vs. geometry for digital transmission line, filled in with Si, $a/h = 0.25$ [10].

The obtained results have been used to compute the transmission line parameters of the bus [7, 8, 10, 18]. The bus model is based on multiconductor TEM transmission line theory [5, pp. 356–363]. In case of the TEM transmission line the inductance per unit of length matrix follows directly from the capacitance per unit of length matrix and the material [18]. in case of small conductor cross sections the resistance per unit of length becomes such high that the inductance per unit of length matrix can be neglected in comparison with the resistantce per unit of length. In this case the impedance per unit of length matrix becomes diagonal [8]. The ohmic losses in the conductors are modeled by resistance per unit length R'. These parameters determine the lumped element equivalent circuit of the bus shown in Fig. 1 b.

The crosstalk between the conductors of the bus has been investigated in [18]. Figure 3 a shows the response of the crosstalk voltage at the end of the line. The results, obtained by solving the transmission line equations have been compared with numerical Method of Moment (MoM) full-wave simulations. The analytic model exhibits good accuracy up to frequencies beyond 10 GHz. Figure 3 b shows the pulse distortion at the end of the transmission line. Analytical data computed with the transmission line model have been compared with numerical data obtained from SPICE simulation.

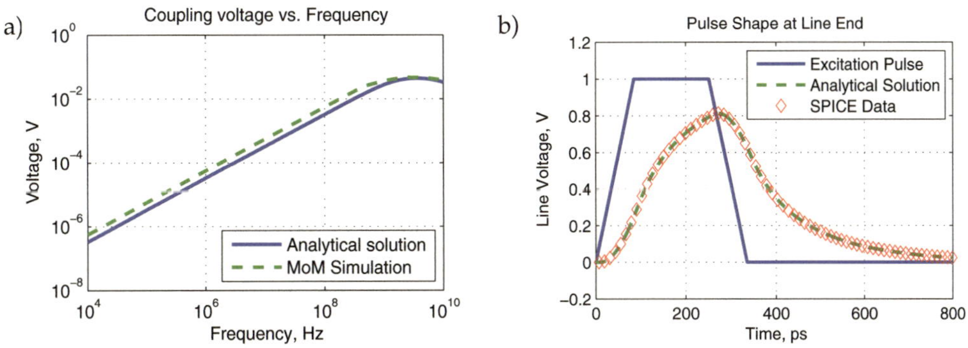

Figure 3. a) Frequency response of the crosstalk voltage at the end of the line, b) pulse response at the end of the line [18].

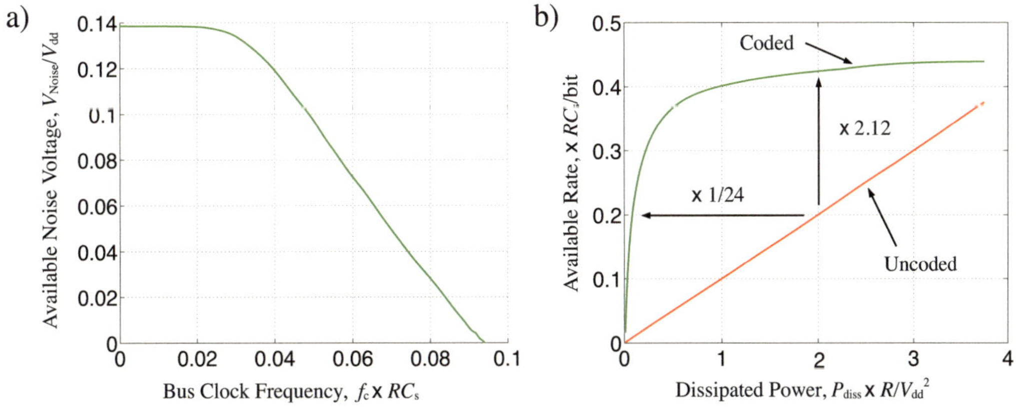

Figure 4. a) Maximum permissible noise-voltage at the receiver of a four-conductor bus as function of the bus clock frequency, b) achievable information rates of coded and uncoded transmission as function of the dissipated power [7].

The space-time intersymbol interference present in on-chip interconnection buses is a limiting factor of the performance of digital integrated circuits. This effect has greater influence as the transfer data rate increases and the circuit dimensions decrease. In order to be able to develop coding techniques for reducing the detrimental effects of intersymbol interference, an efficient and precise method for calculating the impulse response of the interconnect is required [31]. In [7] a quasi-analytical method was applied for computing the impulse response of a digital interconnection bus. The fundamental performance limits of bus systems due to information theory have been analyzed. Figure 3 shows the maximum permissible noise-voltage V_{Noise} at the receiver of a four-conductor bus as function of the bus clock frequency f_c and the achievable information rates of coded and uncoded transmission as function of the dissipated power P_{diss}. The clock frequency for the coded transmission is set to $0.11/(RC_s)$, which is above the cutoff of the uncoded bus and proves to work well with the coded system. Here V_{dd}, R and C_s are the magnitude of the signal voltage at the input, the total resistance, and the total average substrate capacitance, respectively. Here, $C_c = 6C_s$ is assumed, but similar results are obtained for other ratios.

Conclusion

The developed methods allow to compute the impulse response of the multi-conductor bus, and – building on this ground – to compute information theoretic measures, like mutual information. Those measures allow to quantify the possible gains in performance that can be achieved by employing suitable coding schemes to the multi-conductor interconnection bus. The obtained results reveal a huge potential of coded transmission both in terms of increasing the data rate and in decreasing the dissipated power.

2.2. On-chip antennas

An interesting future possibility for handling Gbit/s data streams on chip and from chip to chip will be wireless intra-chip and inter-chip communication. This section describes investigations of integrated on-chip antennas for broad-band intra-chip and inter-chip communications. At frequencies of 60 GHz and beyond antennas can be made sufficiently small to be integrated on monolithic circuits [1, 19]. However, there are still problems when integrating millimeterwave antennas on CMOS circuits. The integration of millimeterwave antennas on silicon requires a high resistivity substrate in order to achieve low losses, whereas for CMOS circuits the substrate resistivity has to be low in order to provide isolation of the circuit elements. Furthermore, chip surface is a cost factor and should not be wasted for antennas.

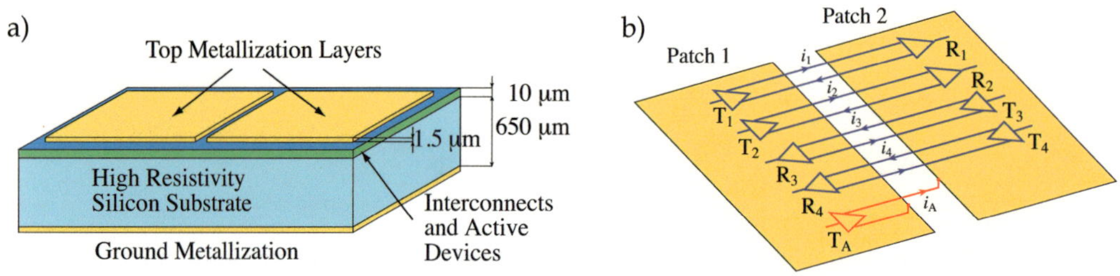

Figure 5. a) Schematic drawing of a chip with an integrated antenna, b) Differential lines, connecting the digital circuits under the separate antenna patches.

An integrated on-chip antenna for chip-to-chip wireless communication, based on the usage of the digital circuits' ground planes as radiating elements was presented in [12–17]. Figure 5 a shows schematically the realization of this principle in slicon technology. The integrated circuit is fabricated on a high resistivity silicon substrate ($\geq$ 1kΩ·cm) with a thickness in the order of of 650 μm. The substrate is backed by a metallic layer. On top of the substrate a low-resistivity layer ($\approx$ 5Ω·cm) of few micrometer thickness is grown. A homogeneous low-resistivity layer of 3 μm to 5 μm thickness is followed by a top with embedded CMOS circuitry and the interconnects. A low resistivity top layer is required for the circuit insulation. The electromagnetic field of the circuits is mainly confined in this top layer. The antenna field is spreading over the whole thickness of the substrate. Due to the high resistivity of the substrate the antenna losses are low. Since only a small fraction of the antenna near-field energy is stored in the low-resistivity layer, the coupling between the antenna near-field and the circuit field is weak. Furthermore, the interference between the CMOS circuits and the antenna field can be reduced when the main part of the circuit is operating in a frequency band distinct from the frequency band used for the wireless transmission.

The utilization of the electronic circuit ground planes as radiating elements for the integrated antennas allows for optimal usage of chip area, as the antennas share the chip area with the circuits. It has to be taken care that the interference between the antenna field and the field propagating in the circuit structures stays within tolerable limits. Consider the structure represented schematically in Fig. 5 b. The structure contains two antenna patches 1 and 2. Both antenna patches serve as the ground planes of circuits. These circuits contain line drivers $T_1 \ldots T_4$ driving over symmetrical interconnection lines the line receivers $R_1 \ldots R_4$. Furthermore there is a driver T_A, the output of which is connected to both patches, however only one conductor bridges the gap between the patches. The currents $i_1 \ldots i_4$ all are flowing back over the symmetric lines. The sum of the currents $i_1 \ldots i_4$ flowing in both directions through the transmission line modes vanishes and is not exciting the antenna. Different from this, the current i_A excites an antenna radiation mode. The circuit for this current is closed via the displacement current in the near-field of the antenna. By exciting the interconnection structures in transmission line modes and the antennas in antenna modes the interference between circuit and antennas can be minimized. We need not to use differential lines between the patches. In general a interconnection structure consisting of N conductors can guide up to $N-1$ quasi-TEM transmission line modes and one antenna mode.

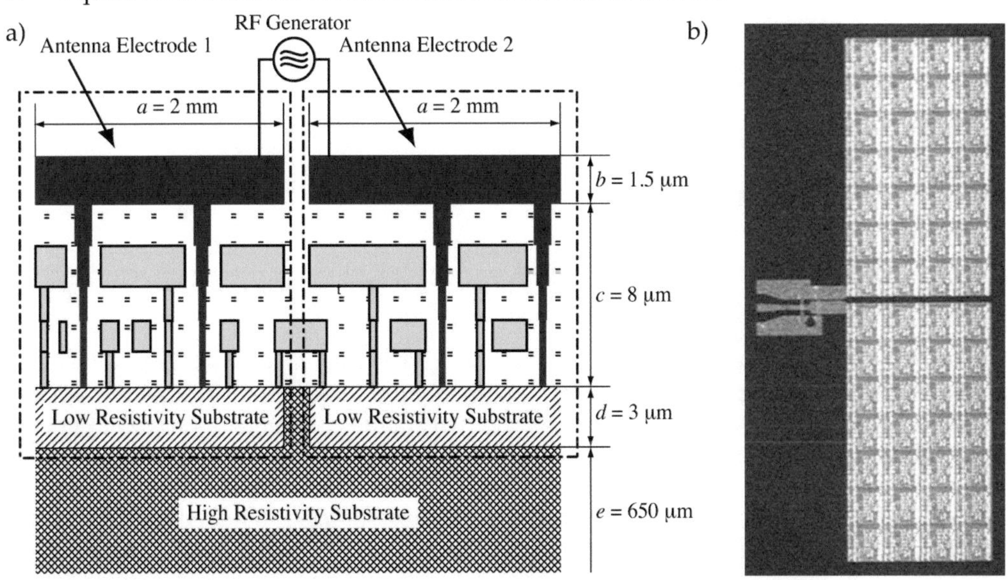

Figure 6. Cross-sectional view of integrated on-chip antenna, using the ground planes as antenna electrodes [11, 14].

Figure 6 a shows a cross-sectional view of the integrated on-chip antenna, using the ground planes in layer b on top of the integrated circuit as the antenna electrodes. Layer c with a total thickenss of 8 μm contains the active and passive circuit elements and the interconnect wiring. The low-resistivity layer d provides isolation of the circuit elements. The thick substrate layer e is of high resistivity. Figure 6 b shows a photograph of the fabricated open-circuit slot antenna with CMOS circuits under the antenna electrode [15]. Figure 7 a shows the simulated current distribution of a two-patch V-band antenna [14, 15]. The current distribution in both patches mainly is concentrated in the neighborhood of the slots. The antenna behaves as an open-circuited slot antenna. The guided wavelength is in the range of a millimeter. The open-circuited slot with a length of about one millimeter is a transmission line resonator with a resonance frequency in the V-band. The standing wave in the slot excites the radiation field.

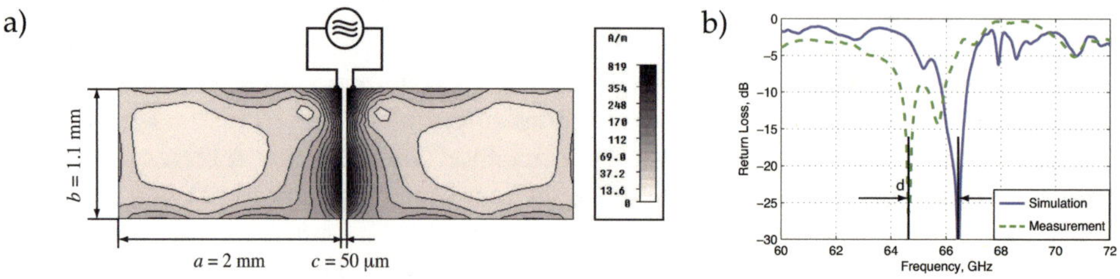

Figure 7. (a) Top view and current distribution of a two-patch dipole antenna, operating at 66 GHz, (b) Measured return loss of the on-chip open slot antenna [14, 15].

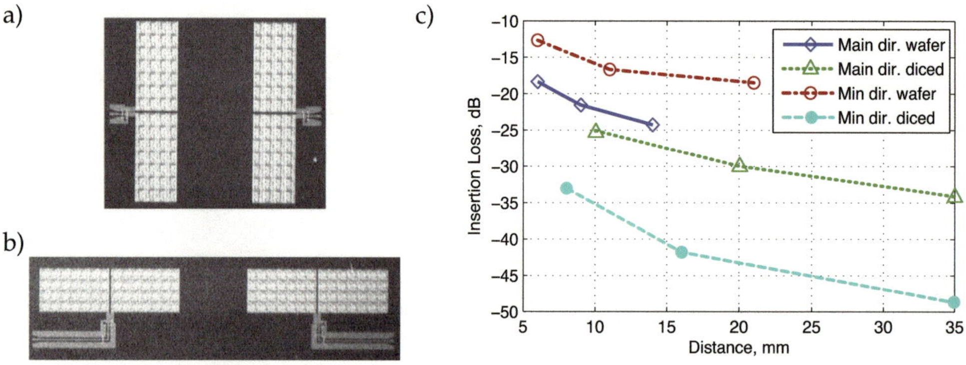

Figure 8. Antenna orientation with a) collinear and b) parallel slots, c) Measured insertion loss of a wireless chip-to-chip link depending [14, 15].

The antennas have been measured on-wafer and diced. The measured return loss of the diced slot antenna from Fig. 7 a is compared with the simulation results in Fig. 7 b. The insertion loss of a transmission link has been measured for the two antenna alignments shown in Figs. 8 a and b, where the antennas were positioned in each other radiation minima and maxima. Figure 8 s c shows the measured insertion loss of a wireless links formed by two antennas. When the antennas are oriented such that their slots are collinear, they are in each other's direction of minimum radiation. When they are oriented such that their slots are parallel, they are in each other's direction of maximum radiation. Both cases were investigated for on-wafer and for diced chips. The chip-to-chip links with both antennas on different chips exhibit higher insertion loss. The lower insertion loss of links between antennas on the same chip is due to the contribution of surface waves. The worst-case transmission link (gain-chip-to-chip link) in the direction of minimum radiation shows an insertion loss of -47dB, which is sufficient for high-rate data links.

Lumped element circuit models can provide a compact description of wireless transmission links [20–22, 25]. Distributed circuits can be modeled also in a broad frequency band with arbitrary accuracy using lumped element network models. A general way to establish network models is based on modal analysis and similar techniques [2–5]. In the case of wireless transmission links, high insertion losses have to be considered. Therefore methods for the synthesis of lossy multiports have to be applied. In [23, 24] a lumped-element two-port antenna model is presented where the antenna near-field is modeled by a reactive two-port and the real resistor R_r terminating the two-port models the energy dissipation in the far-field.

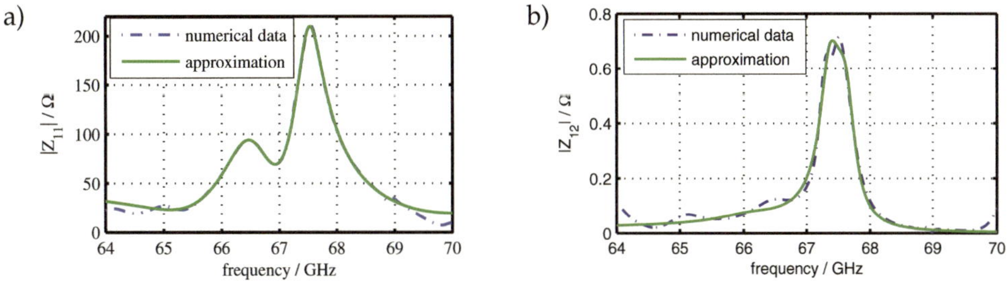

Figure 9. Comparison of the numerical data of (a) $|Z_{11}|$ and (b) $|Z_{12}|$ obtained from the full-wave simulation of the wireless transmission link with the data computed from lumped element model

Figure 9 shows the comparison of the numerical data of the magnitudes of the two-port impedance parameters $|Z_{11}|$ and $|Z_{12}|$ obtained from the full-wave simulation of the wireless transmission link with the data computed from lumped element model. For details of the model see [3]. The numerical full-wave simulations have been performed using CST software. An accurate model of Z_{11} is achieved for the frequency band from 65 GHz to 69 GHz based on four pairs of poles and two single poles at zero and infinity. The frequency range may be extended by increasing the number of poles.

Conclusion

We have investigated methods for an area-efficient design of on-chip integrated antennas, based on the utilization of the same metallization structures both as a CMOS circuit ground plane and as antenna electrodes. An experimental setup has been designed for validating the computed antenna parameters, as well as the interference between the CMOS interconnects and the antenna. Equivalent circuits have been established to model integrated antennas and wireless intra-chip and inter-chip transmission links.

3. Communication theoretical limits, coding and signal processing

Both multiconductor interconnects and wireless multiantenna interconnects can be interpreted as discrete-time, multi-input-multi-output (MIMO) systems. Such systems have been subject to extensive study in the recent past in the field of digital, especially mobile communications. Starting from the analysis of their promising information theoretic capabilities (e.g., [41]), a large amount of signal processing and coding techniques have been developed, that aim at achieving the information theoretic bounds (e.g., [42–44, 46]).

The common approach to handle spatio-temporal interference in MIMO systems, involves either linear or non-linear transmit and receive signal processing, which job is to transform the original MIMO system into a »virtual« MIMO system, where large amounts of spatio-temporal interference have been removed [48]. All state of the art MIMO signal processing techniques have in common that they assume that either the receiver, the transmitter, or both, have access to, or can generate signals with arbitrary precision. This implies, in practice, the existence of ADC and DAC components with a large enough resolution such that the non-linear effects of signal quantization can be neglected. However, in multiconductor, or wireless

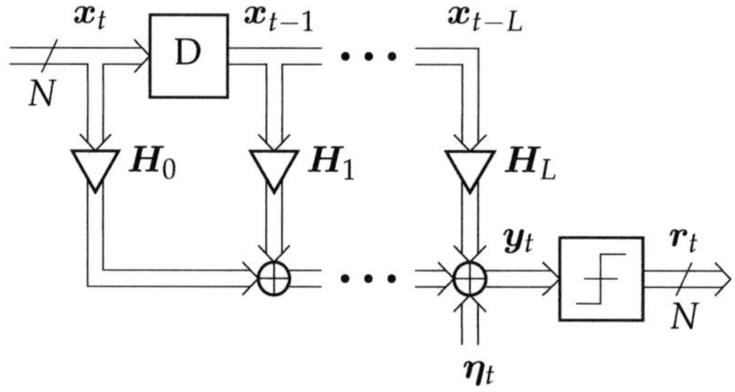

Figure 10. MIMO channel with ISI and single-bit outputs modeling wired as well as the wireless interconnects.

multiantenna interconnects, used for high-speed on-chip or chip-to-chip communication, such an assumption of having available high-resolution ADC and DAC components, cannot be made.

In case of on-chip multiconductor interconnects, the DAC and the ADC components are formed by the output or the input of a logic CMOS inverter, respectively. Hence, the ADC and DAC components, perform a *single-bit* conversion between the analog and the digital domain. With such coarse quantization, all state of the art techniques for MIMO signal processing fail.

In the case of wireless multiantenna interconnects for chip-to-chip communication the situation can be expected to be slightly better. However, because of the huge bandwidth, the requirements on conversion time are extremely high, such that only moderate resolution (4–5 bits) ADC and DAC components are reasonable. As it turns out, such a moderately high resolution is still too low for reliable operation of state of the art MIMO signal processing.

In this section, we treat the ADC and DAC components as an integral part of the MIMO system. We develop signal processing and coding techniques, which utilize the information theoretic gains of MIMO systems with very-low to moderately-low resolution signal quantization. We first provide suitable design principles for low-latency channel-matched codes applied on general frequency selective MIMO channels, which are based on an information theoretic ground.

3.1. Single-bit ADC/DAC: Coding and performance limits

Consider the MIMO channel with inter-symbol interference (ISI) and single-bit output quantization shown in Fig. 10. The channel has a memory of length L and it is governed by the channel law

$$\boldsymbol{r}_t = \mathcal{Q}\left\{\boldsymbol{y}_t\right\} = \mathcal{Q}\left\{\sum_{k=0}^{L} \boldsymbol{H}_k \boldsymbol{x}_{t-k} + \boldsymbol{\eta}_t\right\}. \tag{1}$$

Here, $\boldsymbol{H}_k \in \mathbb{C}^{N\times N}$ is the k-th channel matrix. $\boldsymbol{x}_k \in \mathcal{X}$, $\boldsymbol{\eta}_k \in \mathbb{C}^N$, $\boldsymbol{y}_k \in \mathbb{C}^N$ and $\boldsymbol{r}_k \in \mathcal{Y} = \{\alpha + \mathrm{j}\beta | \alpha, \beta \in \{+1, -1\}\}^N$ denote the channel input vector, the noise vector, the unquantized receive vector and the channel output vector, at the k-th time instant, respectively. The single-bit quantization operator $\mathcal{Q}$ returns the sign of the real and imaginary part of each component of the unquantized received signal $\boldsymbol{r}_t$, i.e.,

$$\mathcal{Q}\left\{\boldsymbol{y}_t\right\} = \mathrm{sign}(\mathrm{real}\{\boldsymbol{y}_t\}) + \mathrm{j} \cdot \mathrm{sign}(\mathrm{imag}\{\boldsymbol{y}_t\}). \tag{2}$$

The conditional probability of the channel output satisfies

$$\Pr(\boldsymbol{r}_t | \boldsymbol{x}_{-L}^{\infty}, \boldsymbol{r}_1^{t-1}, \boldsymbol{r}_{t+1}^{\infty}) = \Pr(\boldsymbol{r}_t | \boldsymbol{x}_{t-L}^{t}),\ t \geq 0. \tag{3}$$

Here, $\boldsymbol{x}_0^\infty$ and $\boldsymbol{r}_1^{t-1}$ stand for the sequences $[\boldsymbol{x}_{-L}, x_{-L+1}, \ldots]$ and $[\boldsymbol{r}_1, \boldsymbol{r}_2 \ldots, \boldsymbol{r}_{t-1}]$, respectively. The noise is additive white Gaussian with covariance matrix $\mathrm{E}[\boldsymbol{n}_t \boldsymbol{n}_t^{\mathrm{H}}] = \sigma_\eta^2 \boldsymbol{I}_N$. The transmit signal energy is normalized to 1, that is $\|\boldsymbol{x}_t\|_2 = 1$. The signal-to-noise ratio is accordingly defined as

$$\mathrm{SNR} = 1/\sigma_\eta^2. \tag{4}$$

The channel transition probabilities can be calculated via

$$\Pr(\boldsymbol{r}_t | \boldsymbol{x}_{t-L}^t) = \prod_{c \in \{\mathbb{R}, \mathbb{I}\}} \prod_{i=1}^{N} \Phi\left(\frac{[\boldsymbol{y}_t]_{c,i} \cdot [\bar{\boldsymbol{y}}]_{c,i}}{\sigma_\eta / \sqrt{2}}\right). \tag{5}$$

Here, $[\boldsymbol{x}_t]_{\mathbb{R},i}$ ($[\boldsymbol{x}_t]_{\mathbb{I},i}$) denotes the i-th real (imaginary) component of the input vector $\boldsymbol{x}_t$, $\bar{\boldsymbol{y}} = \sum_{k=0}^{L} \boldsymbol{H}_k \boldsymbol{x}_{t-k}$ is the noise-free unquantized receive vector and $\Phi(x) = \frac{1}{\sqrt{2\pi}} \int_{-\infty}^{x} e^{-\frac{t^2}{2}} dt$ is the cumulative normal distribution function. The input symbols are modulated using QPSK (or BPSK with 1-bit DAC). Consequently, we consider ISI channels with the input and output cardinality $|\mathcal{X}| = |\mathcal{R}| = 4^N$.

3.1.1. Code-design

We are looking for codes which maximally increase throughput and that allow fast decoding with good performance. In this way, the code design has to focus on finding low-complexity codes providing good coding gain while having low overhead both with respect to circuit complexity and power dissipation.

Even though linear block and convolution coding schemes are favorable candidates for error correction, they are not able to decrease power consumption and eliminate the residual error floor caused by the crosstalk even in the noiseless case [31]. This is due to their structural properties (linear codes) and the coarse quantization of the channel.

Therefore, the coding schemes which are needed are non-linear and – for having good performance and, at the same time, a low complexity – for instance have memory of order one. In [51–53], an information-theoretic framework was developed as a practical design guideline for novel codes. To this end, the following optimization has been considered in [52]

$$\begin{aligned} C_{M_s}^{\text{uniform}} = \max_{P_{ij}:(i,j) \in \mathcal{T}} \mathcal{I}(\boldsymbol{x}; \boldsymbol{y}) \quad \text{s.t.} \quad & P_{ij} = \Pr(S_k = j | S_{k-1} = i) \in \{0, 2^{-K}\}, \\ & S_k = \boldsymbol{x}_{k-M_s+1}^k \in \mathcal{X}^{M_s}, \\ & M_s \geq L, \end{aligned} \tag{6}$$

where K defines the rate $R = K/(2N)$ of the code (for QPSK modulation) and $C_{M_s}^{\text{uniform}}$ is the maximum channel capacity that can be attained with a homogeneous Markov source of order M_s. In general, the capacity of an unconstrained Markov source of order M_s [30] is higher than this *uniform capacity* ($P_{i,j}$ takes 0 or 2^{-K}), i.e., $C_{M_s}^{\text{uniform}} \leq C_{M_s}$.
This coding approach incorporates the following four ideas:

1. In order to avoid the complexity of maximizing an arbitrary Markov source, we restrict the optimization to homogeneous Markov sources.

2. We choose the memory length M_s of the source to be roughly as long as the number of channel taps L but not shorter. The reason is that the information rate of a Markov source of order $M_s = L$ is noticeably larger compared to the i.u.d. capacity, but memory lengths above L yield only a small additional gain in information rate.
3. As we want to avoid the use of distribution shapers, the number of transmit symbols is fixed at 2^K (irrespective of the current state). Thus, the encoder can be realized as a look-up table and we obtain the data encoding rule

$$\boldsymbol{x}_n = \text{ENC}(\boldsymbol{d}_n, [\boldsymbol{x}_{n-1}, \dots, \boldsymbol{x}_{n-M_s}]),$$

where $\boldsymbol{d}_n \in \{0,1\}^K$ is the data vector of the source. Using QPSK modulation the realizable code rates R are

$$\frac{K}{2N} \in \left\{ \frac{1}{2N}, \frac{2}{2N}, \cdots, 1 \right\}.$$

4. The optimized transition probabilities are uniformly distributed and at the same time they approximate the capacity-achieving input distribution. Hence, the optimized transition probability matrix P_{ij} serves as an inner code that can be concatenated with an outer Turbo-like code, e.g. *low density parity check code* (c.f. section 4), in order to reach information rates (well) above the i.u.d. capacity [53].

Fig. 11 illustrates the channel model together with the encoder and decoder. The optimization

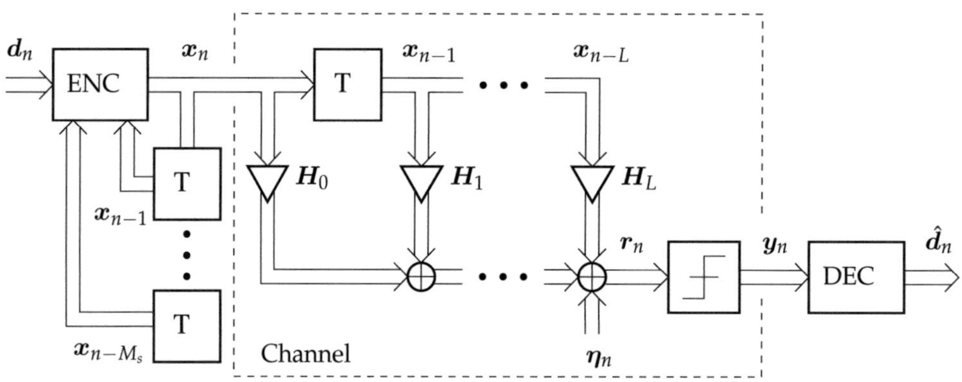

Figure 11. Encoding and Decoding for MIMO Channels with ISI and Single-Bit Output Quantization.

(6) is non-convex but can be solved by an efficient greedy algorithm [52] that delivers an optimized transition probability matrix $\boldsymbol{P} = [P_{i,j}]$ that maximizes the mutual information between the input and the output.

In Figure 12, a coded bus system employing a memory-based code with a code rate of K/N is shown. A fixed bus access time $T_{\text{CU}}^{\text{cod}}$ is chosen such that two channel temporal taps are significant ($L = 1$). The encoding scheme is time-invariant and has the property that the data vector, $\boldsymbol{d}_n = [d_1[n], \dots, d_K[n]]^{\text{T}} \in \{0,1\}^{K\times 1}$, is encoded and decoded instantaneously (without latency). The actual code vector $\boldsymbol{x}_n$ depends on the input data vector $\boldsymbol{d}_n$ and the previous transmitted vector $\boldsymbol{x}_{n-1}$.

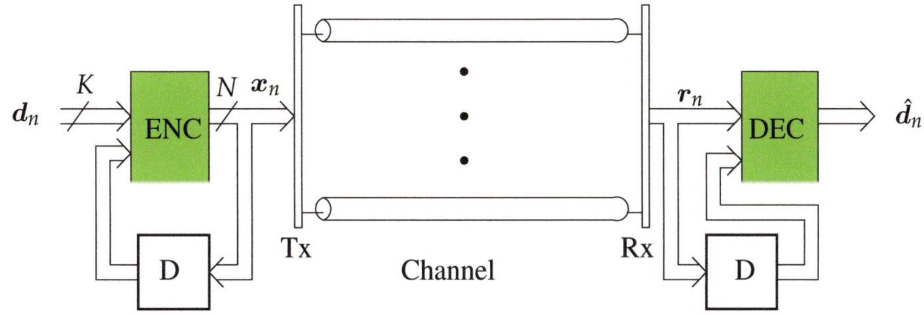

Figure 12. Noisy bus communication with a memory-based Code.

$x_{n-1} \setminus$ Data	000	001	010	011	100	101	110	111
0000	0000	0001	0011	0111	1000	1100	1110	1111
0001	0000	0001	0111	1000	1001	1100	1101	1111
0010	0000	0010	0011	0100	1010	1011	1110	1111
0011	0000	0010	0011	0100	1010	1011	1110	1111
0100	0000	0001	0100	0101	0111	1100	1101	1111
0101	0000	0001	0100	0101	0111	1100	1101	1111
0110	0000	0011	0100	0110	0111	1100	1110	1111
0111	0000	0001	0100	0110	0111	1100	1110	1111
1000	0000	0001	0011	1000	1001	1011	1110	1111
1001	0000	0001	0011	1000	1001	1011	1100	1111
1010	0000	0010	0011	0100	1010	1011	1110	1111
1011	0000	0010	0011	0100	1010	1011	1110	1111
1100	0000	0001	0100	0101	0111	1100	1101	1111
1101	0000	0001	0100	0101	0111	1100	1101	1111
1110	0000	0010	0011	0110	0111	1000	1110	1111
1111	0000	0001	0011	0111	1000	1100	1110	1111

Table 1. The mapping function of a 3/4 optimized code: $\boldsymbol{x}_n = \mathrm{ENC}(\boldsymbol{d}_n, \boldsymbol{x}_{n-1})$.

At the receiver side, the decoder uses the value of the current and the previous channel outputs to reconstruct an estimate of the data vector $\hat{\boldsymbol{d}}_n$ as

$$\hat{\boldsymbol{d}}_n = g(\boldsymbol{y}_n, \boldsymbol{y}_{n-1}) = \arg\max_{\boldsymbol{d}} \Pr\left(\boldsymbol{d}_n = \boldsymbol{d} \mid \boldsymbol{y}_n, \boldsymbol{y}_{n-1}\right). \tag{7}$$

Obviously, the mapping done by this function performs a maximum-likelihood estimation of $\boldsymbol{d}_n$ based on $\boldsymbol{y}_n$ and $\boldsymbol{y}_{n-1}$ [1].

Although this approach seems quite heuristic, its usefulness can be demonstrated by simulation. Table 1 lists the mapping function of a code designed for a bus with $N = 4$ mutually coupled, tapped RC lines as shown in Fig. 1 used at $T_{\mathrm{CU}} = 9RC_{\mathrm{s}} = 1.5RC_{\mathrm{c}}$ symbol time, where R is the serial resistance, C_{s} is the ground capacitance and C_{c} is the coupling capacitance (c.f. section 2.1).

[1] The reliability of the decoding can be improved if earlier or later outputs are considered using the BCJR algorithm (forward-backward algorithm), at the cost of some latency.

Its performance in terms of symbol error rate (SER) when applied to a noisy bus system, compared to uncoded transmission, is shown in Figure 13. The uncoded transmission reveals an error floor (a residual SER at vanishing noise variance) due to signaling belong the RC–specific time. However, as we see in Figure 13, the optimized code does not see any error floor. Besides, it turns out that the achievable power savings of this code (in terms of energy per transmitted information bit) is 40%, without taking into account the power overhead of the codec circuit. The SER curve of a space-only code, which has been optimized by exhaustive search, is also plotted. Due to its simplicity, this code performs inherently worse than the discussed memory-based code. Although several coding schemes can be found in the literature [29, 32], such a unified framework that jointly address power, rate, and reliability aspects, simultaneously is new.

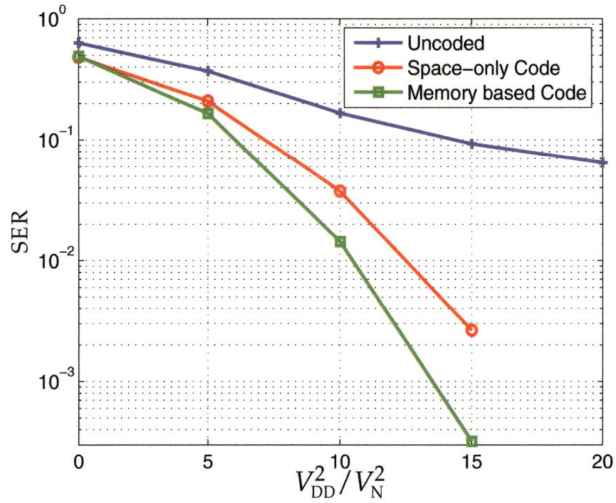

Figure 13. Symbol error rate (SER) as function of V_{DD}^2/V_N^2 for an interconnect with four lines, employing a memory-based code of rate 3/4, compared to the uncoded case. The performance of a space-only code is also plotted.

We note that, for large buses, it is impractical to encode all bits at once because of the large complexity in the design and the implementation of the codec circuit. Therefore, partial coding can be employed in which the bus is partitioned into sub-buses of smaller width, which are encoded separately. The partitioning requires some additional wires since a shielding wire has to be placed between every two adjacent sub-buses.

3.2. Low-resolution ADC: Linear signal-processing

In the following, we concentrated on receive signal processing and our aim is to study the applicability of standard equalization techniques for our application, where the receiver is equipped with a low to moderate ADC for each antenna or port. A modified version of the standard linear receiver designs is presented in the context of MIMO communication with quantized output, taking into account the presence of the quantizer. An essential aspect of our analysis is that no assumption of uncorrelated white quantization errors is made. The performance of the modified receiver designs as well as the effects of quantization are studied theoretically and experimentally. Thereby, perfect channel state information (CSI) at the receiver is assumed, which can be obtained even with coarse quantization as discussed in Section 3.3.

In [33], the joint optimization of the linear receiver and the quantizer in a MIMO system is addressed. The figure of merit that has been used for the design of the optimum quantizer and receiver is the *mean square error* (MSE). Based on this MSE approach, the communication performance (in terms of channel capacity) of the quantized MIMO channel is studied. Our work [34] generalizes this modified MMSE filter to frequency selective channels. Motivated by the same approach, the authors of [36] optimized the Decision Feedback Equalizer (DFE) for the flat MIMO channel with quantized outputs.

In this and the following Section, we provide a summary of these works. Throughout these sections, $r_{\alpha\beta}$ denotes $\mathrm{E}[\alpha\beta^*]$. The operators $(\bullet)^{\mathrm{T}}$, $(\bullet)^{\mathrm{H}}$, $(\bullet)^*$, $\mathrm{tr}[\bullet]$ stand for transpose, Hermitian transpose, complex conjugate, and trace of a matrix, respectively.

3.2.1. System model

Let us now consider a point to point MIMO Gaussian channel, where the transmitter operates M antennas and the receiver employs N antennas. Figure 14 shows the general form of a quantized MIMO system, where $\boldsymbol{H} \in \mathbb{C}^{N\times M}$ is the channel matrix. For simplicity, inter-symbol interference (ISI) is ignored, even though considering it would be straightforward. The vector $\boldsymbol{x} \in \mathbb{C}^M$ comprises the M transmitted symbols with zero-mean and covariance $\boldsymbol{R}_{xx} = \mathrm{E}[\boldsymbol{x}\boldsymbol{x}^{\mathrm{H}}]$. The vector $\boldsymbol{\eta}$ refers to zero-mean complex circularly symmetric Gaussian noise with a covariance matrix $\boldsymbol{R}_{\eta\eta} = \mathrm{E}[\boldsymbol{\eta}\boldsymbol{\eta}^{\mathrm{H}}]$, while $\boldsymbol{y} \in \mathbb{C}^N$ is the unquantized channel output:

$$\boldsymbol{y} = \boldsymbol{H}\boldsymbol{x} + \boldsymbol{\eta}. \tag{8}$$

In our system, the real parts $y_{i,\mathrm{R}}$ and the imaginary parts $y_{i,\mathrm{I}}$ of the receive signals y_i, $1 \le i \le N$, are each quantized by a b-bit resolution uniform/non-uniform scalar quantizer. Thus, the resulting quantized signals are given by

$$r_{i,l} = Q(y_{i,l}) = y_{i,l} + q_{i,l},\ l \in \{\mathrm{R},\mathrm{I}\},\ 1 \le i \le N, \tag{9}$$

where $Q(\cdot)$ denotes the quantization operation and $q_{i,l}$ is the resulting quantization error. The matrix $\boldsymbol{G} \in \mathbb{C}^{M\times N}$ represents the receive filter, which delivers the estimate $\hat{\boldsymbol{x}}$

$$\hat{\boldsymbol{x}} = \boldsymbol{G}\boldsymbol{r}. \tag{10}$$

Our aim is to choose the quantizer and the receive matrix $\boldsymbol{G}$ minimizing the MSE $= \mathrm{E}[\|\hat{\boldsymbol{x}} - \boldsymbol{x}\|_2^2]$, taking into account the quantization effect. Since the ADC can drastically affect the performance of the system, it should be also designed carefully.

3.2.2. Quantizer characterization

Each quantization process can be given a distortion factor $\rho_q^{(i,l)}$ to indicate the relative amount of quantization noise generated, which is defined as follows

$$\rho_q^{(i,l)} = \frac{\mathrm{E}[q_{i,l}^2]}{r_{y_{i,l}y_{i,l}}}, \tag{11}$$

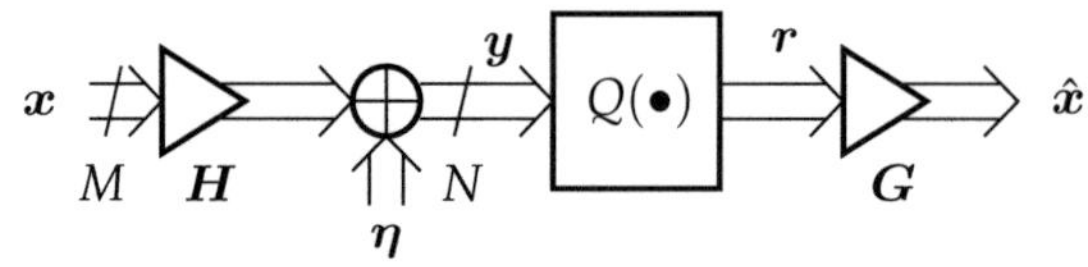

Figure 14. Quantized MIMO System.

where $r_{y_{i,l}y_{i,l}} = \mathrm{E}[y_{i,l}^2]$ is the variance of $y_{i,l}$ and the distortion factor $\rho_q^{(i,l)}$ depends on the number of quantization bits b, the quantizer type (uniform or non-uniform) and the probability density function of $y_{i,l}$. Note that the signal-to-quantization noise ratio (SQNR) has an inverse relationship with regard to the distortion factor. The uniform/non-uniform quantizer design is based on minimizing the *mean square error* (distortion) between the input $y_{i,l}$ and the output $r_{i,l}$ of each quantizer. In other words, the SQNR values are maximized. With this optimal design of the scalar finite resolution quantizer, whether uniform or not, the following equations hold for all $0 \le i \le N, l \in \{R, I\}$ [35, 37, 38]

$$\mathrm{E}[r_{i,l}q_{i,l}] = 0 \tag{12}$$

$$\mathrm{E}[y_{i,l}q_{i,l}] = -\rho_q^{(i,l)} r_{y_{i,l}y_{i,l}}. \tag{13}$$

Obviously, (13) follows from (11) and (12). Under multipath propagation conditions and for large number of antennas, the quantizer input signals $y_{i,l}$ will be approximately Gaussian distributed and thus, they undergo nearly the same distortion factor ρ_q, i.e., $\rho_q^{(i,l)} = \rho_q \ \forall i \forall l$. Furthermore, the optimal parameters of the uniform as well as the non-uniform quantizer and the resulting distortion factor ρ_q for Gaussian distributed signal are tabulated in [35] for different resolutions b.

Now, let $q_i = q_{i,\mathrm{R}} + \mathrm{j}q_{i,\mathrm{I}}$ be the complex quantization error. Under the assumption of uncorrelated real and imaginary part of y_i, the following relations are obtained

$$r_{q_iq_i} = \mathrm{E}[q_iq_i^*] = \rho_q r_{y_iy_i}, \text{ and } r_{y_iq_i} = \mathrm{E}[y_iq_i^*] = -\rho_q r_{y_iy_i}. \tag{14}$$

This particular choice of the (non-)uniform scalar quantizer minimizing the distortion between $\boldsymbol{r}$ and $\boldsymbol{y}$, combined with the receiver developed in the next Section, is also optimal with respect to the total MSE between the transmitted symbol vector $\boldsymbol{x}$ and the estimated symbol vector $\hat{\boldsymbol{x}}$, as we will see later.

3.2.3. Nearly optimal linear receiver

The linear receiver $\boldsymbol{G}$ that minimizes the MSE, $\mathrm{E}[\|\boldsymbol{\varepsilon}\|_2^2] = \mathrm{E}[\|\boldsymbol{x} - \hat{\boldsymbol{x}}\|_2^2] = \mathrm{E}[\|\boldsymbol{x} - \boldsymbol{G}\boldsymbol{r}\|_2^2]$, can be written as:

$$\boldsymbol{G} = \boldsymbol{R}_{xr}\boldsymbol{R}_{rr}^{-1}, \tag{15}$$

and the resulting MSE equals

$$\mathrm{MSE} = \mathrm{tr}\left(\boldsymbol{R}_{\varepsilon\varepsilon}\right) = \mathrm{tr}\left(\boldsymbol{R}_{xx} - \boldsymbol{R}_{xr}\boldsymbol{R}_{rr}^{-1}\boldsymbol{R}_{xr}^{\mathrm{H}}\right), \tag{16}$$

where $\boldsymbol{R}_{xr}$ equals

$$\boldsymbol{R}_{xr} = \mathrm{E}[\boldsymbol{x}\boldsymbol{r}^{\mathrm{H}}] = \mathrm{E}[\boldsymbol{x}(\boldsymbol{y}+\boldsymbol{q})^{\mathrm{H}}] = \boldsymbol{R}_{xy} + \boldsymbol{R}_{xq}, \tag{17}$$

and $\boldsymbol{R}_{rr}$ can be expressed as

$$\boldsymbol{R}_{rr} = \mathrm{E}[(\boldsymbol{y}+\boldsymbol{q})(\boldsymbol{y}+\boldsymbol{q})^{\mathrm{H}}] = \boldsymbol{R}_{yy} + \boldsymbol{R}_{yq} + \boldsymbol{R}_{yq}^{\mathrm{H}} + \boldsymbol{R}_{qq}. \tag{18}$$

We have to determine the linear filter $\boldsymbol{G}$ as a function of the channel parameters and the quantization distortion factor ρ_q. To this end, we derive all needed covariance matrices by using the fact that the quantization error q_i, conditioned on y_i, is statistically independent from all other random variables of the system. First we calculate $r_{y_i q_j} = \mathrm{E}[y_i q_j^*]$ for $i \neq j$

$$\begin{aligned}
\mathrm{E}[y_i q_j^*] &= \mathrm{E}_{y_j}\Big[\mathrm{E}[y_i q_j^* | y_j]\Big] \\
&= \mathrm{E}_{y_j}\Big[\mathrm{E}[y_i | y_j]\mathrm{E}[q_j^* | y_j]\Big] \\
&\approx \mathrm{E}_{y_j}\Big[r_{y_i y_j} r_{y_j y_j}^{-1} y_j \mathrm{E}[q_j^* | y_j]\Big] \qquad (19) \\
&= r_{y_i y_j} r_{y_j y_j}^{-1} \mathrm{E}[y_j q_j^*] \\
&= -\rho_q r_{y_i y_j}. \qquad (20)
\end{aligned}$$

Note that, in (19), we approximate the Bayesian estimator $\mathrm{E}[y_i|y_j]$ with the linear estimator $r_{y_i y_j} r_{y_j y_j}^{-1} y_j$, which holds with equality if the vector $\boldsymbol{y}$ is jointly Gaussian distributed. Eq. (20) follows from (14). Summarizing the results of (14) and (20), we obtain

$$\boldsymbol{R}_{yq} \approx -\rho_q \boldsymbol{R}_{yy}. \tag{21}$$

Similarly, we evaluate $r_{q_i q_j}$ for $i \neq j$ using (21), and with (14) we arrive at

$$\boldsymbol{R}_{qq} \approx \rho_q \boldsymbol{R}_{yy} - (1-\rho_q)\rho_q \mathrm{nondiag}(\boldsymbol{R}_{yy}), \tag{22}$$

where nondiag$(\boldsymbol{A})$ obtained from a matrix $\boldsymbol{A}$ by setting its diagonal elements to zero. Inserting the expressions (21) and (22) into (18), we obtain

$$\boldsymbol{R}_{rr} \approx (1-\rho_q)(\boldsymbol{R}_{yy} - \rho_q \mathrm{nondiag}(\boldsymbol{R}_{yy})). \tag{23}$$

Also in a similar way, we get $\boldsymbol{R}_{xq} = \mathrm{E}[\boldsymbol{x}\boldsymbol{q}^{\mathrm{H}}] \approx -\rho_q \boldsymbol{R}_{xy}$, and (17) becomes

$$\boldsymbol{R}_{xr} \approx (1-\rho_q)\boldsymbol{R}_{xy}. \tag{24}$$

In summary, we get from (23) and (24) the following expression for the Wiener filter from (15) operating on quantized data

$$\boldsymbol{G}_{\mathrm{WFQ}} \approx \boldsymbol{R}_{xy}(\boldsymbol{R}_{yy} - \rho_q \mathrm{nondiag}(\boldsymbol{R}_{yy}))^{-1}, \tag{25}$$

and for the resulting MSE, we obtain using (16)

$$\mathrm{MSE}_{\mathrm{WFQ}} \approx \mathrm{tr}\Big[\boldsymbol{R}_{xx} - (1-\rho_q)\boldsymbol{R}_{xy}(\boldsymbol{R}_{yy} - \rho_q \mathrm{nondiag}(\boldsymbol{R}_{yy}))^{-1}\boldsymbol{R}_{xy}^{\mathrm{H}}\Big]. \tag{26}$$

We obtain $\boldsymbol{R}_{yy}$ and $\boldsymbol{R}_{xy}$ easily from our system model

$$\boldsymbol{R}_{yy} = \boldsymbol{R}_{\eta\eta} + \boldsymbol{H}\boldsymbol{R}_{xx}\boldsymbol{H}^{\mathrm{H}}, \tag{27}$$

$$\boldsymbol{R}_{xy} = \boldsymbol{R}_{xx}\boldsymbol{H}^{\mathrm{H}}. \tag{28}$$

Let us examine the first derivative of the MSE_{WFQ} in (26) with respect to ρ_q

$$\frac{\partial \text{MSE}_{\text{WFQ}}}{\partial \rho_q} = \text{tr}\left[\boldsymbol{G}_{\text{WFQ}}\text{diag}(\boldsymbol{R}_{yy})\boldsymbol{G}_{\text{WFQ}}^{\text{H}}\right] > 0, \tag{29}$$

where $\boldsymbol{G}_{\text{WFQ}}$ is given in (25). Therefore the MSE_{WFQ} is monotonically increasing in ρ_q. Since we choose the quantizer to minimize the distortion factor ρ_q, our receiver and quantizer designs are jointly optimum with respect to the total MSE.

3.2.4. *Lower bound on the mutual information and the capacity*

In this section, we develop a lower bound on the mutual information rate between the input sequence $\boldsymbol{x}$ and the quantized output sequence $\boldsymbol{r}$ of the system in Figure 14, based on our MSE approach. Generally, the mutual information of this channel can be expressed as [26]

$$I(\boldsymbol{x},\boldsymbol{r}) = H(\boldsymbol{x}) - H(\boldsymbol{x}|\boldsymbol{r}). \tag{30}$$

Given $\boldsymbol{R}_{xx}$ under a power constraint $\text{tr}(\boldsymbol{R}_{xx}) \leq P_{\text{Tr}}$, we choose $\boldsymbol{x}$ to be Gaussian, which is not necessarily the capacity achieving distribution for our quantized system. Then, we can obtain a lower bound for $I(\boldsymbol{x},\boldsymbol{r})$ (in bit/transmission) as

$$\begin{aligned} I(\boldsymbol{x},\boldsymbol{r}) &= \log_2|\boldsymbol{R}_{xx}| - h(\boldsymbol{x}|\boldsymbol{r}) \\ &= \log_2|\boldsymbol{R}_{xx}| - h(\boldsymbol{x}-\hat{\boldsymbol{x}}|\boldsymbol{r}) \\ &\geq \log_2|\boldsymbol{R}_{xx}| - h(\boldsymbol{x}-\hat{\boldsymbol{x}}) \qquad (31) \\ &\geq \log_2\frac{|\boldsymbol{R}_{xx}|}{|\boldsymbol{R}_{\varepsilon\varepsilon}|}. \qquad (32) \end{aligned}$$

Since conditioning reduces entropy, we obtain inequality (31). On the other hand, the second term in (31) is upper bounded by the entropy of a Gaussian random variable whose covariance is equal to the error covariance matrix $\boldsymbol{R}_{\varepsilon\varepsilon}$ of the linear MMSE estimate of $\boldsymbol{x}$. Finally, we get using (26) and (28)

$$I(\boldsymbol{x},\boldsymbol{r}) \gtrapprox -\log_2\left|\mathbf{I}-(1-\rho_q)\boldsymbol{R}_{xy}(\boldsymbol{R}_{yy}-\rho_q\text{nondiag}(\boldsymbol{R}_{yy}))^{-1}\boldsymbol{H}\right|. \tag{33}$$

Considering the case of low SNR values, we get easily with $\boldsymbol{R}_{yy} \approx \boldsymbol{R}_{\eta\eta}$, (33) and (28), the following first order approximation of the mutual information[2 3]

$$I(\boldsymbol{x},\boldsymbol{r}) \gtrapprox (1-\rho_q)\text{tr}[\boldsymbol{R}_{xx}\boldsymbol{H}^{\text{H}}\boldsymbol{R}_{\eta\eta}^{-1}\boldsymbol{H}]/\log(2). \tag{34}$$

Compared with the mutual information $I(\boldsymbol{x},\boldsymbol{y})$ for the unquantized case, also at low SNR [39], the mutual information for the quantized channel degrades only by the factor $(1-\rho_q)$. For the spacial case $b = 1$, we have $\rho_q|_{b=1} = 1 - \frac{2}{\pi}$ (see [35]) and the degradation of the mutual

[2] We assume also that $\rho_q \ll 1$ (or $\boldsymbol{R}_{\eta\eta}$ is diagonal).
[3] Note that $\log|\mathbf{I}+\Delta\boldsymbol{X}| \approx \text{tr}(\Delta\boldsymbol{X})$.

information becomes

$$\lim_{\text{SNR}\to 0} \frac{I(\boldsymbol{x},\boldsymbol{r})}{I(\boldsymbol{x},\boldsymbol{y})}\bigg|_{b=1} \approx \frac{2}{\pi}. \tag{35}$$

In other words, the power penalty due to the 1-bit quantization is approximately equal $\frac{\pi}{2}$ (1.96 dB) at low SNR. This shows that mono-bit ADCs may be used to save system power without an excessive degradation in performance, and confirms the significant potential of the coarsely quantized UWB MIMO channel. Using a different approach, [40] presented a similar result, and showed that the above approximation is asymptotically exact.

3.2.5. Simulation results

The performance of the modified Wiener filter for a 4- and 5-bit quantized output MIMO system (WFQ), in terms of BER averaged over 1000 channel realizations, is shown in Figure 15 for a 10×10 MIMO system (QPSK), compared with the conventional Wiener filter (WF) and Zero-forcing filter (ZF). The symbols and the noise samples are assumed to be uncorrelated, that is, $\boldsymbol{R}_{xx} = \sigma_x^2\mathbf{I}$ and $\boldsymbol{R}_{\eta\eta} = \sigma_\eta^2\mathbf{I}$. Hereby, the SNR (in dB) is defined as

$$\text{SNR} = 10 \cdot \log_{10}\left(\frac{\sigma_x^2}{\sigma_\eta^2}\right). \tag{36}$$

Furthermore, we used a generic channel model, where the entries of $\boldsymbol{H}$ are complex-valued realizations of independent zero-mean Gaussian random variables with unit variance. Clearly, the modified Wiener filter outperforms the conventional Wiener filter at high SNR. This is because the effect of the quantization error is more pronounced at higher SNR values when compared to the additive Gaussian noise variance. Since the conventional Wiener filter converges to the ZF-filter at high SNR values and loses its regularized structure, its performance degrades asymptotically to the performance of the ZF-filter, when operating on quantized data. For comparison, we also plot the BER curves for the WF and ZF filter, for the case when no quantization is applied.

3.3. Channel estimation

Because in general, the MIMO channel cannot be assumed known a-priori, a channel estimation has to be performed. In practice, it is highly desirable that the channel is estimated directly by the communication device – in our case by on-chip digital circuitry. However, this implies that the channel estimator is restricted to use received signal samples of a pilot sequence after single-bit quantization in the extreme case. This motivates investigation of channel estimation with coarse quantization. This problem was first addressed by [27], where a maximum likelihood (ML) channel estimation with quantized observation is presented. In general, the solution cannot be given in closed form, but requires an iterative numerical approach, which hampers the analysis of performance.

In [28], it has been shown that – in contrast to unquantized channel estimation – different orthogonal pilot sequences (with same average total transmit power and same length) yield different performances. Especially, orthogonality in the time-domain (time-multiplexed pilots) can be preferable to orthogonality in space. With orthogonal pilots

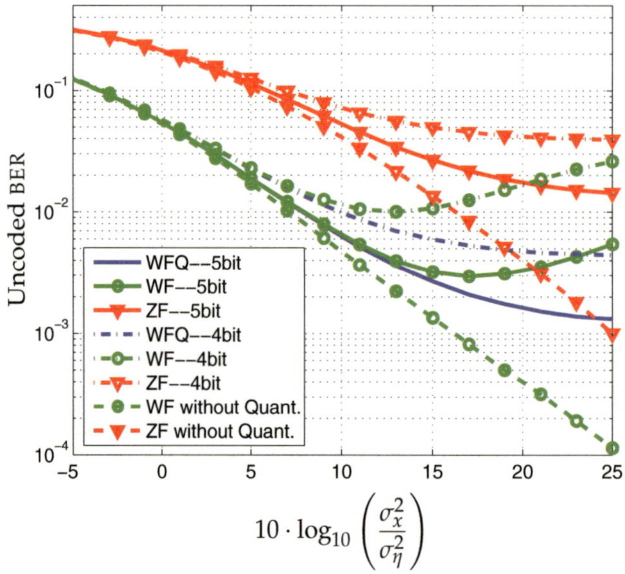

Figure 15. The WFQ vs. the conventional WF and ZF receivers, QPSK modulation with $M = 10$, $N = 10$, 4- ($\rho_q = 0.01154$) and 5- ($\rho_q = 0.00349$) bit uniform quantizer.

that are multiplexed in time, the problem can be reduced from the MIMO to the SIMO (single-input-multiple-output) case, because each line of the multiconductor interconnect is excited separately for time-multiplexed pilots. Finally, the problem can be reduced to the SISO (single-input-single-output) case, when the channel estimation is performed separately in parallel at each receiving end of the multiconductor interconnect. For this case, in [28], a closed-form solution can be found for the maximum likelihood channel estimation problem, which makes performance analysis possible in an analytical fashion.

In [50], a more general setting for parameter estimation based on quantized observations was studied, which covers many processing tasks, e.g. channel estimation, synchronization, delay estimation, Direction Of Arrival (DOA) estimation, etc. An Expectation Maximization (EM) based algorithm is proposed to solve the Maximum a Posteriori Probability (MAP) estimation problem. Besides, the Cramér-Rao Bound (CRB) has been derived to analyze the estimation performance and its behavior with respect to the signal-to-noise ratio (SNR). The presented results treat both cases: pilot aided and non-pilot aided estimation. The paper extensively dealt with the extreme case of single bit quantization (comparator) which simplifies the sampling hardware considerably. It also focused on MIMO channel estimation and delay estimation as application area of the presented approach. Among others, a 2×2 channel estimation using 1-bit ADC is considered, which shows that reliable estimation may still be possible even when the quantization is very coarse, with any desired accuracy, provided the pilot sequence is long enough. Since in on-chip and chip-to-chip communications, the channel almost does not change in time, it is possible to use very long pilot sequences, and run the channel estimation only once, or once in a while.

4. Efficient digital hardware architecture

Sole optimization of transmitting power in the standardization and conception phase of communication channels results in highly complex and energy-intensive receivers with a complex channel decoder as one of its key components. Neglecting the energy dissipation of the integrated decoder in this early phase results in suboptimal and, thus, costly communication systems in terms of manufacturing and usage costs. In the previous part of this chapter approaches to reduce the ADC complexity and, thus, the complexity of the subsequent digital components by using single-bit or medium-low resolution quantizations have been discussed. A quantitative comparison of these new approaches to standard receivers requires accurate cost models of the digital components. Quite accurate cost models are available for most of the communication system components except for channel decoders. While such cost models can be easily derived for Viterbi, Reed-Solomon, and Turbo Decoders, an estimation of the silicon area and the energy dissipation of LDPC decoders is challenging due to the high internal communication effort between the basic components.

Although LDPC codes have already been introduced by Gallager in 1962 [55], up to now they are known to achieve the best decoding performance [56] and are adopted in various communication standards (e.g. [57],[58],[59]) and other applications such as hard-disk drives [60]. They belong to the class of block codes and, thus, can be defined by a parity-check matrix H with m rows and n columns or by the corresponding Tanner Graph. Both are shown in Figure 16 for a very simplified LDPC code. Each row of the parity-check matrix represents one parity check wherein a '1'-entry in column i and row j indicates, that the received symbol i takes part in parity check number j. In the Tanner Graph such a parity check is represented by one so called check node and each column by one bit node. Furthermore, the number of one entries per row d_C (column d_V) defines the number of connected bit (check) nodes per check (bit) node.

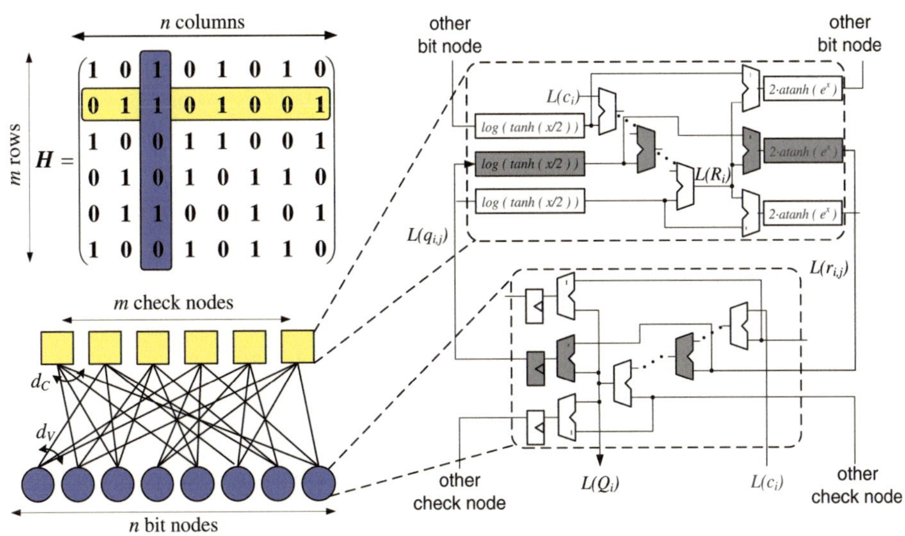

Figure 16. Parity-check matrix, Tanner-Graph and non-linear recursive decoder loop

Figure 16 also illustrates one of the $n \cdot d_V$ non-linear recursive loops of the resulting decoder. In each iteration the extrinsic information $L(q_{i,j})$ on the received symbol i is sent to check node j. Here, new A-posteriori information $L(r_{i,j})$ is derived. The sign of $L(r_{i,j})$ is chosen in such a way, that the confidence in the received symbol i indicated by the magnitude of $L(q_{i,j})$ increases. For the sake of clarity only the magnitude calculation is illustrated in Figure 16. Considering the original Sum-Product decoding algorithm [55] the check node consists of transcendent functions and a multi-operand adder with subsequent subtractor stages. Here, the basic idea is, whenever all participating symbols in that parity check feature a high confidence in their current estimation, the magnitude of the A-posteriori information is high. The A-posteriori information $L(r_{i,j})$ is then sent back to the bit node. Here, all information of symbol i, namely the d_V A-posteriori information and the received information $L(c_i)$, are combined using a multi-operand adder resulting in a new estimation $L(Q_i)$ of symbol i. To avoid decoding-performance-demoting cycles, in the next decoding iteration only the extrinsic information $L(q_{i,j}) = L(Q_i) - L(r_{i,j})$ is used instead of $L(Q_i)$. For more information on the decoding algorithm and possible fix-point realizations refer to [61].

A metric for the code's and, thus, the decoder's complexity is the number of '1'-entries in the matrix $n \cdot d_V = m \cdot d_C$. Each '1'-entry can be assigned to a part of the bit- and check-node logic as highlighted in gray in Figure 16. Thereby, each '1'-entry leads to four two-operand adders/subtractors, a block for the calculation of $log\left(tanh\left({}^{x}/_{2}\right)\right)$, a block for the calculation of $2 \cdot atanh\left(e^{x}\right)$ and a register stage at the output of the bit node. Additionally, $n \cdot d_V$ is a measure for the communication between the nodes as $2 \cdot w \cdot n \cdot d_V$ bits are exchanged between the bit and the check nodes in each decoding iteration with w being the word length of the exchanged messages.

In high-throughput applications with a time-invariant parity-check matrix all bit and check nodes are typically instantiated in parallel as in the first integrated LDPC decoder [66]. Here, typically the m check nodes are realized in the center of the decoder floorplan surrounded by the n bit-node instances. The communication between the nodes is then realized by $2 \cdot w \cdot n \cdot d_V$ dedicated interconnect lines. In [66] the logic area, which is the accumulated silicon area of all logic gates, is approximately $25\,mm^2$. However, the total of 26,624 interconnect lines can not be realized on this area. The silicon area needs to be artificially expanded until a successful routing of all interconnect lines could be established. The resulting global interconnect has a length of $80\,m$ on a macro size of $52.5\,mm^2$. Thus, only 50% of the active silicon area is utilized in the final decoder. The impact of the complex global interconnect complicates the derivation of accurate area, timing and energy cost models which might be the reason why no cost models are available in literature so far. However, such models are necessary to avoid costly wrong decisions in early design phases, for example when choosing a certain LDPC code in the system-conception phase. Also in later design phases those models are indispensable, for example for a quantitative exploration of the architecture design space.

4.1. Accurate area, timing, and energy cost models

In general the silicon area of a high-throughput LDPC decoders can be estimated using

$$A_{DEC\_P} = max(A_L, A_R), \tag{37}$$

with A_L being the logic area and A_R the required area to realize the global interconnect. To reduce the logic area typically the approximative Min-Sum algorithm [70] is used which estimates the magnitude of $L(r_{i,j})$ using the minimal and second minimal magnitude of $L(q_{i,j})$ (e.g. [62],[63],[64],[65]). The derivation of A_L for this decoding algorithm as the accumulated silicon area of all logic gates has been presented in [67]. The resulting total logic area can be estimated using

$$A_L = l_L^2 = 1000 \cdot n \cdot d_V \cdot (11.5 \cdot w + 2 \cdot ld\,(d_V)) \cdot \lambda^2. \quad (38)$$

This equation reveals a linear dependency between the code complexity $n \cdot d_V$ and the accumulated gate area.

The major challenge in deriving an accurate routing-area model is the adaptability to different LDPC codes. It is possible to divide the problem into two parts: an estimation of the available and the required Manhattan length. Considering a certain logic area, the available Manhattan length is a measure for the routing resources above the decoder's node logic. Considering that the node layouts require M_L of the total M metal layers in the CMOS stack for the local interconnect, $M_R = M - M_L$ metal layers are available for the realization of the global bit- and check-node communication. The required routing area A_R can then be determined by equalizing the available and the required Manhattan lengths. This means, that the available Manhattan length allows the realization of the required Manhattan length. Thereby, the available Manhattan length can be derived as

$$l_{AVAIL} = \frac{u}{p} \cdot \left(l_L^2 \cdot M_R + min\left(l_{DEC}^2 - l_L^2, 0\right) \cdot M\right), \quad (39)$$

with the routing pitch p, an utilization factor u for each metal layer, and a decoder macro side length of l_{DEC}. Considering that no artificial increase of the decoder is required ($l_{DEC} \leq l_L$) the second term is zero and the available routing resources are on top of the node logic. If the decoder needs to be expanded, the whole metal stack is available between the node instances for the realization of the global interconnect. Therefore, this part is weighted with M.

The estimation of the required Manhattan length is more challenging as it depends on code characteristics as for example the number of interconnect lines and the average length of one interconnect line. An upper bound estimation of the required Manhattan length could be derived by using the maximum possible length l_{MAX} of one bit- and check-node connection which is shown in Figure 17(a). In a typical placement with the bit nodes surrounding the check-node array the longest possible connection runs from one corner of the decoder macro to the opposing corner of the check-node array. An analysis of the logic model [67] shows that the check nodes occupy about 60% of the complete decoder macro leading to a maximum Manhattan length of

$$l_{MAX} = 2 \cdot \left(l_{CNA} + \frac{l_{DEC} - l_{CNA}}{2}\right) = l_{DEC} + l_{CNA} = 1.77 \cdot l_{DEC}. \quad (40)$$

When looking at the wire-length histogram for an exemplary code (see Figure 17(b)) the average Manhattan length is significantly smaller than the maximum length leading to an overestimation of the required Manhattan length and, thus, of the required routing area. An analysis of various LDPC codes showed, that the shape of the wire-length histogram is always similar. Especially, the ratio between the average and the maximum Manhattan length was found to be almost constant as can be seen from Table 2. For the derivation of the average Manhattan lengths all placements have been optimized using a custom simulated annealing process [62]. While code no. 11 is the code adopted in [57], the other codes are taken from

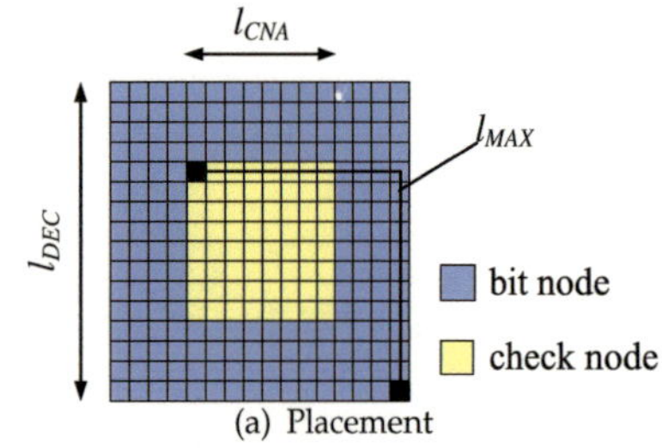

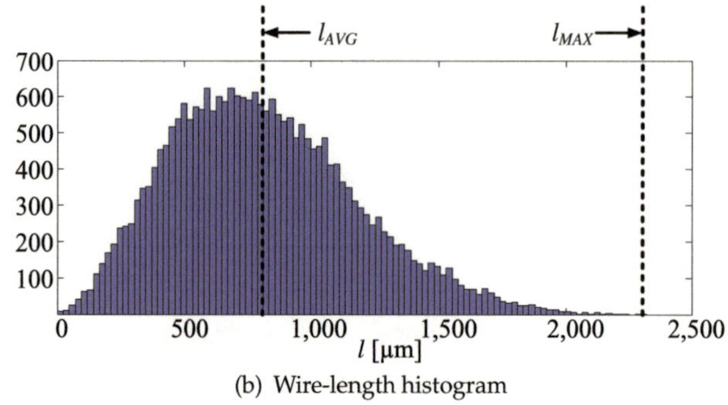

Figure 17. Bit- and check-node architecture

Code nr.	n	m	d_V	d_C	$n{\cdot}d_V$	l_{AVG}/l_{MAX}	ρ_{AVR}/ρ_{MAX} vertical	ρ_{AVR}/ρ_{MAX} horizontal
1	96	48	3	6	288	0.33	0.58	0.59
2	408	204	3	6	1224	0.31	0.56	0.55
3	408	204	3	6	1224	0.30	0.55	0.56
4	408	204	3	6	1224	0.31	0.54	0.50
5	816	408	3	6	2448	0.31	0.52	0.57
6	816	408	5	10	4080	0.34	0.52	0.58
7	816	408	5	10	4080	0.34	0.52	0.57
8	816	408	5	10	4080	0.34	0.50	0.55
9	999	111	3	27	2997	0.37	0.36	0.33
10	1008	504	3	6	3024	0.32	0.54	0.46
11	2048	384	6	32	12288	0.37	0.42	0.40
12	4000	2000	3	6	12000	0.34	0.57	0.54
13	4000	2000	4	8	16000	0.35	0.55	0.55
14	8000	4000	3	6	24000	0.35	0.55	0.45

Table 2. Interconnect properties of various LDPC codes

[68]. For a wide range of LDPC codes with code complexities $n \cdot d_V$ between 300 and 24,000 the ratio varies only between 0.30 and 0.37. Approximating the ratio of the average to the maximum Manhattan length with 0.35 and using (40), the required Manhattan length can be estimated based on the decoder side length as

$$l_{REQ} = 1.2 \cdot n \cdot d_V \cdot w \cdot l_{DEC}. \tag{41}$$

Additionally, an estimation of the achievable utilization is possible based on the comparison of the average routing density ρ_{AVG} and the maximum routing density ρ_{MAX}. The ratios of these values for vertical and horizontal interconnect lines are also given in Table 2. Although there are exceptions (e.g. code no. 9) the utilization $u = \rho_{AVG}/\rho_{MAX}$ is almost constant and will be chosen to $u = 0.5$ in the following.

Considering that the decoder area needs to be expanded and assuming a uniform stretch (39) and (41) still hold. Then, the minimal required decoder area to realize the global interconnect A_R can be calculated by equating (39) and (41) and solving for l_R as

$$A_R = l_R^2 = \left(1.2 \cdot \frac{n \cdot d_V \cdot w}{M} \cdot p + \sqrt{\left(1.2 \cdot \frac{n \cdot d_V \cdot w}{M} \cdot p\right)^2 - l_L^2 \cdot \frac{M_R - M}{M}}\right)^2. \tag{42}$$

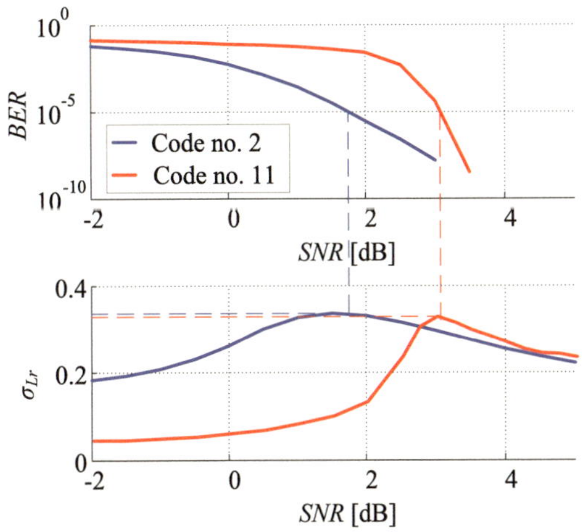

(a) SNR-dependent switching activity

Code nr.	*SNR [dB]*	$\sigma_{L(q)}$	$\sigma_{L(r)}$
1	2.6	0.14	0.30
2	1.7	0.17	0.34
3	1.7	0.17	0.33
4	2.9	0.14	0.30
5	1.5	0.18	0.34
6	0.5	0.16	0.33
7	0.5	0.16	0.33
8	0.5	0.16	0.33
9	9.8	0.41	0.29
10	1.4	0.19	0.34
11	3.0	0.20	0.33
12	1.2	0.19	0.34
13	0.1	0.20	0.35
14	1.2	0.19	0.34

(b) Table of switching activities ($BER = 10^{-5}$)

Figure 18. Switching activity

In contrast to the logic area, the routing area shows a quadratic dependence on code complexity. By comparing (38) and (42) it can be shown that the bit-parallel decoder is routing dominated as soon as

$$n \cdot d_V \geq 500 \cdot \frac{M_R^2}{w}. \tag{43}$$

The required artificial increase of the silicon area also impacts the other two decoder features: the energy per iteration E_{IT} and the iteration period, which is the required time for one decoding iteration and the inverse of the block throughput [69]. Here, only the interconnect fraction of the decoder energy will be discussed in detail. For more information on the derivation of the iteration period and the total decoder energy refer to [67]. The dynamic energy dissipation of the global interconnect can be estimated using

$$E_{INT} = \frac{1}{2} \cdot \left(\sigma_{Lq}\,(BER) + \sigma_{Lr}\,(BER)\right) \cdot \alpha \cdot C' \cdot \frac{l_{REQ}}{2} \cdot V_{DD}^2, \tag{44}$$

with V_{DD} being the supply voltage, C' the capacitive load per unit length of a minimum-spaced interconnect line and α a fitting factor to cover the fact, that on average the global interconnect lines are not minimum spaced [67]. Furthermore, the switching activities on the interconnect lines from bit to check nodes (σ_{Lq}) and vice versa (σ_{Lr}) need to be considered. In Figure 18(a) the BER and the switching activity σ_{Lr} for two codes from Table 2 and different signal-to-noise ratios are illustrated. The switching activity highly depends on the considered SNR and is especially high in the so-called waterfall region when the BER starts to get significantly smaller. Furthermore, the two codes strongly differ when it comes to comparing the switching activities for a given SNR (e.g. $1dB$) . But, considering a specific BER (indicated by the dashed lines) an almost equal switching activity for the two codes can be observed (approx. 0.33 for a BER of 10^{-5}). The comparison of the switching activities σ_{Lq} and σ_{Lr} for all codes listed in Figure 18(b) shows, that this behavior is common for almost all other codes, as well. Therefore, a quite accurate estimation of the decoder energy based on the code parameters n and d_V is possible without knowledge of the actual LDPC code.

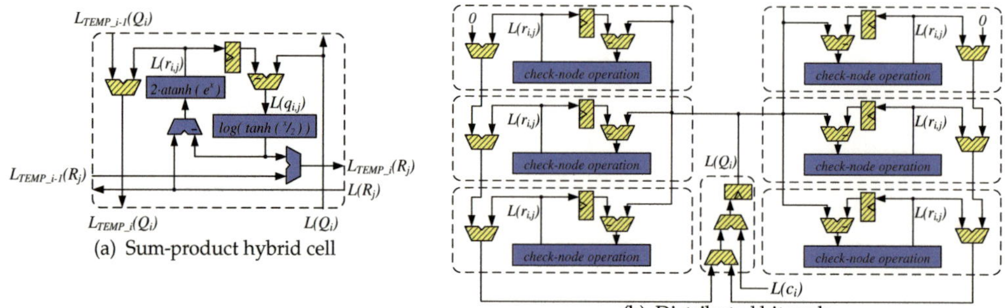

(a) Sum-product hybrid cell

(b) Distributed bit node

Figure 19. Hybrid-cell decoder architecture

4.1.1. Hybrid-cell decoder architecture

The main routing problem of the bit- and check-node architecture arises from the high routing density at the border of the check-node array as it can be seen in the interconnect-density chart in Figure 20(a) for an exemplary code [57]. To overcome this drawback it is possible to break up the bit- and check-node clustering of the logic and rearrange it. The new idea is based on the observation, that each '1'-entry in the parity-check matrix can be assigned to certain parts of the decoder loop. Then, the decoder consists of $n \cdot d_V$ small, equal basic components. A combination of the logic for one '1'-entry (see grey blocks in Figure 16) leads to the block diagram of one hybrid cell, as it is shown in Figure 19(a). This hybrid cell gets the accumulated information $L_{TEMP\_i-1}(Q_i)$ of the received A-priori information $L(c_i)$ and of all A-posteriori information of the previous hybrid cells and adds the A-posteriori information $L(r_{i,j})$ of check node j. The resulting information $L_{TEMP\_i}(Q_i)$ is forwarded to the next hybrid cell. The last hybrid cell in that column calculates $L(Q_i)$ and sends this value back to all participating hybrid cells. A similar structure is used in the check-node part of the hybrid-cell where the calculation of $L(R_i)$ is distributed over d_C hybrid cells. Although, here, the hybrid-cell approach considers a Sum-Product algorithm, it is also applicable to a Min-Sum based decoder. Therefore, the Φ function and the multi-operand adder have to be replaced with basic compare-and-swap cells.

In contrast to the bit- and check-node architecture, in which the $(d_V + 1)$-operand adder in the bit node and the d_C-operand adder in the check node would be realized using a tree topology, the hybrid-cell architecture is based on an adder chain topology. However, it is possible to introduce tree-stages for the bit-node operation as illustrated in Figure 19(b). The $L(r_{i,j})$ values are accumulated in two branches and the intermediate results are added to the channel information $L(c_i)$ in an additional IO cell. A similar topology is possible for the check-node operation. The global interconnect of the hybrid-cell architecture has been realized in a 90-nm CMOS technology using five metal layers for the same code as used for the bit- and check-node architecture in Figure 20(a). In a first step, the placement of the nodes has been optimized using a custom simulated annealing process. Thereby, a placement scheme as depicted in Figure 20(b) with the hybrid cells surrounded by the io cells has been assumed. The advantage of the hybrid-cell architecture becomes obvious when comparing the two interconnect densities. The routing density of the hybrid-cell architecture is distributed more uniformly especially without high density peaks at the border of the bit- and check-node array. Thus, the average routing density of the hybrid-cell interconnect is higher than of the bit- and check-node architecture, promising a smaller silicon area.

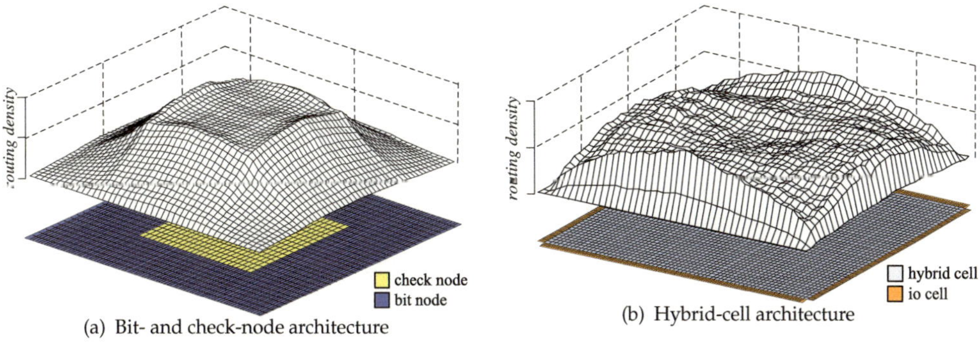

Figure 20. Routing density

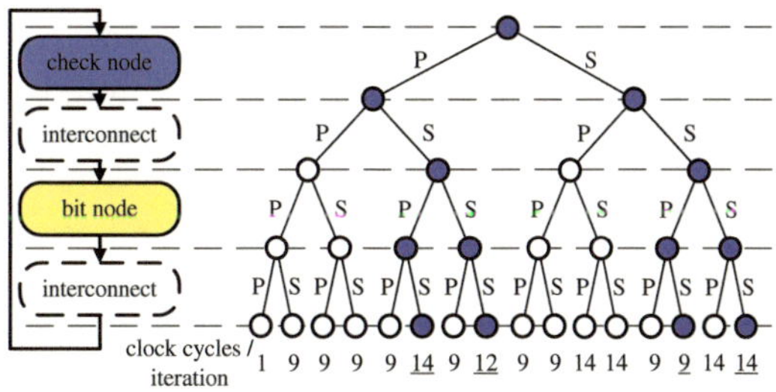

Figure 21. Bit- and check-node architecture design space.

4.1.2. Hardware-efficient partially bit-serial decoder architecture

Another promising approach to reduce the decoder's silicon area is the introduction of a bit-serial interconnect as proposed in [65]. The number of interconnect lines can be reduced by a factor of w resulting in a significant reduction in decoder area because of the quadratic dependency in (42). While the realized minimum search in the check node requires a most-significant-bit-first data flow in the check node the multi-operand adder in the bit node has to be realized using a least-significant-bit-first data flow. Therefore, the order of the bits needs to be flipped twice per iteration resulting in a high number of clock cycles. Although the clock frequency of the decoder is higher due to the bit-serial node logic, the high number of clock cycles per iteration limit the achievable decoder throughput and block latency. However, it is possible to introduce a bit-serial data flow in a more fine-grained way. A systematic architecture analysis is possible by breaking the decoder loop into four parts as shown in Figure 21, namely the bit and check node and the communication between the nodes in both directions. Now, possible architectures can be distinguished by assuming either a bit-serial or a bit-parallel approach in each of the four parts. Obviously, also a digit-serial approach is possible as discussed in [69]. Considering only a bit-serial or bit-parallel data flow, in total 16 different architectures are possible. As a first order metric of the decoder throughput, the number of clock cycles per iteration considering a message word length of $w = 6$ is given. To avoid extensive routing-induced extensions of the silicon area, especially the highlighted architectures with a bit-serial communication in both directions should be taken into account. When comparing the number of clock cycles per iteration for these four architectures, the

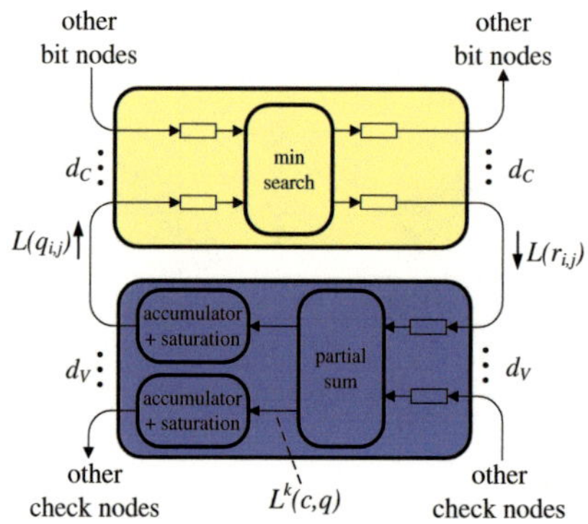

Figure 22. Partially bit-serial architecture

architecture with a bit-parallel bit node allows for the smallest number of clock cycles per iteration and, thus, promises the highest decoder throughput. As the bit-parallel realization of the bit node would result in a large silicon area and a long critical path, further optimizations on arithmetic level have to be done. Here, it is possible to gain from the bit-serial input data stream by realizing the multi-operand adder in the bit node bit-serially using an MSB-first data flow. Within each clock cycle a partial sum $L^k(c, q)$ for the received bit-weight is generated which is accumulated subsequently to derive the new estimation $L(Q_i)$ as is shown in the decoder loop in Figure 22. The long ripple path in the accumulator unit running over the complete word length can be reduced using a carry-select principle. For further details of the realization on arithmetic and circuit level refer to [62].

4.1.3. Quantitative architecture comparison

The cost models have been adapted to the new architecture concepts to allow for a quantitative evaluation of the architecture design space. Figure 23(a) illustrates the resulting silicon area A and iteration period T_{IT} of the fully bit-parallel, fully bit-serial, hybrid-cell and partially bit-serial decoder architecture for three different code complexities $n \cdot d_V = 5,000$, $10,000$ and $15,000$. For all code complexities the new architecture concepts are Pareto optimal as they allow for a trade-off between silicon area and iteration period in comparison to the bit-parallel and bit-serial architectures. Considering small code complexities the decoder architectures with a bit-parallel interconnect show the smallest area-time (AT) product and, therefore, are most AT-efficient. Considering a specified decoder throughput, the hybrid-cell architecture is promising whenever the timing constraints cannot be met by using bit-serial approaches, as it reduces the silicon area significantly in comparison to the bit-parallel bit- and check-node architecture. The new partially bit-serial architecture features the smallest area-time product for all code complexities larger than $9,000$. In comparison to the bit-serial architecture a significantly smaller iteration period with only a slightly increased area is achieved. The architectures with a bit-parallel interconnect are located further and further away from the curve representing the smallest achievable area-time product. Here, the timing advantage of the bit-parallel architectures vanishes for large code complexities. Figure 23(b) depicts the energy per decoding iteration E_{IT} of the four decoder architectures for different code complexities. The advantage with respect to energy of the decoder architectures with a

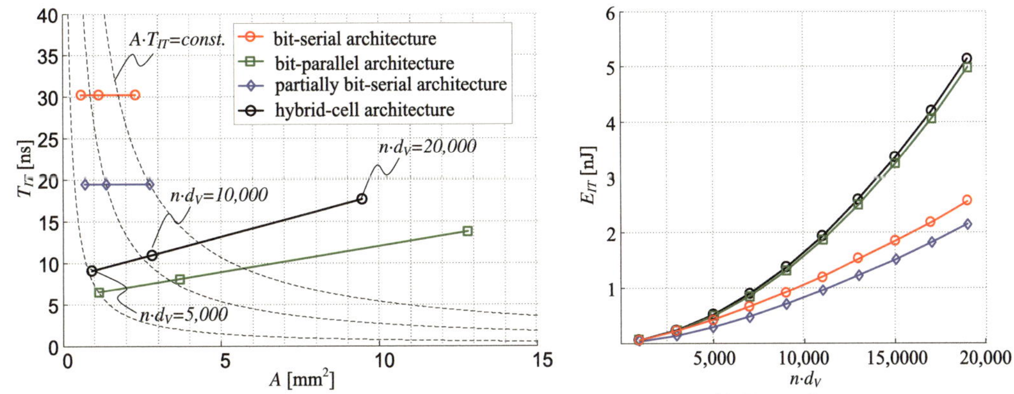

(a) Area and timing features (b) Energy features

Figure 23. Comparison of area, timing and energy features ($d_V = 6, d_C = 32, M_R = 4, w = 6, \lambda = 40nm$)

bit-serial interconnect becomes apparent. Considering code complexities larger than 10,000, the energy per iteration of the bit-parallel decoder becomes more than twice as high as the partially bit-serial architecture. The latter allows for the smallest energy per iteration in the complete code complexity range. This emphasizes the efficiency of the new partially bit-serial architecture which allows for the smallest area-time product in a wide range of code complexities and the smallest decoding energy, simultaneously. This work has been supported by the German Research Foundation (DFG) under the priority program UKoLoS (SPP1202).

5. Conclusion

This chapter presented results, accomplished within the frame of the DFG priority program »Ultrabreitband Funktechniken für Kommunikation, Lokalisierung und Sensorik«. Focus was put primarily on the analysis and optimization of on-chip and chip-to-chip multi-conductor/multi-antenna interconnects. While we could show that special techniques of physical optimization, coding and signal processing can improve interconnect performance to a remarkable degree, it is expected that even higher performance is achievable in chip-to-chip communication, when multi-conductor interconnects are replaced by wireless ultra-wideband multi-antenna interconnects. Hereby, the signal pulses do not necessarily increasingly disperse as they travel along their way to the receiving end of the interconnect. The propagating nature of the wireless interconnect can make for a much more attractive channel for chip-to-chip communications. The primary goal has been the development of both theoretical and empirical foundations for the application of ultra-wideband multi-antenna wireless interconnects for chip-to-chip communication. Suitable structures for integrated ultra-wideband antennas have been developed, their properties theoretically analyzed and verified against measurements performed on manufactured prototypes. Qualified coding and signal processing techniques, which aim at efficient use of available resources of bandwidth, power, and chip area has been proposed. In addition, attention was given to the implementation of iterative decoding structure for LDPC codes. Detailed cost-models, which are based on signal flow charts and VLSI implementations of dedicated functional blocks have be developed, which allow for an informative analysis of elementary trade-offs between throughput, required chip area, and power consumption. This work has been supported by the German Research Foundation (DFG) under the priority program UKoLoS (SPP1202).

Author details

Nossek Josef A., Mezghani Amine and Michel T. Ivrlač
Institute for Circuit Theory and Signal Processing, Technische Universität München, Germany

Russer Peter, Mukhtar Farooq, Russer Johannes A. and Yordanov Hristomir
Institute for Nanoelectronics, Technische Universität München, Germany

Noll Tobias and Korb Matthias
Chair of Electrical Engineering and Computer Systems, RWTH Aachen University, Germany

6. References

[1] Russer P (1998) Si and SiGe millimeter-wave integrated circuits, Si and SiGe based monolithic integrated antennas for electromagnetic sensors and for wireless communications. IEEE Transactions on Microwave Theory and Techniques 1998: 590-603.

[2] Johannes A. R, Kuznetsov Y, Russer P, Russer J.A, Kuznetsov K, Russer P (2010) Discrete-time network and state equation methods applied to computational electro magnetics. Mikrotalasma revija: 2-14.

[3] Johannes A.R, Baev A, Kuznetsov Y, Mukhtar F, Yordanov H, Russer P (2011) Combined lumped element network and transmission line model for wireless transmission links. German Microwave Conference (GeMIC): 1-4.

[4] Felsen L.B, Mongiardo M, Russer P (2009) Electromagnetic Field Computation by Network Methods. Heidelberg: Springer-Verlag.

[5] Russer P (2006) Electromagnetics, Microwave Circuit and Antenna Design for Communications Engineering. Boston: Artech House.

[6] Russer J.A, Mukhtar F, Wane S, Bajon D, Russer P (2012) Broad-Band Modeling of Bond Wire Antenna Structures. German Microwave Conference (GeMIC): 1-4.

[7] Yordanov H, Ivrlač M.T, Mezghani A, Nossek J.A, Russer P (2008) Computation of the impulse response and coding gain of a digital interconnection bus. 23rd Annual Review of Progress in Applied Computational Electromagnetics: 494-499.

[8] Yordanov H, Russer P (2008) Computing the Transmission Line Parameters of an On-chip Multiconductor Digital Bus. Time Domain Methods in Electrodynamics: 69-78.

[9] Yordanov H, Ivrlač M.T, Nossek J.A, Russer P (2007) Field modelling of a multiconductor digital bus. Microwave integrated circuit conference, 2007: 579-582.

[10] Yordanov H, Russer P (2008) Computation of the Electrostatic Parameters of a Multiconductor Digital Bus. International Conference on Electromagnetics in Advanced Applications 2007: 856-859.

[11] Yordanov H, Russer P (2010) Antennas embedded in CMOS integrated circuits. Proceedings of the 10th Topical Meeting on Silicon Monolithic Integrated Circuits in RF Systems: Electronics and Energetics 2010: 53-56.

[12] Yordanov H, Russer P (2009) Wireless Inter-Chip and Intra-Chip Communication. Proceedings of the 39th European Microwave Conference: 145-148.
[13] Yordanov H, Russer P (2009) On-Chip Integrated Antennas for Wireless Interconnects. Semiconductor Conference Dresden (SCD) 2009.
[14] Yordanov H, Russer P (2010) Integrated On-Chip Antennas Using CMOS Ground Planes. Proceedings of the 10th Topical Meeting on Silicon Monolithic Integrated Circuits in RF Systems: 53-56.
[15] Yordanov H, Russer P (2010) Area-Efficient Integrated Antennas for Inter-Chip Communication. Proceedings of the 10th Topical Meeting on Silicon Monolithic Integrated Circuits in RF Systems: 401-404.
[16] Yordanov H, Russer P (2008) Chip-to-chip interconnects using integrated antennas. Proceedings of the 38th European Microwave Conference, EuMC 2008: 777-780.
[17] Yordanov H, Russer P (2008) Integrated on-chip antennas for chip-to-chip communication. IEEE Antennas and Propagation Society International Symposium Digest 2008: 1-4.
[18] Yordanov H, Ivrlač M.T, Nossek J.A, Russer P (2007) Field modelling of a multiconductor digital bus. Microwave Conference, 2007. European. 37th European: 1377-1380.
[19] Russer P, Fichtner N, Lugli P, Porod W, Russer A.J, Yordanov H (2010) Nanoelectronics Based Monolithic Integrated Antennas for Electromagnetic Sensors and for Wireless Communications. IEEE Microwave Magazine: 58-71.
[20] Mukhtar F, Yordanov H, Russer P (2010) Network Model of On-Chip Antennas. URSI Conference 2010.
[21] Mukhtar F, Yordanov H, Russer P (2011) Network model of on-chip antennas. Advances in Radio Science: 237-239.
[22] Mukhtar F, Kuznetsov Y, Russer P (2011) Network Modelling with Brune's Synthesis. Advances in Radio Science: 91-94.
[23] Russer J.A, Dončov N, Mukhtar F, Stošić B, Asenov T, Milovanović B, Russer P (2011) Equivalent lumped element network synthesis for distributed passive microwave circuits. Mikrotalasna revija: 23-28.
[24] Russer J.A, Gorbunova A, Mukhtar F, Yordanov H, Baev A, Kuznetsov Y, Russer P (2011) Equivalent circuit models for linear reciprocal lossy distributed microwave two-ports. Microwave Symposium Digest: 1-4.
[25] Russer J.A, Dončov N, Mukhtar F, Stošić B, Milovanović B, Russer P (2011) Compact equivalent network synthesis for double-symmetric four-ports. Telecommunication in Modern Satellite Cable and Broadcasting Services (TELSIKS), 2011 10th International Conference on: 383-386.
[26] Cover T.M, Thomas J.A (1991) Elements of Information Theory. New York: John Wiley and Son.
[27] Lok T.M, Wei V.K.W (1998) Channel Estimation with Quantized Observations. IEEE International Symposium on Information Theory 1998: 333.
[28] Ivrlač M.T, Nossek J.A (2007) On MIMO Channel Estimation with Single-Bit Signal-Quantization. ITG Smart Antenna Workshop 2007.
[29] Sotiriadis P.P, Wang A, Chandrakasan A (2000) Transition Pattern Coding: An Approach to Reduce Energy in Interconnect. ESSCIRC 2000: 320-323.

[30] Kavčić A (2001) On the capacity of markov sources over noisy channels. IEEE Global Communications Conference 2001: 2997-3001.
[31] Ivrlac M.T, Nossek J.A (2006) Chalanges in Coding for Quantized MIMO Systems. IEEE International Symposium on Information Theory 2006: 2114-2118.
[32] Lyuh G, Kim T (2002) Low Power Bus Encoding with Crosstalk Delay Elimination. 15th Annual IEEE International ASIC/SOC Conference 2002: 389-393.
[33] Mezghani A, Khoufi M.S, Nossek J.A (2007) A Modified MMSE Receiver for Quantized MIMO. ITG/IEEE WSA 2007.
[34] Mezghani A, Nossek J.A (2007) Wiener Filtering for Frequency Selective Channels with Quantized Outputs. IEEE SSD 2007.
[35] Max J, Quantizing for Minimum Distortion (1960) IEEE Trans. Inf. Theory 1960: 7-12.
[36] Mezghani A, Khoufi M.S, Nossek J.A (2008) Spatial MIMO Decision Feedback equalizer Operating on Quantized Data. IEEE ICASSP 2008.
[37] Jayant N.S, Noll P (1984) Digital Coding of Waveforms. New Jersey: Prentice-Hall 1984.
[38] Proakis J.G (1995) Digital Communications. New York: McGraw Hill 1995.
[39] Verdú S (2002) Spectral Efficiency in the Wideband Regime - IEEE Trans. Inform. Theory 2002: 1319-1343.
[40] Mezghani A, Nossek J.A (2007) On Ultra-Wideband MIMO Systems with 1-bit Quantized Outputs: Performance Analysis and Input Optimization - IEEE International Symposium on Information Theory 2007.
[41] Teletar E (1999) Capacity of Multi-Antenna Gaussian Channels. European Transactions on Telecommunications 1999: 585-595.
[42] Schubert M, Boche H (2005) Iterative multiuser uplink and downlink beamforming under SINR constraints. IEEE Transactions on Signal Processing: 2324-2334.
[43] Mezghani A, Joham M, Huger R, Utschick W (2006) Transceiver Design for Multi-user MIMO Systems. Proc. ITG Workshop on Smart Antennas 2006.
[44] Joham M, Kusume K, Gzara M.H, Utschick W, Nossek J.A (2002) Transmit Wiener Filter for the Downlink of TDD DS-CDMA Systems. ISSSTA 2002: 9-13.
[45] Ivrlač M.T, Utschick W, Nossek J.A (2002) On Time-Switched Space-Time Transmit Diversity. Proceedings of the 56th IEEE Vehicular Technology Conference 2002: 710-714.
[46] Ivrlač M.T, Choi R.L.U, Murch R, Nossek J.A (2003) Effective use of Long-Term Transmit Channel State Information in Multi-user MIMO Communication Systems. Proceedings of the IEEE Vehicular Technology Conference 2003.
[47] Widrow B, Kollár I (2008) Quantization Noise. Cambridge University Press.
[48] Joham M, Utschick W, Nossek J.A (2003) Linear Transmit Processing in MIMO Communication Systems. IEEE Transactions on Signal Processing 2003.
[49] Mezghani A, Khoufi M.S, Nossek J.A (2008) Maximum Likelihood Detection For Quantized MIMO Systems. In Proc. ITG/IEEE WSA 2008.
[50] Mezghani A, Antreich F, Nossek J.A (2010) Multiple Parameter Estimation With Quantized Channel Output. In Proc. ITG/IEEE WSA 2010.
[51] Lang T, Mezghani A, Nossek J.A (2010) Channel adaptive coding for coarsely quantized MIMO systems. In Proc. ITG/IEEE WSA 2010.
[52] Lang T, Mezghani A, Nossek J.A (2010) Channel-matched trellis codes for finite-state intersymbol-interference channels. In Proc. SPAWC 2010.

[53] Mezghani A, Ivrlac M, Nossek J.A (2008) On Ultra-Wideband MIMO Systems with 1-bit Quantized Outputs: Performance Analysis and Input Optimization. IEEE International Symposium on Information Theory and its Applications 2008.

[54] Gray R.M (1990) Quantization noise spectra. IEEE Trans. Inform. Theory 1990: 1220-1244.

[55] Gallager R. (1962) Low-density parity-check codes. Information Theory, IRE Transactions on, vol. 8, pp. 21-28.

[56] Sae-Young C., Forney, Jr. G. D. , et al. (2001) On the design of low-density parity-check codes within 0.0045 dB of the Shannon limit. Communications Letters, IEEE, vol. 5, pp. 58-60.

[57] IEEE (2006) IEEE Standard for Information Technology-Telecommunications and Information Exchange Between Systems-Local and Metropolitan Area Networks-Specific Requirements Part 3: Carrier Sense Multiple Access With Collision Detection (CSMA/CD) Access Method and Physical Layer Specifications. IEEE Std 802.3an-2006 (Amendment to IEEE Std 802.3-2005), pp. 1-167.

[58] IEEE (2005) 802.16e. Air Interface for Fixed and Mobile Broadband Wireless Access Systems. IEEE P802.16e/D12 Draft, oct 2005.

[59] IEEE (2006) 802.11n. Wireless LAN Medium Access Control and Physical Layer specifications: Enhancements for Higher Throughput. IEEE P802.16n/D1.0, Mar 2006.

[60] Galbraith R. L., Oenning T., et al. (2010) Architecture and Implementation of a First-Generation Iterative Detection Read Channel. Magnetics, IEEE Transactions on, vol. 46, pp. 837-843.

[61] Korb M., Noll T. G. (2012) A Quantitative Analysis of Fixed-Point LDPC-Decoder Implementations using Hardware-Accelerated HDL Emulations. to appear in Embedded Computer Systems (SAMOS) International Conference.

[62] Korb M. Noll T. G. (2011) Area- and energy-efficient high-throughput LDPC decoders with low block latency. ESSCIRC, Proceedings of the, 2011, pp. 75-78.

[63] Mohsenin T., Baas B. M. (2006) Split-Row: A Reduced Complexity, High Throughput LDPC Decoder Architecture. Computer Design, International Conference on, 2006, pp. 320-325.

[64] Zhang Z., Anantharam V., et al. (2009) A 47 Gb/s LDPC Decoder With Improved Low Error Rate Performance. VLSI Circuits, Symposium on, 2009, pp. 286-287.

[65] Darabiha A., Carusone A. C., et al. (2007) A 3.3-Gbps Bit-Serial Block-Interlaced Min-Sum LDPC Decoder in 0.13-ţm CMOS. Custom Integrated Circuits Conference, 2007. CICC '07. IEEE, 2007, pp. 459-462.

[66] Blanksby A. J., Howland C. J. (2002) A 690-mW 1-Gb/s 1024-b, Rate-1/2 Low-Density Parity-Check Code Decoder. Solid-State Circuits, IEEE Journal of, vol. 37, pp. 404-412.

[67] Korb M. Noll T. G. (2010) LDPC decoder area, timing, and energy models for early quantitative hardware cost estimates. System on Chip (SoC), 2010 International Symposium on, 2010, pp. 169-172.

[68] MacKay D. http://www.inference.phy.cam.ac.uk/mackay/codes/data.html

[69] Korb M., Noll T. G. (2009) Area and Latency Optimized High-Throughput Min-Sum Based LDPC Decoder Architectures. ESSCIRC, Proceedings of the, 2009, pp. 408-411.

[70] Fossorier M. P. C., Mihaljevic M., et al. (1999) Reduced complexity iterative decoding of low-density parity check codes based on belief propagation. Communications, IEEE Transactions on, vol. 47, pp. 673-680, 1999.

Chapter 4

Non-Coherent UWB Communications

Nuan Song, Mike Wolf and Martin Haardt

Additional information is available at the end of the chapter

http://dx.doi.org/10.5772/53219

1. Introduction

The use of ultra wide band (UWB) signals can offer many advantages for communications. It can provide a very robust performance even under harsh multipath and interference conditions, the capability of precision ranging and a reduced power consumption. Since the power spectral density is very low, it is possible to overlay UWB networks with already existing non-UWB emissions.

Early UWB concepts for communications have almost exclusively relied on impulse radio, where the whole available bandwidth, i.e., up to 7.5 GHz, is covered at once by means of very short pulses which are generated with a low duty cycle. Meanwhile, a bandwidth of 7.5 GHz is only available in the US [1, 3]. In Europe, the spectrum which is available with the same transmit power spectral density of -41.3 dBm/MHz ranges only from 6.0 to 8.5 GHz [4], if no detect and avoid techniques are applied[1]. A potential UWB system has therefore to be able to 'live' with a mean transmit power of less than -7.3 dBm.

This is a small value, but fortunately UWB systems may exploit the signal energy very efficiently because firstly, even at data rates in the Gbps range, it is not required to use bandwidth efficient (but energy inefficient) modulation schemes like a 1024-QAM. Secondly, UWB transmission benefits from a good fading resistance.

For a measured indoor channel [7], Fig. 1 shows that even a bandwidth of 'only' 500 MHz ensures a very good fading resistance: If the receiver is moved over all x-y-positions in the non-LOS case, the smallest power value at the receive antenna lies less than 3 dB below the mean power, averaged over all positions. Thus the fading margin could be chosen in the order of 3-4 dB — even for indoor channels which exhibit the largest coherence bandwidth.

The second energy efficiency argument claimed above is underpinned in Fig. 2. It shows the channel capacity depending on the bandwidth, where additive white Gaussian noise (AWGN) is assumed. A value of 83 MHz just corresponds to the total bandwidth available in the 2.4 GHz ISM band, which is chosen for comparison. At 1 Gbps, a 2.5 GHz bandwidth promises an advantage of 25 dB with respect to the required receive-power. Furthermore, even binary modulation (on the inphase and quadrature components) promises high data rates.

[1] With detect and avoid techniques, -41.3 dBm/MHz is also permitted between 4.2 and 4.8 GHz.

Unfortunately, a very large signal bandwidth is also associated with some serious problems. These problems are related to the transceiver components itself (availability of broadband antennas, amplifiers etc.), and to the technical effort which is required for synchronization, channel estimation and interference rejection. Since UWB networks operate in frequency bands already assigned to other RF-systems, the probability that narrowband interference occurs at all increases with the bandwidth, too.

Furthermore, by increasing the bandwidth, more and more multipath arrivals with different path gains and delays are resolvable at the receiver, which makes it more difficult to collect the multipath energy coherently — although the power at the receive antenna does not suffer from the fading effect. Fig. 1 shows an example that one may lose 10 dB and more, if an UWB-receiver uses only the strongest signal echo. Thus, especially in non-LOS scenarios, a coherent RAKE receiver requires a very large number of RAKE fingers and a precise channel knowledge to efficiently capture the multipath energy. Such a coherent RAKE receiver will be very complex and costly, such that the hardware itself may consume a lot of power. This fact is the major motivation for systems using non-coherent detection, which are discussed in this chapter.

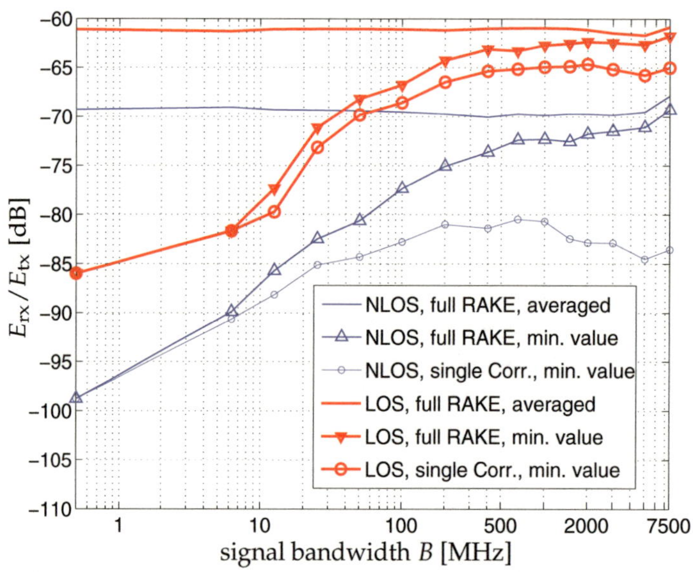

Figure 1. Received energy E_{rx} normalized by the transmitted symbol's energy E_{tx} in dB, versus different signal bandwidths. The thickness of the curves indicates LOS or non-LOS regimes. The curves without markers show E_{rx}/E_{tx} averaged across all x-y-positions within an rectangular area of 30 cm × 40 cm (1 cm grid, data from [7]), if an ideal full RAKE-receiver is used. The curves marked with triangles show the minimum value of E_{rx}/E_{tx} which occurs within these positions, again assuming a full RAKE. Thus the small-scale fading effect becomes visible. The curves with circles depict the normalized receive energy for a receiver which exploits only the strongest propagation path, i.e., a single correlator is applied. Transmitters-receiver separation is about 3 m, the carrier frequency is always set to 6.85 GHz.

Non-coherent UWB transmission is an attractive approach especially if simple and robust implementations with a small power consumption are required. Main application fields are low data rate sensor or personal area networks, which require low cost devices and a long battery life time. It should be noted that the current IEEE802.15.4a UWB-PHY for low data rate communications enables non-coherent detection, too [18]. The main advantage of a

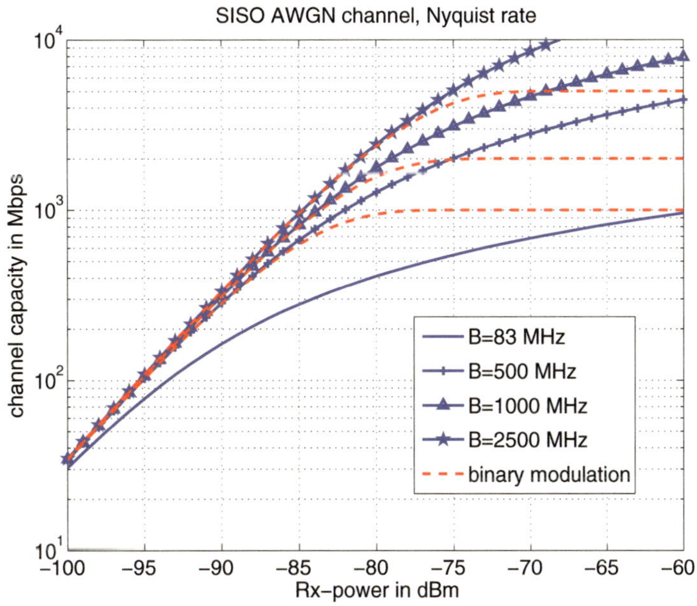

Figure 2. Capacity of an AWGN channel as a function of the receive-power for different signal bandwidths. The 2.4 GHz ISM band offers a bandwidth of 83 MHz.

non-coherent receiver is clearly the dramatically reduced effort which is required for channel estimation, synchronization, and multipath diversity combining. This advantage is, however, bought by a serious drawback: non-coherent detectors are more susceptible to narrowband interference (NBI), multi-user interference (MUI), and inter-symbol interference (ISI).

Non-coherent detection can either rely on envelope detection or on differential detection. In the simplest case, path-diversity combining is carried out via an analog integration device. However, the change from analog to digital combining stimulates new perspectives. Since a digital code matched filter can be applied prior to the non-coherent part of the receiver, the capability to distinguish users (or networks) by means of code division multiple access is improved. We show that digital receiver implementations with user specific filtering have also an enhanced interference rejection capability and energy efficiency. Moreover, we present well suited solutions for the analog-to-digital conversion, the spread-spectrum code sequences, and the modulation format.

2. Non-coherent detection in multipath AWGN

Although non-coherent detection is not restricted to low data rates — even orthogonal frequency division multiplex (OFDM) can be combined with non-coherent modulation and detection [19] — we focus our attention on low data rate single carrier transmission.

Non-coherent detection can either be based on envelope detection or on differential detection. In the simplest case, path-diversity combining can be achieved by means of a single, analog integrate and dump filter, see Fig. 3 and Fig. 4. The integration effectively provides a binary weighting of the multipath arrivals: all components inside the integration window of size T_{int} are weighted with "1", while all the others are weighted with "0". Regardless of whether envelope or differential detection is chosen, we assume that the receiver uses a quadrature

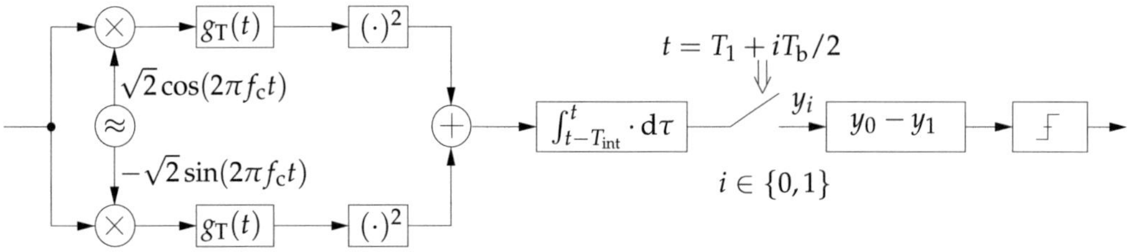

Figure 3. Envelope detection of a 2-PPM signal using a quadrature down-conversion stage. The analog integrate and dump filter (integrator) is required to capture the multipath energy.

down-conversion stage, since an ECC-conform UWB-signal is a 'truth' band-pass signal with a maximum bandwidth of 2.5 GHz and a lower cut-on frequency of 6 GHz. It is also assumed that each the inphase and the quadrature branch contain a low-pass filter, whose impulse response $g_T(t)$ is matched to the transmitted pulse $\psi_1(t)$. Without multipath, this ensures that the energy of the received signal is focused at the sampling times. For example, according to the IEEE 802.15.4a UWB PHY standard a receiver needs to perform a matched de-chirp operation, if the optional "chirp on UWB" pulse is used.

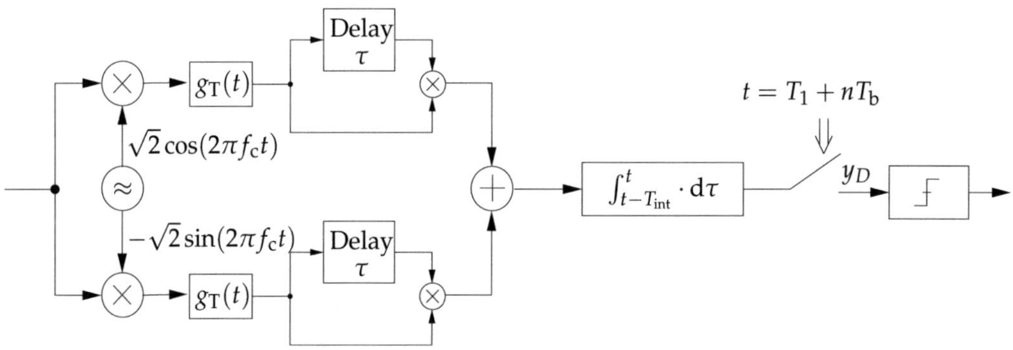

Figure 4. Differential detection for DPSK ($\tau = T_b$) or transmitted-reference PSK ($\tau < T_b$).

In the following, we consider binary pulse-position modulation (2-PPM) and differential phase-shift keying (DPSK) as straightforward modulation schemes to be combined with either envelope detection or with differential detection.

If E_b denotes the mean energy per bit, a 2-PPM signal (single carrier with fixed carrier frequency) given in the complex baseband can be written as

$$s(t) = \sqrt{E_b} \sum_{n=-\infty}^{\infty} (1 - b_n)\psi_1(t - nT_b) + b_n\psi_1(t - nT_b - T_b/2), \tag{1}$$

where $\psi_1(t)$ needs to be orthogonal to $\psi_1(t - T_b/2)$. T_b is the bit interval, which is split into two subintervals each of length $T_b/2$. Depending on the binary information b_n, $b_n \in \{0,1\}$, to be transmitted, the waveform $\sqrt{E_b}\psi_1(t)$ is generated either at the time nT_b or $T_b/2$ seconds later. For single carrier transmission with a fixed carrier frequency, the unit energy basis function $\psi_1(t)$ needs to exhibit a bandwidth of at least 500 MHz, i.e., it is a spread spectrum waveform. For example, according to the IEEE 802.15.4a UWB PHY, $\psi_1(t)$ consists of a single pulse with a duration of 2 ns (or less) or a burst of up to 128 such pulses with a scrambled polarity.

A (single carrier) DPSK signal given in the complex baseband can be written as

$$s(t) = \sqrt{E_b} \sum_{n=-\infty}^{\infty} (2\tilde{b}_n - 1)\psi_1(t - nT_b), \tag{2}$$

where $\tilde{b}_n = b_n \oplus \tilde{b}_{n-1}$, $\tilde{b}_n \in \{0, 1\}$, denotes differentially encoded bits.

2.1. Performance estimation for single-window combining

By increasing the transmission bandwidth B, more and more multipath arrivals are resolvable at the receiver. For example, with $B = 500$ MHz and an assumed excess delay of 50 ns, about 25 individual paths are resolvable in the time domain. In contrast to a channel matched filter (or its RAKE receiver equivalent), which combines all these arrivals coherently and with an appropriate weighting, the integrate and dump filter shown in Fig. 3 and Fig. 4 clearly allows only a non-coherent combining. This suboptimal combining leads to a performance loss, which increases with the product $B \cdot T_{int}$, where it is assumed that the whole signal energy is contained within the integration interval T_{int}. If it is further assumed that $B \cdot T_{int}$ is an integer $N \geq 1$, the BER p_b for 2-PPM (energy detection) can be estimated as [5]

$$\begin{aligned} p_b &= \frac{1}{2^N} \exp\left(-\frac{E_b}{2N_0}\right) \sum_{i=0}^{N-1} \frac{1}{2^i} \mathcal{L}_i^{N-1}\left(-\frac{E_b}{2N_0}\right) \\ &\approx \frac{1}{2}\text{erfc}\left(\frac{E_b/N_0}{2\sqrt{E_b/N_0 + N}}\right), \end{aligned} \tag{3}$$

where $\mathcal{L}_i^{N-1}$ is a generalized Laguerre polynomial. The second expression corresponds to a Gaussian approximation which can be used for $N > 15$. Eqn. (3) is only valid if the integration interval T_{int} contains the whole bit energy E_b, and if no ISI occurs, which is the case if the channel excess delay is small compared to T_b. The compact solution (3) has its origin in the fact that the samples y_0 and y_1 (see Fig. 3) are χ^2-distributed with $2N$ degrees of freedom [10].

For DPSK and differential detection, the statistical description is very similar. Thus, it is only required to substitute N_0 by $N_0/2$ in (3), i.e., the E_b/N_0-performance differs by exactly 3 dB in favor of DPSK.

Fig. 5 shows the penalty with respect to the required E_b/N_0, if we switch from — rather hypothetical — coherent channel matched filter detection to non-coherent detection with equal gain single-window combining (SinW-C). A value of 1.2 dB at $N = 1$ just corresponds to the E_b/N_0-penalty in AWGN ($p_b = 10^{-3}$).

The results must be interpreted very carefully, since it is assumed that the whole bit energy is concentrated within the interval T_{int}. In reality, this is surely not the case and an appropriate T_{int} must be found. If T_{int} is increased, more and more noise is integrated which leads to the loss shown in Fig. 5, but more signal energy may be collected as well.

From the energy efficiency point of view, any non-coherent combining should take place with respect to the multipath energy only. If chirp or direct sequence spread spectrum (DSSS) signals are used, where the signal energy is spread over time even at the transmitter, it is preferable to equip the receiver with a matched filter $g_T(t)$ which focuses the energy of the chirp or DSSS-waveform before the non-coherent processing takes place, as it was assumed in Fig. 3 and Fig. 4.

Fig. 6 shows the E_b/N_0-performance of 2-PPM in the case of a non-LOS indoor channel, where E_b is interpreted as the bit energy available at the receiver's antenna output. For the results shown, we have used measured UWB channels (5m×5m×2.6m office) including the antennas. The measurements were carried out by the IMST GmbH [6, 7]. B_6 is chosen to be 500 MHz.

The results show that the BER strongly depends on T_{int}, whereas the position of the integration window is always chosen optimally. For the reference indoor channel considered here [10], the optimal value of T_{int} is about 20 ns. At $T_{int} = 20$ ns, SinW-C loses additionally 1.5 dB compared to an optimal non-coherent detector, cf. Section 2.3.

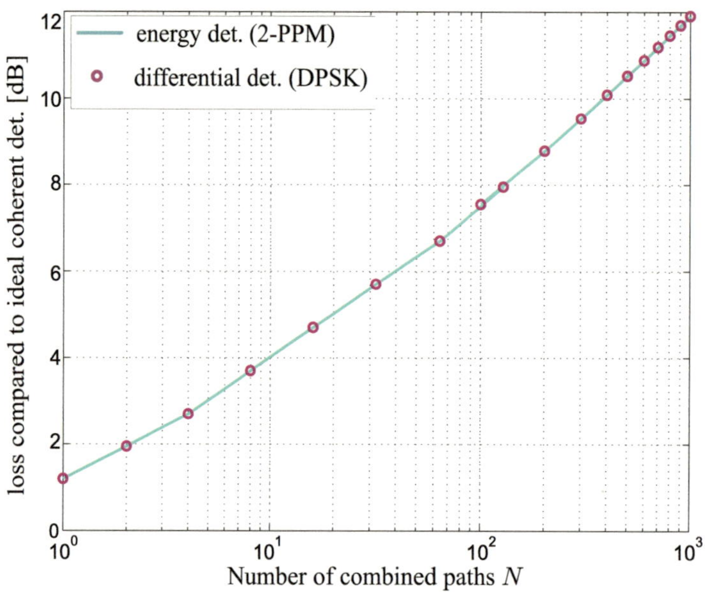

Figure 5. Power penalty due to non-coherent combining as a function of the time bandwidth product N ($p_b = 10^{-3}$). Coherent matched filter detection acts as the reference.

2.2. Weighted sub-window combining

In the case of SinW-C, at least two parameters need to be adjusted, the window size T_{int} and its position T_1 (synchronization). Since the BER over T_{int} performance of SinW-C may also exhibit several local minima, the practical determination of appropriate T_{int} and T_1 values may be more difficult than it seems. These problems are reduced, if weighted sub-window combining (WSubW-C) is used, where the whole integration window is divided into a number of N_{sub} sub-windows of size T_{sub}. From the E_b/N_0 performance point of view, it is preferable to choose the N_{sub} weighting coefficients according to the sub-window energies. In [10] we have shown that WSubW-C with $T_{sub} = 4$ ns (which corresponds to a sampling rate of 250 MHz) outperforms SinW-C (with optimum synchronization) by about 0.5 dB, if indoor channels are considered.

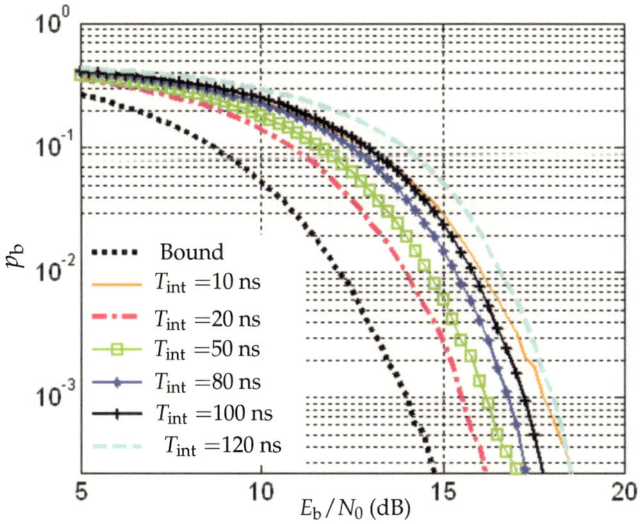

Figure 6. BER performance of non-coherent 2-PPM detection for SinW-C and a 6 dB signal bandwidth of 500 MHz. A measured indoor non-LOS channel is used to obtain the results.

2.3. Performance limit

If the sampling rate of the WSubW-C is equal to the signal bandwidth B, i.e., $T_{\mathrm{sub}} = 1/B$, each resolvable multipath component[2] can be weighted according to its energy. In [10] we have shown that this approach ensures the benchmark performance, if non-coherent combining is used. The advantage compared to perfectly synchronized SinW-C is about 1.5 - 2 dB for indoor channels, cf. Fig. 6 (black, dotted curve).

3. Analog receiver implementations

3.1. Feasibility of analog differential detection

ISI will degrade the BER performance of DPSK systems, if the symbol interval is considerably smaller than the channel excess delay. However, it is crucial to realize delays on the order of 50 ns or more in the analog domain, if the ultra-wideband nature of the signals is taken into account. Fig. 7 shows the normalized group delay of a Bessel-Thomson all-pass filter with a maximum flat group delay $t_g(f)$. If a 5 % group delay error is chosen to define the cut-off frequency, it is clear that a 5th order filter can provide a usable frequency range of $\approx 1/t_g(0)$, i.e., for a desired cut-off frequency of $f_g = B_6/2 = 250$ MHz, the delay is only 4 ns. Even a huge and completely unrealistic filter order of 20 could only provide a delay of $5.6 \cdot 4 = 22$ ns, if $f_g = 250$ MHz. It should be noted that two of these analog delay lines have to be implemented, if a quadrature down-conversion stage as shown in Fig. 4 is used.

A basic motivation of impulse radio based on transmitted reference (TR) signaling is that shorter delays can be used. This is possible, since the autocorrelation does not take place with the previous modulated symbol but rather with an additional reference pulse. Our results show that the performance of TR-signaling varies extremely from channel realization

---

[2] At a total transmission bandwidth B, multipath components can be resolved down to $1/B$ in the time domain.

to channel realization, since the autocorrelation process is disturbed by intra-symbol interference. Additionally, if a reference pulse is periodically inserted prior to each modulated pulse, a 3 dB loss occurs. Delay hopping techniques or reference symbol averaging may reduce this 3 dB loss, but require even more (and longer) delay lines.

It is more than unlikely that analog implementations of differential receivers will have a chance to be applied in low cost products. For the multi-user case with analog multipath combining (next section), we have therefore focused on energy detection combined with time-hopping (TH) impulse radio.

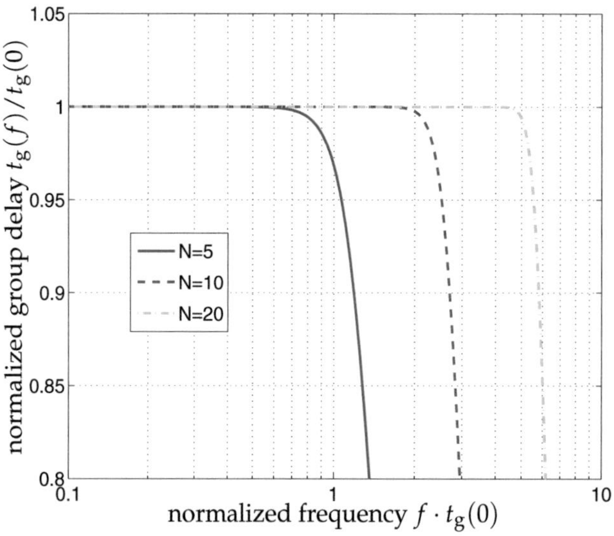

Figure 7. Normalized group delay of a Bessel-Thomson all-pass filter of order N.

3.2. Multiple access for analog multipath combining

As mentioned above, the transmitted pulse $\psi_1(t)$ may be also a chirp or direct-sequence spread spectrum waveform. To realize the filter $g_T(t)$ in the analog domain, SAW-filters (SAW: surface acoustic wave) could be a solution, but only if $\psi_1(t)$ is a fixed, user independent waveform. Therefore, as long as the non-coherent signal processing (or a part of it) takes place in the analog domain, we assume a user independent $\psi_1(t)$.

To still enable multiple access (MA) communications, we assume that each symbol to be transmitted is represented by several short pulses which are generated at distinct times according to a user specific TH-code. The decoding will be carried out digitally, i.e., after the non-coherent signal processing took place, cf. Fig. 8.

Compared to direct sequence MA codes which contain a large number of (non-zero) chips, the sparseness of TH codes facilitates the receiver processing, reduces the complexity and keeps the additional loss due to the non-coherent combining of the code elements within limits. A TH code is determined by two parameters: the number of pulse repetitions N_s, which is equivalent to the code weight, and the number of hopping positions[3] N_h. In [12], we have presented a semi-analytical method to assess the multiple access performance. It

[3] One symbol interval can be divided into an integer number N_s of equally spaced intervals (named frames), where each frame contains one pulse. Within each frame, N_h positions are possible.

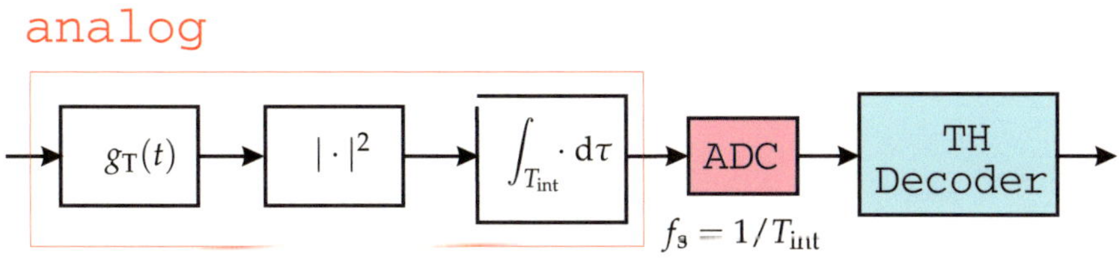

Figure 8. Block diagram of a TH code division multiple access receiver with analog multipath combining. Differential detection is not considered since analog wide band delays are difficult to realize.

is based on the statistics of the total code collisions (or "hits", determined by N_s and N_h) as well as the first- and second-order moments of the multipath channel's energy within the integration window. For non-coherent TH-PPM systems, the proposed method provides a more accurate and comprehensive evaluation of the multiple access performance than the existing code correlation function based analysis.

In [12], we have investigated various MA codes to be applied for a non-coherent TH-PPM system. It can be concluded that for a moderate number of users, optical orthogonal codes (truncated Costas codes, prime codes) with low code weights ensure a good multiple access performance while adding only a very small additional non-coherent combining loss.

4. Digital receiver implementations

Digital receiver implementations according to Fig. 9 have several advantages. First of all, they offer a superior interference rejection capability [11, 16], since user specific filtering can take place prior to the non-coherent signal processing. This restricts the non-coherent combining loss to the multipath arrivals (which exhibit stochastic path weights), whereas that part of the signal energy, which is already spread by a user-specific code at the transmitter, is coherently summed up. Furthermore, digital receiver implementations enable advanced modulations such as Walsh-modulation [15, 17] or advanced NBI-suppression strategies based on soft-limiting [13].

The block diagram of a receiver with a "digital code matched filter" (DCMF) is shown in Fig. 9. The ADC operates with a sampling rate, which is not smaller than the UWB signal bandwidth, where the ADC resolution has been chosen between 1 and 4 bits. Regarding the following results, we have always assumed TH impulse radio transmission.

Fig. 10 shows the E_b/N_0 improvement of a DCMF-based receiver as a function of N_s, where N_s depicts the number of non-zero elements of the user-specific code. As the DCMF combines the corresponding pulses coherently, the benefit compared to an analog receiver increases with N_s.

4.1. Applicability of low-resolution ADCs (single user case)

In [14] we have shown that in the 2-PPM case and under certain conditions, low-resolution ADCs can almost achieve the full resolution E_b/N_0-performance. One important condition is the number of pulse repetitions N_s within one modulated symbol, which should not be too small. For the one bit ADC case, $N_s = 8$ and $N_s = 20$ just correspond to quantization penalties of 2 dB and 1.5 dB, respectively, cf. Fig. 11(a). If the resolution is increased from 1 bit to 2 or 4 bits, the penalty may decrease, but only if the input level of the ADC is well controlled by an additional gain-control circuit. In [14] we have also proven that a 1 bit ADC with its inherent clipping characteristic offers a superior interference rejection capability.

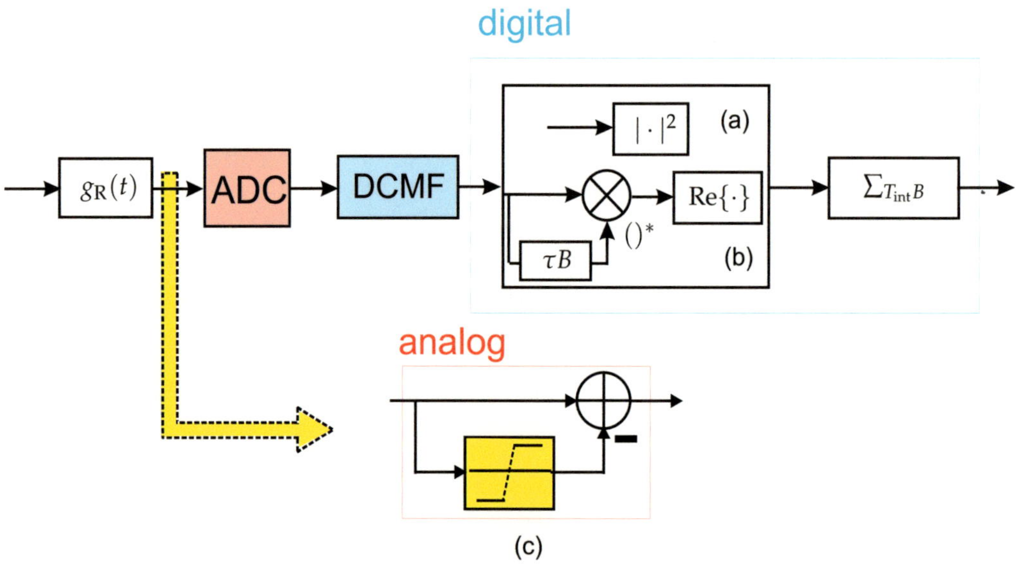

Figure 9. Block diagram of the DCMF-based non-coherent receiver shown in the complex baseband for (a) energy detection and (b) differential detection. (c) A NBI suppression scheme using a soft-limiter.

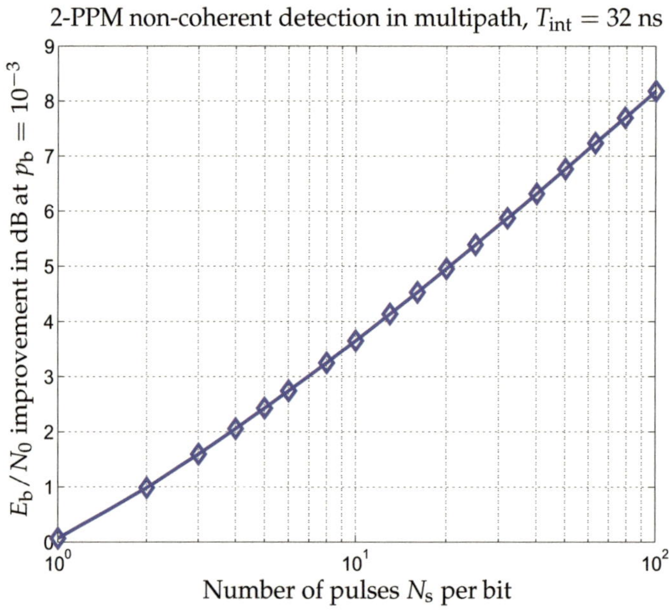

Figure 10. E_b/N_0 performance improvement of a fully digital 2-PPM non-coherent receiver compared to its analog counterpart for the multipath single-user case. The integration window extents over $T_{int} = 32$ ns. A full-resolution ADC is assumed.

We have also investigated the applicability of Sigma-Delta ADCs, especially if M-ary Walsh modulation is used, cf. Section 4.4. The results show that the full resolution performance can be obtained for an oversampling rate of 4. Since the power consumption of an ADC depends linearly on the sampling rate, but exponentially on the resolution [8], Sigma-Delta ADCs can thus be considered as attractive candidates.

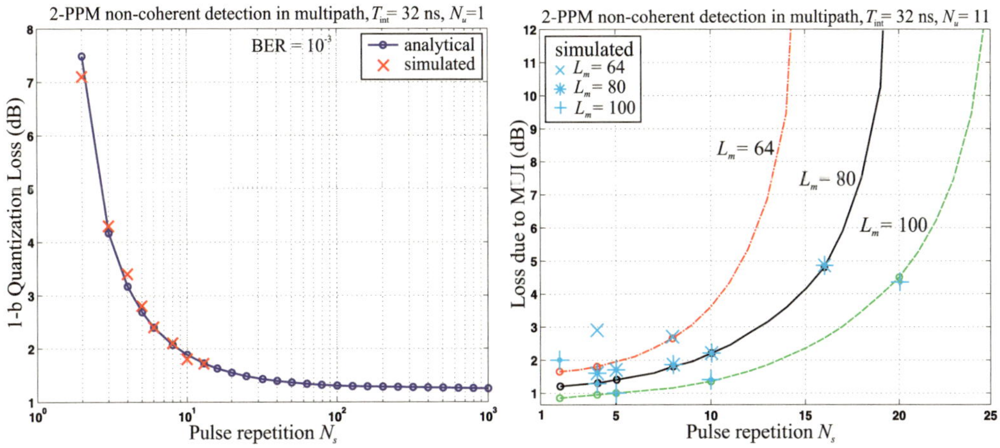

(a) (b)

Figure 11. Power penalty due to 1-bit quantization (a) and multi-user interference (b), where random-codes are assumed. BER is 10^{-3}.

4.2. Multiple access codes for time hopping PPM

Fig. 11(b) shows the power penalty due to MUI for a network with 11 users, where perfect transmit power control and a full resolution ADC was assumed. Random codes were applied. For a given processing gain $N_s \cdot N_h$, the penalty depends strongly on the ratio of the parameters N_s (number of non-zero pulses) and N_h (number of hopping positions). Since N_s determines the ADC quantization induced penalty, too, we conclude that N_s on the order of 8 leads to a good trade-off between the quantization loss and the MUI penalty. In [14] we have shown that this rule does not only apply to random codes but also to optical orthogonal codes as suggested for analog receiver implementations.

4.3. Performance of simultaneously operating piconets

A test geometry of simultaneously operating piconets (SOP) is shown in Fig. 12, where a single co-channel interference is considered. The reference distance d_{ref} (desired piconet 1) is chosen such that the power at the receiver is 6 dB above the receiver sensitivity threshold. The interfering transmitter (uncoordinated piconet 2) operates at the same power as the transmitter of piconet 1, but at a distance d_{int} to the reference receiver. We have considered the IEEE 802.15.4a channel model 3 (indoor LOS) and the channel model 4 (indoor non-LOS) [9], where random TH codes with $N_s = 10$ and $N_h = 8$ are applied altogether with forward error correction (Reed Solomon code with a rate of 0.87). The results shown in Table 1 prove clearly that digital receiver implementations outperform analog ones. Furthermore, 1-bit ADCs are desirable.

Channels	Analog	DCMF (full)	DCMF (1 bit)
CM3	1.53	0.64	0.32
CM4	1.05	0.71	0.45

Table 1. d_{int}/d_{ref} at 1% packet error ratio for SOP tests

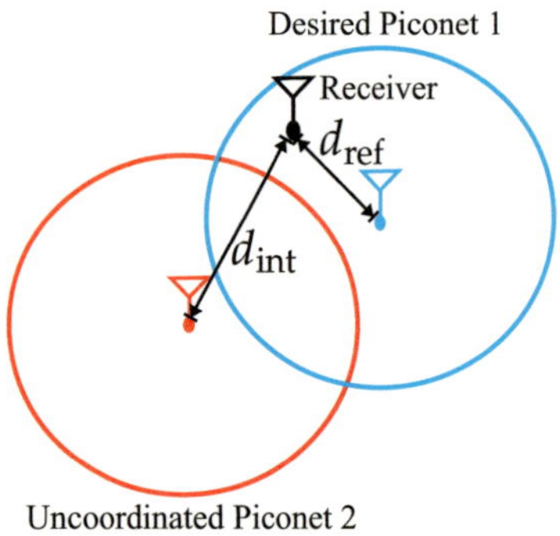

Figure 12. The SOP test geometry with a single co-channel interference

4.4. Power efficient Walsh-modulation

In [17], we have proposed two advanced (low data rate) transmission schemes based on M-ary Walsh-modulation, namely repeated Walsh (R-Walsh) and spread Walsh (S-Walsh). For both schemes, the fast Walsh Hadamard transformation can be used to efficiently implement the demodulator. Whereas the more implementation friendly R-Walsh transmission is favorable for data rates of up to 180 kbps ($M = 8$ or $M = 16$), S-Walsh transmission with $M \geq 32$ is an option for higher data rates.

In [15], we have compared R-Walsh transmission with M-PPM. It has been shown that R-Walsh works well with a 1-bit quantization. Fig. 13 shows that in the case of Walsh modulation, the quantization loss (compared to the full resolution case) is only about 1.5 dB — independently of M and N_s. However, if a strong near-far effect is present, M-PPM outperforms R-Walsh with respect to the MA-performance.

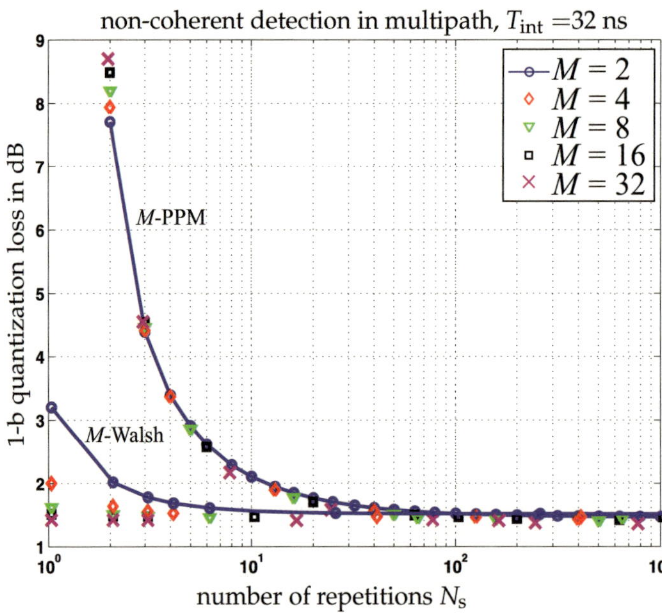

Figure 13. Quantization loss due to a one-bit ADC versus the number of repetitions N_s at $p_b = 10^{-3}$ for M-Walsh and M-PPM in multipath channels (analytical estimation).

4.5. Advanced narrowband interference suppression schemes

In [13] we have presented a new NBI-mitigation technique, which is shown in Fig. 9c). It is based on a soft limiter, where the soft limiter itself was originally proposed to suppress impulse interference [2]. The thresholds of the soft limiters are adjusted according to NBI power.

We have shown that the proposed receiver can effectively mitigate the NBI, if the threshold factor and the input level of the subsequent ADC are chosen appropriately. Furthermore, the performance improves if the ADC resolution is increased. In the presence of the OFDM interference, the proposed scheme could also be used, but it is required to adjust the threshold dynamically. It should be noted that the performance also depends on the frequency of the interference, since the DCMF has a frequency dependent transfer function.

5. Conclusions

We have derived concepts for energy efficient impulse radio UWB systems with a low transceiver complexity. These concepts are especially suitable for wireless sensor networks operating at low data rates. The E_b/N_0-performance of non-coherently detected 2-PPM and DPSK is very similar. It differs by 3 dB in favour of DPSK. However, if the multipath combining should take place in the analog domain, i.e., by means a simple integrate and dump filter, the difficulty to realize analog broadband delays makes it almost impossible to use differential detection and thus DPSK. On the contrary, digital receiver implementations enable advanced modulation schemes and offer superior interference rejection capabilities. With low-resolution ADCs, only a small quantization loss is observed. Compared to the full-resolution case, a one-bit receiver shows a higher MUI suppressing capability. Sigma-Delta ADCs can be considered as attractive candidates for the analog to digital conversion. Our results show that the full resolution performance can be obtained for an oversampling rate of 4.

Author details

Nuan Song, Mike Wolf and Martin Haardt
Ilmenau University of Technology, Germany

6. References

[1] 05-58, F. [2005]. Petition for Waiver of the Part 15 UWB Regulations Filed by the Multi-band OFDM Alliance Special Interest Group, ET Docket 04-352.

[2] Beaulieu, N. C. & Hu, B. [2008]. Soft-Limiting Receiver Structures for Time-Hopping UWB in Multiple Access Interference, *IEEE Transactions on Vehicular Technology* 57.

[3] Commission, F. C. [2002]. First Report and Order: Revision of Part 15 of the Commission's Rules Regarding Ultra-Wideband Transmission Systems, ET Docker 98-153.

[4] ECC [2007]. ECC Decision of 24 March 2006 amended 6 July 2007 at Constanta on the harmonised conditions for devices using Ultra-Wideband (UWB) technology in bands below 10.6 GHz, amended ECC/DEC/(06)04.

[5] Hao, M. & Wicker, S. [1995]. Performance evaluation of FSK and CPFSK optical communication systems: a stable and accurate method, *Journal of lightwave technology* (13): 1613–1623.

[6] Kunisch, J. & Pamp, J. [2002]. Measurement Results and Modeling Aspects for the UWB Radio Channel, *IEEE Conference on Ultra Wide-Band Systems and Technologies* .

[7] Kunisch, J. & Pamp, J. [n.d.]. *IMST-UWBW: 1-11 GHz UWB Indoor Radio Channel Measurement Data*, IMST GmbH. available at http://www.imst.com/imst/en/research/radio-networks/funk_wel_dow.php.

[8] Le, B., Rondeau, T. W., Reed, J. H. & Bostian, C. W. [2005]. Analog-to-digital converters, *IEEE Signal Processing Magazine* 22(6): 69–77.

[9] Molisch, A. F. et al. [2005]. IEEE 802.15.4a channel model - final report, *Tech. Rep. Document IEEE 802.15-04-0662- 02-004a.*

[10] Song, N., Wolf, M. & Haardt, M. [2007]. Low-complexity and Energy Efficient Non-coherent Receivers for UWB Communications, *Proc. 18-th Annual IEEE International Symposium on Personal Indoor and Mobile Radio Communications (PIMRC07)*, Greece.

[11] Song, N., Wolf, M. & Haardt, M. [2009a]. A Digital Code Matched Filter-based Non-Coherent Receiver for Low Data Rate TH-PPM-UWB Systems in the Presence of MUI, *IEEE International Conference on Ultra Wideband (ICUWB 2009)*, Vancouver, Canada.

[12] Song, N., Wolf, M. & Haardt, M. [2009c]. Performance of PPM-Based Non-Coherent Impulse Radio UWB Systems using Sparse Codes in the Presence of Multi-User Interference, *Proc. of IEEE Wireless Communications and Networking Conference (WCNC 2009)*, Budapest, Hungary.

[13] Song, N., Wolf, M. & Haardt, M. [2010a]. A Digital Non-Coherent Ultra-Wideband Receiver using a Soft-Limiter for Narrowband Interference Suppression, *Proc. 7-th International Symposium on Wireless Communications Systems (ISWCS 2010)*, York, United Kingdom.

[14] Song, N., Wolf, M. & Haardt, M. [2010b]. A *b*-bit Non-Coherent Receiver based on a Digital Code Matched Filter for Low Data Rate TH-PPM-UWB Systems in the Presence of MUI, *Proc. of IEEE International Conference on Communications (ICC 2010)*, Cape Town, South Africa.

[15] Song, N., Wolf, M. & Haardt, M. [2010d]. Time-Hopping M-Walsh UWB Transmission Scheme for a One-Bit Non-Coherent Receiver, *Proc. 7-th International Symposium on Wireless Communications Systems (ISWCS 2010)*, York, United Kingdom.

[16] Song, N., Wolf, M. & Haardt, M. [Feb., 2009b]. Non-coherent Receivers for Energy Efficient UWB Transmission, *UKoLoS Annual Colloquium*, Erlangen, Germany.

[17] Song, N., Wolf, M. & Haardt, M. [Mar., 2010c]. Digital non-coherent UWB receiver based on TH-Walsh transmission schemes, *UKoLoS Annual Colloquium*, Günzburg , Germany.

[18] TG4a, I. . [2007]. Part 15.4: Wireless Medium Access Control (MAC) and Physical Layer (PHY) Specifications for Low-Rate Wireless Personal Area Networks (WPANs), *IEEE Standard for Information Technology* .

[19] Wolf, M., Song, N. & Haardt, M. [2009]. Non-Coherent UWB Communications, *FREQUENZ Journal of RF-Engineering and Telecommunications* . special issue on Ultra-Wideband Radio Technologies for Communications, Localisation and Sensor applications.

Chapter 5

Coding, Modulation, and Detection for Power-Efficient Low-Complexity Receivers in Impulse-Radio Ultra-Wideband Transmission Systems

Andreas Schenk and Robert F.H. Fischer

Additional information is available at the end of the chapter

http://dx.doi.org/10.5772/53261

1. Introduction

Impulse-radio ultra-wideband (IR-UWB) is a promising transmission scheme, especially for short-range low-data-rate communications, as, e.g., in wireless-sensor networks. One of the main reasons for this is its potential to employ noncoherent, hence low-complexity, receivers even in dense multipath propagation scenarios, where channel estimation required for coherent detection would be overly complex due to the large signal bandwidth and hence rich multipath propagation [23].

Differential pulse-amplitude-modulated IR-UWB in combination with autocorrelation-based detection constitutes an attractive variant of noncoherent detection schemes [11]. The inherent loss in performance of traditional noncoherent autocorrelation-based differential detection (DD), as compared to coherent detection based on explicit channel estimation, can be alleviated by advanced autocorrelation-based detection schemes operating on the output of an extended autocorrelation receiver (ACR). This ACR delivers correlation coefficients of symbols separated by several symbol durations. In this context, block-based detection schemes, which partition the receive symbol stream into (possibly overlapping) blocks and thus process multiple symbols jointly, have proven to enable power-efficient, yet low-complexity detection in both uncoded and coded IR-UWB transmission systems [3, 6, 11, 12, 15–18].

In this chapter, a comprehensive review of block-based detection schemes is presented. Starting with an exposition of the operation in uncoded schemes, we discuss the generation of soft output, required in coded IR-UWB systems employing autocorrelation-based detection. For the design of such systems, an information theoretic performance analysis of IR-UWB transmission with autocorrelation-based detection delivers design rules for coded IR-UWB systems. In particular, optimum rates for the applied channel code are derived, which improve the overall power efficiency (i.e., required signal-to-noise ratio to guarantee a desired error rate) of the system. The chapter concludes with a brief summary.

2. IR-UWB system model with autocorrelation-based detection

2.1. IR-UWB system model

Throughout this chapter binary pulse-amplitude-modulated IR-UWB transmission in combination with bit-interleaved coded modulation (BICM), as shown in Fig. 1, is considered. Avoiding up-/downconversion due to operation at a carrier frequency, transmission takes place in the baseband; hence, all signals are real valued. The sequence of information bits (assumed to be equiprobable and independent, i.e., with maximum entropy) is encoded with a channel code of rate R_c. After symbolwise mapping from (interleaved) codeword bits c_k to binary information symbols $a_k \in \{\pm 1\}$, differential encoding is performed, yielding the transmit symbols $b_k \in \{\pm 1\}$, where $b_k = b_{k-1} a_k$ and $b_0 = 1$. The IR-UWB receive signal, after propagation through an UWB multipath channel, is given by [23]

$$r(t) = \sum_{k=0}^{+\infty} b_k p(t - kT) + n(t) \tag{1}$$

where T is the symbol duration and $p(t)$ denotes the overall receive pulse shape, resulting from the convolution of transmit (TX) pulse, receive (RX) filter, and channel (CH) impulse response; its energy is normalized to one, thus, the energy per information symbol[1] is given as $E_s = 1$. We assume the channel to remain constant within one codeword. $n(t)$ results from white Gaussian noise of two-sided power-spectral density $N_0/2$ passed through the RX filter. To preclude inter-symbol interference, the symbol duration T is chosen sufficiently large, such that each pulse has decayed before the next pulse is received. For clarity, we do not explicitly consider the typically applied frame structure used for time-hopping and code-division multiple access, as it can be averaged out prior to further receive signal processing [6].

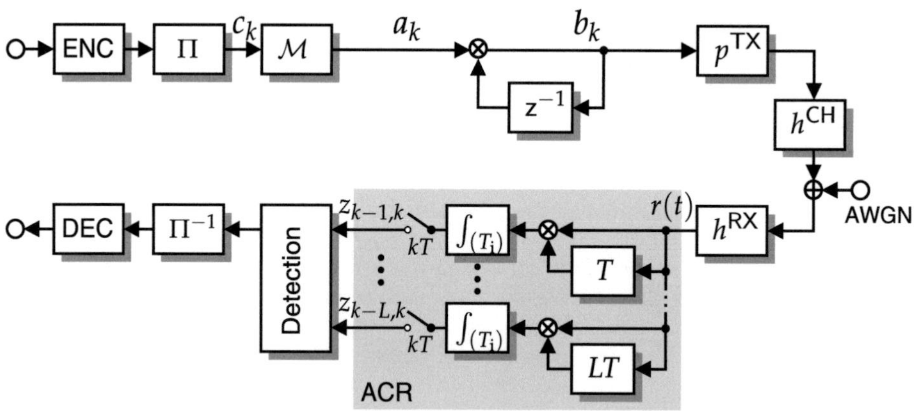

Figure 1. System model of coded IR-UWB transmission with autocorrelation-based detection.

For convenient representation and in view of an all-digital implementation of the receiver, we define the sampled receive signal of the kth symbol interval as $\bar{r}_k = [r(kT), r(kT + T_s), \ldots, r(kT + (N_s - 1)T_s)]^\mathsf{T}$, where $N_s = f_s T$ is the number of samples per symbol interval, and $f_s = 1/T_s$ is the sampling rate (greater than or equal to the Nyquist rate). With respective

[1] For long bursts, the energy for the reference symbol may be neglected.

definitions for the end-to-end impulse and the noise, assuming no inter-symbol interference, we compactly write

$$\bar{\boldsymbol{r}}_k = b_k \bar{\boldsymbol{p}} + \bar{\boldsymbol{n}}_k \,. \tag{2}$$

The noise components in $\bar{\boldsymbol{n}}_k$ are modeled as uncorrelated Gaussian random variables with variance $\sigma_n^2 = f_s N_0/2$, which is the case for a square-root Nyquist low-pass receiver front-end filter with two-sided bandwidth f_s.

2.2. Autocorrelation-based detection

Autocorrelation-based noncoherent detection of IR-UWB, cf., e.g., [2, 3, 6, 9, 15], requires to compute the correlation of the current symbol with up to L preceeding symbols, as shown in Fig. 1. Significant gains are achieved by adopting the integration interval to the channel characteristics at hand [23], i.e., choosing $T_i < T$ in the order of the expected channel delay spread. Simplified, larger T_i lead to decreased performance, but become inevitable in case of only coarse synchronization or insufficient knowledge of the channel characteristics[2]. Defining the time-bandwidth product $N = f_s T_i$ and $\boldsymbol{r}_k$ as the part of $\bar{\boldsymbol{r}}_k$ relevant for the ACR integration, i.e., (typically the first) N successive components out of N_s, in an digital implementation we have, for $l = 1, \ldots, L$,

$$z_{k-l,k} = \boldsymbol{r}_{k-l}^{\mathsf{T}} \boldsymbol{r}_k \,. \tag{3}$$

The correlation coefficients serve as input for various detection schemes, cf., Sec. 3 and [3, 6, 9, 10, 12, 15, 23]. E.g., symbolwise differential detection (DD) utilizes only the correlation coefficient of the current symbol and its predecessor, i.e., $L = 1$, and, since $b_{k-1} b_k = a_k$, the decision rule for the information symbols reads $a_k^{\mathrm{DD}} = \mathrm{sign}(z_{k-1,k})$.

We explicitly point out the major drawback of an autocorrelation-based receiver, namely the required accurate analog delay lines in an analog implementation, or the large sampling rate[3] in an all-digital implementation. Especially approaches based on the principle of compressed sensing (CS) promise to circumvent these problems [14, 24]. These approaches avoid sampling the receive signal at the (possibly prohibitively) large Nyquist rate by taking fewer measurements in a different domain (e.g., frequency or some transform domain). In [14] it has been shown, that a CS-front-end can readily be applied prior to an ACR, i.e., via direct correlation of the measurements, thus also avoiding the need for computationally complex CS-reconstruction algorithms. In combination with autocorrelation-based DD the inherent loss in performance of CS/ACR-based detection is proportional only to the square root of the compression ratio (number of measurements over N_s) [14].

2.3. Equivalent discrete-time system model

Based on the all-digital implementation, we introduce an equivalent discrete-time system model of ACR-based detection. The ACR-output can be written as

$$z_{k-l,k} = E_i x_{k-l,k} + \eta_{k-l,l} \tag{4}$$

---

[2] A typical setting for realistic IR-UWB scenarios, e.g., modelled by the IEEE channel models [7, 8] is $T_i = 33$ ns, whereas $T = 75$ ns to avoid inter-symbol interference. With $f_s = 12$ GHz, we have $N_s = 900$ and $N \approx 400$ [23].

[3] With the advance in micro electronics, one can expect that an all-digital implementation becomes realistic within no later than the next decade.

where $E_\text{i} = \boldsymbol{p}^\mathsf{T}\boldsymbol{p}$ denotes the captured pulse energy. It is composed of the phase transition from b_{k-l} to b_k, i.e., $x_{k-l,k} = b_{k-l}b_k$, and "information × noise" and "noise × noise" terms, summarized in the equivalent noise term

$$\eta_{k-l,l} = b_{k-l}\boldsymbol{p}^\mathsf{T}\boldsymbol{n}_k + b_k\boldsymbol{n}_{k-l}^\mathsf{T}\boldsymbol{p} + \boldsymbol{n}_{k-l}^\mathsf{T}\boldsymbol{n}_k\,. \tag{5}$$

A detailled analysis of the components of the equivalent noise term in (5) shows that already for moderate time-bandwidth products N it is reasonable to approximate the respective terms as uncorrelated Gaussian random variables [9, 10, 14]. In particular, the "information × noise" terms are zero-mean with variance σ_n^2, and the "noise × noise" term, as the sum of N products of independent Gaussian random variables, is zero-mean with variance $N(\sigma_n^2)^2$. Consequently, $\eta_{k-l,k}$ may be modeled as a zero-mean Gaussian random variable with variance $\sigma_\eta^2 = 2\sigma_n^2 + N(\sigma_n^2)^2$. Since each $\eta_{k-l,k}$ results from the multipication of different parts of noise and symbols, the equivalent noise samples at different time instances and ACR branches are uncorrelated.

This approximation is only valid under the following prerequisites, which typically are fulfilled in common IR-UWB systems: i) the symbol duration is chosen sufficiently large, such that no inter-symbol interference is present, ii) the integration interval of the ACR and the time-bandwidth product N are chosen sufficiently large, such that the Gaussian approximation holds, iii) the receiver front-end filter is a square-root Nyquist low-pass with two-sided bandwidth f_s to avoid correlations of the noise samples, and iv) the channel remains constant over the block of at least $L+1$ symbols. We emphasize that this model not only enables the subsequent information theoretic analysis of ACR-based detection of IR-UWB, but also serves as a tool for efficient numerical simulations of the IR-UWB transmission chain.

3. Advanced detection schemes for IR-UWB

3.1. Multiple-symbol differential detection

One of the most powerful detection schemes is based on the principle of multiple-symbol differential detection (MSDD), cf., [1] and its modifications for IR-UWB detection [3, 6, 15]. In MSDD the stream of receive symbols is decomposed into blocks of $L+1$ symbols (note that the blocks have to overlap by at least one symbol), and for each block the blockwise-optimal sequence of L information symbols is decided jointly based on the correlation coefficients corresponding to this block. The decision metric given a hypothesis of information symbols grouped into a vector $\tilde{\boldsymbol{a}}$ and the corresponding hypothesis of the ACR output $\tilde{\boldsymbol{x}}$, reads

$$\Lambda(\tilde{\boldsymbol{a}}) = \sum_{k=1}^{L}\left(\sum_{l=0}^{k-1}\left(|z_{l,k}| - \tilde{x}_{l,k}\,z_{l,k}\right)\right)\,. \tag{6}$$

The blockwise-optimal sequence $\boldsymbol{a}^\mathsf{MSDD}$ of hard-output MSDD is given as the sequence with minimum decision metric.

To fully exploit the benefits of channel coding, reliability information on the estimated codeword bits should be delivered to the subsequent channel decoder, i.e., so-called soft-output MSDD (SO-MSDD) should be performed. Sticking to the so-called max-log approximation, in terms of log-likelihood ratios (LLRs) reliability information corresponds to the (scaled) difference of the decision metric of the optimum sequence [12], i.e.,

$$\Lambda^{\mathsf{MSDD}} = \Lambda(\boldsymbol{a}^{\mathsf{MSDD}}) = \min_{\tilde{\boldsymbol{a}}\in\{\pm1\}^L} \Lambda(\tilde{\boldsymbol{a}}) \tag{7}$$

and the decision metric of the corresponding counter hypothesis, i.e., the minimum metric with the restriction $\tilde{a}_k = -a_k^{\mathsf{MSDD}}$, i.e., for $k = 1, \ldots, L$,

$$\Lambda_k^{\overline{\mathsf{MSDD}}} = \min_{\tilde{\boldsymbol{a}}\in\{\pm1\}^L, \tilde{a}_k = -a_k^{\mathsf{MSDD}}} \Lambda(\tilde{\boldsymbol{a}}) \,. \tag{8}$$

Finally, the reliablitiy of the kth symbol/codeword bit is proportional to

$$\mathrm{LLR}_k \sim a_k^{\mathsf{MSDD}} \left(\Lambda_k^{\overline{\mathsf{MSDD}}} - \Lambda^{\mathsf{MSDD}}\right) \,. \tag{9}$$

In the case of SO-DD ($L = 1$), the LLRs are directly given as the (scaled) ACR output, i.e., $\mathrm{LLR}_k^{\mathsf{DD}} \sim z_{k-1,k}$ [12].

Utilizing the triangular structure of the decision metric, an efficient solution to the MSDD search problem (7) is obtained by employing the sphere decoder algorithm [6, 18, 19]. In the case of SO-MSDD, incorporating modifications in the sphere decoder algorithm proposed for efficient soft-output detection in multi-antenna systems [21], the $L + 1$ search problems per block, (7) and (8), can be solved in a single sphere decoder run per block using the single-tree-search soft-output sphere decoder [12, 21]. Thus, SO-MSDD can be realized at only moderate complexity increase compared to hard-output MSDD.

3.2. Decision-feedback differential detection

A closely-related detection scheme is blockwise decision-feedback differential detection (DF-DD), cf., [5] and its modifications for IR-UWB detection [15], which decides the symbols within each block in a successive manner taking into account the feedback from already decided symbols within the block. The blockwise processing of the receive signal enables to optimize the decision order, such that in each step the most reliable symbol is decided next, resulting in almost the performance of MSDD at lower and in particular constant complexity.

Briefly sketched, following [15] and focusing on the first block, with $\hat{k}_0 = 0$, $b_0^{\mathsf{DF\text{-}DD}} = 1$, the optimized decision order and the estimates are given by

$$\hat{k}_i = \operatorname*{argmax}_{k\in\{1,\ldots,L\}/\{\hat{k}_1,\ldots,\hat{k}_{i-1}\}} \left|\sum_{l=0}^{i-1} z_{\hat{k}_l,k}\, b_{\hat{k}_l}^{\mathsf{DF\text{-}DD}}\right| \tag{10}$$

$$b_{\hat{k}_i}^{\mathsf{DF\text{-}DD}} = \operatorname{sign} \sum_{l=0}^{i-1} z_{\hat{k}_l,\hat{k}_i}\, b_{\hat{k}_l}^{\mathsf{DF\text{-}DD}} \,. \tag{11}$$

Basically, the optimized decision order forces reliable decisions for the first decided symbols, which then strongly influence the upcoming decisions. In contrast to the related detection scheme BLAST in multiple-antenna systems, sorting is done per block based on the actual receive symbols and previous decisions, rather than on the channel realization.

3.3. Low-complexity soft-output detection via combining multiple observations

The blockwise processing of the receive symbol stream enables a further possibility to improve the performance without increase of the maximum delay of the ACR [17]. This method utilizes an overlapping block-structure. Since multiple blocks thus contain the same symbol, processing of each block delivers (possibly different) beliefs on the same symbol, i.e., multiple observations are available. Suitably combining the observations obtained from processing of each block, results in a (possibly more reliable) final decision. Depending on the applied blockwise decision scheme (here SO-MSDD and DF-DD are considered), there are different options how to combine multiple soft/hard observations to deliver a final hard and/or soft decision for the respective symbol [17]. The most interesting option is to combine multiple hard decisions, e.g., obtained from DF-DD, of the same symbol to form a single soft decision. This method can be implemented by using the sum of the individual hard-decisions as (quantized and scaled) "soft-output"; it preserves the low complexity of blockwise DF-DD, yet enables to exploit the additional gain of soft- vs. hard-decision channel decoding.

3.4. Performance of advanced detection schemes for uncoded IR-UWB transmission

In Fig. 2 the presented ACR-based detection schemes are compared with respect to bit error rate of uncoded IR-UWB transmission and a time-bandwidth product of $N = 400$. This parameter setting is based on the reasoning in Footnote 2; the Gaussian approximation as described in Sec. 2.3 is employed assuming that the integration interval captures the entire pulse energy, i.e., $E_{\mathrm{i}} = E_{\mathrm{s}}$). It can be observed, that i) with increasing blocksize performance improves over traditional DD (the significant loss compared to coherent detection is mainly caused by the squared original noise variance σ_n^2 in the equivalent noise variance σ_η^2) and approaches coherent detection with perfect channel estimation, ii) DF-DD with optimized decision order (dashed lines) achieves almost the performance of MSDD (solid lines, exactly the same performance for $L = 2$ with minimum overlap, and iii) combining multiple observations obtained by introducing a maximum block-overlap, but using the same ACR front-end (right hand side of Fig. 2) leads to significant gains over traditional blockwise processing without overlapping blocks (left hand side of Fig. 2), for both soft-output MSDD and hard-output DF-DD (except for $L = 2$) as blockwise detection scheme. This gain comes at the cost of an increased computational complexity (roughly proportional to L).

4. Design rules for coded IR-UWB systems

Based on an information theoretic performance analysis of IR-UWB in combination with ACR-based detection [16], in this section design rules for coded IR-UWB transmission systems are derived and verified by means of numerical results employing convolutional codes.

4.1. Capacity of IR-UWB with MSDD

In contrast to coded modulation using multi-level codes [22], common IR-UWB systems adopt the conventional serial concatenation of coding and modulation at transmitter, and detection and decoding at receiver side, as shown in Fig. 1, i.e., restrain to the BICM philosophy. This approach offers increased flexibility and robustness in fading scenarios. The BICM capacity of the overall transmission chain composed of mapping, differential encoding, and ACR-based

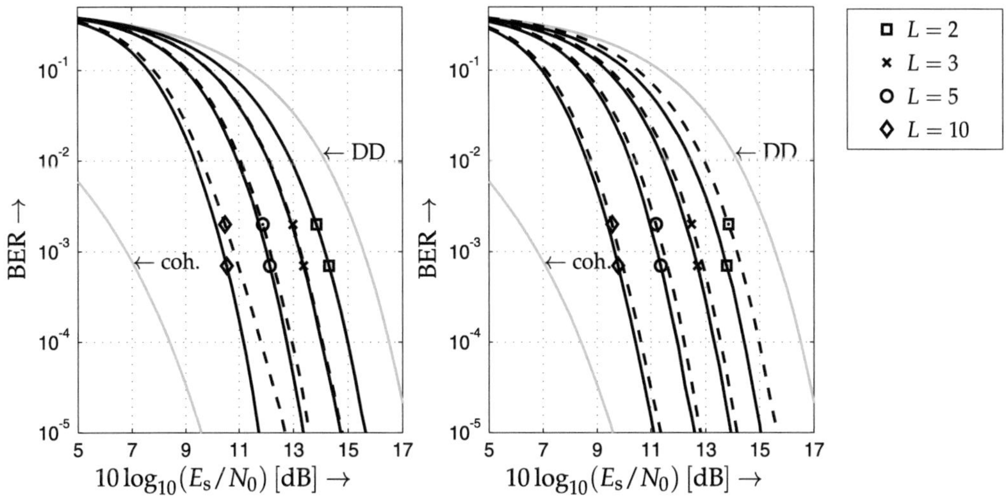

Figure 2. BER of uncoded BPSK IR-UWB transmission with autocorrelation-based detection with $L = 2$, 3, 5, and 10 (right to left). Solid: MSDD, dashed: DF-DD, left: processing of non-overlapping blocks, right: combining of multiple observations obtained by processing of maximum-overlapping blocks. Gaussian approximation with time-bandwidth product $N = 400$.

detection, is depicted in Fig. 3 (using the same parameter setup as in Sec. 3.4). Since an exact evaluation of the BICM capacity of the IR-UWB system at hand is overly complex, the equivalent discrete-time channel model and the Gaussian approximation, as derived in Sec. 2.3, have been applied [16]. Soft-output MSDD with $L = 2$, 5, and 10 (solid black), DF-DD with $L = 2$ and 5 (dashed black), and soft- and hard-output DD (solid gray/dashed gray) are shown; for comparison the capacity of BPSK with coherent detection is included.

In line with the BER results, the ACR operation causes a significant gap compared to coherent detection; the capacity improves with increasing blocksize. As expected for noncoherent detection schemes, cf., e.g., [20], the capacity curves of IR-UWB with ACR-based detection plotted vs. E_b/N_0, with E_b denoting the energy per information bit, have a C-like shape. Thus, as opposed to coherent detection, the minimum ratio E_b/N_0, which still guarantees reliable transmission, is obtained at non-zero rates (indicated with markers). At the operating point of minimum E_b/N_0 and optimum rate, both options, decreasing and increasing the code rate, lead to operating points which do not allow reliable transmission. Consequently, as known from other noncoherent detection schemes [20], also in realistic BICM IR-UWB systems the code rate should be carefully selected. Especially for increasing L this minimum gets more and more pronounced, and higher code rates should be favored compared to the probably more common choice of $R_c = 0.5$ [4]. In all cases, the optimum rate for the hard-output schemes DD and DF-DD is larger than the respective optimum rate of soft-output MSDD.

These effects are also observed for noncoherent detection (energy detection) of pulse-position modulation [20]. However, in this case already the application of BICM in combination with coherent detection leads to optimum operating points at non-zero code rates [13].

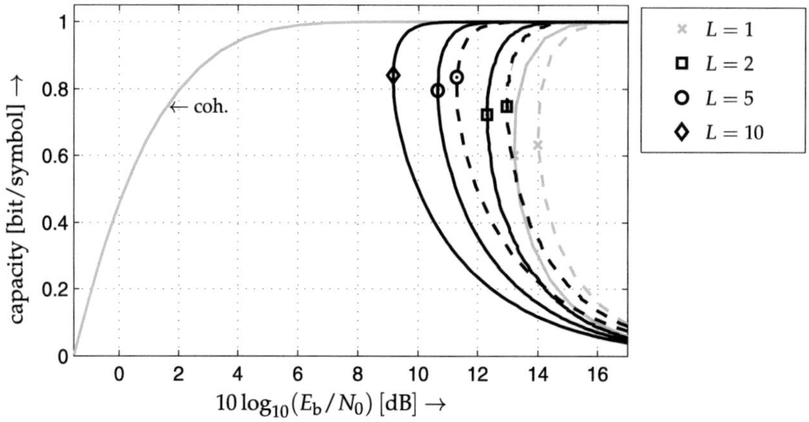

Figure 3. Capacity of BICM BPSK IR-UWB for soft/hard-output DD (solid gray/dashed gray), soft-output MSDD with $L = 2, 5$, and 10 (solid black), and DF-DD with $L = 2$ and 5 (dashed black). Gaussian approximation with time-bandwidth product $N = 400$.

In addition, a more detailled analysis shows that in non-fading scenarios an interleaver is not required for BICM IR-UWB [16].

4.2. Performance of advanced detection schemes for coded IR-UWB transmission

Finally, the design rules derived above are verified by means of numerical simulations. Fig. 4 depicts the BER of coded IR-UWB transmission using convolutional codes with optimized code rate compared to the default rate choice of $R_c = 0.5$. We apply the same channel model as in Sec. 3.4), and nonrecursively nonsystematically encoded maximum-free-distance convolutional codes with constraint length $\nu = 4$. For soft-output DD, the optimum rate

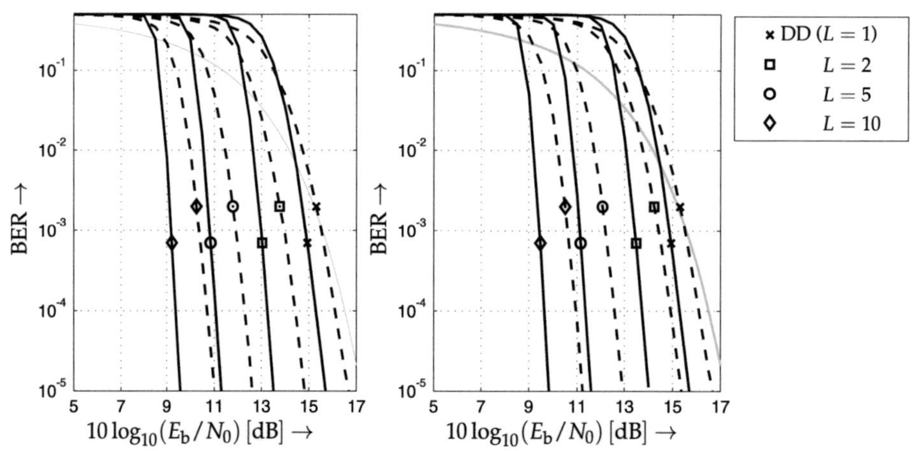

Figure 4. BER of convolutionally-coded BICM BPSK IR-UWB with autocorrelation-based detection with $L = 1, 2, 5$, and 10 (right to left). Solid: optimum code rate ($R_c = 2/3$ for DD with $L = 1$ and $R_c = 3/4$ for $L = 2, 5$ and 10), dashed: $R_c = 1/2$, gray: DD uncoded, left: soft-output MSDD, right: DF-DD, both using multiple-observations combining with maximum overlap. Gaussian approximation with time-bandwidth product $N = 400$.

is quantized to $R_c = 2/3$. Note that due to the increased decoder complexity of high-rate convolutional codes, for MSDD/DF-DD $R_c = 3/4$ is selected for all L, although higher rates are suggested by Fig. 3. ACR-based detection using soft-output MSDD (left) and DF-DD (right) with multiple observations combining is applied. It can clearly be observed that the performance is significantly improved with an optimized choice of the code rate, although the optimum code rates are larger than the default setting of $R_c = 0.5$ for all L—of course the relations are exactly opposite for coherent detection. As expected from the shape of the curves in Fig. 3, this effect is emphasized for larger blocksizes, yielding gains of almost 1 dB for $L = 10$ compared to $R_c = 0.5$.

Similar results are obtained for different coding schemes, such as LDPC codes with belief-propagation decoding [16], and also for coded IR-UWB pulse-position modulation in combination with energy detection.

5. Summary and conclusions

In this chapter we have presented a comprehensive review of coding, modulation, and detection for IR-UWB binary phase-shift keying. We conclude that noncoherent autocorrelation-based receivers in combination with blockwise detection constitute a power-efficient low-complexity reference for uncoded, as well as coded transmission. We derived and verified design rules for coded IR-UWB systems, in particular optimum code rates, which take into account the noncoherent detection at the receiver side.

Author details

Andreas Schenk
Lehrstuhl für Informationsübertragung, Friedrich-Alexander-Universität Erlangen-Nürnberg, Germany

Robert F.H. Fischer
Institut für Nachrichtentechnik, Universität Ulm, Germany

6. References

[1] Divsalar, D. & Simon, M. K. [1990]. Multiple-Symbol Differential Detection of MPSK, *IEEE Trans. Commun.* 38(3): 300–308.
[2] Farhang, M. & Salehi, J. A. [2010]. Optimum Receiver Design for Transmitted-Reference Signaling, *IEEE Trans. Commun.* 58(5): 1589–1598.
[3] Guo, N. & Qiu, R. C. [2006]. Improved Autocorrelation Demodulation Receivers Based on Multiple-Symbol Detection for UWB Communications, *IEEE Trans. Wireless Commun.* 5(8): 2026–2031.
[4] *IEEE Std 802.15.4a-2007, IEEE Standard for PART 15.4: Wireless MAC and PHY Specifications for Low-Rate Wireless Personal Area Networks* [2007].
[5] Leib, H. & Pasupathy, S. [1988]. The Phase of a Vector Perturbed by Gaussian Noise and Differentially Coherent Receivers, *IEEE Trans. Inf. Theory* 34(6): 1491–1501.
[6] Lottici, V. & Tian, Z. [2008]. Multiple Symbol Differential Detection for UWB Communications, *IEEE Trans. Wireless Commun.* 7(5): 1656–1666.
[7] Molisch, A. F., Cassioli, D., Chong, C.-C., Emami, S., Fort, A., Kannan, B., Karedal, J., Kunisch, J., Schantz, H. G., Siwiak, K. & Win, M. Z. [2006]. A Comprehensive Standardized Model for Ultrawideband Propagation Channels, *IEEE Trans. Antennas Propag.* 54(11): 3151–3166.

[8] Molisch, A. F., Foerster, J. R. & Pendergrass, M. [2003]. Channel Models for Ultrawideband Personal Area Networks, *IEEE Wireless Commun. Mag.* 10(6): 14–21.

[9] Pausini, M. & Janssen, G. J. M. [2007]. Performance Analysis of UWB Autocorrelation Receivers Over Nakagami-Fading Channels, *IEEE J. Sel. Topics Signal Process.* 1(3): 443–455.

[10] Quek, T., Win, M. Z. & Dardari, D. [2007]. Unified Analysis of UWB Transmitted-Reference Schemes in the Presence of Narrowband Interference, *IEEE Trans. Wireless Commun.* 6(6): 2126–2139.

[11] Schenk, A. & Fischer, R. F. H. [2010a]. Multiple-Symbol-Detection-Based Noncoherent Receivers for Impulse-Radio Ultra-Wideband, *2010 International Zürich Seminar on Communications (IZS)*, Zürich, Switzerland, pp. 70–73.

[12] Schenk, A. & Fischer, R. F. H. [2010b]. Soft-Output Sphere Decoder for Multiple-Symbol Differential Detection of Impulse-Radio Ultra-Wideband, *2010 IEEE International Symposium on Information Theory (ISIT)*, Austin (TX), U.S.A., pp. 2258–2262.

[13] Schenk, A. & Fischer, R. F. H. [2011a]. Capacity of BICM Using (Bi-)Orthogonal Signal Constellations in the Wideband Regime, *2011 IEEE International Conference on Ultra-Wideband (ICUWB)*, Bologna, Italy.

[14] Schenk, A. & Fischer, R. F. H. [2011b]. Compressed-Sensing (Decision-Feedback) Differential Detection in Impulse-Radio Ultra-Wideband Systems, *2011 IEEE International Conference on Ultra-Wideband (ICUWB)*, Bologna, Italy.

[15] Schenk, A. & Fischer, R. F. H. [2011c]. Decision-Feedback Differential Detection in Impulse-Radio Ultra-Wideband Systems, *IEEE Trans. Commun.* 59(6): 1604–1611.

[16] Schenk, A. & Fischer, R. F. H. [2011d]. Design Rules for Bit-Interleaved Coded Impulse-Radio Ultra-Wideband Modulation with Autocorrelation-Based Detection, *2011 International Symposium on Wireless Communication Systems (ISWCS)*, Aachen, Germany, pp. 146–150.

[17] Schenk, A. & Fischer, R. F. H. [2012]. Low-Complexity Soft-Output Detection for Impulse-Radio Ultra-Wideband Systems Via Combining Multiple Observations, *2012 International Zürich Seminar on Communications (IZS)*, Zurich, Switzerland, pp. 119–122.

[18] Schenk, A., Fischer, R. F. H. & Lampe, L. [2009a]. A New Stopping Criterion for the Sphere Decoder in UWB Impulse-Radio Multiple-Symbol Differential Detection, *2009 IEEE International Conference on Ultra-Wideband (ICUWB)*, Vancouver, Canada, pp. 606–611.

[19] Schenk, A., Fischer, R. F. H. & Lampe, L. [2009b]. A Stopping Radius for the Sphere Decoder and its Application to MSDD of DPSK, *IEEE Commun. Lett.* 13(7): 465–467.

[20] Stark, W. [1985]. Capacity and Cutoff Rate of Noncoherent FSK with Nonselective Rician Fading, *IEEE Trans. Commun.* 33(11): 1153 – 1159.

[21] Studer, C., Burg, A. & Boelcskei, H. [2008]. Soft-Output Sphere Decoding: Algorithms and VLSI Implementation, *IEEE J. Sel. Areas Commun.* 26(2): 290–300.

[22] Wachsmann, U., Fischer, R. F. H. & Huber, J. B. [1999]. Multilevel Codes: Theoretical Concepts and Practical Design Rules, *IEEE Trans. Inf. Theory* 45(5): 1361–1391.

[23] Witrisal, K., Leus, G., Janssen, G., Pausini, M., Troesch, F., Zasowski, T. & Romme, J. [2009]. Noncoherent Ultra-Wideband Systems, *IEEE Signal Process. Mag.* 26(4): 48–66.

[24] Zhongmin W., Arce, G. R., Paredes, J. L. & Sadler, B. M. [2007]. Compressed Detection for Ultra-Wideband Impulse Radio, *2007 IEEE 8th Workshop Signal Processing Advances in Wireless Communications (SPAWC)*, pp. 1–5.

Chapter 6

Interference Alignment for UWB-MIMO Communication Systems

Mohamed El-Hadidy, Mohammed El-Absi, Yoke Leen Sit, Markus Kock, Thomas Zwick, Holger Blume and Thomas Kaiser

Additional information is available at the end of the chapter

http://dx.doi.org/10.5772/55083

1. Introduction

Due to the enormous occupied bandwidth even by a single pair of UWB users, one major scientific challenge in UWB communications is interference management. Recently, Interference Alignment (IA) has become popular not only to well manage the interference, but also to optimally exploit the possible capacity gain caused by multiple pairs of transmitters and receivers. Theoretically, IA scales the channel capacity by $K/2$, where K is the number of user pairs. This fact makes IA highly attractive for future communication systems with numerous pairs of users. However, in the literature it has been reported that IA is not robust [7] against imperfections such as channel estimation errors. Thanks to the interdisciplinary research of antenna, communication and hardware engineers within the UKoLoS project *Decimus*, we were able to jointly search for solutions to improve IA robustness while not suffering from too perfect simulation idealities or too unrealistic hardware requirements. We find that *antenna* or *pattern selection* is a promising approach to improving the robustness of IA while keeping the underlying algorithms at a reasonable complexity and feasibility for implementation. Our contribution is structured as follows: first, a brief introduction of IA tailored to UWB is given. Then, we propose an antenna selection algorithm with low complexity and demonstrate its performance. In the third chapter a general methodology of MIMO UWB antenna design for orthogonal channels maximizing the channel capacity is presented. A first outcome of this methodology is a multi-mode orthogonal antenna which has been used for the investigated antenna selection approach. At last, the hardware requirements of IA systems with the proposed antenna selection method are studied. A conclusion summarizes the outcomes of our contribution.

2. MIMO-UWB interference mitigation by interference alignment

IA is a promising technique which achieves the maximum degrees of freedom (DoF) for K users in interference channels [4]. This can be achieved by a combination of linear precoding at the transmitters and interference suppression at the receivers. IA permits to force interfering signals at each receiver in one subspace and the desired signal in another orthogonal subspace [5].

Consider a K-user UWB Multi Band Orthogonal Frequency Division Multiplexing (MB-OFDM) interference channel with M_j transmit antennas at transmitter j and N_i receive antennas at receiver i. All users transmit d_s streams using N sub-carriers. Every transmitter communicates with his desired receiver and causes interference to other pairs of transmitter and receiver. The discrete-time complex received signal over the nth subcarrier at the ith receiver over a flat channel is represented as[21],[28]:

$$\mathbf{y}_i^n = \sum_{j=1}^{K} \mathbf{H}_{ij}^n \mathbf{V}_j^n \mathbf{x}_j^n + \mathbf{z}_i^n = \mathbf{H}_{ii}^n \mathbf{V}_i^n \mathbf{x}_i^n + \sum_{j=1, j\neq i}^{K} \mathbf{H}_{ij}^n \mathbf{V}_j^n \mathbf{x}_j^n + \mathbf{z}_i^n \tag{1}$$

where $\mathbf{y}_i^n$ is the $N_i \times 1$ received vector at receiver i , $\mathbf{H}_{ij}^n$ is the $N_i \times M_j$ flat frequency domain channel matrix over nth subcarrier between jth transmitter and ith receiver, $\mathbf{V}_j^n$ is the $M_j \times d_s$ unitary precoding matrix which is applied for the transmitted $M_j \times 1$ vector $\mathbf{x}_j^n$ from the jth transmitter, and $\mathbf{z}_i^n$ is the $N_i \times 1$ zero mean unit variance circularly symmetric additive white Gaussian noise vector at receiver i. The Channel State Information (CSI) is assumed to be perfectly known at each node. To reconstruct the transmitted d_s signal at the ith receiver, the received signal is decoded using a unitary linear suppression interference matrix $\mathbf{U}_i^n$. The reconstructed data $\widehat{\mathbf{y}}$ at receiver i is defined as:

$$\widehat{\mathbf{y}}_i^n = \mathbf{U}_i^{nH} \mathbf{H}_{ii}^n \mathbf{V}_i^n \mathbf{x}_i^n + \sum_{j=1, j\neq i}^{K} \mathbf{U}_i^{nH} \mathbf{H}_{ij}^n \mathbf{V}_j^n x_j^n + \mathbf{U}_i^{nH} \mathbf{z}_i^n \tag{2}$$

For perfect interference alignment, the following conditions need to be fulfilled [1]:

$$\text{rank}\left(\mathbf{U}_i^{nH} \mathbf{H}_{ii}^n \mathbf{V}_i^n\right) = d_s \qquad \forall i \tag{3}$$

and

$$\mathbf{U}_i^{nH} \mathbf{H}_{ij}^n \mathbf{V}_j^n = 0 \qquad \forall j \neq i \tag{4}$$

According to (3) and (4), the received signal after processed by the linear suppression interference matrix is:

$$\widehat{\mathbf{y}}_i^n = \mathbf{U}_i^{nH} \mathbf{H}_{ii}^n \mathbf{V}_i^n \mathbf{x}_i^n + \mathbf{U}_i^{nH} \mathbf{z}_i^n \tag{5}$$

2.1. Closed-form interference alignment

In order to achieve a closed-form IA solution, 3 users interference channel (K=3) has been considered, where each node has $M = 2d$ antennas, and each user wishes to achieve d degrees of freedom by applying the IA principles. The conditions of IA given in (3) and (4) are obtained by setting the precoding matrices as [4]:

$$\mathbf{V}_1 = \text{eign}(\mathbf{H}_{31}^{-1} \mathbf{H}_{32} \mathbf{H}_{12}^{-1} \mathbf{H}_{13} \mathbf{H}_{23}^{-1} \mathbf{H}_{21}) \tag{6}$$

$$\mathbf{V}_2 = (\mathbf{H}_{32}^{-1} \mathbf{H}_{31} \mathbf{V}_1) \tag{7}$$

$$\mathbf{V}_3 = (\mathbf{H}_{23}^{-1} \mathbf{H}_{21} \mathbf{V}_1) \tag{8}$$

and the interference suppression matrix for receiver i is given by:

$$\mathbf{U}_i = \mathrm{null}([\mathbf{H}_{ij}\mathbf{V}_j]) \qquad \forall j \neq i \tag{9}$$

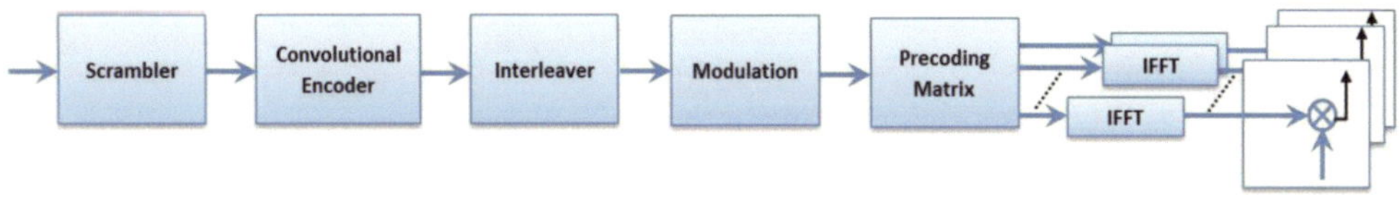

Figure 1. Transmitter block diagram of a MIMO UWB MB-OFDM communication system.

2.2. Artificial channel diversity algorithm

For successful applying the IA principle a sufficient orthogonality between all channels is required. In real-world indoor environments and MIMO UWB systems such orthogonality is not guaranteed; instead, the small distances between neighboring antennas and a possible low scattering could lead to high correlation among the several channels of the communication system. Therefore, the orthogonal component of the desired signal to the plane of the aligned undesired signals would be less pronounced, leading to worse overall system performance and robustness.

To overcome this problem, an artificial channel diversity technique is applied utilizing an antenna selection algorithm. The goal of this algorithm is to maximize the *orthogonality* of the desired signal on the plane of the aligned undesired ones. Here each transmitting node has Q_{i} antennas and only the best M_{i} antennas will be selected for maximizing the orthogonality of the desired component on the undesired signals plane. A brut force iterative process is carried out for all available combinations to choose the best selection that realizes this maximum orthogonality [8]. In the next section, the antenna selection criteria are illustrated in more detail.

3. Low complexity signal processing antenna selection algorithm

As mentioned before we propose an antenna selection algorithm in order to increase the required orthogonality directly leading to a more robust communication system in terms of minimum Bit Error Rate (BER). Note that antenna selection is a widely known approach in order to capture diversity and to improve the SNR of the communication systems [19],[14].

3.1. Antenna selection criterion

In the following we consider a K-user MIMO system with perfect IA. The selection algorithm consists of choosing the best M out of the L available transmit antennas. Denote by S_k the selected subset indices of the transmit antennas of k users [6]. The goal of the selection is to find S_k for all users $\{k = 1 : K\}$ which maximizes the average SNR_S for the multi-user system by increasing the projected desired signal into the interference-free space. This can be achieved by minimizing the principal angles between the desired signal subspace and the interference-free subspace. Minimizing the principal angles is equivalent to maximizing so-called *canonical correlations*.

Let ζ_1 and ζ_2 be subspaces in the complex plane $\mathbb{C}$. Considering the dimension of ζ_1 is smaller than or equal to the dimension of ζ_2 ($\dim \zeta_1 = d_1 \leqslant \dim \zeta_2 = d_2$). The canonical correlations are defined as the cosines of the principal angles between any two linear subspaces, which can uniquely defined as [3]

$$\cos\theta_i = \max_{\mathbf{a}_i \in \zeta_1} \max_{\mathbf{b}_i \in \zeta_2} \mathbf{a}_i^H \mathbf{b}_i, \qquad i = 1, ..., d_1, \tag{10}$$

where $\mathbf{a}_i$ and $\mathbf{b}_i$ are principal vectors of ζ_1 and ζ_2 respectively, subject to $\mathbf{a}_i^H \cdot \mathbf{a}_i = \mathbf{b}_i^H \cdot \mathbf{b}_i = 1$ and $\mathbf{a}_i^H \cdot \mathbf{a}_j = \mathbf{b}_i^H \cdot \mathbf{b}_j = 0$, $i \neq j$.

If $\mathbf{Q}_1$ and $\mathbf{Q}_2$ are orthonormal bases of the two subspaces ζ_1 and ζ_2 respectively, the canonical correlations are obtained as singular values of $\mathbf{Q}_1^H \mathbf{Q}_2 \in \mathbb{C}^{d_1 \times d_2}$ as follows [3]

$$\mathbf{Q}_1^H \mathbf{Q}_2 = \mathbf{P}_1 \Lambda \mathbf{P}_2^H \quad , \tag{11}$$

where $\mathbf{P}_1$ is a $d_1 \times d_1$ unitary matrix and $\mathbf{P}_2$ is a $d_2 \times d_2$ unitary matrix, Λ is is $d_1 \times d_2$ diagonal matrix with nonnegative real numbers on the diagonal. Therefore, $\Lambda = \mathrm{diag}(\alpha_1, .., \alpha_{d_1})$ and $\alpha_1, .., \alpha_{d_1}$ are the canonical correlations of the subspaces.

Observe that the principal angles are given by

$$\theta_i = \cos^{-1}(\alpha_i), \qquad i = 1, .., d_1. \tag{12}$$

In order to maximize the SNR at the receiver, the antenna selection criterion relies on maximizing the canonical correlations between $\mathbf{U}_k$ and $\mathbf{H}_{kk}\mathbf{V}_k$ as follows

$$S_k = \arg\min \angle\left(\mathbf{U}_k; \mathbf{H}_{kk}\mathbf{V}_k\right) \qquad ; k = 1, .., K \tag{13}$$

$$S_k = \arg\max \cos\left(\angle\left(\mathbf{U}_k; \mathbf{H}_{kk}\mathbf{V}_k\right)\right) \quad ; k = 1, .., K \tag{14}$$

$$S_k = \arg\max \left(\alpha_1, ..., \alpha_{d_s}\right) \qquad ; k = 1, 2, .., K, \tag{15}$$

where $(\alpha_1, ..., \alpha_{d_s})$ are the canonical correlations between subspace $\mathbf{U}_k$ and subspace $\mathbf{H}_{kk}\mathbf{V}_k$.

3.2. Relation between sum-rate and canonical correlations

The impact of the canonical correlations on the sum rate of a K-user MIMO system is given by [6]:

$$C = \sum_{k=1}^{K} \log \left| \mathbf{I}_N + \left(\sigma^2 \mathbf{I}_N + \sum_{l \neq k} \mathbf{W}_{kl} \right)^{-1} \mathbf{W}_{kk} \right| , \tag{16}$$

where C is the sum-rate, $\mathbf{W}_{kl} = \mathbf{H}_{kl}\mathbf{V}_l\mathbf{V}_l^H\mathbf{H}_{kl}^H$ denotes the $N \times N$ interference covariance matrix of the signal from the l-th transmitter to the k-th receiver, σ^2 is the variance of the additive white Gaussian noise, and $\mathbf{W}_{kk} = \mathbf{H}_{kk}\mathbf{V}_k\mathbf{V}_k^H\mathbf{H}_{kk}^H$ denotes the $N \times N$ covariance matrix of the desired signal. While perfect IA is assumed according to (3) and (4), the interference channel is equivalent to a set of parallel Gaussian MIMO channels, where the MIMO channel transfer function is given by $\overline{\mathbf{H}}_k = \mathbf{U}_k^H \mathbf{H}_{kk} \mathbf{V}_k$, for $k = 1, .., K$. Then the sumrate equation in (16) reduces to

$$C = \sum_{k=1}^{K} \log \left| \mathbf{I}_N + \frac{1}{\sigma^2} \mathbf{U}_k^H \mathbf{H}_{kk} \mathbf{V}_k \mathbf{V}_k^H \mathbf{H}_{kk}^H \mathbf{U}_k \right|. \tag{17}$$

Note that at high SNR, (17) can be approximated as

$$C \simeq \sum_{k=1}^{K} \log \left| \frac{1}{\sigma^2} \mathbf{U}_k^H \mathbf{H}_{kk} \mathbf{V}_k \mathbf{V}_k^H \mathbf{H}_{kk}^H \mathbf{U}_k \right|, \tag{18}$$

and by applying a thin QR decomposition

$$\mathbf{U}_k = \mathbf{Q}_{U_k} \mathbf{R}_{U_k}, \tag{19}$$

where

$$\mathbf{H}_{kk} \mathbf{V}_k = \mathbf{Q}_{V_k} \mathbf{R}_{V_k} \tag{20}$$

and $\mathbf{Q}_{U_k}$, $\mathbf{Q}_{V_k}$ are orthonormal $N \times d_s$ matrix and $\mathbf{R}_{U_k}$, $\mathbf{R}_{V_k}$ are $d_s \times d_s$ upper triangle matrix it follows

$$C = \sum_{k=1}^{K} \log \left| \frac{1}{\sigma^2} \left(\mathbf{Q}_{U_k} \mathbf{R}_{U_k} \right)^H \left(\mathbf{Q}_{V_k} \mathbf{R}_{V_k} \right) \left(\mathbf{Q}_{V_k} \mathbf{R}_{V_k} \right)^H \left(\mathbf{Q}_{U_k} \mathbf{R}_{U_k} \right) \right|. \tag{21}$$

Since $\mathbf{U}_k$ is a unitary matrix, meaning $\left| \mathbf{R}_{U_k} \mathbf{R}_{U_k}^H \right| = 1$ it furthermore follows

$$C = \sum_{k=1}^{K} \log \left(\left(\frac{1}{\sigma^2} \right)^2 \left| \mathbf{Q}_{U_k}^H \mathbf{Q}_{V_k} \right| \left| \mathbf{Q}_{V_k}^H \mathbf{Q}_{U_k} \right| \left| \mathbf{R}_{V_k} \mathbf{R}_{V_k}^H \right| \right). \tag{22}$$

Since $\mathbf{Q}_{U_k}$ and $\mathbf{Q}_{V_k}$ are the orthonormal basis of the two subspaces $\mathbf{U}_k$ and $\mathbf{H}_{kk} \mathbf{V}_k$ respectively, (22) can be linked to the principal angles between the two subspace using (11). Therefore, (22) can be written as

$$C = \sum_{k=1}^{K} \log \left(\left(\frac{1}{\sigma^2} \right)^2 \left| \mathbf{P}_{k1} \Lambda \mathbf{P}_{k2}^H \right| \left| \mathbf{P}_{k2} \Lambda \mathbf{P}_{k1}^H \right| \left| \mathbf{R}_{V_k} \mathbf{R}_{V_k}^H \right| \right) \tag{23}$$

such that

$$\mathbf{Q}_{U_k}^H \mathbf{Q}_{V_k} = \mathbf{P}_{k1} \Lambda \mathbf{P}_{k2}^H,$$

where $\mathbf{P}_{k1}$ and $\mathbf{P}_{k2}$ are $d_s \times d_s$ unitary matrices and Λ is $d_s \times d_s$ diagonal matrix equals $\text{diag}(\alpha_1, .., \alpha_{d_s})$.

Thereafter, (22) can be re-formulated as

$$C = \sum_{k=1}^{K} \log \left(\left(\frac{1}{\sigma^2} \right)^2 \left(\prod_{i=1}^{d_s} \alpha_i \right)^2 \left| \mathbf{R}_{V_k} \mathbf{R}_{V_k}^H \right| \right), \tag{24}$$

where $(\alpha_1, ..., \alpha_{d_s})$ are the canonical correlations between subspace $\mathbf{U}_k$ and subspace $\mathbf{H}_{kk} \mathbf{V}_k$. From (23) it is shown that maximizing the canonical correlations increases C, but still does not result in the maximum C because the term $\left| \mathbf{R}_{V_k} \mathbf{R}_{V_k}^H \right|$ is linked to the matrix of coefficients participating in the linear combinations yielding the columns of $\mathbf{H}_{kk}$.

3.3. Simulation results analysis

All the following simulation results have been obtained based on real-world deterministic scenarios. The deterministic hybrid EM ray-tracing channel model was considered for the MIMO UWB channel [9]. This model considers the spatial channel and the environmental effects such as path-loss, frequency dependence, reflections, transmissions, and also diffractions. It considers as well the characteristics of the antennas as part of the effective channel such as directional gain, matching and polarization. A fair comparison has been carried out among three systems: the first uses two omnidirectional Half-Wave Dipole (HWD) antennas at each node, the second uses three directional antennas (horn antenna) at each transmitter node and two directional antennas (horn antenna) at each receiver node, in this system antenna selection (AS) technique is applied to select two antennas from the three at each transmitter. The third system uses two directional antennas at each node without AS (we choose the worst case in this manner). Fig. 2 shows a comparison between the three systems using the *average* BER vs. E_b/N_o for the whole multiuser system. As shown in Fig. 2 the artificial diversity technique improves the performance of the system significantly compared to the system that uses a HWD omnidirectional antennas and the other which uses the directional antennas without using AS. This figure proves that AS is a powerful technique to improve the BER performance of the system.

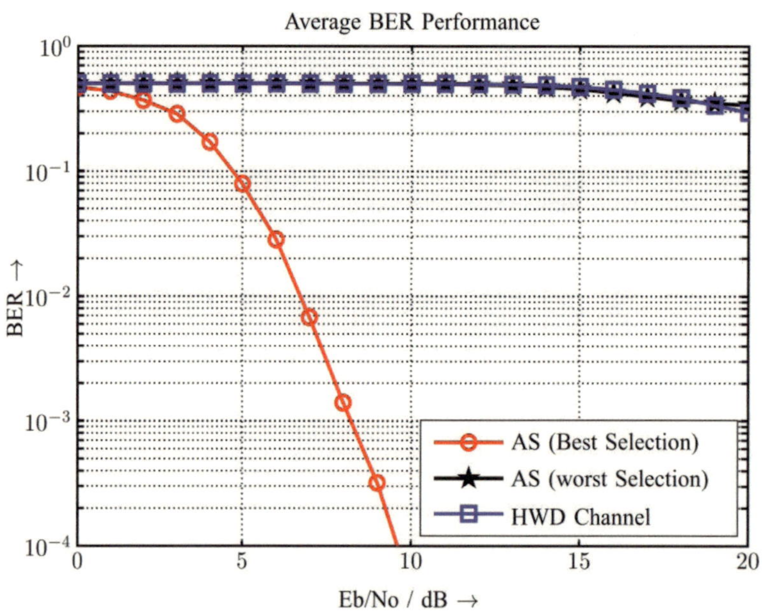

Figure 2. Comparison of average BER vs. Eb/No for a 3-user system by applying antenna selection (best and worst scenarios) and using omni-directional HWD antennas.

MIMO systems typically use antenna arrays and beamforming and spatial multiplexing. These beamforming methods do not result in (1) orthogonal channels nor (2) guarantee optimum data rates within the predefined environment e.g. indoors, outdoors, urban, suburban, etc. In designing optimal antenna systems, which fulfill the previously stated conditions, up to now, only heuristic methods have been employed. Presented here is a systematic framework to fulfill the above 2 conditions, that aids in reducing the number of real world antennas to be used. The goal is to synthesize an optimal UWB antenna system which can be used by both the transmitter and receiver at any location within the investigated scenario. This synthesis method is based on [24] and has been used to design antennas for narrowband systems in [25]. The method is now extended to the design of UWB antenna systems. In the following sections, the theory and concepts behind this systematic synthesis will be given, followed by the methodology of realization and the results.

4. Spatial sampling with sampling antennas

The concept of 'spatial sampling' is presented in greater detail in [24, 30], but can be simplified to this: 'Given a predefined overall antenna aperture confined to a limited volume V, there exists a maximum spatial capacity limit with transmission system parameters i.e. antenna aperture size and element spacing. This can be determined by sampling the transmit and receive volumes with a set of ideal sampling antennas'. In a realistic case sampling antennas possess an overall aperture size and occupy a certain spatial volume. Therefore three parameters have to be considered for their design. First, the antenna aperture size, which will approximate the size of the real world antennas. The larger the aperture size, the higher the capacity, but this size is limited by the physical size of the desired real world antennas. Second, the minimum distance between the antenna elements in order to decrease their correlation [[18, 22]]. This parameter also influences the number of sampling antenna elements within the selected aperture size. It should be noted that the more elements the aperture contains, the more time is needed for the synthesis algorithm. Third, the frequency dependency of the previously mentioned parameters.

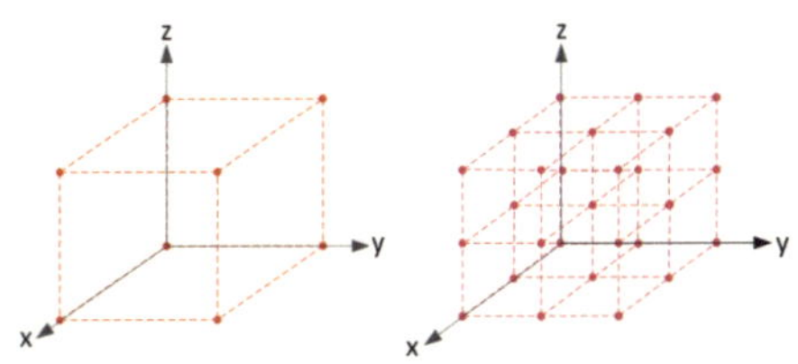

Figure 3. Sampling antenna configurations for spatial sampling, (left) minimum elements of 2×2×2, (right) more elements added, 3×3×3.

Here, a 5×5×5 cm^3 aperture based on the configuration in Fig. 3 has been chosen to illustrate the sampling antenna design and antenna synthesis algorithm in section 4.1. This configuration is then used in the subsequent antenna synthesis steps and the changes to the resulting radiation pattern over frequency is noted. If the resulting radiation pattern varies too much over frequency, the number of sampling antenna elements is then increased and the synthesis algorithm is repeated. This procedure is iterated until the resulting radiation pattern over frequency appear similar.

4.1. Channel diagonalization

For a time-invariant system, the transfer function is a transfer coefficient for all transmit and receive antenna pairs and can be expressed in the form of an $N \times M$-dimentional matrix $\mathbf{H}$. Letting $\mathbf{y}_{\text{out}}$ be an $N \times 1$ matrix and $\mathbf{x}_{\text{in}}$ be an $M \times 1$ matrix with $\mathbf{n}$ as the noise vector, the received signal vector of a communication system can then be described in the frequency domain as $\mathbf{y}_{\text{out}} = \mathbf{H}\mathbf{x}_{\text{in}} + \mathbf{n}$. Using singular value decomposition (SVD) $\mathbf{H}$ can be decomposed into $\mathbf{H} = \mathbf{U}\mathbf{S}\mathbf{V}^{\dagger}$, where $\mathbf{S} = \mathbf{U}^{\dagger}\mathbf{H}\mathbf{V}$ is a diagonal matrix whose elements are non-negative square roots of the eigenvalues λ_i of the matrix $\mathbf{H}\mathbf{H}^{\dagger}$. $\mathbf{U}$ and $\mathbf{V}$ are unitary matrices, which fulfill the condition $(\mathbf{X}^{-1})^{\dagger} = \mathbf{X}$. Multiply the input vector $\mathbf{x}_{\text{in}}$ and the output vector $\mathbf{y}_{\text{out}}$ with the matrices $\mathbf{U}^{\dagger}$ and $\mathbf{V}$ respectively, and the original channel becomes and equivalent channel,

$$\hat{\mathbf{y}}_{\mathbf{out}} = \mathbf{U}^{\dagger}\mathbf{y}_{\mathbf{out}} = \mathbf{U}^{\dagger}(\mathbf{H}\mathbf{x}_{\mathbf{in}} + \mathbf{n}) = \mathbf{U}^{\dagger}(\mathbf{H}\mathbf{V}\hat{\mathbf{x}}_{\mathbf{in}} + \mathbf{n}) = \mathbf{S}\hat{\mathbf{x}}_{\mathbf{in}} + \hat{\mathbf{n}} \tag{25}$$

where $\hat{\mathbf{x}}_{\mathbf{in}}$, $\hat{\mathbf{y}}_{\mathbf{out}}$ and $\hat{\mathbf{n}}$ are the equivalent input, output and noise vectors respectively. The diagonal matrix $\mathbf{S}$ now becomes the channel matrix of the equivalent channel where each Eigenmode is interpreted as an independent SISO (single-input-single-output) subchannel and the capacity of the system becomes a sum over these SISO capacities as expressed by [12]

$$C = \sum_{i=1}^{K} \log_2 \left(1 + \frac{p_i \lambda_i}{\sigma^2_{\text{noise}}}\right) \tag{26}$$

with $K = \min(M, N)$, which is the rank of the matrix $\mathbf{H}\mathbf{R}_{\mathbf{xx}}\mathbf{H}^{\dagger}$ with $\mathbf{R}_{\mathbf{xx}}$ being the covariance matrix of the transmit signal, its Eigenvalues $\lambda_i(i = 1, 2, ..., K)$ and power coefficients $p_i(i = 1, 2, ..., K)$. From (26) the capacity of a MIMO system can be seen as a sum of independent K SISO subchannels (Eigenmodes) represented by the Eigenvalues λ_i, where each Eigenmode corresponds to one orthogonal subchannel. More explanation can be found in [16, 24].

Employing the waterfilling algorithm in the case of channel state information (CSI) known to the transmitter will result in an optimum capacity solution for such a MIMO channel. The $\mathbf{U}$ and $\mathbf{V}$ matrices are called the 'beamforming matrices' as they determine the mapping and weighting of all the signals onto the antenna elements.

5. Scenario-based MIMO antenna synthesis

In order to apply the SVD technique to obtain parallel subchannels, the channel matrix $\mathbf{H}$ of the intended scenario must be provided. One of the most reliable and repeatable way of obtaining the SISO channel matrix $\mathbf{H}$ is through ray-tracing with the software developed by [10]. The ray-tracing simulations are done in parallel with the design of the sampling antenna configuration.

A typical indoor scenario (with glass windows, furniture, ceiling and floor) was built for simulation as shown in Fig. 4. The size of the room is $10 \times 10 \times 3$ m and simulations were done for transmitters and receivers at randomized positions in the room with an antenna height 1.5 m over the frequency band of 3.1 GHz to 10.6 GHz. Omni-directional antennas (dipoles) are used, along with the option of using both vertical and horizontal polarizations for an added degree of freedom for the design of the real world antenna. Around 45600 random transmitter-receiver points were simulated in order to acquire a synthesis result, that when the averaging strategies in section 6 have been applied, will be applicable from virtually any point in the intended scenario.

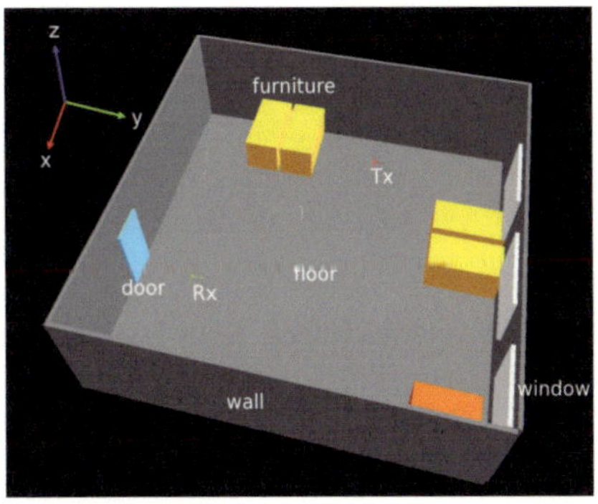

Figure 4. Indoor scenario for ray-tracing

5.1. SISO to MIMO extrapolation

A complete characterization of the MIMO channel matrix with ray-tracing requires $N \times M$ runs taking into account the sampling antenna configuration, which is computationally inefficient. [11] presents a method to reduce the calculation effort by assumming that the same plane wave impinges on all sampling antenna elements. Since the sampling antenna elements spacing is small and fixed, the difference of the incident wave at the origin of the sampling antenna configuration shown in Fig. 3 with the other antenna elements is only the phase difference expressed as

$$\Delta\varphi_i = -\beta\left(\Delta x_i \sin\vartheta\cos\psi + \Delta y_i \sin\vartheta\sin\psi + \Delta z_i\cos\vartheta\right) \tag{27}$$

where ϑ and ψ are the angles of arrival or departure of the incident wave in elevation and azimuth respectively. Hence the SISO to MIMO extrapolation reduces the computation of the MIMO **H** matrix to only one SISO run.

5.2. Antenna system simplification

In order to simplify the system, a plot of the eigenvalues (obtained after the SVD) versus the frequency is used to identify the channels with the strongest power. For instance, if only the first two subchannels were identified as having significant power as compared to the rest, the beamforming matrices **U** and **V** can be modified to contain only those two subchannels. With this, the system will now comprise only 2 inputs and 2 outputs.

6. Synthesis results

The resulting synthesized antenna radiation patterns for both the transmitter and receiver at one point for several different frequencies are as shown in Fig. 5. The figure shows the 3D plot of the radiation patterns for two subchannels and is computed using:

$$\vec{E}(d,\vartheta,\psi) = \vec{E}_{\text{single}}(d,\vartheta,\psi)\cdot\frac{e^{-j\beta d}}{d}\cdot\sum_{i=1}^{N_{\text{ant}}} a_i\, e^{-j(\beta(d_i-d)+\zeta_i)} \tag{28}$$

where $\vec{E}_{\text{single}}$ is the electric field of the sampling antenna used (a dipole in this case), N_{ant} is the total number of transmitter or receiver antennas (since they both use the same sampling antenna configuration), β is the wave number, d is the distance from the origin of the sampling antenna to a far-field observation point, $a_i \angle \zeta_i$ is the weighting from the **U** and **V** beamforming matrices, $d_i - d = \Delta\varphi$ in (27) and $\Delta x_i, \Delta y_i, \Delta z_i$ are the position of the individual elements in the array according to the Cartesian coordinate system.

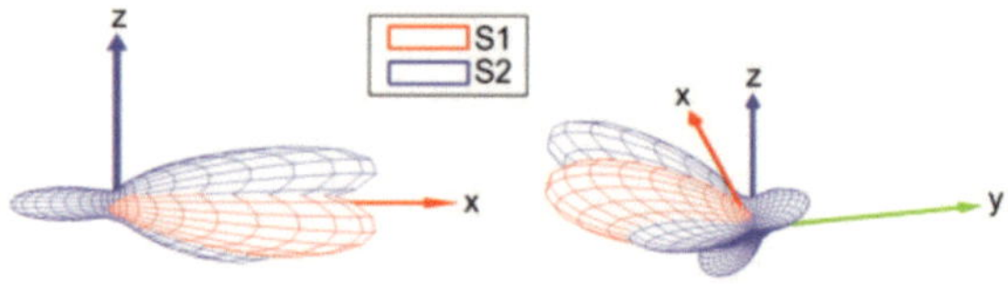

Figure 5. Resulting synthesized antenna radiation pattern for a transmit-receive pair (horizontal polarization) for subchannel 1 (S1), with line-of-sight propagation, and subchannel 2 (S2), with propagation paths reflected from the ceiling and floor, for the aperture size 5×5×5 cm^3 with 5×5×5 elements, (left) side view and (right) bottom view. (Image taken from [29])

Averaging strategy

Three averaging strategies were used, namely averaging over frequency, over location, and of transmitter and receiver radiation patterns. The first averages all radiation patterns obtained at frequency points between 3.1 to 10.6 GHz to obtain a pattern which is valid for the UWB. The second averages the radiation pattern obtained from random points around the scenario so that the resulting radiation pattern is valid for use in the whole scenario. The third averaging is done if the resulting transmitter and receiver radiation patterns look qualitatively similar, so that both can use the same antennas.

6.1. Capacity analysis

The capacity for the averaged synthesized patterns according to the number of sampling antennas across the ECC (Electronic Communications Committee) standard's UWB band was analyzed using (26). The term p_i is taken from the power spectral density levels of the ECC UWB spectral mask, λ_i is the Eigenvalue of the subchannel from the matrix **S** and $\sigma^2_{\text{noise}} = kTB$, where k is the Boltzmann constant, $T = 297$ K and $B = 100$ MHz. Fig. 6 shows the capacity of the synthesized radiation pattern using 2×2×2 till 5×5×5 sampling antenna elements within the defined 5×5×5 cm^3 physical space. It can be seen that the higher the number of sampling antennas, the more the capacity increases, agreeing with the theory in [18]. Noting that the rise in the capacity is decreasing with the higher element configuration used, we conclude that the 5×5×5 configuration is nearing the capacity saturation limit.

7. Real world antennas

The real world antennas which match the elevation characteristics over frequency of the optimized synthesized antennas can be found in [1]. Two dual orthogonal polarized antennas are used along with a 180° hybrid coupler to form the two subchannels. The comparison of the radiation pattern over frequency of the synthesized antennas and the real world antennas is shown in Fig. 7 .

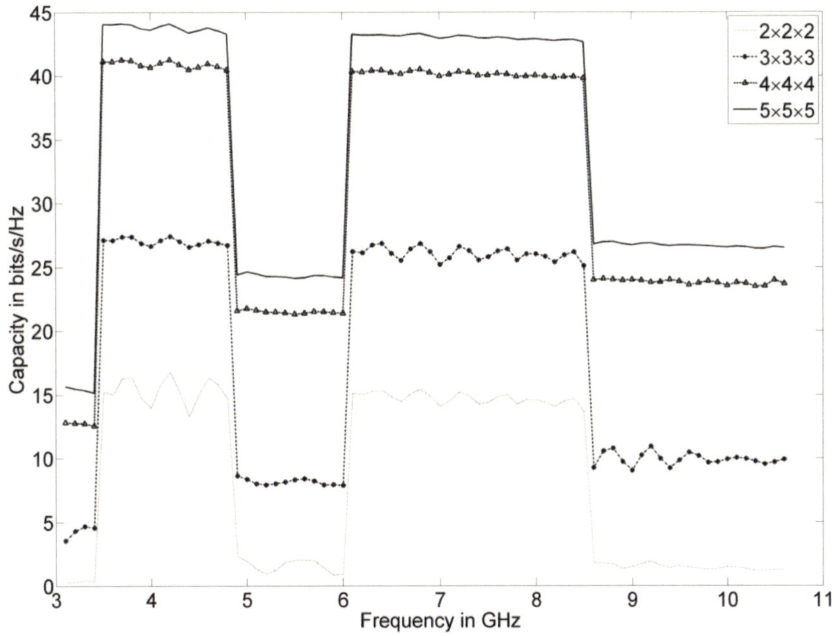

Figure 6. Capacity analysis for the aperture size $5 \times 5 \times 5\,\text{cm}^3$. (Image taken from [29])

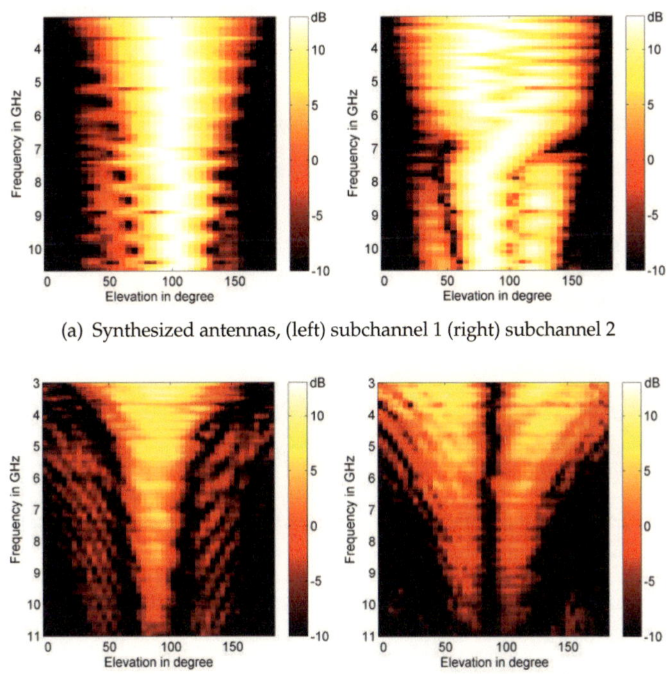

(a) Synthesized antennas, (left) subchannel 1 (right) subchannel 2

(b) Real world antennas (left) subchannel 1 (right) subchannel 2

Figure 7. Synthesized antennas vs. real world antennas: radiation pattern (elevation) over frequency

7.1. Orthogonal channels multimode antenna selection criteria

Before the above mentioned real world UWB antennas are used, the IA algorithm is tested with a narrowband multimode antenna with orthogonal channels. The narrowband antenna system is designed for use at 5.9 GHz to 6.15 GHz, consisting of four monopoles built on a finite ground plane as shown in Fig. 8(a). This antenna system is capable of radiating four different orthogonal modes based on the amplitude and phase of the excitation signals to the antenna ports. More details about this antenna can be found in [15].

Two modes as shown in Fig. 8(c) and (d) were chosen and the antenna system was simulated with a ray-tracing software within the scenario shown in Fig. 8(b). The simulation has been performed for 1000 different transmit and receive nodes locations. Simulation results shown in Fig. 9 illustrates that the overall BER system performance has been significantly improved with the multimode antenna system compared to the half wave dipoles (HWD). That is due to the additional path diversity to the communication system.

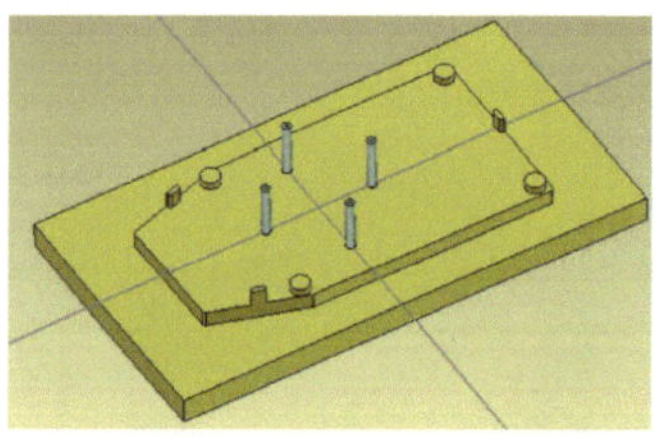

(a) Multimode narrowband antennas

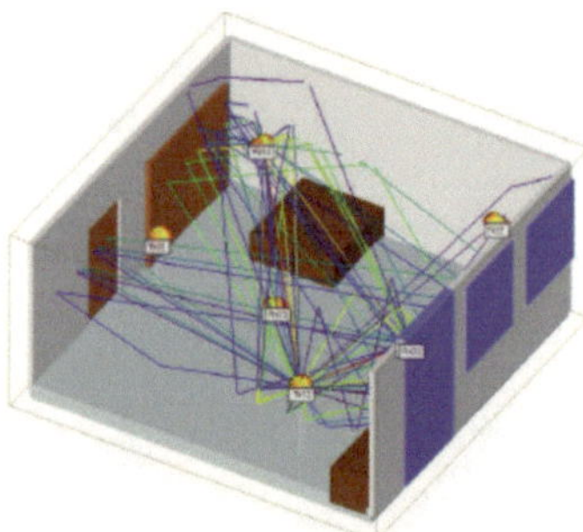

(b) Ray-tracing scenario for Interference Alignment test using the narrowband multimode antennas

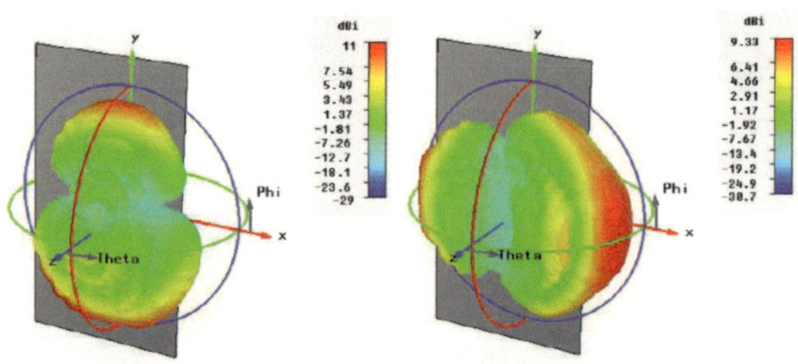

(c) Orthogonal radiation patterns of the multimode antennas (left) mode 1, (right) mode 2

Figure 8. Multimode narrowband antennas for preliminary Interference Alignment algorithm analysis

8. Hardware requirements of IA systems

High throughput wireless communication systems including LTE, ECMA-368 (WiMedia) and IEEE 802.11ac are built around sophisticated digital signal processing algorithms. Among the research goals for future communication standards are higher spectral efficiency, higher

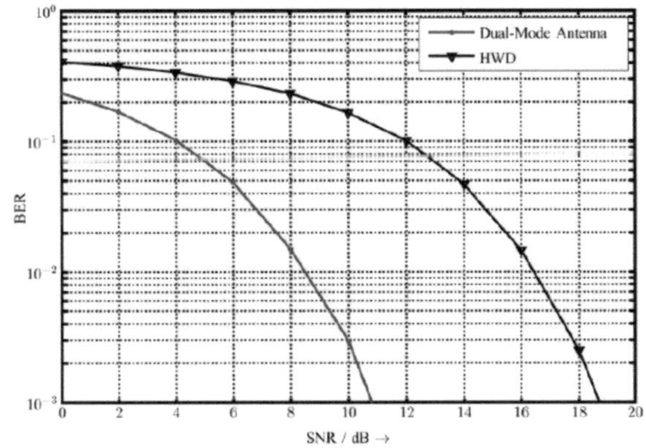

Figure 9. Comparison of the average BER vs. E_b/N_0 of the IA system by applying multimode antenna systems and comparing to HWDs.

energy efficiencies towards the Shannon limit and increased data rates. Naturally, these benefits come at the price of higher computational complexity. The demand for flexible realtime hardware platforms capable of delivering the required huge number of operations per second at a severely limited power and silicon area budget has led to the development of specialized hardware platforms for software defined radio (SDR) applications. Techniques from FPGA-based ASIC verification and rapid prototyping are combined in this project for the design space exploration of highly optimized hardware architectures.

In a typical implementation scenario for complex designs, high level reference models are used. The choice of optimization-blocks is often based on profiling results, with those blocks contributing significantly to the overall resource requirements being chosen for optimization. This leads to a hybrid design consisting of a mixture of high level blocks and highly optimized blocks, running on hardware ranging from general purpose processors (GPP), application specific instruction set processors, FPGA-based rapid prototyping systems and dedicated hardware accelerators. The presented design space exploration framework reflects this structure and allows the designer to freely move processing blocks between the different layers of optimization.

The design space exploration framework created within this project is presented in Section 8.2. A hardware implementation case study of a closed-form 3-user IA algorithm has been selected for presentation in Section 8.3. Cost functions for an iterative IA algorithm are given in Section 8.4.

8.1. Wireless communication systems design space exploration

The process of designing complex digital electronic circuits offers a large variety of options to the designer. There are many valid possible implementations that fulfill the specification, but they differ in certain properties, e.g. silicon area, power efficiency, flexibility, testability and design effort. These properties span the so-called design space. A design space exploration establishes relations between possible points in the design space, ultimately leading to cost functions modeling the relation between the design properties and parameters. These models serve as a quantitative basis to make important design decisions in an early design phase.

Certain parameters are of special interest in the domain of wireless communication platforms. The limited power budget in mobile devices puts hard constraints on the power efficiency, requiring power optimization across all layers of algorithm development, design implementation and semiconductor technology.

Deriving comprehensive cost models using Monte-Carlo methods requires visiting a significantly larger number of points in the design space compared to existing heuristically driven parameter optimization approaches covered by existing FPGA-based simulation acceleration systems. The achievable simulation speedup is a key factor enabling the characterization and optimization of complex communication systems using Monte-Carlo approaches which are infeasible for pure software simulation due to the large required stimuli sets.

8.2. Development framework

The FPGA-based hybrid hardware-in-the-loop research and design space exploration (DSE) framework created in this work combines high-level tools (e.g. MATLAB/Simulink) and optimized hardware blocks [17]. Its application domain ranges from the design, optimization and verification of efficient and optimized signal processing blocks for computationally demanding next-generation wireless communication systems to system characterization and DSE.

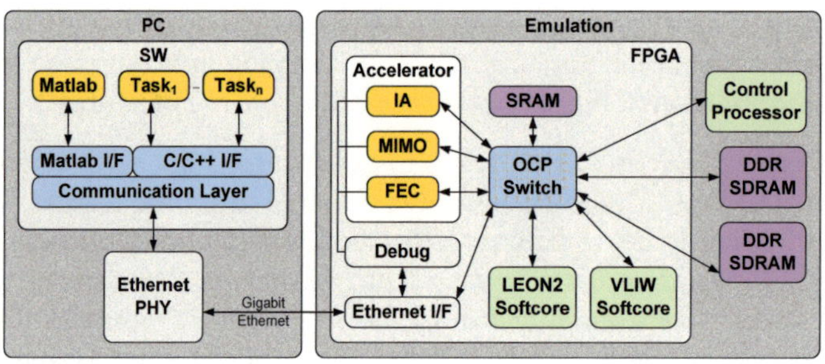

Figure 10. Emulation framework block diagram

The framework consists of a host PC, FPGA-based emulation systems, a generic fully synthesizeable VHDL SoC infrastructure, dedicated processors, processor softcores and a software library providing a transparent communication application programming interface (API). This allows signal processing blocks to be split and run distributed on a highly heterogeneous signal processing system. Software API libraries provide unified transparent communication between MATLAB, C/C+, embedded software and the hardware on-chip multilayer bus system. The same resources are accessible from all components, enabling a flexible partitioning and migration of processing task between high-level software, embedded software and dedicated hardware modules. The framework block diagram is shown in Figure 10. The properties of the optimized on-chip infrastructure template make it suitable for usage in final ASIC targets and thus enable the test, debugging and characterization of signal processing blocks in their target environment. Using standard FPGA design flows, new computationally intensive processing cores are directly implemented as the optimized hardware target modules. Instrumentation is used to enable dynamic, software controlled parameter adjustment. The remaining blocks may continue to run as high-level models, enabling a divide-and-conquer implementation and verification approach. The framework provides transparent data transport between the substituted MATLAB modules and multiple parallel instances of their FPGA hardware counterparts. The same interfaces are available for hardware simulation via the Modelsim foreign language interface (FLI), effectively also providing a verification and debugging environment at minimal extra effort.

The PC is connected to the emulation systems via Gigabit Ethernet. The generic FPGA infrastructure template comprises an OCP multilayer bus, the ethernet DMA interface, SDRAM controllers, on-chip memories and massively parallel parameterized softcore processors [20]. It has been adapted to and tested on a Xilinx Virtex-6 LX550T based BEE4 rapid prototyping system, the Xilinx Virtex-6 ML605 Evaluation Kit and the Virtex-5 LX220 based MCPA board [2] developed at IMS, see Fig. 11.

Figure 11. FPGA-based emulation system developed at IMS

8.3. Case study: 3-user antenna selection interference alignment

An implementation of the antenna selection interference alignment algorithm presented in Section 3 has been chosen as a case study using the development framework presented in Section 8.2. The proposed 3-user 2x2 MIMO zero-forcing IA antenna selection algorithm computes precoding matrices $\mathbf{V}$, decoding matrices $\mathbf{U}$ and a metric η based on [4]. Compared to the experimental testbed for fixed antenna patterns presented in [13], our implementation also chooses a subset of channels (i.e. antennas or radiation patterns) from the available channels. This leads to an increased channel orthogonality for the chosen channels at a reduced number of RF front ends.

The problem of finding the optimum antenna combination $\hat{i}$ from a set of I combinations can be formulated as

$$\hat{i} = \arg\max_{i=1...I} \sum_{k=1}^{K} \eta(\mathbf{V}_{k,i}, \mathbf{U}_{k,i}) \tag{29}$$

where $\eta(\mathbf{V}, \mathbf{U})$ is a function of the resulting SNR according to Section 3.1. $\mathbf{V}_{k,i}$ and $\mathbf{U}_{k,i}$ are the precoding and decoding matrices of user k for a given antenna combination i. Equation 29 is solved by visiting all I antenna combinations.

8.3.1. Computational complexity

The resource requirements of an optimized efficient integer implementation of the proposed novel antenna-selection IA algorithm is presented in this section, based on FPGA implementation results. Target systems include SDR platforms, FPGAs and ASICs.

This section focuses on the costs of the 3-user 2x2 MIMO processing consisting of matrix inversions, matrix multiplications, eigenvector computation and normalization, see Equations 6 to 9. The metric η is computed for both eigenvectors. All intermediate matrices can be independently scaled by arbitrary scalars without affecting the antenna decision or **V** and **U**. Exploiting this makes the cost of all involved 2x2 matrix inversions negligible and allows intermediate matrices to be block-normalized by shifting, i.e. extract a common power of 2 from all matrix elements. This results in reduced integer word lengths and thus reduced hardware costs. Table 1 summarizes the number of required real-valued mathematical base operations for antenna selection and the computation of **V** and **U** per antenna combination and subcarrier, without a final normalization step of **V** and **U**. Complex multiplications are composed of three real multiplications, three additions and two subtractions, `INVSQRT` denotes the reciprocal square root [26].

OP	ADD	MUL	SQRT	INVSQRT
Matrix mult.	696	348	0	0
Eigenvectors	15	8	3	0
Metric score	46	82	6	2
#OPC	757	438	9	2

Table 1. Operation counts for the computation of η per antenna combination i and subcarrier

To keep the total transmit power constant, the chosen antenna combination's precoding matrices V need to be normalized, resulting in 3 `ADD`, 8 `MUL` and 1 `INVSQRT` additional operations #OPN per transmitter and subcarrier. The above analysis implies that in general, the implementation cost is dominated by the multiplications in terms of silicon area and power consumption.

For the case of $K = 3$ users with $M = 2$ active transmit antennas used out of $L = 3$ physical antennas per transmitter and $N = 2$ antennas per receiver, there are 27 antenna combinations to be visited per subcarrier.

For realtime operation, the maximum allowable latency is defined to be T_0. Assigning relative operation costs α_i to each operation type OP_i, the total computational cost C for S subcarriers becomes

$$C = \frac{S}{T_0} \cdot \left(n \cdot \sum_{i \in \text{OP}} \alpha_i \cdot \#\text{OPC}_\text{i} + K \cdot \sum_{i \in \text{OP}} \alpha_i \cdot \#\text{OPN}_\text{i} \right) \tag{30}$$

8.3.2. Hardware cost estimation

Using α as relative silicon area costs, the total silicon area implementation cost of an architecture without resource sharing can be estimated from Eq. (30). The relative area α of 16-bit arithmetic operations for an ASIC implementation based on [23] results in the values given in Table 2. The relative costs α_{MUL} of a multiplier are defined to be 1.

OP	ADD	MUL	SQRT	INVSQRT
α	0.108	1	1.73	3

Table 2. Relative silicon area costs of 16-bit arithmetic operations

For a system using antenna selection at the transmitter only with $L = 3$ antennas, $S = 128$ subcarriers and $T_0 = 1$ ms, the total IA costs are estimated to be $C = 1.875$ GOPS.

For the configuration above, the original MATLAB algorithm takes 3.63 s on an Intel Xeon 2.4 GHz CPU running MATLAB R2012a for the computation of the optimal antenna combination $\hat{i}$ and its corresponding precoding and decoding matrices V_k and U_k from a set of channel information H. The FPGA implementation created in this case study achieves realtime operation, requiring 380 μs at 100 MHz clock frequency on a Xilinx Virtex-6 LX550T FPGA in a BEE4 emulation system. Thus, the achieved speedup is 9553.

8.4. Cost functions for K-user IA

The implementation presented in the previous section is a based on a closed-form $K = 3$ user IA algorithm. There is no known closed-form solution for $K > 3$ users, but iterative algorithms exist. In this section, we present implementation complexity estimates of the minimum mean square error (MMSE) IA algorithm presented in [27].

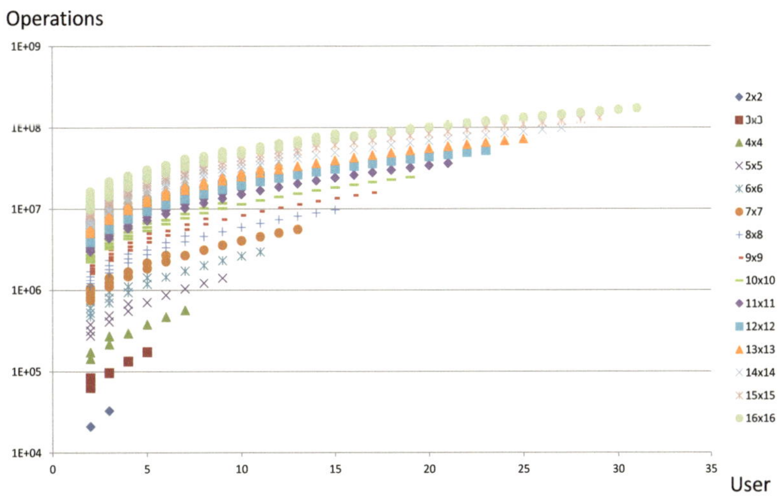

Figure 12. Number of operations for iterative MMSE interference alignment (4 iterations)

The MMSE-IA algorithm starts with arbitrary precoding matrices $\mathbf{V}_k$, then iteratively updates the decoding and precoding matrices $\mathbf{U}_k$ and $\mathbf{V}_k$ according to Eq. (31) and (32) until convergence. The Lagrange multiplier $\lambda_k \geq 0$ is computed to satisfy $\|\mathbf{V}_k\|_2^2 \leq 1$ by Newton iteration.

$$\mathbf{U}_k = \left(\sum_{j=1}^{K} \mathbf{H}_{kj}\mathbf{V}_j\mathbf{V}_j^H\mathbf{H}_{kj}^H + \sigma^2\mathbf{I} \right)^{-1} \mathbf{H}_{kk}\mathbf{V}_k \tag{31}$$

$$\mathbf{V}_k = \left(\sum_{j=1}^{K} \mathbf{H}_{jk}^H\mathbf{U}_j\mathbf{U}_j^H\mathbf{H}_{jk} + \lambda_k\mathbf{I} \right)^{-1} \mathbf{H}_{kk}^H\mathbf{U}_k \tag{32}$$

The number of required iterations is data-dependent. Each iteration step requires the following operations to be executed: matrix multiplication, pseudo-inverse, Newton iterations. Figure 12 summarizes the estimated number of operations for the computation of a set of **V** and **U** matrices, based on well-known optimized hardware implementations. Comparing the iterative approach computational complexity to the the closed-form 2x2 IA implementation presented in Section 8.3, the number of operations is increased by a factor of approximately 60.8.

9. Conclusion

IA is a promising approach for communications with numerous pairs of users. In our contribution we have investigated the usefulness of IA for MIMO UWB communication systems. Beside apart of the significant power processing needed for the high data rate applications, the MIMO UWB antenna design remains a challenge. The antenna synthesis presented here can be viewed as synthesizing an antenna system with optimal radiation pattern catered towards an intended scenario. This antenna system radiates orthogonalized channels (after the averaging strategies) with sufficient power and has fixed beamforming (direction optimized according to the scenario and with averaging over various positions) at the transmitter and receiver antenna systems. Also, the resulting system has been simplified to 2 inputs and 2 outputs based on the subchannels with the strongest power. The whole system has been simulated by an indoor ray tracing tool and the corresponding MIMO UWB base band modulation schemes and detection techniques. Moreover, an antenna selection method is proposed in order to increase the robustness of IA in real environments. It is demonstrated that by using orthogonal multimode antennas a significant gain can be obtained. In the third and last part of our contribution the hardware efforts of IA algorithms are studied in more detail. It is worked out that highly challenging system blocks like IA can be elaborated today only by the help of suitable hardware emulation platforms which are typically FPGA-based. Therefore, a generic methodology has been elaborated and implemented which allows to explore the implementation design space. The hybrid hardware-in-the-loop research and design space exploration (DSE) framework created in this work combines high-level tools (e.g. Matlab/Simulink) and optimized hardware blocks. The properties of the elaborated optimized on-chip infrastructure template make it suitable for usage in final ASIC targets and thus enable the test, debugging and characterization of signal processing blocks in their target environment. This DSE framework has been used to derive cost models for K-user iterative IA algorithms. Estimates for the implementation effort (e.g. in terms of operation counts in dependency of the number of users) have been derived. Because of this project, a generic DSE framework is available and can be used to work out suitable architectures for further challenging building blocks.

Author details

Mohamed El-Hadidy, Mohammed El-Absi and Thomas Kaiser
Duisburg-Essen University, Institute of Digital Signal Processing (DSV), Germany

Yoke Leen Sit and Thomas Zwick
Karlsruhe Institute of Technology, Institut für Hochfrequenztechnik und Elektronik (IHE), Germany

Markus Kock and Holger Blume
Leibniz Universität Hannover, Institute of Microelectronic Systems (IMS), Germany

10. References

[1] Adamiuk, G. [2010]. *Methoden zur Realisierung von Dual-orthogonal, Linear Polarisierten Antennen für die UWB-Technik*, PhD thesis, Karlsruhe.
URL: *http://digbib.ubka.uni-karlsruhe.de/volltexte/1000019874*

[2] Banz, C., Hesselbarth, S., Flatt, H., Blume, H. & Pirsch, P. [2012]. Real-Time Stereo Vision System using Semi-Global Matching Disparity Estimation: Architecture and FPGA-Implementation, *Transactions on High-Performance Embedded Architectures and Compilers, Springer* .

[3] Björck, Å. & Golub, G. H. [1973]. Numerical methods for computing angles between linear subspaces, *Math. Comp.* 27: 579–594.

[4] Cadambe, V. & Jafar, S. [2008]. Interference alignment and degrees of freedom of the k -user interference channel, *Information Theory, IEEE Transactions on* 54(8): 3425 –3441.

[5] Cadambe, V. & Jafar, S. [2009]. Reflections on interference alignment and the degrees of freedom of the k-user mimo interference channel, *IEEE Information Theory Society Newsletter* 54(4): 5 –8.

[6] El-Absi, M., El-Hadidy, M. & Kaiser, T. [2012]. Antenna selection for interference alignment based on subspace canonical correlation, *2012 International Symposium on Communications and Information Technologies (ISCIT)* .

[7] El Ayach, O., Peters, S. & Heath, R. [2010]. The feasibility of interference alignment over measured mimo-ofdm channels, *Vehicular Technology, IEEE Transactions on* 59(9): 4309 –4321.

[8] El-Hadidy, M., El-Absi, M. & Kaiser, T. [2012]. Articial diversity for uwb mb-ofdm interference alignment based on real-world channel models and antenna selection techniques, *2012 IEEE International Conference on Ultra-Wideband (ICUWB)* .

[9] El-Hadidy, M., Mohamed, T., Zheng, F. & Kaiser, T. [2008]. 3d hybrid em ray-tracing deterministic uwb channel model, simulations and measurements, 2: 1 –4.

[10] Fügen, T., Maurer, J., Kayser, T. & Wiesbeck, W. [2006]. Capability of 3-D Ray Tracing for Defining Parameter Sets for the Specification of Future Mobile Communications Systems, *Antennas and Propagation, IEEE Transactions on* 54(11): 3125 –3137.

[11] Fügen, T., Waldschmidt, C., Maurer, J. & Wiesbeck, W. [2003]. MIMO capacity of bridge access points based on measurements and simulations for arbitrary arrays, *5th European Personal Mobile Communications Conference*, pp. 467–471.

[12] Gesbert, G., Shafi, M., Shiu, D., Smith, P. J. & Naguib, A. [2003]. From Theory to Practice: An Overview of MIMO Space-Time Coded Wireless Systems, *IEEE Journal on Selected Areas in Communications* 21: 281–302.

[13] González, O., Ramírez, D., Santamaría, I., García-Naya, J. & Castedo, L. [2011]. Experimental validation of interference alignment techniques using a multiuser MIMO testbed, *Smart Antennas (WSA), 2011 International ITG Workshop on*, pp. 1 –8.

[14] Heath, R., Sandhu, S. & Paulraj, A. [2001]. Antenna selection for spatial multiplexing systems with linear receivers, *IEEE Commun. Letters* 5(4): 142 –144.

[15] Jereczek, G. [2010]. *Design of Capacity Maximizing MIMO Antenna Systems for Car-2-Car Communication*, Master's thesis, Karlsruhe Institute of Technology (KIT), Karlsruhe, Germany.

[16] Khalighi, M. I., Brossier, J., Jurdain, G. & Raoof, K. [2001]. Water Filling Capacity of Rayleigh MIMO Channels, *IEEE Transactions on Antennas and Propagation* 1: A155–A158.

[17] Kock, M., Hesselbarth, S. & Blume, H. [2013]. Hardware-accelerated design space exploration framework for communication systems, *Wireless Innovation Forum Conference*

on Wireless Communications Technologies and Software Defined Radio (SDR-WInnComm 2012), Washington, DC, USA. Accepted for publication.

[18] Loyka, S. & Mosig, J. [2006]. *Information Theory and Electromagnetism: Are They Related?*, CRC Press.

[19] Molisch, A., Win, M. & Winters, J. [2001]. Capacity of mimo systems with antenna selection, *IEEE International Conference on Communications (ICC)* (4): 570 –574.

[20] Paya-Vaya, G. & Blume, H. [2012]. TUKUTURI: A dynamically reconfigurable multimedia soft-processor for video processing applications, *Poster presentation at the 7th International Conference on High-Performance and Embedded Architectures and Compilers, HiPEAC'12, Paris, France*, Vol. USB Proceedings.

[21] Peters, S. & Heath, J. [2009]. Interference alignment via alternating minimization, *Int. Conf. on Acoust. Speech and Signal Processing,(ICASSP)* .

[22] Petersen, D. P. & Middleton, D. [1962]. Sampling and Reconstruction of Wave-limited functions in N-dimensional Euclidean Spaces, *Information and Control* 5: 279 – 323.

[23] Pirsch, P. [1998]. *Architectures for Digital Signal Processing*, John Wiley & Sons, Inc.

[24] Pontes, J. [2010]. *Optimized Analysis and Design of Multiple Element Antenna for Urban Communication*, Karlsruher Institut fuer Technologie (KIT).

[25] Reichardt, L., Pontes, J., Sit, Y. & Zwick, T. [2011]. Antenna optimization for time-variant mimo systems, *Antennas and Propagation (EUCAP), Proceedings of the 5th European Conference on*, pp. 2569 –2573.

[26] Salmela, P., Burian, A., Järvinen, T., Happonen, A. & Takala, J. H. [2011]. Low-complexity inverse square root approximation for baseband matrix operations, *ISRN Signal Processing* vol. 2011.

[27] Schmidt, D., Shi, C., Berry, R., Honig, M. & Utschick, W. [2009]. Minimum mean squared error interference alignment, *Signals, Systems and Computers, 2009 Conference Record of the Forty-Third Asilomar Conference on*, pp. 1106 –1110.

[28] Shen, M., Zhao, C., Liang, X. & Ding, Z. [2011]. Best-effort interference alignment in ofdm systems with finite snr, *IEEE International Conference on Communications (ICC)* 2(4): 5 –9.

[29] Sit, Y. L., Reichardt, L., Liu, R., Liu, H. & Zwick, T. [2012]. Maximum Capacity Antenna Design for an Indoor MIMO-UWB Communication System, *10th International Symposium on Antennas, Propagation and EM Theory, Xian, China.*

[30] Wallace, J. W. & Jensen, M. A. [2002]. Intrinsic Capacity of the MIMO Wireless Channel, *IEEE Proceedings of the 56th Vehicular Technology Conference VTC 2002-Fall*, pp. 701–705.

Chapter 7

Antennas and Propagation for On-, Off- and In-Body Communications

Markus Grimm and Dirk Manteuffel

Additional information is available at the end of the chapter

http://dx.doi.org/10.5772/55080

1. Introduction

The ultra-wideband technology seems very attractive to be transferred to the challenging field of body centric communications. This technology involves the potential to establish robust communication links or high resolution localization systems. All these applications require a characterization of the propagation channel and the influence of the corresponding user to the system performance. Due to the inevitable interaction between the antenna and the related propagation channel a separation of both characteristics via traditional antenna theory methods is hardly applicable. The scope of this study is to establish a so called antenna de-embedding i.e. to separate the antenna form the underlying channel.

Traditional antenna parameters (e.g. directivity, gain, effective area) are based on free space propagation conditions. Underlying is the well known model of an isotropic radiator which enables the separation of channel, transmitter and receiver. It will be shown that this theory can be adapted to deduce approximations of equivalent antenna parameters for body centric communications. Key factor of this approach is the development of equivalent far field models of the corresponding in- and on-body scenarios. For off-body scenarios the propagation direction points away from the human body. The matching and the radiation pattern of the respective antennas may change due to the interaction with the human body but in general no modifications of the far field model are necessary. Therefore, the traditional theory is applicable with just minor restrictions and will not be discussed in further detail here.

The study is structured in two sections. The first part focuses on an in-body link i.e. the main propagation path of the electromagnetic wave leads through the tissue of the human body. Typical applications for this scenario are medical implants like wireless endoscopy or the RF breast cancer detection systems. The second part characterizes an on-body link. This means that the propagation path is defined along the body surface and the antenna is located in close proximity of the human body. The universality of this theory will be shown for the characterization of an UWB teardrop antenna.

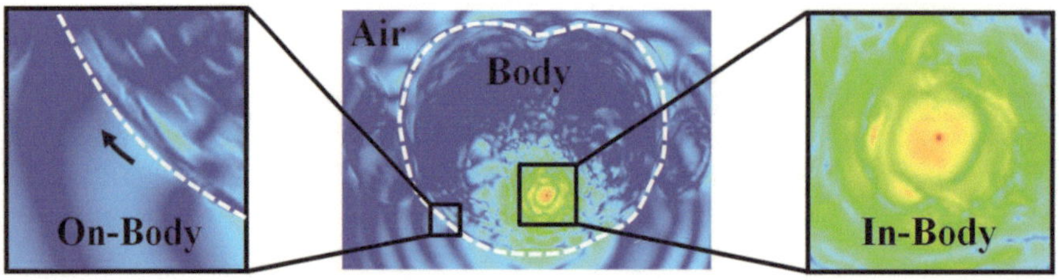

Figure 1. Distribution of the electric field of an implant located within the human abdomen; Left: On-body scenario showing surface waves guided by the body curvature; Right: In-body scenario characterized by circular shaped attenuation within the body.

2. Antenna de-embedding for in-body applications

From an electromagnetic point of view the human body consists of a large number of lossy dielectric materials of various combinations and spacial arrangements. In case of an in-body scenario the antenna is integrated into this complex dielectric structure and along an arbitrary propagation path various electromagnetic propagation effects occur. However, the most dominant effect is the attenuation of the propagating wave due to the lossy character of human tissues, see Figure 1. Furthermore, it has been shown that the average attenuation through the inhomogeneous tissue structure can be described for some scenarios by an equivalent homogeneous medium with appropriate properties [1]. In this case the analogy between the in-body and free space scenario enables an evident description of the related antenna far field. For this purpose the basic assumption of an isotropic point source is generalized for a lossy dielectric medium.

2.1. Model of an infinite homogeneous lossy dielectric media

The description of this model is based on the solution of the Helmholtz equation for a spherical wave which propagates in a homogeneous dissipative dielectric media. Due to this approach the absolute value of the electric field decreases proportionally to the reciprocal distance and shows an additional exponential attenuation with progressing distance compared to a lossless media. Therefore, the absolute value of the electric field E can be approximated by

$$E(d) \propto \frac{U_{\mathrm{n}}}{d} \mathrm{e}^{-\alpha d}, \tag{1}$$

where d, α and U_{n} denote the distance to the antenna, the attenuation constant of the media and a normalization factor related to the equivalent source of the spherical wave. For radiating elements, other than point sources, the accuracy of this model depends on the spacial current distribution of the source and is therefore a function of the distance. Due to the analogy of this dependency to the free space scenario the standard far field criteria seems applicable with the usual phase restrictions [2]. Related to the design of antennas which operates within the UWB frequency range [3] the radiating component is electrically large at the upper band edge frequency. In this case the appropriate far field criterium leads to following formulation:

$$d_{\mathrm{ff}} = \frac{2{D_{\mathrm{max}}}^2}{\lambda_{\mathrm{i}}}, \tag{2}$$

where d_{ff} denotes the approximated far field distance, λ_i the wave length of a TEM wave in the dielectric media at the band edge frequency and D_{max} the largest diameter of the antenna. Note that D_{max} may be increased by passively excited components in the near field of the antenna which may contribute to the radiation, like the PCB of an implant or its encapsulation. Based on this far field model it is possible to deduce the equivalent gain of an in-body antenna. Doing this it seems logically consistent to refer the normalization constant U_n to an isotropic radiator. In this case the general definition of the gain G [2] is altered to

$$G = \frac{S\left(d \geq d_{ff}\right)}{S_{iso,lossy}\left(d \geq d_{ff}\right)}, \tag{3}$$

where S denotes the power density of the antenna and $S_{iso,lossy}$ denotes the power density of an isotropic source in a lossy medium at the same distance. Please note, due to the fact that the propagation medium itself contains losses and the directivity is by definition a lossless quantity, the definition of the directivity is not appropriate in this case. In order to achieve a constant normalization ratio versus the distance, the losses expressed by the exponential term of equation 1 have to be taken into account. Therefore, the absolute value of the lossy isotropic power density is given by

$$S_{iso,lossy} = \frac{P_{rad}}{4\pi d^2} e^{-2\alpha d}. \tag{4}$$

In equation 4, P_{rad} denotes the radiated power of the antenna. For an antenna whose losses are restricted to the surrounding medium P_{rad} is equal to the power on the antenna P_{ant}. Due to the dissipative nature of the tissue the antenna efficiency η decreases exponentially with progressing far field distance and can be calculated by

$$\eta\left(d_{ff}\right) = \frac{1}{P_{ant}} \oiint_A \mathbf{S} \cdot \mathrm{d}\mathbf{\Lambda}, \tag{5}$$

where A is the enclosure of the antenna at the distance d_{ff}. In order to calculate the path loss between two in-body antennas, the receive properties of such antennas have to be characterized as well. Due to the fact that the equivalent tissue medium is source free, linear and isotropic the definition of the effective antenna area A_{eff} [2] is also applicable for the in-body scenario. With respect to the definition of the gain in equation 3, the effective antenna area yields

$$A_{eff} = G \frac{\lambda_i^2}{4\pi}. \tag{6}$$

Note, that the theory given above is based on an intrinsic far field model. Aim of this theory is an approach to give an intuitive formulation for the antenna design and handy path loss estimations. It raises no claim to give a closed analytical solution of the given problem. Despite this fact the model enables even estimations of theoretic problematic scenarios, such as a totally immersed antenna which is not insulated from the surrounding tissue media. As shown in [4], a theoretical formulation of this problem would lead to an inexpressible formulation. Nevertheless, to find a description between source and far field it is suggested to use an antenna which is electrically insulated from the surrounding media. Moreover the applicability of the model depends on the specific in-body scenario. The following chapter addresses the quality of the proposed model.

2.1.1. Validation of the in-body model on the example of an UWB teardrop antenna

Using the example of [1] the validity of the proposed in-body far field approach has been discussed by the evaluation of UWB localization of deep brain implants. As result it has been shown that the attenuation of an electromagnetic pulse within the inner human brain structure can be modeled by a homogeneous tissue with dielectric properties equivalent to grey matter. Despite the complex structure of the focused brain region time domain analyses indicate a radial wave conservation within the inner brain structure. These investigations indicate an average propagation velocity in arbitrary directions of the head which emphasize to the applicability of the suggested model.

The following analysis summarizes the results of an in-body scenario within the human trunk. In this case the representative media parameters are set to homogeneous muscle tissue, calculated by [5], and the validity of the approach is shown on the example of an UWB teardrop antenna. The antenna has been designed for the ultra wideband frequency range from 3.1 GHz to 10.6 GHz which is specified by [3]. Within this frequency range the return loss of the immersed teardrop antenna has been optimized to be lower than 10 dB. The capsulation of the antenna has been designed for the center frequency of 6.85 GHz by a lossless dielectric cylindrical insulation. The related permittivity of the capsulation has been set to $\varepsilon_r = 49.9$ to achieve an impedance matching between the insulation and the tissue. The validation of the model has been performed by numerical calculations with the FDTD simulation software EMPIRE XCcelTM [6].

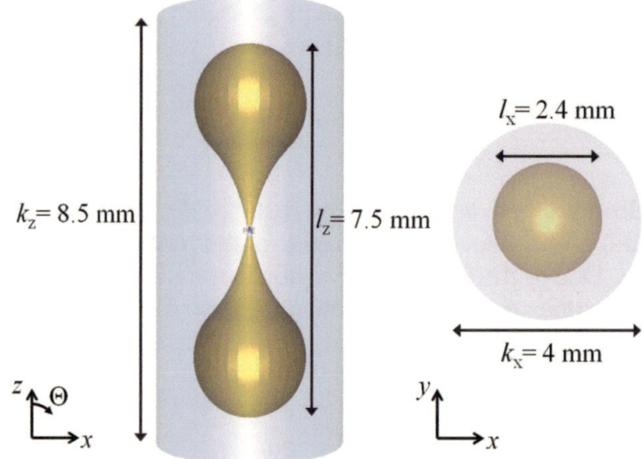

Figure 2. Insulated teardrop antenna designed for the focused UWB frequency band.

According to equation 2, the minimum far field distance for the upper band edge frequency ($f_u = 10.6$ GHz) is $d_{ff} = 3$ cm. Here, due to the insulation, the maximum antenna size D_{max} has been set to the maximal capsule dimension. The quality of the proposed far field model has been verified by the absolute difference ΔS between the calculated power density values S_{FDTD} and the far field model S_{ff}. For the angle $\Theta = 90°$ the difference ΔS has been calculated at the far field distance d_{ff} by the following equation:

$$\Delta S = \frac{|S_{FDTD} - S_{ff}|}{S_{FDTD}}. \tag{7}$$

Table 1 shows the calculated values for the lower, upper and center frequency of the UWB frequency range investigated. The error rises with increasing frequency and is still within the typical range of the equivalent free space far field considerations.

	$f = 3.1$ GHz	$f = 6.85$ GHz	$f = 10.6$ GHz
$\Delta S[\%]$	1.1	3.7	6.9
$G(\Theta = 90°)$	1.9	2.3	3.3
$\eta_{\mathrm{eff}}(d_{\mathrm{ff}})$	$3.8 \cdot 10^{-2}$	$8.4 \cdot 10^{-5}$	$1.5 \cdot 10^{-8}$
$A_{\mathrm{eff}}[m^2]$	$2.75 \cdot 10^{-5}$	$7.22 \cdot 10^{-6}$	$4.74 \cdot 10^{-6}$

Table 1. Calculated equivalent in-body antenna parameters of the UWB teardrop antenna at $d_{\mathrm{ff}} = 3$ cm.

Table 1 also shows the gain calculated by the equations 3 and 4. As stated, additional derivations have shown that the value is nearly constant for distances greater than d_{ff}. Nevertheless, the value is not constant within the observed frequency range and rises with increasing frequency due to the variation of the electrical length of the antenna. To characterize the losses within the near field of the antenna the efficiency has been calculated by equation 5 at the minimum far field distance d_{ff}, see Table 1. As it might be expected, the efficiency decreases drastically with increased frequency due to the higher power consumption of the tissue medium. For greater distances the efficiency decreases exponentially with increasing distance. The effective antenna area, as shown in Table 1, is calculated using equation 6. The derived values of the gain and the effective antenna area enable the approximation of the path gain for arbitrary distances greater than the minimum far field distance. This ratio of transmitted to received power (path gain) is shown for the related frequencies in Figure 3.

As shown above the model of an homogeneous dissipative medium enables the definition of antenna parameters to establish an antenna de-embedding for in-body scenarios. Therefore, the consideration of the whole system is not necessary to achieve a path loss estimation. Moreover this assumption enables a basis for a purposeful antenna development.

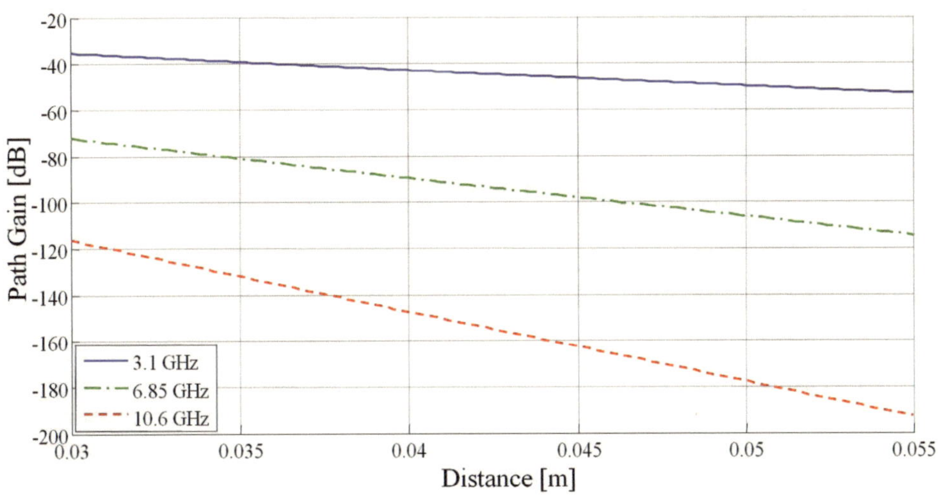

Figure 3. Calculated path gain of the UWB teardrop antenna.

2.1.2. Limitations of the in-body model

As shown in [1], boundaries between high water content tissues (muscle tissue) and air filled regions (paranasal sinus, frontal sinus) or low water content tissues (fat tissue, bone tissue) lead to various electromagnetic interactions which reduce the accuracy of the approach presented above. The average attenuation of the electromagnetic field may still be characterized by the proposed model but especially with respect to time domain analysis the multi path behavior of these structures leads to insufficient results. Moreover, the specific anatomical location of the RF application may have a strong influence on the corresponding radiation characteristic of the antenna which cannot be adequately described by a homogeneous tissue model. Despite this fact the generality of the approach enables an extension of the theory by including anatomic realistic human models in the antenna design process to derive the resulting radiation characteristic. In this case the corresponding far field distance has to be enlarged related to the anatomical structure but the average description of the far field may still be given by the proposed homogeneous model.

3. Antenna de-embedding for on-body applications

Encouraged by [7], the didactic first step to deduce a de-embedding for a wide class of on-body applications is the deduction of an adequate far field propagation model of a radiating source near a planar tissue like surface.

3.1. Model of an infinite homogeneous lossy plane surface

The propagation mechanisms of an electric doublet near a dissipative infinite homogenous plane has been investigated in [8]. These results provide a description of the electric field components as a function of the given geometry. Therefore the absolute value of the electric field E can be calculated depending on the dipole current distribution $\mathbf{i}$, its effective height h, the frequency f, the complex dielectric parameters of the media $\underline{\varepsilon}$ and a tangential to the surface defined distance d, see Figure 4.

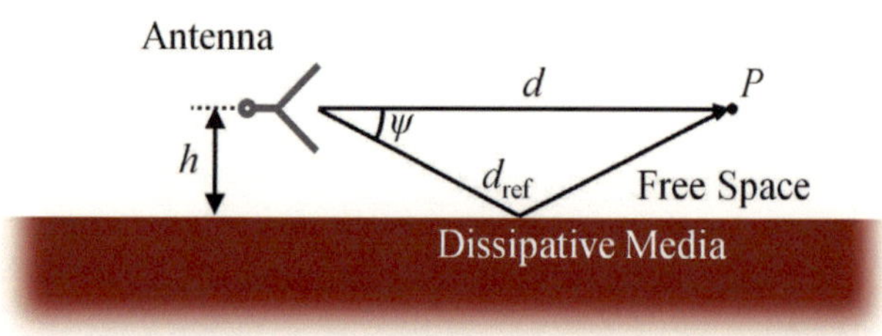

Figure 4. Principle geometry of the on-body scenario.

Included in this theory is the separation of the total electric field in its space and surface wave. As defined by Figure 4 the space wave consists of the wave components which propagates along the direct path d and the ground reflected path d_{ref}. A surface wave component which is excited by the dissipative nature of the tissue is guided by the air-media boundary. It only contributes to the related electromagnetic field if the antenna is located in close proximity to the surface. This means that the effective antenna high h is typically lower than a few wave lengths. Otherwise, this wave component is negligibly small compared to the space wave. The electric field of any observation point P along a parallel surface path at the effective antenna high h can be described by Norton's formulation N as follows:

$$E \propto \frac{U_n}{d} N\left(\mathbf{i}_{\mathrm{v,h}}, d, h, f, \underline{\varepsilon}\right), \tag{8}$$

where the normalization value U_n depends on the antenna excitation. The term N is a function of the current distribution of the source $\mathbf{i}_{\mathrm{v,h}}$, the distance d, the frequency f and the complex tissue permittivity $\underline{\varepsilon}$. The absolute value of the electric field is given by the superposition of the space wave E_{sp} and surface wave E_{sw} as follows:

$$E(d) = E_{\mathrm{sp}} + E_{\mathrm{sw}}, \tag{9}$$

where E is given by

$$E(d) \propto \frac{U_n}{d}\left[\underbrace{e^{j\frac{2\pi d}{\lambda}}}_{\text{direct wave}} \underbrace{+ R_{\mathrm{v,h}}\cos^3(\psi)\, e^{j\frac{2\pi d_{\mathrm{ref}}}{\lambda}}}_{\text{ground-reflected wave}} + \underbrace{\left(1 - R_{\mathrm{v,h}}\right) F_{\mathrm{v,h}}\cos^2(\psi)\, e^{j\frac{2\pi d_{\mathrm{ref}}}{\lambda}}}_{\text{surface wave}}\right]. \tag{10}$$

In equation 10, λ denotes the free space wave length, $R_{\mathrm{v,h}}$ the plane-wave reflection coefficient of the ground, $F_{\mathrm{v,h}}$ the surface-wave attenuation function and d_{ref} the reflected path at angle ψ. Both, the reflection and the attenuation functions depend on the current distribution of the antenna and are given by [9] for vertical and horizontal antenna orientations. The related far field solution is valid for sufficient great distances d and depends in first place on the mathematical description of the surface-wave attenuation function. If an adequate description can be assumed the minimal valid distance, referred by Norton in [8], has to be greater than one free space wave length. An additional limitation is the assumption of locally plane waves for the derivation of the reflection coefficient of the ground. Under the assumption that the surface acts as a perfect mirror the electrical antenna size may be enlarged by the mirror image. Analog to equation 2 the far field distance d_{ff} is defined by:

$$d_{\mathrm{ff}} = \frac{2D^2_{\mathrm{max,eff}}}{\lambda}, \tag{11}$$

where the modified maximum antenna dimension is denoted by $D_{\mathrm{max,eff}}$. For an adequate derivation of the far field distance the quantity $D_{\mathrm{max,eff}}$ has to be chosen appropriate under the aspect that the enlargement of the antenna by the ground acts primarily in normal direction to the surface. For distances greater than the minimum far field distance, given by equation 11, the formulation of equation 10 enables the definition of the directivity for on-body scenarios. Analog to equation 3 the directivity D is defined by the normalization of the power density to the far field model:

$$D = \frac{|S|}{|S_{\mathrm{Norton}}|}. \tag{12}$$

The related electromagnetic field is in general a superposition of TE-, TM- and TEM-wave components and therefore a function of the parameters given in equation 8. Despite this fact, the TEM-wave component contributes the most significant part to the power flux density. Even if a strong surface wave is excited, the resulting TM-wave component is comparatively low. This fact enables a simple approximation of the power density by the given expression of electric field:

$$D \approx \frac{|E|^2}{|E_{\text{Norton}}|^2} = \frac{|E|^2}{\left|\frac{U_n}{r} N\right|^2}. \quad (13)$$

Considering the free space and in-body definition of the directivity it is consistent to define the normalization value U_n related to a isotropic source which is modified by the function N:

$$D = \frac{|E|^2}{\frac{\eta_0 P_{\text{rad}}}{4\pi d^2} |N|^2}. \quad (14)$$

Due to the fact that the effective area of an antenna is defined for the condition in which the antenna receives a locally plane electromagnetic wave [10] the received power of an antenna inserted in the far field of the transmitting on-body antenna cannot be calculated directly by equation 6. Nevertheless, as shown by a preceding study, see [11], even for body worn antennas a derivation of the received power as a function of the incident power density is possible. The results imply a nearly constant ratio between the received power P_{out} and the incident power density S if the corresponding antennas are farther than the minimum far field distance apart. The constant ratio A'_{eff} can be calculated by

$$A'_{\text{eff}} = \frac{|P_{\text{out}}|}{|S|}. \quad (15)$$

Note, that the ratio defined by equation 15 is a function of the parameters given above and is therefore limited in its applicability to the specific setup. Apart from these aspects the equation enables the opportunity to calculate the received power as a function of arbitrary far field distances.

3.1.1. *Validation of the on-body model on the example of an UWB teardrop antenna*

The verification of the suggested model has been done for a vertically and horizontally orientated UWB teardrop antenna for the frequency range defined by [3]. The effective height of the antenna has been set to a quarter free space wave length at $f = 3.1$ GHz to avoid intersections between the antenna and the tissue medium. In contrast to the in-body design an encapsulation of the radiating antenna elements is not necessary to obtain an adequate matching of the antenna. The key parameters of the antenna geometry are set to $l_z = 39$ mm and $l_x = 18.5$ mm, see Figure 2, to achieve a return loss below 10 dB for the desired frequency range. The tissue properties have been set to muscle tissue, given by [5], and analog to [7] the geometry has been numerically calculated by the FDTD method presented in [6]. The minimum far field distance d_{ff} has been calculated along equation 11 and is shown in Table 2 for a vertical and horizontal orientated teardrop antenna.

Table 2 also shows the quantities D and A'_{eff} which are calculated for the minimum far field distance $d_{\text{ff,v}} = 0.54$ m and $d_{\text{ff,h}} = 0.43$ m. Analog to the in-body scenario the quality of the suggested model has been verified by the absolute difference ΔS defined by equation 7. The absolute difference shows a sufficient applicability of the suggested on-body model. In contradiction to the in-body scenario it describes a non monotone behavior for the target frequencies due to the inverse frequency dependence of the reflection coefficient of the ground and the surface-wave attenuation function defined by equation 10. As shown in Table 2, the

	$\mathbf{i}_{v,h}$	$f = 3.1$ GHz	$f = 6.85$ GHz	$f = 10.6$ GHz
$d_{ff}[\mathrm{m}]$	v	0.16	0.35	0.54
	h	0.13	0.28	0.43
$\Delta S[\%]$	v	10.38	4.83	5.56
	h	6.64	3.15	3.48
$D[\mathrm{lin}]$	v	1.37	1.07	0.99
	h	1.34	1.23	0.47
$A'_{\mathrm{eff}}[\mathrm{m}^2]$	v	$8.7 \cdot 10^{-4}$	$1.7 \cdot 10^{-4}$	$4.4 \cdot 10^{-5}$
	h	$10.3 \cdot 10^{-4}$	$1.8 \cdot 10^{-4}$	$3.4 \cdot 10^{-5}$

Table 2. Calculated on-body antenna parameters of the UWB teardrop antenna. The quantities ΔS, G and A'_{eff} are calculated for the maximum far field distance of the considered frequency range with $d_{\mathrm{ff,v}} = 0.54$ m and $d_{\mathrm{ff,h}} = 0.43$ m.

directivity of the evaluated antennas decreases with increased frequencies. This behavior implies a reduced excitation of surface waves in the upper UWB frequency range due to the greater effective height of the antenna. Note, the derived gain is not directly comparable to the free space or in-body values. Due to the dependency of the channel model to the antenna polarization even the vertical and horizontal quantities are incomparable to each other. Despite this restriction the formulation of the directivity defines a quantity which enables an adequate discussion of various on-body antenna types and to enhance the corresponding design process.

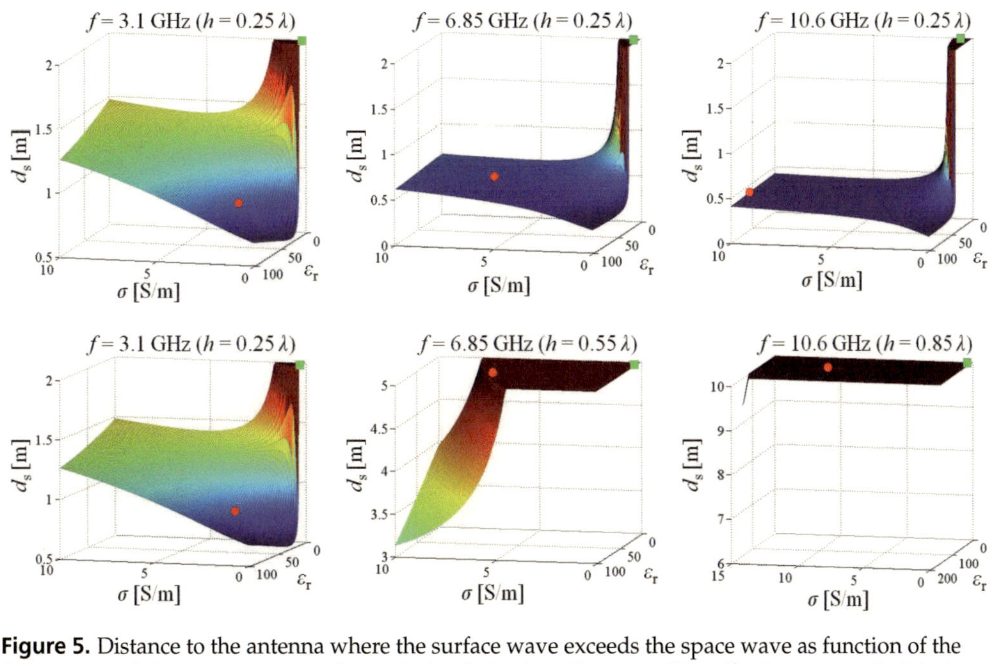

Figure 5. Distance to the antenna where the surface wave exceeds the space wave as function of the frequency and tissue parameters of a vertical polarization; First row: The effective antenna height is set to a quarter wave length of the respective frequency; Second row: Effective antenna height is set to a quarter wave length at $f = 3.1$ GHz; Green square marker: Fat tissue; Red round marker: Muscle tissue.

Figure 5 shows an analysis of the effective antenna height in relation to the surface wave excitation. It discusses the distance d_s where the surface wave exceeds the space wave as function of the tissue parameters. The first row shows the intersect point for effective antenna heights of a quarter wave length of the respective frequency. The second row shows the intersection point for a fixed effective antenna height which has been set to a quarter wave length of the lower UWB edge frequency ($f = 3.1$ GHz). The comparison of the results shows that with increasing frequency even at distances greater than 10 m the surface wave component is lower than the corresponding space wave. This fact implies a relatively weak far field and causes a reduction of the directivity at high frequencies. With respect to the design of future UWB on-body antennas this circumstance has to be considered. Additional investigations have also shown that vertical polarized antennas excite a much more dominant surface wave than equivalent horizontal orientated antenna configurations, see [7]. These results are in accordance to the theory given by [8] and should be considered to optimize UWB applications for given propagation scenarios.

3.1.2. Limitations of the on-body model

The validation of the suggested model with respect to the anatomical structure of the human body, with its numerous tissue types and curved surfaces, is done by a path gain calculation of a complete human body voxel model. Basis for this derivation is the numerical IT'IS virtual family Duke model [12]. The selected scenario consists of a transmitting antenna TX which is located at the right shoulder front. The corresponding receiver RX is shifted along the front side of the trunk above the right leg to the right foot. Figure 6 shows the path gain along the chosen path d. In addition, the path gain of the suggested on-body model has been calculated for homogeneous muscle and fat tissues. As seen in Figure 6, the calculated path gain of the voxel model lies between the graphs of the theoretical models.

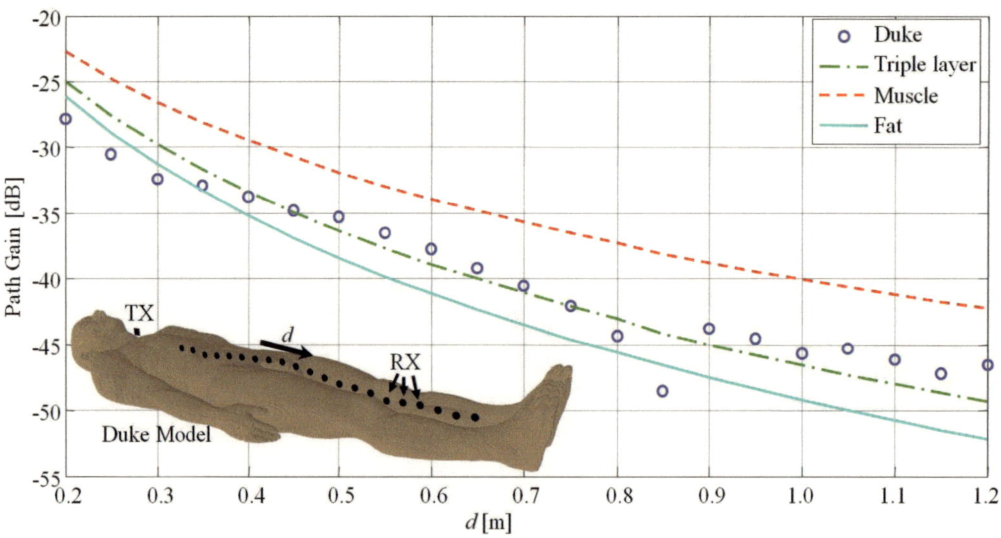

Figure 6. Path gain of the Duke voxel model in comparison to the homogeneous model of fat and muscle; Additional included is a numerical validation graph of a layered model analog to [7].

Analog to [7] a numerical model of a layered plane surface has been implemented to realize a more realistic representation of the human tissue structure. The suggested simulation model consists of 2 mm skin and 5 mm fat tissues which are positioned on an infinite muscle tissue.

As shown in Figure 6 the modified model is capable to give an adequate description of the detailed voxel simulation. In future approaches this fact can be used to enhance the presented on-body far field model by a modification of the surface-wave attenuation function $F_{v,h}$, see equation 10, by an adaption of the numeric distance as function of the surface impedance [13].

An additional effect, which is not included in the model presented, is the propagation in shadowed regions of the human body. While locally small shadowed regions are still covered by the model, large shadowed regions seem not to be described by this theory [7]. Nevertheless, these aspects have also been discussed in the last century with respects to the wave propagation above a spherical earth [9] and may also be transferred to the field of body centric communications.

4. Conclusion

The study has shown that an antenna de-embedding for in- and on-body applications can be realized by the derivation of corresponding far field models with reasonable accuracy for practical applications. Related to this theory, quantities as the directivity and the effective antenna area have been defined to derive good approximations of propagation models. Especially for on-body applications the suggested model gives a detailed insight by the separation of the electromagnetic field in its space and surface wave components.

Moreover, the presented theory enables the calculation of average path gain models of arbitrary antennas which can be reduced to a source of vertical and horizontal orientated current distributions. By this, the numerical calculation space can be reduced to the minimum far field distance of the corresponding model. Additionally, an insulated UWB tear drop antenna design has been presented for in-body communication applications to give an adequate validation example. For the on-body scenario the UWB teardrop antenna has been modified and also been discussed.

In future studies the in-body approach has to be modified to a multipath channel model to include additional propagation effects like surface waves. In addition, the on-body model has to be extended to give a wider applicability with respects to the complex structure of the human body. Moreover, the effective antenna area for on-body applications has to be described as function of the given model. With these improvements a structured combination of the on-, off- and in-body scenarios seems realizable to develop an optimized antenna theory for body centric communications.

Author details

Markus Grimm and Dirk Manteuffel
University of Kiel, Germany

5. References

[1] Grimm, M. & Manteuffel, D. (2011). Characterization of electromagnetic propagation effects in the human head and its application to Deep Brain Implants, *IEEE-APS Topical Conference on Antennas and Propagation in Wireless Communications (APWC)*, pp. 674-677, ISBN 978-1-4577-0046-0

[2] Balanis, C. A. (2005). *Antenna Theory*, John Wiley & Sons., ISBN 978-0-4716-6782-X

[3] Federal Communication Commission (2002). Revision of Part 15 of the communication's rules regarding ultra wideband transmission systems. First report and order, ET Docket 98-153, FCC 02-48

[4] Tai, C. T. & Collin, R. E. (2000). Radiation of a Hertzian Dipole Immersed in a Dissipative Medium, *IEEE Transactions on antennas and propagations*, Vol. 48, No. 10, pp. 1501-1506

[5] Gabriel, C. & Gabriel, S. et al. (2012). An Internet Resource for the Calculation of the Dielectric Properties of Body Tissues, Italian National Research Council website, Available from: http://niremf.ifac.cnr.it/docs/DIELECTRIC/home.html

[6] IMST (2012). EMPIRE XCcelTM, URL: http://www.empire.de

[7] Grimm, M. & Manteuffel, D. (2010). Electromagnetic Wave Propagation on Human Trunk Models excited by Half-Wavelength Dipoles, *Antennas and Propagation Conference (LAPC)*, pp. 493-496, ISBN 978-1-4244-7304-5

[8] Norton, K. A. (1937). The Propagation of Radio Waves over the Surface of the Earth and in the Upper Atmosphere, *Proceedings of the Institute of Radio Engineers*, , Vol. 24, No. 10, pp. 1367-1387

[9] Norton, K. A. (1941). The Calculation of Ground-Wave Field Intensity over a Finitely Conducting Spherical Earth, *Proceedings of the Institute of Radio Engineers*, Vol. 29, No. 12, pp. 623-639

[10] Friis, H. (1946). A Note on a Simple Transmission Formula, *Proceedings of the IRE*, Vol. 34, pp. 254-256

[11] Grimm, M. & Manteuffel, D. (2012). Evaluation of the Norton Equations for the Development of Body-Centric Propagation Models, *European Conference on Antennas and Propagations (EUCAP)*

[12] IT'IS Foundation (2012). Whole-Body Human Models, Enhanced Anatomical Models, URL: http://www.itis.ethz.ch

[13] Wait, J.R. (1996). *Electromagnetic Waves in Stratified Media*, IEEE Press, ISBN 0-7803-1124-8

Chapter 8

Power Allocation Procedure for Wireless Sensor Networks with Integrated Ultra-Wide Bandwidth Communications and Radar Capabilities

Gholamreza Alirezaei, Rudolf Mathar and Daniel Bielefeld

Additional information is available at the end of the chapter

http://dx.doi.org/10.5772/53005

1. Introduction

In this chapter, we analyze the problem of power allocation for a distributed wireless sensor network with sensor nodes based entirely on ultra-wide bandwidth (UWB) technology. The network is used to perform object detection as well as object classification, where the absence, the presence, or the type of an object is observed by the sensors independently. UWB signals can be used for data communication between the sensor nodes as well as for radar applications. The approach of misemploying the communication sensors as radar sensors, such that the data transmission is misused as a radar beam in order to detect or to classify a target object, helps in realizing an energy-efficient radar system with compact and cheap sensor nodes. A further advantage of such radar systems is the fulfillment of major requirements of wireless sensor networks. This exploitation presupposes that the integration of sensing functionality into usual UWB sensors is implementable easily without the usage of any additional hardware units. Since the compact and low complexity UWB sensors are limited in power and communication capabilities, the detection and classification performance of a single sensor is restricted compared to that of a common complex radar system. To obtain an appropriate overall system performance, we consider the case of distributed detection and classification, where the local observations of the sensors are fused into a reliable global decision. Due to noisy communication channels and differences in distances between the object and the sensor nodes, both, the observations and their transmissions are unequally interfered. One simple way to suppress noise interference is to increase the power of each sensor node. But if the total power of the entire network is limited, then power allocation procedures are needed in order to increase the overall detection and classification probabilities. In general, it is challenging to evaluate the detection and the classification probabilities analytically, if possible at all. This particularly holds for the detection probability under a Neyman-Pearson-hypotheses-test criterion as well as for the classification probability under a Bayesian-hypotheses-test criterion [5]. This limits the usability of these criterions for analytical optimization of power allocation. Bounds, such as the Bhattacharyya bound [8], are also difficult to use for optimizing multidimensional problems. Therefore, we employ an information theoretic approaches [3], which help to solve the power allocation problem with a lower mathematical complexity. This approach yields a simple however suboptimal analytical solution for the power allocation problem. Furthermore, the proposed technique enables the consideration of object detection and classification at the same time. This is a further advantage of this method, which enables the usage of the same allocation algorithm in both cases. Hence a sensor network, which is used to classify target objects, can also be used to detect the absence or the presence of a target object with equal system settings. Therefore we only describe the case of object classification, which also includes the case of object detection, in the following sections.

The origin of research on distributed detection has been the attempt to fuse signals of different radar devices [10]. Currently, distributed detection is usually discussed in the context of wireless sensor networks, where the sensor unit of the nodes might be based on radar technology [7, 9, 14]. Other applications for UWB radar systems, which require or benefit from the detection and classification capabilities, are for example localization and tracking [6] or through-wall surveillance [4]. The physical layer design for an integrated UWB radar network that utilizes OFDM technology was analyzed in [11]. In [2] the case of object detection is considered, where for the problem of power allocation an approach based on the maximization of the Kullback-Leibler distance is used. In a recent publication [1] another approach is discussed, where the bit-error probability of data communication is used in order to allocate the transmission power and to increase the overall detection probability.

This chapter is divided into the following three sections except the introduction. First, the system model of the wireless network including sensor nodes and the fusion center is described. Here, all system parameters and assumptions with detailed mathematical formulations are introduced. Furthermore, the global classification rule in the fusion center as well as the local decision rules in the nodes are motivated. In the second section, we present a novel approach for power allocation in order to increase the overall classification probability, following which, the solution of this optimization approach is briefly discussed. The last section shows some results and demonstrates the feasibility of object classification by using the proposed power allocation method in UWB signaling systems. This chapter concludes with an interpretation of the achieved system performance.

2. System model

Throughout this chapter we denote the set of natural, real, and complex numbers by $\mathbb{N}$, $\mathbb{R}$, and $\mathbb{C}$, respectively. Note that the set of natural numbers does not include the element zero. Furthermore, we use the subset $\mathbb{F}_N \subseteq \mathbb{N}$ which is defined as $\mathbb{F}_N := \{1, \ldots, N\}$ for any given natural number N. The mathematical operations $|z|$ and $|\mathbf{z}|$ denote the absolute value of a real or complex-valued number z and the Euclidian length of a real or complex vector $\mathbf{z}$, respectively.

Distributed *target object* classification can be formally modeled by a multiple hypotheses testing problem with hypotheses $H_k \forall k \in \mathbb{F}_K$ for a specified number $K \in \mathbb{N}$ of different objects. We assume that all objects have the same size, shape, alignment, and position. They only differ in material and are classified by their complex-valued reflection coefficients $r_k \in \mathbb{C}$, which are ordered in a strictly increasing manner $0 \leq |r_1| < \cdots < |r_K| \leq 1$. Therefore the reflection coefficients are the only recognition features in this work. Generally, this assumption is not realistic, but, this case describes an ideal scenario for increasing the classification probability by performing a power allocation and is not really suitable for analyzing the problems of manifoldness.

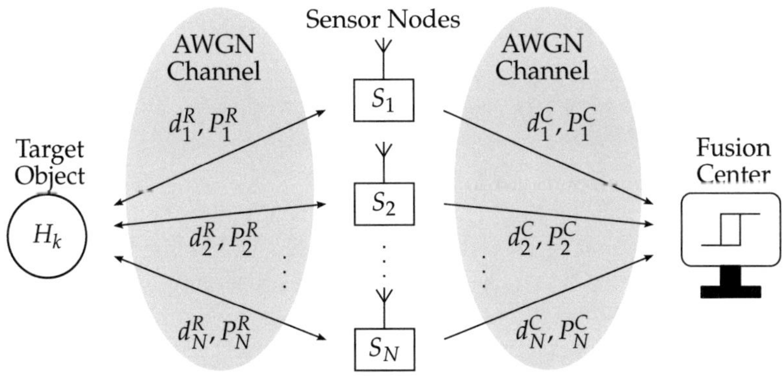

Figure 1. System model of the distributed wireless sensor network.

At any instance of time, a network of $N \in \mathbb{N}$ independent and spatially distributed sensors, as shown in Figure 1, obtains random observations $X_1, \ldots, X_N \in \mathbb{R}$. In the case of energy classification, X_n models the received signal at the receiver of the nth sensor. If a target object is present, then the received energy is a part of the radiated energy of the same sensor, which is reflected from the object's surface and is weighted by its reflection coefficient. We refer to this communication channel, between the sensors and the target object, as the *first* communication link and denote all dedicated parameters by the superscript R. The random observations $X_1, \ldots, X_N$ are assumed to be conditionally independent for each of the underlying hypotheses, i.e., the joint conditional probability density function of all the observations factorizes according to

$$f^R(X \mid H_k) := \prod_{n=1}^{N} f_n^R(X_n \mid H_k), \; \forall k \in \mathbb{F}_K, \tag{1}$$

where X denotes the sequence of random variables $X_1, \ldots, X_N$. In general, the observations are not identically distributed because the sensor nodes have different distances d_n^R from the target object and their radiated powers P_n^R are also different. Therefore, the signal-to-noise ratio (SNR) varies between the sensor nodes. Due to the distributed nature of the problem, the nth sensor S_n performs independent measurements and processes its respective observation X_n by generating a local decision $U_n := \theta_n(X_n) \forall n \in \mathbb{F}_N$, which depends only on its own observation and not on the observations of other sensor nodes. After deciding locally, each sensor transmits its decision to a fusion center located at a remote location. The communication between the sensor node and the fusion center is determined by the corresponding distance d_n^C as well as by the transmission power P_n^C of the same sensor node. We refer to this communication channel, between the sensor nodes and the fusion center, as the *second* communication link and denote all dedicated parameters by the superscript C. Furthermore, we assume that both communication channels are non-fading channels and that all data transmissions are affected only by additive white Gaussian noise (AWGN). We disregard time delays within all transmissions and assume synchronized data communication. We use two distinct pulse-shift patterns for each sensor node in order to distinguish its first and second communication link from the communication links of other sensor nodes as described in [13]. Each pattern has to be suitably chosen in order to suppress inter-user interference at each receiver. Hence the N received signals at the fusion center are uncorrelated and are assumed to be conditionally independent for each of the underlying hypotheses. These received random signals correspond to the local decisions $U_1, \ldots, U_N$ and are mapped to $\tilde{\mathbf{X}}_1, \ldots, \tilde{\mathbf{X}}_N \in \mathbb{R}^K$. Their joint conditional probability density function factorizes according to

$$f^C(\tilde{\mathbf{X}} \mid H_k) := \prod_{n=1}^{N} f_n^C(\tilde{\mathbf{X}}_n \mid H_k), \ \forall k \in \mathbb{F}_K, \tag{2}$$

where $\tilde{\mathbf{X}}$ denotes the sequence of random vectors $\tilde{\mathbf{X}}_1, \ldots, \tilde{\mathbf{X}}_N$. In general, these observations are – similar to the observations $X_1, \ldots, X_N$ – not identically distributed, because of variation in distances d_n^C as well as that of the radiated powers P_n^C. Unlike the local decision rules, the global decision rule $U_0 := \theta_0(\tilde{\mathbf{X}}_1, \ldots, \tilde{\mathbf{X}}_N)$ depends on all observations in order to increase the overall classification probability.

All described assumptions are necessary in order to obtain an ideal framework suited for analyzing the power allocation problem without studying problems of different classification methods in specific systems and their settings.

2.1. Local classification rules

The local decision and classification rules θ_n are mappings of the kind $\theta_n\colon \mathbb{R} \to \mathbb{F}_K, \ \forall n \in \mathbb{F}_N$. In this work, hard-decision rules are used for performing local classification given by

$$\theta_n(X_n = x_n) = k, \text{ if } \tau_{n,k} < x_n \leq \tau_{n,k+1}, \ k \in \mathbb{F}_K, \forall n \in \mathbb{F}_N, \tag{3}$$

where the thresholds $\tau_{n,k} \in \mathbb{R}$ are suitably chosen. The thresholds must be calculated separately for every sensor node in order to perform optimal classification. They depend on the prior probabilities of the hypotheses. Their values can be calculated by a suboptimal approach which is described in Section 3.1. In this way, every sensor node has a local probability of correct decision given by

$$\text{Prob}(U_n = k \mid H_k) = \text{Prob}(\tau_{n,k} < X_n \leq \tau_{n,k+1} \mid H_k), \ \forall k \in \mathbb{F}_K, \forall n \in \mathbb{F}_N \tag{4}$$

and a local probability of false decision given by

$$\text{Prob}(U_n \neq k \mid H_k) = 1 - \text{Prob}(U_n = k \mid H_k), \ \forall k \in \mathbb{F}_K, \forall n \in \mathbb{F}_N. \tag{5}$$

2.2. Fusion of local decisions and the global classification rule

The local decisions $U_1, \ldots, U_N$ at the sensor nodes are conditionally independent due to uncorrelated and independent noisy communication channels. By applying the Bayesian-hypotheses-test criterion the optimal fusion rule at the fusion center is given by

$$U_0 = \theta_0(\tilde{\mathbf{X}} = \tilde{\mathbf{x}}) = \underset{k \in \mathbb{F}_K}{\text{argmax}} \big(\pi_k f^C(\tilde{\mathbf{x}} \mid H_k)\big), \tag{6}$$

where $\pi_k := \text{Prob}(H_k)$ with $\sum_{k=1}^{K} \pi_k = 1$ denotes the prior probability of hypothesis H_k. We use this formula to classify the target object. However, in order to optimize the allocation of the total power to the sensor nodes, we have to consider the overall classification probability. Therefore, we consider K pairwise disjoint regions $\mathcal{R}_1, \ldots, \mathcal{R}_K$ with

$$\mathcal{R}_k := \big\{\tilde{\mathbf{x}} \in \mathbb{R}^{K \times N} \mid \pi_k f^C(\tilde{\mathbf{x}} \mid H_k) \geq \pi_l f^C(\tilde{\mathbf{x}} \mid H_l), \forall l \in \mathbb{F}_K, l \neq k\big\}, \ \forall k \in \mathbb{F}_K. \tag{7}$$

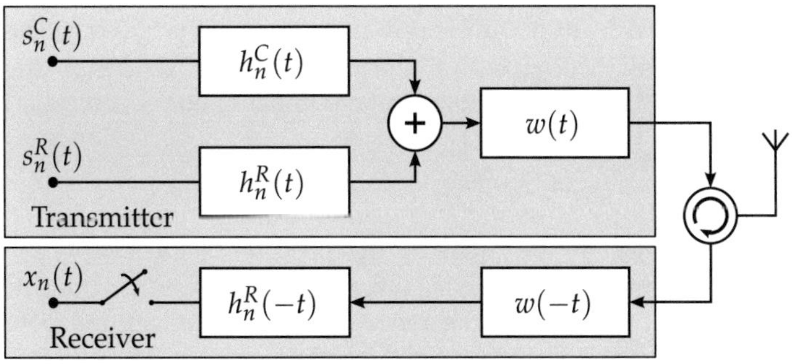

Figure 2. System model of the nth sensor node with circulator and antenna.

According to [5], the expected value of correct classification is given by

$$P_c := \sum_{k=1}^{K} \text{Prob}(\tilde{\mathbf{x}} \in \mathcal{R}_k, H_k), \tag{8}$$

which in general cannot be analytically evaluated. Therefore, the previous formula cannot be used to optimize the allocation of the total power analytically. Consequently, we choose a different approach for the optimization, which is described in Section 3.3.

2.3. Ultra-wide bandwidth sensor nodes

In Figure 2 the system model of the considered impulse-radio UWB (IR-UWB) sensor nodes with pulse position modulation (PPM) is shown. The transmitter generates two streams of data symbols $s_n^C(t)$ and $s_n^R(t)$.

The symbol stream s_n^C is used to transmit the local decisions $u_n(i) \in \mathbb{F}_K$ at the time index i to the fusion center, which are generated by the algorithm defined in (3). We describe the data symbols by Dirac delta functions $\delta(t - [u_n(i) - 1]\Delta)$, which are shifted pulses on the time axis. Their alignment is determined by the modulation index Δ. We assume that the product $K\Delta$ is much smaller than the symbol duration. Thus K different data symbols can be transmitted to the fusion center. The transmission power P_n^C of this stream is variable in order to adjust transmission power and to enable distributed power allocation.

The symbol stream s_n^R establishes the radiation to the target object and uses always the same data symbol. Its transmission power P_n^R is also variable. In order to increase the available power range at every sensor node, time-division multiple-access (TDMA) method is used to separate both streams into different time slots and to periodically share the same power amplifier.

In order to eliminate collisions due to multiple access, each user stream is assigned to a distinctive time-shift pattern after passing through the blocks $h_n^C(t)$ and $h_n^R(t)$. Their transfer functions are based on time-hopping sequences [13].

After superposition of both streams, a monocyclic pulse shape filter $w(t)$ limits the bandwidth of the signal. This filter has to fulfill the Nyquist intersymbol interference (ISI) criterion in order to avoid the intersymbol interferences.

When this superposition is transmitted, a part of the radiated signal s_n^R will be reflected from the target surface back to the antenna. The received signal will pass through the matched-filter

$w(-t)$ and will be decoded from its time-hopping sequence by $h_n^R(-t)$. The additive noise signal $b_n^R(t)$ will pass as well through both filters at the receiver. We denote the corresponding noise power by P_{noise}. If all receiver components are linear, then we can describe the received power by

$$\tilde{P}_{n|k}^R := P_n^R \frac{\alpha_n^R |r_k|^2}{g^2(2d_n^R)}, \; \forall k \in \mathbb{F}_K, \forall n \in \mathbb{F}_N, \tag{9}$$

where the transmitted power is weighted by the product of the factors $\alpha_n^R > 0$, $|r_k|^2$, and $g^{-2}(2d_n^R)$. The factor α_n^R includes the radar cross section, the influence of the antenna, the impacts of the filters, and all additional attenuation of the transmitted power. Due to the reflection coefficient r_k of the target object the received power depends on the underlying hypothesis. The path loss function g depends on the assumed multipath propagation channel and is usually an increasing function of the distance between transmitter and receiver. Here, the factor of two in the distance results from that back and forth transmission between the transceiver and the object. The ratio of $\tilde{P}_{n|k}^R$ and P_{noise} is the observed conditional SNR at the receiver and is given by

$$\gamma_{n|k}^R := \frac{P_n^R}{P_{\text{noise}}} \cdot \frac{\alpha_n^R |r_k|^2}{g^2(2d_n^R)}, \; \forall k \in \mathbb{F}_K, \forall n \in \mathbb{F}_N. \tag{10}$$

Due to the Gaussian distribution of the noise, each sample is also a Gaussian random variable, which is conditionally distributed according to

$$f_n^R(X_n = x_n \mid H_k) := \frac{1}{\sqrt{2\pi P_{\text{noise}}}} \exp\left(-\frac{\left(x_n - \sqrt{\tilde{P}_{n|k}^R}\right)^2}{2P_{\text{noise}}}\right), \; \forall k \in \mathbb{F}_K, \forall n \in \mathbb{F}_N. \tag{11}$$

The local decision probabilities $\text{Prob}(U_n = l \mid H_k)$, see (4) and (5), can be computed by solving the integral

$$\begin{aligned}\tilde{\pi}_{n,l|k} := \text{Prob}(U_n = l \mid H_k) &= \int_{\tau_{n,l}}^{\tau_{n,l+1}} f_n^R(x_n \mid H_k)\, \mathrm{d}x_n \\ &= \frac{1}{2}\left[\text{erf}\left(\frac{\sqrt{\tilde{P}_{n|k}^R} - \tau_{n,l}}{\sqrt{2P_{\text{noise}}}}\right) + \text{erf}\left(\frac{\tau_{n,l+1} - \sqrt{\tilde{P}_{n|k}^R}}{\sqrt{2P_{\text{noise}}}}\right)\right]\end{aligned} \tag{12}$$

for all $k, l \in \mathbb{F}_K$ and for all $n \in \mathbb{F}_N$. Here, the mapping $\text{erf}(z)$ denotes the error function of z.

2.4. Fusion center

After radiation of the stream s_n^C by the sensor node S_n, the signal is attenuated depending on the distance and it reaches the antenna at the fusion center as depicted in Figure 3. The received signal is matched-filtered and decoded from its time-hopping sequence. Then a data splitter $\mathbf{v}(t)$ is used to split the received signal into a K-dimensional vector space. This is necessary in order to retain the Euclidian distances between all transmitted symbols and achieve a higher classification probability. This filter is mathematically implemented as $\sum_{k=1}^{K} \mathbf{e}_k\, \delta(t - (k-1)\Delta)$, where $\mathbf{e}_k$ is the standard basis vector of the K-dimensional space that points in the kth direction. Therefore the received signals $\tilde{\mathbf{X}}_1, \ldots, \tilde{\mathbf{X}}_N \in \mathbb{R}^K$ are K-dimensional vectors. This new approach extends the method given by [13].

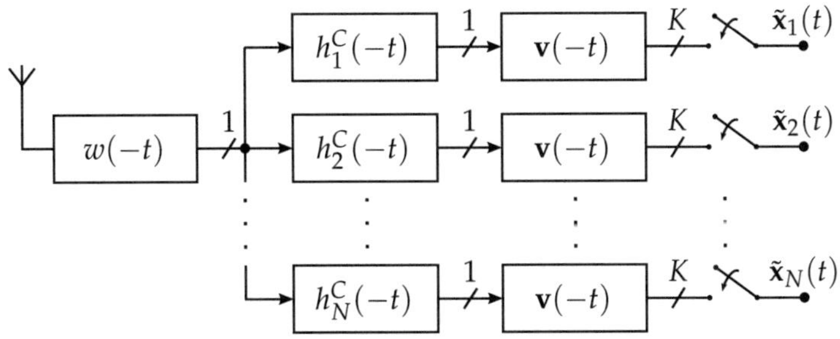

Figure 3. System model of the fusion center.

In case of additive zero-mean noise and due to the assumptions of $w(t)$, each vector sample of the received signal has the expected value of

$$\mathbf{m}_{n|l} := \mathrm{E}\big(\tilde{\mathbf{X}}_n \mid U_n = l\big) = \sqrt{P_n^{\mathrm{C}} \frac{\alpha_n^{\mathrm{C}}}{g^2(d_n^{\mathrm{C}})}} \cdot \mathbf{e}_l,\ \forall l \in \mathbb{F}_K, \forall n \in \mathbb{F}_N, \tag{13}$$

which depends on the transmitted symbol $U_n = l$. Thus the received power from the nth sensor node is given by

$$\tilde{P}_n^{\mathrm{C}} := P_n^{\mathrm{C}} \frac{\alpha_n^{\mathrm{C}}}{g^2(d_n^{\mathrm{C}})},\ \forall n \in \mathbb{F}_N, \tag{14}$$

where we assume that the path loss function is the same as for the first communication link. The power $\tilde{P}_n^{\mathrm{C}}$ is independent of the underlying hypothesis because the data stream s_n^{C} has the same power for all kinds of transmitted data symbols.

The additive noise signal $b_n^{\mathrm{C}}(t)$ will also pass through all the filters. We assume that the noise spectral density at the fusion center is the same as at the sensor nodes. Due to similarity in architecture of the fusion center and the sensor nodes, the noise power in each dimension of each stream is equal to P_{noise}. Because of the whiteness of noise, the interferences are uncorrelated in each dimension of each stream. Therefore the noise covariance matrix is determined by the product $P_{\text{noise}} \cdot \mathbf{I}_K$. Here $\mathbf{I}_K$ denotes the identity matrix of size K.

Similar to (10) we define an observed SNR for each data stream at the fusion center and denote it by

$$\gamma_n^{\mathrm{C}} := \frac{P_n^{\mathrm{C}}}{P_{\text{noise}}} \cdot \frac{\alpha_n^{\mathrm{C}}}{g^2(d_n^{\mathrm{C}})},\quad \forall n \in \mathbb{F}_N. \tag{15}$$

Due to the Gaussian distribution of noise, each vector sample is a Gaussian random vector, which is conditionally distributed according to

$$f_n^{\mathrm{C}}(\tilde{\mathbf{X}}_n \mid H_k) := \sum_{l=1}^{K} \frac{\tilde{\pi}_{n,l|k}}{(2\pi P_{\text{noise}})^{K/2}} \exp\Big(-\frac{(\tilde{\mathbf{x}}_n - \mathbf{m}_{n|l})^{\mathrm{T}}(\tilde{\mathbf{x}}_n - \mathbf{m}_{n|l})}{2P_{\text{noise}}}\Big),\ \forall k \in \mathbb{F}_K, \forall n \in \mathbb{F}_N, \tag{16}$$

where the operator $\mathbf{z}^{\mathrm{T}}$ denotes the transpose of any vector $\mathbf{z}$.

Because of the convex superposition of multivariate Gaussian distributions, it is difficult to use (16) with the properties of (2) to optimize the distributed power allocation. Bounds, such as the Bhattacharyya bound [8], are also difficult to use due to multidimensional nature of (2) and (16). Therefore we propose an applicable technique which is motivated by concepts of information theory and is described in the next section.

3. Allocating power to the radar and to the communication task

In this section, we motivate and present an approach to suboptimally allocate transmission power to the radar and to the communication task. The objective is to maximize the overall classification probability, given a limited total transmission power P_{tot} that can be arbitrarily allocated to the radar task as well as to the communication task. A direct solution to this problem does not exist, since no analytical expression for the overall classification probability (8) is available. Instead, we independently maximize the mutual information of both communication channels to increase the information flow and in order to determine the power allocation. The motivation for this approach is the separation of the power allocation problem from the object classification procedure. Because in this case the data communication does not affect the classification of the target object.

Note that this theoretical concept is not realistic. However, we apply this concept as a heuristical method in this work.

3.1. Threshold calculation

For the optimization of the thresholds in Section 2.1, in order to increase the overall classification probability, the analytic evaluation of (8) is needed. Due to the fact that this explicit form for the overall classification probability is unknown and due to the separation of the data communication from the classification task we propose the following simple approach to calculate the thresholds.

We increase the probability of correct decision of each sensor node independently to achieve suboptimal values for the thresholds. Thus, the overall classification probability should be increased as well. According to equations (4) and (12) the local probability of correct decision, which has to be maximized, is given by

$$\sum_{k=1}^{K} \mathrm{Prob}(H_k)\,\mathrm{Prob}(U_n = k \mid H_k) = \sum_{k=1}^{K} \frac{\pi_k}{2}\left[\mathrm{erf}\left(\frac{\sqrt{\tilde{P}^R_{n|k}}-\tau_{n,k}}{\sqrt{2P_{\mathrm{noise}}}}\right) + \mathrm{erf}\left(\frac{\tau_{n,k+1}-\sqrt{\tilde{P}^R_{n|k}}}{\sqrt{2P_{\mathrm{noise}}}}\right)\right]. \quad (17)$$

Its solution can be found explicitly by using differential calculus. The corresponding result is identical to the one obtained by using the Bayesian-hypotheses-test criterion. It is given by

$$\tau_{n,k} = \begin{cases} \inf(\mathbb{I}_{n,k}) & \text{if } \mathbb{I}_{n,k} \neq \{\},\ k \in \mathbb{F}_K, \\ \tau_{n,k+1} & \text{if } \mathbb{I}_{n,k} = \{\},\ k \in \mathbb{F}_K, \\ \infty & \text{if } k = K+1, \end{cases} \quad (18)$$

for all $n \in \mathbb{F}_N$, where the function $\inf(\mathbb{I}_{n,k})$ is the infimum of the interval $\mathbb{I}_{n,k}$ that is defined by

$$\mathbb{I}_{n,k} := \left\{x \in \mathbb{R} \mid \pi_k f_n^R(x \mid H_k) > \pi_l f_n^R(x \mid H_l), \forall l \in \mathbb{F}_K, l \neq k\right\},\ \forall k \in \mathbb{F}_K, \forall n \in \mathbb{F}_N. \quad (19)$$

3.2. Limitation of transmission power

We assume that both the radar and the communication signal use the same bandwidth and are uncorrelated to each other, due to separation of the sensing task and the communication task into different time slots (see Section 2.3). In this case and for each new classification process, the limited total transmission power P_{tot} is an upper bound for the sum

$$\sum_{n=1}^{N} \underbrace{\underbrace{P_n^R}_{\text{Radar sensing}} + \underbrace{P_n^C}_{\text{Data communication}}}_{\text{Total transmission power of one sensor for a single observation}} \leq P_{\text{tot}} \,. \tag{20}$$

By using this restriction, we present the power allocation procedure in the next section. But, we will have a look at some special cases previously.

In real applications the transmission power of each sensor node is also limited. Consider the case in which all sensor nodes have the same power limitation P_{max} with $\frac{P_{\text{tot}}}{N} \leq P_{\text{max}} < P_{\text{tot}}$. If the power regulation, which is described in the next section, wants to allocate a higher power to $P_n^R > P_n^C$ of the nth sensor node than its limitation, then we set the transmission power P_n^R equal to its highest possible limitation given by P_{max}, recalculate P_n^C which is given in terms of $P_n^R = P_{\text{max}}$, discard this n^{th} sensor node from the list of unallocated sensor nodes, decrease the given total transmission power P_{tot} by $P_{\text{max}} + P_n^C(P_{\text{max}})$, and reallocate the remaining total power $P_{\text{tot}} - P_{\text{max}} - P_n^C(P_{\text{max}})$ recursively to the remaining sensor nodes by the same procedure described in the next section. In a case, where the power P_n^C instead of $P_n^R > P_n^C$ will be regulated higher than P_{max}, we can reverse the roles of both transmission powers and repeat this reallocation method until no more sensor nodes are left which exceed their power limitation. Therefore, the described limitation of the total transmission power is the generalized case which includes the limitation of the transmission power of each sensor node.

Note that this procedure is applicable for individual power constraints per node as well. Furthermore, note that in each iteration more than one node can be discarded from the list of unallocated sensor nodes in order to decrease the computation complexity.

3.3. Mutual information-based power allocation

For the maximization of the information flow we set the mutual information of both communication channels equal. This leads to the same symbol error probabilities on both sides for low SNR values. For each sensor node an upper bound for the mutual information of its first and second link can simply be calculated. The identity of obtained bounds

$$\frac{1}{2}\log\left[1 + \frac{P_n^R \alpha_n^R (|r_K| - |r_1|)^2}{4P_{\text{noise}} g^2(2d_n^R)}\right] = \frac{K}{2}\log\left[1 + \frac{P_n^C \alpha_n^C (K-1)}{P_{\text{noise}} g^2(d_n^C) K^2}\right] \tag{21}$$

has to be computed in order to find the relationship between the powers for all $n \in \mathbb{F}_N$. After calculation and usage of the simple approximation

$$\sqrt[K]{1+x} \approx 1 + \frac{x}{K} \tag{22}$$

for any small values of x we obtain the analytical relationship

$$P_n^C = P_n^R \cdot \frac{\alpha_n^R}{\alpha_n^C} \frac{g^2(d_n^C)}{g^2(2d_n^R)} \frac{K}{K-1} \frac{(|r_K| - |r_1|)^2}{4}, \; \forall n \in \mathbb{F}_N. \tag{23}$$

In the next step, we increase the overall information flow by maximization of the cumulative mutual information subject to the given total power of the sensor network. Then the optimization problem is given by

$$\underset{P_1^R,\dots,P_N^R}{\text{maximize}} \quad \sum_{n=1}^{N} \frac{1}{2} \log\left[1 + \frac{P_n^R \alpha_n^R \left(|r_K| - |r_1|\right)^2}{4 P_{\text{noise}} g^2(2d_n^R)}\right] \quad \text{subject to} \quad \sum_{n=1}^{N} P_n^C + P_n^R \leq P_{\text{tot}}. \tag{24}$$

It has to be considered that the sum of concave functions is also concave and that the arguments of the logarithms are linear functions of the powers. Furthermore, the domain of the feasible set is a closed convex set and, therefore, only one global maximum of the problem exists. This maximum can be explicitly calculated by using the method of Lagrange multipliers which is equivalent to the water-filling power allocation result [3]. The result is given by

$$P_n^R = P_{\text{noise}} \frac{g^2(2d_n^R)}{\alpha_n^R} \frac{4}{\left(|r_K| - |r_1|\right)^2} \cdot \max\left(0, \frac{\lambda}{\beta_n} - 1\right), \forall n \in \mathbb{F}_N, \tag{25}$$

where the factor β_n is defined by

$$\beta_n := \frac{g^2(2d_n^R)}{\alpha_n^R} \frac{4}{\left(|r_K| - |r_1|\right)^2} + \frac{g^2(d_n^C)}{\alpha_n^C} \frac{K}{K-1}, \forall n \in \mathbb{F}_N. \tag{26}$$

For the following equations, we assume that the factors β_n are ordered in an increasing manner. Then the water-filling level λ is a value specified by the inequality

$$\beta_{\tilde{N}} < \lambda \leq \frac{1}{\tilde{N}} \left[\frac{P_{\text{tot}}}{P_{\text{noise}}} + \sum_{n=1}^{\tilde{N}} \beta_n \right], \tag{27}$$

where the number $\tilde{N}$ with $1 \leq \tilde{N} \leq N$ is a suitably chosen integer value for which the inequality

$$\sum_{n=1}^{\tilde{N}} \left(\beta_{\tilde{N}} - \beta_n\right) < \frac{P_{\text{tot}}}{P_{\text{noise}}} \tag{28}$$

holds. From (23) and (25) the allocated power for the second channel is determined as

$$P_n^C = P_{\text{noise}} \frac{g^2(d_n^C)}{\alpha_n^C} \frac{K}{K-1} \cdot \max\left(0, \frac{\lambda}{\beta_n} - 1\right), \forall n \in \mathbb{F}_N. \tag{29}$$

This allocation has the following interpretation. The sensor node S_n with the lowest β_n gets the largest part of the total power because its communication channels are possibly the best due to the low distances. Therefore the observation of the target object is less interfered by noise and consequently results in better data communication. Sensor nodes with higher distances get smaller parts of the total power and some of them do not get any power at all. The last ones participate neither in the data communication nor in the classification of the target object. Their information reliability is too poor to be considered for data fusion. More and more sensor nodes will become active by increasing the total power. Then the overall classification probability increases because more correct information is provided by the observations.

Note that we have used the approximation (22) in order to simplify the maximization problem and to find analytical solutions for all equations. Without any approximation the maximization problem yields the *Lambert's trinomial equation*, which still does not have any analytical solutions. Although the above approximation is only valid for low transmission powers, we use the same solution for high transmission powers, too. If instead another approximation is used, the results are indeed different, but the behavior of solutions remains generally valid. However, this study is not the subject of this work.

3.4. Computational effort

In order to calculate the transmission powers (25) and (29) the computation of β_n, λ, and $\tilde{N}$ is necessary. The parameters K, N, P_{tot}, P_{noise}, r_k, α_n^R, and α_n^C are fixed system parameters which are known to the computation unit. The distances d_n^R and d_n^C depend on the position of the target object and are therefore unknown. They can be estimated for example by a tracking algorithm. If these values are also determined, then the equations (25) to (29) can be calculated with little effort, because of simple mathematical operations such as summation and multiplication. The only difficulty is the evaluation of the path loss function g, which can include complex mathematical operations. Its complexity depends on the given multipath channel.

However, the computation effort of the equations (25) to (29) is less complex than the evaluation of the classification algorithm such as (6). If one can find simpler algorithms than (6) (see, for example [12]), then the assessment of the calculation effort becomes important and should be considered in detail.

4. Numerical results and conclusions

In this section we present some numerical results obtained by applying the proposed optimization method from Section 3. We simulate target objects with equal prior probabilities $\pi_k = \frac{1}{K}\ \forall k \in \mathbb{F}_K$ in sensor networks with different settings as described in Section 2. In all results, we consider three different kinds of target objects with reflection coefficients chosen as $|r_1| = 0$, $|r_2| = \frac{1}{2}$, and $|r_3| = 1$. Furthermore, the path loss function is modeled as line-of-sight propagation. The ratio $\text{SNR} = 10dB\log\left(\frac{P_{\text{tot}}}{P_{\text{noise}}}\right)$, instead of *received* SNRs, is depicted on the abscissa of all figures.

The verification of the proposed power allocation between both communication links of a single sensor node is shown in Figure 4. The overall error probability of the classification increases for higher SNR values for the case where the allocated power of one link is reduced by 10% and at the same time the power of the other link is stepped up by this 10%. When we reallocate a power amount of $10\% - 30\%$ to both links in an inverse manner, then the classification probability remains almost valid. This result shows that the proposed method allocates the given total power nearly optimal to both communication links, especially for higher SNR values.

In Figure 5 another verification of the proposed power allocation is shown, where a network of two sensor nodes is considered. The overall error probability of the classification decreases if we decrease the allocated power of the sensor node, which has the smallest part of the total power, by 10% and allocate this amount of power to the other sensor node. This result shows that the proposed method assigns the given total power suboptimal to the sensor nodes. The curves disperse, because of the approximation (22) which has been used for the equation (23).

As shown in Figure 6 the proposed method yields a better classification probability in comparison to a uniform power allocation where a network of ten sensor nodes is considered.

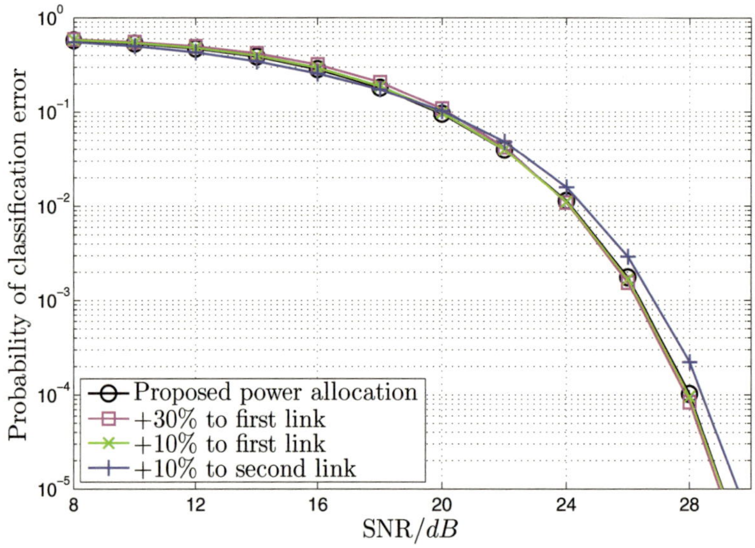

Figure 4. Verification of proposed power allocation between the two communication links of a single sensor node network.

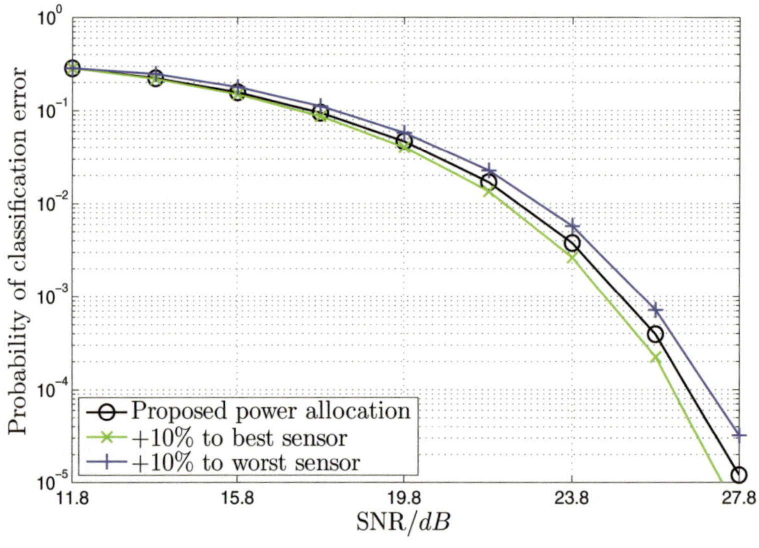

Figure 5. Verification of proposed power allocation between two sensor nodes.

In particular, it is shown that the same overall classification probability can be achieved with much lower transmission power, especially for low SNR values, by using an efficient power allocation method. Furthermore, the symbol-error probability of the sensor node with the highest part of the total power is also shown. The classification accuracy is better than the best symbol-error probability for higher SNR values, which affirms the gain of data fusion and illustrates the feasibility of object classification in this kind of distributed sensor networks.

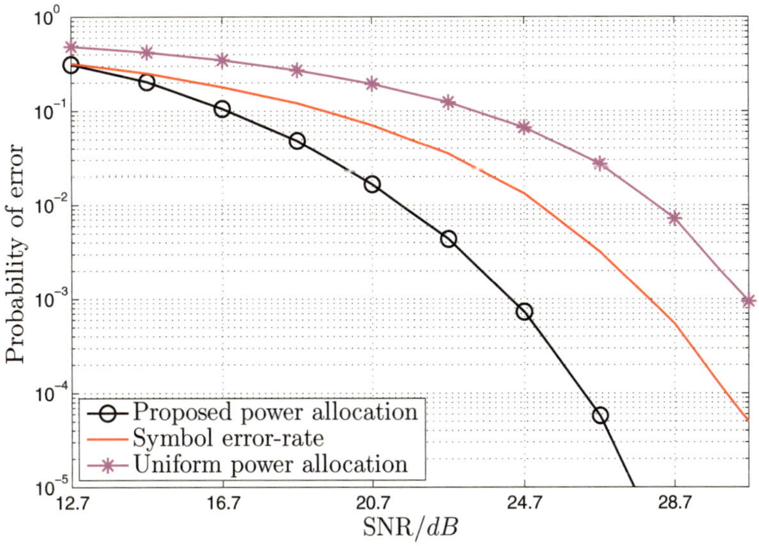

Figure 6. Comparison of proposed power allocation to a uniform power allocation in a network of ten sensor nodes.

Author details

Gholamreza Alirezaei, Rudolf Mathar and Daniel Bielefeld
Institute for Theoretical Information Technology, RWTH Aachen University, D-52056 Aachen, Germany

5. References

[1] Alirezaei, G. [2012]. Channel capacity related power allocation for ultra-wide bandwidth sensor networks with application in object detection, *IEEE ICUWB 2012 - International Conference on Ultra-Wideband*, Syracuse, NY, USA.

[2] Bielefeld, D., Fabeck, G., Zivkovic, M. & Mathar, R. [2010]. Optimization of cooperative spectrum sensing and implementation on software defined radios, *Proc. Int. Workshop Cogn. Radio Advanced Spectr. Management CogART*, pp. 1–5.

[3] Cover, T. M. & Thomas, J. A. [2006]. *Elements of Information Theory*, 2nd edn, John Wiley & Sons, Inc.

[4] Debes, C., Riedler, J., Zoubir, A. M. & Amin, M. G. [2010]. Adaptive target detection with application to through-the-wall radar imaging, *IEEE Trans. Signal Process.* Vol. 58(No. 11): 5572–5583.

[5] Duda, R. O., Hart, P. E. & Stork, D. G. [2000]. *Pattern Classification*, 2nd edn, John Wiley & Sons, Inc.

[6] Gezici, S., Tian, Z., Giannakis, G. B., Kobayashi, H., Molisch, A. F., Poor, H. V. & Sahinoglu, Z. [2005]. Localization via ultra-wideband radios: A look at positioning aspects for future sensor networks, *IEEE Signal Process. Mag.* Vol. 22: 70–84.

[7] Hume, A. L. & Baker, C. J. [2001]. Netted radar sensing, *Proc. IEEE Int. Radar Conf.*, pp. 23–26.

[8] Lapidoth, A. [2009]. *A Foundation in Digital Communication*, Cambridge University Press.

[9] Pescosolido, L., Barbarossa, S. & Scutari, G. [2008]. Radar sensor networks with distributed detection capabilities, *Proc. IEEE Int. Radar Conf.*, pp. 1–6.

[10] Srinivasan, R. [1986]. Distributed radar detection theory, *IEE Proceedings-F* Vol. 133(No. 1): 55–60.

[11] Surender, S. C. & Narayanan, R. M. [2011]. Uwb noise-ofdm netted radar: Physical layer design and analysis, *IEEE Trans. Aerosp. Electron. Syst.* Vol. 47(No. 2): 1380–1400.

[12] Varshney, P. K. [1997]. *Distributed Detection and Data Fusion*, 1st edn, Springer-Verlag

[13] Win, M. Z. & Scholtz, R. A. [2000]. Ultra-wide bandwidth time-hopping spread-spectrum impulse radio for wireless multiple-access communications, *IEEE Trans. on Communications* Vol. 48(No. 4): 679–691.

[14] Yang, Y., Blum, R. S. & Sadler, B. M. [2010]. Distributed energy-efficient scheduling for radar signal detection in sensor networks, *Proc. IEEE Int. Radar Conf.*, pp. 1094–1099.

Chapter 9

Cooperative Localization and Object Recognition in Autonomous UWB Sensor Networks

Rudolf Zetik, Honghui Yan, Elke Malz, Snezhana Jovanoska, Guowei Shen, Reiner S. Thomä, Rahmi Salman, Thorsten Schultze, Robert Tobera, Hans-Ingolf Willms, Lars Reichardt, Malgorzata Janson, Thomas Zwick, Werner Wiesbeck, Tobias Deißler and Jörn Thielecke

Additional information is available at the end of the chapter

http://dx.doi.org/10.5772/55077

1. Introduction

Ultra-wideband (UWB) radio sensor networks promise interesting perspectives for emitter and object position localization, object identification and imaging of environments in short range scenarios. Their fundamental advantage comes from the huge bandwidth which could be up to several GHz depending on the national regulation rules. Consequently, UWB technology allows unprecedented spatial resolution in the geo-localization of active UWB radio devices and high resolution in the detection, localization and tracking of passive objects.

With the lower frequencies (< 100 Hz) involved in the UWB spectrum, looking into or through non-metallic materials and objects becomes feasible. This is of major importance for applications like indoor navigation and surveillance, object recognition and imaging, through wall detection and tracking of persons, ground penetrating reconnaissance, wall structure analysis, etc. UWB sensors preserve their advantages -high accuracy and robust operation- even in complicated, multipath rich propagation environments. Compared to optical sensors, UWB radar sensors maintain their capability and performance in situation where optical sensors collapse. They can even produce useful results in non-LOS (non-Line of Sight) situations by taking advantage of multipath.

Despite the excellent range resolution capabilities of UWB radar sensors, detection and localization performance can be significantly improved by the cooperation between spatially distributed nodes of a sensor network. This allows robust localization even in the case of partly obscured links. Moreover, distributed sensor nodes can acquire comprehensive knowledge of the structure of an unknown environment and construct an electromagnetic image which is related to the relative sensor-to-sensor node coordinate system. Distributed observation allows the robust detection and localization of passive objects and the identification of certain features of objects such as shape, material composition, dynamic parameters, and time-variant behavior. This all makes UWB a promising basis for the autonomous navigation of mobile sensor nodes, such as maneuverable robots- in an unknown environment that may arise as a result of an emergency situation.

The objective of the CoLOR project (Cooperative Localization and Object Recognition in Autonomous UWB Sensor Networks) was to develop and demonstrate new principles for localization, navigation and object recognition in distributed sensor networks based on UWB radio technology. The application scenario of the CoLOR project is described by mobile and deployable sensor nodes cooperating in an unknown or even hostile indoor environment without any supporting infrastructure as it may occur in emergency situations such as fire disasters, earthquakes or terror attacks. In this case, UWB can be used to identify hazardous situations such as broken walls, locate buried-alive persons, roughly check the integrity of building constructions, detect and track victims, etc. In this scenario, it is assumed that optical cameras and other sensors cannot be used. Data fusion of optical image information and UWB radar was not in the scope of this project.

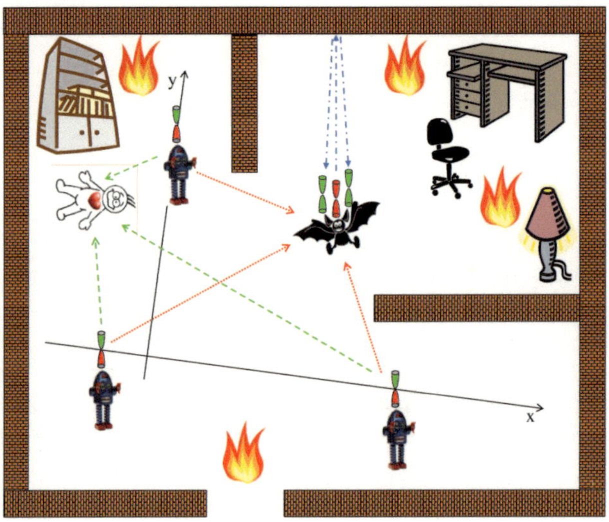

Figure 1. CoLOR scenario.

A possible scenario is shown in Fig. 1. The unknown environment is first explored by autonomous robots that deploy fixed nodes at certain positions. Those nodes being able of transmitting and receiving shall play the role as anchor nodes. They span a local coordinate system and should be placed at "strategic" positions, i.e. they span a large volume and ensure a complete illumination of the environment. Moving nodes will localize themselves relative to this anchor node reference. Moreover, when moving they collect information about the structure of the environment by receiving reflected waves. This way, they build an "electromagnetic image" of the environment and recognize basic building structures ("landmarks"). Step by step a map of the environment is built which can be used as another reference for navigation. This procedure is already well known from autonomous robot navigation as SLAM (Simultaneous Localization and Mapping). However, here we do not consider optical methods but UWB to recognize the feature vector. If there are solitary objects, the moving sensor may scrutinize shape and material composition by circling around. Other objects like humans may walk around and create time-variant reflections that identify their moving trajectory. Humans and animals may also reveal themselves by time variant features resulting from vital functions such as breathing.

The organization of this chapter is as follows. Section 2.1 describes the architecture of the sensor network and basic parameters of UWB sensor nodes that we used to achieve our objectives. Section 2.2 specifies the simulated test scenario that was applied for the development of the localization and imaging algorithms. Simulated data allowed us to develop algorithms in parallel to the demonstrator, which is presented in Section 2.3. The demonstrator was used to assess performance and to evaluate the developed data extraction algorithms in realistic scenarios. The algorithms were developed and evaluated using data obtained from the UWB wave propagation simulator described in Section 3. Within the CoLOR project algorithms were developed for: the cooperative localization of sensor nodes, see Section 4, the evaluation of sensor network topology, see Section 5, simultaneous localization and map building, see Section 6, object parameter estimation, see Section 7, imaging of environment, see Section 8 and the detection and localization of moving objects, see Section 9. Special attention was given to algorithms that promise real-time capability.

2. System architecture

2.1. Sensor network architecture

To accomplish the tasks described above, the UWB sensor nodes used can be heterogeneous in terms of their sensing capabilities and mobility. Most simple nodes may act just as illuminators of the environment. This requires only transmitting operation, but no sensing and processing capability. However, multiple transmit signal access has to be organized. This could be CDMA (Code Division Multiple Access) or TDMA (Time Division Multiple Access). TDMA requires some time frame synchronization. CDMA, on the other hand, would need multiple orthogonal codes which would complicate transmitter circuit design and increase self-interference because of non-perfect orthogonality. We preferred TDMA switching.

With its unprecedented temporal resolution, ToA (Time of Arrival) based localization methods are the natural choice for localization in the case of UWB. ToA, however, requires temporal synchronization between the nodes. This can be achieved by the RTToA (Round Trip Time of Arrival) approach which means that sensors involved must be able to retransmit received signals.

Deployable nodes, placed at verified positions may act as location reference or anchor nodes for the localization of roaming nodes. Other nodes are spread around as static or moving observers. Static observers are well suited to observe moving objects since they can most easily distinguish between desired information and reflections from the static environment (clutter) by exploiting time variance. On the other hand, moving observers (or illuminators, since propagation phenomena are reciprocal) can collect information about static objects and environments (e.g. structure of walls). By applying coherent data fusion methods, moving nodes will create an image of the static propagation environment. This is a full multi-static approach which requires a number of widely distributed cooperating sensor nodes having precise (relative) location information and synchronization at least between subsets of sensors (e.g. between receivers or transmitters or both).

Synchronization issues can be relaxed if we construct a more complex node containing, e.g. one transmitting antenna (Tx) and two receiving (Rx) antennas. Such a sensor already constitutes a small-baseline bi-static radar, which somehow resembles the sensing capability of a bat and allows the estimation of both object range and direction (by using time difference of arrival, TDoA). We will also refer to these nodes as "bat-type sensors". Their advantage is that mutual coherent processing is restricted to one platform. So, if several of those nodes are cooperating, they can directly exchange range and DoA (direction of arrival) information, which we call "non-coherent cooperation" since no exact phase synchronization between nodes is required. There is, of course, also a mixed approach which may consist of non-synchronized illuminators and locally coherent (differential) observers. So, with regard to the capabilities of sensor nodes involved and to their mutual cooperation, we distinguish between three basic structures of sensor networks:

- The multi-static approach, which assumes full coherent cooperation. The position of those nodes is estimated and tracked w.r.t. the anchor nodes.
- The "bat-type" approach consisting of self-contained sensor nodes which are able to detect and recognize characteristic features of the propagation environment on their own. This allows the building up of partial maps of the environment, the investigation of unknown objects in more detail, the identification of object features, etc.
- The mixed approach is characterized by a cooperation of all types of sensors. For example, the multi-static sensors will support the localization of the scout sensors, and these will deliver additional reference information that relates the local sensor coordinate system to the structural details of the environment.

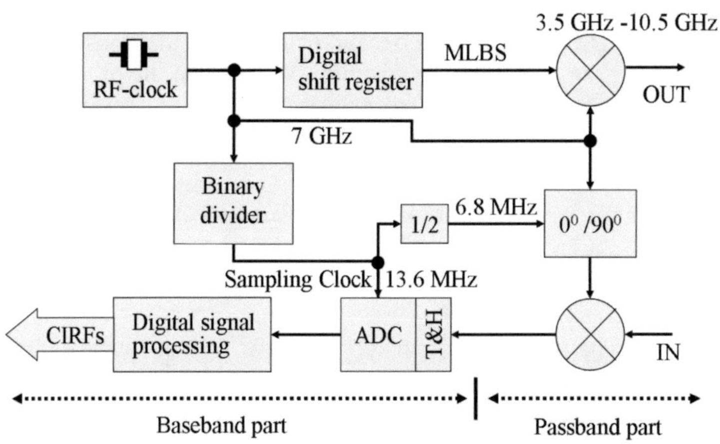

Figure 2. Basic architecture of UWB sensors node.

The basic architecture of a bat-type sensor is described in Fig. 2, see [47], [29], and see Chapter 14. The transmit signal is a periodic M-sequence spread spectrum signal which is generated by a high speed 12-stage digital shift register. It is clocked by a stable RF-clock generator - a dielectric resonance oscillator operating at 7 GHz which allows a usable frequency band up to 3.5 GHz (baseband). The sequence contains 4095 chips corresponding to 585 ns duration and about 175 m of usable distance. In the passband mode the signal is up-/down-converted by the same frequency which results in usable frequency band from 3.5 GHz to 10.5 GHz, which corresponds to the well-known FCC-specification. The receiver applies 512 time subsampling which relaxes ADC (analog-to-digital converter) bandwidth requirements and still allows real-time data recording with a measurement rate of 50 impulse responses per second.

Data measured in real-time by static anchor nodes allow the extraction of information about time-varying objects and people within the inspected scenario. Real-time data measured by moving nodes of the network allow the extraction of information about the geometric structure of the environment. In order to develop data extraction algorithms in parallel with the development of the demonstrator, we simulated a test scenario in terms of electromagnetic waves propagation. The simulated scenario is described in the following Section 2.2.

2.2. Simulated scenario

As we did not have everything ready for the demonstration scenario, e.g. the robot was not available, a scenario was simulated using the ray tracing tool 3, in order to develop and test the algorithms for map building, localization and imaging independently.

For the final demonstration, the scenario described in Section 2.3 was used.

The simulated scenario is shown in Fig. 3. It consists of a room of the size 9x8x4 m with a pillar in its center. There are six different objects (shown in Fig 3 on the right side) distributed in the room. Their positions are marked in the figure. Metal is chosen as material for the walls and for the six objects.

The bat-type sensor follows the track indicated by the dotted line. The sensor is equipped with one transmit and two receive antennas. The transmit antenna is located at the point (0,0,1) of the local Cartesian coordinate system of the robot (with x being the moving direction, y, and z the height) and the receive antennas at (0,0.5,1) and (0,-0.5,1). At 78 positions, indicated by the red circles on the track, the robot stops and takes measurements (the channel is simulated accordingly). Therefore, the transceiver array rotates in the azimuth plane (x-y-plane) with an angle resolution of 3°, so that at each position 120 simulations are performed. A horn antenna with a 3 dB beam-width of around 60° at 9 GHz was used as the radiation pattern for the transmit and receive antennas. The channel was simulated from 4.5 to 13.4 GHz in steps of 6.25 MHz in order to describe the frequency selective behavior realistically.

The inclusion of high-frequency aspects in terms of the transmission channel is particularly important for the accuracy of the simulation results. Otherwise, unrealistic conclusions will result from the simulation. In order to achieve a realistic evaluation of the efficiency of sensing and imaging systems, the channel model must describe the multi-path propagation realistically. A description of the used channel simulator as well as some verification measurements are given in Section 3.

2.3. Demonstrator

To prove robustness and applicability, research results obtained within CoLOR were experimentally validated in an extensive measurement campaign. The main focus was placed on a robust transfer of the algorithms from laboratory conditions to a more realistic indoor propagation scenario. The goals of these extensive multi-disciplinary measurements were

- to verify UWB sensing algorithms in real indoor scenarios,

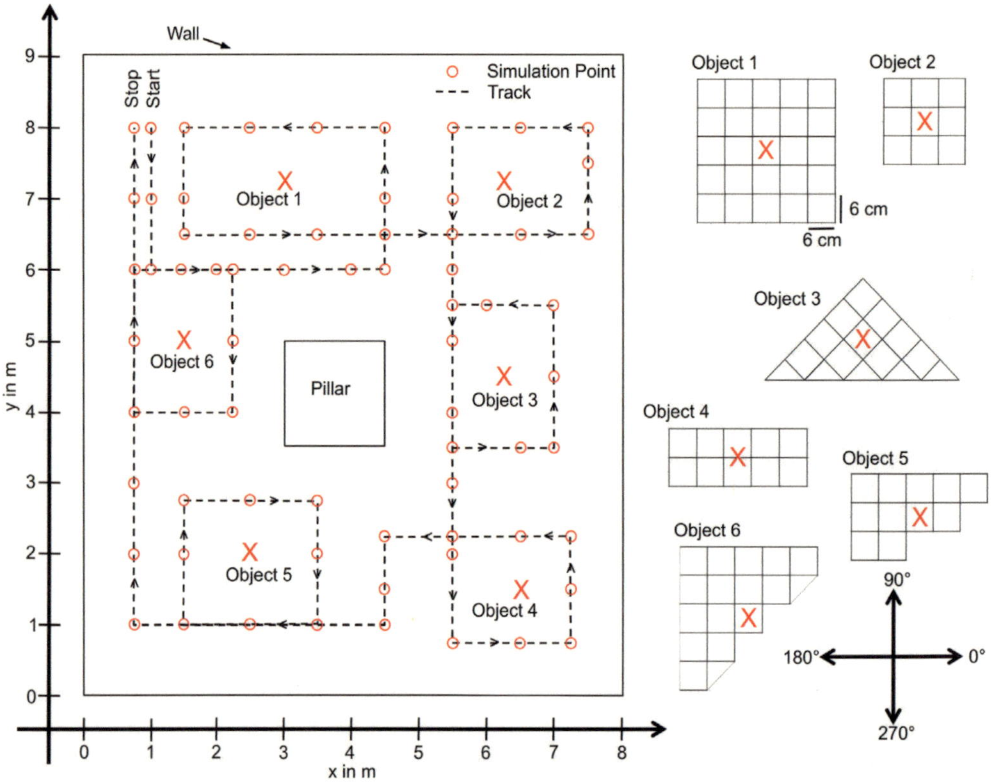

Figure 3. Simulated scenario.

- to perform and evaluate the simultaneous working algorithm for the cooperative approach,
- to determine to what extent previously achieved research results are applicable under these conditions.

With regard to a final experimental validation, the investigations in CoLOR were accompanied by experimental practice throughout the project, i.e. algorithms were always validated with regard to real world conditions.

The measurements made so far were simplified to limited sensor and/or object motions, to ideal movements by means of motorized linear arrays or by reducing the number of simultaneously performing cooperative algorithms. Therewith, only hardware complexity and mechanical effort could be reduced and meaningful validations of the investigated algorithms could be obtained. However, measurement scenarios and instrumental investments were extensively expanded within the scope of a demonstrator to fully meet the demands of a realistic indoor propagating scenario. Hence, an autonomous mobile security robot with professional motion units was fully equipped with UWB-devices, RF components, a power supply unit and a laptop for data acquisition and communication with the data fusion computer. This security robot serves as the previously mentioned mobile bat-type sensor.

Most algorithms of previous investigations could be adopted with manageable complexity to the mobile bat-type scenario. However, some challenges came up unexpectedly, e.g. erroneous robot motion due to uneven floor and slippage of robot tires, potentially more clutter in a realistic scenario, handling and transformation between a global coordinate system of the static environment and an additional local coordinate system of the dynamic robot.

To provide a realistic indoor scenario with corners as well as edges and dimensions like that of a larger office room (56 m^2), the fire detection laboratory of the University Duisburg-Essen was modified and used as the location for the measurement campaign. The modification consists of partly installed portable metallic walls to give the room a more complicated shape. The ground plane of the designed indoor scenario is depicted in Fig. 4 .

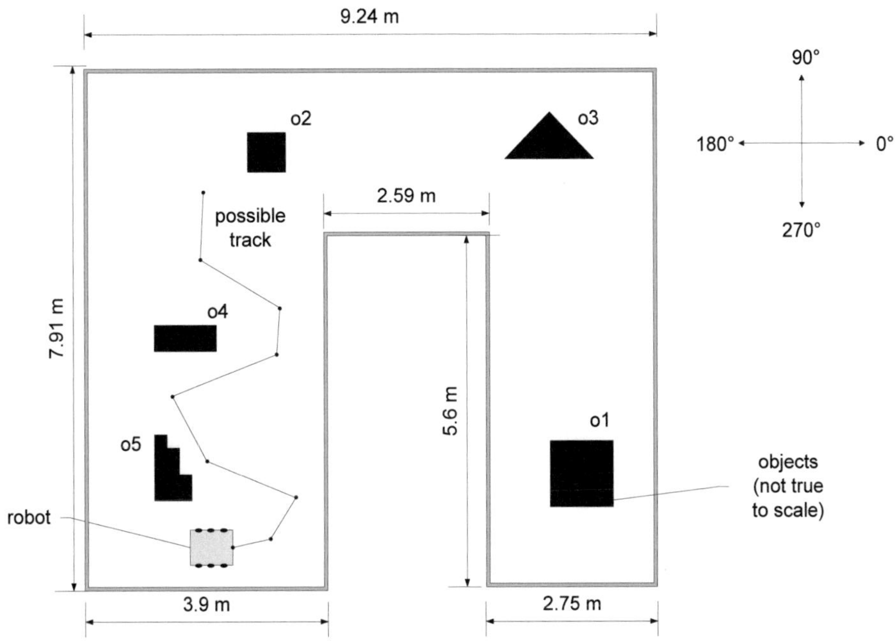

Figure 4. Ground plane of measurement location and robot with possible track.

As previously mentioned, the bat-type sensor is realized by a mobile security robot which is schematically represented in Fig. 5

As super-resolution techniques in short-range UWB sensing have to be performed, certain demands on the motion accuracy result are necessary. However, localization accuracy is predominantly achieved by advanced algorithms described later in chapter 7. Additionally, some assisting robot specifications have also been taken into account to improve accuracy. There are three actuators in the robot, two in the motion unit at the bottom, and one at the top which rotates the antenna array. The actuators are all hollow-shaft servo motors, which offer unique features unsurpassed by conventionally geared drives. Used in highly demanding industrial and medical servo systems, they provide outstanding precision motion

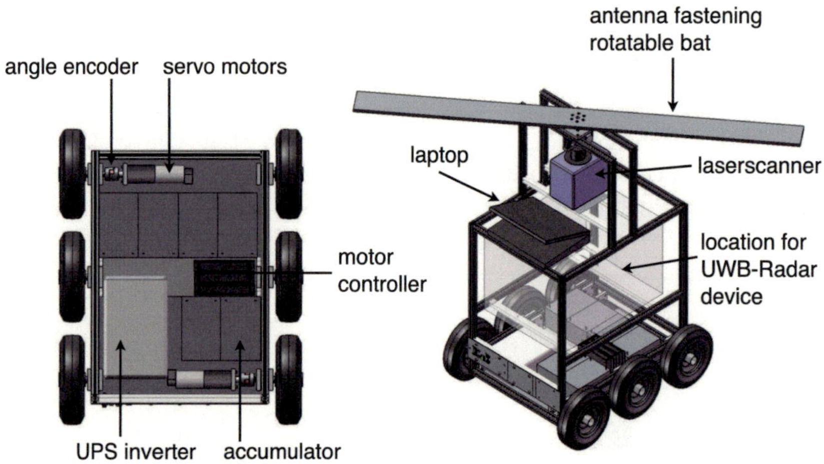

Figure 5. Schematic drawing of the motion unit (left) and of the robot (right).

control in sub-mm range and high torque capacity in a very compact package. The robot has 3 solid rubber tires on both sides which are connected by a chain-drive. Actually, both sides could be used autonomously with different acceleration, deceleration and speed, which results in curved tracks. However, for further accuracy different triggering is avoided so that the servo motors are driven synchronously. Evidently, the robot moves straight forward when the motors drive into the same direction and rotation is performed when the motors drive in the opposite direction. To maintain a more gliding rotation of the robot with reduced positioning errors, the circumference of the middle tire is minimally higher than those of the other ones. The dimensions of the robot as well as the tire position maintain a rotation center in the middle of the robot which also equals the middle of the bat antenna platform at the top. Hence, the movement of the robot was entirely restricted to translations and rotations, strictly avoiding curvature paths. Because of that, a track is split into several straight segments which are separated by a change of orientation. A resulting possible track is shown in Fig. 4. As mentioned previously, to further minimize erroneous robot motions, the bat is equipped with its own rotational unit. The orientation change can be performed by just rotating the bat, which is preferred compared to rotating the whole robot. This is more sensitive to errors due to an uneven or slippery floor. The robot is also equipped with a laser-based indoor navigation system. This highly accurate and well-proven localization system shall provide reference data for subsequent performance analysis. It has neither assisting nor guiding functionality in the localization process of CoLOR. The localization process is first and foremost handled by UWB-Radar technology.

Due to the different localization and imaging applications in this project, the requirements placed on the antenna characteristics differ. Fig. 6 gives an image and a photograph of the antenna array used. It consists of three different antenna types. A broadband monopole antenna (Tx_{2_1}), a dual-polarized broadside radiating antenna ($Tx_{2\_2}, Rx_{2\_1}, Rx_{2\_2}$) and an end-fire radiating antenna ($Tx_{1\_1/1\_2}Rx_{1\_1/1\_2}$). The requirements as well as the design of the antennas itself are described in detail in subsections 7 and 6 .

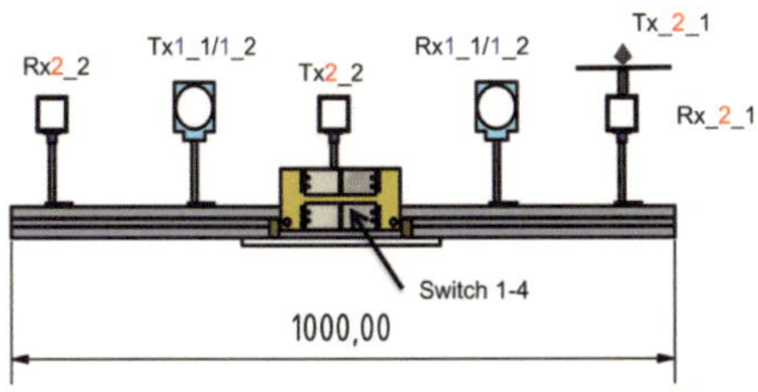

Figure 6. Schematic diagram (left) and picture (right) of the antenna array used

An array of switches, as shown in Fig. 6, allows the change between the different receiver and transmitter configurations.

3. Hybrid deterministic-stochastic channel simulation

In the framework of this project, a realistic UWB multi-path propagation simulation tool was developed in order to test and compare different algorithms and antenna arrangements for indoor UWB sensing and imaging. Multi-path propagation implies that the transmitted signal does not only arrive over the direct propagation path at the receiver, but also over paths which are dependent on the propagation environment in a complex manner. The received signal is then a combination of a multiplicity of reflected, diffracted and scattered electromagnetic waves. Wave propagation models, in general, can be classified into deterministic and stochastic ones. Deterministic models are based on the physical propagation characteristics of electromagnetic waves in a model of the propagation scenario. In contrast, stochastic models describe the behavior of the channel through stochastic processes.

By now, some statistical channel models have been established for the early design phase and for testing the ideas for possible applications. Statistical models randomly generate channel impulse responses of a channel based on the probability functions, which are usually obtained from measurements. However, if a system has to be tested in a specific environment, deterministic channel models are required, which approximate real physical phenomena.

One of the most popular deterministic channel modeling approaches is ray tracing (RT), based on geometrical optics and the uniform theory of diffraction. In outdoor areas ray tracing simulations emulate the propagation conditions very well [18]. Furthermore, it has been shown that ray tracing can be also easily extended to simulate ultra-wideband channels.

However, comparisons between the measurements and simulations with respect to UWB indoor channels show that the ray tracing results are often underestimated in terms of received power, mean delay and delay spread [26, 34, 42]. This is due to insufficient modeling or the complete neglect of diffuse scattering in the ray tracing model.

Diffuse scattering causes contributions to the power delay profile, which are not resolvable (dense multipath components). Through these contributions the power delay profile is smoother than in the scenarios with reflection and diffraction only.

In the model described in this section, a simple approach is proposed which combines the ray tracing method with statistically distributed scatterers. The approach is inspired by the diffuse scattering model for UWB channels presented in [33], by the spatiotemporal model for urban scenarios presented in [46] and by the geometrically-based stochastic channel model [40]. The scatterer placement and properties are bound to the geometry of the considered scenario. The parameters for the stochastic part of the model are derived from measurements. As only few additional scattering contributions per surface are added, the increase in computational effort is very little. The placement of scatterers ensures that part of the contributions are resolvable for the UWB system. Some preliminary results of this model were previously presented in [23], [24] and [25].

To get an impression of the sources of dense multipath components, the spatial behavior of the channel a stationary office scenario (scenario A), shown in Fig. 7, is analyzed.

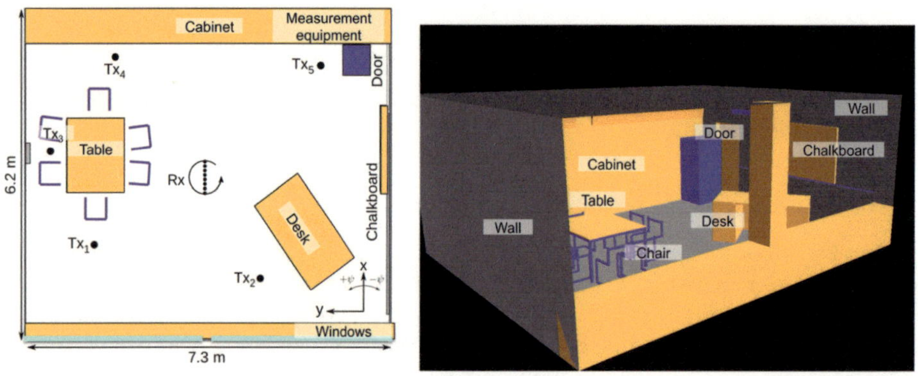

(a) Schematic diagram of the measurement scenario

(b) Simulation environment

Figure 7. Schematic diagram of the measurement (left) and simulation scenario (right, view from outside through the windows) for the DOA analysis of dense multipath components.

The measurement setup used consists of a vector network analyzer, low noise amplifier and of a set of step motor controlled positioners. The frequency range sampled by the analyzer is 2.5 - 12.5 GHz. For the measurements in scenario A, the motor controlled turntable is used to rotate the strongly directive antenna array described in [2] around its z-axis. This antenna is used as a receiver (Rx) and placed approximately in the middle of the room. The transmitter (Tx) is equipped with a UWB omnidirectional monocone antenna and placed at 5 positions marked in Fig. 7 (left).

In the same scenario, ray tracing simulations, as shown in Fig. 7 (right), with up to 5 reflections and up to 3 diffractions have been performed. At this point, no scattering is considered in the simulations. The transmission has not been considered here, earlier radar measurements in comparable rooms, [23], showed that no significant paths are to be expected from their side of the walls. On the other hand, objects inside the wooden cabinets may cause significant dense contributions. The patterns of the antennas used in the measurements have been measured in an anechoic chamber for the considered frequency band, and are considered in the simulations.

In Fig. 8 the measured and simulated power delay profiles (PDP) are depicted for the transmitter position 1, see Fig. 7 (left). For each rotation angle indicated on the x-axis, the time dependent PDP is plotted along the y-axis. As the radiation pattern of the antenna array has a narrow main beam the directions of arrival of the propagation path, impinging on the antenna can be directly identified in the picture.

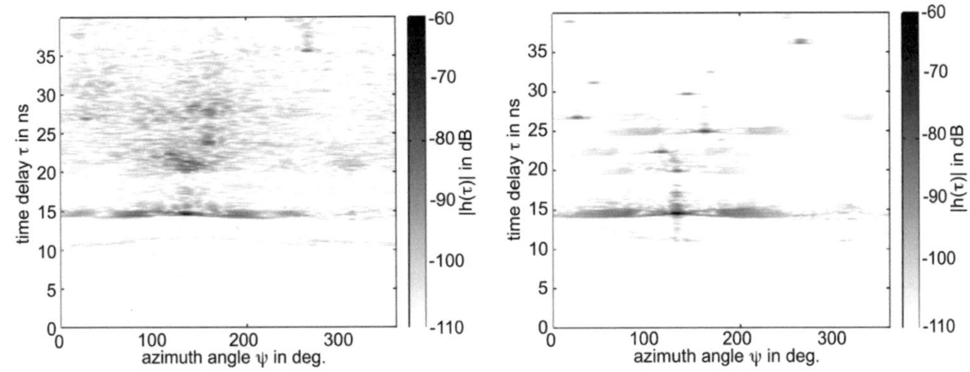

(a) Tx_1 – Measurement (b) Tx_1 – Simulation

Figure 8. Measured and simulated angle-dependent power delay profiles for the transmitter position Tx_1.

The comparison shows that most of the strong contributions are present in both measurements and simulations. Moreover, the amplitudes of the measured and simulated reflection contributions have been captured with good accuracy.

Nevertheless, a significant amount of power is missing in the simulations. Figure 9 shows the power delay profiles of Fig. 8 averaged over the delay time and over the angle. It can be observed that although some strong contributions such as the direct path and a reflection at τ = 25 ns is at the same level as that from the measurements, for most delay times the simulated power is considerably lower. From the PDP values averaged over the angle we may conclude that this effect is present for all observation angles. The dense components are not distributed evenly over the angle, but create *clusters* around the strong contributions. The observed dense contributions are not an effect of rough surfaces, as all surfaces in the room can be considered to be smooth within the frequency range used. Some contributions may arise from small objects such as doorknobs present in this scenario. Other contributions are most likely due to the scattering from the inhomogeneities within the walls or cabinets. In [44] it was shown that typical inhomogeneous building materials distort the transmitted signal significantly. Such distortions are expected to be present also in the reflected signal and are likely to cause dense components with delay times slightly larger than the delay time of the reflected signal. The amplitudes of such components are decreasing almost exponentially with the delay time.

This means that if the dense contributions are to be modeled, their delay times and angles of arrival should be grouped around the significant contributions.

3.1. Scattering model

To achieve the clustering effect, additional contributions are generated by placing point scatterers around the reflection points calculated by the ray tracing model, see Fig. 10. These scatterers represent small structures on the surface, which have not been considered in the

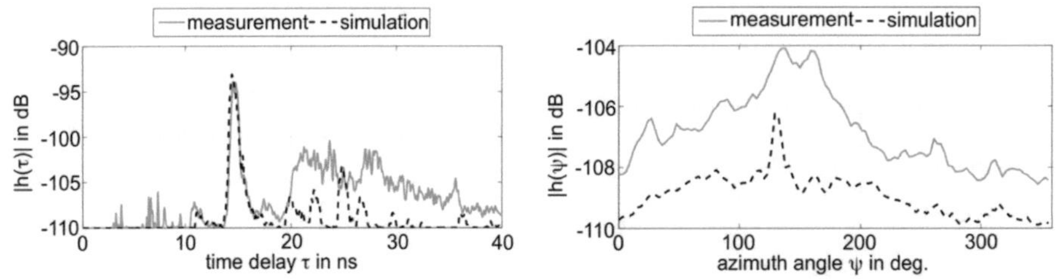

(a) Tx_1 – PDP averaged over the angle (b) Tx_1 – PDP averaged over the delay time

Figure 9. Averaged angle-dependent power delay profiles for the transmitter position Tx_1.

scenario data so far, as well as interactions with inhomogeneities inside the objects and with objects behind the walls. The delay time of the additional multi-path contributions due to the scattering points is approximately equal to the delay time of the reflected path. Their scattering coefficients are adjusted so that the resulting amplitudes of these multi-path components are slightly below the amplitude of the reflected path.

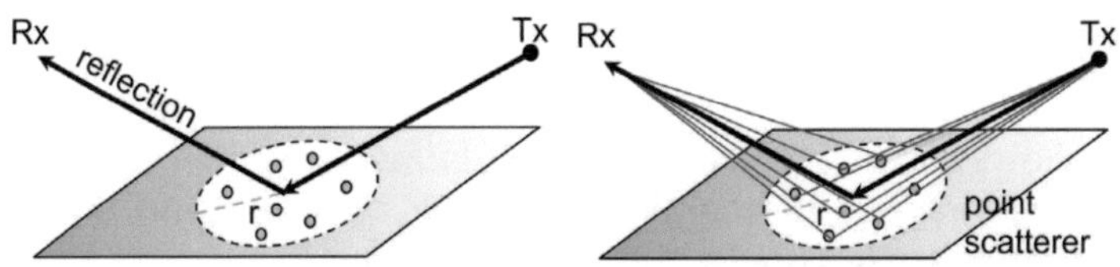

Figure 10. Modeling approach for single reflections.

The number of scatterers $n = 1 \ldots N$ is a model parameter and is assumed to be constant for all clusters. The scatterers are distributed uniformly on the objects' surface within a radius r around the reflection point. The scatterers whose position is outside the considered area are discarded.

In order to keep the model as general as possible, each scatterer is characterized by the complex full polarimetric scattering matrix $\underline{\mathbf{S}}$. The field scattered from the n-th scatterer $\underline{\mathbf{E}}^{\mathrm{s}}$ is described in the frequency domain by:

$$\underline{\mathbf{E}}^{\mathrm{s}} = \frac{e^{-jk_0 d}}{d} \cdot \underline{\mathbf{S}} \cdot \underline{\mathbf{E}}^{\mathrm{i}} \tag{1}$$

where $\underline{\mathbf{E}}^{\mathrm{i}}$ is the incident field, k_0 is the wave number, and d is the distance between the scatterer and the observation point.

In this subsection, only the vertical co-polarized element $\underline{S}_{\mathrm{vv}}$ is considered. The parameterization of other scattering matrix components can be done in the same way. As the frequency band used is very wide, at least some of the contributions can be resolved by the system. Therefore, the scattering contributions in the model are coherently summed at the receiver.

To obtain the amplitudes of scattered contributions in the same order of magnitude as the amplitudes of the reflected path, their scattering factors are related to the reflection coefficient $\underline{\Gamma}$ by a proportionality factor a, which is derived from the measurements as well. The reflection factor $\underline{\Gamma}$ is calculated at the position of the scatterer using the material parameters of the corresponding surface. Depending on the polarization of the impinging wave, the reflection coefficient either for parallel or vertical case is used. Thus, in the case of single reflection paths and assuming vertical polarization, the resulting field at the receiver consists of the reflection contribution and of the sum of scattered contributions $\underline{E}^{\mathrm{s}}_{\mathrm{v,tot}}$ given by:

$$\underline{E}^{\mathrm{s}}_{\mathrm{v,tot}} = \sum_{n=1}^{N} \frac{e^{-jk_0 d_n}}{d_n} \cdot a\underline{\Gamma}_{\mathrm{v},n} \cdot \underline{E}^{\mathrm{i}}_{\mathrm{v},n} \tag{2}$$

Due to the single scattering approach, the model covers only the part of the power delay profile with relatively short excess delay times. For the reliable simulation of delay spreads longer multiple reflected propagation paths have to be considered. The intuitive approach would be to place additional scatterers around the higher order reflections points and to use the impinging reflected wave as an excitation, see Fig. 11. This would then have to be incorporated into the reflection path search algorithm of the ray tracing approach, which would require much computational effort. To keep the excess simulation time of the hybrid part of the model as short as possible, the multiple scattering processes are replaced here with "virtual" single bounce scatterers. These scatterers are placed at the point of the multiple reflections. Their scattering factors contain an additional term $e^{-jk_0\delta}$, where δ corresponds to the path length between the point of the first interaction and the considered higher order interaction. This term adds δ/c_0 to the delay corresponding to the distance between the transmitter and the scatterer d_{geom}.

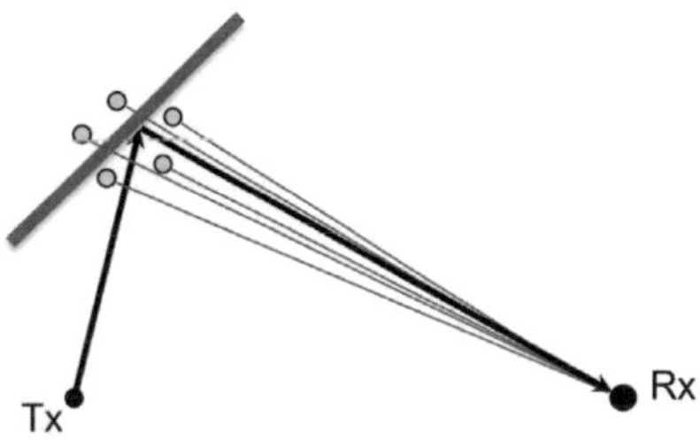

(a) 1st order reflection

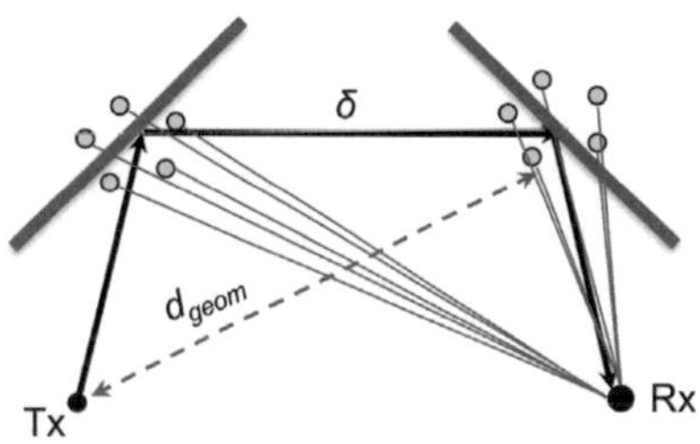

(b) 2nd order reflection - simplified approach

Figure 11. Modeling approach for multiple reflections

Thus, the delay time similar to the delay of a propagation path containing one or more reflections can be realized. However, the amplitude of such a path decreases proportionally to $\delta \cdot d_{\mathrm{geom}}$ whereas the amplitude of a path containing the reflection would experience a slower decay. To counterbalance this effect, an additional term $p \cdot \delta$ is included in the scattering coefficients of the higher order scatterers resulting in the following expression for all scatterers:

$$\underline{S} = (a + p \cdot \delta) \cdot \underline{\Gamma} \cdot e^{-jk_0\delta} \tag{3}$$

For the scatterers placed around single reflection, this expression reduces to $\underline{S} = a \cdot \underline{\Gamma}$ since in this case $\delta = 0$. Thus, the resulting model is characterized by 4 parameters:

- N - number of scatterers
- r - placement radius

- a and p - proportionality factors.

These parameters are estimated from the measurements. The derivation approach is described in the following Sections.

3.2. Derivation of the model parameters

For the derivation of the model parameters, a series of measurements with synthetic arrays at both transmitter (Tx) and receiver (Rx) have been conducted in three different office and lab scenarios in the IHE building of the Karlsruhe Institute of Technology. Two are office scenarios (scenario B and D), where the number of details in both rooms is small. The third scenario (scenario C) is a cluttered lab scenario [25]. Here, a large number of small details such as cables, tools, books etc. is distributed over the tables and shelves. These small objects were neglected in the scenario model. The transmitter is placed within a 0.12 m long linear positioner, and the receiver is moved on a 1.2 m by 0.6 m rectangle. Thus, a linear and a rectangular virtual array are obtained. The spacing between two consecutive antenna positions in both Tx and Rx arrays is 3 cm. With the exception of the antennas, the measurement setup used is identical with the setup described before. The simulation settings are also the same.

For the derivation of the model parameters, the behavior of the channel characteristics (path loss L and delay spread σ_{D}) in the measurements and the simulations are analyzed and compared [25]. To find adequate model parameters, simulations with different parameter sets are conducted and compared with the measurements. As the test of all possible parameter combinations would be computationally prohibitive, an initial parameter set has been chosen based on previous work findings in [23] and the parameters have been varied one by one.

The scatterer generation is done only once in each realization for Tx position in the middle of the Tx array and for Rx position in the middle of the Rx arrays. For each other Tx/Rx configuration the same scatterers are used.

Due to the statistical nature of the model, some variation of the simulated channel parameters for consecutive simulations with the same model parameter set is to be expected. Hence, for each parameter set 5 realizations are then simulated and the channel parameters derived from them are averaged. This number is small enough to be simulated quickly, and large enough to give approximate mean values for a given parameter set.

To derive the model parameters, their influence on the chosen channel characteristics is analyzed. It can be observed that:

- the mean relative error in path loss has a minimum at $a = 0.175$, $N = 22$ and $p = 0.075$.
- the mean error in delay spread rises with rising a and N. It changes also very quickly with p. The error minimum is at $p = 0.025$.
- for $r <= 0.5$ m all error values rise. For $r > 0.5$ m and $r < 1.5$ m, the error values are stable.

Considering this observation, first p is set to 0.03 because it has the strongest influence on the error. Thus, values of $a = 0.2$ and $N = 16$ are chosen which give a good tradeoff between path loss and delay spread errors. Finally, the scattering radius is set to $r = 1$ m.

Another indirect model parameter is the order of reflection which is considered in the scatterer placement. The influence of the considered reflection order on the delay spread is shown for a single position in the middle of the Rx and Tx array of an office scenario. The measured delay spread for this point is 3.8 ns.

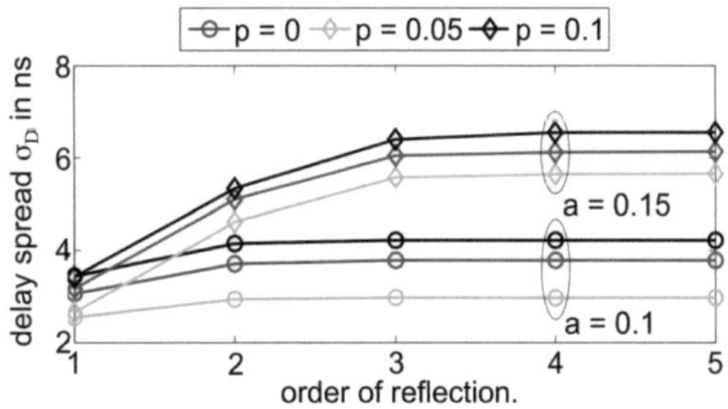

Figure 12. Influence of the considered reflection order on the delay spread in scenario 2.

The curves show that depending on the chosen model parameter (a and p have the strongest influence here) the inclusion of reflections of up to 3rd order influences the delay spread. The same has been observed in other scenarios and for the path loss. Thus, in the following the scatterers will be placed around the reflection points of up to the 3rd order.

3.3. Model performance

To test the parameterized model, it is compared with measurements with respect to path loss, delay spread, azimuth spread, power delay profiles and azimuth spectra at the receiver [25]. For the estimation of the power delay profiles and azimuth spectra, the first Tx position and a rectangular track along the edges of the positioning table at Rx is considered. Each edge of the rectangular track is placed 9 cm (3 Rx positions) away from the edge of the positioning table. The estimation of azimuth angle is done using the sensor-CLEAN algorithm [8] using 4×4 elements with a spacing of 6 cm, with the midpoint at each comparison-track point. In contrast to the measurement the simulations can provide also the angle of arrival of individual paths. However, due to the enormous amount of data obtained if the properties of each individual path are recorded, it is more convenient to apply the estimation also to the simulation data. In this case, only the coherent sum of all paths for a particular Tx/Rx position has to be recorded.

The placement of the comparison points and of the arrays used for calculating the angles of arrival (AoA) at the receiver is shown in Fig. 13 .

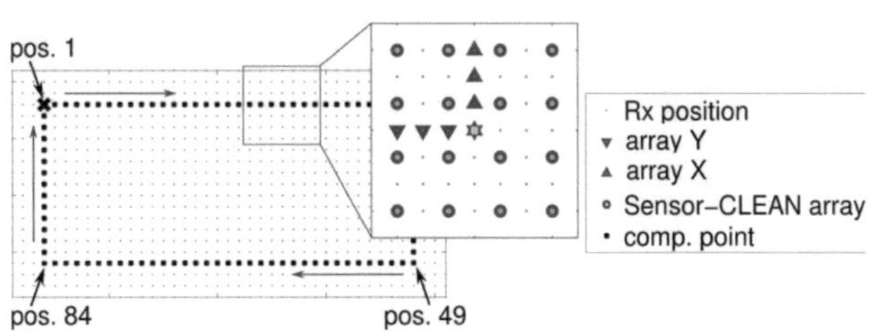

Figure 13. Placement of the comparison points and of the arrays used for the calculation of AoA at the rectangular positioner.

The "array X" configuration is used for positions along the shorter edge, whereas the "array Y" configuration is used for positions along the longer edge of the rectangular positioner. The elevation is neglected here as the measurements with a 2-D array do not allow for resolution of paths impinging from below and above the array.

The analysis of the corresponding PDPs shows that the impulse responses simulated with the hybrid approach bear much more similarity to the measurements. Although the scatterers are generated in a statistical way, their properties are tightly bound to the properties of underlying reflections so that their contributions do not dominate in the channel impulse response but fill the missing dense components of the impulse responses and angular spectra.

In the next step, the mean error μ_e and the standard deviation of the error between the measurement and the hybrid model σ_e is calculated for the path loss L, delay spread σ_D and the angular spread $\sigma_{\psi R}$. For this purpose, all possible Tx/Rx positions as described in Subsection 3.2 are used. These values and the corresponding values of the error between the measurement and conventional ray tracing are shown in Table 1 .

	Scenario B		Scenario C		Scenario D.	
	RT	Hyb.	RT	Hyb.	RT	Hyb.
μ_{e_L} in dB	4.34	1.64	1.98	-0.07	3.27	1.68
σ_{e_L} in dB	0.57	0.77	1.25	1.03	0.73	0.75
$\mu_{e_{\sigma_D}}$ in ns	1.70	0.83	1.23	0.04	1.81	-0.76
$\sigma_{e_{\sigma_D}}$ in ns	0.78	0.59	0.60	0.56	0.66	0.67
$\mu_{e_{\psi R}}$ in deg	22.35	10.26	19.85	7.15	4.33	-1.19
σ_{ψ_D} in deg	9.00	9.86	7.10	8.01	5.16	6.03

Table 1. Mean values and standard deviations of the error between the measurement and ray tracing simulation (RT) and between the measurement and hybrid simulation (Hyb.) .

Except for azimuth spread, the standard deviation values are very small. In the case of azimuth spread, however, additional errors are imposed due to path estimation. In a few cases, an insignificant rise is observed. The mean values are improved simultaneously for all considered channel characteristics.

The spread of the error values of path loss, delay spread and capacity resulting from the statistical nature of the model is analyzed also. For this purpose 40 realizations of the channel with the same parameter set are generated. For each realization, the mean error of each channel parameter is calculated. To describe the spread, the standard deviation over all mean values is adopted.

Finally, the derived model is applied to scenario A from Subsection 6.4 to prove the space-time distribution of the additional contributions. The angle dependent PDP simulated with the hybrid method is shown in Fig. 14 . The comparison with Fig. 8 shows that the additional contributions are properly placed in the azimuth-delay space, thus, depicting better the clustering effects in the scenario.

With this, a simple and effective modeling approach for directional UWB channels is proposed. The ray tracing method is combined with a simple geometric-stochastic model which represents the dense part of the channel.

The parameters of the stochastic model are connected to the properties of reflected paths so that they form a cluster with a certain delay and angle range around the reflected contribution. The stochastic clusters are also implemented around the points of multiple reflections. The

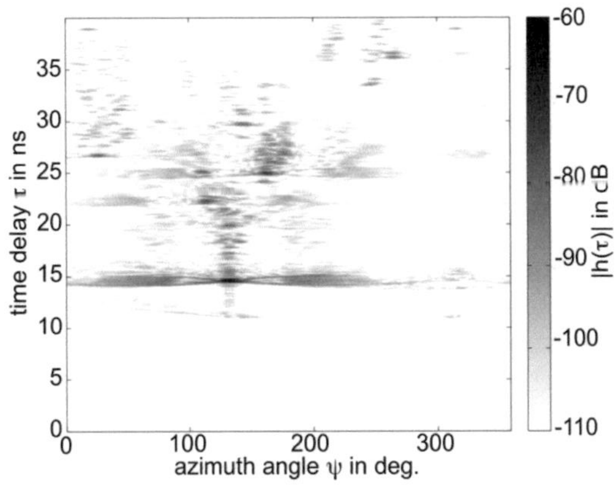

Figure 14. Angle-dependent power delay profile for the transmitter position Tx_1 in Scenario A simulated with the hybrid method.

model delivers very realistic channel impulse responses, azimuth spectra, and resulting channel parameters. The mean error between measurement and simulation is considerably improved in comparison to conventional ray tracing. This includes also the geometrical structure of the channel. Moreover, the deviation of the values of the simulated channel parameters due to the random placement of the scatterers is very small. Thus, a good reproducibility of the results is given.

4. Cooperative localization of mobile sensor nodes

In the application scenario envisaged in the introduction, an unknown environment is inspected by a UWB sensor network. Static anchor nodes of the network are placed at strategic positions. They span a local coordinate system and passively localize people or other moving objects just by electromagnetic waves scattered from them. "Electromagnetic images" of the environment are provided by moving nodes of the network. All data extraction algorithms that evaluate data measured by the sensor network require a priori information about the position of corresponding sensor nodes. In this section, basic principles of the cooperative localization of sensor nodes are described.

UWB localization is usually achieved in two steps, parameter extraction and data fusion, [20, 57]. The parameter extraction estimates parameters of signals received by sensor nodes that are required in the data fusion step. Typical parameters that are used in radio based localization systems are time of arrival (ToA), time difference of arrival (TDoA), angle of arrival (AoA) and/or received signal strength (RSS). The range-based schemes, ToA and TDoA, are shown to yield the best localization accuracy due to the excellent time resolution of UWB signals [19]. The range based ToA approach appears to be the most suitable approach for localization in UWB sensor networks. However, there are still many challenges in developing a real-time ToA based indoor UWB localization system. Due to the number of error sources, such as thermal noise, multipath propagation, direct path (DP) blockage and DP excess delay, the accuracy of the range estimation may get worse. In indoor environments, it is proven that the major sources of errors are multipath components (MPCs) and the NLOS situation [54, 56] that strongly influence the parameter estimation step - the range estimation. The quality of the range estimation is related to the SNR (or distance between Tx and Rx) and the LOS/NLOS situation. It could be improved if suitable a priori information is available. This information is usually obtained from subsequent location estimations. In what follows, we propose a novel UWB localization approach which does not require such a priori information and, instead, is based on the NLOS identification and mitigation.

4.1. UWB localization in realistic environments

The first step in our approach is high precision ToA estimation. Conventionally, ToA estimation for UWB localization is performed via a correlator or equivalently, via a matched filter (MF) receiver, [19]. However, it is difficult and not practical to implement this estimator since the received waveform with many unknown parameters must be estimated. This is almost impossible especially in realistic indoor scenarios. Another approach is the maximum likelihood (ML) based method for joint estimation of path amplitudes and ToAs described e.g. in [31, 67, 76]. This is, however, a computationally extensive method that is not suitable for real-time operations. Although various low-complexity ranging algorithms exist, their performance is not sufficient for high precision ToA estimation. Examples of low-complexity threshold-based methods such as the peak detection method, the fixed threshold method, or the adaptive threshold approach are given e.g. in [11], [17] and [22]. In these approaches, the received signal is compared to an appropriate threshold δ, and the first threshold-exceeding sample index corresponds to the ToA estimate, i.e.,

$$\hat{t}_{ToA} = t_n, \; n = \min\{i | z[i] \geq \delta\}. \tag{4}$$

For the high precision ToA estimation we proposed an adaptive threshold-based ToA estimation algorithm, the maximum probability of detection (MPD) method, in [56]. It aims at improving the robustness in multipath and NLOS situations. The main idea is to compare the probabilities for a number of possible peaks in the obtained CIR of being the ToA estimates. The probability that a certain sample, e.g. the ith sample, is determined as the ToA estimate when its amplitude, $z[i]$, is equal to or greater than the threshold and the samples before are smaller than it, i.e.,

$$P_{\mathrm{d}}(i) = \mathrm{P}(\hat{n}_{\mathrm{ToA}} = i) = \left[\prod_{n=1}^{i-1} \mathrm{P}\left(z[n] < \delta\right)\right] \cdot \mathrm{P}\left(z[i] \geq \delta\right), \tag{5}$$

where, $\hat{n}_{ToA}$ denotes the estimated index, and $i = 1, 2, \ldots, N$ are the sample indices. The one which has the highest probability leads to the final ToA estimate.

The next step in our localization approach is the NLOS identification and mitigation. The advantage of this approach is that, if the identification is correct, the accuracy of the localization can be considerably improved. Several attempts to cope with the NLOS identification problem have been proposed, such as, methods based on the sudden decrease of the SNR or on the multipath channel statistics, or method by comparing statistics of the estimated distances with a threshold in [7, 68]. However, these methods usually need to record a history of channel statistics. The advantage of our approach based on a hypothesis test proposed in [54] is that it could also be applied in cases, when the target node is static or within the halting period of a moving node. The algorithm compares the mean squared error (MSE) of the estimated range estimates with known variance of the LOS range estimates. The two hypotheses are:

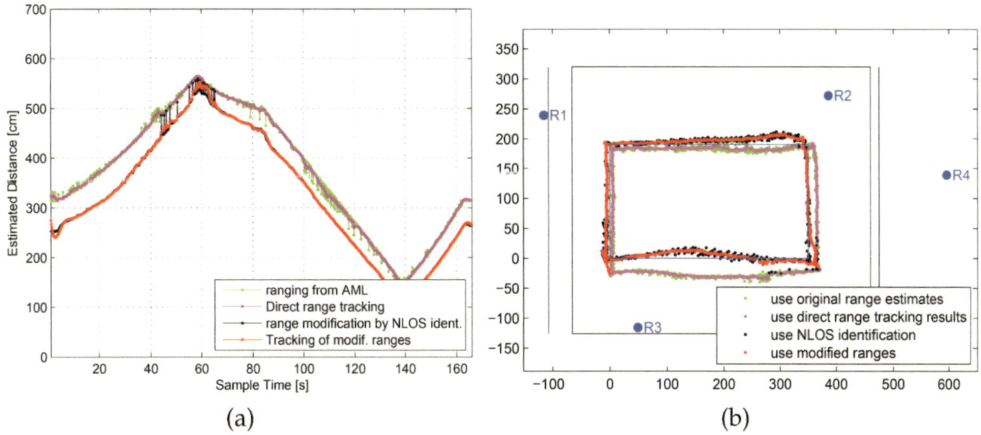

Figure 15. Data processing results of scenario with 2 LOS nodes and 2 NLOS nodes. (a) Ranging results by using different approaches for one NLOS channel; (b) the localization results by using different approaches before. location tracking.

$$\begin{cases} H_0 : M \leqslant \sigma^2_{LOS}, & \text{no NLOS node exists,} \\ H_1 : M > \sigma^2_{LOS}, & \text{NLOS nodes exist.} \end{cases} \quad (6)$$

where M is the MSE of the estimated range estimates.

After the NLOS identification, the location estimation is performed by using the identified LOS nodes only. For the implementation of the location estimation, trilateration systems are widely used. Many range-based location estimation methods with different complexity and restrictions have been proposed in the literature. All of them try to acquire a high precision of the location estimate from the range estimates. Different location estimation algorithms, which aim to find the closest position to the current coordinate of the target node, offer different accuracies and complexities. In [55], performances of a number of location estimation algorithms are compared, such as the least squares method, the Taylor series method and the approximate Maximum likelihood method.

4.2. Measurement-based verification

In order to verify our localization approach described above, a measurement was performed in a radar laboratory environment. Two Rx antennas were situated in one room, another two antennas were situated in the neighboring room and in the corridor. The Tx antenna was mounted on a positioning unit and moved along a predefined rectangular track. The MPD-based algorithm was used for range estimation. The hypothesis test-based NLOS identification and mitigation algorithm, which compares the MSEs of range estimates with the variance of the LOS range estimates, was used for location estimation. For comparison, the approximately Maximum likelihood method was applied for the location estimation, too. For both, range tracking and location tracking, the Kalman filter was applied.

The ranging results obtained for a sensor network containing one NLOS node is shown in Fig. 15(a). The result of the localization is displayed in Fig. 15(b). Both figures illustrate the feasibility of the proposed localization approach and its better performance in most cases compared with a number of other approaches.

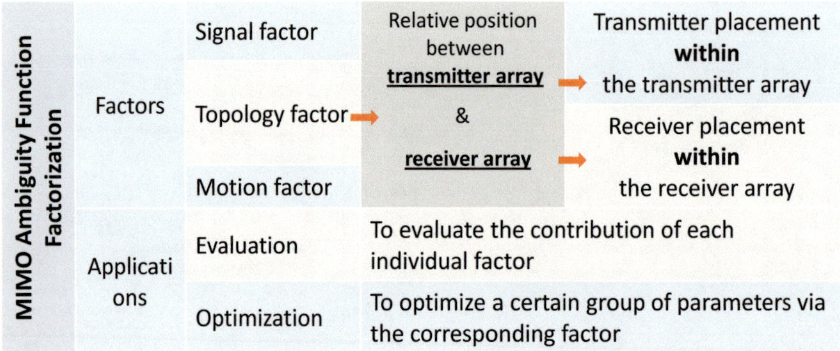

Figure 16. Factorization of MIMO ambiguity function and its potential applications.

5. Evaluation and optimization of the topology via ambiguity function analysis

As discussed in the chapter, sensor network imaging is one of the important applications of UWB sensors. In the UWB sensor network, there are stationary sensors (e.g. anchors), and mobile sensors. In the perspective of radar imaging, the spatial distribution of the stationary sensors would form a "real array", while the movement of the mobile sensors would generate a "virtual array" (i.e. synthetic aperture). The beam patterns of this "real array" and "virtual array" highly depend on their spatial configurations (topologies). In other words, the resolving performance of the real/virtual array highly depends on the topology itself.

Beyond the topologies of the real and virtual arrays, the signal parameters such as the waveform, bandwidth, etc., could also impact the resolving performance of the system. That is, the overall resolving performance of the system is jointly decided by the parameters, including the topologies of the real and virtual arrays, as well as the signal parameters. This makes the analysis of the topology even more challenging.

Sensor networks are designed to be highly accurate for their intended purpose. Always, designers and engineers are required to know the level of resolution expected from a particular sensor configuration. In order to evaluate and optimize the topologies, ambiguity function analysis is introduced in this section. Via ambiguity functions, we could know how the topology of the real/virtual array (i.e. the array formed by stationary/mobile sensors) contribute to the resolving performance of the system, and then further optimize it [30, 69].

Generally, in the far field, the ambiguity function can be factorized into several factors such as the signal related factor, the topology factor (associated with the "real array"), and the motion factor (associated with the "virtual array") [69], as shown in Fig. 16. The combination of these factors results in the overall resolving ability of the system. Theoretically, each individual factor can be used to evaluate a certain aspect of the resolution characteristics or optimize certain parameters instead of using the complicated ambiguity function of the system as a whole.

As described in the scenario, a number of UWB sensors are deployed to image the environment in order to provide necessary information for further applications. As shown in Fig. 17, it can consist of a number of moving transmitters ($T_i \in \{T_1, T_2, .., T_M\}$) and a number of stationary receivers ($R_j \in \{R_1, R_2, .., R_N\}$) . In this way, a UWB sensor network is constructed. The transmitters can move along predefined tracks (e.g. Track 1 & 2) to probe

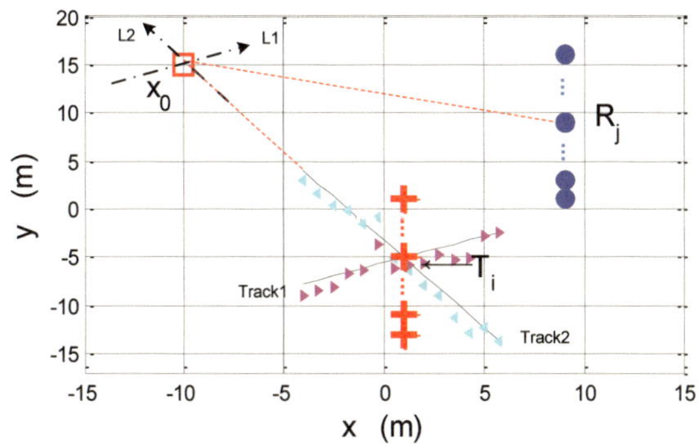

Figure 17. UWB MIMO imaging scenario. "Track 1 & 2" are the transmitter tracks; Triangles: nonlinear tracks.

the environment. The receivers collect the backscattered probing signals to produce an image of the environment. Meanwhile, they could also serve as anchor sensors to support other applications, such as localizing and tracking the position of the moving transmitters [61].

The sensor motion factor is shown in Fig. 18 (a) and (b), with respect to different tracks (Track1 and Track2, as defined in Fig. 17). In the figures, apparently, the ripples are narrower in the direction of L1 compared to the direction of L2. It indicates that the resolving performance in the direction of L1 is better than that of L2, due to the total angular rotation in the direction of L1 is far greater than the angular rotation in the direction of L2 with respect to the reference $\mathbf{x}_0$. For similar reasons, the resolving performance of "Track1" is better than the resolving performance of "Track2" in the corresponding directions. In addition, it is shown in Fig. 18 (a) that a "ghost" object occurs in the direction of L2, due to an insufficient illumination of the object. Generally, it would generate a false object image, and consequently worsen the quality of the image.

In Fig. 18 (a) and (b), the motion factors are given with respect to linear tracks. However, in practice, the sensors are not necessarily moving along linear tracks. There may be more practical irregular tracks as shown in Fig. 17 where the triangles indicate the transmission positions. The irregular movement of the sensors could improve the performance of ghost suppression, since the irregular tracks can provide a more sufficient illumination of the environment compared to the linear tracks.

According to the sensor topology in Fig. 17, the topology factor is given in Fig. 18(c). In the figure, the ghost image is partially suppressed. As shown in the figure, the suppression residuals exist at the ghost image position. However, they are not as strong as the real object. Theoretically, the ghost image can be further suppressed by optimizing the sensor spatial placement.

Figure 18 (a), (b) and (c) indicate the resolution contribution of the sensor motions and the sensor placement topology to the overall resolution. As given in Fig. 17, the overall performance of the system is the combination of all involved individual factors. It implies that we can try to realize a better overall resolving performance by (i) optimizing each individual factor, or (ii) trading-off between related factors. For example, in order to suppress the "ghost" image, on the one hand, we can optimize the movement tracks via the motion factor and the sensor placement topology via the topology factor. On the other hand, a compromise can be made between the motion factor and the topology factor. In this sense, due to the interaction

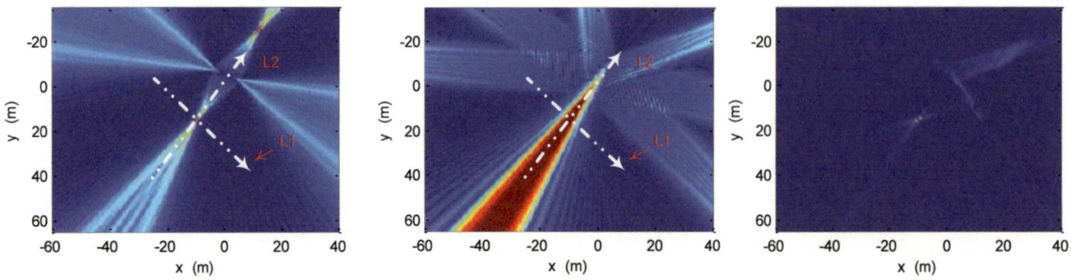

(a) Motion factor: Track 1 (b) Motion factor: Track 2 (c) Topology factor

Figure 18. The motion and topology factors given at a certain frequency f. For the motion factor, $v \cdot PRT = 5c/f$, where v is the sensor speed, PRF is the pulse repetition time, and c is the signal propagation speed. For the topology factor, the sensor element interval is $5c/f$, and the number of sensor elements on each array is 30.

of the factors, the sensor network provides more degrees of freedom for the system designer, compared with a single sensor system.

6. UWB for map building and localization

This section deals with the problem of building a map of the surrounding area using the bat-type scenario introduced in Section 2.1 . This scenario is characterized by the fact that no supporting infrastructure is used and no external information about the location of the mobile, robot-like sensor is needed. Our goal is to build a map of the surrounding for the robot, while at the same time the robot localizes itself relative to the map. In the field of robotics, this problem is known as simultaneous localization and mapping (SLAM).

To solve the complete SLAM problem, many different approaches have been presented e.g. in [62], but there well established sensor technologies like LASERs or optical cameras are used that would not work in the envisaged scenario and cannot make use of the unique capabilities of UWB radar, see Section 2.1. Other solutions are based on WLAN [53], RFID [32] or other external sources of information and, thus, must also be discarded.

There are other approaches for indoor localization and/or map building using UWB technology, but they are restricted to estimate the 2D dimensions of a strictly rectangular room [14] or need a priori information about the positions of walls to calculate virtual anchor nodes [39]. The solution presented here is more general and copes with arbitrary room shapes as long as adjacent walls are straight and orthogonal to each other.

The main advantage of this approach in comparison with the object recognition or the imaging in Section 7 and 8, is the fact that it is able to deliver a solution with a far lower number of measurements.

In the following section, a solution to the SLAM problem using a UWB radar in the bat-type scenario is described. It uses measurement models incorporating three different typical room characteristics: straight walls, corners and edges and a state-space description of the room and the robot. Algorithms for dynamic state estimation are used to calculate the desired states. Data association of measured propagation times and room features is vital here and is dealt with in great detail. Results using simulated and measured data then show the feasibility of the concept. Special requirements for the antennas are also discussed.

6.1. Measurement model

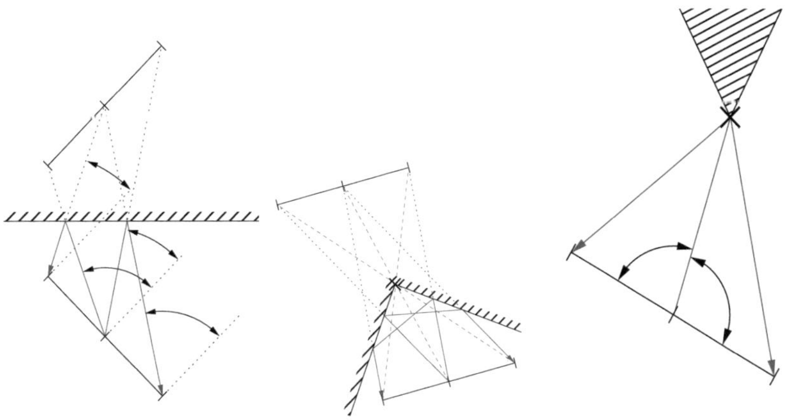

(a) single reflection (b) double reflection (c) scatterer

Figure 19. Three room features used in the state-space measurement equation.

To detect and localize the features, the bat-type UWB radar is used to measure the round-trip-times between the transmitting antenna, features of the surroundings, and the receiving antennas, which are extracted from UWB impulse responses and stored in the measurement vector **z**. A two-dimensional geometrical model of the real world is used in the algorithm, similar to [4]. Walls and corners are represented as single and double reflections, respectively. Edges and small objects are represented as scatterers. Schematic illustrations of the propagation models are shown in Fig. 19 .

Using an estimated initial position and orientation of the antenna array and an estimated initial position of a feature, an expected time-of-flight between transmitter, feature and receiver can be calculated. This can be used for dynamic state estimation, for example in an Extended Kalman filter or a particle filter, to iteratively improve the estimate of the positions. By measuring at different positions or rotating the antenna array, it is possible to distinguish the different features and calculate an initial estimate of their positions. To do this, a state-space description of the room and the robot is needed.

6.2. State-space description of the room and the robot

To solve the SLAM problem, a state-space description is used. The state vector **x** to be estimated consists of three different parts.

$$\mathbf{x} = \left[\mathbf{x}_{robot}, \mathbf{x}_{sensor}, \mathbf{x}_{map}\right]^T \tag{7}$$

$\mathbf{x}_{robot}$ contains the information about the robot position in x and y direction, p_x and p_y, as well as the speed of the robot, represented as movement angle p_ϕ and the absolute value of the speed v. All values are in relation to the local coordinate system.

$$\mathbf{x}_{\mathbf{robot}} = \left[p_x, p_y, p_\phi, v\right]^T \tag{8}$$

In $\mathbf{x}_{sensor}$ relevant information about the sensors like biases, vector $\mathbf{b}$, or the orientation ϕ_{array} of the antenna array relative to the robot is stored.

$$\mathbf{x}_{sensor} = \left[\mathbf{b}, \phi_{array}\right]^T \tag{9}$$

The map consists of the coordinates of recognized features of the surroundings, called landmarks, and are stored in $\mathbf{x}_{map}$.

$$\mathbf{x}_{map} = \left[x_{landmark_1}, y_{landmark_1}, x_{landmark_2}, y_{landmark_2}, ...\right]^T \tag{10}$$

This is the largest part of the state vector.

To estimate the current state $\mathbf{x}_k$ of the system in time step k, first the a priori state estimate $\hat{\mathbf{x}}_k^-$ is calculated using the previous state $\hat{\mathbf{x}}_{k-1}$ by using the system transition function g

$$\hat{\mathbf{x}}_k^- = g\left(\mathbf{u}, \hat{\mathbf{x}}_{k-1}\right) \tag{11}$$

where $\mathbf{u}$ is a control vector used to model external influences like movement commands to the robot.

In a second step, the state estimate $\hat{\mathbf{x}}_k$ is updated from $\hat{\mathbf{x}}_k^-$ using the measurements from the UWB radar. In what follows, the index k is often discarded to make the text better readable.

The remaining problem is that of data association discussed in the next subsection.

6.3. Data association

A major task in employing the UWB radar for SLAM is data association, in this case the task of assigning the time-of-flight measurements of the radar to corresponding features of the surroundings. The solution presented in this subsection uses two different grouping algorithm, one working in the state space, the other working in the measurement space. In both cases, the estimation of the map is done using a Rao-Blackwellized particle filter, as presented in [13].

The principal challenge we are facing is the fact that in indoor environments there is always an abundance of echoes to deal with. It is not always obvious which particular echoes belong together comparing the two impulse responses from the left and the right receiver channel. It is even harder to identify the feature which caused a particular pair of echoes. In order to improve the current state estimate, the task of determining which pair of impulses belongs to which already identified landmark has to be performed. This is achieved by applying different data association algorithms.

6.3.1. Data association in state space

The first method is a probabilistic method in state space. For a given data association vector $\mathbf{c}$ an importance distribution $\pi_j(i)$ is calculated for all impulses i and landmarks j, using the a priori state estimate $\hat{\mathbf{x}}^-$ and the time of flights z_i.

$$\pi_j(i) = p\left(z_i|\hat{\mathbf{x}}^-, c_i = j\right) p\left(c_i = j\right) \tag{12}$$

This is possible because the a priori state estimate $\hat{\mathbf{x}}^-$ contains the position and orientation of the antenna array as well as the position of the landmarks, so an expected time of flight can be calculated using the measurement functions presented earlier. Here, a Gaussian distribution with known covariance is assumed. $p(c_i = j)$ represents the probability of impulse z_i being associated with landmark j.

From the normalized importance distribution π the data association vector $\mathbf{c}$ is drawn by means of the Monte Carlo method. In this way, even in the case of false measurements being closer to the predicted measurement than the correct measurement, there is a chance that the right one is chosen. It is important to note that the opposite case, a false measurement chosen over a correct one, is also possible.

So, at first glance, this method has no advantage over a simple Nearest Neighbor method, where only the measurement closest to the prediction is used. The Monte Carlo method tends to produce slightly worse results than simply choosing the association with the highest probability. This method makes sense if not only one state is estimated but many hypotheses of possible states. That is what the particle filter can handle.

The particle filter tracks many hypotheses. These hypotheses are depicted as points, or particles, in the multi-dimensional state space. Each particle s^l is composed of the estimated state $\hat{\mathbf{x}}^l$, the covarianz matrix $\mathbf{P}^l$ of the state, a data association variable $\mathbf{c}^l$ where the association between measurements and landmarks is stored, and a weight w^l.

$$s^l = \left[\hat{\mathbf{x}}^l, \mathbf{P}^l, \mathbf{c}^l, w^l\right] \tag{13}$$

The index l denotes the lth particle. The weight w^l is an indicator of the likelihood that a certain hypothesis holds true. It is calculated using the weight from the previous iteration. Thus, the weight serves as a memory. In the long run, the hypothesis with the highest weight will be the one that best approximates the real world.

$$w_k^l = w_{k-1}^l p\left(\mathbf{z}_k | \mathbf{x}_k^l, \mathbf{c}_k^l\right) \tag{14}$$

6.3.2. Data association in measurement space

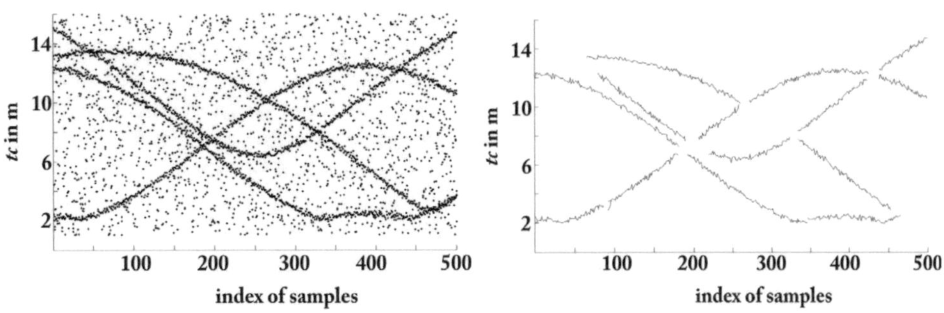

(a) Cluttered measurements

(b) Associated measurements

Figure 20. Left: Cluttered sample measurements over time, right: Associated measurement curves.

One problem that arises in using a particle filter is that due to the probabilistic nature of the algorithm the number of particles can grow considerably high. This is because the number of possible data associations increases with every measurement step. Using more particles or discarding unlikely hypotheses by resampling can only partially solve this problem.

To reduce the necessary number of particles, we use a second approach: Measurements are not directly associated to landmarks. Instead, they are first grouped in the measurement space. To do this, the fact is used that the antenna array moves only in small steps between consecutive measurements, so measurements originating from the same feature also change only slightly. This correlation can be exploited. By employing a simple Kalman Filter in the measurement space, it is possible to predict and group measurements that belong to the same feature. Only whole groups of measurements are passed to the particle filter, which greatly reduces the number of hypotheses needed and therefore the number of particles necessary. Figure 20 shows a simulation of this process. In the left figure, measurements are taken as the robot travels through the environment. Dots indicate extracted time-of-flights. The measurements are cluttered, but almost continuous echoes originating from room features can be made out. The right figure shows the result of the grouping algorithm.

The disadvantage of this procedure is that the grouping introduces a time delay in the system. Moreover, it requires measurements to be made more frequently, and so partly weakens one advantage of the room reconstruction algorithm.

6.4. Simulations

The algorithm was first tested with simulated data. The ray tracing algorithm described in Section 3 was used to calculate the impulse response function of a room. The outline of the room is shown in Fig. 3, alongside with 78 measurement points. At each point, 120 measurements were made by turning the bat-type antenna array in steps of 3 degrees. The simulated environment consisted of a rectangular room with the size of 8 m by 9 m, with a rectangular column the size of 1.5 m by 1.5 m roughly in the middle. Between the walls and the column, six complex objects used to test the object recognition algorithm from Section 7 were placed. Walls, objects and the column were assumed to be of metal. The frequency response function was calculated from 4.5 GHz to 13.4 GHz for an antenna array with three double-ridged horn antennas in the bat-type configuration. The distance between the antennas was set to 0.5 m.

Figure 21 shows the complete radargram for one whole rotation of the antenna array at point 16. In this example, it can be seen that for every angle, there are clear peaks in the impulse response that connect to peaks in the next measurement step, so data association in measurement space is possible. Mapping the whole room just from this position is not possible, because some features simply cannot be seen from there. To map the whole room, measurements from all 78 points were used. The result can be seen in Fig. 23. Walls, shown as solid and dotted lines, and corners, shown as triangles, are mapped at an accuracy of approximately 20 cm. Due to their small size, the placed objects are detected as point scatteres which are depicted as stars. Their positions correspond to the object locations shown in Fig. 3. At this point, a separate object recognition algorithm as described in Section 7 could be used to identify and distinguish them. Note that the origin of the coordinate system is set arbitrarily at the point of the first measurements.

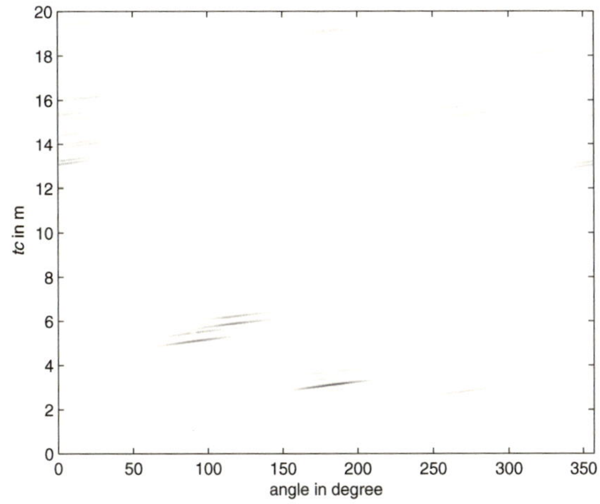

Figure 21. Radargram at position 16 of the simulated room.

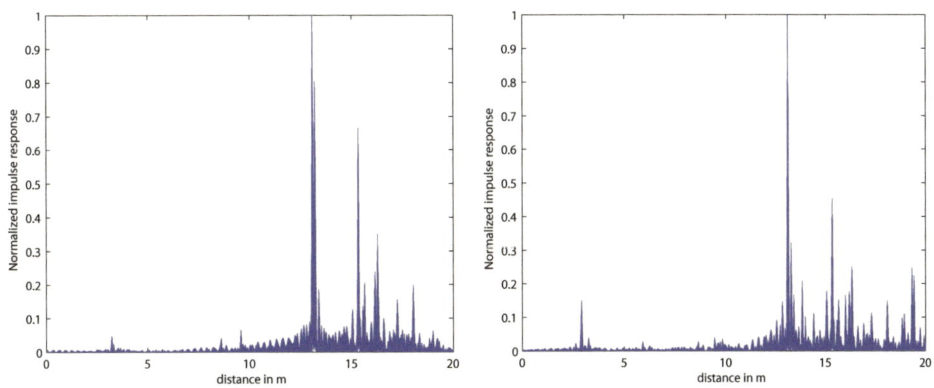

(a) Left receiver channel

(b) Right receiver channel

Figure 22. Impulse responses at position 16 of the simulated room, facing 0 degrees.

Figure 22 shows the impulse responses for the left and right channel at point 16 of the scenario described in Section 2.2. On the x-axis, the time t times the speed of light c indicates the distance the pulse has travelled. Peaks reflected or scattered from different room characteristics can easily be separated.

6.5. Measurements

To further verify the results, measurements were made in a laboratory room. The room included furniture, some metal pipes on the walls, and was filled with assorted laboratory equipment at one end of the room. The sensor array consisted of three double ridged horn antennas 0.46 m apart, similar to the simulation.

The antenna array was placed in the middle of the room and rotated manually. Pictures of the room can be seen in Fig. 25. As in the simulation, no information about the current angle of

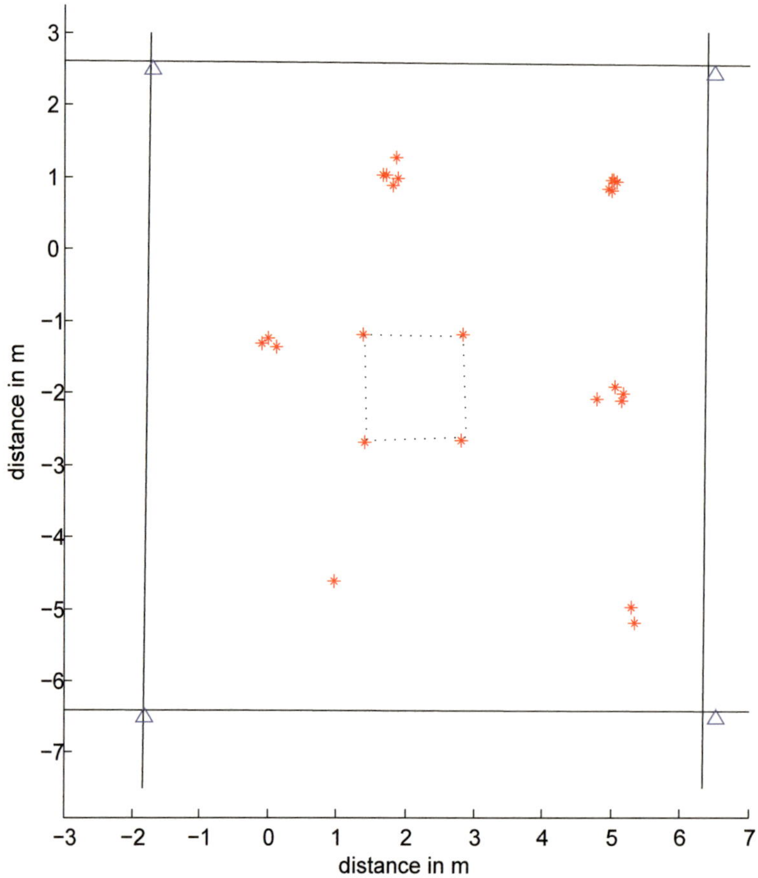

Figure 23. Reconstructed room.

the array was passed to the algorithm. The algorithm only used the UWB measurements to reconstruct the room. Figure 24 shows a sample of a recorded impulse response.

In a first test, the array was rotated only by 180°, illuminating the tidy side of the room. In this case, results similar to those of the simulation could be produced; walls and corners could be mapped with 10-20 cm accuracy. A reconstruction of the whole room was not possible. Many objects on the other, chaotic side of the room produced a large number of echoes and made it impossible to associate the measurements reliably. Here, the algorithm reached its limits.

In a second scenario, measurements were made at 15 positions in an L-shaped corridor, as depicted in Fig. 26. The array was rotated in 3 degree steps at every position, resulting in 1800 measurements.

To test the ability of the algorithm to cope with sparse measurements, only 24 measurements at 5 positions (station 1, 4, 6, 8 and 12 of the scenario) were used, resulting in a total of only 120 measurements. Here, additional information about the position of the robot had to be used, in this case odometry data about the way the robot traveled and the direction the antenna faced. The use of inertial measurement units is also possible.

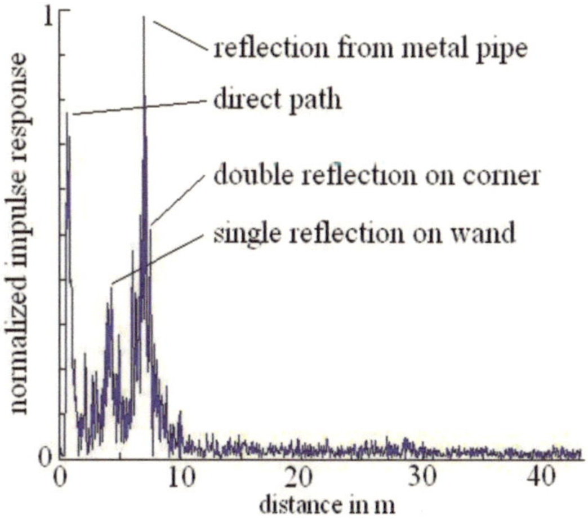

Figure 24. Impulse response showing the features of the room

(a) Tidy side of the room

(b) Chaotic side of the room

Figure 25. Room used for measurements

The results in Fig. 27 show that the algorithm is able to recognize the outline of the room using only these few measurements, although a higher number of measurements still improves the quality of the reconstruction. There is also a trade-off between the different data association methods. While grouping in measurement space is only possible if the measurements are taken frequently, data association in state space can cope with few measurements, but rely on additional sensor data.

6.6. Optimized antenna design for SLAM

To further optimize the results of the SLAM algorithm described in this section, antennas with a broader 3 dB beam-width (>60°) than for the object recognition in section 7 are needed.

Apart from the broad frequency band of 3.5 to 10.5 GHz in order to meet further conditions the antenna also has to be dual-orthogonally polarized. The radiation phase center should be constant over frequency, and the two polarizations should have identical radiation conditions. In literature several types of UWB antennas can be found. Most of them are either biconical

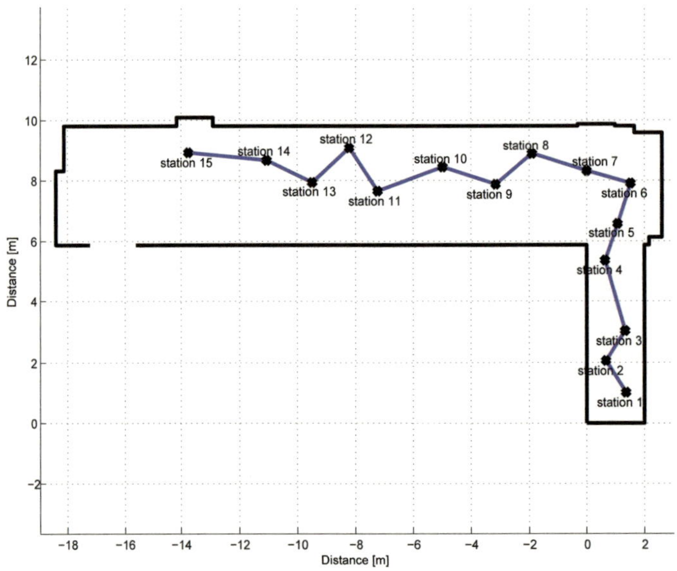

Figure 26. Outline of the measurement scenario.

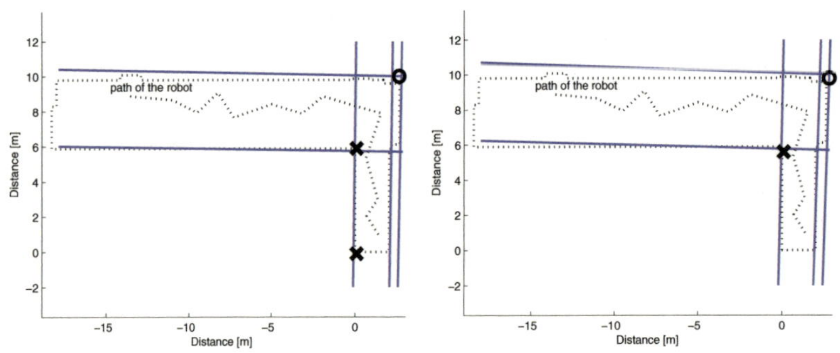

(a) Reconstruction using 1800 measurements (b) Reconstruction using 120 measurements

Figure 27. Reconstructions of an L-shaped corridor using different numbers of measurements. Dotted lines show actual room outline.

structures [5], traveling wave radiators [16], or even a combination of both [6]. The requirement of a common phase center (over frequency) limits the possible solutions. For this purpose a planar solution, a broad-side radiating antenna is chosen and optimized.

The antenna consists of two elliptically shaped dipoles surrounded by a metallic ground plane as shown in Fig. refschematic . The ellipses for each polarization (vertical and horizontal) are orthogonal to each other. Contrary to normal dipoles the feeding is separated and placed between the ground plane and the single ellipses. This is outlined by the arrows in the radiating element shown in Fig. 28. This type of feeding allows a separate feeding for each polarization and helps to keep the current distribution in the radiation zone symmetrical resulting in a constant phase center (of each polarization) exactly in the middle of the elliptical dipoles (two monopoles) [1].

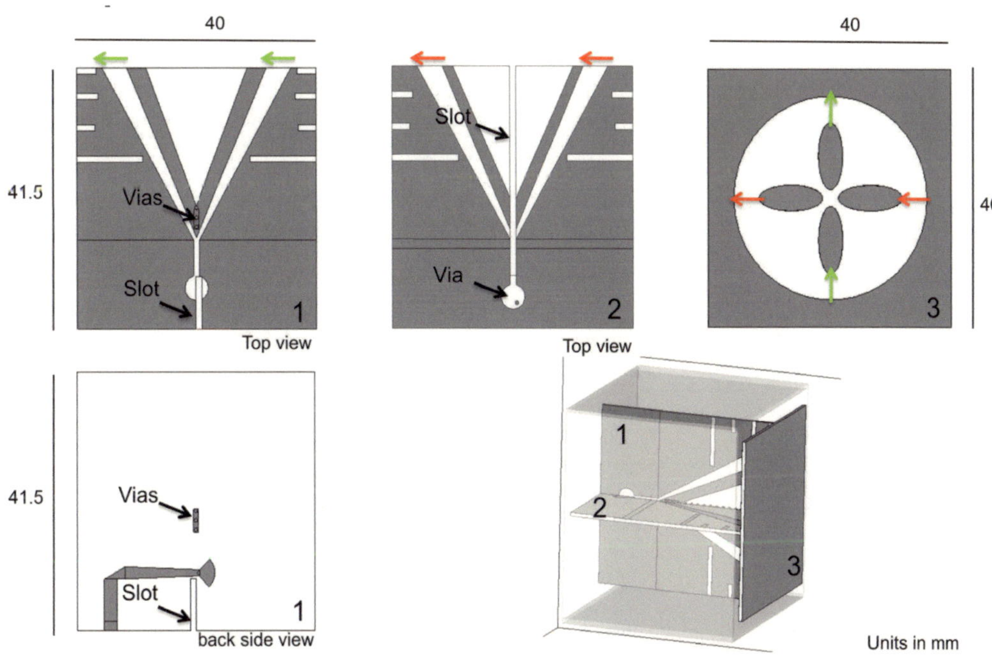

Figure 28. Schematic illustration of the dual polarized antenna element and the feeding networks, all units in mm).

The feeding networks themselves are placed orthogonally to the radiating element, see Fig. 29 (left). Similar to the Vivaldi structure, Fig. 32, a balun is used for microstrip to slotline transition. The slotline is then split up and used to feed the monopoles. The two polarizations are realized by shifting the two orthogonal feeding elements into each other as shown in Fig. 28. Therefore, a slot has to be cut into both (feeding) networks. The gaps in the metallic structures have to be closed again. This is realized by soldering through vias in the respectively orthogonal feeding network.

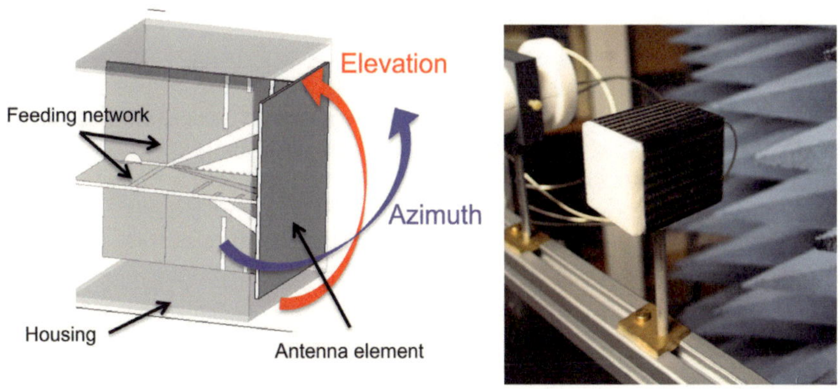

Figure 29. Schematic illustration (left) and photograph (right) of the 4-elliptical antenna

As this type of antenna is radiating broad-side in both directions (forward and backward) and should be used as (mono-) directional radiating antenna, the backward radiation (illuminating towards the feeding network) must be absorbed by a carbon fiber housing as shown in Fig. 29. This results in a reduction of the radiation efficiency. An alternative would be to use a reflector, which, however, would limit the bandwidth of the antenna.

The antenna characteristics are measured with a vector network analyzer in an anechoic chamber. The input impedance matching is around -10 dB between 3.5 and 10.5 GHz and the decoupling of the two ports is approximately 20 dB.

Figure 30 shows the 2D gain over frequency and angle for both planes (E- and H-plane) of one polarization in co-polarization arrangement. The measured gain of the second polarization is very similar and not specifically shown.

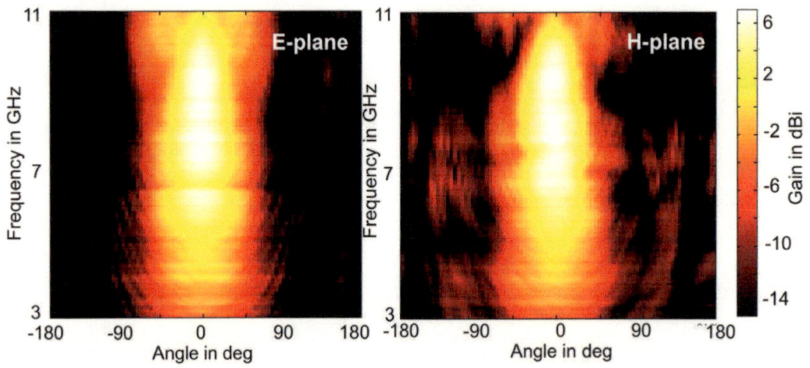

Figure 30. Measured gain over frequency and angle in the E-plane (left) and H-plane (right).

This solution of a planar, dual-polarized UWB antenna covers the frequency range between 3.5 and 10.5 GHz. In this frequency range, the radiation pattern of the antenna remains stable and directive. Both polarizations have the same radiation phase center, which is frequency-independent.

7. Short range super-resolution UWB-radar sensing

In recent years short range UWB radar sensing and imaging has gained steadily increasing interest in research. The demand for a wide absolute bandwidth results from the smallest dimensions to be resolved. However, the request for increased resolution capabilities strove for innovative algorithms, new hardware equipment, and for performance which is not restricted by the bandwidth defined by the hardware. In this chapter novel and pioneering methods, algorithms and antennas are presented which were investigated within CoLOR for UWB radar applications especially in short range UWB radar applications.

7.1. Antenna design and measurement results

As the use of polarization diversity allows further information about the object characteristics to be obtained, for instant about the surface structure as it is shown in this section, the antenna has to be orthogonally polarized. Apart from orthogonal polarizations, further conditions for the (object recognition) antenna design are a high gain and a common phase center (for both polarizations).

A common antenna for such an application is the so-called Vivaldi antenna. The radiation mechanism is based on exponentially tapered slots and the traveling wave principle [3]. This type of antenna has a convenient time domain behavior as shown in [59] and a relatively stable radiation pattern in the whole frequency range. In CoLOR the frequency range for the final demonstrator covers 3.5 to 10.5 GHz. A second band from 4.5 to 13.5 GHz was also used during the development process. Therefore, the objective of the antenna design is to cover the hole frequency range from 3.5 GHz to 13.5 GHz.

A 3D illustration (left) and a photograph (right) of the fabricated antenna is shown in Fig. 31, see also [45]. Combining the integration of two tapered slot line antennas on a single substrate with embedding them into Polytetrafluoroethylene (PTFE) allows the total bandwidth to be covered. The integration of two radiation elements per polarization yields a higher gain and saves space compared with an array of two separated antennas. Furthermore, it is less cost intensive and easier to manufacture. The possibility of varying the tapering of the inner and outer structure, see Fig. 32, can be used to focus the beam.

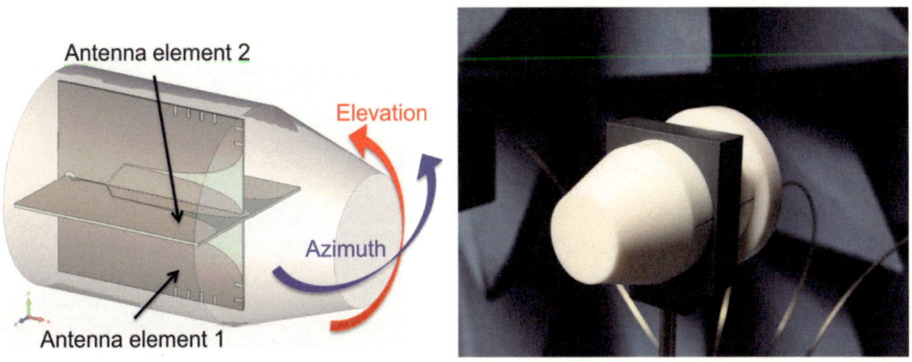

Figure 31. Antenna, embedded in the dielectric.

Both radiating structures are fed by a network shown in Fig. 32 (left), and for the second polarization in Fig. 32 (right), respectively. Starting with microstrip, where the connector is soldered on, an aperture coupling transforms to slotline which finally divides the power and feeds it to the two elements.

As already mentioned, the integration of the antenna elements into a dielectric reduces the effective wavelength. This affects several advantages compared to an antenna in free space. The antenna is capable of radiating a lower frequency, the far-field conditions are fulfilled in a closer distance and the shaping of the dielectric can (also) be used to focus the beam and for sidelobe suppression. For this work PTFE is chosen as dielectric. Its permittivity of $\epsilon_r = 2.1$ is similar to that of the substrate used (Rogers Duroid 5880 with $\epsilon_r = 2.2$). Furthermore, PTFE has low losses and can be easily shaped to adapt to the antenna design. The shaping and the dimensions of the PTFE structure are given in [45]. The conically shaped rod (in radiation direction) allows a smoother transition of the guided wave into free space.

The two polarizations are realized by shifting two orthogonal elements into each other as shown in Fig. 31 left. For doing this, a slot has to be cut into both elements. Thus, the metalized structures are interrupted, see Fig. 32. They have to be galvanically connected again. This is realized by introducing vias in the orthogonal antenna.

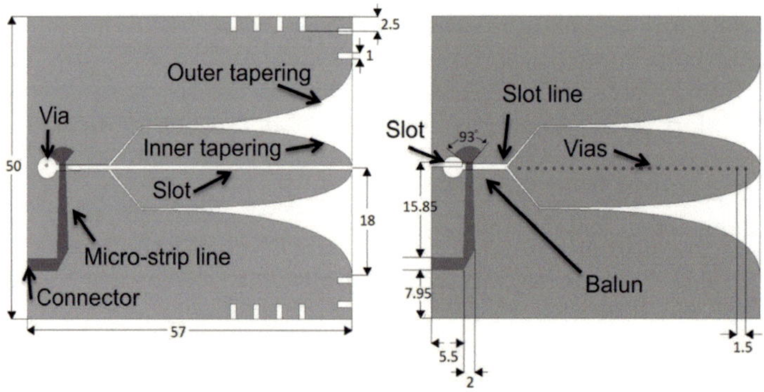

Figure 32. Schematic illustration of the antenna elements 1 (left) and 2 (right), in [mm].

The antennas are manufactured with aid of a circuit board plotter on a Duroid RT5880 substrate of a thickness of 0.79 mm. The measured S-parameter, see Fig. 33 left, show a good impedance matching for both polarizations and antenna elements, respectively, between 3.5 GHz and 13.5 GHz (and even higher). The decoupling (S_{21}, S_{12}) between the two elements is over the biggest portion of the bandwidth better than -25 dB.

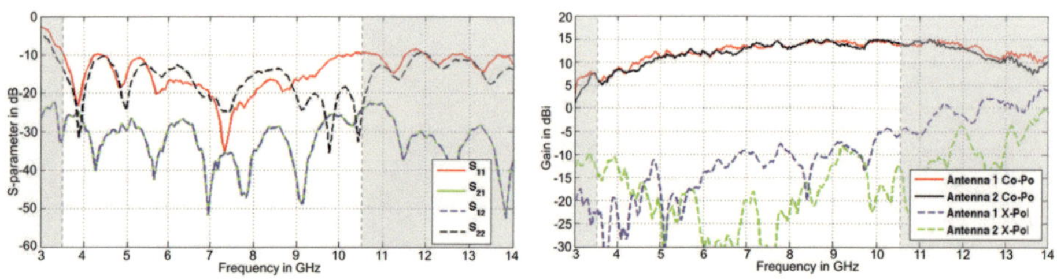

Figure 33. Measured S-parameter (left) and antenna gain in the main beam direction (right).

The gain and the pattern of the antenna were measured in an anechoic chamber. The results for antenna element 2 are presented in Fig. 34. The E-plane corresponds in this case to the azimuth direction, see Fig. 34 left, the H-plane to the elevation one. Antenna element 1 shows a similar characteristic. The maximum gain measured is around 15 dBi at 9 GHz.

To evaluate the polarization properties, the gain for both co-polarizations (Co-Pol) and both cross-polarizations (X-Pol) in the main beam direction was measured, see Fig. 33 right. The difference (between Co-Pol and X-Pol regarding the antenna gain) provides the information about the polarization purity. The cross-polarization suppression is better than 20 dB at the low frequencies up to 10.5 GHz. Starting from 10.5 GHz, the values of the X-Pol of antenna 2 are increasing (deteriorating). This is due to the current distribution of higher modes which cannot be avoided for higher frequencies. Nevertheless the measured performance allows a successful use in polarization diversity systems even above 10.5 GHz, see [45].

7.2. Data pre-processing

The raw radar data provided by the M-sequence radar needs some form of data pre-processing to smooth pulse shape, improve dynamic range, minimize the signal to interference plus noise ratio (SINR) by reducing range sidelobes and finally to enhance the

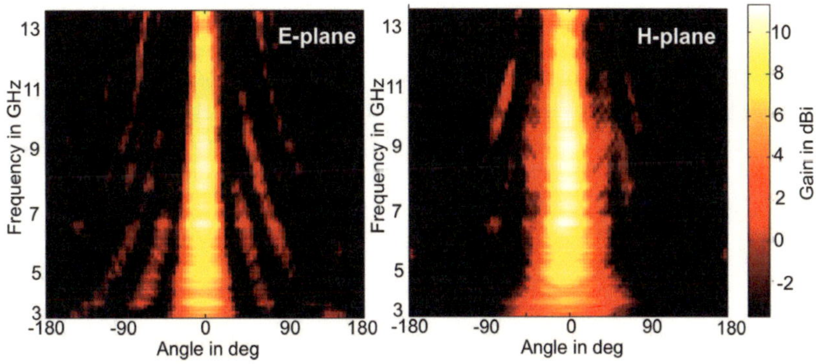

Figure 34. Measured gain [dBi] and pattern of antenna element 2 in the E-plane (left) and H-plane (right) .

temporal resolution. The channel impulse response of the radar link can be extracted by deconvoluting with a reference pulse, as we assume the link as an LTI-system. However, it is well known that classical deconvolution by spectral division may drastically distort the result especially at low SNR values. A highly efficient method with low complexity to perform the deconvolution is to apply a simplified Wiener filter with the transfer function

$$H_{wiener}(f) = \frac{1}{H_{ref}(f)} \frac{|H_{ref}(f)|^2}{|H_{ref}(f)|^2 + 1} = \frac{H_{ref}(f)^*}{|H_{ref}(f)|^2 + 1} \tag{15}$$

where $H_{ref}(f)$ is the Fourier transform of a previously measured offline reference pulse. Hence, the estimate of the deconvoluted channel impulse response $h_{deconv}(t)$ is then obtained as $h_{deconv}(t) = h_{wiener}(t) * h_{measured}(t)$, with $h_{measured}(t)$ being the measured impulse response under test. Depending on the power level, the Wiener filter either acts as an inverse or matched filter for the deconvolution. In Fig. 35 an example of the channel impulse extraction is shown. Note that both pulses are normalized to the same power to enable visual comparison in the plot.

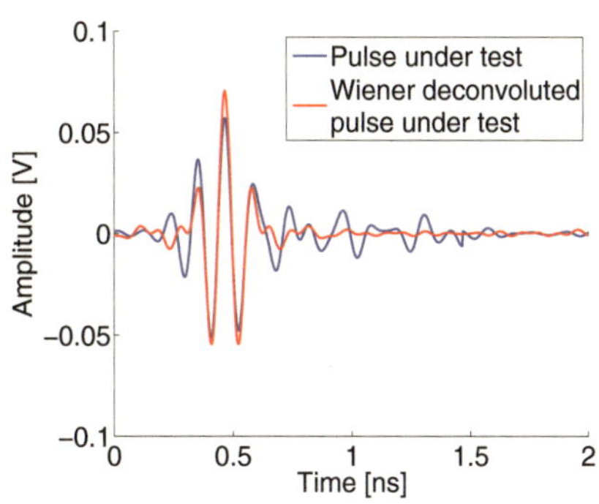

Figure 35. Example of deconvoluted pulses normalized to the same power

7.3. Material characterization

A material characterization in hostile and pathless scenarios requires a remote measurement. Hence, a method known from optics, the ellipsometry has been adapted to the UWB microwave range. The estimation of the dielectric characteristics, especially the permittivity and the emissivity are based on the ratio of the reflected power measured at two orthogonal polarizations. The orientation of the polarization is defined with regard to the plane of incidence. The plane of incidence is orthogonal to the surface of the object and is spanned by the incoming and the reflected ray. In Fig. 36 a drawing of the functional principle is given.

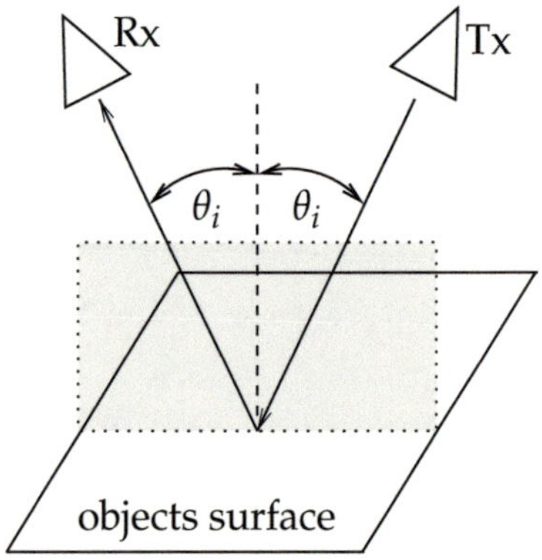

Figure 36. Schematic representation of the functional principle.

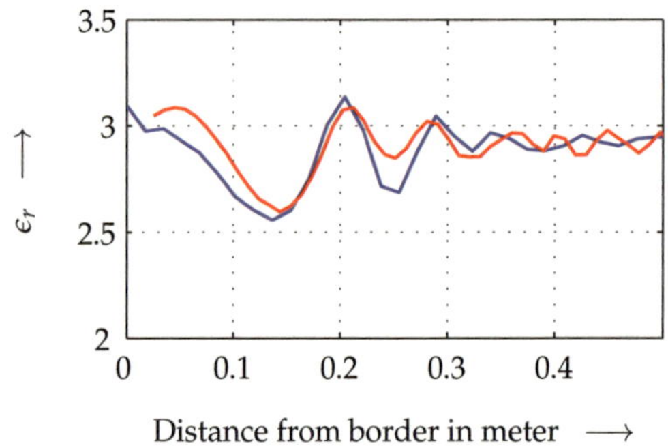

Figure 37. Permittivity (blue) and simulation values (red) as a function from the distance from the border.

The calculation of the permittivity is performed by the inverse application of the Fresnel-formulas. Assuming a material with a relative permeability $\mu_r = 1$, the expressions for the calculation of the relative permittivity ϵ_r and the emissivity e can be written as follows, with E as the received electric field strength for each polarization:

$$\epsilon_r = \left(\frac{\sin^2 \theta_i \left(\frac{E_\perp}{E_\parallel} - 1 \right)}{\cos \theta_i \left(\frac{E_\perp}{E_\parallel} + 1 \right)} \right)^2 + \sin^2 \theta_i \quad \text{and} \quad e = 1 - \left| \frac{1 - \sqrt{\epsilon_r}}{1 + \sqrt{\epsilon_r}} \right|^2 . \tag{16}$$

Here, it is important to note that the given expression for the emissivity e is valid for a straight monitoring of the hot spot. The additional information about the hot spot dimension and distance e.g. to a radiometer can be supplied by the UWB radar. So, an estimation of the hot spot temperature is possible.

The ellipsometry method allows an accurate characterization of plane surfaces with the main restriction to measure at a distance of at least 25 cm to any corner. The estimation of the permittivity of a small object or of objects with a complex shape is significantly influenced by effects of diffraction and scattering. As an example in Fig. 37, the deviation of the estimated relative permittivity of a MDF wall (with relative permittivity of approximately 2.9) is plotted as blue line over the distance from the antennas to the corner of the wall. In order to overcome this restriction, the UWB ellipsometry method is used in a combination with the object recognition process using the imaging methods as described later in this chapter. The distortions of the estimated permittivity values, which arise due to the diffraction effects, are then simulated (red curve in Fig. 37) and the calculated patterns of the permittivity curve are then compared with the corresponding measured pattern. For a first investigation, a simplified simulation algorithm (designed for online measurements) to consider the effects of reflection and diffraction for canonical 2 D objects was implemented. The results show that an accurate estimation of the dielectric characteristics of small objects is possible, with an accuracy of about $\pm 3\,\%$ for typical indoor objects (e.g. composed of fiberboards or bricks) with dimensions greater than 10 cm.

The effect of the object surface structure on the material estimation was analyzed by measurements of bulk materials. For slight roughness, i.e. height deviations much smaller than the wavelength, there is almost no influence on the estimation of the permittivity. For surfaces with a roughness in the order of the wavelength, the estimation of the permittivity has an uncertainty of less than 20 %. The surface roughness can be estimated by the analysis of the depolarization, i.e. measuring cross-polarized to the transmitted polarization. For the measured indoor materials with rough surfaces, the cross-polarized power is at least about 15 dB higher than for flat surfaces.

7.4. Pulse separation

The fundamental problem common to all super-resolution approaches is the precise extraction of the round trip time of UWB pulses. While this approach can easily be performed for single reflection measurements, things become challenging when the distance between multiple scattering centers drops below the range resolution. Constructive and destructive interferences are caused, and the shape of the resulting superposed pulses is distorted massively. Common algorithms for this purpose were analyzed, evaluated and extensively tested under various circumstances. In most cases, they can hardly resolve richly interfered pulses which overlap almost the whole pulse width or have vast computational load. Often, to some extent a priori information is necessary (e.g. the number of pulses to be separated), otherwise these algorithms suffer from inflexible termination conditions or need huge post-processing.

Within the CoLOR project a novel wavefront extraction algorithm called Dynamic Correlation Method (DCM) was proposed, [51]. The DCM is based on a correlation search using a set of two differently shifted reference pulses. Thus, the resulting correlation coefficients are no more just a function of one temporal parameter but rather of two parameters which result in a matrix of correlation coefficients. DCM does not ignore the interfering signature of backscattered pulses and provides a pair of pulses taking the interference pattern of them into account. Additionally, it terminates adaptively to different power levels which enables the detection of weak reflections and avoids post-processing. For a further detailed description and a comparison with alternative methods, see [51].

7.5. Short range imaging algorithm

In [28] , an imaging algorithm called Range Point Migration (RPM) was proposed that utilizes fuzzy estimation for the Direction Of Arrival (DoA). It extracts a direct mapping by combining the measured distance of the wavefront with its DoA. It realizes a stable imaging of even complex objects and requires no pre-processing like clustering or connecting discontinuous wavefronts. The 2D RPM was introduced in [28] for a planar sensor track nearby the object which allows either only a limited image of the lateral region of the object or requires huge scan distances. Consistently, the back region of the object is not scanned and, hence, not imaged. In [28] the RPM was extended to 3D where the sensors are placed on a planar surface in front of the 3D object. However, this 2D sensor track, too, is not capable of a full perspective of all stereoscopically distributed voxels (volume pixel) of 3D objects.

In CoLOR, the RPM was extensively analyzed, validated with numerous measurements in different scenarios and further improved [49]. For a full perspective of the object, the sensor track and the antenna alignment need circumnavigation to extract entirely all stereoscopically distributed voxels. In order to reconstruct a full 3D object contour, the scan pattern of the sensors was modified and extended to a spatial scanning including the z-axis. Experimental validations were carried out based on complex test objects with small shape variations relative to the wavelength used (for results see Fig. 40).

The main principle of the improved RPM, which is called Fuzzy imaging henceforth, shall briefly be described on the basis of the following illustration of a simple scenario with 3 measurements.

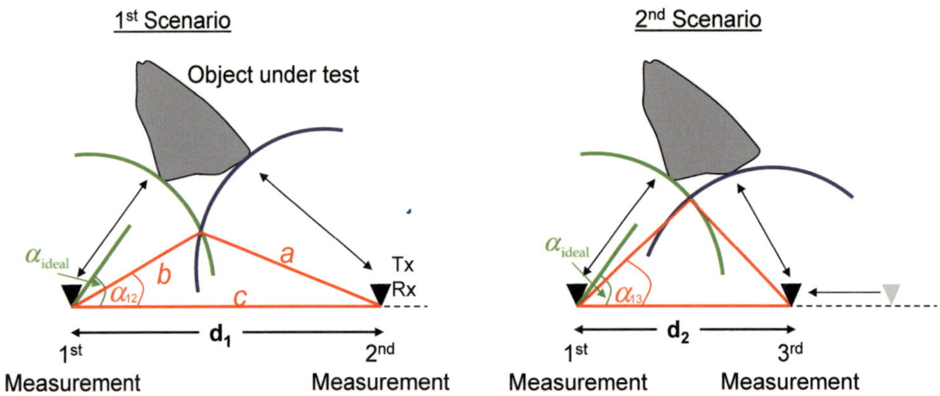

Figure 38. Measurement scenario to illustrate the principle of Fuzzy imaging

In the 1st scenario on the left side of Fig. 38, an object under test is measured at two antenna positions which are separated from each other by a distance d_1 which equals c. Both positions provide the time of flight and the distances a and b, respectively. With a geometric approach (law of cosines), the intersection points of the wavefronts as well as the angle α_{12} are extracted. Afterwards, shown on the right side of Fig. 38 a 3rd measurement is performed at a distance of $d_2 < d_1$ and provides the new angle α_{13}. Evidently, for $d_2 \to 0$, the antenna configuration converges to a mono-static configuration. In that case, $\alpha_{13} \to \alpha_{ideal}$ which would result in an exact image point when combined with the corresponding time of flight measurement.

The set of angles α_{1n} with its n neighboring wavefronts are called crisp set in terms of Fuzzy logic. Each of these scalar values could be regarded as one Dirac delta function at the corresponding angle value and is used for further processing. However, this would only make sense in the case of ideal point scatterers. Once the dimensions of the object are expanded or if the object consists of additional complex structures (e.g. edges and corners), it would result in erroneous image points. Fuzzy technology is applied here to compensate such influences and to still use a kind of convergence of nearby wavefronts. Therefore, each angle of the discrete crisp set is Fuzzyfied by a Gaussian membership function. Hence, the result does not only depend on one scalar but on a Fuzzy set around each scalar.

A scaling/clipping operation of the amplitude of the neighboring reflection point is performed to focus on strong reflections and scattering. Additional weighting is performed which scales the fuzzy sets as a function of the distance between the neighboring position and the one under test. Thus, it is ensured that the influence of sensors being further away is minimized. Afterwards, an accumulation of these differently weighted fuzzy sets is performed. The DoA can be estimated by a maximum defuzzyfication operator.

Within CoLOR it was investigated whether the standard deviation of the Gaussian membership functions are crucial parameters for the image processing. Depending on these parameters, the algorithm either extracts straight parts of the contour of the object, if it is larger than several wavelengths, or it extracts object features, i.e. scattering centers (edges and corners). A strong echo, i.e. a specular reflection is received only when the main lobe of the antenna is aligned to the normal of a smooth surface of the object. However, in the case of a circular track this occurs very rarely, if the cross-section of the object is not a circle. Scattering and diffraction effects overbalance immensely within a circular track, even more if the scan track is spread over a large circular arc. To overcome this problem, weak echoes which spread spherically from the edges can be recorded from any line of sight position. For detailed information, an extensive discussion, as well as the extension to 3D imaging see [49].

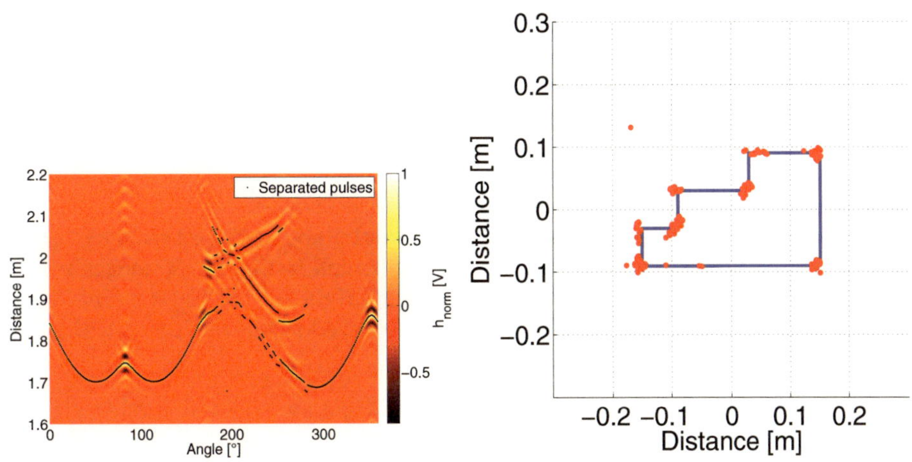

Figure 39. Radargram with extracted wavefronts on the left side and corresponding object on the right with extracted image points.

On the right side of Fig. 39 the object under test is depicted in blue. This object was scanned on a circular track with 1° grid resulting in the radargram shown on the left side. The wavefronts which are extracted with the DCM are also plotted in the radargram. With the previously discussed Fuzzy imaging the red image points shown on the right side are extracted.

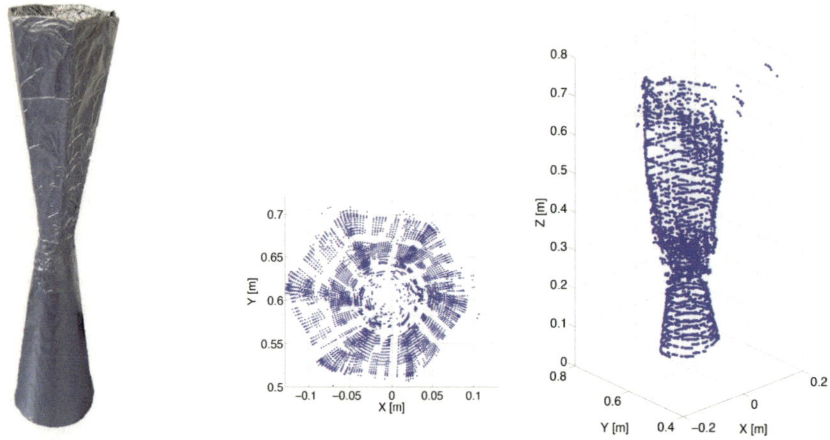

Figure 40. Photograph of an 3D test object on the left. Extracted 3D Radar image with Fuzzy imaging in the middle and top view on the right.

7.6. Exploitation of polarimetric diversity gain

The exploitation of polarization diversity in radar application provides additional information compared to mono-polarized sensing techniques. This polarization diversity gain enhances the efficiency of object classification according to the information contained in the backscattered signal. Hence, additional characteristics of objects such as shape, details of surface structure and orientation are gathered which may remain invisible for mono-polarized systems. However, in the literature polarization characteristics are rarely considered and most approaches use mono-polarized EM-waves. By the exploitation of polarization diversity the performance can be increased significantly.

Unpolarized electromagnetic incident waves on an object are diffracted or scattered in all directions. The spatial distribution of scattered energy depends on the object geometry, material composition, the operating frequency and polarization of the incident wave.

7.6.1. Polarization diversity gain in short-range UWB radar object imaging

Investigations and results shall be demonstrated by a complex edged 2D object with 6 corners. This object is measured on a circular track at a 1° grid which results in 360 measurements. The contour of the object can easily be recognized on the right side of Fig. 41. The sensors are the previously introduced crossed Vivaldi antennas embedded in PTFE which allow full polarimetric measurements.

The imaging algorithm used in this work is Kirchhoff Migration (KM). KM relies on some form of coherent summation, which means that a pixel of the radar image is produced by integrating the phase-shifted radar data of the field amplitude measurements at each antenna position. KM image spots of high intensity correspond to the scattering centers of the object. The image contrast is higher with increasing number of recorded impulse responses at different antenna positions. Here, the object is of 1 m height with about 1 m distance to the object in a bi-static configuration with 0.25 m distance between the transmitter and receiver. Actually, the object is a column with no variation in the height. Thus, it has vertical predominant directions causing stronger reflection in co-polarization or VV, respectively (notation: the first index indicates the polarization of the transmitter, the second index the one of the receiver).

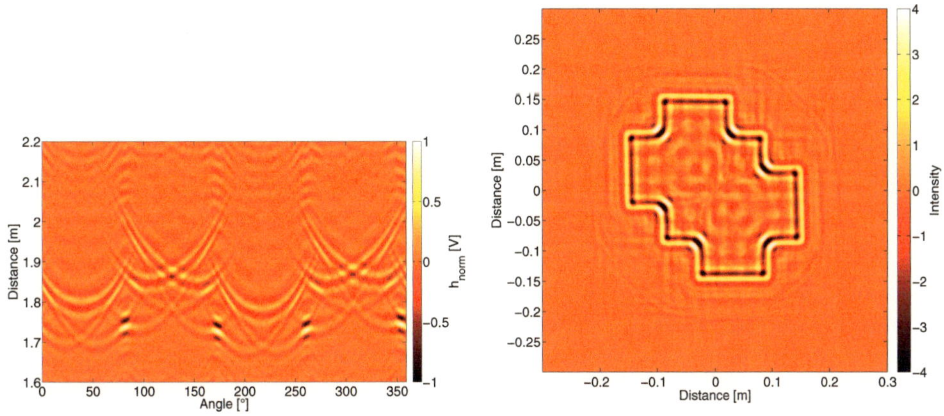

Figure 41. Radargram of the object under test with cross-polarization and 45° rotated antennas on the left and the extracted KM image from this radar data.

However, this object has more or less parts which imitate the scattering and reflection characteristic of flat plates or 0°- dihedrals. In the field of polarization research it is well known that dihedrals have strong polarizing effects. For example, a 0°-dihedral (the angle between the fold line of the dihedral and the vertical axis) has only co-polarized components, whereas a 45°-dihedral has only cross-polarized components. Therefore, dihedrals are especially suitable for calibration in polarimetric measurements.

In order to exploit polarimetric diversity gain a 45° shift is missing in the radar link [52]. As mentioned before, this would actually depolarize the wave. However, by rotating both antennas by 45°, the scattering characteristic of the object edges are comparable to 45°-dihedrals. Hence, in Fig. 42 both sensors are positioned diagonally. Using such a rotated configuration a cross-polarized measurement was performed with cross polarization.

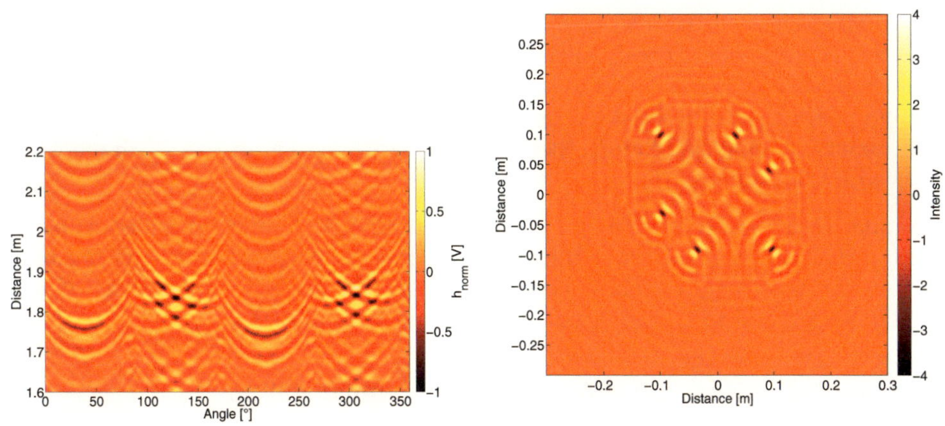

Figure 42. Radargram of the object under test with cross-polarization and 45° rotated antennas on the left and the extracted KM image from this radar data.

In Fig. 42, it can clearly be seen that only the corners are focused in the resulting image but not the flat structures. The reason for this effect is that a flat plate does not depolarize but a 45°-dihedral does.

Thus, the detection capability of a UWB radar system can be improved by exploiting polarization diversity. Under certain circumstances, the radar images detect object features which would have remained invisible in mono polarized radar systems. Supplementary information about the contour, orientation and dimensions of the object can thus be obtained, which upgrades super-resolution UWB radar significantly.

7.6.2. Polarimetric investigation of bulk goods with rough surfaces

One possible discrimination criterion between smooth and rough surfaces is to exploit specific depolarization effects. Rough in this context means that the standard deviation of the outer surface height distribution is in the range of some wavelengths of the operating carrier frequency, i.e. 9 GHz in this case. To obtain results which are independent both of material and shape, extensive measurements were performed.

Four metallic objects with smooth surfaces were used for the investigations, 2 objects with a square cross section of different size, an object with isosceles triangular cross section, and an object with a rectangular cross section. Details are shown in Fig. 44. For comparison with rough surfaces 4 polystyrene bins were built which were filled with bulk goods made of chunky gas concrete, chunky sand-lime brick, medium density fiberboard (MDF) blocks with 0.01 m edge length and M16 × 25 mm screws. Due to the material composition, the polystyrene bin itself has a vanishing radar cross section so that reflections caused by the bin are negligible.

For depolarization investigations, these 8 objects under test were scanned in both co- as well as in both cross-polarized configurations on a circular track with a radius of 1 m at a 1° grid. The depolarization is expressed by the relation of signal powers P_{VH}/P_{VV}.

In Fig. 43 the results for an object with a square cross section are compared with those of bulk good objects. Expectedly, objects with a smooth surface depolarize least, i.e. about -20 dB in the mean of all measurements. This complies with the 20 dB cross-polarization suppression of the antenna characteristic. The depolarization is least when an edge with predominantly vertical orientation is illuminated. Gas concrete and MDF are the bulk goods which depolarize most. The higher the permittivity is the more the bulky material depolarizes, i.e. sand-lime brick with about 8 dB more and the screws which depolarize most with over 10 dB more than the objects with a flat surface.

So UWB radar seems to be capable of discriminating objects of different materials by the roughness of their surface, subject to the condition that the height deviation is not much smaller than the wavelength. These results highlight the superior capabilities of fully polarimetric systems and recommend their use in future radar systems.

7.7. Object recognition for full and restricted circumnavigation

The object recognition (OR) method proposed in this work is part of the previously introduced super-resolution radar imaging system. The investigated objects and the reference alphabet derived from these consist of simple canonical and some polygonal complex objects shown in Fig. 44.

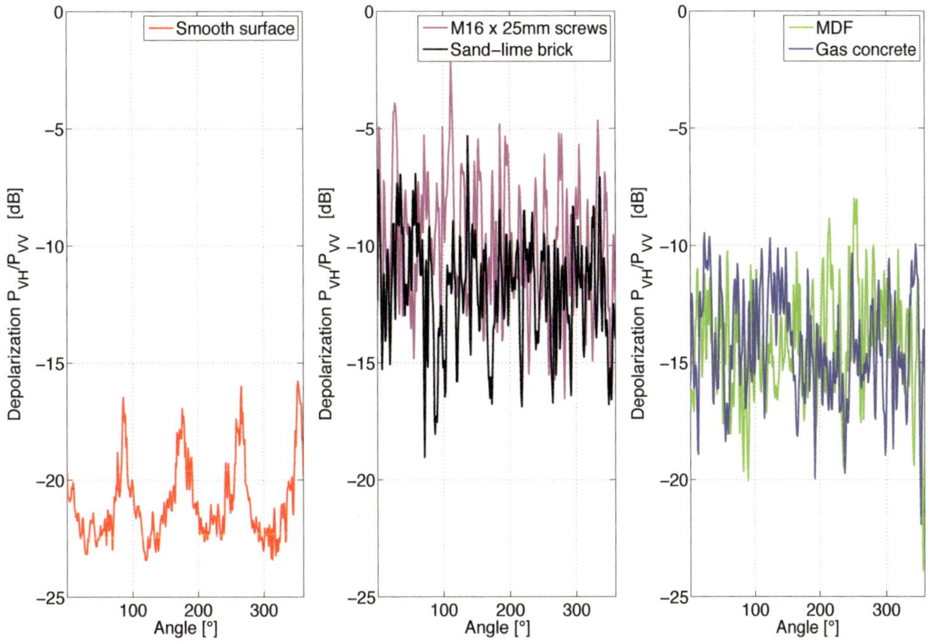

Figure 43. Results of the depolarization investigation of the object with square cross section, with 0.3 m edge length and the bin with same inner dimension, respectively. The depolarization coefficient P_{VH}/P_{VV} of the smooth surface is depicted on the left. The same coefficient is shown of the bulk goods in the middle and on the right side.

As mentioned above, the OR algorithms in [50] and [48] yield very robust results with the method of moment invariants and Fourier descriptors, which was proven for images obtained on complete circular tracks around the objects. However, in many cases such complete tracks are not possible as they lead to unfinished object images. To perform an OR also in such situations the method of Curvature Scale Space (CSS) was applied due its ability to robustly recognize contour parts [35].

7.7.1. Object recognition algorithm with the curvature scale space

The CSS representation is invariant against rotation of objects as this causes a circular shift of the CSS which has no effect on the recognition process, since the CSS of an object under test is compared by a correlation to the CSS of all reference objects. Moreover, CSS is highly robust against noise as most of its influence is compensated for to some degree by a smoothing Gaussian filter. Another property of the CSS is that it retains the local properties of shapes. Each peak of the CSS corresponds to a concavity or a convexity. A local deformation of the shape causes a change just in the corresponding local contour of the CSS image. Thus, a restricted curve can exactly be found in the CSS of the whole curve. Moreover, the absolute value of a CSS peak indicates the curvature radius, and the algebraic sign of the peak indicates whether the curve is concave or convex.

In Fig. 45 a plane curve with 8 convex or concave parts is drawn in blue. The Gaussian smoothed curve is drawn in red. On the right side of Fig. 45 , the corresponding curvature of the smoothed red curve is plotted.

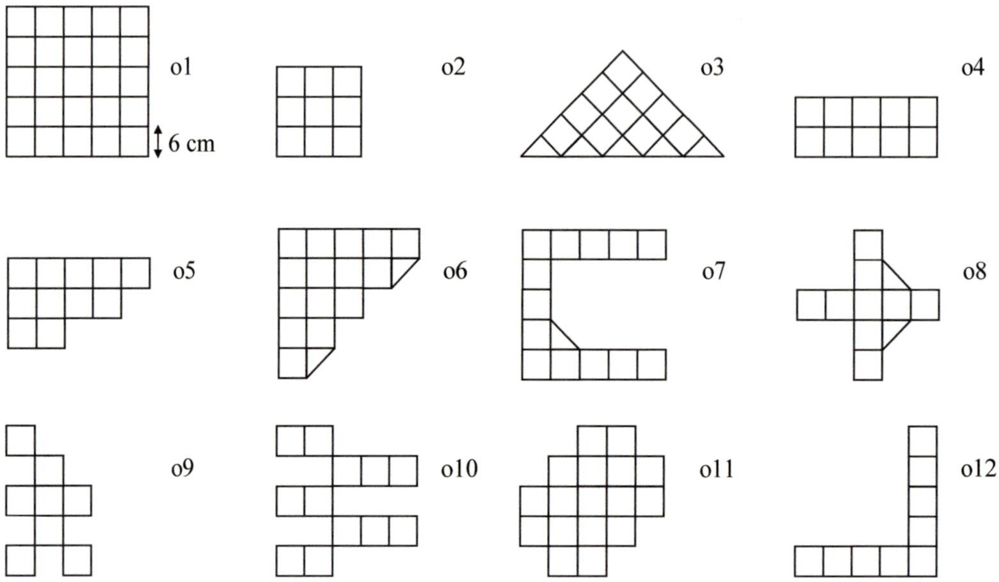

Figure 44. Cross-section of the 12 reference objects. The first 6 objects o1 - o6 were used for experimental OR investigations.

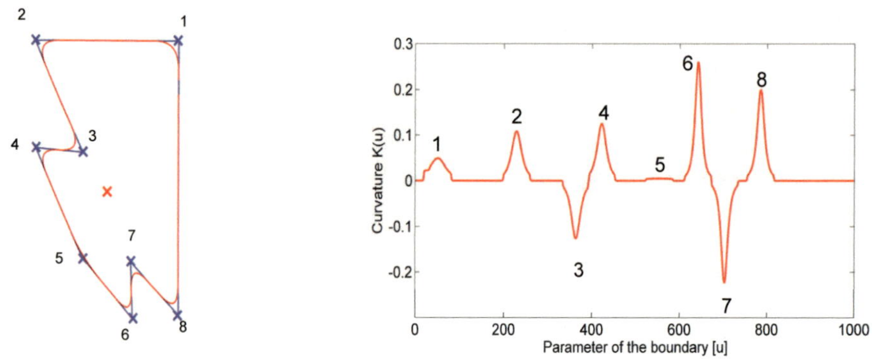

Figure 45. A plane curve in blue with its Gaussian smoothed version on the left, and the curvature of the smoothed curve on the right.

7.7.2. Object recognition performance test

The previously discussed OR algorithm based on the comparison of CSS was applied to Objects o1 - o6 in a vast measurement series. Every object under test was measured for 400 different orientations and different locations with an offset up to 0.2 m for a full circular track, a restricted track of 270° and 180°, respectively.

In the evaluations, every CSS of an object under test was compared with all 12 reference CSS data by a correlation function. The global maximum of the 12 correlation functions then indicated the recognized object.

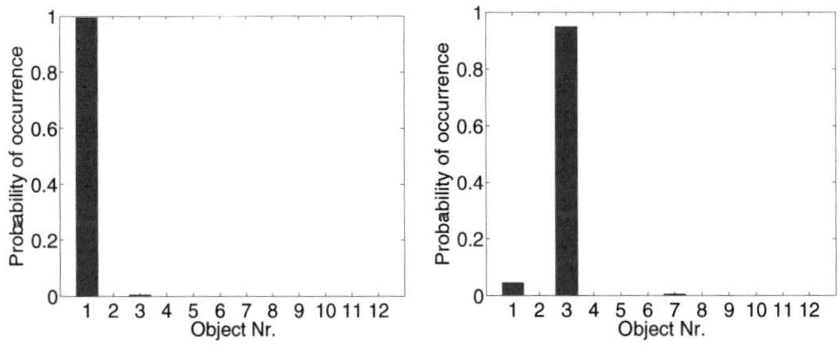

Figure 46. Discrete probability density function of the recognition process of object o1 for 270° track (left) and 180° track (right).

Figure 46 shows that object o1 can clearly be recognized even with the restricted track of 270°. Only for the 180° angle restriction, object o3 is recognized instead of o1 which is not unexpected since o3 equals a part of o1.

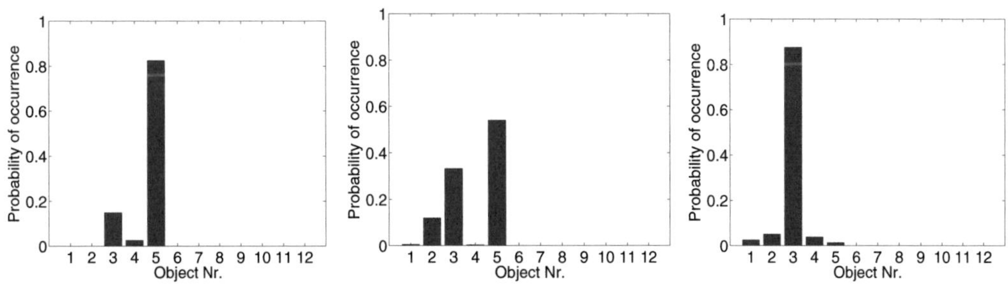

Figure 47. Discrete probability density function of the recognition process of object o5 for a full track (left), 270° track (middle) and 180° track (right).

Figure 47 shows that object o5 can be recognized in about 80% of all cases for a full track. This value drops to about 60% when the track is restricted to 270°. For 180° track o5 cannot be recognized anymore and is confused with o3.

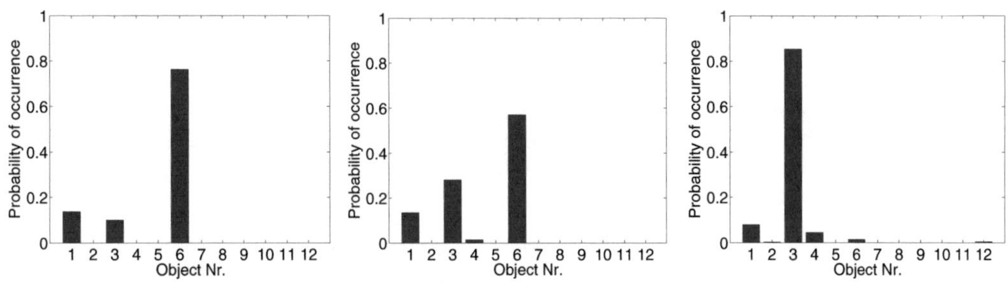

Figure 48. Discrete probability density function of the recognition process of object o6 for a full track (left), 270° track (middle) and 180° track (right).

Object o6 has approximately the same recognition characteristic as o5, as shown in Fig. 48. As both edge lengths of o6 are 0.3 m, o1 is falsely recognized as a certain part of its cross section has the same form as o5 or o6.

7.8. Transfer onto the mobile robot

Research investigations and algorithms had to be transferred onto a new mobile platform to demonstrate the essential results obtained in CoLOR. Many challenges in the area of robotics, localization and robot motion resulted. Additionally, an optimal alignment of the sensors and robust UWB radar sensing conditions had to be taken into account for the motion of the robot. The motion algorithm for the robot when moving around the objects are designed to be contour adaptive, exhibit collision avoidance features, and use the shortest tracks through a room while maintaining optimal antenna alignments.

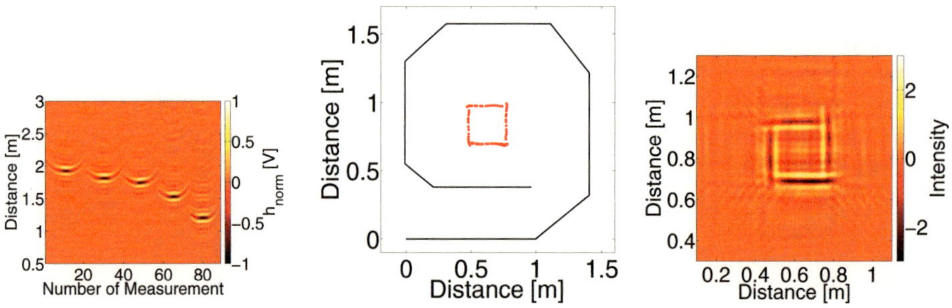

Figure 49. Example of the inspected object o1. The obtained radargram (left), the wavefront-based imaging of the object with robot track (middle) and the same object with a migrated image (right).

8. UWB Imaging

8.1. Imaging in distributed multi-static sensor networks

Although imaging in distributed multi-static sensor networks results in a rough image of the environment, its quality is usually better than the quality of the image obtained by a single bat-type sensor autonomously operating in an unknown environment. Sensor networks offer more diverse information for imaging algorithms. The resulting quality of the image greatly depends on the number and the positions of the illuminating and observing nodes. If there are enough illuminating and observing nodes available simultaneously, then the instantaneous information can be used even for imaging of time variant scenarios with moving objects. For imaging of static environments, one moving observer is enough. The moving observer collects data, and the image is built sequentially and improved gradually by an imaging algorithm. The time domain imaging algorithms are referred to as Kirchhoff migration. They rely on Born's approximation, which presumes undisturbed ray-optical propagation [9]. Their basic principle is depicted in Fig. 50 (a).

The illuminator transmits a signal at the fixed point $[x_T, y_T]$. At the variable position $[x_{Ri}, y_{Ri}]$ the receiver collects the impulse responses. Assuming single bounce reflection and the propagation velocity v, the echo reflected from an object situated at the position $[x_o, y_o]$ can be found at the time delay $\tau_i = (r_T + r_{Ri})/v$ in the measured impulse response $R_i(\tau)$. One observation determines an ellipse around the transmitter-receiver pair. In order to focus the image, observations from different positions must be fused. Conventional migrations propose to fuse the observations by a simple summation

$$o(x_o, y_o) = \frac{1}{N} \sum_{i=1}^{N} R_i(\tau_i) \tag{17}$$

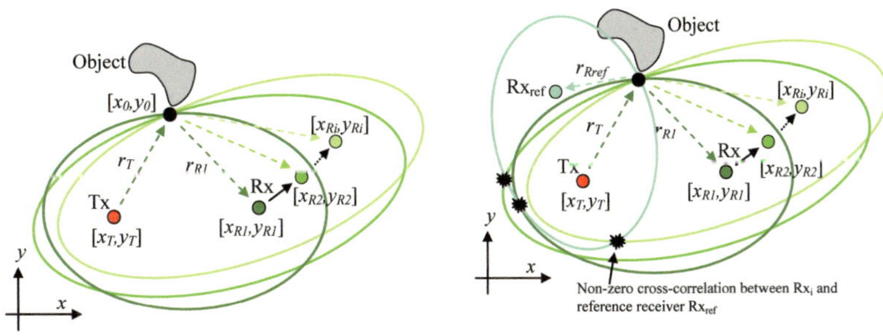

(a) Kirchhoff migration (b) Cross-correlated imaging

Figure 50. Principle of imaging algorithms.

where N is the number of observations available for the data fusion. A disadvantage of this fusion is indicated in Fig. 50 (a). The signal energy does not only cumulate at the desired pixel. Other local maxima arise at every position where ellipses are crossing. Moreover, the elliptical traces are added to the focused image as well. Both effects decrease the image quality. In [73] we have proposed a data fusion method which reduces these artifacts and enhances the image quality. The algorithm is based on the cross-correlation of multiple observations. The spatial scheme for the cross-correlation algorithm is indicated in Fig. 50 (b). The observation from the reference receiver Rx_{ref} is cross-correlated with each observation of the moving receiver. The cross-correlation results in non-zero values only at positions where the ellipse related to the reference receiver crosses the ellipse related to the moving receiver. The non-zero values which result from the cross-correlation are summed by the data fusion algorithm according to

$$o(x_o, y_o) = \frac{1}{N}\sum_{i=1}^{N}\int_{-T/2}^{T/2} R_i(\tau_i + \zeta)R_{ref}(\tau_{ref} + \zeta)d\zeta \tag{18}$$

where $\tau_{ref} = (r_T + r_{Rref})/v$ is the time delay which is related to the echo measured by the reference receiver, and T is the pulse duration of the autocorrelation function of the stimulation signal. In order to improve image quality, the number of reference receivers can be increased. The performance of this imaging algorithm strongly depends on the spatial arrangement of the cross-correlated receivers. If the arrangement of the receivers is disadvantageously chosen, the performance of this algorithm tends toward the conventional migration described by (17). If the spatial arrangement is selected in an optimum way the performance of the cross-correlated imaging outperforms the conventional algorithm. In order to illustrate the differences between the conventional and the cross-correlated algorithm, data simulated by the ray tracing algorithm [60] were used for the data fusion. The simulated scenario consisted of a sensor node which moved through a rectangular room of size 8 m by 9 m. The inspected room contained seven objects, namely, one large distributed object (some meters) and six smaller objects (decimeters). The sensor node had one transmitting and two receiving antennas and operated in the frequency band from 4.5 GHz to 13.4 GHz. The antennas were directional, therefore, the moving sensor scanned the room by a full turn at selected positions. Almost 10 thousand impulse responses were simulated for this scenario. Figure 51 (a) shows the result obtained by the conventional migration algorithm. The focused image contains intersecting ellipses. However, as indicated before, the intersections are not exclusively at the target positions. The ellipses also intersect at positions where no target is situated and even traces of these ellipses make the interpretation of the image difficult. Figure 51 (b) illustrates the image of the same scenario obtained by the cross-correlated algorithm. The color scaling of both images in Fig. 51 (a) and (b) is the same. The reduction of image artifacts by means of the cross-correlated algorithm is evident. Elliptical traces were significantly reduced. This helps to identify the inspected environment in this figure more clearly. A more detailed discussion on the cross-correlated imaging with measurement examples is given in [74] and [75].

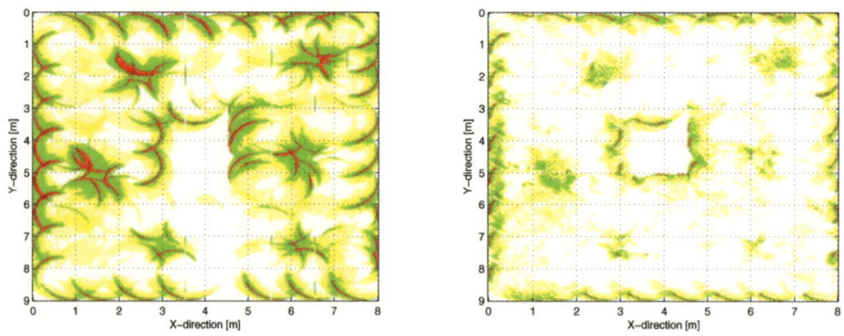

(a) Range migration (b) Cross-correlated imaging

Figure 51. Imaging in sensor networks.

8.2. Stationary target detection in an unknown indoor environment

Indoor stationary object detection in an unknown environment is important for many civilian and military applications, such as indoor surveillance, search and rescue operations, logistics, security and so on. Compared with target detection in a known environment, it presents some challenges:

- *The detection takes place in an unknown environment (e.g. an environment after disaster).* Thus, it is not possible to probe the environment before the presence of target. A range of techniques based on *"a priori"* information of the background cannot be utilized (e.g. [41, 64]). Additionally, the statistical distributions of clutter and noise are sometimes unknown, leading to further complications in the detector design.
- *The targets concerned are stationary with respect to the background.* There is no distinct speed difference between the targets and the background. Hence, it prevents the application of the motion-based detection techniques, such as Doppler based approaches, subtraction (or cancellation) between sequence snapshots, etc [58].
- *It is a highly cluttered environment.* The targets (objects of interest) are typically surrounded by clutter (objects, which are not of interest, as shown in Fig. 52). In this case, the response of targets is not always stronger than clutter. In other words, the Signal to Interference and Noise Ratio (SINR) of the system is not always high enough to ensure a reliable detection.

8.2.1. Signal enhancement

Due to the challenges mentioned above, a possible detection scheme is given in Fig. 53. In the scheme, the objective of the first step, "signal enhancement" which is realized by a *"time-shift & accumulation"* operation, is to raise the SINR of system, and to transform the unknown environment into a Gaussian clutter and noise environment so that the detector could be

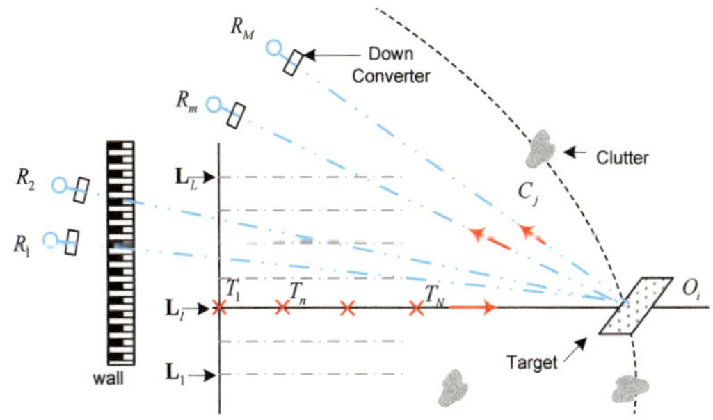

Figure 52. Measurement configuration. $T_1...T_n...T_N$ are different transmission positions on the track $\mathbf{L}_l$. "$R_1...R_M$" are sparsely spaced receivers with different reception angles. The dashed curve is an ellipse segment with the foci R_m and T_n.

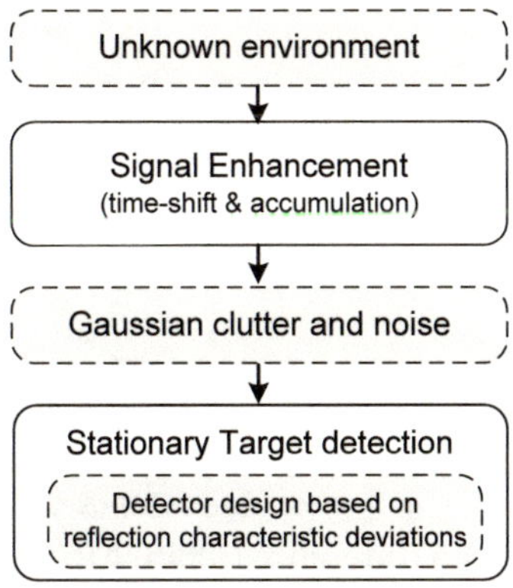

Figure 53. Flow chart of the algorithm .

designed based on Gaussian clutter and noise. Mathematically, the *"time-shift & accumulation"* operation is described as

$$\sum_{n=1}^{N} s_{Rm,Tn}(t + t_{Tn,T1}^{diff}) \tag{19}$$

where $s_{Rm,Tn}(t)$ is the signal received by the receiver R_m with respect to the transmission position T_n, and $t_{Tn,T1}^{diff} = |\mathbf{T}_n - \mathbf{T}_1| / c$ is the time-delay to be compensated for. $\mathbf{T}_n$ and $\mathbf{T}_1$ are the position vectors of T_n and reference transmission position T_1 , respectively. In the operation,

- the responses of the objects (O_i) in the direction of interest (for example the direction of $\mathbf{L}_l$ in Fig. 52, are time-shifted, aligned and then accumulated. It is a coherent operation in which the parameter $t_{Tn,T1}^{diff}$ is used for time-delay correction. Hence, responses are enhanced by N times compared to the case of single channel data.
- The responses of objects (C_j) out of the direction of interest, are non-uniformly time-shifted, disturbed and then accumulated. It is an incoherent operation. As a consequence, for a certain time-slot, the responses from objects located on different ellipses (or ellipsoids) are non-uniformly time-shifted and accumulated. Hence, these responses are attenuated compared to the output of the coherent operation above, [70].

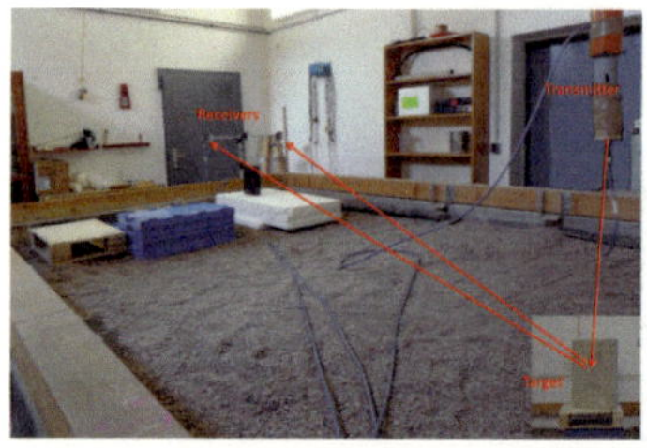

Figure 54. Measurement environment. R_3 and R_4 are placed behind the wall.

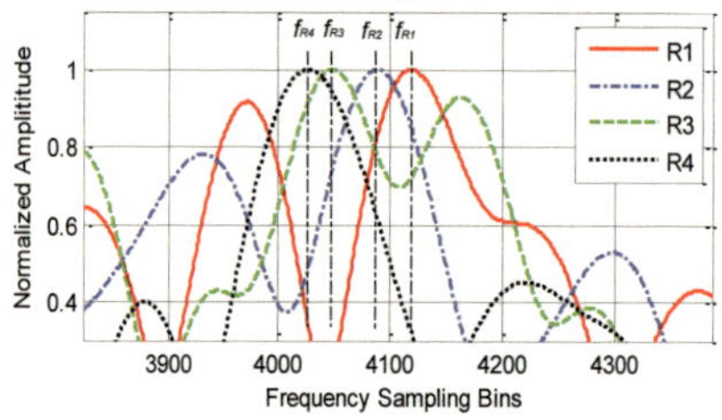

Figure 55. Measured spectrum-shift

8.2.2. The detection principle

For a detection problem, generally, detection algorithms are designed based on the differences (or deviations) between the targets and the background (clutter, noise). The ability to distinguish objects depends on how much their properties (e.g. electromagnetic properties, motion properties) deviate from the properties of the background. For the concerned stationary target which has no speed differences relative to the environment it can be detected based on the reflection characteristic deviations between the target and the background. Here, we take a planar-surface target as an example to illustrate the principle of indoor stationary target detection.

Consider a detection scenario as shown in Fig. 52 and Fig. 54. The transmitter moves along track $\mathbf{L}_l$ to probe the environment, and to enhance the signal. The target is a concrete brick-like object with a square plane-surface ($50cm \times 50cm$). It is located at $x = 24cm$ on the track $\mathbf{L}_l$, with an orientation angle of 60°. The receivers R_1, R_2, R_3 and R_4 receive the scattered signals from the surroundings. They are located at different directions with respect to the target. The transmitter-target-receiver angles are 13.9°, 20.3°, 26.3° and 31.7°, respectively. The antennas of the transmitter and receivers are omnidirectional and horn antennas, respectively. The receivers R_1 and R_2 are in the same room with the target and the transmitter, while the receivers R_3 and R_4 are placed behind a 20cm-thick concrete wall.

Theoretically, it can be proved that if the target is a planar-surface object with diffuse reflections, after passing through the down-converter and certain mathematical transforms, the received signals from different directions would have a spectrum-shift, which is given by $\Omega_{j,i} = \omega_c(\gamma_{j,i} - 1)$ under the illuminating geometry in Fig. 52. Here, the subscripts i and j are the indices of the receivers located at the different directions, ω_c is the angular central frequency, and $\gamma_{j,i}$ is a parameter associated with the reflection angles,

$$\gamma_{j,i} \approx 1 + \left(\sin\theta_{Rj} - \sin\theta_{Ri}\right) / \left(|\mathbf{V}_{bi}| \sin\theta_{bi}\right) \tag{20}$$

where, θ_{Ri} and θ_{Rj} are the reflection angles of the receiver R_i and R_j,respectively. $\mathbf{V}_{bi}$ is the bisector of the angle $\measuredangle R_i X T_1$. θ_{bi} is the reflection angle of $\mathbf{V}_{bi}$. According to this principle, stationary plane-surface objects can be detected.

In practice, (20) is demonstrated in Fig. 55, where the spectra of receivers R_1-$R4$ are measured based on the scenario of Fig. 54. It can be shown that $\frac{f_{Ri}-f_{R1}}{f_{Rj}-f_{R1}} \approx \frac{\gamma_{i,1}-1}{\gamma_{j,1}-1}$, where $f_{R_i}, f_{R_j} \in [f_{R_1}, f_{R_2}, f_{R_3}, f_{R_4}]$ are the frequencies at which the maximum amplitudes are located. The differences between f_{R1}-f_{R4} indicate the spectrum shifts between R_1, R_2, R_3 and R_4, correspondingly.

8.2.3. Detection

According to the principle discussed above, if we take the observation of R_1, $\bar{s}_{R1,Tn}(\tau)$, as a reference signal, the observation of receiver R_j can be given by $\bar{s}_{Rj,Tn}(\tau) \approx \bar{s}_{R1,Tn}(\tau)\exp(-j\gamma_{j,1}\tau)$, where $\gamma_{j,1}$ is determined by the illuminating geometry according to (20). Based on this relationship, if we consider the effects of unwanted contributions due to clutter $\mathbf{c}$ and noise $\mathbf{n}$, the signal model for detection can be given as

$$\mathbf{y} = \bar{\mathbf{s}}^{ref} \otimes \mathbf{K} + \mathbf{c} + \mathbf{n} = \mathbf{x} + \mathbf{c} + \mathbf{n} \tag{21}$$

where $\mathbf{y}$ is an $N_{MN} \times 1$ vector. $N_{MN} = M \times N$, where M and N are the numbers of receivers and transmission positions, respectively. $\mathbf{x} = \bar{\mathbf{s}}^{ref} \otimes \mathbf{K}$, where $\bar{\mathbf{s}}^{ref} = [\bar{s}^{ref}_{T1}, \bar{s}^{ref}_{Tn}, ..., \bar{s}^{ref}_{TN}]^T$, $\bar{s}^{ref}_{Tn} = \bar{s}_{R1,Tn}(\tau)$, and $\mathbf{K} = [1, \exp(-j\gamma_{2,1}\tau), ..., \exp(-j\gamma_{M,1}\tau]^T$. The symbol $\otimes$ denotes the Kronecker product.

According to the Central Limit Theorem (CLT) of the probability theory, if a large number of clutter from different sources (scattered from different objects) is accumulated, the statistical distribution of the sum will approach a Gaussian distribution. As the scenario concerned takes place in a cluttered indoor environment which has many scatterers, we assume that the clutter $\mathbf{c}$ and the noise $\mathbf{n}$ can approach a Gaussian distribution due to the "*time-delay & accumulation*" operations given by (19). Hence, our detection problem simplifies to searching targets in Gaussian clutter and noise. We assume that the noise $\mathbf{n}$ and the clutter $\mathbf{c}$ are $N_{MN} \times 1$ independent zero-mean complex Gaussians with $N_{MN} \times N_{MN}$ known covariance matrices $\mathbf{M}_{c+n} = E[(\mathbf{c}+\mathbf{n})(\mathbf{c}+\mathbf{n})^H]$. $\mathbf{M}_{c+n}$ is a positive semidefinite and Hermitian symmetric matrix [63]. The superscript H indicates conjugate transpose of a matrix. Based on the signal model above, a matched filter detector is given as

$$\mathbf{h} = \mathbf{M}^{-1}_{c+n}\mathbf{x} = \mathbf{M}^{-1}_{c+n}(\bar{\mathbf{s}}^{ref} \otimes \mathbf{K}) \tag{22}$$

The probability of detection can be given as $P_d = 1 - \Phi(\frac{k_{th} - \mathbf{h}^H \mathbf{x}}{D})$, and the probability of false alarms can be given as $P_f = 1 - \Phi(\frac{k_{th}}{D})$. Here, k_{th} is the threshold for a likelihood test and $\Phi(x) = \frac{1}{\sqrt{2\pi}} \int_{-\infty}^{x} e^{-t^2/2} dt$. The parameter D is defined as $D^2 = \mathbf{x}^H \mathbf{M}^{-1}_{c+n} \mathbf{x} = N_{MN} \frac{x^2_{av}}{E[(\mathbf{y}-\mathbf{x})^H(\mathbf{y}-\mathbf{x})]}$, where x^2_{av} is the average signal power of $\mathbf{x}$. Generally, $\frac{x^2_{av}}{E[(y-x)^H(y-x)]}$ can be regarded as the SINR at the output of the detector [43].

In terms of the Neyman-Pearson criterion, the threshold k_{th} could be set as $k_{th} = D\Phi(1 - P_f)$, and the detection probability can be given as $P_d = 1 - \Phi(\Phi^{-1}(1 - P_f) - D)$.

9. Detection and localization of moving objects

In this section we describe methods for the detection and localization of several moving persons who do not have a tag or device attached to them. This is useful for applications where the targets being tracked are not expected to cooperate with the system.

A single UWB sensor suffers from narrow-band interferences as well as shadowing when detecting multiple targets. Often, the closest target can be observed best. Due to shadowing caused by the closest target, the other targets are usually invisible, although they lie within the coverage of the sensor. The closest target attenuates the electromagnetic waves that propagate toward the targets located behind it. Thus, the shadowed targets are almost impossible to detect by a single sensor node. Using a distributed network of UWB radars, the estimated target positions are refined by fusing the information available from the multiple sensors which are able to detect the targets. It also allows for the observation of the targets from different viewing angles.

A network of multiple static bat-type sensors distributed around the inspected area is used. Each bat-type sensor node is capable of autonomously detecting and localizing the present targets. The weak echoes of the targets are first detected in the backscattered signal, after which the detections are correctly assigned to the targets and finally, the target information from all sensors is fused together resulting in location estimates of the targets in the scenario. Here we describe two methods that can cope with this challenging task. The first one uses simplification assumptions of one target detection per sensor to combat the data association problem. It is described in Section 9.1. The second one is based on the probabilistic hypothesis density (PHD) filter for range estimation and position tracking and is described in more detail in Section 9.2.

In both methods, background subtraction is used for target detection [9, 21, 38, 71]. The echoes evoked by the moving targets are usually weak and must be detected in the presence of other strong multipath signals. The disturbing signals are usually time invariant and overlay echoes from the target. Therefore, one received impulse response is insufficient to separate them. However, due to their time-invariance these background signals can be estimated by a suitable estimation technique from a sequence of received impulse responses. The subtraction of the estimated background from the measured data leads to a signal where the weak target echo can be recognized (see Fig. 56(b)) and its range can be estimated more easily.

For the verification of the methods, a sensor network constellation as in Fig. 56(a) is used. Five UWB bat-type sensors are placed around a large foyer, and one bat-type sensor is placed behind one of the walls, directed towards the area of interest. Directional horn antennas with different size and quality are used, resulting in variable target detection performance between the sensors. A scenario with three moving persons was used. All three persons move in the area of interest during the whole measurement time. They move in a straight line from one wall to the opposite and back. The starting position of the targets is shown by a star in Fig. 56(a). The arrow signifies the direction of movement at the start of the measurement.

9.1. Low complexity method for the localization of multiple targets in sensor networks

The method presented in this section was proposed as a computationally efficient method for the localization of multiple targets in dense sensor networks. Each target can be usually observed by at least one sensor node. As can be seen in Fig. 56(b), the closest target

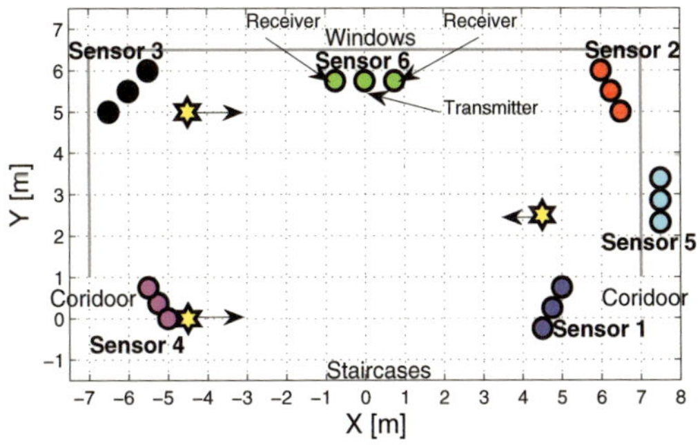

(a)

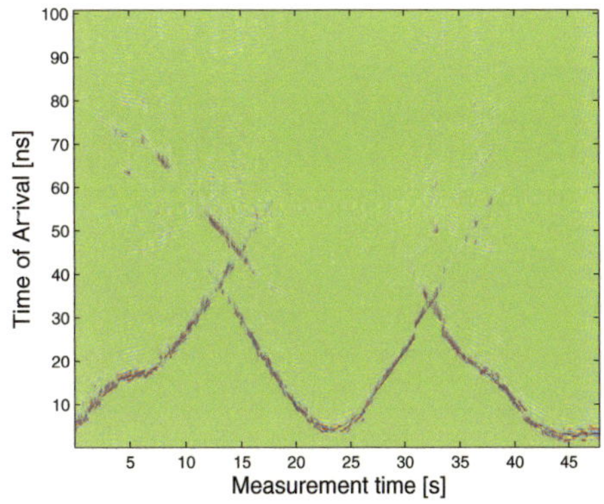

(b)

Figure 56. a) The verification scenario and b) the received impulse responses after background subtraction

to the sensor evokes the strongest echo, and almost completely shadows the other two targets when it is very close to the sensor. In order to simplify the localization algorithm, we make the assumption that each sensor node can observe only the closest target. This assumption allows the algorithms for multi-target detection such as the constant false alarm rate (CFAR) techniques [15], algorithms for data association, or multi-hypotheses tests to be skipped. This makes the proposed algorithm computationally efficient and suitable for real-time applications. On the other hand, this simplification restricts the number of targets that can be localized by the approach. A sensor network with N sensor nodes can localize up to N targets in real time. A more detailed description of this method is presented in [72].

9.1.1. Single target range estimation

After the time-invariant background signals have been removed from the received impulse response, the targets' ranges need to be estimated. The distance from the transmitter to the target and back to the receiver is considered as target range. The target range of the closest target can be estimated by using one of the various threshold-based approaches where the leading edge of the received signal is detected [10–12, 22, 56]. Threshold-based approaches offer simple techniques which detect the leading edge of a received signal by comparing the received signal magnitude or energy with a predefined threshold. The choice of an appropriate threshold is mandatory for a good performance of this estimator.

9.1.2. Localization

As in our scenario each sensor has two receivers, from each sensor node we have the estimated range of the closest target by the two receivers. The estimated ranges determine two ellipses whose focal points are determined by the locus of the transmitting antenna and the locus of the respective receiving antenna. The target position is determined as the intersection of the two ellipses within the area of detection of the respective sensor. The location estimates with respect to each of the six sensor nodes is shown in Fig. 57(a). Here, the estimates are represented by the same color as the respective sensor node.

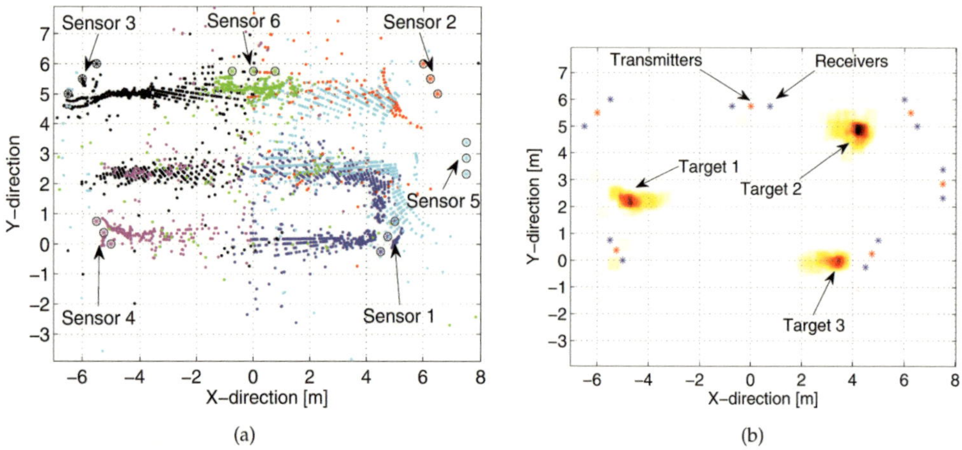

Figure 57. The target position estimates a) by each sensor and b) after data fusion

9.1.3. Data fusion

The location estimates from all sensor nodes must be fused together by an appropriate algorithm. We propose an imaging algorithm that does not require any data association. It creates a sampled image of the inspected area at a given time i which is stored in a matrix $\mathbf{P_i}$. Each element of the matrix corresponds to the spatial coordinates within the inspected area. The image is updated according to

$$\mathbf{P_i} = \alpha \mathbf{P_{i-1}} + (1 - \alpha)\mathbf{U_i} \tag{23}$$

where $\mathbf{U_i}$ is the innovation matrix and α is the forgetting factor. Thus, the new image estimate $\mathbf{P_i}$ takes a fraction of the previous estimate $\mathbf{P_{i-1}}$ and a fraction from the innovation matrix $\mathbf{U_i}$. The innovation matrix $\mathbf{U_i}$ maps the location estimates of all sensor nodes. The matrix $\mathbf{P_i}$ indicates moving targets in the environment as hot spots within the image.

An example of the data fusion is given in Fig. 57(b). It depicts one snapshot of the inspected area computed for a certain measurement time. Due to the exponential averaging each target appears in the snapshot like a "comet" with a tail. This tail indicates previous locations of the objects. Its length can be adjusted by adjusting the weighting factor α in Eqn. 23.

9.2. Localization and tracking of multiple targets in sensor networks based on the PHD filter

Unlike the method presented in Section 9.1, this method uses the range information related to all targets detected by each sensor. Traditionally, multiple targets are tracked using two-step approaches: data association of the observations to targets followed by single target tracking on the associated data. The traditional approaches are generally computationally expensive, thus, we use a less computationally expensive alternative based on random finite sets (RFS) [36, 37]. It allows for tracking of time-variant number of unknown moving targets in the presence of false alarms, miss-detections and clutter. An approximation of the Bayesian multi-target tracking represented by RFSs, by its first order moment leads to the PHD filter [27]. There are two implementations of the PHD filter, one based on a sequential Monte Carlo approach [65], and the other based on Gaussian mixtures [66]. Here, we use the

Gaussian mixture (GM) approach. We first use it for estimating the target ranges with respect to each sensor, and later for fusion of the target location estimates by all sensors. The method is explained in greater detail in [27].

9.2.1. Multiple target range estimation

In Fig. 58 we describe the processing done on a measured impulse response from raw measurement to range estimation. The impulse response presented is from a scenario with three moving targets. First the measured impulse response is shown in dark blue. As can be seen, the target echoes are non-detectable. After subtracting the time-invariant background (resulting signal shown in green) the echoes of the moving targets are detectable more easily. A Gaussian adaptive threshold constant false alarm rate (CFAR) method as in [15] is used for extracting the ranges from the resulting signal. The adaptive threshold calculated is shown in cyan. A Neyman-Pearson detector is used to discriminate between the noise and the target echo. The ranges extracted using the CFAR approach are shown as points in magenta.

The CFAR detector is not immune to clutter noise and false detections. As we can see in Fig. 58, four targets are detected by the CFAR method although there are only three targets in the scenario. In addition, there are also multiple detections per target, making it difficult to decide if there are multiple targets in the vicinity of each other, or it is only a single target. By using a GM-PHD filter we improve the target range estimates. The ranges extracted by the CFAR method are used as observations for the PHD filter, and a linear Gaussian model is assumed. The target detection and survival probability are assumed to be state independent. To estimate the number of targets and their states, first the Gaussian terms with low weights are pruned and the Gaussian terms that are within a small distance of each other are merged together. The number of targets is estimated by the number of Gaussian terms with a weight above a predefined threshold, and their state is represented by the mean terms of these Gaussian mixtures. The surviving Gaussian terms for the impulse response given in Fig. 58 are shown in black.

9.2.2. Target localization

As in Section 9.1.2, the target locations are computed using the ranges estimated by the two receivers of the sensor. As multiple targets are detected by each receiver, the range estimates from both receivers corresponding to the same target need to be associated. An intersection threshold T_s is defined for each sensor. The ranges of a target in the inspected area with respect to both receivers of a sensor s are first calculated as $r_k^{s,1}$ and $r_k^{s,2}$. The intersection threshold is then defined as the maximum possible absolute difference between these ranges such that the target lies in the area of interest.

$$T_s = \max_{k \in A} |r_k^{s,1} - r_k^{s,2}| \tag{24}$$

As the size of the inspected area is known or approximated by the sensor detection range, the intersection threshold is calculated prior to scanning the environment. The range association is done such that for each range estimate from the first receiver, we associate the range estimate from the second receivers which satisfies

$$|r_k^{s,1} - r_k^{s,2}| < T_s \tag{25}$$

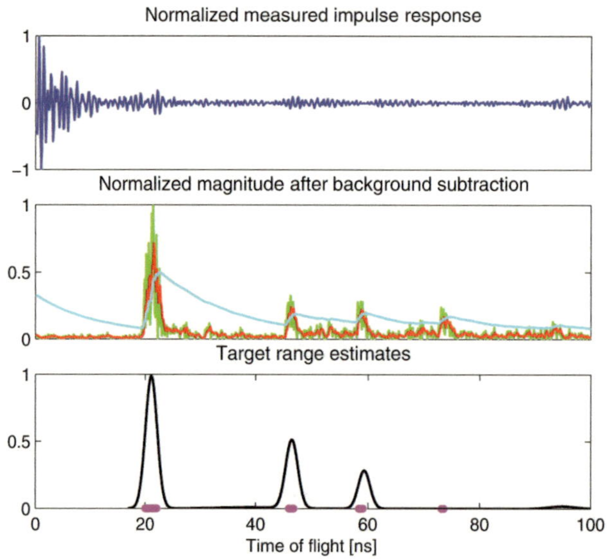

Figure 58. Target echo detection - measured impulse response (blue), normalized signal magnitude after background subtraction (green), CFAR test statistic (red), CFAR adaptive threshold (cyan), indices of detected targets by CFAR (magenta) and Gaussian mixtures representing the estimated target ranges (black) are shown

as a range estimate that corresponds to the same target. When multiple range estimates comply with this rule, the range estimate which results in the smallest absolute difference is chosen.

The target location is analytically calculated as the intersection of the ellipses defined by the associated range estimates. The location estimates with respect to each of the six sensor nodes is shown in Fig. 59(a), where the estimates are represented by the same color as the respective sensor node.

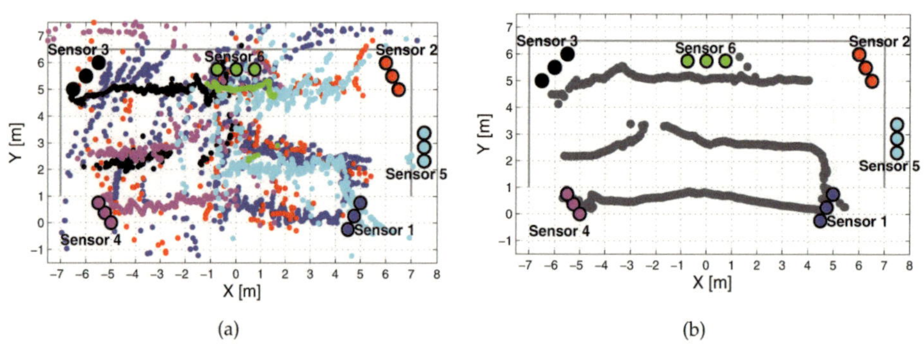

Figure 59. a)The target position estimates by each sensor and b) the target tracks after data fusion

9.2.3. Data fusion and target tracking

The location estimates from all sensors are fused together resulting in a single target location per target. The estimated target locations are not in a track form and contain a significant amount of noise due to small errors in the range estimation. A simplified GM-PHD filter is applied for fusing the target location estimates from all sensors. The location estimates are used to form the observation RFS for the GM-PHD filter. The target state is defined by the 2D target coordinates and velocity vector, $\mathbf{x} = [x\ \dot{x}\ y\ \dot{y}]^T$. The targets are identified using nearest neighbor association, and the results for the three-target scenario can be seen in Fig. 59(b).

Both methods presented in this section can be used for the localization of multiple non-cooperative targets using distributed UWB sensor network with static sensors as in the scenario described in Section 2.1 .

10. Conclusion

The CoLOR project was devoted to the recognition of unknown environments using UWB technology. This topic encompasses a number of partial challenges. In order to obtain a complex picture of some catastrophic scenario, like the detection of victims after some natural disasters, their location in collapsed buildings, the geometrical information and the status of the buildings, we derived new detection, localization and imaging algorithms. Their performance was analyzed on simulated data and data measured in realistic scenarios using UWB sensors. These UWB sensors are capable of real-time operation in MIMO configuration. This allows us to analyze the application of UWB sensor networks and cooperative approaches for the localization of sensor nodes within the network, for the localization of people, for the detection and monitoring of their live signs, and for the imaging of their surroundings.

It was shown that by using a mobile UWB radar with multiple antennas, it is possible in an efficient way to reconstruct the basic layout of rooms and the position of freestanding objects. The detected features are added to a map while at the same time the own position is estimated (SLAM). To minimize the computational cost and the number of measurements needed, simplified models for wave propagation and stochastic, dynamic state space estimators where enhanced. The method of data association proved to be most critical regarding the precision and reliability of the map.

Using this map as a-priori information, the detection, localization and the imaging of the objects within an indoor scenario can be performed using the developed localization and imaging algorithms. By knowing the location of the individual objects, the potential of UWB radar was fully tapped by obtaining super-resoluted local information about 2D as well as 3D complex objects (concerning the outer contour). The interior of objects was gathered by novel algorithms which are based on exact radiation patterns depending on the permittivity of the medium while showing low computational load. The obtained radar images are post-processed by means of object recognition algorithms designed for full, fragmented or restricted illumination to provide recognition of the object under test from a finite alphabet. By the adaption of classical ellipsometry to the UWB-range, an estimation of dielectric surface properties can robustly be performed even for small dimensioned objects with a size of a couple of wavelengths. In addition polarimetric measurements as well as polarimetric data processing were taken into account to obtain object features which may remain invisible in mono polarized systems.

In order to test and compare different algorithms and antenna arrangements for indoor UWB sensing and imaging, a realistic UWB multi-path propagation simulation tool was developed. The propagation model is based on a hybrid approach which combines the deterministic ray tracing method (based on geometrical optics and the uniform theory of diffraction) with statistically distributed scatterers. Verification measurements show that the new model delivers very realistic channel parameters like channel impulse response, azimuth spectra, and path loss. Thus, it is suitable for an application in UWB system simulations.

Further within CoLOR a flexible and dual-polarized UWB antenna array has been developed. The major challenges beside the huge bandwidth itself were to design antenna elements which are able to meet the requirements regarding size, pattern, beam-width, polarization and the location of the phase center (over frequency). Via switches it is possible to select and control the single elements of this array, which allows its adaption to the different localization, imaging and object recognition algorithms and applications in this project. Due to the dual polarized antenna elements, the possibility to take advantage of polarization diversity is given and demonstrated.

Our results show that the UWB technology and especially the cooperative approach that fuses diverse information from multiple sensors provide a big potential for safety, security and emergency applications.

Author details

Rudolf Zetik, Honghui Yan, Elke Malz, Snezhana Jovanoska,
Guowei Shen and Reiner S. Thomä
Ilmenau University of Technology, Germany

Rahmi Salman, Thorsten Schultze, Robert Tobera and Hans-Ingolf Willms
University Essen-Duisburg, Germany

Lars Reichardt, Malgorzata Janson, Thomas Zwick and Werner Wiesbeck
Karlsruhe Institute of Technology (KIT), Germany

Tobias Deißler and Jörn Thielecke
Friedrich-Alexander University Erlangen-Nürnberg, Germany

11. References

[1] Adamiuk, G., Beer, S., Wiesbeck, W. & Zwick, T. [2009]. Dual-orthogonal polarized antenna for UWB-IR technology, *Antennas and Wireless Propagation Letters, IEEE* 8: 981 –984.

[2] Adamiuk, G., Janson, M., Wiesbeck, W. & Zwick, T. [2009]. Dual-polarized UWB antenna array, *IEEE International Conference on Ultra-Wideband, ICUWB'09*, Vancouver, Canada, pp. 159–163.

[3] Adamiuk, G., Zwick, T. & Wiesbeck, W. [2008]. Dual-orthogonal polarized vivaldi antenna for ultra wideband applications, *17th International Conference on Microwaves, Radar and Wireless Communications, 2008. MIKON.*

[4] Araujo, E. G. & Grupen, R. [1998]. Feature detection and identification using a sonar-array, *In Proceedings of the IEEE International Conference on Robotics and Automation*, IEEE Computer Society Press, pp. 1584–1589.

[5] Balanis, C. A. [1996]. *Antenna Theory: Analysis and Design*, New York: Wiley.

[6] Blech, M. & Eibert, T. [2007]. A dipole excited ultrawideband dielectric rod antenna with reflector, *Antennas and Propagation, IEEE Transactions on* 55(7): 1948 –1954.

[7] Borras, J., Hatrack, P. & Mandayam, N. B. [1998]. Decision theoretic framework for NLOS identification, *Proc. 48th IEEE Vehicular Technology Conf. VTC 98*, Vol. 2, pp. 1583–1587.

[8] Cramer, R.-M., Scholtz, R. & Win, M. [2002]. Evaluation of an ultra-wide-band propagation channel, *IEEE Transactions on Antennas and Propagation* 50(5): 561–570.

[9] Daniels, D. J. (ed.) [2004]. *Ground penetrating radar (2nd Edition)*, Vol. 15 of *IEE radar, sonar, navigation, and avionics series*, Institution of Electrical Engineers, London, UK.

[10] Dardari, D., Chong, C.-C. & Win, M. Z. [2006]. Analysis of threshold-based TOA estimator in UWB channels, *Proc. Eur. Signal Process. Conf. (EUSIPCO)*, Florence, Italy.

[11] Dardari, D., Chong, C.-C. & Win, M. Z. [2008]. Threshold-based time-of-arrival estimators in UWB dense multipath channels, *IEEE Transactions on Communications* 56(8): 1366–1378.

[12] Dardari, D. & Win, M. Z. [2006]. Threshold-based time-of-arrival estimators in UWB dense multipath channels, *Proc. IEEE International Conference on Communications*, Vol. 10, pp. 4723–4728.

[13] Deissler, T., Salman, R., Schultze, T., Thielecke, J. & Willms, I. [2010]. UWB radar object recognition for slam, *Proceedings of the 11th International Radar Symposium*, pp. 378–381.

[14] des Noes, M. & Denis, B. [2012]. Benefits from cooperation in simultaneous anchor-less tracking and room mapping based on impulse radio - ultra wideband devices, *19th International Conference on Systems, Signals and Image Processing, IWSSIP 2012*.

[15] Dutta, P., Arora, A. & Bibyk, S. [2006]. Towards radar-enabled sensor networks, *Information Processing in Sensor Networks, 2006. IPSN 2006. The Fifth International Conference on*, pp. 467 –474.

[16] Elsherbini, A., Zhang, C., Lin, S., Kuhn, M., Kamel, A., Fathy, A. & Elhennawy, H. [2007]. UWB antipodal vivaldi antennas with protruded dielectric rods for higher gain, symmetric patterns and minimal phase center variations, *Antennas and Propagation Society International Symposium, 2007 IEEE*, pp. 1973 –1976.

[17] Falsi, C., Dardari, D., Mucchi, L. & Win, M. Z. [2006]. Time of arrival estimation for UWB localizers in realistic environments, *EURASIP Journal on Applied Signal Processing* 2006: 1–13.
URL: *http://www.hindawi.com/journals/asp/2006/032082.abs.html*

[18] Fügen, T., Maurer, J., Kayser, T. & Wiesbeck, W. [2006]. Capability of 3-D ray tracing for defining parameter sets for the specification of future mobile communications systems, *IEEE Transactions on Antennas and Propagation* 54(11): 3125–3137.

[19] Gezici, S. [2008]. A survey on wireless position estimation, *Wireless Personal Communications* 44: 263–282.
URL: *http://www.springerlink.com/content/w1481x761266m417*

[20] Gezici, S. & Poor, H. [2009]. Position estimation via ultra-wide-band signals, *Proceedings of the IEEE* 97(2): 386 –403.

[21] Gezici, S., Tian, Z., Giannakis, G., Kobayashi, H., Molisch, A., Poor, H. & Sahinoglu, Z. [2005]. Localization via ultra-wideband radios: a look at positioning aspects for future sensor networks, 22(4): 70 – 84.

[22] Guvenc, I. & Sahinoglu, Z. [2005]. Threshold-based TOA estimation for impulse radio UWB systems, *Proc. IEEE International Conference on Ultra-Wideband ICU 2005*, pp. 420–425.

[23] Janson, M., Fügen, T., Zwick, T. & Wiesbeck, W. [2009]. Directional channel model for ultra-wideband indoor applications, *IEEE International Conference on Ultra-Wideband, ICUWB'09*, Vancouver, Canada.

[24] Janson, M., P., J., Zwick, T. & Wiesbeck, W. [2010]. Directional hybrid channel model for ultra-wideband MIMO systems, *4th European Conference on Antennas and Propagation, EuCAP'10*.

[25] Janson, M., Pontes, J., Fügen, T. & Zwick, T. [2012]. A hybrid deterministic-stochastic propagation model for short-range MIMO-UWB communication systems, *accepted for publication in Frequenz, Journal of RF-Engineering and Telecommunications* .

[26] Jemai, J., Eggers, P., Pedersen, G. & Kurner, T. [2009]. Calibration of a UWB sub-band channel model using simulated annealing, *IEEE Transactions on Antennas and Propagation* 57(10): 3439 –3443.

[27] Jovanoska, S. & Thomä, R. [2012]. Multiple target tracking by a distributed UWB sensor network based on the PHD filter, *Proc. 15th Int. Conf. Information Fusion (FUSION)*, Singapore, Singapore. submitted for publication.

[28] Kidera, S., Sakamoto, T. & Sato, T. [2010]. Accurate UWB radar three-dimensional imaging algorithm for a complex boundary without range point connections, *Geoscience and Remote Sensing, IEEE Transactions on* 48(4): 1993 –2004.

[29] Kmec, M., Sachs, J., Peyerl, P., Rauschenbach, P., Thomä, R. & Zetik, R. [2005]. A novel Ultra-wideband real-time MIMO channel sounder architecture, *XXVIIIth URSI General Assembly*, New Delhi, India.

[30] Li, J. & Stoica, P. (eds) [2008]. *MIMO radar signal processing*, Hoboken, NJ: Wiley.

[31] Li, X. & Pahlavan, K. [2004]. Super-resolution TOA estimation with diversity for indoor geolocation, *IEEE Transactions on Wireless Communications* 3(1): 224–234.

[32] Loeffler, A., Wissendheit, U., Gerhaeuser, H. & Kuznetsova, D. [2010]. A multi-purpose rfid reader supporting indoor navigation systems, *IEEE International Conference on RFID-Technology and Applications (RFID-TA)* 1: 43–48.

[33] Lostanlen, Y. & Gougeon, G. [2007]. Introduction of diffuse scattering to enhance ray-tracing methods for the analysis of deterministic indoor UWB radio channels, *International Conference on Electromagnetics in Advanced Applications, ICEAA'07*, pp. 903–906.

[34] Lostanlen, Y., Gougeon, G., Bories, S. & Sibille, A. [2006]. A deterministic indoor UWB space-variant multipath radio channel modelling compared to measurements on basic configurations, *First European Conference on Antennas and Propagation, EuCAP'06*, pp. 1–8.

[35] Mackworth, S. & Mokhtarian, F. [1988]. The renormalized curvature scale space and the evolution properties of planar curves, *Computer Vision and Pattern Recognition, 1988. Proceedings CVPR '88., Computer Society Conference on*, pp. 318 –326.

[36] Mahler, R. [2003]. Multitarget Bayes filtering via first-order multitarget moments, 39(4): 1152–1178.

[37] Mahler, R. P. S. [2001]. *Random set theory for target tracking and identification*, Handbook of Mutlisensor Data Fusion, CRC Press, London, p. 14.

[38] McIvor, A. M., Zang, Q. & Klette, R. [2001]. The background subtraction problem for video surveillance systems, *RobVis*, pp. 176–183.

[39] Meissner, P. & Witrisal, K. [2012]. Multipath-assisted single-anchor indoor localization in an office environment, *19th International Conference on Systems, Signals and Image Processing, IWSSIP 2012*.

[40] Molisch, A., Kuchar, A., Laurila, J., Hugl, K. & Bonek, E. [1999]. Efficient implementation of a geometry-based directional model for mobile radio channels, *IEEE VTS 50th Vehicular Technology Conference, 1999. VTC 1999 - Fall* 3: 1449–1453.

[41] Moura, J. M. F. & Jin, Y. [2008]. Time reversal imaging by adaptive interference canceling, *IEEE Transactions on Signal Processing* 56(1): 233–247.

[42] Nasr, K. [2008]. Hybrid channel modelling for ultra-wideband portable multimedia applications, *Microwaves, Antennas & Propagation, IET* 2(3): 229–235.

[43] Poor, H. V. [1994]. *An introduction to signal detection and estimation, Chapter III*, 2nd edn, Springer-Verlag.

[44] Porebska, M., Adamiuk, G., Sturm, C. & Wiesbeck, W. [2007]. Accuracy of algorithms for UWB localization in NLOS scenarios containing arbitrary walls, *2nd European Conference on Antennas and Propagation, EuCAP'07*, Edinburgh, UK.

[45] Reichardt, L., Kowalevski, J., Zwirello, L. & Zwick, T. [2012]. Compact, teflon embedded, dual-polarized ultra wideband (UWB) antenna, *accepted for pulicatiob on IEEE, Antennas and Propagation Society International Symposium*.

[46] Ren, Y.-J. & Tarng, J.-H. [2007]. A hybrid spatio-temporal model for radio propagation in urban environment, *2nd European Conference on Antennas and Propagation, EuCAP'07*, Edinburgh, UK.

[47] Sachs, J., Peyerl, P., Zetik, R. & Crabbe, S. [2003]. M-sequence ultra-wideband-radar: state of development and applications, *International Radar Conference*, Adelaide, Australia, pp. 224 – 229.

[48] Salman, R. & Willms, I. [2010]. A novel UWB radar super-resolution object recognition approach for complex edged objects, *Ultra-Wideband (ICUWB), 2010 IEEE International Conference on*, Vol. 2, pp. 1 –4.

[49] Salman, R. & Willms, I. [2011a]. 3d UWB radar super-resolution imaging for complex objects with discontinous wavefronts, *Ultra-Wideband (ICUWB), 2011 IEEE International Conference on*, pp. 346 –350.

[50] Salman, R. & Willms, I. [2011b]. Super-resolution object recognition approach for complex edged objects by UWB radar, *Object Registration*, InTech.

[51] Salman, R. & Willms, I. [2012]. Joint efficiency and performance enhancement of wavefront extraction algorithms for short-range super-resolution UWB radar, *Microwave Conference (GeMiC), 2012 The 7th German*, pp. 1 –4.

[52] Salman, R., Willms, I., Reichardt, L., Zwick, T. & Wiesbeck, W. [2012]. On polarization diversity gain in short range UWB-radar object imaging, *2012 IEEE International Conference on Ultra-Wideband (ICUWB)*.

[53] Seitz, J., Vaupel, T., Jahn, J., Meyer, S., Boronat, J. G. & Thielecke, J. [2010]. A hidden markov model for urban navigation based on fingerprinting and pedestrian dead reckoning, *Information Fusion (FUSION), 2010 13th Conference on*, IEEE, pp. 1–8.

[54] Shen, G., Zetik, R., Hirsch, O. & Thomä, R. S. [2010]. Range-based localization for UWB sensor networks in realistic environments, *EURASIP Journal on Wireless Communications and Networking* pp. 1–9.
URL: *http://www.hindawi.com/journals/wcn/2010/476598.html*

[55] Shen, G., Zetik, R. & Thomä, R. S. [2008]. Performance comparison of TOA and TDOA based location estimation algorithms in LOS environment, *Proc. 5th Workshop on Positioning, Navigation and Communication (WPNC 2008)*, Hannover, pp. 71–78.

[56] Shen, G., Zetik, R., Yan, H., Hirsch, O. & Thomä, R. S. [2010]. Time of arrival estimation for range-based localization in UWB sensor networks, *2010 IEEE International Conference on Ultra-Wideband (ICUWB 2010)*, Nanjing, China.

[57] Shen, G., Zetik, R., Yan, H., Jovanoska, S. & Thomä, R. S. [2011]. Localization of active UWB sensor nodes in multipath and NLOS environments, *the 5th European Conference on Antennas and Propagation (EUCAP 2011)*, Rome, Italy.

[58] Skolnik, M. (ed.) [2002]. *Introduction to Radar Systems*, 3rd ed edn, Mc Graw-Hill.

[59] Sörgel, W., Waldschmidt, C. & Wiesbeck, W. [2003]. Transient responses of a vivaldi antenna and a logarithmic periodic dipole array for ultra wideband communication, *IEEE, Antennas and Propagation Society International Symposium*.

[60] Sturm, C., Sörgel, W., Knorzer, S. & Wiesbeck, W. [2006]. An adequate channel model for ultra wideband localization applications, *Antennas and Propagation Society International Symposium 2006, IEEE*, pp. 2140 –2143.

[61] Thomä, R. S., Hirsch, O., Sachs, J. & Zetik, R. [2007]. UWB sensor networks for position location and imaging of objects and environments, *Proc. Second European Conf. Antennas and Propagation EuCAP 2007*, pp. 1–9.

[62] Thrun, S., Burgard, W. & Fox, D. [2001]. *Probabilistic Robotics (Intelligent Robotics and Autonomous Agents)*, The Mit Press.

[63] van den Bos, A. [1995]. The multivariate complex normal distribution-a generalization, *IEEE Transactions on Information Theory* 41(2): 537–539.

[64] Varslot, T., Yazici, B., Yarman, C.-E., Cheney, M. & Scharf, L. [2007]. Time-reversal waveform preconditioning for clutter rejection, *Proc. Int. Waveform Diversity and Design Conf*, pp. 330–334.

[65] Vo, B.-N., Singh, S. & Doucet, A. [2003]. Sequential Monte Carlo Implementation of the PHD filter for Multi-target Tracking, *Proc. 6th Int. Conf. Information Fusion (FUSION)*, Vol. 2, Cairns, Australia, pp. 792–799.

[66] Vo, B.-N. V. & Ma, W.-K. [2006]. The Gaussian Mixture Probability Hypothesis Density Filter, 54(11): 4091–4104.

[67] Win, M. & Scholtz, R. [2002]. Characterization of ultra-wide bandwidth wireless indoor channels: a communication-theoretic view, *Selected Areas in Communications, IEEE Journal on* 20(9): 1613 – 1627.

[68] Wylie, M. P. & Holtzman, J. [1996]. The non-line of sight problem in mobile location estimation, *Proc. 5th IEEE Int Universal Personal Communications Record. Conf*, Vol. 2, pp. 827–831.

[69] Yan, H., Shen, G., Zetik, R., Hirsch, O. & Thomä, R. S. [2012]. Ultra-wideband MIMO ambiguity function and its factorability. IEEE Trans. Geosci. Remote Sensing, to be published. DOI: 10.1109/TGRS.2012.2201486.

[70] Yan, H., Shen, G., Zetik, R., Malz, E., Jovanoska, S. & Thomä, R. S. [2011]. Stationary symmetric object detection in unknown indoor environments, *Loughborough Antennas & Propagation Conference 2011, LAPC 2011)*.

[71] Yarovoy, A. G., van Genderen, P. & Ligthart, L. P. [2000]. Ground penetrating impulse radar for land mine detection, *in* D. A. Noon, G. F. Stickley, & D. Longstaff (ed.), *Society of Photo-Optical Instrumentation Engineers (SPIE) Conference Series*, Vol. 4084 of *Society of Photo-Optical Instrumentation Engineers (SPIE) Conference Series*, pp. 856–860.

[72] Zetik, R., Jovanoska, S. & Thomä, R. [2011]. Simple method for localisation of multiple tag-free targets using UWB sensor network, *2011 IEEE Int. Conf. Ultra-Wideband (ICUWB)*, pp. 268 –272.

[73] Zetik, R., Sachs, J. & Thomä, R. [2005]. Modified cross-correlation back projection for UWB imaging: numerical examples, *Ultra-Wideband, 2005. IEEE International Conference on*, p. 5.

[74] Zetik, R. & Thomä, R. [2008]. Monostatic imaging of small objects in UWB sensor networks, *Ultra-Wideband, 2008. IEEE International Conference on*, Vol. 2, pp. 191 –194.

[75] Zetik, R. & Thomä, R. S. [2010]. Imaging of Distributed Objects in UWB Sensor Networks, *in* Sabath, F., Giri, D. V., Rachidi, F., & Kaelin, A. (ed.), *Ultra-Wideband, Short Pulse Electromagnetics 9*, Springer New York, pp. 97–104.

[76] Zhan, H., Ayadi, J., Farserotu, J. & Le Boudec, J.-Y. [2009]. Impulse radio ultra-wideband ranging based on maximum likelihood estimation, *IEEE Transactions on Wireless Communications* 8(12): 5852–5861.

Chapter 10

Pedestrian Recognition Based on 24 GHz Radar Sensors

Steffen Heuel and Hermann Rohling

Additional information is available at the end of the chapter

http://dx.doi.org/10.5772/53007

1. Introduction

Radar sensors offer in general the capability to measure extremely accurately target range, radial velocity, and azimuth angle for all objects inside the observation area. These target parameters can be measured simultaneously even in multiple target situations, which is a technical challenge for the waveform design and signal processing procedure. Furthermore, radar systems fulfil these requirements in all weather conditions, even in rain and fog, which is important for all automotive applications, [1], [2]. Advanced driver assistant systems (ADAS) are currently under investigation to increase comfort and safety in general. For Adaptive Cruise Control (ACC) applications a single 77 GHz radar sensor is used, which has a maximum range of 200 m and covers a narrow azimuth angle area of 15 degree for example. Many other and additional automotive applications, like Stop & Go, Pre-Crash or Parking Aid, consider a completely different observation area [3]. In this case a maximum range of 50 m, but a wide azimuth angle area of 120 degrees is required. For these applications 24 GHz radar sensors are used. Besides the range and velocity parameters, additional information concerning the target type are of great interest, as one of the main objectives of future safety systems will be the increased protection of all pedestrians and other vulnerable road users.

By extending the radar signal processing part of a 24 GHz radar sensor with a pedestrian recognition scheme, the same radar sensor which is used for the mentioned applications can be applied additionally for pedestrian recognition and allows the design of pedestrian safety systems. Therefore, the radar signal processing part has to be adapted to the assumption of extended targets with a characteristic range profile and a velocity profile (e.g. based on the Doppler Spectrum) in general [4]. The detailed analysis of the resulting range profile and target's velocities is possible and can be used to recognize pedestrians in urban areas with conventional 24 GHz radar sensors.

2. Radar sensor and measurements

Several proposals for pedestrian recognition schemes have been described, which are based on video cameras and computer vision systems [7], [8]. But automotive radar sensors in the

24 and 77 GHz band are also strong candidates for automotive safety systems. Compared with vision systems, they have some additional important advantages of robustness in all weather conditions, simultaneous target range and radial velocity measurement and a high update rate. These properties are especially important for pedestrian recognition, as the object classification should be available immediately and at any time.

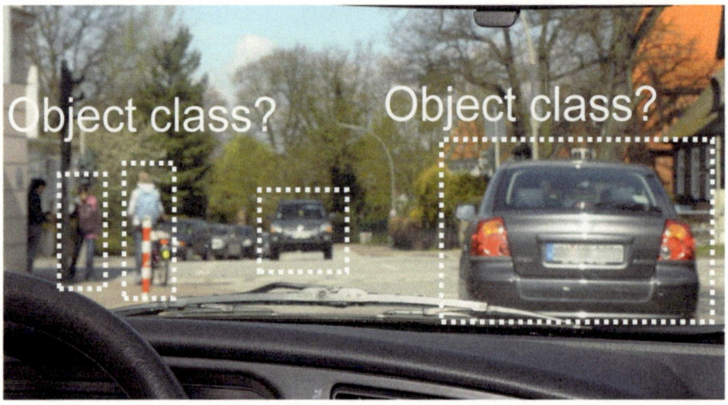

Figure 1. Daily traffic situation in an urban area with an oncoming vehicle and pedestrians walking on the sidewalk.

This chapter presents the modulation scheme of an automotive radar sensor and explains the features of pedestrians and vehicles by which a robust classification is possible in an urban area from a moving vehicle with a mounted 24 GHz radar sensor, see Figure 1.

2.1. Modulation scheme

The automotive 24 GHz radar sensor allows a simultaneous and unambiguous measurement of target range R and radial velocity v_r even in multiple target situations. This is achieved by combining the advantages of the Frequency Shift Keying (FSK) waveform and the Frequency Modulated Continuous Waveform (FMCW) in a so called Multi Frequency Shift Keying (MFSK) waveform [19], which is already used in commercial automotive radar sensors to enable Adaptive Cruise Control (ACC) or Blindspot Detection (BSD) [20], [21]. Applying an FSK waveform, the target range R and radial velocity v_r can be measured. However, there is no range resolution. Multiple objects measured at the same spectral line in the Doppler spectrum result in an unusable range information, as the determination procedure assumes a single target. To mitigate this drawback, the FMCW waveform resolves targets in range R and velocity v_r. Limitations will occur in this case in multi target situations due to ambiguous measurements. The specific MFSK waveform is applied in the 24 GHz Radar sensor for a range and Doppler frequency measurement even in multi target situations with a bandwidth of $f_{\text{sweep}} = 150\,\text{MHz}$ and a resulting range resolution of $\Delta R = 1.0\,\text{m}$. It is a classical step-wise frequency modulated signal with a second linear frequency modulated signal in the same slope but with a certain frequency shift f_{step} integrated into this waveform in an intertwined way. The chirp duration is denoted by $T_{\text{CPI}} = 39\,\text{ms}$ which results in a velocity resolution of $\Delta v = 0.6\,\text{km/h}$.

It is important to notice that this waveform is not processed by a matched filter or analyzed by an ambiguity function. Instead it is processed in a non-matched filter form to get an unambiguous and simultaneous target range and Doppler frequency measurement with high resolution and accuracy. The echo signal of the stepwise and intertwined waveform is downconverted by the corresponding instantaneous transmit frequency into baseband and sampled at the end of each short frequency step. This time discrete signal is Fourier transformed separately for the two intertwined signals to measure the beat frequency f_B which is simultaneously influenced by the target range R and radial velocity v_r.

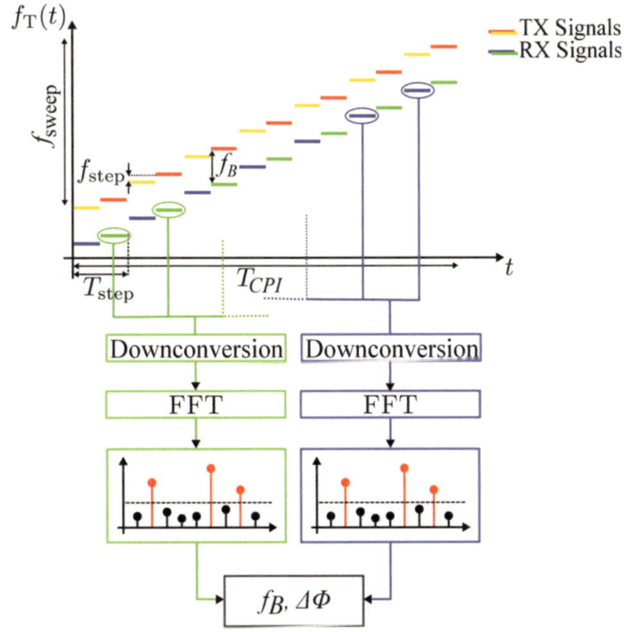

Figure 2. MFSK waveform principle with two intertwined transmit signals.

$$f_B = -\frac{2v_r}{\lambda} - \frac{2R \cdot f_{sweep}}{c} \cdot \frac{1}{T_{CPI}} \tag{1}$$

In any case, a single target will be measured and will be detected on the same spectral line at position f_B for the two intertwined signals. Therefore, after the detection procedure the phase difference $\Delta\Phi$ between the two complex-valued signals on the spectral line f_B will be calculated. The step frequency f_{step} between the intertwined transmit signals determines the unambiguous phase measurement $\Delta\Phi$ in the interval $[-\pi; \pi)$. This phase difference $\Delta\Phi$ again is influenced by the target range R and radial velocity v_r described in Equation (2).

$$\Delta\Phi = -\frac{2\pi}{f_{sample}} \cdot \frac{2v_r}{\lambda} - \frac{4\pi R \cdot f_{step}}{c} \tag{2}$$

The target range R and radial velocity v_r can be determined by solving the linear equation described in Equation (1) and (2) in an unambiguous way. In this case, ghost targets are completely avoided since this waveform and signal processing combines the benefits of linear FMCW and FSK technology. The system design and the sensor parameters can be determined like in a linear FMCW radar system. The range and velocity resolution ΔR and Δv are determined by the bandwidth f_{sweep} of the radar sensor and the chirp duration T_{CPI} as described in Equation (3) and (4), respectively.

$$\Delta R = \frac{c}{2} \cdot \frac{1}{f_{\text{sweep}}} \tag{3}$$

$$\Delta v = -\frac{\lambda}{2} \cdot \frac{1}{T_{\text{CPI}}} \tag{4}$$

The table below shows the system parameters of the automotive radar sensor in detail.

Carrier Frequency	$f_T = 24\,\text{GHz}$
Sweep Bandwdith	$f_{\text{sweep}} = 150\,\text{MHz}$
Maximum Range	$R_{\text{max}} = 200\,\text{m}$
Range Resolution	$\Delta R = 1\,\text{m}$
Chirp Length	$T_{\text{CPI}} = 39\,\text{ms}$
Maximum Velocity	$v_{\text{max}} = 250\,\text{km/h}$
Velocity Resolution	$\Delta v = 0.6\,\text{km/h}$

Table 1. 24 GHz Radar Sensor Parameters.

Classical UWB-Radar Sensors have a sweep bandwidth of $f_{\text{sweep}} = 2\,\text{GHz}$. Using such a bandwidth, a high range resolution is determined, which allows also pedestrian classification. The technical challenge in this chapter is to realize pedestrian recognition based on a 24 GHz radar sensor with a bandwidth of only 150 MHz. This sensor is used in automotive applications, therefore an extension of the signal processing in terms of pedestrian classification is desirable.

2.2. Radar echo signal measurements

The possibility to recognize pedestrians with a static radar sensor using the Doppler effect has been shown in [15]. A moving vehicle is equipped with an automotive radar sensor with a built-in feature extraction and classification to recognize pedestrians. The feature extraction in the backscattered radar echo signals resulting from superposition of the reflection points of an object is done automatically in the radar sensor signal processing. Detected targets are therefore tracked in the environment and an additional feature extraction and classification is performed.

To distinguish between the echo signal characteristics of pedestrians and vehicles, a target recognition model is described which is based on the specific **velocity profile** and **range profile** for each object separately [4]. The velocity profile describes the extension of the different velocities of an object measured by the radar sensor, while the range profile shows the physical expansion of a target.

In case of a longitudinally moving pedestrian, different reflection points at the trunk, arms and legs with different velocities are characteristic in radar propagation. Therefore an *extended* velocity profile will be observed in a single radar measurement of a pedestrian as the velocity resolution Δv of the radar sensor is higher than the occuring velocities. Carrying out several measurements with a time duration of 39 ms each, a sinusoidal spreading and contraction of the velocity profile can be observed in the case of a pedestrian, due to the movement of arms and legs for example in the swing and stand-phase of the legs. For a laterally moving pedestrian, the velocity profile is less extended due to the moving direction of the pedestrian. Furthermore, the extension depends mainly on the azimuth angle under which the pedestrian is measured. In contrast, the radar echo signal in case of a vehicle shows a very narrow (*point shaped*) velocity profile due to a uniform motion.

Additionally, a *point shaped* range profile will occur in the case of a longitudinally or laterally moving pedestrian as the physical expansion is small compared to the range resolution of $\Delta R = 1.0\,\text{m}$. In contrast, a vehicle shows an *extended* range profile, due to several reflection points spaced in several range cells. The measurement result of a single observation is shown in the range Doppler diagram in Figure 4.

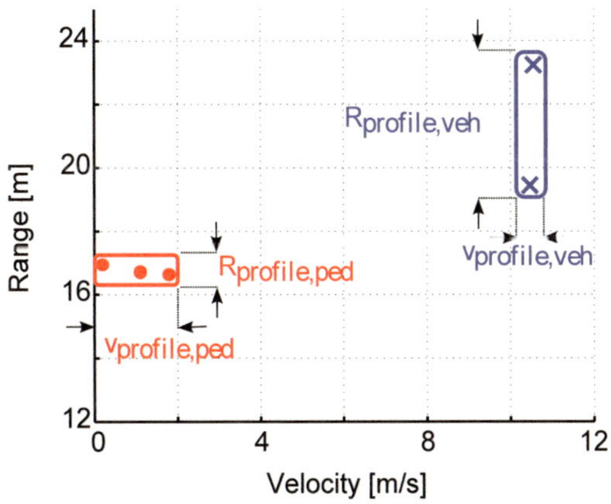

Figure 3. Range profile and velocity profile of a single measurement.

Under the use of an MFSK modulation signal, a range profile and the velocity profile can be extracted from a series of received signals as shown in Figure 4. As an example, four consecutive range and velocity measurements are depicted in a range Doppler diagram. The red dots show a longitudinally walking pedestrian, the blue crosses an in front moving vehicle. The figures depicted are based on radar measurements taken in an urban area with an ego speed of 50 km/h. It can be observed that neither velocity profile nor a range profile can be seen in the first measurement, consequently, those feature values are zero. In the second measurement, however, several range and velocity measurements allow to calculate an extended range profile for the vehicle and an extended velocity profile for the pedestrian.

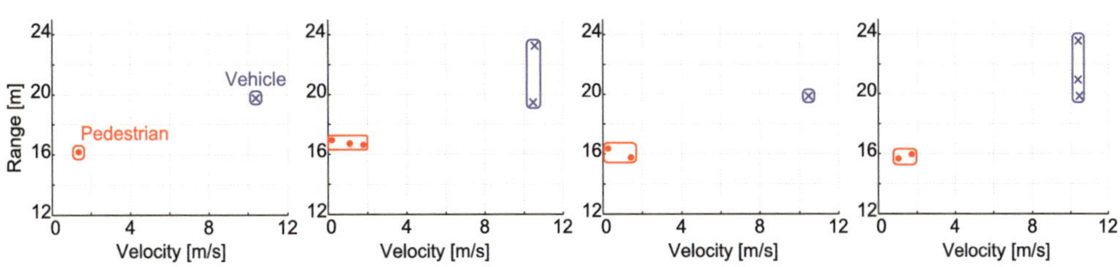

(a) Measurement 1 (b) Measurement 2 (c) Measurement 3 (d) Measurement 4

Figure 4. Sequence of range and velocity measurements.

The range profile and velocity profile features do not depend on the modulation signal. Solely, the range and velocity resolution must be smaller than the expected extension. For example, in the case of a continous wave modulation signal, the range profile can be read directly from the Fourier transformed radar echo signal and the velocity profile can be evaluated from the Doppler spectrum. Also, instead of these spectra or the frequency spectrum and phase difference analysis, it is possible to calculate the extension of an object in range and velocity on the basis of target lists by applying a detection algorithm. On this basis, an extended range profile with a point-shaped velocity profile can also be measured for a vehicle. For a pedestrian, the profiles remain vice versa. Figure 5 depicts this context.

Figure 5. Range profile and velocity profile of a pedestrian and a vehicle.

The longitudinally and laterally moving pedestrians are classified as **pedestrians**, longitudinally and laterally moving vehicles as **vehicles**, all other signals received from objects such as parked cars, poles, trees and traffic signs are classified as **other** objects.

3. Classification

Target recognition is a challenge for each radar engineer. A reliable feature extraction and classification process has to be implemented. To describe the characteristics of pedestrians and vehicles, the velocity profile and range profile signal features have been introduced. These are the basis in the feature extraction and target recognition system based on a single radar measurement (single look of 39 ms duration) as described in subsection 3.1. An extended extraction based on the spreading and contraction of the spectra by observing several measurements is considered in subsection 3.2. Finally in subsection 3.3, a tracker feedback is calculated where additional features based on the Kalman gain and innovation are extracted. In the next step, the classification process is performed, which maps the extracted features into classes. An evaluation of the classification results is shown by means of a confusion matrix for the case of a single measurement- and multiple measurements-feature extraction.

3.1. Classification based on a single radar measurement

Radar sensors provide continuously available measurement results in an interval of a few milliseconds. This interval is determined by the duration of the transmission signal $T_{CPI} = 39\,\text{ms}$ in which a single MFSK signal is transmitted. The echo signal is downconverted and Fourier-transformed, which allows the described features to be extracted continuously. Rather than examinating a received sequence of radar echoes, this subchapter will initially focus on a single radar measurement of $T_{CPI} = 39\,\text{ms}$.

3.1.1. Feature extraction

Automotive radar sensors are an important source of information for security and comfort systems. The information is measured in terms of range, radial velocity and signal level. However, information about the object types do not exist. To fill this gap, features from the available information are extracted, which describe the object types and allow a decision of the related class on the basis of measured sensor data. To describe a detected object, this signal processing step calculates a number of features, which are discriminant for measurements containing different object types and match for objects from the same type. Thereby, moderately separated features achieve even in a perfect classification algorithm only moderate or even poor results ([16], [17]). An ideal feature extractor on the other hand shows good classification performance by using simple linear classifiers. This is why the feature extraction is so important. For a distinct classification, transformation-invariant features are sought. Still, there is no recipe to determine a feature set and since each sensor type describes an object specifically and each task is different, the feature set for pedestrian recognition based on an automotive radar sensor is explained shortly. Figure 6 shows the feature extraction with the specific object description in the context of the signal processing chain.

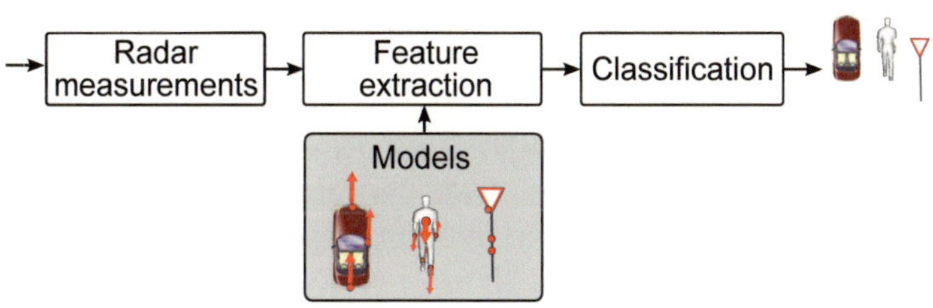

Figure 6. Context of the feature extraction in the signal processing chain and the object description using the range profile and velocity profile.

The basis for feature extraction are the velocity profile and the range profile of a detected object which has been described previously. The term of the profile describes the physical and kinematic dimensions of an object in the distance, angle and velocity. This can be measured in the case of multiple reflection points with different velocities greater than zero for any object. It can involve an extended or a point-shaped profile for the range and velocity depending on the type of expansion. On this basis, a number n of features can be calculated which describe the object in terms of a radar measurement. All n calculated real valued features $x_1, ..., x_n$ are saved in a feature vector $\vec{x}$ and build the basis for further signal processing steps.

$$\vec{x} = (x_1, x_2, ..., x_n)\; n \in \mathbf{N}, x_i \in \mathbf{R} \tag{5}$$

Exemplarily, the calculation of the range profile R_{profile} is given in Equation (6). Analogously, the velocity profile v_{profile} of the spectrum can be calculated, Equation (7).

$$x_1 = R_{\text{profile}} = R_{\max} - R_{\min} \tag{6}$$

$$x_5 = v_{\text{profile}} = v_{\max} - v_{\min} \tag{7}$$

The approach in feature extraction, using stochastic features, assumes that the measured data are random variables with independent and identical distribution. From this data within a single measurement cycle the variance and the standard deviation is estimated. To support the classification process, the number of scatterers is extracted, which describes the number of detected reflection points of an object. This approach allows a classification of the object type within a single measurement. The entire feature set for $n = 8$ features is shown in Table 2 below.

Feature	Annotation	Description
x_1	R_{profile}	Extension in range
x_2	$std(R)$	Standard deviation in range
x_3	$var(R)$	Variance in range
x_4	v_{r}	Radial Velocity
x_5	v_{profile}	Extension in velocity
x_6	$std(v_r)$	Standard deviation in velocity
x_7	$var(v_r)$	Variance in velocity
x_8	scatterer	Number of scatterers

Table 2. Feature Set of each object in a single measurement.

To determine the quality of a feature, the common area index (CAI) of two histograms is considered. While a common area index of 0 describes a complete overlapping of the feature space, a CAI of 1 describes an absolutely separable feature.

Several urban measurement scenarios of longitudinally moving vehicles and pedestrians were measured with an automotive radar sensor. From the detections of each single measurement cycle the features are extracted. Exemplarily, the velocity profile of a vehicle and a pedestrian is depicted in Figure 7 as a histogram. It shows a strong overlap of the area with a point shaped extension. This results directly from the model. A pedestrian is not extended at all times, because the arms and legs move sinusoidally. In addition, the echo signal fluctuates which causes fewer detections in a measurement. The vehicle equipped with the radar sensor moves also. The quality of the feature is calculated to CAI = 0.57.

3.1.2. Classification

The assignment of a measured object to a class is performed by a subjective decision algorithm based on the extracted characteristics. This process is called classification. The features therefore have been described previously, and are extracted within a single radar measurement of $T_{\text{CPI}} = 39$ ms. In supervised classifiers, the model of the classifier is generated in a training phase by using a training data set. The verification is performed in an evaluation phase with a test data set. The training data and test data consist of randomly selected feature vectors $\vec{x}$ of the radar measurements and corresponding assigned class labels. In the training and evaluation phase the classification result can be compared to the class labels and make a

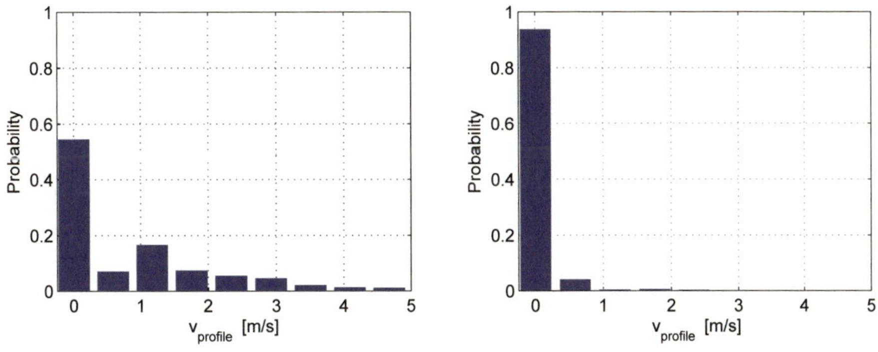

Figure 7. Feature histogram of the velocity profile using single radar measurements as a basis for feature extraction. The common area index is calculated to 0.57.

statement about the performance of the algorithm and designed model. Figure 8 depicts this process.

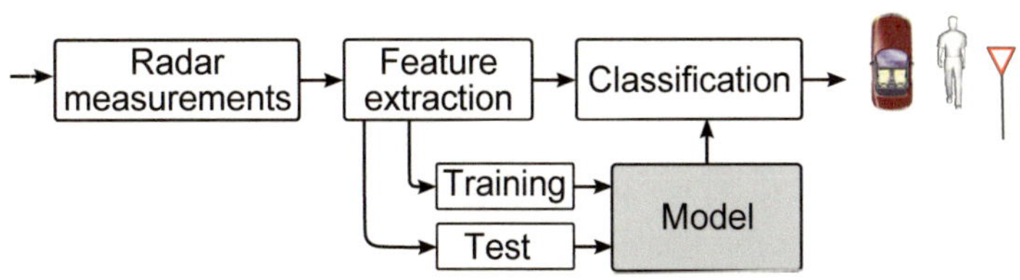

Figure 8. Signal flow graph of the classification process. Using a training data-set a model can be evaluated which performance is measured by a labelled test data set. This model is used for the classification process.

A classifier based on statistical learning theory is the support vector machine (SVM), introduced by Boser et al. in 1992 [24]. The SVM became very famous as studies about classification algorithms show good performance [25]. The classification process has low complexity and is very effective for high dimensional feature vectors. An SVM separates a set of training data by calculating a hyperplane $h(x)$ with maximum margin between the two classes ± 1 in a higher dimensional space in order to find the best classification function.

In this classification process, the SVM is able to map the extracted feature set into three different classes by using a majority voting algorithm. The verification of the previous training is conducted with the help of test data sets recorded from real urban measurements.

Table 3 shows the classification results. A trained and tested SVM was applied to different data sets of the extracted features from single measurements. All measurements were taken in an urban area with an ego velocity of 50 km/h. Applying new test data results in 71.32% true positive for a vehicle and 45.20% true positive for a pedestrian.

These quantitative results show already a possibility to distinguish between vehicles, pedestrians and other objects. However the performance is not good enough. Therefore

	Vehicle	Pedestrian	Other
Vehicle	71.32	5.87	22.81
Pedestrian	10.29	45.20	44.52
Other	23.56	26.65	49.78

Table 3. Confusion matrix: classification applied to a single measurement test data set containing 8000 data samples.

multiple radar measurements are considered to extract a more significant features set for the classification process.

3.2. Classification based on multiple radar measurements

From the continuously available radar measurements, a single measurement can be used to extract a feature set on the basis of range profile and velocity profile and estimated stochastic features. In this section, several range and velocity measurements are buffered and build the basis for the additional extraction process. From these buffered measurements a second **multiple measurement feature set** is extracted. This extends the classification process, which was previously based on a single measurement only.

3.2.1. Feature extraction

To gain performance, the choice of the measurement buffer dimension is cruical. A long measurement buffer builds the basis for a more successful feature extraction, however results in a long classification time, as the classifier has to wait for the buffer to be filled. A short buffer, on the other hand is not always able to build the basis for separable features, as shown in the previous section where a single measurement is used. In this section, the dimension of the buffer is explained by means of "probability of maximum velocity profile" and fast availability deduced from the step frequency of a moving pedestrian.

An ideal measurement of a moving pedestrian allows to extract the step frequency from the spreading and contraction of the velocity profile [9]. This step frequency of $f_{\text{ped}} = 1.4 - 1.8\,\text{Hz}$ can be used to determine a necessary buffer dimension. Every 1.4 Hz the maximum velocity profile can be observed considering a moving pedestrian, which allows to extract the maximum velocity profile and range profile. At all other times, the expansion in velocity is lower or even zero. To detect at least one expansion in velocity, the number of measurements should therefore span a period of $T_{\text{ped}} = \frac{1}{f_{\text{ped}}}$. However, it can be assumed that all measurements are independent of each another, due to the ego motion of the radar, vibrations and fewer detections. Applying the feature extraction process using a single measurement, the probability P to extract an extended velocity is then given by:

$$P_{v_{\text{profile}},\text{single}} = \frac{T_{\text{CPI}}}{1/f_{\text{ped}}} \tag{8}$$

Assuming an equal distribution, multiple measurements increase the probability to extract a velocity profile by the factor $\frac{T_{\text{Buffer}}}{T_{\text{CPI}}}$.

$$P_{v_{\text{profile}},\text{multiple}} = \frac{T_{\text{Buffer}}}{T_{\text{CPI}}} \cdot \frac{T_{\text{CPI}}}{1/f_{\text{ped}}} \tag{9}$$

It can be seen that a long measurement buffer increases the probability to detect the maximum velocity profile. But even a smaller velocity profile can be detected and fulfils the requirements. In these measurements, a buffer of $T_{\text{Buffer}} = 150\,\text{ms}$ is applied.

Using current and time delayed range measurements leads to incorrect range profiles, due to the movement of the objects during the elapsed buffer time. To cope this, the corresponding range measurements inside the buffer must therefore be predicted in range. Each stored measurement is predicted in range by the elapsed time ΔT_{Buffer} and the velocity $\vec{v} = (v_x, v_y)$ during the measurements to compensate the movement. A new estimated range $\widehat{R} = |\vec{R}^*|$ can be calculated by a cartesian representation using velocity and elapsed time:

$$\vec{R}^* = \vec{R} + \Delta T_{\text{Buffer}} \cdot \vec{v} \text{ with } \vec{R} = (X, Y) \text{ and } \vec{v} = (v_x, v_y) \tag{10}$$

The multiple measurement feature set is shown in Table 4. It consists of the same characteristics as the single measurement feature set, but is calculated from a basis of several buffered velocity and predicted range measurements.

Feature	Annotation	Description
x_9	$\widehat{R}_{\text{profile,buf}}$	Extension in range
x_{10}	$std(\widehat{R},\text{buf})$	Standard deviation in range
x_{11}	$var(\widehat{R},\text{buf})$	Variance in range
x_{12}	$v_{\text{r,buf}}$	Radial Velocity
x_{13}	$v_{\text{profile,buf}}$	Extension in velocity
x_{14}	$std(v_{\text{r,buf}})$	Standard deviation in velocity
x_{15}	$var(v_{\text{r,buf}})$	Variance in velocity
x_{16}	scatterer,buf	Number of scatterers

Table 4. An additional feature set extracted from multiple measurements. To cover multiple measurements, each single measurement R, v_r is stored in a buffer of several milliseconds. This ensures a quick availability of an additional feature set for pedestrian classification.

In the previous section the characteristic velocity profile of a vehicle and a pedestrian was depicted as a histogram. On the basis of a single radar measurement a quality of CAI = 0.57 was determined. Using multiple radar measurements the feature extraction is based on a larger number of measurement values. This leads to a higher separability of the features as shown exemplarily in Figures 9(a), 9(b). The common area index has a value of CAI = 0.88 and thus increases by 31% compared to the single measurement feature extraction.

3.2.2. Classification

In the single measurement it is described how range and velocity measurements build the basis for the feature vector and the classification process. In the multiple measurement, the basis is an extended feature vector based on several range and velocity measurements of an object stored inside a buffer. Instead of using 8 features from a single measurement, additional 8 features are available for the first time with a filled buffer. For a successful classification using a SVM, a new model is built, which is also trained/tested with in total 16 features and is a basis for the following classification process. As shown in the confusion matrix in Table 5 by the additional features, the correct classification and the overall performance achieved, increases by using a total number of 16 features from single and multiple measurements.

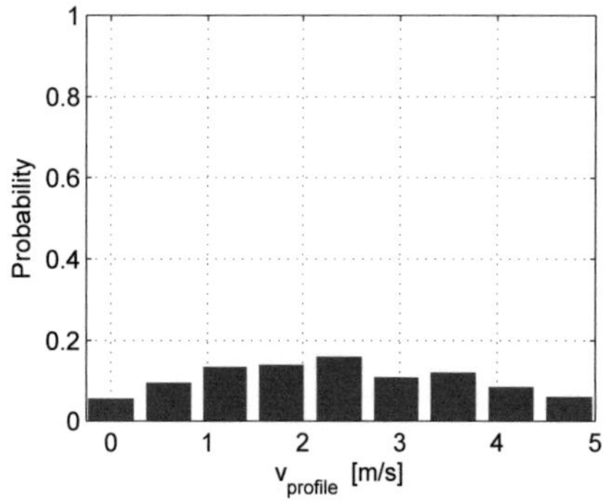

(a) $v_{\text{profile,buf}}$, Pedestrian

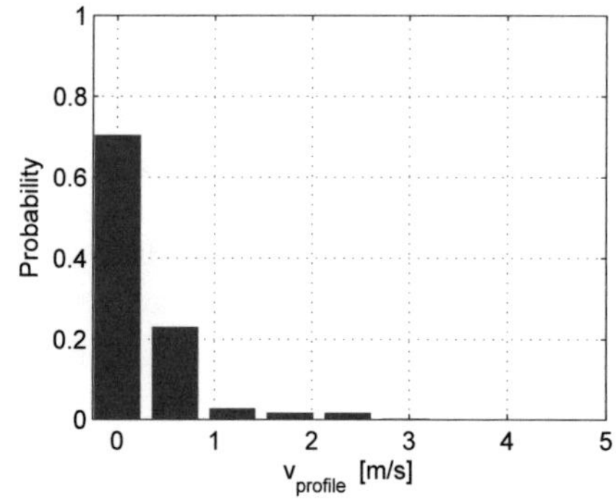

(b) $v_{\text{profile,buf}}$, Vehicle

Figure 9. Feature histogram of the velocity profile using multiple radar measurements as a basis for feature extraction. The common area index is calculated to 0.88.

Additionally, classifying feature vectors from single and multiple measurements results in fewer false positives.

	Vehicle	Pedestrian	Other
Vehicle	90.71	0.58	8.71
Pedestrian	4.94	53.94	41.12
Other	16.28	16.64	67.08

Table 5. Confusion matrix: classification applied to a single and multiple measurement test data set containing 8000 data samples.

Due to additionally multiple measurements as a basis for feature extraction, an improvement in the classification result is shown. Especially in terms of correct classification of a pedestrian and false alarms in which a pedestrian was classified as a vehicle, a significant enhancement is seen.

3.3. Classification based on the tracker feedback

A single radar measurement and multiple radar measurements were considered for feature extraction. These features are already a good basis for the classification process. In this subchapter, a third, additional feature set is described. This set can be extracted from the tracking algorithm. In this adaptive algorithm, different process noise of pedestrian radar measurements and vehicle radar measurements result in different gains for the track. On this basis, the process noise Q and the calculated gain K are additional features and are added to single and multiple radar measurement features in the classification process.

3.3.1. *Feature extraction*

Tracking is defined by a state estimation of moving targets. This state estimation is determined from the state parameters such as position, velocity and acceleration from a detected target. Known tracking methods are for example the alpha-beta filter or Kalman filter [26] which estimate a new state using a well-known prior state (e.g. position, velocity, acceleration). This reduces false alarms and smoothes movements of the objects.

The Kalman filter is a linear, recursive filter, whose goal is to determine an optimal estimate of the state parameters. The optimal estimate is based on available measurements and the models which describe the observed objects. In the equation of the motion model and observation model, the measurement noise is assumed to be average free, white Gaussian noise with the known covariance Q_{k-1} and R_k respectively. Under the given conditions, i.e., linear models and Gaussian statistics, the Kalman filter provides the optimal solution for the estimation of the state in the sense of minimizing the mean squared error, as described in [26].

The tracking for the object described by the motion model of the Kalman filter works fine as long as the motion models fit to the object. Pedestrians, vehicles, and static objects have different motion, which makes the tracking more difficult. Instead of creating a different motion and observation model for each object, it is proposed to determine the covariance of process noise Q_k and the measurement noise R_k adaptively. The process noise considers a non-modeled behavior in the motion model, while the measurement noise consideres uncertainty in the measurement. The original Kalman filter is not adaptive, which is why deviations from the model can not be handled. The gain matrix K, which is calculated from the process and measurement noise, reaches a stable condition after a short measurement time. An increase in the covariance Q_k leads to a larger value for K, so that the measured values are weighted more strongly, a decrease in Q_k relies more on the estimation.

In addition to an improved tracking effect, additional features can be extracted from the adaptive adjustment of the process noise, as pedestrian measurements in range and velocity differ from those of vehicles. Next to the process noise $Q_{k,v}$ of the velocity, the Kalman gain K_v (velocity component of the matrix K) is a good feature as measurements show. Anyhow, in an adaptive adjustment of the process noise, a compromise between the compensation of non-modeled movements and the filtering effect to reduce noise must be found, even though features are extracted.

The process noise matrix Q describes object-specific measurement properties that are initially set and are readjusted during operation of the tracker. For example, the readjustment of a single coefficient $Q_{k,v}$ in the Q matrix at the measurement k in respect to the velocity v is based on the actual target range R, the velocity v, the parameters a and b in an alpha-beta filter. Equation (11) shows the relation. The velocity of a pedestrian deviates between consecutive measurements, while the velocity deviation of a car within consecutive measurements is small or even zero. Consequently, the velocity v can be used to update the matrix Q.

$$Q_{k,v} = (1-\beta)Q_{k-1,v} + \beta \frac{|z_{k,v} - v_{k|k-1}|}{R_{k|k} \cdot a + b} \tag{11}$$

The predicted covariance matrix $P_{k|k-1}$ in the tracking process depends on the motion model F_{k-1}, the currently measured covariance $P_{k-1|k-1}$ and the process noise Q_{k-1} as shown in Equation (12).

$$P_{k|k-1} = F_{k-1} P_{k-1|k-1} F_{k-1}^{T} Q_{k-1} \tag{12}$$

Under the use of this covariance matrix $P_{k|k-1}$ and the innovation covariance S_k, the gain K_k can be calculated.

$$K_k = P_{k|k-1} S_k^{-1} \tag{13}$$

Both, matrix Q_k and gain K_k are used as additional features in the classification process. The separation of the feature space is depicted in Figure 10. These histograms show that a vehicle (Figure 10(c), 10(d)) has lower process noise Q_v and thus results in a smaller gain K compared to a pedestrian (Figure 10(a), 10(b)). This is due to a mostly linear trajectory of a vehicle with one main reflection point. A pedestrian echo fluctuates, which is reflected in a greater process noise and thus larger gain K.

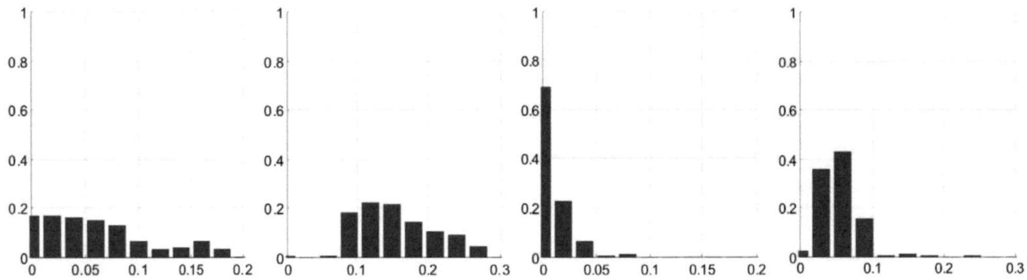

(a) Q_v, Pedestrian (b) K, Pedestrian (c) Q_v, Vehicle (d) K, Vehicle

Figure 10. Feature Histogram of Process Noise Q_v and Gain K.

The additional features Q and K are available for a created and active track. For each detected object a track is created, which is however not activated until several measurements can be associated with the track. A track must be confirmed by subsequent measurements, otherwise the track remains semi-active. In any case, additional features are available for classification.

Feature	Annotation	Description
x_{17}	Q_v	Velocity component of the process noise matrix Q
x_{18}	K_v	Veleocity component of the Kalman gain matrix K

Table 6. Additional Feature Set extracted from the tracker using an adaptive process noise.

3.3.2. *Classification*

An additional feature extraction based on the process noise and Kalman gain has been described. These features are added to the prior feature set. Table 7 shows the results of the classifier using all proposed features, single measurement, multiple measurements and tracker feedback. Even in an urban area with a high density of static targets pedestrians could be detected, tracked and classified with a true positive rate of 61.22% in the test data set. These results outperform prior outcomes using single- and the multiple measurements as a feature basis.

The classification results show an increasing performance in terms of correct classification and misclassification of pedestrians.

	Vehicle	Pedestrian	Other
Vehicle	92.84	0.50	6.66
Pedestrian	5.71	61.22	33.07
Other	10.06	13.30	76.64

Table 7. Confusion matrix: classification applied to a single, multiple and tracker feedback measurement test data set containing 8000 data samples.

4. Summary and conclusions

This chapter described a pedestrian classification algorithm for automotive applications using an automotive 24 GHz radar sensor with a bandwidth of 150 MHz as a measuring device. Three different systems for pedestrian recognition have been considered. The first system was based on a single radar measurement. The second system extracted a feature set on the basis of multiple radar measurements. Finally a tracking procedure was adapted to extract an additional feature set. The results show an increasing performance in the classification accuracy by using single-, multiple- and tracker feedback features. It is also pointed out that is not necessary to equip radar sensors with large bandwidths in order to classify pedestrians in urban areas.

Author details

Hermann Rohling and Steffen Heuel
Department of Telecommunications, Hamburg University of Technology, Hamburg, Germany

5. References

[1] Foelster, F.; Oprisan, D.; Rohling, H., (2004) Detection and Tracking of extended targets for a 24GHz automotive radar network, International Radar Symposium IRS 2004, Warsaw, Poland; pp 75-80

[2] Meinecke, M.-M.; Obojski, M.; Toens, M.; et.al, (2003) Approach For Protection Of Vulnerable Road Users Using Sensor Fusion Techniques, International Radar Symposium IRS 2003, Dresden, Germany, pp 125-130

[3] Schiementz, M.; Foelster, F. (2003) Angle Estimation Techniques for different 24 GHz Radar Networks, International Radar Symposium IRS 2003, Dresden, Germany, 2003

[4] Rohling, H.; Heuel, S.; Ritter, H. (2010) Pedestrian Detection Procedure integrated into an 24 GHz Automotive Radar, IEEE RADAR 2010, Washington D.C.

[5] Schmid, V.; Lauer, W.; Rollmann, G. (2003) Ultra Wide Band 24 GHz Radar Sensors for Innovative Automotive Applications, International Radar Symposium IRS 2003, Dresden, Germany, pp 119-124

[6] Oprisan, D.; Rohling, H.,(2002) Tracking Systems for Automotive Radar Networks, IEE Radar 2002, Edinburgh

[7] Philomin, V.; Duraiswani, R.; Davis, L. (2000) Pedestrian tracking from a moving vehicle, Proceedings of the IEEE Intelligent Vehicle Symposium 2000, Dearborn (MI), USA

[8] Ran, Y. et al (2005) Pedestrian classification from moving platforms using cyclic motion pattern, IEEE International Conference on Image Processing 2005, Volume 2, pages: II 854-7

[9] Van Dorp, P.; Groen, F. (2003) Real-Time Human Walking Estimation with Radar, International Radar Symposium IRS 2003, Dresden, Germany, pp 645-650

[10] Lei, J.; Lu, C. (2005) Target Classification Based on Micro-Doppler Signatures, IEEE International Radar Conference 2005; Arlington, USA; pp 179-183

[11] Klotz, M. (2002) An Automotive Short Range High Resolution Pulse Radar Network, Phd. Thesis, Hamburg University of Technology, Hamburg, Germany

[12] Luebbert, U. (2005) Target Position Estimation with a Continuous Wave Radar Network, Phd. Thesis, Hamburg University of Technology, Hamburg, Germany

[13] Foelster, F., Rohling, H. and Meinecke, M. M. (2005) Pedestrian recognition based on automotive radar sensors, 5th European Congress on Intelligent Transportation Systems and Services 2005, Hannover, Germany

[14] Heuel, S.; Rohling, H. (2011) Two-Stage Pedestrian Classification in Automotive Radar Systems, International Radar Symposium IRS 2011, Leipzig, Germany

[15] Nalecz, M.; Rytel-Andrianik, R.; Wojtkiewicz, A. (2003) Micro-doppler analysis of signals received by FMCW radar, Proceedings of International Radar Symposium 2003, Dresden, Germany, pp. 651-656

[16] Duda, R. O.; Hart, P. E.; Stork, D. G. (2001) Pattern classification, Wiley, New York

[17] Schuermann, J. (1996) Pattern Classification - A Unified View of Statistical and Neural Approaches, Wiley, New York

[18] Schiementz, M. (2005) Postprocessing Architecture for an Automotive Radar Network, Phd. Thesis, Hamburg University of Technology, Hamburg, Germany

[19] Meinecke, M.-M.; Rohling, H. (2000) Combination of LFMCW and FSK modulation principles for automotive, German Radar Symposium GRS 2000, Berlin, Germany

[20] Winner, H., Hakuli, S., Wolf, G., (2009), Handbuch Fahrerassistenzsysteme, Vieweg+Teubner Verlag / GWV Fachverlage GmbH, Wiesbaden, Germany

[21] Smart Microwave Sensors GmbH, (2012), UMRR | LCA BSD Technical Information Sheet, available at http://www.smartmicro.de/images/stories/contentimage/automotive/LCA and BSD Technical Information.pdf, Accessed: 30 August 2012

[22] Perry, J. (1992) Gait Analysis - Normal and Pathological Function, SLACK Incorporated, ISBN 1556421923

[23] Foelster, F.; Ritter, H., Rohling, H. (2007) Lateral Velocity Estimation for Automotive Radar Applications, The IET International Conference on Radar Systems, Edinburgh, Great Britain

[24] Boser, B. E.; Guyon, I. M.; Vapnik, V. N. (1992) A training algorithm for optimal margin classifiers, In D. Haussler, editor, 5th Annual ACM Workshop on COLT, Pittsburgh, PA, pp. 144-152

[25] Wu, X.; Kumar, V.; Ross Quinlan, J.; Ghosh, J.; Yang, Q.; Motoda, H.; McLachlan, G. J.; Ng, A.; Liu, B., Yu, P. S. and others (2008) Top 10 algorithms in data mining, Knowledge and Information Systems, Vol. 14, No. 1., pp. 1-37

[26] Branko, R.; Sanjeev, A.; Nei, G. (2004) Beyond the Kalman Filter: Particle Filters For Tracking Applications, Artech House Inc.

ultraMEDIS – Ultra-Wideband Sensing in Medicine

Ingrid Hilger, Katja Dahlke, Gabriella Rimkus, Christiane Geyer, Frank Seifert, Olaf Kosch, Florian Thiel, Matthias Hein, Francesco Scotto di Clemente, Ulrich Schwarz, Marko Helbig and Jürgen Sachs

Additional information is available at the end of the chapter

http://dx.doi.org/10.5772/54987

1. Introduction

The exploitation of electromagnetic interaction with matter specifically with organic tissues is a powerful method to extract information about the state of biological objects in a fast, continuous and non-destructive (i.e. painless) way. These interactions are mainly based on two groups of phenomena.

One proceeds on an atomic and molecular level, which is typically described by the macroscopic quantities permittivity ε, permeability μ and conductivity σ. The physical reasons of possible interactions may be quite manifold. Here, in connection with ultra-wideband sounding, we restrict ourselves to pure electric interactions which affect the permittivity and conductivity via the motion of free charge carriers (free electrons and ions), the Maxwell-Wagner polarization (also Maxwell-Wagner-Sillars polarization) at boundaries, reordering of dipolar molecules or oscillations on an atomic or nuclear level. We assume that all involved substances have the permeability of vacuum $\mu = \mu_0$. An overview of relevant interaction mechanisms for biological tissue is given in [1], and sub-chapter 3 deals with some selected examples. The related effects are scattered over a huge frequency band covering 15…18 decades. In this paper, we limit ourselves to RF and lower microwave frequencies. Water – the key building block of life –shows dipole relaxation within the considered frequency band. Additionally, it has a very high permittivity in comparison with other natural substances. Hence, water will play an important role for UWB-sounding of biological tissue or human and animal subjects. Examples exploiting this fact are discussed in sub-chapter 5 dealing with breast cancer detection or in [2], which refers to lung edema. The frequency bands of our experiments were selected depending on physical requirements (propagation attenuation, relaxation time) and implementation issues of the sensor electrodes (e.g. antennas).

The second group of phenomena refers to macroscopic effects like reflection and refraction of electromagnetic waves. These effects appear at boundaries between substances of different permittivity or conductivity. Thus, a human body illuminated by radio waves will generate new waves which may be registered by an UWB radar sensor. The strongest waves are provoked by the skin reflection due to the large contrast between air and skin. But also inner organs will leave a trace in the scattered waves since firstly, electromagnetic waves within the lower GHz range may penetrate the body, and secondly, the various organs have different permittivity (e.g. due to different water content) leading to reflections at the organ boundaries. These waves can be used to reconstruct high resolution 3D microwave images of external or internal body structures and also to track their shape variation and motion.

It should be emphasized that motion is a strong indicator of vital activities like breathing, heartbeat or walking which can be registered remotely via UWB-radar sensing. This opens up new approaches of medical supervision as exemplified in sub-chapter 4, rescue of people in dangerous situation [3], [4] or supervision of people in need [5], [6].

In what follows, we like to review first some important requirements and technical solutions of high-resolution short-range UWB-sensor aimed at medical applications before we discuss a couple of selected aspects of medical ultra-wideband sensing in greater detail as for example:

a. Impedance (or dielectric) spectroscopy: It is performed to quantify and qualify biological tissue and cells. Here, we have to deal with reflection measurements at an open ended coaxial probe which is in direct contact to the material under test.
b. Organ motion tracking: It is aimed to observe temporal shape variations of the heart and the lung in order to trigger a magnetic resonance (MRI)-tomography. This task requires a remotely operating MIMO-antenna array with an up-date rate which is sufficiently high to follow mechanical motions up to 200 Hz.
c. Remote microwave imaging for surface reconstruction: It may be used as first step in a chain of further UWB-investigations based on remote sensing for inner organ evaluation. In the scope of this work, the data capture was implemented by scanning a torso. Under real conditions, such measurements must be made in real time using a large MIMO array (large in the sense of the number of antennas) in order to avoid artifacts due to body motions during the scan time.
d. Contact-based microwave imaging: In this case, the antennas are placed onto the skin either directly or through a thin dielectric layer in order to emphasize the radiation into the body. Applications are cancer detection or organ supervision and monitoring requiring densely occupied MIMO-arrays based on small radiators.

2. ultraMedis sensing devices

2.1. Requirements

The following overview summarizes some technical key features and requirements to be satisfied by the sensor electronics corresponding to the application types a) – d).

Bandwidth: UWB sensing is an indirect measurement method. As a general rule of thumb, one can state that the quantity respectively the reliability of the gathered information increases with the bandwidth of the sounding signal. It is predetermined and limited by the physical effects involved as well as technical implementation issues. In the case of impedance spectroscopy (application type a)), we applied Network Analyzers or M-sequence devices (see below) whose operational band was spanned from several hundred KHz to some GHz. For UWB-radar experiments, the frequency band was typically limited to 1-13 GHz or to 1-4…8 GHz. The lower cut-off frequency is typically determined by the size of the antennas while wave penetration into the body limits the upper frequencies. Correspondingly, the first frequency band was applied for application type c) which involves only propagation in air. The sensor device was a modified M-Sequence radar [7], [8]. If the sounding field must penetrate the body (applications b) and d)), the upper frequency may be reduced since wave attenuation dominates the other effects. Some details concerning the sensor structure are summarized in the next sub-chapter.

Field exposition: The strength of field exposition appearing in connection with UWB-sensing is usually harmless for biological tissue. Nevertheless, one should distinguish between an average charge and a short impact. A certain average charge of the test objects is required in order to achieve a given quality (in terms of signal-to-noise ratio) of the captured signals. The strength of the maximum impact is related to the type of sounding signals applied by the sensor. Sine waves and M-sequences cause maximum impacts of about the same strength as their average exposition is. In contrast, pulse systems lead to high-peak impacts even if their average charge keeps the same value as for sine waves or M-sequences. Hence, some attention should be paid to the selection of the sensor principle if applicators in direct contact with tissue and short electrode distances are involved (applications of type a) and d)) since this may lead to high field strengths within the test objects causing non-linear effects or even local damages.

Time stability: Here, the term 'time stability' refers to a summary of several facets of sensor performance like precision of equidistant sampling (i.e. linearity of time axis), long-term stability (drift), and short-term stability (jitter). These aspects pertain to all applications. They strongly affect the quality of the captured signals and, hence, the achievable results of the signal processing. In detail, the following items are concerned:

- the quality of time-frequency conversions via FFT, which is an important tool for signal processing
- the quality and durability of sensor calibration (3- or 8-term calibration),
- the limits of super-resolution techniques and the quality of background removal,
- the capability to detect weak targets in the neighborhood of strong ones, and
- the micro-Doppler sensitivity with respect to weak target detection and slow motions tracking.

Some additional aspects of this topic are summarized in Chapter 14. A thorough discussion of related problems and their linkage to the sensor principles is given in [9].

Measurement rate, channel number, data handling: Except for impedance spectroscopy, the applications mentioned above require MIMO-sensor arrays which have to run at a reasonable update rate. On the one hand, this assumes cascadable sensors in order to build multi-channel systems, and on the other it poses some challenges for the data handling resulting from the large number of measurement channels and the high measurement rate. Chapter 14 (section 2.1) adverts to some measures which avoid redundant and inefficient data. Irrespective of these measures, the data throughput will be still quite high so that standard PCs and interfaces quickly reach their capacity limits.

Radiators: The radiators represent the interface between sensor electronics and test object for applications b) – d). They have to convert guided signals into free waves and vice versa. As they are linear and time-invariant devices, they may be operated with any type of signals. Certainly, their major features are the bandwidth and the beam width which should be as large as possible if they are applied for UWB imaging. However, these characteristics describe their performance only insufficiently particularly for UWB short-range applications. Ideal UWB antennas for our purposes should provide a short and angular independent impulse response (time shape and wave front), they should convert the incoming signal completely into a free wave, and the incident fields should be converted into voltage signals avoiding any re-radiation or scattering by the antenna. These conditions are contradictory and cannot be met by a physically realizable antenna.

A short impulse response is needed for high range and image resolution as well as the ability to recover weak targets closely behind surfaces. Otherwise, we risk the loss of the target since a slowly decaying surface reflex distorts the target response. If that signal is too abundant, even sophisticated background removal strategies will not be able to dig it out.

The angular independent impulse response is important for the imaging algorithm. For every image pixel or voxel, it has to coherently integrate signals which are captured from different positions. In order to ensure the coherence of this integration, the propagation time to the considered pixel (voxel) must exactly be known. This involves the knowledge of the propagation speed as well as the knowledge of the deviation from a spherical wave front created by the antennas. In order to achieve a simple and manageable imaging algorithms, the involved antennas should avoid such distortions, hence they should be (electrically) small [9].

However, this contradicts the physical conditions for an efficient conversion between guided signals and freely propagating waves (see Bode-Fano limit and Chu-Wheeler limit [10]). Additionally, efficient antennas backscatter (re-radiate) half of the incident power in the ideal case. For targets in close proximity of the antennas, this leads to multiple reflections which are hardly to remove by signal processing. As we saw for the impulse behavior, the inefficient antennas behave again best regarding their re-radiations (structural antenna reflections are omitted here for shortness). Hence, one has to find a reasonable compromise between efficiency and impulse as well as scattering performance. Antenna efficiency is an important issue in connection with noise suppression and high path losses. For imaging at very close distances, noise induced measurement errors are falling below the strength of clutter and systematic deviations. Here, efficiency should take a back seat in antenna design in favor of a clear impulse response and low self-reflections. The sensitivity of the sensor electronics should compensate for the efficiency degradation of the antennas.

Furthermore, radiator related items concern array aspects such as the geometric shape of the array, radiator density (depending on antenna size and acceptable cross talk) and distribution within the array as well as polarimetric issues.

In the context of this chapter, we distinguish two types of antenna modes. For the first one, the antenna radiates in air, whereas the other mode refers to interfacial antennas which are in contact with the test object. In both cases, due to the short target distance, we have to deal with spherical waves and their specific reflection and refraction behavior which are accompanied by wave front deformations as well as the generation of evanescent and head waves [9].

Device miniaturization: The application of unusual radiators and the operation of dense MIMO-arrays require new sensor concepts avoiding long RF-cables (which have to be matched at both sides) as well as large and heavy measurement devices as network analyzers. Future MIMO-array implementations for medical microwave imaging should have jointly integrated radiator and sensor electronics in order to permit the operation of mismatched antennas, to increase the stability of the system, to reduce its susceptibility to environmental conditions (e.g. temperature variation or strong magnetic fields) and to simplify its handling. The project HaLos (Chapter 14) was addressing related questions of sensor integration.

2.2. ultraMEDIS sensor electronics

In view of the previous discussion, we mostly abstain from the use of network analyzers since they will not meet the requirements of future developments of the sensing technology even if they best fulfill the demands with respect to sensitivity, bandwidth and reliability of measurement data. A new sensor concept with comparable performance but higher measurement speed, better MIMO capability and integration friendly device layout exploits ultra-wideband pseudo-noise sequences (namely M-sequences) for the target stimulation instead of the sine waves of a network analyzer. This measurement approach was favored for our investigations. Device concepts applying sub-nanosecond pulses were rejected due to their inherent weakness concerning noise and jitter robustness. The interested reader is referred to Chapter 14 and [9] for further discussions of the pros and cons of various sensor principles.

The block schematics of the M-sequence prototype devices applied by ultraMEDIS are depicted in Figs. 4 and 6 in Chapter 14. The integrated RF key components were provided by the project HaloS while the implementation of prototype devices was performed by MEODAT GmbH and later on by ILMSENS. A special issue of an M-sequence device provides 12 GHz bandwidth. Its implementation is based on [8].

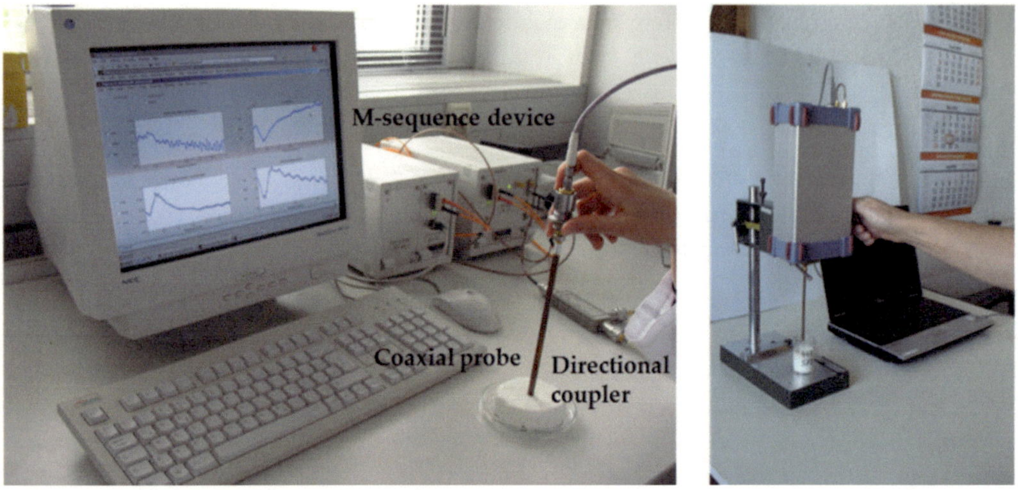

Figure 1. M-sequence based impedance spectroscopy (bandwidth 17 MHz – 4 GHz; 9th order M-sequence). Left: Device implementation with external coupler. Right: M-sequence device with internal coupler and rigid probe connection to improve measurement reliability.

Figure 4 of Chapter 14 (HaLoS-project) relates to the basic structure which can be found in all device modifications. Such device configurations were applied in an early project state for microwave imaging and organ motion tracking experiments. Involving a directional coupler, it is further used for impedance spectroscopy as exemplified in Fig. 1. Multi-channel systems and MIMO-arrays are based on the device philosophy as illustrated in Fig. 6 of Chapter 14. Implemented examples are depicted in Fig. 2 to Fig. 4.

Figure 2. M-sequence two-port network analyzer (operational band 40 MHz – 8 GHz, 9th order M-sequence, USB2 interface). It can be extended by an RF-switch matrix for MIMO-radar imaging.

Figure 3. 4Tx-8 Rx MIMO device for organ motion tracking in MRI tomographs. (operational band 17 MHz – 4.5 GHz; 9th order M-sequence; maximum up data rate 530.4 Hz, Ethernet data link, data acquisition on Linux PC)

Figure 4. 8 Tx-16 Rx MIMO radar for microwave breast imaging (operational band 20 MHz – 6 GHz; 9th order M-sequence, USB2 interface). M-sequence units (as shown in Fig. 6 of Chapter 14) and RF-front ends are separated to get more flexibility for experimental purposes.

2.3. Antennas and sensor elements

2.3.1. Introduction

The exploitation of UWB microwave technologies for biomedical diagnostics requires the development of antennas and sensors tailored to this application. The integration of the antennas as a part of a complex system leads to serious compatibility and functionality constraints, which must be properly addressed for high system performance. Within ultraMEDIS, two goals were pursued: Firstly, the detection of early stage breast cancer and secondly, the development of a magnetic resonance imaging (MRI) compatible navigator system (Section 4). These two goals provide different challenges in terms of antenna design, implementation, and experimental evaluation, both with respect to mechanical and electrical constraints [10]. As both applications involve different approaches, they will be treated separately.

2.3.2. Dielectrically scaled antennas

For the process of detecting early stage breast cancer by means of microwave imaging (Section 5), the antenna size, the effective radiation of electromagnetic energy into the body, and the operational bandwidth turn out to be the main constraints regarding the design of the antenna.

The miniaturization of the antenna is of main concern to meet the requirements of the devised imaging technique (Section 5) of placing an array of many antennas surrounding the target under investigation (i.e. the human breast), considering also the small anatomic dimensions on the scale of the wavelengths of operation. In general, electrically small antennas are mismatched or narrowband [10], [11]. One possibility to overcome these obstacles is to use the antenna in *contact mode*, i.e., placing the antennas in contact with the target under investigation (e.g., the human body). With this *modus operandi* the antenna will radiate into a dielectric material (the human body), and it can be geometrically scaled by a factor of about $\sqrt{\varepsilon}$, where ε represents the dielectric permittivity of the target, without changing its electrical dimensions and, therefore, its radiative properties [12].

The *contact mode* presents advantages also with respect to the constraint of the effective radiation of electromagnetic energy into the body. In fact, it will not suffer from reflections occurring at the air-skin interface, due to the dielectric mismatch between the two grossly different media. This will also simplify the imaging processing since it prevents the need of surface reconstruction [10], [12]. Though, for practical and hygienic reasons, it is less convenient to put the array of antennas in direct contact with the patient's skin. However, the addition of a further layer, e.g. a disposable thin dielectric film, could spoil the effective radiation into the body[1]. Electromagnetic simulations (confirmed by measurements, Section 5.4) have shown that even the addition of a thin layer (~ 0.5 mm) can reduce the radiated power to less than half compared to the direct contact case. The implementation of a matching layer (with similar permittivity to human body tissues) can cure this effect (Fig. 5).

Eventually, particular attention has to be paid to the operational bandwidth of the antenna, especially to the lower cut-off frequency, which limits the penetration depth into the target. Based on a specific 14-layer model mimicking a trans-thoratic slice from the visual human data set, we have computed the penetration of electromagnetic waves into a human body, as shown in Fig. 6 [13], [14] and [15]. A strong increase of the signal attenuation with increasing frequency is clearly seen. Therefore, the lower cut-off frequency has to be set between 1 GHz and 2 GHz.

---

[1] The relevance of this phenomenon depends also on the antenna type used

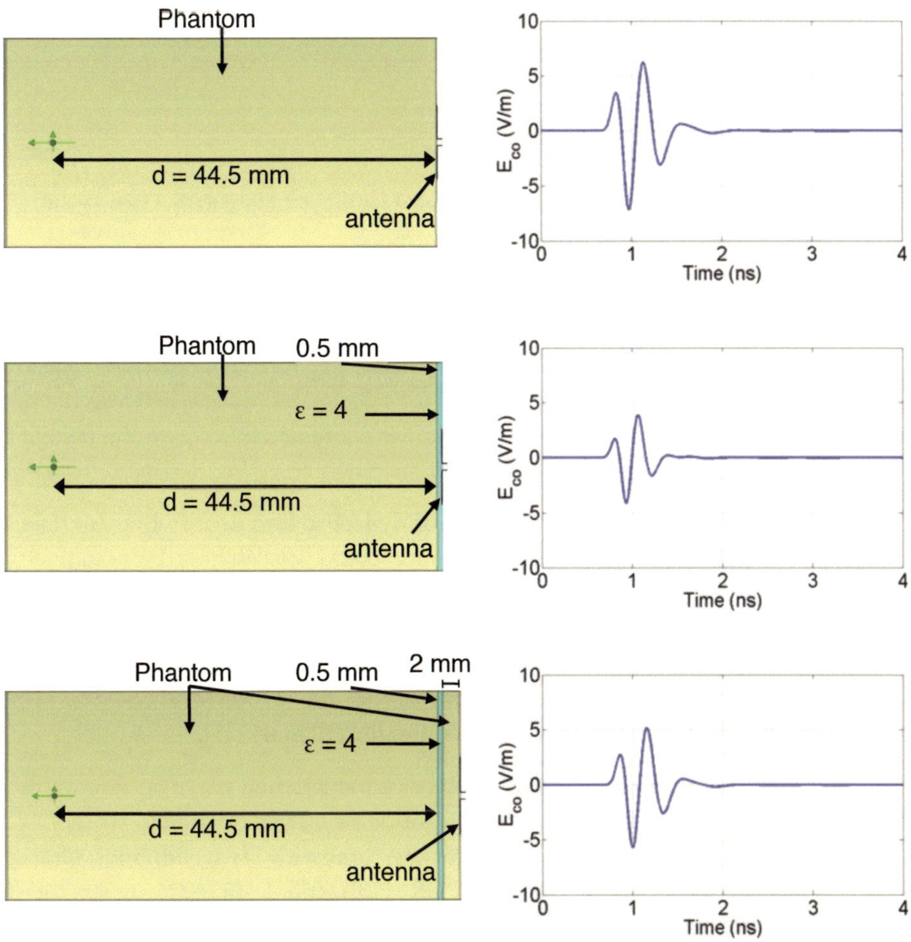

Figure 5. Simulated scenarios to investigate the effective radiation of electromagnetic energy into the body (on the left). The antenna used is a bow-tie excited by a Gaussian pulse of a duration of around 80 ps FWHM. The *Phantom* material is a homogenous dispersive material simulating the dielectric behavior of the human body tissues. The results (on the right) represent the time-dependent co-polar component of the electric field evaluated at a distance of 44.5 mm from the phantom interface (the green spot in the figure). The examined cases are, from top to bottom: the antenna in direct contact with the phantom; with the implementation of a thin dielectric layer; with the implementation of a matching layer plus a thin dielectric layer, respectively.

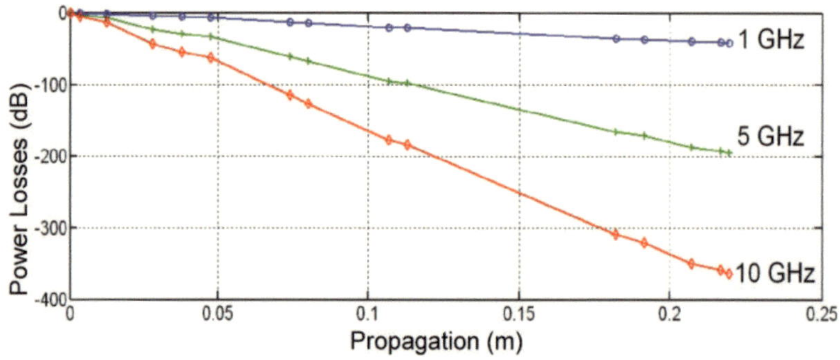

Figure 6. Penetration depth into a 14-layer model computed for different frequencies (see legend) [13].

The dielectric scaling of traveling-wave antennas, like tapered slot antennas and horn antennas, has to consider all three factors - the phase constant, the wave impedance, and the feeding structure [16]. Accordingly, extended iterative full-wave simulations were performed. The key parameters of the wideband radiation properties of double-ridged horn (DRH) antennas turned out to be the curvature of the ridges, the spacing between them, and the geometry of the pyramidal casing itself. Traveling-wave antennas have the benefit that the dielectric medium used to scale the antenna will serve as an embedded *matching medium*.

Our initial design was based on the immersion of dielectric liquids into the sinkhole of a dielectrically scaled DRH antenna [15]. The antenna could successfully be manufactured, using acetone as dielectric medium, with a scaling factor of about 4, but it was still insufficient to obtain a sufficiently compact antenna. The straight-forward approach to solve this problem was to replace the acetone by alternative high-permittivity dielectric materials, like a solid sintered ceramic. The ceramic powder is the commercial product LF-085 manufactured by MRA Laboratories based on neodymium titanate [17].

As the complex permittivity ε of the ceramic body of the antenna plays an important role not only in terms of antenna design but also in terms of *matching medium*, we performed measurements to access the complex permittivity following two different strategies: employing a split-post dielectric resonator (SPDR, [18]) and a dielectric resonator (DR) technique [19][20]. Both techniques are resonant methods and, hence, limit the experimental studies to a small set of discrete frequencies, because specific sample geometries are required for each measured frequency. The results showed that the sintered ceramic presents low frequency dispersion with a mean value of the real part of the permittivity $\varepsilon' \cong 72$ [21], offering the potential for a scale factor of around 8.

The exploitation of the full potential of dielectric scaling leads to an aperture size of only 11 mm × 16 mm, but also to the reduction of the input impedance by the same scaling factor as by which the geometrical dimensions are scaled, resulting in a low value below 10 Ω. This value implies a large mismatch in terms of standard electronic equipment, which is usually designed for a characteristic impedance of 50 Ω. In order to maintain the compatibility with standard electronic equipment, the antenna retains an aperture size of 24 mm × 24 mm, and a frequency bandwidth ranging from 1.5 to 5.5 GHz (Fig. 7).

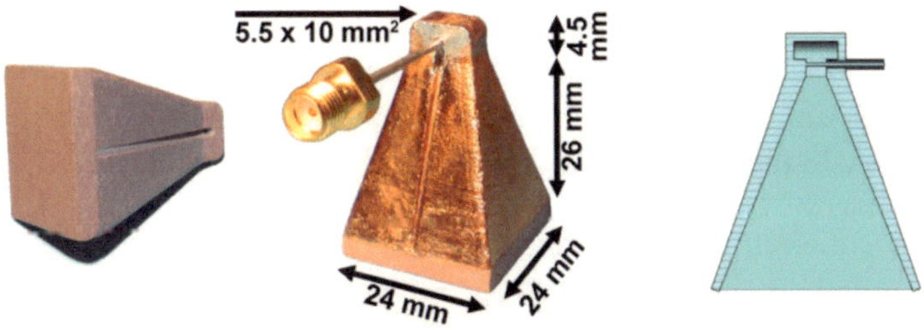

Figure 7. The left-hand panel shows the completely processed ceramic body of the antenna. The center panel depicts the final antenna including metallization, feed line and dimensions. The right-hand panel displays the profile of the ridges.

In order to properly manufacture the antenna and obtain good yield (of around 75 %) and reproducibility, specific manufacturing steps have to be devised, as indicated in the right image of Fig. 8 [22]. First, the white-colored ceramic powder (Fig. 8a) is pressed into the specific pyramidal shape; a cubic base accommodates the asymmetric feed (Fig. 8b). The dimensions of the raw body are slightly enlarged in order to respect the shrinkage upon sintering. The sintered body is complemented by grooves, which form the ridges (Fig. 8c), and is subsequently galvanically metalized with copper or gold (Fig. 8d). Due to the high permittivity of the ceramic body and a feed impedance of 50 Ω, the ridges are nearly linear in geometry, in contrast to the markedly curved shapes found in antennas for operation in air [12][15]. The feed is provided by a coaxial cable whose center conductor is fed through a small bore to the narrow end (diameter about 1.2 mm) of the ridged waveguide. A plastic housing and epoxy fixture provide a compact and mechanically rigid construction, to protect the ceramic body and the coaxial feed against torque and damages due to improper handling. It also provides a mechanical fixture to mount the antennas in an array of complex geometry (Fig. 8e). Further details of the manufacturing processes are given in [23].

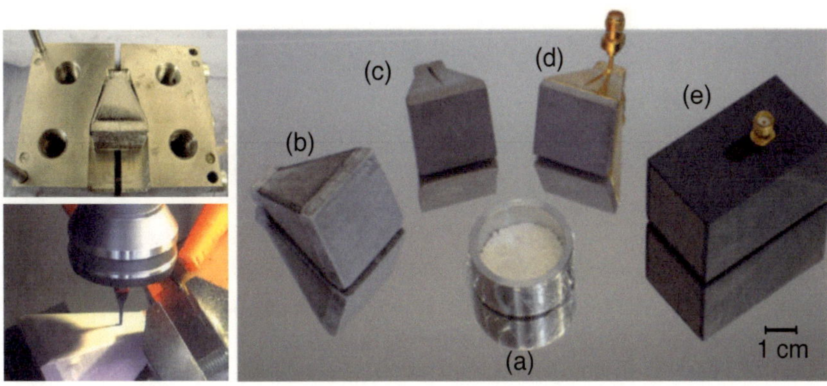

Figure 8. The left-top picture shows a dry pressed green body of the antenna with some lubricant on it inside the dismantled mold. The left-bottom picture depicts the milling process to work in the ridges into the sintered ceramic body. The right picture illustrates the manufacturing steps for the sintered horn antenna: powder raw material (a), pressed raw (b), sintered (c), metallized (d), and fully packaged versions (e).

A further issue of dielectrically scaled antennas is related to their measurement and characterization. As common measurement techniques and equipment cannot be applied, we followed three different strategies: measurements made in the frequency domain, measurements made in the time domain, and basic tests with volunteers.

In order to provide dielectrically matched surrounding conditions for the antenna body, the antennas were tested in de-ionized water. The results were then compared with data obtained in a more realistic environment, i.e. with the antenna put into contact with phantoms mimicking human tissues. The phantoms consisted of oil, water and some additives [24]. The dielectric permittivity ε' and the loss tangent ε''/ε' of the phantoms can be controlled by changing the percentage of oil [24], [25] (Section 5.2).

The frequency behavior of the reflection coefficient is shown in Fig. 9. It can be observed that while the reflection coefficient for the test against the phantom (with 40% oil, Section 5.2) approaches levels around –8 dB, the antenna is even better matched in water, leading to a further decrease of the reflection coefficient by 4 dB in the frequency range of interest. The compromise between input matching to a certain medium and the geometrical dimensions of the antenna denotes the key trade-off exploited for our design. In order to study the reflection occurring at the aperture plane, which is influenced by the permittivity matching between the dielectric medium composing the antenna body and the human skin, we performed time domain reflectometry (TDR) measurements by having the antenna radiate into different media [22].

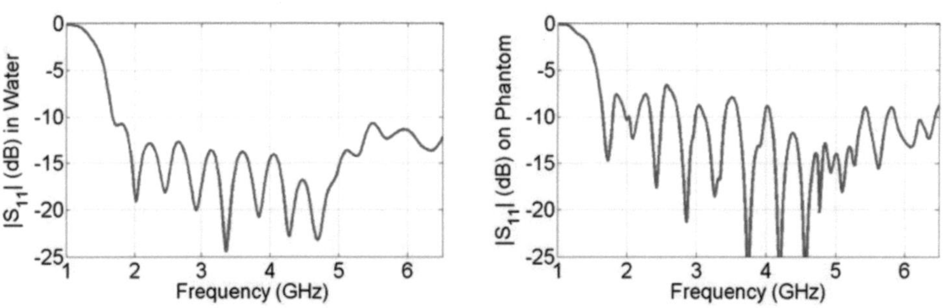

Figure 9. Reflection coefficient measured in water (left) and on a skin mimicking phantom (right).

The results for water and skin (in-vivo) are shown in Fig. 10. The amplitude of the reflected signal with the antenna operating in water is significantly smaller than the one with the antenna operating on skin. This result indicates, in agreement with the frequency domain measurements, that the antenna is better matched to water than to skin. We note from Fig. 10 that the reflection occurring at the aperture due to impedance mismatch results in a signal with a longer decaying time. The larger the impedance mismatch is the longer the decaying time is. This feature is due to the fact that part of the reflected energy does not leave the antenna through the well-matched feed towards the signal source but remains within the antenna body.

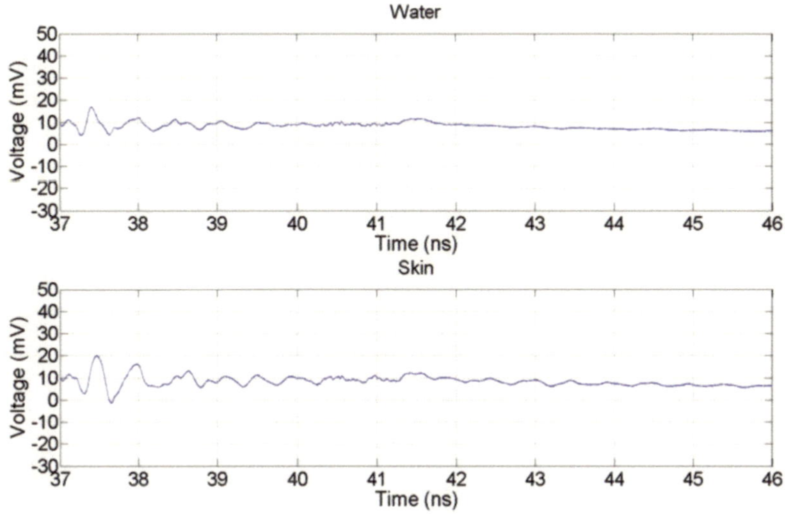

Figure 10. Time domain reflections caused by the aperture plane due to dielectric mismatch between the antenna and water and skin, respectively (top and bottom).

The measurement of the radiating behavior is more complicated. To evaluate standard antenna parameters (e.g. gain, radiation pattern, etc.), the antennas should be placed in the Fraunhofer region. However, due to the high dielectric losses of water, the antenna could be placed at a maximum distance of 10 cm, which is not sufficient to meet the Fraunhofer region (starting from around 35 cm), but still is large enough to let the antennas operate in the Fresnel region. Near-field measurements are of main concern since the antennas are designed to be used in contact mode for biomedical imaging applications. The results show that the antennas present a flat frequency response (measured along the boresight direction), after the compensation of the frequency dispersion of the water (left diagram in Fig. 11), and 3 dB beam widths of nearly 20° for the E-Plane, and nearly 28° for the H-Plane (right diagram in Fig. 11).

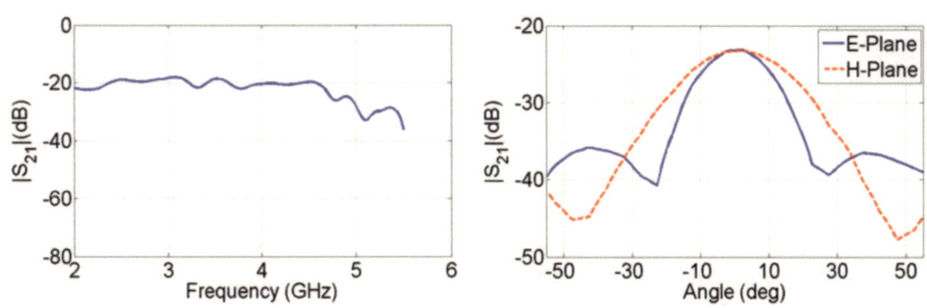

Figure 11. Transmission behavior between two identical antennas operating in de-ionized water at a distance of 10 cm. The frequency response, compensated for water frequency dispersion is shown in the left diagram, while the near-field pattern is represented on the right. The blue and red curves illustrate a cut along the E-Plane and the H-Plane, respectively.

Eventually, in order to demonstrate the functionality of the ceramic antennas under realistic conditions, we have performed preliminary transmission measurements through a breast of a female volunteer, and the monitoring of the heartbeat, as illustrated in Figs. 12 and 13. The dynamic range of the achievable signal can be determined from Fig. 12 in comparison with the ideal transmission through a 4 cm path inside distilled water by a face-to-face arrangement of the antennas. The monitoring of the heartbeat was performed on a 35 years old healthy male volunteer. During measurement, the volunteer was sitting still and was holding breath in the state of maximal breathing in. The measurement was performed with an M-sequence radar in a bi-static configuration [Section 2.2]. Figure 13 shows the heartbeat signals as monitored. Upon Fourier transformation, we extracted a beat rate of nearly 75 beats per minute, which is considered normal for an adult. The results display very clearly the characteristic feature of heartbeat, thus manifesting a favorable dynamic range. This opens up promising applications for realistic monitoring and imaging tasks. Further details of these tests can be found in [26] and [22].

The full dielectric scaling, as previously stated, offers the potential for a further size reduction of the antenna. Accordingly, we continued our research and succeeded in developing a ceramic DRH antenna with an aperture of only 16 mm × 11 mm and a lower cut-off frequency around 1.5 GHz. Due to the input impedance of the antenna below 10 Ω, active receive and transmit versions are under development in the framework of HALOS (Chapter 14), employing an UWB low-noise subtraction circuitry and power amplification [27]. The manufacture of the tiny antenna followed similar production steps as for the previous version. It proved quite challenging because of the reduced size, requiring additional specific production steps and iterative testing procedures [28].

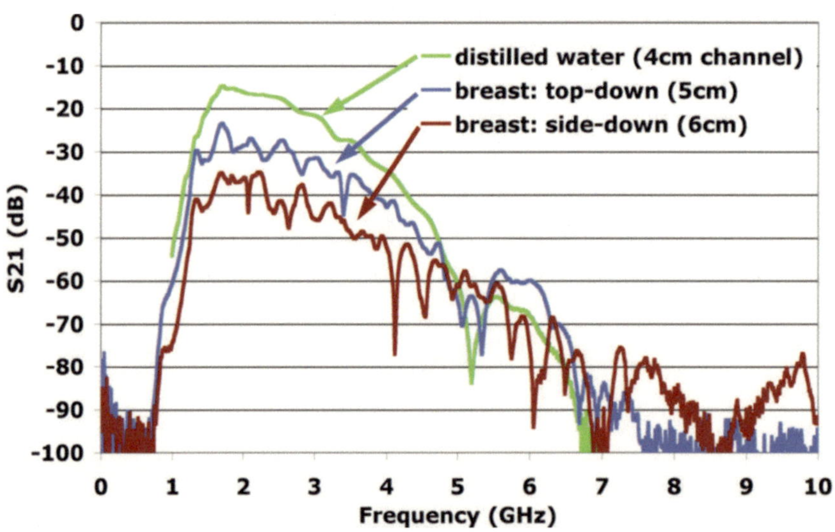

Figure 12. Transmission measurements through a female breast performed with the ceramic double-ridged horn antennas in comparison with a reference measurement of 4 cm distilled water (upper curve).

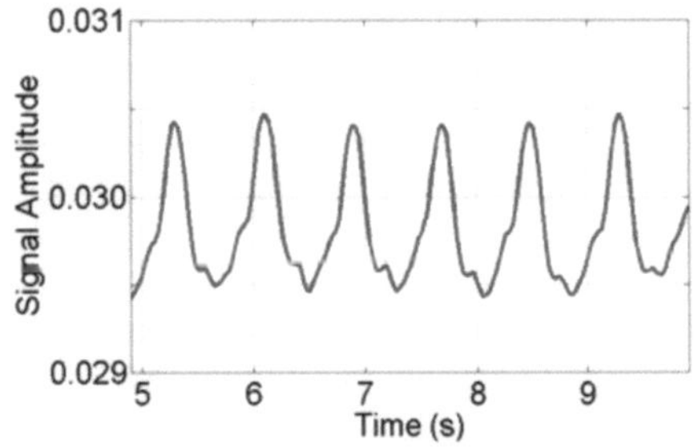

Figure 13. Monitored heartbeat signal of a 35 years old healthy male volunteer. Upon Fourier transformation, we extracted a beat rate of nearly 75 beats per minute.

2.3.3. *MRI compatible antennas*

Magnetic resonance imaging (MRI) systems are among the most sensitive diagnostic methods in medicine for the visualization of soft tissue [10]. At present, more than ten million MRI examinations of patients are performed per year worldwide. Given such a progressive development, further improvements of this diagnostic technique are under way. However, MRI systems are not *per se* capable of creating focused images of moving objects like the human heart or the thorax of the patient while breathing. Instead, additional techniques like breath holding, ECG triggering, or MR navigation methods are required. Such techniques either cause some inconvenience for the patient, or they are even not applicable for upcoming generations of MR scanners. A novel approach which overcomes these obstacles is the use of low-power multi-static UWB radar as a contactless navigator technology for MR tomography [29], [30]. To devise such navigators, the design of antennas compatible with MRI systems, i.e. antennas which do not interact neither electrically nor mechanically with the operation of the MR scan, is needed. The strategy to follow when designing an MRI compatible antenna is the minimization of mutual interaction between the metallized antennas and the strong static and gradient magnetic fields. Several experiments with conventional wideband antennas showed strong mechanical interactions, pointing out the need for special antenna designs [31]. The operational conditions inside an MR scanner are determined by three different types of fields. First, a static magnetic field B_{stat} = 1.5 to 7 T provides a reference orientation of the nuclear spins of the regions under inspection. Furthermore, gradient magnetic fields with a slope of dB_{grad}/dt = 50 T/s at the rising edge are switched during diagnostic measurements to allow for spatial (depth) information of the acquired molecular information. Finally, the nuclear spins are set into precession by a strong (KW range) RF signal at 42.58 MHz/T. The gradient fields induce eddy currents in the metallized sections of the antenna which, in turn, interact with the static magnetic field by exerting a mechanical torque on the antenna structure. The torque can reach peak values of the order of 0.045 Nm for a contiguous metallized area of 20 mm × 30 mm. This value is high enough to result in mechanical amplitudes of several millimeters, deforming or moving the antenna structure, especially in the case of mechanical resonances. Furthermore, the magnetic fields of the eddy currents can lead to artifacts of the MR-image. These interactions inhibit the beneficial application of UWB navigation for magnetic resonance imaging and, therefore, must be avoided. We used a 3-T MR scanner with the resulting RF frequency of 127.8 MHz, which is ten times smaller than the lower cut-off frequency of the UWB antennas employed. As the frequency response of a typical antenna corresponds to a high-pass filter of first order, the stop-band attenuation amounts to 20 dB per decade, indicating the risk of collecting RF power even in the presence of path-loss and shadowing.

The minimization of contiguous metallized area and, hence, eddy currents, is therefore the main issue to be addressed by the antenna design. Additional constraints arise from the intended applications in biomedical diagnostics: weakly frequency-dependent radiation patterns over the entire operational bandwidth, good decoupling between neighboring antennas, and a lower cut-off frequency around 1 GHz. The DRH antenna was identified to be a suitable UWB antenna type to accomplish these requirements. Due to the functional principle of DRH antennas, the minimization of contiguous metallized areas and the realization of a weakly frequency-dependent radiation pattern are in conflict with each other. Horn antennas are typically made entirely out of metallic parts of high electrical conductivity σ, thus suffering from the induction of eddy currents under MR-scanner conditions. Therefore, the major challenge was to modify the double-ridged horn antenna to achieve MR-compatibility, without compromising the favorable radiation properties.

Inspired by commercial counterparts of DRH antennas, we removed the H-plane sidewalls of the pyramidal horn, leaving just a thin wire in the plane of the aperture, as illustrated by the left picture in Fig. 14. As a result, the lower cut-off frequency could be reduced from 2.6 GHz to 1.5 GHz for otherwise unchanged dimensions and operation in air. The comparison with a conventional double-ridged horn antenna with a similar bandwidth revealed that this improvement was achieved at the expense of increased beam width, side-lobes and backward radiation, predominantly at frequencies below 3 GHz, due to the modified aperture distribution and diffraction at the edges of the open construction. The increased beam width led to a slightly increased crosstalk [32]. It can easily be compensated for by re-orienting the antennas relative to each other. While the crosstalk for conventional DRH antennas becomes small for an H-plane alignment, the MR-compatible versions have to be aligned along the E-plane due to the removed H-plane sidewalls and, thus, reduced shielding.

The thickness of the metallization was also reduced in order to exploit the skin effect for a decoupling of the low-frequency eddy current paths. The metal planes were replaced by metallized dielectric boards with a metallization thickness of 12 μm (Fig. 14). This value corresponds to about twice the skin depth at the lowest frequency used. The high-frequency currents determining the radiation of the antenna remain essentially undisturbed while the eddy currents in the KHz range are strongly attenuated by the high sheet resistance. For further optimization of the remaining metallized areas, the distribution of surface currents in the UWB frequency range was inspected by electromagnetic simulations (right image of Fig. 14). Typical results for the normalized surface current are illustrated at 5 GHz (left-most). The surface current is concentrated near the position of the ridge and the edges of the pyramidal frame. According to our expectations, the number of current loops was found to increase with frequency; in contrast, the current distribution across the backward cubical part of the antenna showed little frequency dependence. Based on these observations, a compromise was sought to reduce the plane metallization with the minimal possible distortion of the broadband current distribution. As a result, the conductor faces of the horn sections were separated into strip lines, straight and elliptically shaped, separated by 1 mm, and oriented parallel to the most common current paths, with plain connections at the face edges only. The central part of the right image of Fig. 14 illustrates the resulting geometric arrangement of the slots, while the normalized surface current of the modified antenna at 5 GHz is shown in the right-most part. The main features of the current distribution could be sustained qualitatively both on the pyramidal faces and the backward cubical part of the antenna. Differences occurred mainly for the currents oriented perpendicular to the slots. It is this minor change in current distribution which causes the modified radiation properties discussed above. The ridges themselves required special attention. A grid of holes was eventually identified as the proper solution to reduce the metallization area of the ridges without disturbing the high-frequency current distribution too much. In order to reduce the maximal loop size for low-frequency currents, the outer contour of the horn section was cut and shortened by standard surface-mounted device capacitors.

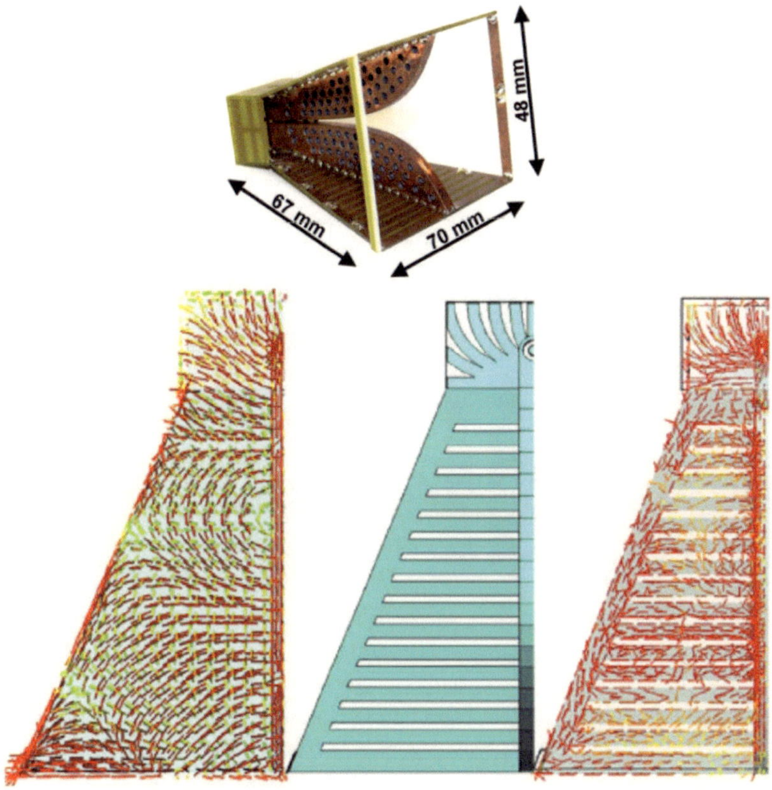

Figure 14. MR-compatible double-ridged horn antenna for a lower cut-off frequency of 1.5 GHz (top). The lower image shows the simulated, normalized current distribution of an unmodified DRH antenna at 5 GHz (left-hand part), the layout of the resulting MR-compatible DRH antenna (center part), and the current distribution of the modified DRH antenna at 5 GHz (right-hand part).

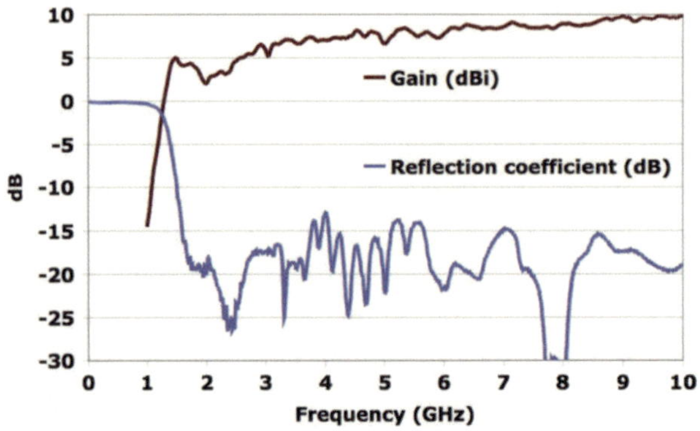

Figure 15. Measured reflection coefficient of the MR-compatible double-ridged horn antenna (lower curve) and the measured antenna gain (upper curve) versus frequency.

Fig. 15 displays measured results for the reflection coefficient and the gain of the modified DRH antenna. A return loss above 10 dB was achieved over the frequency range from 1.5 to 12 GHz. Radiation measurements in an anechoic chamber yielded the radiation patterns illustrated in Fig. 16 for two orthogonal cuts with respect to the plane of the ridges. The half-power beam width, indicated as the black contour line in Fig. 16, was found to vary between 30 and 50 degrees, thus covering a range suitable for the envisaged applications. Except for frequencies around 2 GHz, the main lobe showed little spectral variation. The corresponding frequency variation of the antenna gain is displayed in Fig. 16. These results were found in good agreement with the numerical simulations.

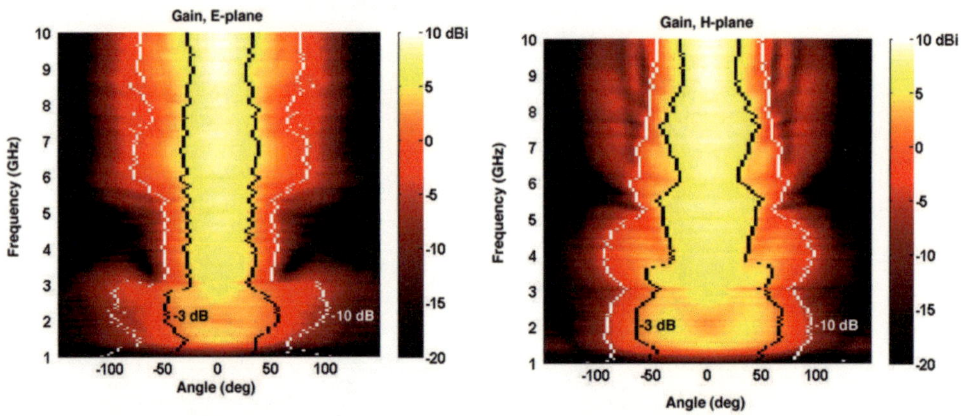

Figure 16. Two-dimensional representation of the measured radiation pattern of the MR-compatible double-ridged horn antenna for the E-plane (left) and the H-plane (right) through the main beam. The scales indicate the antenna gain in dBi. The black and white contour lines illustrate the corresponding beam widths at 3 and 10 dB below the frequency-dependent maximum gain, respectively.

The transient response of the antenna is shown in Fig. 17. Despite the open geometry of the MR-compatible antenna, a low signal distortion could be sustained. The slight angular dependence of the time responses can be attributed to an offset between the phase centers of the antennas and the center of rotation of the antenna positioning system.

The MRI compatible DRH antennas were implemented as part of a UWB MR navigator, by means of which it was possible to take images of the myocardium for the first time without using an ECG as navigator. The quality achieved was comparable with the one achievable with the ECG navigator (see Section 4).

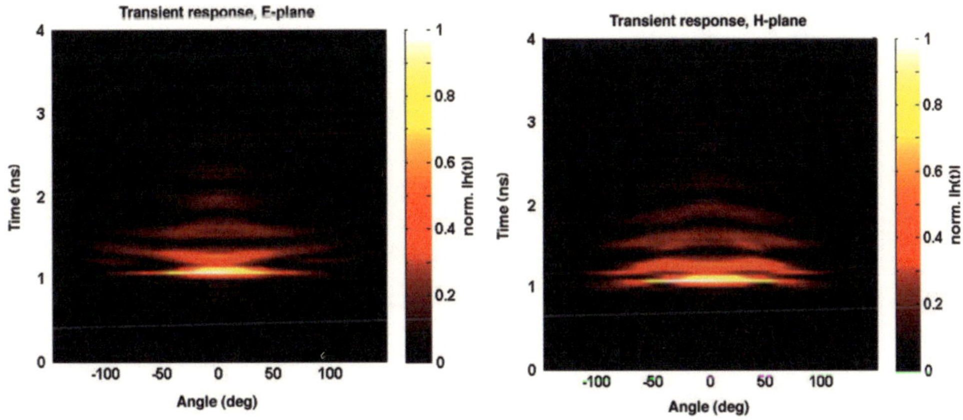

Figure 17. Two-dimensional representation of the measured time domain response of the MR-compatible DRH antenna for the E-plane (left) and the H-plane (right) through the main beam. The scales indicate the normalized impulse response of the antenna.

3. Weak electromagnetic fields and biological tissue

3.1. Impact on living cells

The electrical properties of biological tissues and cell suspensions have been of interest for over a century for many reasons. They determine the pathways of current flow through the body and are very important for the analysis of a wide range of biomedical applications such as functional electrical stimulation and the diagnosis and treatment of various physiological conditions with weak electric currents, radio-frequency hyperthermia, electrocardiography, and body composition. On a more fundamental level, the knowledge of these electrical properties can lead to an understanding of the underlying basic biological processes. Indeed, biological impedance studies have long been an important issue in electrophysiology and biophysics; interestingly, one of the first demonstrations of the existence of the cell membrane was based on dielectric studies on cell suspensions [33].

Biological tissues are a mixture of water, ions, and organic molecules organized in cells, sub-cellular structures, and membranes, and its dielectric properties are highly frequency-dependent in the range from Hz to GHz. The spectrum is characterized by three main dispersion regions referred to as α, β, and γ regions at low, intermediate, and high frequencies [34]. Biological materials can show large dispersions, especially at low frequencies (Fig. 18). Low frequencies are mainly caused by interfacial polarizations at the surfaces between the different materials of which a cell is composed [35]. Reviews of the dielectric properties of cells and the different dispersions are given in the literature [36], [37].

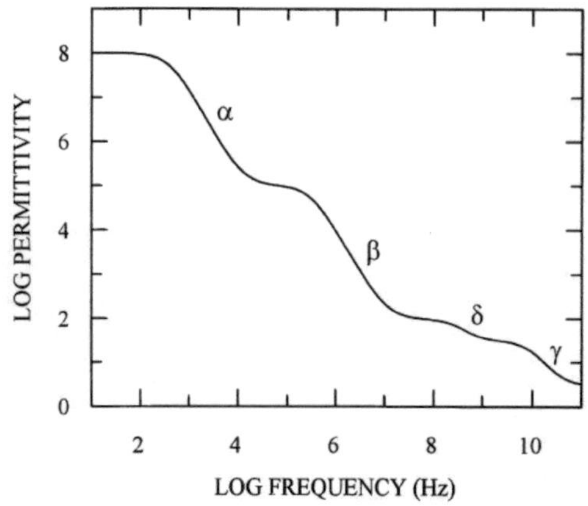

Figure 18. Spectrum of the dielectric properties of cell suspensions and tissues.

The step changes in ε_r are called dispersions and are due to the loss of particular polarization processes as frequency increases. The α-dispersion is due to the flow of ions across cell surfaces, the β-dispersion results from the charge at cell membranes, the δ-dispersion is produced by the rotation of macromolecular side-chains and bound water, and the γ-dispersion is due to the dipolar rotation of small molecules particularly water [35] (figure reproduced with permission from Elsevier).

When exposed to electric fields, living cells behave as tiny capacitors, accumulating charges on the cell surface. The permittivity of living cell suspensions is dependent on the frequency, and falls in a series of the dispersions described above, as frequency increases. The β-dispersion, between 0.1 and 100 MHz, results from the build-up of charges at cell membranes. The difference between permittivity measurements made at two frequencies, on either side of the β-dispersion range, is proportional to the viable biomass concentration. With spherical cells, the permittivity increment is given by equation [38].

$$\Delta\varepsilon = \frac{9\,P\,r\,C_m}{4} \tag{1}$$

As long as there is no change in the cell radius r or the membrane capacitance C_m, the permittivity increment $\Delta\varepsilon$ is proportional to the cell volume fraction P [39].

As a starting point for developing new applications, it is critical to characterize differences in the dielectric properties of the cells, for example human leukocyte subpopulations [40]. Even though, a comparative analysis of the dielectric properties of the cells is necessary, and the UWB radiation on cells itself has to be characterized, too. For this reason, experiments with two different cell lines (tumor cell line BT474 and fibroblasts BJ) were performed. Cell suspensions of these cell lines were disseminated, and the growth rate was determined. Afterwards, the cells were seeded on 96-well plates, cultivated for 24 h and exposed to UWB radiation *via* UWB-M-sequence radar with double-ridged horn antennas of about 10 dBi average gain for 5, 30 or 60 min. As non-treated control, for the same time, plates were placed in a Faraday cage (to avoid any irradiation). After continued incubation for 24, 48 and 72 h, the vitality of cells was determined by colorimetric identification (MTT assay for measuring the activity of enzymes that reduce MTT [3-(4.5-Dimethylthiazol-2-yl)-2.5-diphenyltetrazolium bromide, yellow tetrazole] to formazan, giving a purple color). The measured vitality of control cells was normalized to 100%, and the vitality of exposed cells was put into relation. The vitality of exposed cells was related to non-exposed cells. Due to biological fluctuations, data between 70% and 120% vitality were assessed as not influenced. As depicted in Fig. 19, none of the determined cells was influenced by ultra-wideband electromagnetic waves.

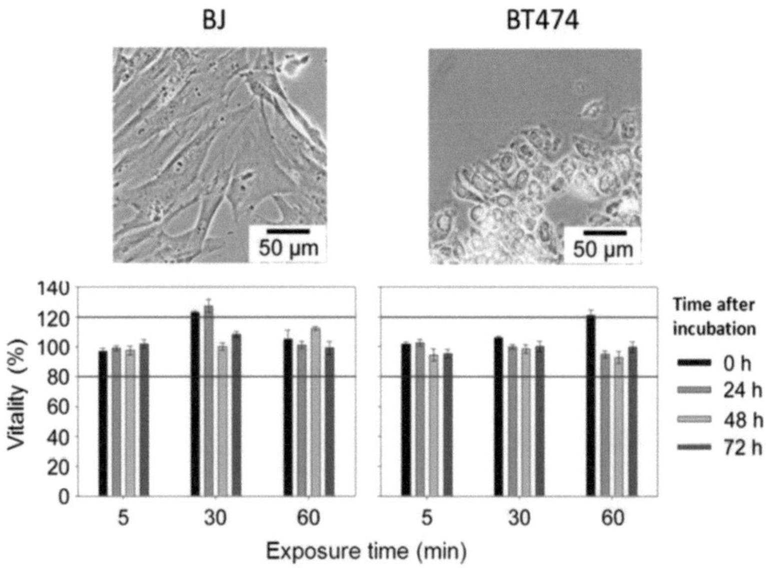

Figure 19. Impact of ultra-wideband electromagnetic waves on the vitality of living cells. The upper part of the figure shows light images of the fibroblast cell line BJ and the cancerous cell line BT474. The lower figure depicts the vitality of the fibroblasts BJ and the cancerous cells BT474 after UWB exposition with 4 mW for 5, 30 or 60 min. The vitality was observed 0, 24, 48 or 72 h after exposure. The depicted vitality of exposed cells is related to non-exposed cells. Due to biological fluctuations, data between 80% and 120% vitality was not considered to be cytotoxic [25].

3.2. Animal tissue

The electrical properties of tissues and cell suspensions are most unusual. They change with frequency in three distinct steps (dispersions as described above) and their dielectric constants reach enormous values at low frequencies. Extensive measurements were carried out over a broad frequency range extending from less than 1 Hz to many GHz. The observed frequency changes of these properties obey causality, i.e., the Kramers-Kronig relationships which relate changes of dielectric constants with conductivity changes. A number of mechanisms were identified which explain the observed data. These mechanisms reflect the various compartments of the biological material. These include membranes and their properties, biological macromolecules and fluid compartments inside and outside membranes [41]. Special topics include a summary of the significant advances in theories on counter ion polarization effects, dielectric properties of cancer *vs.* normal tissues, properties of low-water-content tissues [42], and macroscopic field-coupling considerations. The dielectric properties of tissues are often summarized as empirical correlations with tissue water contents in other compositional variables. The bulk electrical properties of tissues are needed for many bioengineering applications of electric fields or currents, and they provide insight into the basic mechanisms that govern the interaction of electric fields with tissue [43].

Using devices with our own configurations, the dielectric properties of different porcine and bovine tissues were determined [25]. Different measuring points were defined on the surface of udder, fat, liver, muscle, and kidney of porcine and bovine tissue (homogenous structure) and the permittivity of these points was measured three times (selected tissues in Fig. 20, left panel). Afterwards, the tissue under these measuring points was excised and dried to calculate the water content. Water content and permittivity ε' were related to each other, so we could clearly differentiate between fat, low-water-content tissue, with a low permittivity ($\varepsilon' \approx 8$) and liver, muscle or kidney ($\varepsilon' \approx 40$) as high-water-content tissues. The high-water-content tissues show similar permittivity ε' values whereas fat of porcine and bovine origin can be distinguished (Fig. 20, right panel).

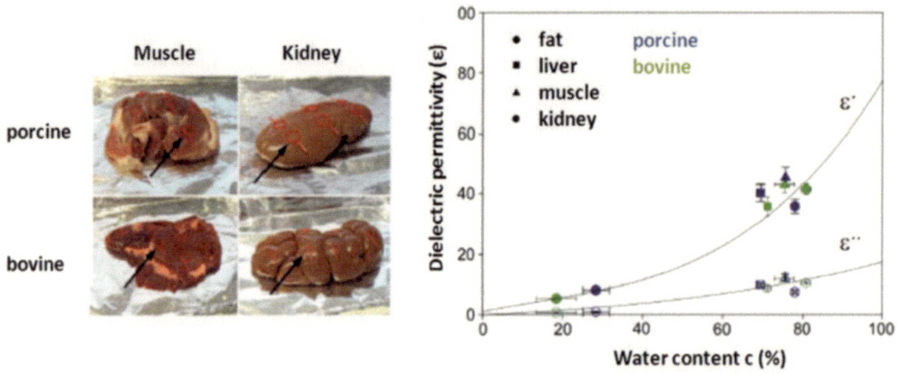

Figure 20. Determination of the dielectric properties of different porcine and bovine tissues at defined measuring points (left panel). Real and imaginary part of permittivity ε' at 2 GHz of porcine and bovine tissue in relation to the water content (right panel). The standard error represents six independent measurements [25].

3.3. Bacterial cell wall identification based on their dielectric properties

The identification of bacterial strains in biological media is a matter of interest in very different fields of modern life. Examples are in food hygiene and food industry, catering and gastronomy [44], [45], in environmental research activities, fermentation processes for the production of medical drugs, such as insulin, antibiotics, and other [46]-[48], and in the diagnosis of infections in clinical and veterinarian applications [49]. Depending on the respective research and application field, bacterial strains are currently detected by complex methods, for example: polymerase chain reaction (technique to amplify a single or a few copies of a piece of DNA), fluorescent *in situ* hybridization, DNA microarray and Raman-spectroscopy, etc.

Different studies have shed some light into the biomass determination of different microbial suspensions *via* dielectric spectroscopy. Mishima *et al.* investigated growth kinetics of bacterial, yeast and animal cells by dielectric monitoring in the frequency range of 10 kHz - 10 MHz [50]. The determination of bacterial growth by dielectric measurements was also shown by Harris *et al.* [51]. Jonsson *et al.* measured the concentration of bacterial cells *via* indirect methods based on the dielectric determination of ions in the suspension, which are released by killed cells [52]. Benoit *et al.* showed that it is possible to discriminate the hydrophobic or hydrophilic features of bacterial suspensions by determining the dielectric permittivity [53]. Nevertheless, no data are available for discrimination on the basis of bacterial structures *per se*, such as the presence of Gram-positive or Gram-negative bacterial strains in biological samples [54].

Therefore, two different Gram-positive bacterial strains (*Micrococcus luteus* and *Bacillus subtilis*) and two Gram-negative bacterial strains (*Escherichia coli* and *Serratia marcescens*) were cultivated under standard conditions using Standard I media and shaking flasks. Bacterial strains were incubated for 24 h at 37°C in an incubation shaker. To assess whether the Gram-status of bacteria could be determined by dielectric spectroscopy, bacterial suspensions were transferred to 50 ml tubes and centrifuged. The supernatant (liquid above precipitate) was removed, the pellet was washed in 0.9% sodium chloride solution and, finally, the dielectric properties of the bacterial biomass (pellet of 10 ml) were determined. Dielectric spectroscopy of bacterial strains and suspensions was performed using a network analyzer in a frequency range from 30 kHz to 6 GHz (HP 8753D) and a coaxial probe (High temperature probe). The real ε' and imaginary ε'' part of permittivity was determined in a frequency range from 50 MHz to 300 MHz [54].

In the frequency range between 50 and 300 MHz, dielectric spectroscopy revealed higher values of the real part of permittivity ($\varepsilon'_{(+)} \approx 160$ Gram-positive) of the Gram-positive bacterial strains *Micrococcus luteus* and *Bacillus subtilis* compared to the Gram-negative strains *Escherichia coli* and *Serratia marcescens* ($\varepsilon'_{(-)} \approx 100$ Gram-negative). From each strain the same cell count and volume was measured. Particularly at a frequency of 50 MHz (maximum of discrimination), the real part of permittivity ε' of both Gram-positive strains was about 60 units higher than of the Gram-negative strains (Fig. 21)

The clear discrimination between the Gram-positive strains *Micrococcus luteus* and *Bacillus subtilis* as well as the Gram-negative strains *Escherichia coli* and *Serratia marcescens* at a frequency up to 100 MHz can be attributed to the β-dispersion. At these frequencies, proteins and other macromolecules of the bacterial cells polarize according to Markx *et al.* [35]. This effect decreases at frequencies above 100 MHz. With increasing frequency the influence of water becomes more prominent.

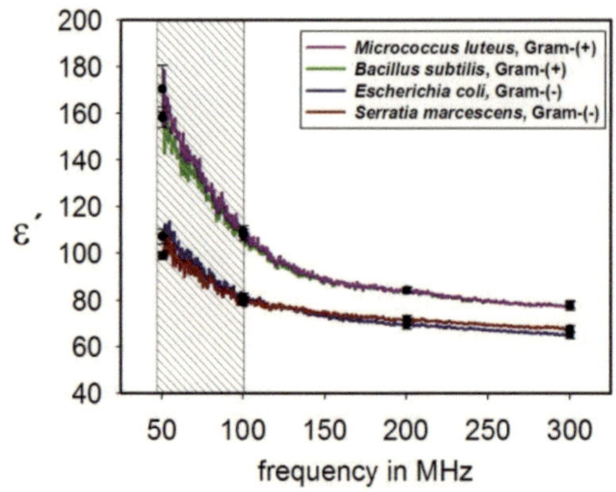

Figure 21. Discrimination of Gram-positive and –negative bacterial strains *via* dielectric spectroscopy. The diagram shows the real part of the permittivity ε' of the biomass of Gram-positive bacterial strains (*Micrococcus luteus* and *Bacillus subtilis* [upper curves]) and Gram-negative bacterial strains (*Escherichia coli* and *Serratia marcescens* [lower curves]) in a frequency range between 50 and 300 MHz. The highlighted area shows the most obvious region of differentiation between Gram-positive and Gram-negative bacterial strains [54].

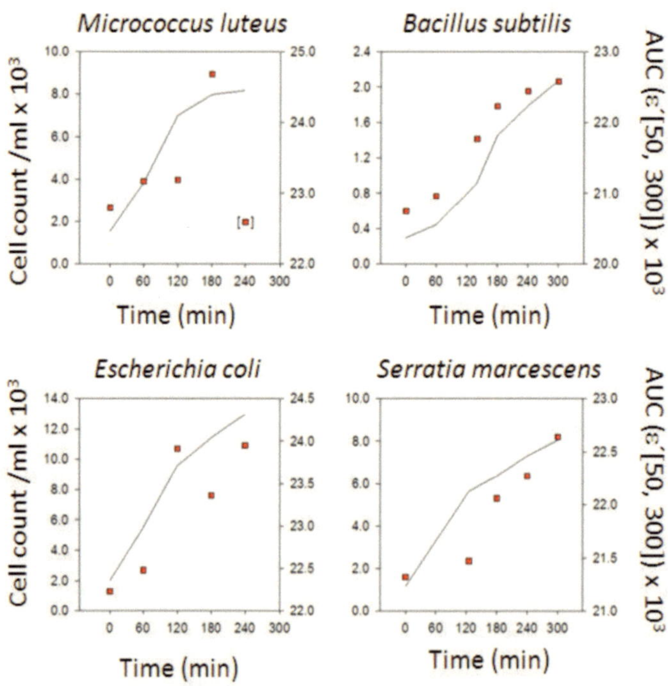

Figure 22. Monitoring of growth kinetics of four bacterial strains (growth phase). Red squares show the area under the curve (AUC) of the real part of permittivity ε' in a frequency range between 50 MHz and 100 MHz derived from measurements during the bacterial growth phase. The permittivity of the cell suspension was taken hourly for 240 or 300 min. Black lines show the cell count per ml taken at the same time as permittivity was measured [54].

All bacterial strains investigated in the present study revealed a characteristic time-dependent correlation between cell counts (black lines in Fig. 22) and ε' (red dots in Fig. 22). The growth kinetics was not influenced by the presence of accumulated metabolites in the culture medium since supernatants (liquid above precipitate) of every bacterial culture showed the same permittivity as the Standard I culture media (Fig. 22; $\varepsilon' = 85 - 78$ and $\varepsilon'' = 600 - 100 \quad @ \quad 50 - 300\,\text{MHz}$) [54].

3.4. Temperature influence on tissue permittivity

In therapeutic or diagnostic applications or biological effects of the electromagnetic field, dosimetric evaluations are greatly dependent on the precise knowledge of the dielectric parameters of biological tissues (relative permittivity ε and electrical conductivity σ). These parameters are sensitive to many influencing factors, which include the temperature of the target organ [55]. During radio-frequency or microwave radiation exposure, the internal temperature of tissue can change, thus influencing the electrical field distribution. For example, the evaluation of the lesion obtained by thermal ablation is a function of the relative permittivity and conductivity at 37°C and also of their evolution during heating. The influence of temperature in dielectric spectroscopy has been studied by several authors [56]-[58]. However, these effects remain misunderstood and the measured values are sparse at various frequencies and exist only for some organs.

To find out in how far temperature-dependent changes in permittivity can result in a parameter identified by ultra-wideband technology, water and different tissues were examined. To assess the basic capability of UWB radar for monitoring local temperatures, dedicated phantom and *in vivo* experiments were performed. Dielectric spectroscopy of water at different temperatures (25 – 80°C in steps of 5°C) and corresponding experiments using porcine and bovine tissue, such as udder, liver, muscle, and kidney revealed a distinct decrease of permittivity with increasing temperature. Nevertheless, heating of tissues to more than 60 °C might also reduce permittivity due to the reduction of water content. No distinct organ-specific differences in the temperature-dependent dielectric properties have been found so far (Fig. 23). Only fat, as low-water-content tissue, exhibited no influence on permittivity at different temperatures [59].

In addition to further studies with improved probes, corresponding analysis were performed using clinically approved temperature-based methods for tumor eradication, such as radio frequency ablation (RFA) or magnetic thermo ablation. For this experiment, a bovine liver was positioned onto a neutral electrode. The second, active electrode was launched into the liver tissue. Both electrodes together create a stress field, and the tissue around the active electrode becomes heated up to 60°C. Bi-static UWB antennas were first positioned in a distance to the region where RFA was thought to detect the signals of liver tissue itself. Then, the antennas were positioned above the region of radio frequency ablation, and changes in impulse response before, while and after radio frequency ablation were detected. The signal analysis displayed an increase of the impulse response during radio frequency ablation (data not shown) [59].

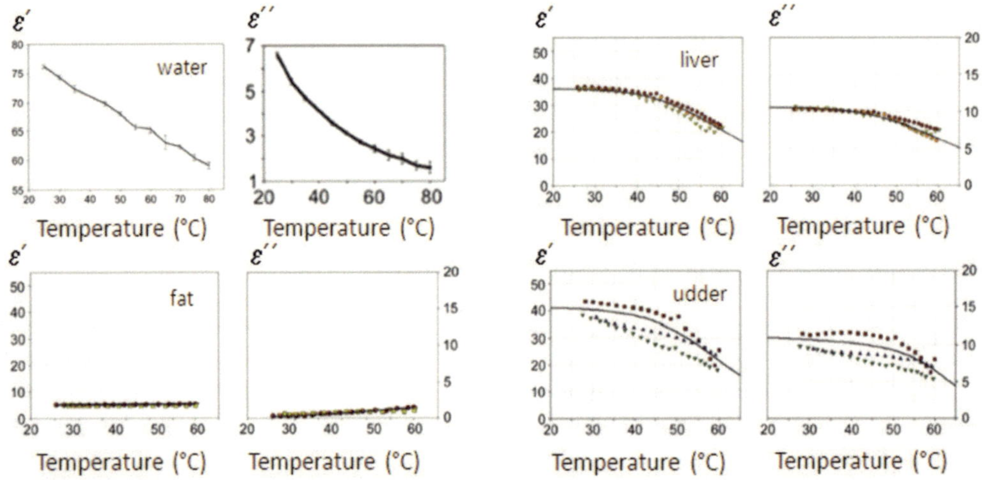

Figure 23. Real part (ε') and imaginary part (ε'') of the permittivity of water, liver, fat, and udder tissue at a frequency of 2 GHz is depicted. Water showed a constant decrease of permittivity in relation to the increase of the temperature. Liver and udder tissue showed a diminished deviation, and in contrast fat showed no change of the permittivity [59] at all.

The applicability of UWB for temperature monitoring was also assessed *in vivo* in mice. Prior to the start of experiments, mice were shaved at the abdominal region. Dielectric spectroscopy of the skin at the animal's liver region before and after euthanasia showed a time-dependent increase of permittivity as a result of decreasing temperature with on-going time after euthanasia. The data provide a good basis for further development of UWB as a non-invasive temperature measurement technology.

3.5. Permittivity variations by contrast media

Microwave-frequency dielectric contrast between malignant and normal tissue in the breast serves as the physical basis for emerging microwave methods of detecting and treating breast cancer. The effective dielectric properties of breast tissue are influenced at microwave frequencies by endogenous polar molecules, such as free and bound water, peptides, and proteins. Consequently, the dielectric properties depend on the type and physiological state of the tissue. The effective dielectric properties - both the dielectric constant and effective conductivity - can also be influenced by exogenous molecules introduced as contrast agents [60].

Detection of dielectric properties of structures and tissues with similar characteristics (such as breast and breast tumor) pose challenges for imaging by ultra-wideband technologies. Therefore, a phantom serving as a model for blood vessels with a constant flow of ethanol (infusion fluid) was created (Fig. 24 left panel) for first trials to test the sensitivity of the measurement apparatus. Additions of contrast agents (in this case a mixture of ethanol and water) were determined [25]. Such basic search is useful for finding suitable contrast agents including feasibilities and limitations regarding the detectability of, for example, concentration variations. The practice of clinical diagnostic radiology has been made possible by advances not only in diagnostic equipment and investigative techniques, but also in the contrast media that permit the visualization of the details of the internal structure of organs, which would not be possible without them. .The remarkably high tolerance of modern contrast media has been achieved through successive developments in chemical pharmacological technology.

The phantom serving as a model for blood vessels with a constant flow of ethanol was arranged. In the first step, the signals of this ethanol flow were received. By using a syringe *via* three-way cock 3 ml of the selected contrast agent (mixture of ethanol and water) were added, and the relative signal change was detected. The results show that with a decrease of water the signals become weaker (Fig. 24, right panel).

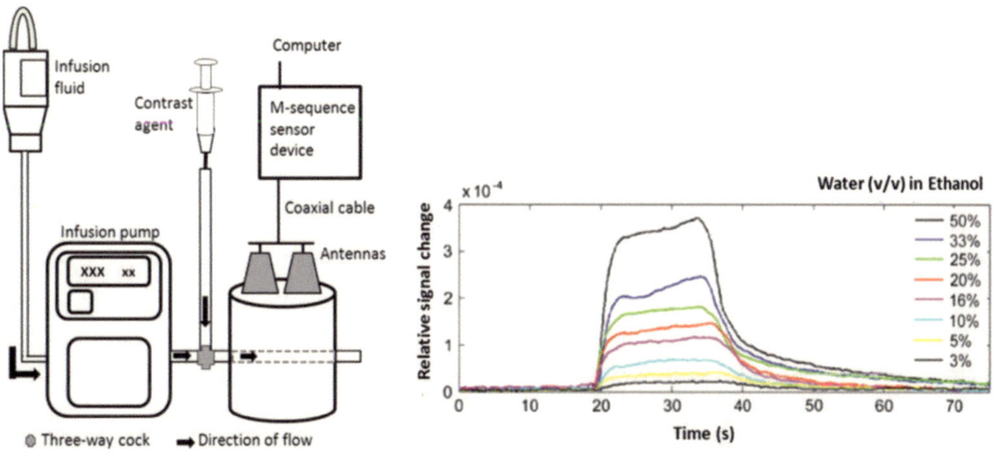

Figure 24. Depiction of the assembly of a phantom serving as a model for blood vessels with a constant flow of ethanol (left panel). The right panel depicts the relative signal variation (change of reflection coefficient) through adding 3 ml of contrast agent in relation to the virgin signal with constant flow of ethanol [43].

Even though dielectric spectroscopy of our group showed promising permittivity values of potential contrast agents such as physiological sodium chloride, the encountered permittivity increases *in vivo* are still to be enhanced to allow for a specific detection *via* UWB radar. One possibility to increase the capability of breast tumor imaging is the application of different clinically approved contrast agents such as ultrasound micro bubbles or iron oxide nanoparticles. Moreover, we expanded our investigations to the assessment of non-clinically approved agents (for example $BaSO_4$) in order to discover potential advantageous mechanistic conditions leading to local signal increase in terms of UWB diagnosis. Experiments will be systematically analyzed using dedicated phantoms, mimicking human tissues and blood flow.

Another challenge is the achievement of a selective accumulation of contrast agents in the target region to be detected by our UWB system. In this regard, a dynamic and transient accumulation *via* the tumor vascularization has been already postulated.

4. Remote organ motion tracking and its application in magnetic resonance imaging

4.1. Cardiac magnetic resonance imaging

Magnetic resonance imaging (MRI) is arguably the most innovative imaging modality in cardiology and neuroscience. It is based on the detection of precessing nuclear spins, mostly from protons of tissue water, in a strong static magnetic field. Using two additional kinds of magnetic fields, the position of the spins inside the human body can be encoded. To this end, the nuclear spin system is excited by resonant RF pulses at the precession frequency of the spin system. After excitation a macroscopic RF signal can be detected by an RF coil providing amplitude and phase information of the precessing nuclear magnetization. Applying additional magnetic field gradients the spin positions can be encoded by generating a well-defined spatial variation of the precession frequencies. Proper sequencing of spin excitation and gradient switching allows the reconstruction of 2D and 3D images from the acquired complex valued MR signals.

MRI data depend crucially on a multitude of physical parameters, e.g. moving spins will cause an additional phase modulation of the signal. One consequence is that MR images of the moving heart or of large vessels with pulsatile blood flow are severely distorted in the whole field of view. Hence, cardiac MRI (CMR) is seriously impaired by cardiac and respiratory motion when no proper gating with respect to both relevant motion types, cardiac and respiratory motion, is applied (Fig. 25). In clinically approved CMR procedures, electrocardiography (ECG) or pulse oximetry are used for cardiac gating and breath holding is applied for freezing respiratory motion [61],[62].

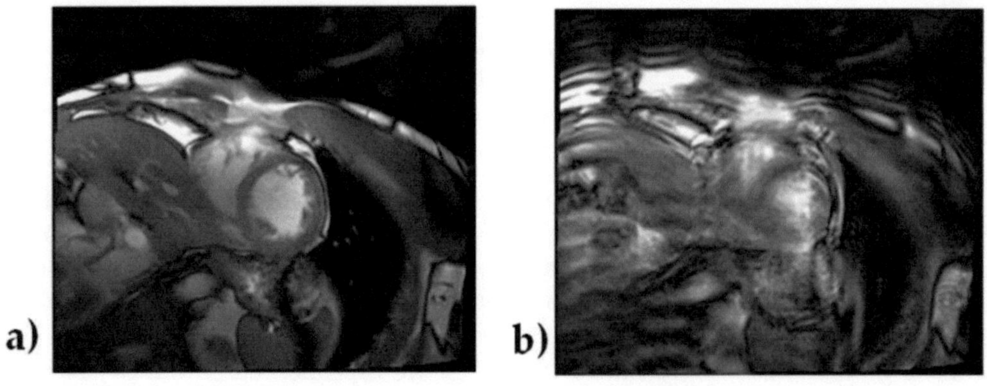

Figure 25. MR image (short axis view) of a human heart. a) Cardiac gating by pulse oximetry and breath hold; b) cardiac gating only, due to free breathing during image acquisition severe image artifacts occur

However, there are unmet needs of clinical CMR, particularly for high (≥ 3 T) and ultra-high (≥ 7 T) field MRI. Higher magnetic fields offer the chance to acquire images of better spatial resolution [63], but on the downside the ECG signal is increasingly perturbed by the magneto-hydrodynamic effect [64] until it becomes effectively useless for cardiac gating at

7 T. Furthermore, ECG electrodes are directly attached to the patient's skin, which may result in local RF burns. In addition, ECG and alternative approaches like pulse oximetric or acoustic cardiac triggering [65] do not provide any information about the respiratory state.

As a cardiac patient's breath hold is typically limited to about 15 s, a 3D whole heart coverage or imaging of the coronaries [66] would require proper respiration gating to acquire MR data under free breathing conditions. A well-established approach for respiration gating is the so-called MR navigator [67]. By means of some extra MR excitations, the momentary position of the diaphragm can be tracked over the respiratory. Unfortunately, these extra excitations interfere with the cardiac imaging sequence itself, making this technique complex and less reliable.

On this background, we propose the simultaneous use of multi-channel UWB radar and MRI to gain complementary information in particular for improving cardiac MRI. The anticipated potentials of this technique are (i) a contactless measuring principle for better patient safety and comfort, thus streamlining the clinical workflow, (ii) concurrent monitoring of a variety of body movements, (iii) direct relation to tissue mechanics [68], (iv) direct tracking of the temporal evolution of inner body landmarks, and (v) absence of any interferences of the UWB radar signals with the MR measurement if MR compatible designed UWB antennas are applied, (s. Section 2.3.3). The decomposition of physiological signatures in UWB radar data is the main challenge of this approach and a prerequisite for a reliable tracking of landmarks within the human body suitable for MRI gating.

Beyond MRI, there are a variety of other possible applications of in vivo UWB radar navigation systems in medical imaging or therapy. Examples are X-ray Computed Tomography (CT), Positron Emission Tomography (PET), Medical Ultrasonography (US), and Radiotherapy using photons or particles, or High Intensity Focused Ultrasound (HIFU). Lessons learned from all these approaches will foster medical applications of standalone UWB radar systems for intensive care monitoring, emergency medical aid, and home-based patient care [70].

4.2. Analytical and numerical modeling of the scenario

4.2.1. Channel-model

For modeling purposes, the human body can be approximated as a multilayered dielectric structure with characteristic reflection coefficients $\Gamma(f)$ (s. Fig. 26) [29], [71], [72]. The UWB signal, which can be a pulse or a pseudo-noise sequence [71] of up to 10 GHz bandwidth, is transmitted utilizing appropriate pulse-radiating antennas T_x (e.g., Double Ridged Horn or Vivaldi antennas). The reflected signal is detected by R_x, and the first step in further signal-processing usually is to calculate the correlation signal $R_{XY}(\tau)$ between received signal S_{Rx} and transmitted signal pulse S_{Tx} is [71], [72]. This represents the impulse response function (IRF) including the transfer functions of the antennas. By UWB measurement of the motion of a multi-layered dielectric phantom [29], the changes of reflections on the single interfaces can be found. Therefore, the signal variance $M(\tau)$ of the correlation signal $R_{XY}(\tau)$ is calculated.

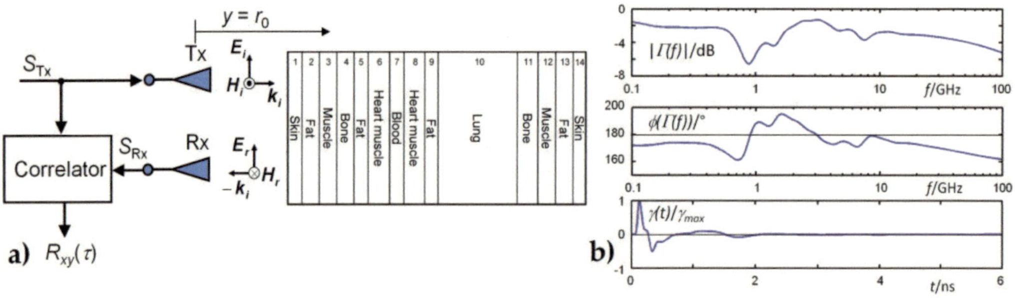

Figure 26. a) 14-layer arrangement to mimic the reflective properties of the human thorax (not to scale). E_i/E_r, H_i/H_r: incident/reflected electric/magnetic field component. k_i: wave vector of incident wave. **b)** Top: calculated magnitude of the reflection response $|\Gamma(f)|$, which is proportional to the frequency response function (FRF) of the object. Middle: unwrapped phase $\phi(f)$ of the reflection response $\Gamma(f)$. Bottom: normalized time domain representation $\gamma(t)$ of $\Gamma(f)$ impulse response function (IRF) of the reflection response.

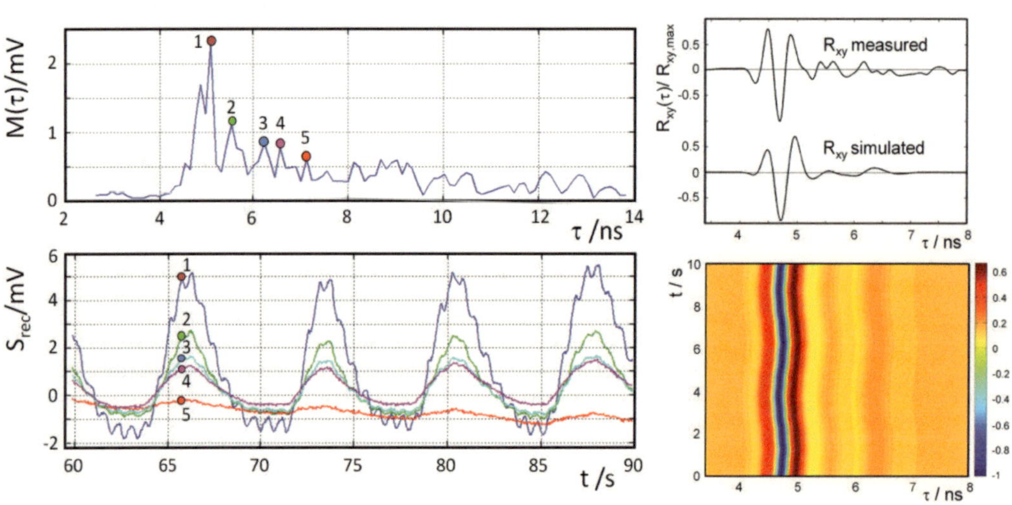

Figure 27. Physiological signatures received by the algorithm described in Ref. [29]. Top: Signal variance M(τ).Bottom: Physiological signatures corresponding to the label local maxima of M(τ).Top right: Measured and simulated correlation signal $R_{xy}(\tau)$.
Bottom right: Radargram of the measured and simulated correlation signal $R_{xy}(\tau)$.

The maxima in M(τ) correspond to the interfaces containing a considerable difference in the permittivity or are close to the illumination side if the transfer functions of the antennas are removed by de-convolution. By these maxima, the time signals corresponding to the interfaces can be extracted [29]. An example of simulated and measured correlation signal $R_{xy}(\tau)$ is given in Fig. 27, top right.

4.2.2. Analytical simulation of the intracranial pulsation detection

It is well known that simultaneously to the head's vibrations intracranial oscillations with spatial varying amplitude occur, induced by physiological sources [73]. Thus, it is only logical to ask whether these oscillations are detectable by UWB radar. Due to the simultaneous occurrence of the intracranial displacement and the vibration of the whole head, decomposing both signals requires sophisticated methods. As an initial step towards the solution to this problem, we need to get a feeling for the change in the acquired UWB reflection signal due to an intracranial oscillation. An analytical approach [71], [72] was applied which models the signal path and the oscillating stratified arrangement of the brain to get signals free of any interfering compositions. Figure 28 schematically depicts the set-up used to probe the human body with a UWB device, where S_{Tx} symbolizes the excitation signal and S'_{Tx} its temporal derivative representing the free space signal E_i in the channel. By the convolution of the impulse response function γ of the multilayered dielectric structure with S'_{Tx}, the reflected electric field component $\gamma * S'_{Tx} = E_r$ is archived and, accordingly, the received current signal $S_{Rx} = (\gamma * S'_{Tx})'$. The * symbol represents the convolution operator.

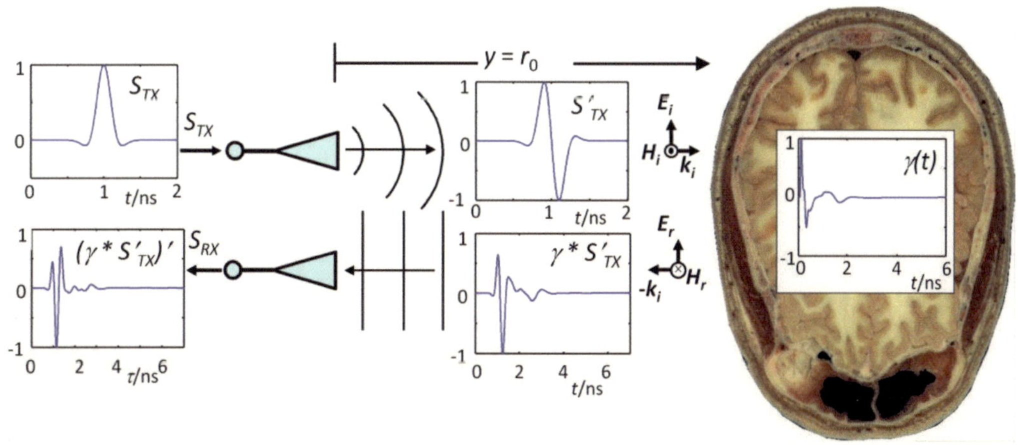

Figure 28. Signal path model for the current transfer function S_{Rx}/S_{Tx}.

Therefore, the spectral response of a dielectric medium is appropriately described in terms of a multiple Cole-Cole dispersion, which – by choosing parameters appropriate to each constituent - can be used to predict the dielectric behavior over the desired frequency range [71]. For such a layered arrangement, the reflection coefficient $\Gamma(\omega)$ can be calculated recursively. In this manner, the response of $\Gamma(\omega, t)$ to the variation of a certain internal interface can be analyzed [68], [73]. We simulated the physiological event by variations of $\Gamma(\omega, t)$, which is done by a sinusoidal oscillation of the white matter. Accordingly, the cerebro spinal fluid varies antipodally [76]. The correlation result $R_{xy}(\tau, t)$ was calculated just as its variation after a certain propagation time. The reconstruction of the intracranial motion applying the reconstruction algorithm proposed in [72] gave us a maximum deviation from the reference oscillation of about 4%. We conclude that the detection of intracranial oscillations using non-contact UWB is indeed feasible [72], [73]. It must be noted that for all real medical applications of this broadband technique trying to monitor variations of the body's interior, sophisticated signal processing techniques must be applied to decompose signals originating from the body's surface and signals originating from deeper sources [74]. The influence of the antenna's transfer function, in contrast, is less of an issue for real applications. For simplicity, we had assumed an ideal transfer function in the above simulation but non-ideal antenna behavior can be extracted from the received signal by using proper de-convolution techniques. Furthermore, the time courses of the ideal channel can be regained [72].

4.2.3. *Full simulation of the electromagnetic field distribution*

Beside the analytical approach, we are interested in the detailed temporal evolution of the electromagnetic fields inside and outside the human thorax. To this end, we investigated complex arrangements mimicking the illumination of a realistic human torso [75] model incorporating the geometry of the antennas by finite-difference time-domain method (FDTD) simulations. By FDTD simulation, we studied, e.g., the dependence of the illumination and detection angles of the transmission and receiving antennas on the quality of the received signal, *i.e.* the correlation result. In this way, an estimate of the optimized antenna placement can be found. Furthermore, by varying organs' boundaries by changing their thickness or/and placement of one or more tissue layers, different functional states can be investigated, e.g. the end-systolic and end-diastolic phase of the myocardium, which consequently determines a characteristic change of the received signals.

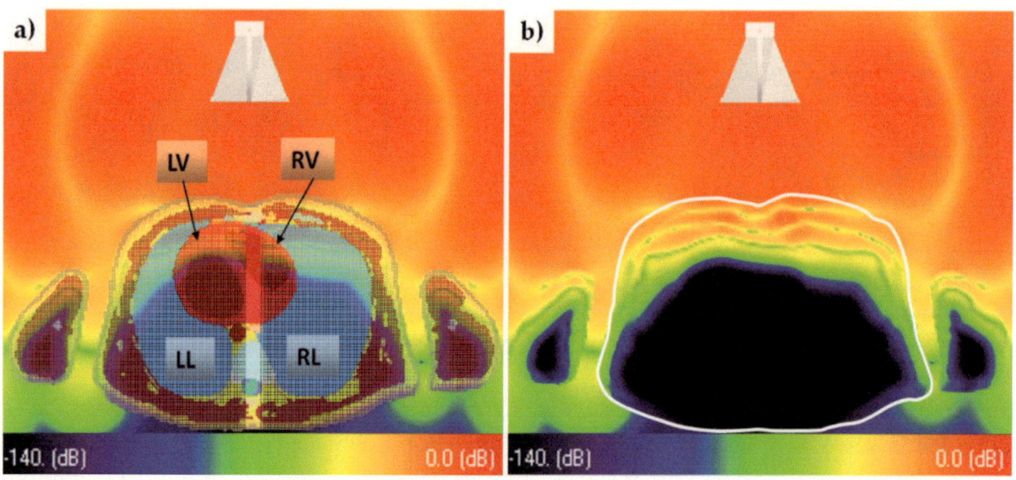

Figure 29. Extra- and intra-corporal electrical field distribution in an axial cross-section of the upper human body **a)** with tissue mesh; **b)** without the mesh showing the wave propagation intra-corporal. The thorax's contour is highlighted by the white line.

An example of the complex wave propagation inside the human torso is shown in Fig. 29. Due to the higher permittivity ε inside the body, the propagation velocity is slowed down according to $c = c_0/\sqrt{\varepsilon}$. Hence, a bending of the extra- and intra-corporal wave fronts results. The transmitted spherical wave fronts are refracted towards the center of the thorax, which is beneficial for the illumination of the myocardial section lying deeper inside the thorax. By these simulations, we achieve an in-depth understanding of the complex electromagnetic field distribution and the dependencies of the resulting output signal of the receiving antenna [73]. Therefore, the results of these simulations are helpful to increase the accuracy of reconstructed physiological signatures from deep sources by finding the optimized antenna position regarding the better penetrability of selected body areas. This, of course, requires the adaptation of the model to the actual thorax geometry of the patient as obtained by MRI scans.

4.3. Detection of motion by UWB radar

4.3.1. Motion detection for a multilayered phantom

We compared the motion detection by variance calculation in a combined MRI/UWB measurement using appropriate MR-compatible phantoms [29]. The dielectric phantoms were arranged in a sandwich structure to mimic the sequence of biological tissue layers of the human thorax.

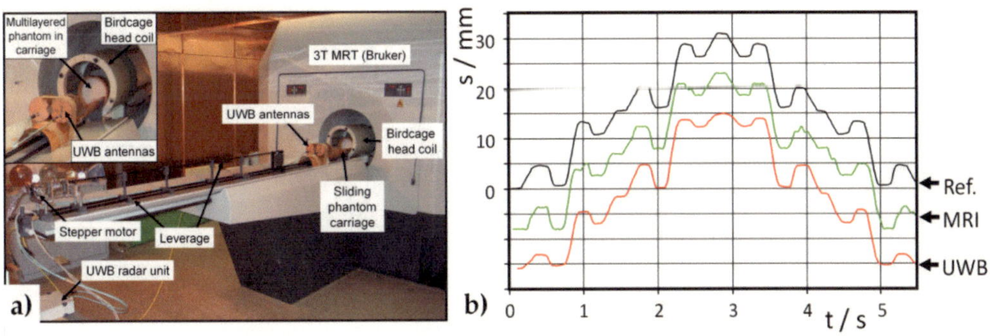

Figure 30. **a)** Set-up of the combined MRI/UWB measurement; **b)** Comparison of the reference profile with the data obtained simultaneously from MR and UWB radar measurements. The profiles are offset for clarity.

Such a sandwich was placed in a moveable sledge-like fixture inside a birdcage MR head coil. The motion profile of the sandwich structure was shaped to approximate respiratory motion of the thorax superimposed by cardiac oscillations (Fig. 30). An M-sequence UWB-Radar system (up to 5 GHz) [76] and MR compatible UWB antennas [10], [32] were utilized to detect the motion of the phantom inside a 3-T MR scanner (Bruker MEDSPEC 30/100). A flow-compensated gradient echo CINE (time resolution 50 ms) sequence was used to reduce artifacts generated by the phantom movements.

Additionally, the physiological signatures monitored by UWB-radar were validated by comparison to simultaneously acquired MR measurements on the same subject [13], [29], [77] (cf. Section 4.5.2 and 4.6).

4.3.2. Detection of micro motion

Subject motion appears to be a limiting factor in numerous MR imaging applications especially in the case of high and ultra-high fields, e.g. high-resolution functional MRI (fMRI). For head imaging the subject's ability to maintain the same head position is limiting the total acquisition time. This period typically does not exceed several minutes and may be considerably reduced in the case of pathologies. Several navigator techniques have been proposed to circumvent the subject motion problem [73]. MR navigators, however, do not only extend the scan because of the time necessary for acquiring the position information, but also require additional excitation pulses affecting the steady-state magnetization. Furthermore, if the very high spatial resolution offered by ultra-high-field MR scanners shall be exploited, the displacements caused by respiration and cardiac activity have to be considered. Thus, we propose to apply an UWB radar technique to monitor such micro motions.

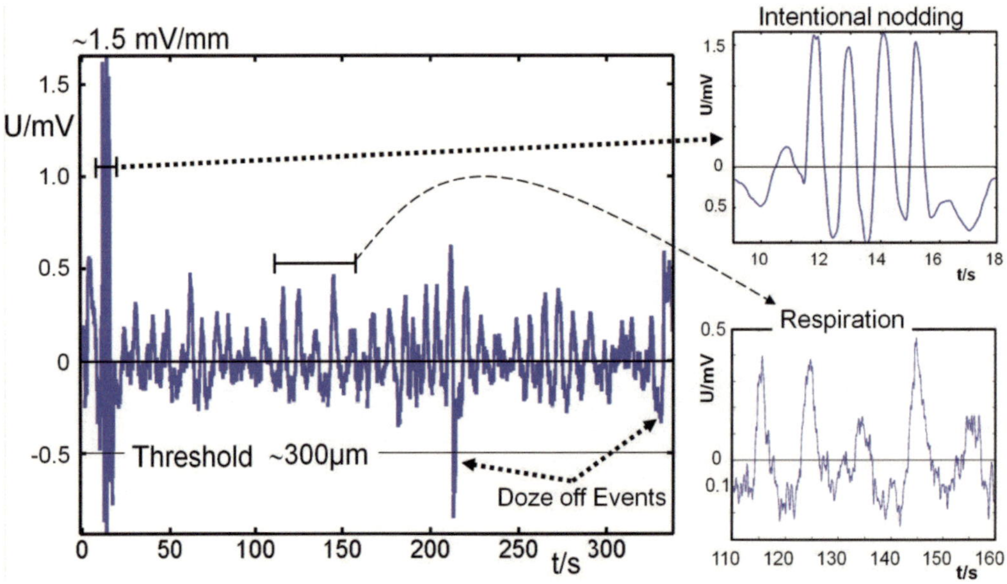

Figure 31. Motion reconstructed from a measured time interval of 350 s. The right inset at the top displays the four nodding events (~1 mm amplitude, episode [t = 10 s,..., t = 18 s]) to localize the surface of the head. Respiratory displacements are clearly visible (right inset bottom, episode [t = 110 s,..., t = 160 s]) and spontaneous twitches are highlighted.

First *in-vivo* motions reconstructed from a measured time interval of 350 s are shown in Fig. 31. By applying appropriate filters in a selected time interval even the cardiac induced displacements were detected with an amplitude of about 40 µm. Thus, we could detect all kinds of involuntary motions (respiratory, cardiac), but also doze-off-events are visible, demonstrating the feasibility of interfacing an MR scanner with an external UWB radar based motion tracking system. Our system is capable of determining the position of interest with sub-millimeter accuracy and an update rate of 44.2 Hz. Using the UWB tracking data of the volunteer's head, the motion artifacts can be compensated for in real time or by post-processing enhancing the actual resolution of the MR scan [73].

4.3.3. Separation of motion components by blind source separation

Monitoring the motion inside the human body, the correlation functions of transmitted and received signals (i.e. the IRF) contain a mixture of all simultaneously occurring motions.

Especially for the human torso where - due to higher harmonics from the highly nonlinear respiratory cycle - the separation of the cardiac cycle by common signal filtering in the frequency domain is limited, another separation of motion components is necessary. For this reason, the separation of motion components based on blind source separation (BSS) was developed.

The IRF from a single UWB shot is a time series of 511 data points with a dwell time of 112 ps. This defines an IRF time scale of 57 ns but is still instantaneous compared to anatomical motions. These shots are then repeated for instance 2000 times at a rate of 44 Hz covering a total time span of 45 s. For further analysis, only the most interesting regime of the IRF data is considered. These are the 100 data points, i.e. a window of 11.2 ns, right after the IRF maximum due to direct cross-talk between *Tx* and *Rx* antenna. Following the temporal evolution of each selected data point over the 2000 repetitions, 100 virtual channels are obtained and subjected to BSS decomposition (ROI, see Fig. 32.a). By removing the mean values in these virtual channels, the changes of the radar signal on the anatomical time scale become visible, see Fig. 32.b. The motion pattern is dominated by respiration; cardiac motion is considerably smaller and not immediately visible in the raw data.

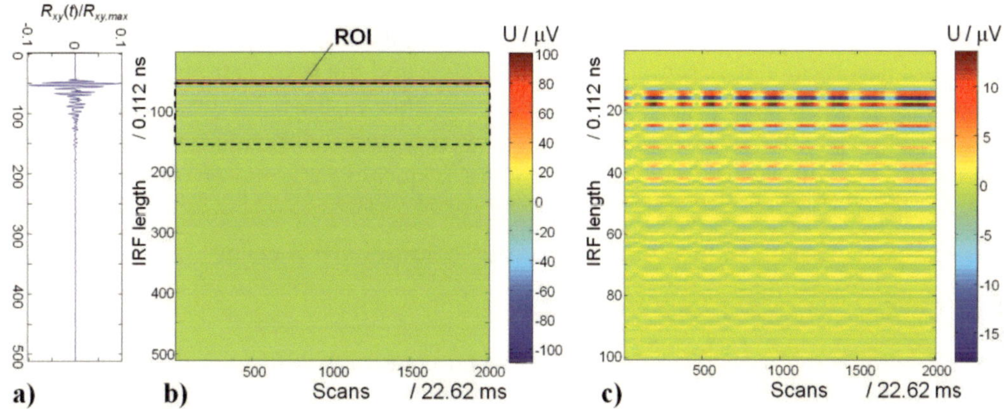

Figure 32. **a)** Single IRF and **b)** radargram of one channel with region of interest and **c)** selected 100 virtual channels, mean value removed

The data analysis is based on the BSS and assumes a measured signal $x(t)$ to be a linear combination of unknown zero-mean source signals $s(t)$ with an unknown mixing matrix **A**:

$$x(t) = \mathbf{A}s(t) \quad x = (x_1, \ldots, x_m)^T \, . \tag{2}$$

The original sources $s(t)$ can be estimated by the components $y(t)$ which can be calculated from the estimation of the de-mixing matrix $\mathbf{A}^* \approx \mathbf{A}^{-1}$:

$$y(t) = \mathbf{A}^* x(t) = \mathbf{A}^* \mathbf{A}s(t) \tag{3}$$

In our analysis, a second-order time-domain algorithm (TDSEP, Temporal Decorrelation source SEParation) was applied which is described in detail in [78]. In TDSEP the unknown mixing matrix **A** is calculated by simultaneous diagonalization of a set of correlation matrices $\mathbf{R}_\tau(x)$ for different choices of τ.

$$\begin{aligned} \mathbf{R}_{\tau(x)} &= \left\langle x(t)x^T(t-\tau)\right\rangle \\ \mathbf{R}_{\tau(x)} &= \left\langle \mathbf{A}s(t)(\mathbf{A}s(t-\tau))^T\right\rangle = \mathbf{A}\mathbf{R}_{\tau(s)}\mathbf{A}^T \end{aligned} \tag{4}$$

where the angular brackets denote time averaging. The quality of signal separation depends strongly on the choice of τ. However, solving $\mathbf{R}_\tau(x) = \mathbf{A}\mathbf{R}_\tau(s)\,\mathbf{A}^T$ for several τ by simultaneous diagonalization eliminates this obstacle. It is recommended by biomagnetic research to choose the number of time shifts τ larger than 40 and to include the time constant of those components which are known a priori, e.g. the range of possible cardiac frequencies $1/\tau_{\text{cardiac}}$ [79]. Additionally, Principal-Component Analysis (PCA) compression was applied to reduce the number of channels used for generating the correlation matrices $\mathbf{R}_\tau(x)$ and reduce computation time for the BSS. The components of the resulting sources are calculated using eq. (3). Automatic identification of the cardiac component was provided by a frequency-domain selection criterion because for non-pathological conditions the main spectral power density of the heart motion falls in a frequency range of 0.5 Hz to 7 Hz. The algorithm searches for the highest ratio between a single narrowband signal (fundamental mode and first harmonic) within this frequency range and the maximum signal outside this range. A high-order zero-phase digital band pass filter of 0.5–5 Hz was applied to the cardiac component of the UWB signal. In a similar way, respiration can be identified by the BSS component with the maximum L2 norm in the frequency range of 0.05 Hz to 0.5 Hz.

4.4. Analyses of cardiac mechanics by multi-channel UWB radar

4.4.1. Compatibility of MRI and UWB radar

Compatibility is the most challenging issue when combining MRI with other modalities. Therefore, the UWB antennas employed are important parts. Eddy currents due to the switching magnetic gradient fields as well as the interference with the powerful RF pulses from the MRI scanner, see Section 2.3.3, were minimized by proper antenna design. The cut-off frequency of the MR-compatible double ridged horn antennas at 1.5 GHz [32] marks the lower limit of the transmitted and received signal frequencies. Coupling to the narrowband MRI frequencies (300 MHz at 7 T, 125MHz at 3T) is thus minimized in both directions. Additionally, the inputs of both our UWB radar systems (MEODAT GmbH, Ilmenau, Germany), one single-module: 1*Tx*-2*Rx*-device and one four-module multi-input-multi-output: 4*Tx*-8*Rx*-MIMO device, were protected by 1.2 GHz high pass filters. In both UWB systems, the transmitted radar signals were generated by a pseudo-random M-sequence. With $m = 9$ it has a length of 2^m-1 = 511 clock signals at $f_0 = 8.95$ GHz [76]. The equivalent UWB power spectrum extends up to $f_0/2$. The impulse response function (IRF) is obtained as mentioned before by correlation of the received signal of the investigated object with the M-sequence [76]. By means of this technique, the signal-to-noise ratio is improved due to the removal of the uncorrelated noise by the correlation of the received signals with the transmitted signal pattern. In this way, even smallest parts of the RF pulses of the MRI were avoided.

4.4.2. Multi-channel UWB radar applying two receiver channels for cardiac trigger events

We started our multi-channel UWB radar development with the single module device enabling us to add a second receiver (*Rx*) antenna, oriented towards the left-anterior oblique direction [68] (Fig. 33.a), to the existing *Rx* and *Tx* antennas facing the antero-posterior direction. The UWB data were recorded at 44.2 Hz. Corresponding to the data selection in Section 4.3.3, we obtained 200 virtual data channels from the IRFs of two UWB measurement channels for the decomposition by blind source separation (BSS).

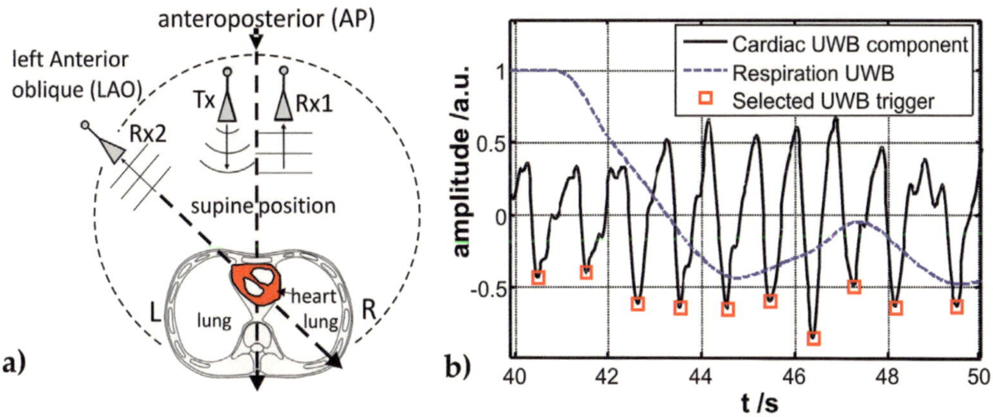

Figure 33. **a)** Scheme of the UWB radar with one transmitter (*Tx*) and two receiver (*Rx*) antennas and measurement set-up; **b)** Cardiac UWB signal applying both *Rx* channels and the calculated trigger events in the signal by combination of low peak and slew rate calculation.

In the cardiac UWB signal, we chose the points of maximum myocardial contraction during the heart cycle. These points are related to the minima of the UWB signal (Fig. 33.b: squares). To increase the robustness of this detection scheme, we combined it with a simple difference calculation at the trailing edge of the minima. Additional consistency checks on the oscillation amplitude were used to suppress double triggering.

By employing two *Rx* channels (Fig. 33.a) the UWB radar detection of the cardiac cycle worked reliably, even in the free breathing mode. In simple cases, e.g. under breath-holding conditions, it is possible to detect cardiac motion with just one *Rx* channel. However, this will not work in general, more complicated situations.

4.4.3. Application of up to 32 receiver channels

By using two *Rx* channels, it was still necessary to align the antennas properly towards the heart. This becomes more critical for measurements during cardiac MRI where the MR coil is placed on the chest of the subject, partly blocking the free line of sight between radar antenna and the heart, see Fig. 38. With our development of a multiple-antenna set-up it is much easier to handle this adjustment by just choosing the 'good channels' in a pool of available channels.

By integration of a MIMO UWB device (MEODAT GmbH, Ilmenau, Germany) containing four modules, each with one *Tx* and two *Rx* channels [76], up to 32 channels became available. In a 1 *Tx* * 8 *Rx* configuration a sampling frequency up to 530.4 Hz can be realized. Using sequentially activated transmitters the set-up can be extended to 32 channels (4 *Tx* * 8 *Rx*) at a reduced sampling frequency of up to 132.6 Hz. For cardiac motion detection, the four *Tx* and eight *RX* antennas are placed over the chest as depicted in Fig. 34 and adjusted to aim for one central point at a distance of 100 cm.

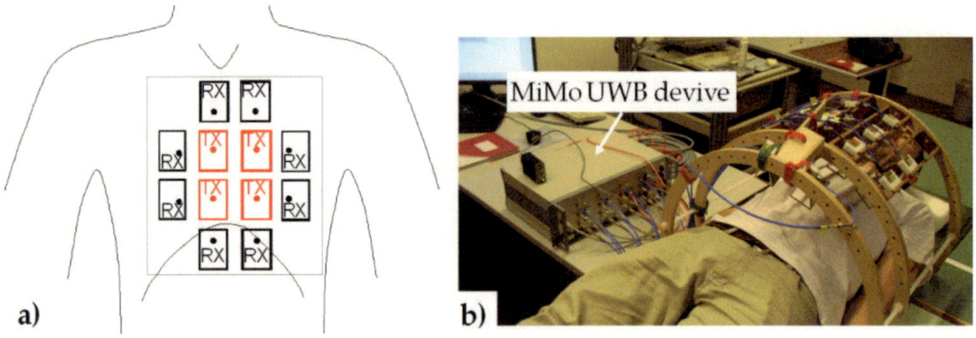

Figure 34. a) Scheme of the UWB radar set-up with 8 *Rx* and 4 *Tx* antennas b) MR compatible measurement set-up.

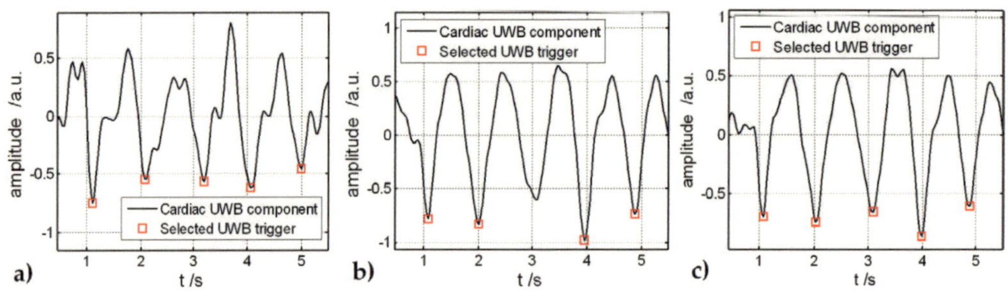

Figure 35. Cardiac signal and detected trigger events for **a)** two hand-picked best channels, **b)** all 32 channels, **c)** the 24 "good channels".

The procedure to identify the most useful channels for triggering starts with a short preparatory measurement, where each channel is analyzed by the BSS to decompose the complex UWB signals [80], extracts the relevant cardiac component and calculates the trigger events as described in Section 4.4.2. The quality of each measurement channel is assessed by calculating the variation of the time interval between trigger events. For comparison, Fig. 35.a depicts the result of the BSS analysis by the best two channels, manually selected for the smallest variation between the trigger events. The cardiac signal based on these two hand-picked channels represents the best achievable result for a set-up like in the section before. By utilizing all 32 channels for the BSS, a smoother cardiac signal is detected, and the motion amplitude shows less variation over the time. However, the sharpness of the trailing slopes is also reduced. Due to this fact, the third trigger event escaped detection (s. Fig. 35.b). Some of the 32 channels contained much noise resulting in a jitter of their trigger contributions and smearing out the sharpness in the combined signal determined by the BSS. By the preparatory check those channels with the highest variation in their 'cardiac' signals were excluded as they were likely contaminated with noise or other motion components. By rejecting the eight noisiest channels and recalculating the BSS with the remaining channels, a cardiac signal is obtained with sharp trailing slopes and well-defined trigger events (s. Fig. 35.c).

The primary goal of this development was to simplify the system handling during cardiac navigation for high-resolution MRI. In addition, the capability of monitoring non-invasively the cardiac activity of a person in an unknown position, e.g. in a patient bed, can be important for a variety of novel medical applications in clinical medicine and biomedical research. As multi-channel UWB radar is unimpeded by bedding or clothing, it is applicable not only in conjunction with MRI. It would also be a valuable stand-alone modality for intensive care monitoring of patient groups not permitting the use of skin contact sensors. Neonates, children at risk of sudden infant death syndrome or burn victims are just a few examples.

4.4.4. Illumination of human thorax by multiple antenna groups

Stand-alone UWB radar enables the detection of cardiac activities by different illumination conditions as shown in [68] for the radiographic standard position. The illumination of the heart from only one side at a time, like the frontal direction for motion detection as depicted in Fig. 33, was extended to the simultaneous illumination of two sides. No averaging was performed to enable the comparison of single heart beats [30]. This approach can open the field for new diagnostic applications by detecting differences and disturbances in comparative measurements of the left and right ventricle, thus recognizing potentially pathological patterns [69]. Two groups of four *Rx* and one *Tx* antennas were applied for this purpose. The first was placed in the left lateral and the second in the right anterior oblique position.

Each antenna group consisted of a single *Tx* antenna surrounded by four *Rx* antennas. All antennas were directed towards the estimated center position of the heart. The challenge was to measure the cardiac motion even from the lateral position, where the attenuation of the reflected signals from the heart is much higher due to the prolonged propagation path in tissue. The data analysis by BSS was applied for both antenna groups separately. For comparison, the data of only two or all four *Rx* channel per group were analyzed.

For lateral position, the UWB signal from the cardiac motion is considerably weaker and much more affected by noise. However, by increasing the number of *Rx*- channels, the signal quality improved substantially, effectively compensating the strong signal attenuation (s. Fig. 36.b). Only healthy volunteers were examined in this particular study but even among them characteristic peculiarities can be found. In both ventricles, the contraction velocity (trailing edge of the UWB motion curve) is higher than the velocity of ventricle dilatation. The duration of the dilation period, on the other hand, is longer for the right ventricle compared to its counterpart on the left. More characteristic features are expected to be visible in patients with cardiac diseases or malfunctions.

4.5. Simultaneous cardiac UWB/ECG, UWB/MRI measurements

4.5.1. UWB radar and high resolution ECG

UWB and ECG were simultaneously acquired. The radar system was equivalent to Section 4.4.2 with one *Tx* and two *Rx* channels. The ECG was recorded with two channels (left arm and left leg against right arm) at a sampling frequency of 8 kHz. For the UWB signals sampled at 44.2 Hz, the same data analysis (see Section 4.4.2) was applied to extract the cardiac signal and determine the trigger events. The usual R-peak detection was applied to trigger on the ECG signal. Cardiac UWB and ECG signals were both re-sampled at 1 kHz to retain more detailed information of the ECG.

The point of maximum mechanical contraction of the heart in the cardiac UWB signals (s. Fig. 37.a) is delayed to the ECG R-peak, indicating the point of the myocardium's peak electrical activity. Therefore, we have to be aware of the difference between detecting cardiac mechanics by UWB radar and the heart's electrical activity by ECG. For the goal of MRI gating, however, the important thing is the existence of a fixed temporal relationship between ECG and UWB signals with as little jitter as possible. For the time lag between ECG and UWB trigger events, we obtained a standard deviation of less than 20 ms which is already smaller than the UWB sampling time of 22.6 ms. This result proves the consistency and robustness of our procedure.

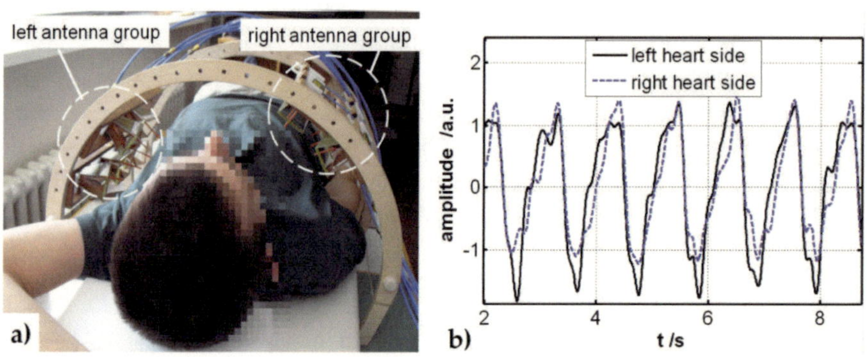

Figure 36. **a)** Measurement set-up with two antenna groups for separate monitoring of the heart's left and right ventricle; **b)** Cardiac signal for left and right ventricle applying four *Rx* channels.

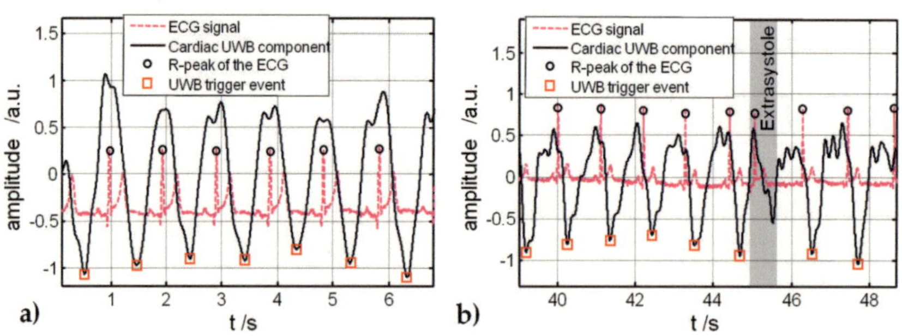

Figure 37. **a)** ECG signal with R-peak trigger events and UWB signal with trigger events located at the maximum of mechanical contraction; **b)** Measurement with an extra-systole.

Another measurement example depicted in Fig. 37.b shows a cardiac sequence containing an extra-systole at 45.5 s. In the ECG signal, this appears as a spontaneous change in the R-R-duration. In the cardiac UWB signal, we observe an incomplete contraction of the myocardium due to the "erroneous" electric excitation. Consequently, no trigger event was generated by this extra-systole. This ability to analyze cardiac mechanics by stand-alone UWB radar or in conjunction with ECG can be exploited, e.g., for infarction detection, as ischemic tissue shows a modified contraction pattern.

4.5.2. Comparison of cardiac UWB Signal and one dimensional MRI

For better understanding the relationship between actual cardiac mechanics and UWB motion-detection signals, a fast MR-sequence was developed with the aim to monitor myocardial landmarks inside the human body in real time. We implemented a very fast 1D gradient echo sequence for low RF power deposition in tissue and high scan repetition frequency on our MR scanner [77]. One dimensional MR profiles and motion sensitive UWB data were acquired simultaneously allowing the comparison of both techniques and hence a verification of the UWB radar navigator. MR compatible UWB antennas [32] mounted above the chest were directed towards the heart (s. Fig. 38). A flexible RF coil with large openings was used to detect the MRI signal. The UWB data were sampled at 132.6 Hz. Using one *Tx* and five *Rx* UWB antennas 500 virtual channels could be constructed from the IRFs.

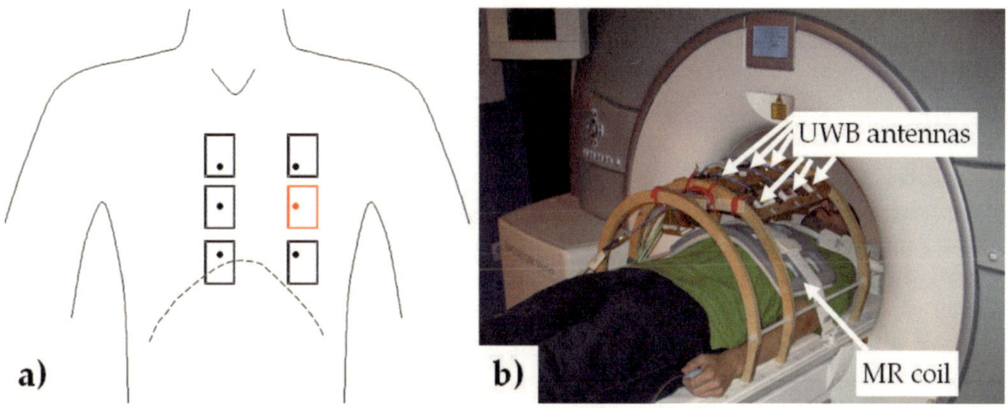

Figure 38. **a)** Scheme of the antenna configuration; **b)** Set-up of simultaneous UWB and MRI measurement.

In the MRI sequence, the one-dimensional 'pencil-like' imaging region is selected by the intersecting volume of two perpendicular slices (s. Fig. 39.a). Both slices are excited in short succession resulting in a saturation effect in the region of the intersection. When the experiment is repeated with a different delay time between both excitation pulses, the two images differ only in the strength of this saturation effect, and subtraction yields the desired 1D image. Placed through the heart in antero-posterior direction, this 'pencil' was scanned at a repetition frequency of 25.4 Hz. The motion components in both data sets, the 500 virtual UWB channels and the MR pencil, were once again separated by applying BSS decomposition.

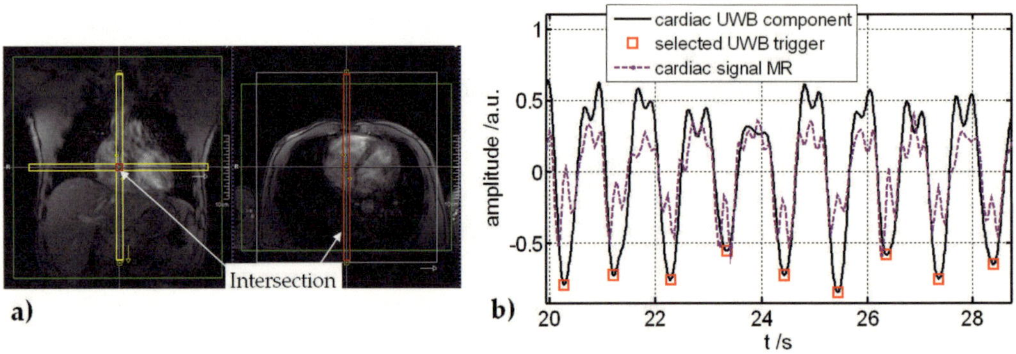

Figure 39. a) Selection of the 'pencil' by two crossing slices in antero-posterior direction through the heart; **b)** Detected cardiac motion component by UWB radar and MR "pencil".

The trigger events (squares in Fig. 39.b) in the UWB cardiac motion data - representing the point of maximum contraction of the myocardium - were determined by applying the algorithm proposed by us. This did not work with the cardiac components of the MR signal due to the pronounced double peaks in this data set. Comparing the cardiac components simultaneously gained by UWB and MR data, we observe perfectly matching slopes of both signals. However, in contrast to UWB radar the MR signal is affected by the blood velocity in the heart producing the double peaks. Keeping this in mind, we can conclude that both modalities render the same motion. Thus, we can assume the cardiac motion detection by UWB radar to be verified.

4.6. Advances for respiratory motions

To establish a UWB navigator for cardiac MRI in free-breathing mode, a landmark tracking of the heart due to the respiration motion is necessary [66]. The time-dependent UWB radar signal contains mainly respiratory motion of the chest, as in Fig. 32.b, which is not necessarily identical to the mechanical displacement of the heart. Therefore, we compared the UWB detected respiratory motion to simultaneously measured 1D MRI as described in Section 4.5.2. Resulting from that comparison, we extended the UWB configuration to allow for the detection of abdominal respiration, too (Fig. 40.a). The antenna configuration applied in the comparison made in Section 4.5.2 was extended by two additional channels above the chest and a second group with one *Tx* and one *Rx* antenna over the abdominal region.

In MR-based navigator techniques [67], the position of the diaphragm is monitored because the shift of the diaphragm is the dominant motion component of the heart due to respiration. The displacement of the diaphragm is mainly orientated in head-foot direction. Hence, the pencil-like one dimensional MRI was placed in head-foot direction across the heart. The UWB data of the first antenna group were decomposed by BSS for detection of breast respiration and cardiac cycle and the second group for the abdominal respiration. In the same way, the motion components were decomposed for "pencil-like" MRI.

The UWB detected breast respiration is not suitable to monitor the mechanical heart shift in head-foot direction. In Fig. 41, a delay between the breast respiration and the heart shift is depicted, and especially in Fig. 41.b it becomes obvious that these are different processes. However, the UWB detected abdominal respiration correlates well with heart motion due to respiration. The correlation factor in measurement a) is 0.932 and 0.81 in measurement b).

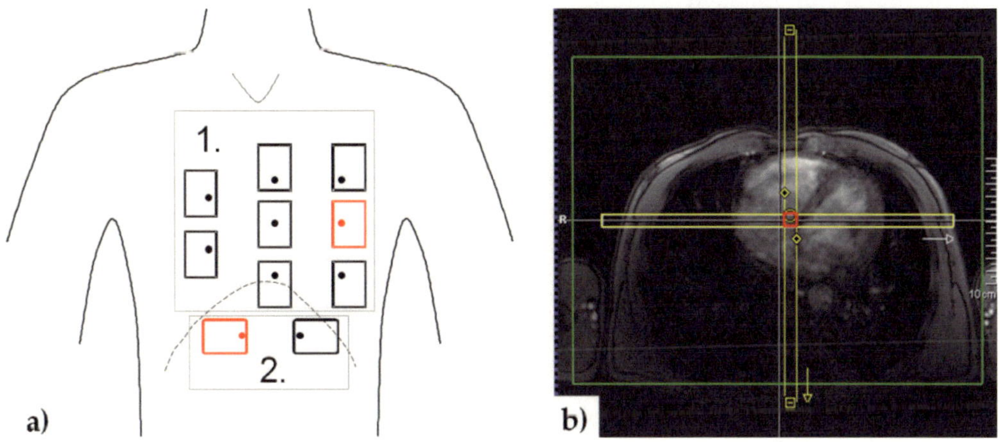

Figure 40. **a)** Extended antenna configuration with second group over the abdominal region; **b)** Placement of the two slices for the 'pencil-like' MRI (Head ⇔ Foot).

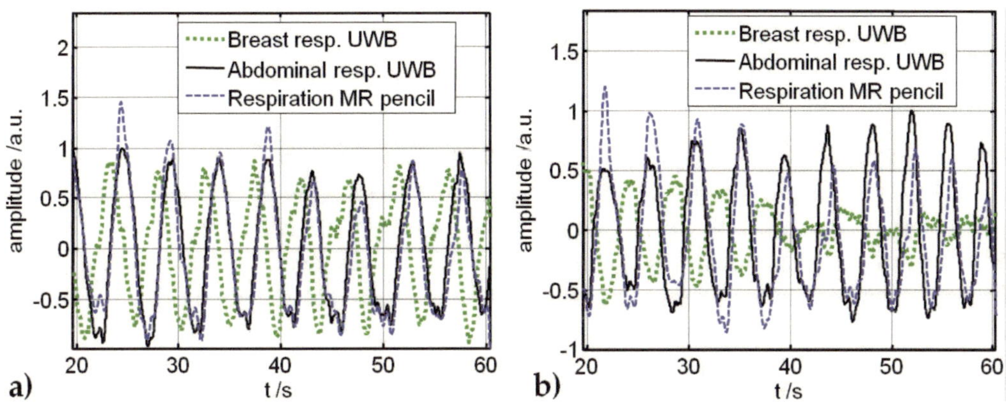

Figure 41. Breast and abdominal respiration by UWB radar and mechanical heart shift in head ⇔ foot direction monitored by MR pencil **a)** with changed breast and abdominal respiration **b)** with fading breast but changed abdominal component.

4.7. UWB triggered cardiac MRI

CMR and UWB signals were acquired simultaneously and synchronously to enable UWB triggering [81]. The UWB antennas were mounted in the same frontal position related to the subject as in Section 4.4.1. Simultaneous pulse oximetry (PO) was applied to compare our approach with another established triggering technique for cardiac MRI.

After acquiring a series of CMR images using a clinical sequence with conventional PO gating, we retrospectively reconstructed the *k*-space data a second time but now using trigger points derived from the simultaneously acquired UWB radar signals [81]. Figure 42.b shows that both methods give virtually undistinguishable results, thus establishing the feasibility of CMR imaging utilizing non-contact UWB radar for triggering. In contrast to established techniques like ECG or PO, however, contact-less UWB-sensing provides cardiac and respiratory information simultaneously and, thus, a sequence-independent external navigator signal.

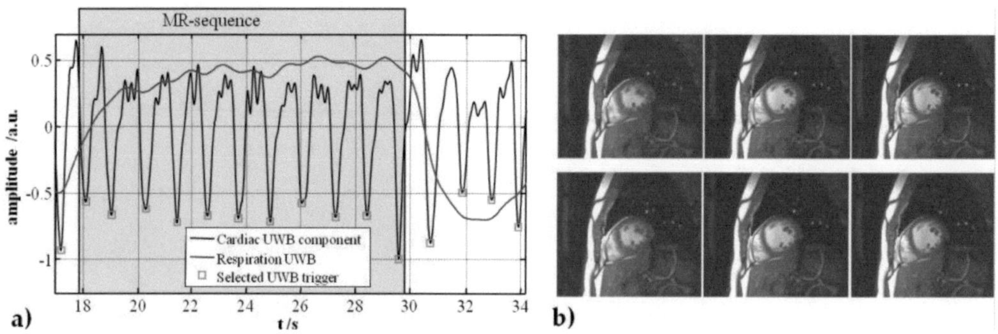

Figure 42. **a)** Cardiac UWB signal with selected trigger events; **b)** Top: Reconstructed images utilizing PO trigger, Bottom: Image reconstruction by UWB trigger events applied.

5. Microwave imaging in medicine

5.1. Introduction

Microwave ultra-wideband (UWB) sensing and imaging represents a promising alternative for the early-stage screening diagnostics of breast cancer. This perspective results from advantageous properties of microwaves: sensitivity of the dielectric properties of human tissue to physiological signatures of clinical interest in this frequency range, especially water content, their non-ionizing nature (compared to X-rays), and the potential of a cost-efficient imaging technology (compared to MRI) [82].

Numerous research groups have been working in this field since the end of the 1990s. Many studies deal with simulations, several groups perform phantom measurements, but only very few have already started some first clinical measurements. The challenges which have to be met concerning real *in vivo* measurements are multifaceted and depend on the conditions of the measurement scenario. The developed strategies and measurement principles of microwave breast imaging can be classified according to various characteristics: active vs. passive vs. heterogeneous microwave imaging systems [83]; microwave tomography (or spectroscopy) imaging [84] vs. UWB radar imaging [85]; examination in prone vs. supine position [83] and some further differentiations. This chapter deals exclusively with active microwave imaging based on the UWB radar principle which can be applied in general in both examination positions.

Figure 43 shows two basic antenna arrangements for the prone examination position. They differ in the antenna-skin distance.

Non-contact breast imaging: The most significant reason for non-contact breast measurements is the size of the antennas compared with the breast size. Thereby, it is impossible to mount a sufficient number of antennas on the breast surface in order to achieve an adequate image quality. The displacement of the antennas from the breast increases the area where additional antennas can be localized. Besides that, it allows mechanical scanning where the antennas can be rotated around the breast in order to create a synthetic aperture. On the other hand, this non-contact strategy is accompanied by a lot of other problems and challenges.

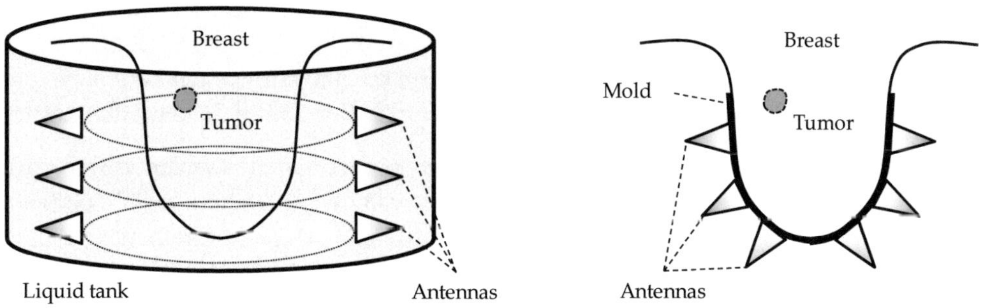

Figure 43. Schematization of non-contact breast imaging using a liquid contact medium (left side) and contact-based breast imaging (right side) in the prone examination position

Depending on the dielectric contrast between the medium surrounding the antennas and the breast tissue, only a fraction of the radiated signal energy will penetrate the breast. The major part will be reflected at the breast surface. It provides clutter which has to be eliminated since it perturbs the signals of interest. In order to reduce the reflection coefficient, several approaches use a liquid coupling medium in which the breast has to be immersed and in which the antennas can surround the breast. The same energy reduction effect appears for reflected components from inside of the breast passing the dielectric boundary in the opposite direction. Furthermore, in the opposite direction (from dielectric dense medium into a less dense medium) waves can only leave the breast below the angle of total reflection which implies an additional reduction of the detectable signal energy outside the breast.

The individual breast shape plays an important role in connection with these effects as well as for image processing. In section 5.3, we describe a method for breast and whole body surface reconstruction based on the reflected UWB signals.

Contact-mode breast imaging: Contact-based breast imaging avoids the disadvantages described above. The antennas are localized directly at the breast surface. Understandably, they have to be small enough in order to arrange a sufficient number of antennas around the breast. The corresponding number of signal channels will be obtained by electronic scanning, that means sequential feeding of all transmitter antennas with simultaneous signal acquisition of all receiving antennas. This strategy involves the problem of individual breast shapes and sizes which influences the contact pressure of the breast skin onto the antenna aperture and, thus, the signal quality [86].

However, we prefer this measurement scenario for our current investigations, and intend to weaken the contact problem in the future by 2 or 3 different array sizes and an additional gentle suction of the breast into the antenna array by a slight underpressure. In section 5.4, we present an experimental measuring set-up where we pursue a strategy of nearly direct contact imaging in order to conjoin the advantages of contact-based imaging with the possibility of mechanical scanning.

5.2. Breast and body phantoms

In the context of UWB tissue sensing, the water content plays a key role as it determines the inherent dielectric properties (ε′ and ε′′) [43]. Moreover, the water content is known to vary among the different human tissues as well as between specific normal and pathologic ones, thus offering a potentially broad spectrum of UWB applications for biomedical diagnostics.

Oil-in-gelatin phantoms, mimicking the dielectric properties of human tissues, were manufactured according to a protocol from [24]. The water concentration varied between 19 and 95% (v/v; ~ 10% water graduation steps), to obtain a set of materials with different permittivity values (ε′ ranging from 8 to 59 and ε′′ ranging from 0.5 to 11, both averaged over frequencies from 1 to 4 GHz). The measurements were carried out by means of the M-sequence devices [76], [87] with HaLoS chipsets and a frequency bandwidth of 4.5 GHz, as well as the radar data acquisition and analysis software "ultraANALYSER" developed for this purpose.

The variation of the oil-water-concentration led to the identification of distinct permittivity values ε′ (Fig. 44, insert) of the different oil-in-gelatin phantoms. The phantom, which was manufactured without oil (95% water), showed values between 53 and 59 for the real part ε′ and between 11 and 10 for the imaginary part ε′′ of the permittivity in the frequency range between 1 and 3.5 GHz (Fig. 44, insert). The results for pure distilled water are also displayed. The real part of permittivity agrees well with literature data [88].

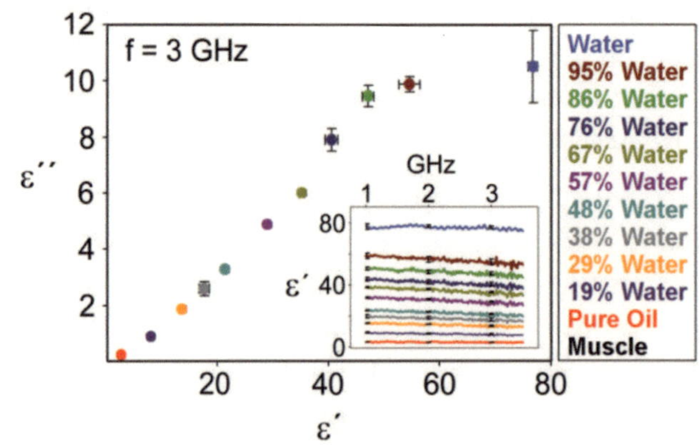

Figure 44. Dielectric properties of nine oil-in-gelatin phantoms with varying percentage of water (from 19% to 95% water (v/v)) and porcine muscle tissue. Depicted is the correlation of the real part ε' and the imaginary part ε'' of the phantoms. Both parts increase with an increasing water-concentration. Error bars represent the standard deviation from an average of three individual measurements on the phantom surface (insert).

5.3. Breast and body surface reconstruction

5.3.1. Method

The benefits of the exact knowledge of the breast surface for non-contact microwave breast imaging are manifold and can improve the results significantly. The inclusion of the breast shape information is essential to calculate the wave traveling path in order to image the interior of the breast based on radar beam-forming techniques. Some approaches use the surface information for initial estimations. Other non-contact measurement approaches strive to illuminate the breast from a specific distance which requires a very fast online surface identification in order to adapt the antenna position during measurement. Furthermore, in the case of varying distances between antenna and breast, the exact knowledge of the breast surface can improve the estimation of the skin reflection component for a better early time artifact removal. In order to reduce the calculation time, the region of interest (i.e. the region for which the image has to be processed) can be restricted based on known surface geometry [89], [90].

Additionally to the significance for breast imaging, UWB microwave radar is suitable for whole body surface reconstruction which can be used in other medical microwave applications as well as in safety-relevant tasks, e.g. under-dress weapon detection.

The Boundary Scattering Transform (BST) represents a powerful approach for surface detection problems. BST and its inverse transform (IBST) were introduced 2004 by Sakamoto and Sato [91] as basic algorithms for high-speed ultra wideband imaging, called SEABED (Shape Estimation Algorithm based on BST and Extraction of Directly scattered waves). Since then, this idea has been extended from mono-static 2D-imaging to the point of bi-static 3D-imaging (IBBST) [92]. The SEABED algorithm represents a high–speed, high-resolution microwave imaging procedure. It does not include the entire radar signal; it uses only wave fronts instead. Furthermore, changes (derivatives) of the propagation time (transmitter → object surface → receiver) depending on the antenna position during the scan process play an important role. SEABED consists of three steps: 1. Detection of the wave fronts and calculation of their derivatives with respect to the coordinates of the scan plane. 2. Inverse Boundary Scattering Transform, which yields spatially distributed points representing the surface of the object. 3. Reconstruction of the surface based on these points.

The practical applicability of the original algorithm to the identification of complex shaped surfaces is limited because of the inherent planar scanning scheme and, therefore, the disadvantage of illuminating only one side of the object. For this reason, we extended the bi-static approach of [92] toward non-planar scanning and a fully three-dimensional antenna movement based on the idea that in the case of arbitrary non-planar scan schemes the current scan plane can be approximated by the tangential plane at each antenna position [93]. An antenna position dependent coordinate transform which ensures that the antenna axis is parallel to the x-axis and the current scan plane is parallel to one plane of the coordinate system allows the application of the IBBST for nearly arbitrary scan surfaces. More precisely, this generalized approach is limited to scenarios where the antennas will be moved orthogonally or parallel to the antenna axis, which is fulfilled in most practical cases. First results of breast shape identification were published in [94], [95].

Based on the following transform equation, the coordinate of the specular point can be calculated

$$\begin{aligned}
\overline{x} &= \overline{X} - \frac{2D^3 D_{\overline{X}}}{D^2 - d^2 + \sqrt{(D^2-d^2)^2 + 4d^2 D^2 D_{\overline{X}}^2}} \\
\overline{y} &= \overline{Y} + \frac{D_{\overline{Y}}}{D^3}\left(d^2(\overline{x}-\overline{X})^2 - D^4\right) \\
\overline{z} &= \overline{Z} + \sqrt{D^2 - d^2 - (\overline{y}-\overline{Y})^2 - \frac{(D^2-d^2)(\overline{x}-\overline{X})^2}{D^2}}
\end{aligned} \tag{5}$$

where $\overline{x},\overline{y},\overline{z}$ are the coordinates of the reflective surface point (specular point), $\overline{X},\overline{Y},\overline{Z}$ are the coordinates of the center between the two antennas, D is the half distance transmitter → reflection point → receiver, d is the half distance between the two antennas, and $D_{\overline{X}} = \frac{dD}{d\overline{X}}$, $D_{\overline{Y}} = \frac{dD}{d\overline{Y}}$ symbolizes the derivatives of the distance with respect to the denoted direction of antenna movement. The bars above the symbols mark the coordinates of the transformed coordinate system [93].

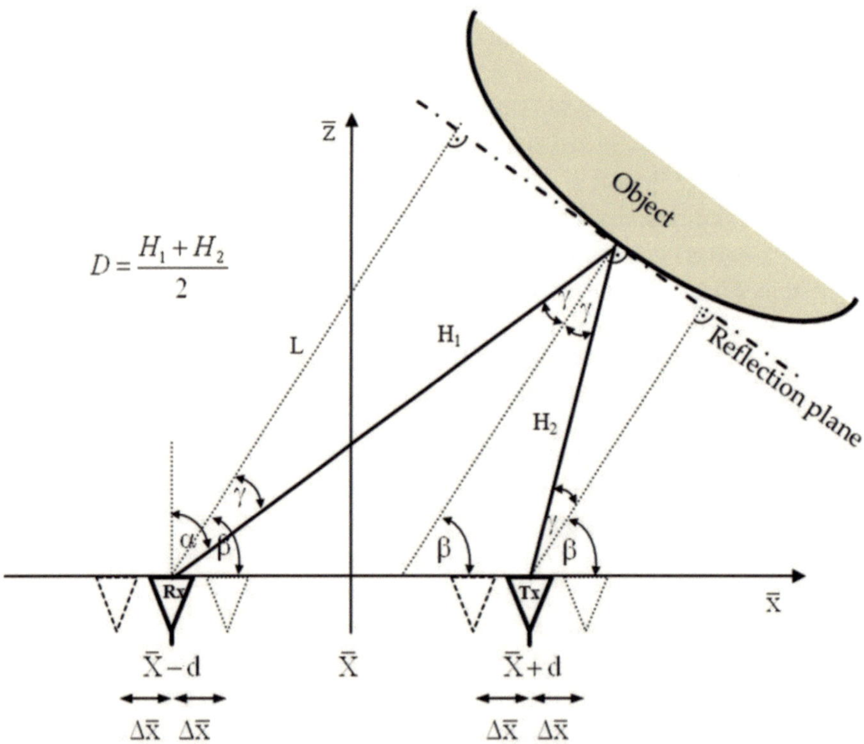

Figure 45. Ray geometry of the inverse bi-static boundary scattering transform (IBBST)

The main challenge is the exact detection of the wave fronts and their proper derivative. For the purpose of wave front detection, we use an iterative correlation-based detection algorithm similar to [96]. In this connection, a short antenna impulse response over a sufficiently wide angular range plays an important role. The difficulties of obtaining appropriate wave front derivatives result from the three-dimensional nature of the problem. The antennas are moved and the transmitted waves are reflected in the three-dimensional space. Especially in the case of wave front crossing and impulse overlapping as well as sparsely detected wave fronts, it is very complicated to recognize which identified wave front at one scan position is related to which wave front at the previous scan position and vice versa. So, it may happen that derivative values are wrongly calculated, which can lead to a spatially false projection of the surface points. In order to avoid such errors, we establish thresholds of feasible derivative values dependent on the antenna beam width.

5.3.2. Detection and elimination of improper wavefronts

General limit values: The range of values of the distance derivatives $D_{\bar{X},\bar{Y}}$ is theoretically bounded between 0 and 1 depending on the slope of the reflection plane (tangent plane of the object surface at the specular point). In the case of parallelism between reflection plane and antenna axis, $D_{\bar{X}} = 0$, whereas in the case of orthogonality, $D_{\bar{X}} = \pm 1$. Thus, calculated values $|D_{\bar{X}}| > 1$ are definitely caused by incorrect wave front detection. Consideration of these general boundaries and exclusion of wave fronts exceeding them yields a significant improvement.

Customized plausibility limit values: The boundary $D_{\bar{X}} = \pm 1$ assumes an antenna radiation angle of 90° or more, which is not given using directive radiators, e.g. horn antennas. In that case, the range of plausible derivative values can further be restricted. Assuming a maximum antenna radiation angle α and a distance between transmitter and receiving antenna of $2d$ the minimum reasonable value $D_{\min}$ can easily be defined by

$$D_{\min} = \frac{d}{\sin\alpha} \tag{6}$$

Wave fronts with lower D values would imply specular points which are located outside the antenna beam and, therefore, can be ignored [98].

Furthermore, a maximum distance derivative $D_{\bar{X}}$ depending on α, d and D can be established:

$$D_{\bar{X}\max} = \frac{\sqrt{(L+\cos\beta\cdot\Delta x)^2 - (L+\cos\beta\cdot\Delta x)\cos\beta\cdot 2d + d^2} - \sqrt{(L-\cos\beta\cdot\Delta x)^2 - (L-\cos\beta\cdot\Delta x)\cos\beta\cdot 2d + d^2}}{2\cdot\Delta x} \tag{7}$$

with the perpendicular from the reflection plane to the distant antenna $L = \cos\gamma\cdot\frac{(D^2-d^2)}{D-d\sin\alpha}$, its perpendicular angle β and the reflection angle γ as depicted in Fig. 45. This value yields

$D_{\bar{X}\max} = \sin\alpha$ for mono-static arrangements $(d = 0)$ and approaches to this value in the case of $L >> d$, respectively. For further details of the derivation of these thresholds and reconstruction examples illustrating the accuracy enhancement due to the application of these thresholds, we refer to [98].

5.3.3. Reconstruction results

For repeatable measurements, we applied a female dressmaker torso which is filled with tissue-equivalent phantom material (Fig. 46). Based on linear and rotational scanners which can move or rotate the object and/or the antennas, several non-planar scan schemes can be realized in order to scan this torso efficiently. In the following, the results of breast shape identification based on a toroidal scan will be shown. The M-sequence radar device used has a bandwidth of 12 GHz [97].

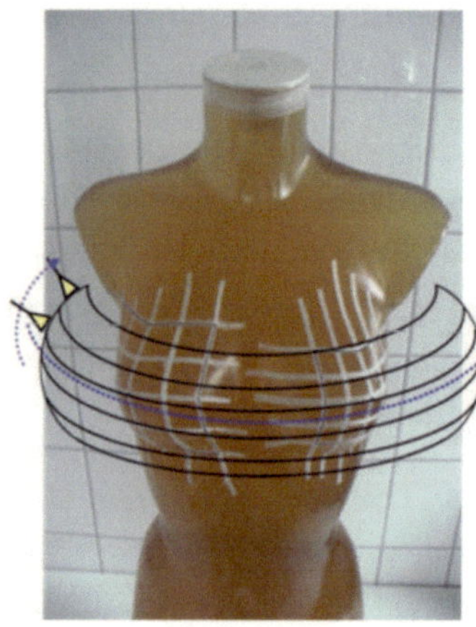

Figure 46. Female torso filled with human tissue mimicking phantom material and delineation of the toroidal scan scheme to reconstruct the chest surface

Numerical problems may arise in the calculation of derivatives from discrete data (discrete time intervals; discrete antenna positions in the space) which have to be considered for setting measurement and processing parameters. The resolutions of spatial scanning and radar signal sampling have to be harmonized carefully with each other in order to avoid derivative artifacts. The maximum possible error of the derivative is $\hat{e}(D_{\bar{x}}) = \frac{\Delta t \cdot v_0}{2\Delta x}$ where Δt is the time resolution of the wave front detection, Δx is the antenna displacement applied for the calculation of $D_{\bar{x}}$ and v_0 is the propagation velocity of the electromagnetic wave. Hence, it will be obvious to meet the requirement of for example $\hat{e}(D_{\bar{x}}) \leq 0.05$ (0.05 is more than 5 percent relative error with respect to $D_{\bar{x}\max}$ for antenna beam widths < 90°!) with an antenna displacement such as $\Delta x = 2.5$ cm in air $(v_0 = c_0)$ the wave front detection has to be realized with a time accuracy of 8.33 ps which has to be provided by interpolation within the wave front detection algorithm. Higher performance requirements presuppose an even more precise wave front identification. Naturally, this is only possible if the radar device fulfills such high time stability requirements.

Figure 47 shows the UWB reconstruction results of the mentioned torso in comparison to a laser reference measurement. In order to quantify the accuracy, the distances between each calculated UWB surface point and the laser-based detected surface is calculated. The resulting mean aberration lower than 1.4 mm underlines the potential of this method. Nevertheless, it is obvious that a further enhancement of the wave front detection represents a residual challenge in order to fill in increasingly the areas of sparsely distributed surface points.

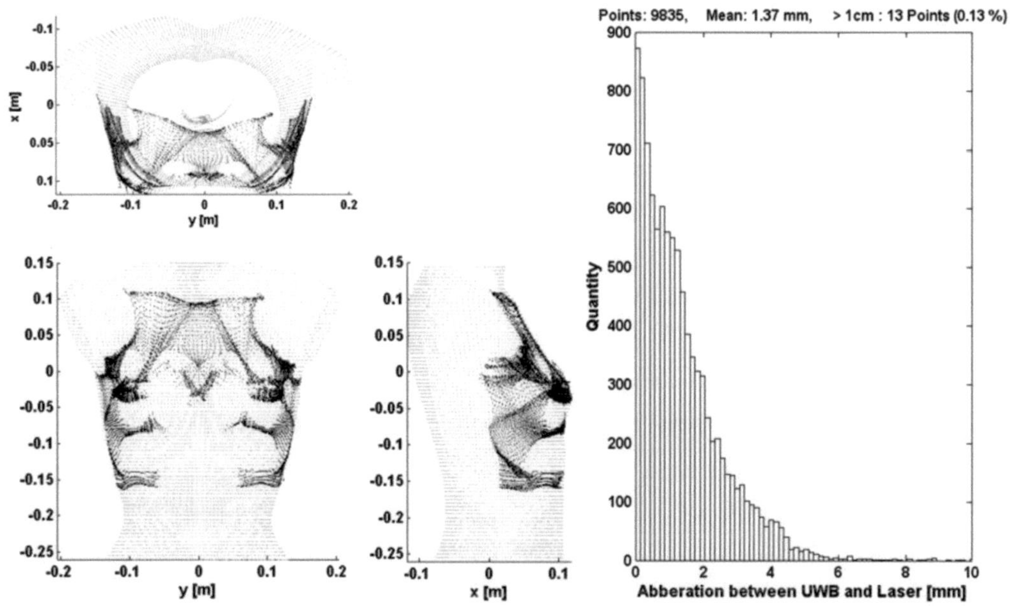

Figure 47. Exact UWB chest surface reconstruction (black) and appraisal of performance values by means of a laser reference measurement (gray) showing a mean aberration lower than 1.4 mm.

Furthermore, the applicability of a 3D-IBBST-based UWB surface reconstruction method for medical applications other than breast imaging as well as for security scenarios (under dress weapon detection) is demonstrated in [98].

5.4. Contact based breast imaging

5.4.1. UWB breast imaging in time domain

The main parts of UWB time domain imaging are the removal of clutter (also referred to as early time artifact removal) and beam-forming (also referred to as migration or back projection). Because the tumor reflections are overlapped by antenna cross-talk and skin reflection, clutter removal is a very important and critical component of signal preprocessing before beam-forming can be carried out. Most clutter removal approaches assume that the clutter appears very similar in each channel and, thus, its estimation improves with increasing channel number. It must be noted that this holds only for channels with comparable clutter parameters. That means clutter estimation and removal has to be done separately for groups consisting of only associated signals (channels with identical antenna distances and boresight angles Tx-Rx), which accomplishes this task. In scientific work on simulation, this circumstance is commonly ignored. For practical applications, however, it has to be taken into consideration.

The simplest approach is to estimate the clutter by means of the average value. Tumor reflections are assumed to appear uncorrelated in the channels and to be negligible in the averaged signal. Even though publications about advanced clutter removal algorithms emphasize the weak points of this self-evident approach, it must be noted that it works relatively robustly in the case of covering tumor response by clutter when some of the proposed alternatives are not applicable.

Image formation algorithms using time domain beam-forming can be included in the following generalized formula:

$$I(\mathbf{r}_0) = \sum_{\tau_h=-T_h/2}^{T_h/2} h(\tau_h,\mathbf{r}_0)\cdot\left(\sum_{n-1}^{N}\sum_{\tau_W=-T_w/2}^{T_w/2} w_n(\tau_w,\mathbf{r}_0)\cdot S_n(t+\tau_n(\mathbf{r}_0)+\tau_w+\tau_h)\right)^2 \tag{8}$$

where N is the number of channels, $S_n(t)$ is the clutter subtracted signal of channel n, $\mathbf{r}_0$ symbolizes the coordinates of the focal point (image position vector), $\tau_n(\mathbf{r}_0)$ is the time delay of channel n related to the focal point at $\mathbf{r}_0$ and $I(\mathbf{r}_0)$ is the back scattered energy which has to be mapped over the region of interest inside the breast. Based on two FIR filters, the different extensions of the common delay-and-sum beam former can be expressed. Path-dependent dispersion and attenuation [99], [100] can be equalized by means of $w_n(\tau_w,\mathbf{r}_0)$ which – in the simplest case - can be only a weight coefficient. Other improvements can also be included by convolution in the time domain, e.g. the cross-correlated back projection algorithm [101]. $h(\tau_h,\mathbf{r}_0)$ represents a smoothing window at the energy level or a scalar weight coefficient [102].

5.4.2. *Measurement setup based on small antennas*

The efficient penetration of the electromagnetic waves into the tissue and the spatial high-resolution registration of the reflected signals are crucial tasks of the antenna array design. In this regard, efficiency is not only a matter of radiation efficiency or antenna return loss, respectively. An efficient antenna array design concerning biomedical UWB imaging purposes comprises also the shape and duration of signal impulses, angle dependence of the impulse characteristics (fidelity), and the physical dimensions of the antenna. These interacting parameters are hardly to accommodate to each other within one antenna design. Generally, compromise solutions have to be found considering basic conditions of scanning, tissue properties and image processing. Here, we pursue the objective of very small antenna dimensions, short impulses and an application in direct or quasi direct contact mode. Therefore, we investigated the usability of small interfacial dipoles.

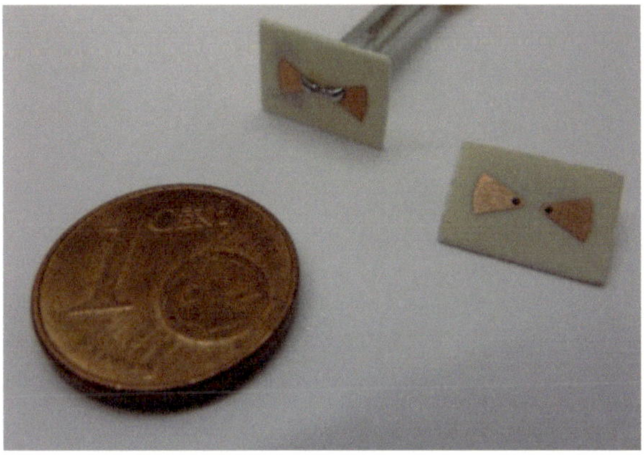

Figure 48. Small bow-ties on Rogers substrate

Initially, we used short bow-ties (Fig. 48) with the dimensions of 8 mm x 3 mm implemented on Rogers® 4003 substrate (0.5 mm) using PCB technology. Dipoles have to be fed differentially. The balanced feeding is realized by differential amplifier circuits [103].

These antennas cannot be matched over a large bandwidth, which leads to unwanted reflections between antenna and amplifier. There are two options concerning the handling of this problem: realization of a sufficient line length between antenna and amplifier (in order to gate out the reflections) or implementation of the amplifier circuits directly at the antenna feed point. On an interim basis, we pursued the first strategy using long cables between antenna and amplifier. Assuming a maximum mean tissue permittivity $\varepsilon' \leq 50$, a 70 cm cable will ensure that any reflections from inside of the breast (diameter ~ 10 cm) and unwanted reflections at the amplifier do not overlap.

As mentioned above, the contact between antennas and breast skin represents a crucial aspect for sufficient signal quality. Regarding clinical requirements (e.g. disinfection) we plan to place the antennas behind a thin examination mold. But this additional interface reduces the signal quality significantly. Therefore, a thin (~2 mm) matching layer consisting of tissue mimicking phantom material was inserted between the examination mold and the antennas in order to increase the signal energy penetrating the tissue and reduce the backward radiation (Fig. 49). The benefit achieved when using a thin contact layer was also investigated and verified by simulations (Fig. 5 in section 2.3.2).

We built up two preliminary array set-ups for phantom measurements, both including eight antennas and distributing them around a circular segment (diameter 9.5 cm) in steps of 22.5°. An array with a horizontal antenna arrangement is shown in Fig. 50. Exemplary phantom measurement results achieved with these prototypes are published in [104] and [105] and will be summarized in section 5.4.3.

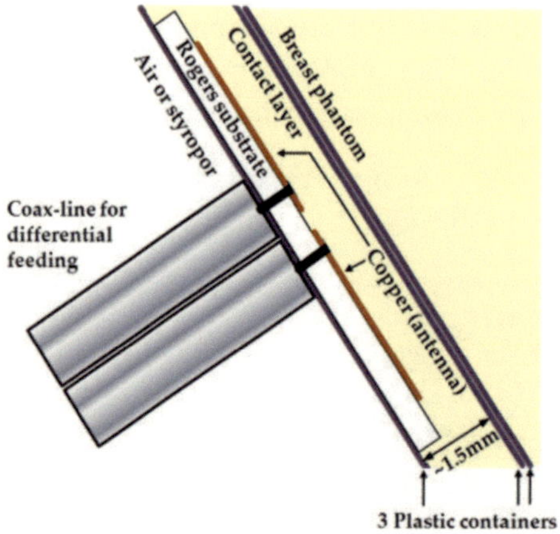

Figure 49. Schematic illustration of the contact layer filled with phantom material and mounted antennas inside

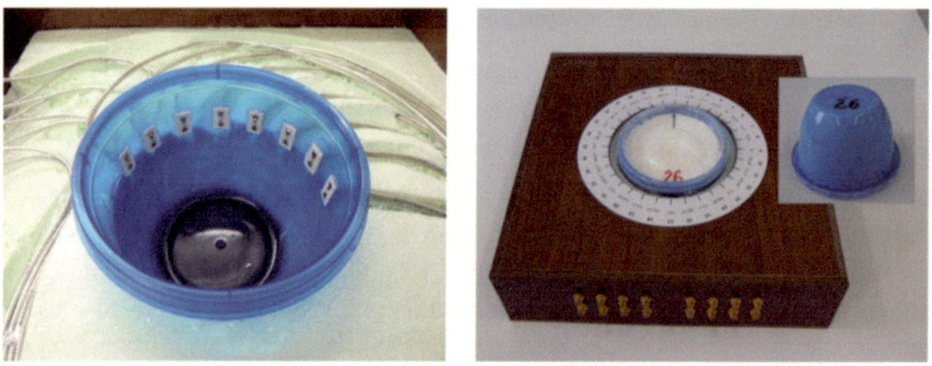

Figure 50. Antenna array: Assembly stage before casting the contact layer. The connected and affixed differential fed antennas and the container for the outer boundary of the contact layer are still visible (left panel). Finished antenna array with inserted rotatable breast phantom (right panel)

After this preliminary development stage, the differential feeding amplifier was relocated into the antenna feed point. By this step, reflections due to antenna mismatch will be avoided, and the quantity of feeding cables will be bisected, because each active antenna element can be fed single-ended (Fig. 51).

In conjunction with this enhancement, the mechanical part of the antenna array was improved. A developed slide-in mounting system (Fig. 51) allows flexible antenna application and replacement and, therefore, facilitates investigations of various Rx-Tx-arrangements without destruction and rebuild of the whole array as it is the case with the preliminary set-up shown in Fig. 50 [106].

Because the contact layer will not be hermetically sealed in this case, the chemical instable oil-gelatin phantom material cannot be used anymore for this task. Thus, investigations of alternative materials have to be considered. We propose polymer-powder composites where dielectric powders (e.g. carbon meal or barium titanate powder) will be admixed to silicone rubber. This special challenging topic is currently under investigation.

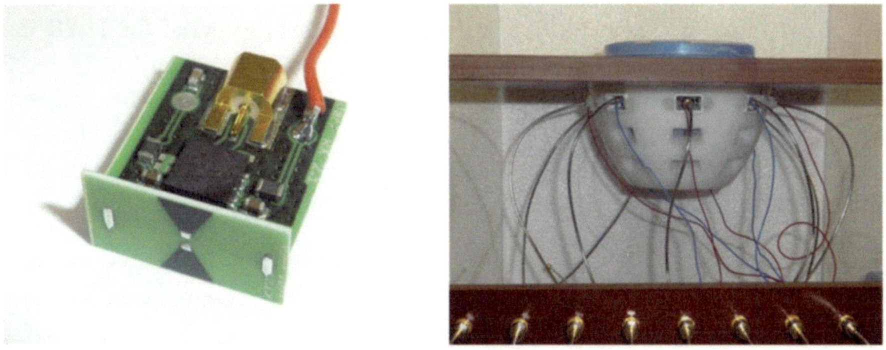

Figure 51. Photographs of an active antenna element (Rx) with 8 mm dipole with amplifier circuit board (left panel) and the slide-on mounting system for phantom measurements as well as *in vivo* measurements

5.4.3. Imaging results of phantom trials

The breast phantoms are tissue mimicking oil-gelatin phantoms according to [24] and described in section 5.2, where the dielectric properties can be adjusted by means of the oil content. For our measurements we used two types of material: 40% oil (57% water) content material mimics healthy tissue which approximately corresponds to group II of adipose-defined tissue (31%-84% adipose tissue) [106]. The 10% oil (85.5% water) content material simulates tumor tissue. Fig. 52 illustrates permittivity, attenuation losses and reflection coefficient between both tissues. In order to realize an optimal contact to the antenna array, the phantom material is filled in identical plastic containers (diameter 9.5cm) as used for the examination mold. The containers are hermetically sealed and stored in the fridge to avoid chemical instability of the phantom material. The phantoms have to be acclimatized at least 3 hours before starting the measurements.

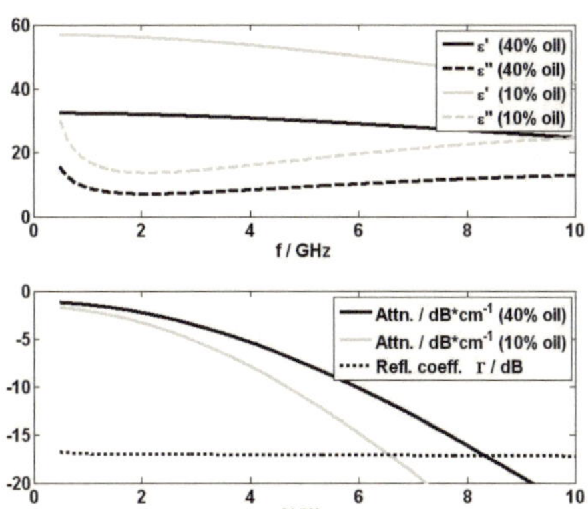

Figure 52. Dielectric values of the tissue mimicking phantom material: Permittivity (above), transmission losses per cm and reflection coefficient between them (below)

Figure 53 shows two measured signals of the proposed antennas which illustrate the appropriate time domain characteristics. The measurement through 6 cm tissue (mimicked by means of phantom material) as well as the cross-talk signal between two antennas show relatively short impulse shapes with low ringing, which is essential for UWB imaging. Including the dispersive tissue impact the spectral bulk ranges between 1 GHz and 3 GHz with a bandwidth greater than 2 GHz for both received impulses. Obviously, because of the dielectric scaling due to the direct contact between tissue and antenna, such small antennas are capable of radiating waves in a frequency range with acceptable attenuation and penetration depth.

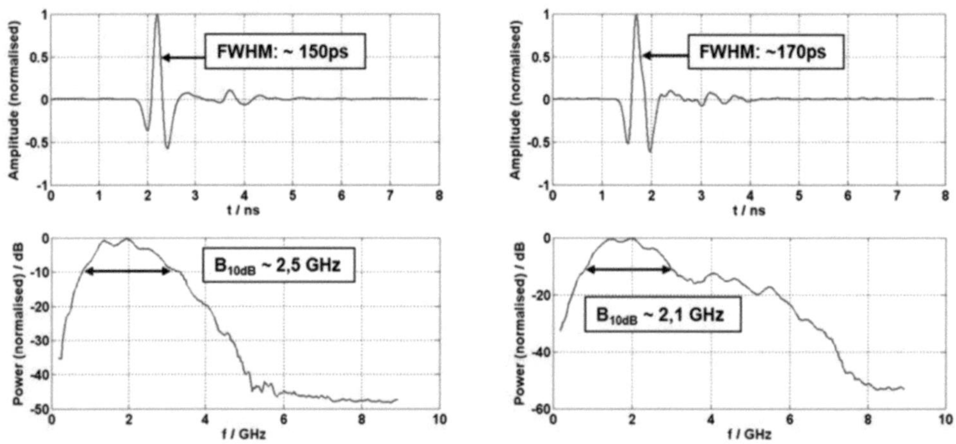

Figure 53. Measurement signals based on the described bow-ties: measurement through 6 cm tissue mimicking phantom material with 40% oil content (left panel) and cross-talk signal between adjacent antennas, separated by 2.5 cm (right panel)

During the phantom measurements, four antennas acted as receivers and are permanently connected with Rx1…Rx4 of the radar device. The transmitter signal was connected to one of 4 transmitter antennas by a coaxial switch matrix. Thus, 16 signal channels could be achieved without rearrangement. Their angles between the boresight directions of Tx and Rx differed in the range 22.5 - 157.5°. Because this amount of signal channels is insufficient for high-resolution imaging, we had to consider robust and reproducible mechanical scanning to achieve a sufficient number of channels. In order to simulate antenna rotation, the phantoms were rotated in steps of 11.25°. This resulted in 512 signals (16 channels x 32 rotations) which could be included into the imaging process of one phantom.

Figure 54 shows exemplary imaging results of the described breast phantoms applying the presented measuring set-up and time domain beam-forming. Despite the relatively low dielectric contrast between both tissue simulations, the tumor inclusions can clearly be identified. The highest interferences (side lobes) are about 11dB (15mm tumor) and around 7dB (10mm tumor) lower than the tumor representation. Additionally, the lower panels of Fig. 54 illustrate the capability of localization and differentiation between multiple tumors, for example two 15 mm tumors with a distance of 30 mm between them. Despite of the relatively low dielectric contrast between healthy and cancerous tissue mimicking phantom material, the tumors can be detected and separated.

The results underline that small dipoles can be profitably applied for UWB breast imaging. The impressive identification of the tumor surrogates promises also the detection of weaker dielectric contrasts. On the other hand, it must be noted that the tumor surrounding tissue imitation is completely homogeneous which does not correspond to reality. Therefore, our breast phantoms must be enhanced in the future toward a better approximation of the breast tissue heterogeneity.

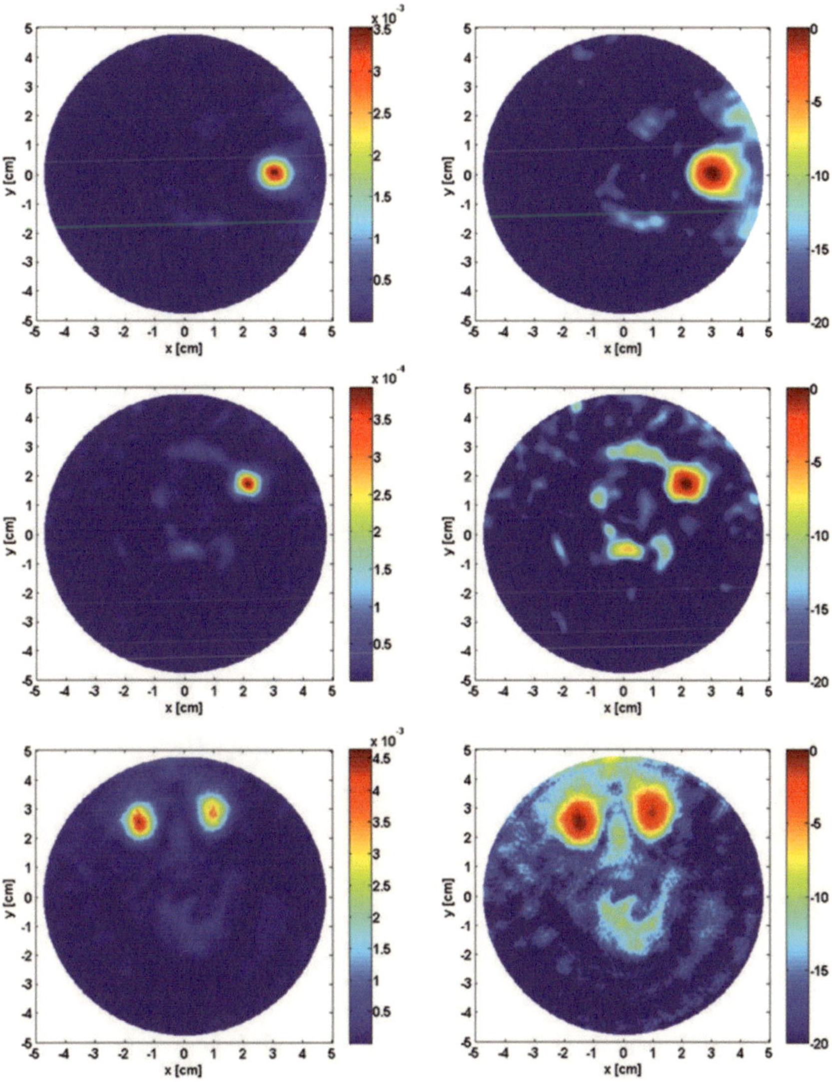

Figure 54. UWB images of phantom trials including a 15mm (top) and a 10 mm (middle) tumor surrogate as well as two 15 mm tumor surrogates, separated by 30 mm (below). Left: linear energy scale; Right: logarithmic scale in dB.

6. Conclusions

In this chapter, we dealt with ultra wideband sensing in medical engineering, i.e. using electromagnetic waves of large bandwidth for probing the human body and biological tissue. Sufficient penetration of the human body combined with antennas of manageable size were our major concern. Also, the frequency band from 1 GHz to 5…8 GHz turned out to be best suited for our purposes. By using active or dielectrically scaled antennas for this frequency range, they can be built sufficiently small. Wave propagation at these frequencies is mostly influenced by water, the most abundant component of biological tissue. The effect of salt becomes less detrimental above 1 GHz. Above 5…8 GHz, however, water absorption will drastically increase the propagation losses. The given frequency band also provides acceptable resolution for microwave imaging and ample micro-Doppler sensitivity.

For experimental investigations, we exploited ultra-wideband pseudo-noise devices. They provide probing signals of very low power, thus avoiding damages to biological tissue. Furthermore, they provide sufficient dynamic range, measurement speed and short term stability for super resolution techniques of microwave imaging and weak-motion tracking.

We demonstrated medical applications of ultra-wideband sensing by three distinctive examples, each standing for a specific class of applications.

1. Contact-based measurements (impedance spectroscopy) aimed to estimate tissue permittivity. This mainly gives some hints on water concentration and water bonds. At lower frequencies, other molecules will also leave their traces in the measured impedance value.
2. Remote motion tracking of organs inside the human body like cardio-pulmonary activity for example of motion correction for magnetic resonance imaging. Remote vital sign detection is a related topic with relaxed conditions referred to tracking precision but increased requirements with respect to area coverage. The analysis of cardiac mechanics for separate heart region accessible by stand-alone UWB radar or in conjunction with the electrical activity from the ECG contains valuable diagnostic information, e.g. for infarction detection, as ischemic tissue shows a modified contraction pattern.

Remote or contact-based microwave imaging of inner organs or malignant tissue, for example the detection of breast tumors.

Author details

Ingrid Hilger, Katja Dahlke, Gabriella Rimkus and Christiane Geyer
Jena University Hospital, Germany

Frank Seifert, Olaf Kosch, Florian Thiel
Physikalisch-Technische Bundesanstalt Berlin, Germany

Matthias Hein, Francesco Scotto di Clemente, Ulrich Schwarz, Marko Helbig, Jürgen Sachs
Ilmenau University of Technology, Germany

Acknowledgement

This work was supported by the German Science Foundation (DFG) in the framework of the priority program UKoLoS (SPP 1202), project acronym ultraMEDIS. The authors appreciate the valuable contributions made by R. Herrmann, P. Rauschenbach and K. Schilling for sensor development, and helpful discussions; K. Borkowski and E. Hamatschek for electronic and mechanical component manufacturing; Ralf Stephan for his support in antenna design and measurement; Hartmut Günther and Stefan Barth for the manufacture of the ceramic antennas; Marina Sieler and Uwe Genatis for the galvanic metallization of the ceramic and MRI compatible antennas.

7. References

[1] U. Pliquett, "Electricity and biology," in 11th International Biennial Baltic Electronics Conference, BEC 2008, pp. 11-20.

[2] K. Nowak, W. Gross, K. Nicksch *et al.*, "Intraoperative lung edema monitoring by microwave reflectometry," *Interactive Cardiovascular and Thoracic Surgery*, pp. 540-544, 2011.

[3] J. Sachs, E. Zaikov, M. Helbig *et al.*, "Trapped Victim Detection by Pseudo-Noise Radar," in 2011 International Conference on Wireless Technologies for Humanitarian Relief (ACWR 2011) Amritapuri, Kollam, Kerala, India, 2011, pp. 265-272.

[4] E. Zaikov, and J. Sachs, "UWB radar for detection and localization of trapped people," *Ultra Wideband*, B. Lembrikov, ed., Rijeka, Croatia: Scivo, 2010.

[5] J. Sachs, M. Helbig, R. Herrmann *et al.*, "Merkmalsextraktion und semantische Integration von Ultrabreitband-Sensoren zur Erkennung von Notfällen," in 3. Deutscher AAL-Kongress 2010 Assistenzsysteme im Dienste des Menschen, Berlin, 2010.

[6] R. Herrmann, J. Sachs, and F. Bonitz, "On benefits and challenges of person localization using ultra-wideband sensors," in Indoor Positioning and Indoor Navigation (IPIN), 2010 International Conference on, 2010, pp. 1-7.

[7] R. Herrmann, J. Sachs, K. Schilling *et al.*, "12-GHz Bandwidth M-Sequence Radar for Crack Detection and High Resolution Imaging," in International Conference on Ground Penetrating Radar (GPR), Birmingham, UK, 2008.

[8] R. Herrmann, "M-sequence based ultra-wideband radar and its application to crack detection in salt mines" PhD-Thesis, Faculty of Electrical Engineering and Information Technology, Ilmenau University of Technology (Germany), Ilmenau, 2011.

[9] J. Sachs, "Ultra-Wideband Short-Range Sensing - Theory, Sensors, Applications", Berlin: Wiley-VCH, 2012.

[10] M. A. Hein, C. Geyer, M. Helbig *et al.*, "Antennas for ultra-wideband medical sensor systems," in 3rd European Conference on Antennas and Propagation, EuCAP 2009, 2009, pp. 1868-1872.

[11] R.C. Hansen, "Fundamental Limitations in Antennas", in Proc. of the IEEE, vol. 69, n. 2, pp. 170-182, Feb. 1981

[12] U. Schwarz, M. Helbig, J. Sachs, R. Stephan, M.A. Hein, "Design and application of dielectrically scaled double-ridged horn antennas for biomedical UWB radar applications", 2009 IEEE International Conference on Ultra-Wideband ICUWB 2009, pp. 150-154

[13] F. Thiel, M.A. Hein, J. Sachs, U. Schwarz and F. Seifert: "Physiological signatures monitored by ultra-wideband-radar validated by magnetic resonance imaging", IEEE International Conference on Ultra-Wideband, ICUWB 2008, pp. 105-108

[14] The Visible Human Project, U.S. National Library of Medicine, 8600 Rockville Pike, Bethesda, MD 20894

[15] U. Schwarz, M. Helbig, J. Sachs, F. Seifert, R. Stephan, F. Thiel, M.A. Hein, "Physically small and adjustable double-ridged horn antenna for biomedical UWB radar applications", IEEE International Conference on Ultra-Wideband, ICUWB 2008, pp. 5-8

[16] H. Schantz, "Ultrawideband Antennas", Artech House, Inc., 2005

[17] MRA Laboratories, Adams, MA, USA. Available online: www.mralabs.com (accessed May 2012)

[18] J. Krupka, R.G. Geyer, J. Baker-Jarvis, J. Ceremuga, "Measurements of the complex permittivity of microwave circuit board substrates using split dielectric resonator and reentrant cavity techniques", 7th International Conference on Dielectric Materials, Measurements and Applications, (Conf. Publ. No. 430), 1996, pp. 21-24

[19] R.E. Collin, "Foundations for microwave engineering", 2nd ed., McGraw-Hill, 1992

[20] J. Krupka, "Computations of Frequencies and Intrinsic Q Factors of TE0nm Modes of Dielectric Resonators", IEEE Transactions on Microwave Theory and Techniques, 1985, pp. 247-277

[21] F. Scotto di Clemente, R. Stephan, and M. Hein, "Compact permittivity-matched ultra-wideband antennas for biomedical imaging", Topical Conference on Antennas and Propagation in Wireless Communications (APWC), 2011 IEEE-APS, pp. 858-860

[22] F. Scotto di Clemente, M. Helbig, J. Sachs, U. Schwarz, R. Stephan, M.A. Hein, "Permittivity-matched compact ceramic ultra-wideband horn antennas for biomedical diagnostics", Proceedings of the 5th European Conference on Antennas and Propagation (EUCAP), pp. 2386-2390

[23] U. Schwarz, R. Stephan, M.A. Hein, "Miniature double-ridged horn antennas composed of solid high-permittivity sintered ceramics for biomedical ultra-wideband radar applications", 2010 IEEE Antennas and Propagation Society International Symposium (APSURSI), 2010, pp. 1-4

[24] M. Lazebnik, E. Madsen EL, G. Frank, S. Hagness, "Tissue-mimicking phantom materials for narrowband and ultra-wideband microwave applications", Physics in Medicine and Biology 2005; 50:4245-4258.

[25] C. Geyer, "Wechselwirkungen von elektromagnetischen Wellen im Radiofrequenzbereich mit biologischen Systemen und lebenden Organismen," Dissertation, Biologisch-Pharmazeutische Fakultät, Friedrich-Schiller-Universität Jena, 2011.

[26] U. Schwarz, R. Stephan, M.A. Hein, "Experimental validation of high-permittivity ceramic double-ridged horn antennas for biomedical ultra-wideband diagnostics", IEEE

International Conference on Wireless Information Technology and Systems (ICWITS), 2010, pp. 1-4.

[27] M. Hein, M. Helbig, M. Kmec, J. Sachs, F. Scotto di Clemente, R. Stephan, M. Hamouda, T. Ussmueller, R. Weigel, M. Robens, R. Wunderlich, S. Heinen, "Ultra-Wideband Active Array Imaging for Biomedical Diagnostics", Topical Conference on Antennas and Propagation in Wireless Communications (APWC), 2012 IEEE-APS (Accepted).

[28] F. Scotto di Clemente, R. Stephan, U. Schwarz, M.A. Hein, " Miniature body-matched double-ridged horn antennas for biomedical UWB imaging", Topical Conference on Antennas and Propagation in Wireless Communications (APWC), 2012 IEEE-APS (Accepted)

[29] F. Thiel, M. Hein, J. Sachs, U. Schwarz, and F. Seifert, "Combining magnetic resonance imaging and ultrawideband radar: A new concept for multimodal biomedical imaging," Rev. Sci. Instrum., vol. 80, no. 1, 2009.

[30] O. Kosch, F. Thiel, F. Scotto di Clemente, M.A. Hein, F. Seifert, "Monitoring of human cardio-pulmonary activity by multi-channel UWB-radar", Topical Conference on Antennas and Propagation in Wireless Communications (APWC), 2011 IEEE-APS, pp.382-385

[31] U. Schwarz, F. Thiel, F. Seifert, R. Stephan, M. Hein, "Magnetic resonance imaging compatible ultra-wideband antennas," in 3rd Eur. Conf. on Antennas and Propagation, 2009, pp. 1102–1105.

[32] U. Schwarz, F. Thiel, F. Seifert, R. Stephan, M.A. Hein, "Ultra-Wideband Antennas for Magnetic Resonance Imaging Navigator Techniques", IEEE Trans. on Antennas and Propag, vol. 58, no. 6, June 2010.

[33] D. Miklavcic, N. Pavselj, and F. X. Hart, "Electrical properties of tissues," *Wiley Encyclopedia of Biomedical Engineering*, M. Akay, ed.: Wiley-Interscience, 2006.

[34] C. Gabriel, "Dielectric properties of biological tissue: Variation with age," *Bioelectromagnetics*, pp. S12-S18, 2005.

[35] G. H. Markx, and C. L. Davey, "The dielectric properties of biological cells at radiofrequencies: Applications in biotechnology," *Enzyme and Microbial Technology*, vol. 25, no. 3-5, pp. 161-171, Aug, 1999.

[36] R. Pethig, "Dielectric-Properties of Biological-Materials - Biophysical and Medical Applications," *IEEE Transactions on Electrical Insulation*, vol. 19, no. 5, pp. 453-474, 1984.

[37] H. P. Schwan, "Analysis of Dielectric Data - Experience Gained with Biological-Materials," *IEEE Transactions on Electrical Insulation*, vol. 20, no. 6, pp. 913-922, Dec, 1985.

[38] H. P. Schwan, and C. F. Kay, "The Conductivity of Living Tissues," *Annals of the New York Academy of Sciences*, vol. 65, no. 6, pp. 1007-1013, 1957.

[39] A. S. Arnoux, L. Preziosi-Belloy, G. Esteban *et al.*, "Lactic acid bacteria biomass monitoring in highly conductive media by permittivity measurements," *Biotechnology Letters*, vol. 27, no. 20, pp. 1551-1557, Oct, 2005.

[40] J. Yang, Y. Huang, X. J. Wang *et al.*, "Dielectric properties of human leukocyte subpopulations determined by electrorotation as a cell separation criterion," *Biophysical Journal*, vol. 76, no. 6, pp. 3307-3314, Jun, 1999.

[41] H. P. Schwan, "Mechanisms responsible for electrical properties of tissues and cell suspensions," *Med Prog Technol,* vol. 19, no. 4, pp. 163-5, 1993.

[42] S. R. Smith, and K. R. Foster, "Dielectric properties of low-water-content tissues," *Phys Med Biol,* vol. 30, no. 9, pp. 965-73, Sep, 1985.

[43] K. R. Foster, and H. P. Schwan, "Dielectric-Properties of Tissues and Biological-Materials - a Critical-Review," *Critical Reviews in Biomedical Engineering,* vol. 17, no. 1, pp. 25-104, 1989.

[44] J. A. Evans, S. L. Russell, C. James *et al.*, "Microbial contamination of food refrigeration equipment," *Journal of Food Engineering,* vol. 62, no. 3, pp. 225-232, May, 2004.

[45] M. Grossi, A. Pompei, M. Lanzoni *et al.*, "Total Bacterial Count in Soft-Frozen Dairy Products by Impedance Biosensor System," *IEEE Sensors Journal,* vol. 9, no. 10, pp. 1270-1276, Oct, 2009.

[46] Y. J. Chang, A. D. Peacock, P. E. Long *et al.*, "Diversity and characterization of sulfate-reducing bacteria in groundwater at a uranium mill tailings site," *Applied and Environmental Microbiology,* vol. 67, no. 7, pp. 3149-3160, Jul, 2001.

[47] S. K. Haack, L. R. Fogarty, T. G. West *et al.*, "Spatial and temporal changes in microbial community structure associated with recharge-influenced chemical gradients in a contaminated aquifer," *Environmental Microbiology,* vol. 6, no. 5, pp. 438-448, May, 2004.

[48] G. Haferburg, D. Merten, G. Buchel *et al.*, "Biosorption of metal and salt tolerant microbial isolates from a former uranium mining area. Their impact on changes in rare earth element patterns in acid mine drainage," *Journal of Basic Microbiology,* vol. 47, no. 6, pp. 474-484, Dec, 2007.

[49] S. E. Beekmann, D. J. Diekema, K. C. Chapin *et al.*, "Effects of rapid detection of bloodstream infections on length of hospitalization and hospital charges," *Journal of Clinical Microbiology,* vol. 41, no. 7, pp. 3119-3125, Jul, 2003.

[50] K. Mishima, A. Mimura, Y. Takahara *et al.*, "Online Monitoring of Cell Concentrations by Dielectric Measurements," *Journal of Fermentation and Bioengineering,* vol. 72, no. 4, pp. 291-295, 1991.

[51] C. M. Harris, R. W. Todd, S. J. Bungard *et al.*, "Dielectric Permittivity of Microbial Suspensions at Radio Frequencies - a Novel Method for the Real-Time Estimation of Microbial Biomass," *Enzyme and Microbial Technology,* vol. 9, no. 3, pp. 181-186, Mar, 1987.

[52] M. Jonsson, K. Welch, S. Hamp *et al.*, "Bacteria counting with impedance spectroscopy in a micro probe station," *Journal of Physical Chemistry B,* vol. 110, no. 20, pp. 10165-10169, May 25, 2006.

[53] E. Benoit, A. Guellil, J. C. Block *et al.*, "Dielectric permittivity measurement of hydrophilic and hydrophobic bacterial suspensions: a comparison with the octane adhesion test," *Journal of Microbiological Methods,* vol. 32, no. 3, pp. 205-211, May, 1998.

[54] Dahlke, K., Geyer, C.; Dees, S.; Helbig, M.; Sachs, J.; Scotto di Clemente, F.; Hein, M.; Kaiser, W.A.; Hilger, I.,"Effects of cell structure of Gram-positive and Gramnegative bacteria based on their dielectric properties," *The 7th German Microwave Conference (GeMiC), 2012,* 12-14 March 2012.

[55] F. Jaspard, and M. Nadi, "Dielectric properties of blood: an investigation of temperature dependence," *Physiological Measurement,* vol. 23, no. 3, pp. 547-554, Aug, 2002.

[56] H. F. Cook, "A Comparison of the Dielectric Behaviour of Pure Water and Human Blood at Microwave Frequencies," *British Journal of Applied Physics,* vol. 3, no. Aug, pp. 249-255, 1952.

[57] H. P. Schwan, and K. Li, "Capacity and Conductivity of Body Tissues at Ultrahigh Frequencies," *Proceedings of the Institute of Radio Engineers,* vol. 41, no. 12, pp. 1735-1740, 1953.

[58] H. P. Schwan, and K. R. Foster, "Rf-Field Interactions with Biological-Systems - Electrical-Properties and Biophysical Mechanisms," *Proceedings of the IEEE,* vol. 68, no. 1, pp. 104-113, 1980.

[59] L. Schmuntzsch, "Thermische Abhängigkeit der Permittivität in biologischem Gewebe," Diploma, Fachbereich Medizintechnik/ Biotechnologie, Fachhochschule Jena, Jena, 2010.

[60] A. Mashal, B. Sitharaman, X. Li *et al.*, "Toward Carbon-Nanotube-Based Theranostic Agents for Microwave Detection and Treatment of Breast Cancer: Enhanced Dielectric and Heating Response of Tissue-Mimicking Materials," *IEEE Transactions on Biomedical Engineering,* vol. 57, no. 8, pp. 1831-1834, Aug, 2010.

[61] Noeske R., Seifert F., Rhein K.H., Rinneberg H., "Human cardiac imaging at 3T using phased array coils", *Magn. Reson. Med.* 44, 978-82 (2000)

[62] Haacke E.M., Li D., Kaushikkar S., "Cardiac MR Imaging: Principles and Techniques". *Top Magn Reson Imaging* 7:200-17, (1995).

[63] Dieringer M.A., Renz W., Lindel T., Seifert F., Frauenrath T., von Knobelsdorff-Brenkenhoff F., Waiczies H., Hoffmann W., Rieger J., Pfeiffer H., Ittermann B., Schulz-Menger J., Niendorf T., "Design and application of a four-channel transmit/receive surface coil for functional cardiac imaging at 7T", *Journal of Magnetic Resonance Imaging* 33 (3): 736-741 (2011-03).

[64] Nijm G.M., Swiryn S., Larson A.C., and Sahakian A.V., "Characterization of the Magnetohydrodynamic Effect as a Signal from the Surface Electrocardiogram during Cardiac Magnetic Resonance Imaging", in *Computers in Cardiology,* Valencia, Spain, p. 269–272, (2006)

[65] Frauenrath T., Hezel F., Renz W., de Geyer T., Dieringer M., von Knobelsdorff-Brenkenhoff F., Prothmann M., Schulz-Menger J., Niendorf T., "Acoustic cardiac triggering: a practical solution for synchronization and gating of cardiovascular magnetic resonance at 7 Tesla", *Journal of Cardiovascular Magnetic Resonance* 12 (1): 67 (2010-11-16).

[66] van Geuns R.J., PA. Wielopolski, HG. de Bruin, et al: "MR coronary angiography with breath-hold targeted volumes: preliminary clinical results", *Radiology* (2000); 217:270–7

[67] Hinks, R.S.. "Monitored echo gating for reduction of motion artifacts". *Magn Reson Imaging,* 6(48), (1988).

[68] Thiel F., Kreiseler D., Seifert F., "Non-contact detection of myocardium's mechanical activity by ultra-wideband RF-radar and interpretation applying electrocardiography", *Rev. Sci. Instrum.,* vol. 80, 11, 114302, ISSN 0034-6748, Melville, NY: American Institute of Physics (AIP), 12 pages, (2009)

[69] Morguet A.J.; Behrens S., Kosch O., Lange C., Zabel M., Selbig D., Munz D.L., Schultheiss H.P., Koch H.: "Myocardial viability evaluation using magnetocardiography in patients with coronary artery disease"; *Coron Artery Dis* 15, 2004: 155 – 162

[70] Muehlsteff J., Thijs J., Pinter R., Morren G., Muesch G., "A handheld device for simultaneous detection of electrical and mechanical cardio-vascular activities with synchronized ECG and CW-Doppler Radar", *IEEE Int. Conf. Engineering in Medicine and Biology Society* (EMBS) 22-26 Aug. (2007), 5758 – 5761

[71] Thiel F. and Seifert F., "Non-invasive probing of the human body with electromagnetic pulses: Modelling of the signal path", *J. Appl. Phys.*, vol.105, issue 4, 044904, ISSN 0021-8979, Melville, NY, American Institute of Physics (AIP), 8 pages, (2009)

[72] Thiel F. and Seifert F., "Physiological signatures reconstructed from a dynamic human model exposed to ultra-wideband microwave signals", *Frequenz, Journal of RF/Microwave-Engineering, Photonics and Communications*, vol. 64, no. 3-4, 34-41, ISSN 0016-1136, (2010).

[73] Thiel F., Kosch O., Seifert F., "Ultra-wideband Sensors for Improved Magnetic Resonance Imaging, Cardiovascular Monitoring and Tumour Diagnostics", Special Issue "*Sensors in Biomechanics and Biomedicine*", 10, 12, 10778-10802, doi:10.3390/s101210778, ISSN 1424-8220, 25 pages, (2010)

[74] Kosch O., Thiel F., Yan D.D., and Seifert F., "Discrimination of respirative and cardiac displacements from ultra-wideband radar data", *Int. Biosignal Processing Conf.* (Biosignal 2010), 14th – 16th July; Berlin, Germany, (2010)

[75] Christ A. et al., "The virtual family – development of surface-based anatomical models of two adults and two children for dosimetric simulations", *Phys. Med. Biol.* 55, N23-N38, (2010)

[76] Sachs J., Peyerl P., Wöckel S., et al., "Liquid and moisture sensing by ultra-wideband pseudo-noise sequence signals", *Meas. Sci. Technol.*, vol. 18, no. 4, pp. 1074-1088, April 2007.

[77] Kosch O., Thiel F., Schneider S., Ittermann B., and Seifert F., "Verification of contactless multi-channel UWB navigator by one dimensional MRT", *Proc. Intl. Soc. Mag. Reson. Med.* 20, (ISMRM), Melbourne, Australia, p.601, (2012)

[78] Ziehe A. and Müller K.R., TDSEP - "An efficient algorithm for blind separation using time structure", *Proc. Int. Conf. Artifical Neural Networks* (ICANN'98), Skövde, Sweden, pp.675-680, (1998).

[79] Sander T.H., Lueschow A., Curio G., Trahms L., "Time delayed decorrelation for the identification of the cardiac artifact in MEG data", *Biomed. Tech.* 2002, 47(1/2), pp. 573-576, (2002).

[80] Kosch O., Schneider S., Ittermann B., and Seifert F., "Self-adjusting multichannel UWB radar for cardiac MRI", *Proc. Intl. Soc. Mag. Reson. Med.* 20, (ISMRM), Melbourne, Australia, p.2467, (2012)

[81] Kosch O., Thiel F., Ittermann B., and Seifert F., "Non-contact cardiac gating with ultra-wideband radar sensors for high field MRI", *Proc. Intl. Soc. Mag. Reson. Med.* 19, (ISMRM), Montreal, Canada, ISSN 1545-4428, p.1804, (2011)

[82] M. Helbig, "Mikrowellen- Ultrabreitband- und THz-Bildgebung," in O. Dössel, T. M. Buzug, (Hrsg.) "Medizinische Bildgebung", Lehrbuchreihe der Biomedizinischen Technik, Bd. 7, Berlin, de Gruyter, 2012

[83] E. C. Fear, S. C. Hagness, P. M. Meaney *et al.*, "Enhancing breast tumor detection with near-field imaging," *Microwave Magazine, IEEE,* vol. 3, no. 1, pp. 48-56, 2002.

[84] P. M. Meaney, M. W. Fanning, D. Li *et al.*, "A clinical prototype for active microwave imaging of the breast," *IEEE Transactions on Microwave Theory and Techniques,* vol. 48, no. 11, pp. 1841-1853, Nov, 2000.

[85] M. Klemm, I. J. Craddock, J. A. Leendertz *et al.*, "Radar-Based Breast Cancer Detection Using a Hemispherical Antenna Array - Experimental Results," *IEEE Transactions on Antennas and Propagation,* vol. 57, no. 6, pp. 1692-1704, 2009.

[86] M. Klemm, I. J. Craddock, J. A. Leendertz *et al.*, "Clinical trials of a UWB imaging radar for breast cancer," *Proceedings of the Fourth European Conference on Antennas and Propagation (EuCAP),* 2010, pp. 1-4.

[87] J. Sachs, "M-sequence radar," *Ground penetrating radar,* D. J. Daniels, ed., pp. 225-237, London: The Institution of Engineering and Technology, 2004.

[88] U. Kaatze, "Complex permittivity of water as a function of frequency and temperature," *J Chem Eng Data,* vol. 34, no. 4, pp. 371-374, Oct, 1989.

[89] T. C. Williams, J. Bourqui, T. R. Cameron *et al.*, "Laser Surface Estimation for Microwave Breast Imaging Systems," *IEEE Transactions on Biomedical Engineering,* vol. 58, no. 5, pp. 1193-1199, 2011.

[90] D. W. Winters, J. D. Shea, P. Kosmas *et al.*, "Three-Dimensional Microwave Breast Imaging: Dispersive Dielectric Properties Estimation Using Patient-Specific Basis Functions," *IEEE Transactions on Medical Imaging,* vol. 28, no. 7, pp. 969-981, 2009.

[91] T. Sakamoto, and T. Sato, "A target shape estimation algorithm for pulse radar systems based on boundary scattering transform," *IEICE Transactions on Communications,* vol. E87b, no. 5, pp. 1357-1365, May, 2004.

[92] S. Kidera, Y. Kani, T. Sakamoto *et al.*, "Fast and accurate 3-D imaging algorithm with linear array antennas for UWB pulse radars," *IEICE Transactions on Communications,* vol. E91b, no. 8, pp. 2683-2691, Aug, 2008.

[93] M. Helbig, M. A. Hein, U. Schwarz *et al.*, "Preliminary investigations of chest surface identification algorithms for breast cancer detection," *IEEE International Conference on Ultra-Wideband, ICUWB 2008,* pp. 195-198.

[94] M. Helbig, C. Geyer, M. Hein *et al.*, "A Breast Surface Estimation Algorithm for UWB Microwave Imaging," IFMBE Proceedings 4th European Conference of the International Federation for Medical and Biological Engineering, J. Sloten, P. Verdonck, M. Nyssen *et al.*, eds., pp. 760-763: Springer Berlin Heidelberg, 2009.

[95] M. Helbig, C. Geyer, M. Hein *et al.*, "Improved Breast Surface Identification for UWB Microwave Imaging," IFMBE Proceedings World Congress on Medical Physics and Biomedical Engineering, September 7 - 12, 2009, Munich, Germany, O. Dössel and W. C. Schlegel, eds., pp. 853-856: Springer Berlin Heidelberg, 2009.

[96] S. Hantscher, B. Etzlinger, A. Reisenzahn *et al.*, "A Wave Front Extraction Algorithm for High-Resolution Pulse Based Radar Systems," *IEEE International Conference on Ultra-Wideband, ICUWB 2007*, pp. 590-595.

[97] J. Sachs, R. Herrmann, M. Kmec *et al.*, "Recent Advances and Applications of M-Sequence based Ultra-Wideband Sensors," *IEEE International Conference on Ultra-Wideband, ICUWB 2007*, pp. 50-55.

[98] M. Helbig, " UWB for Medical Microwave Breast Imaging," in J. Sachs, "*High-Resolution Short-Range Sensing - Basics, Principles and Applications of Ultra Wideband Sensors*, Berlin: Wiley-VCH, 2012

[99] E. J. Bond, X. Li, S. C. Hagness *et al.*, "Microwave imaging via space-time beamforming for early detection of breast cancer," *IEEE Transactions on Antennas and Propagation,* vol. 51, no. 8, pp. 1690-1705, Aug, 2003.

[100] Y. Xie, B. Guo, L. Xu *et al.*, "Multistatic Adaptive Microwave Imaging for Early Breast Cancer Detection," *IEEE Transactions on Biomedical Engineering,* vol. 53, no. 8, pp. 1647-1657, 2006.

[101] R. Zetik, J. Sachs, and R. Thoma, "Modified cross-correlation back projection for UWB imaging: numerical examples," *IEEE International Conference on Ultra-Wideband, ICU 2005,* pp. 650-654.

[102] M. Klemm, I. J. Craddock, J. A. Leendertz *et al.*, "Improved Delay-and-Sum Beamforming Algorithm for Breast Cancer Detection," *International Journal of Antennas and Propagation,* vol. 2008.

[103] M. Kmec, M. Helbig, J. Sachs et al., "Integrated ultra-wideband hardware for MIMO sensing using pn-sequence approach," *IEEE International Conference on Ultra-Wideband,* ICUWB 2012, Syracuse, USA, 17-20 Sept. 2012

[104] M. Helbig, I. Hilger, M. Kmec, *et al.*, "Experimental phantom trials for UWB breast cancer detection", *Proc. German Microwave Conf.*, Ilmenau, Germany, pp. 1-4, 12-14 March 2012.

[105] M. Helbig, M. Kmec, J. Sachs *et al.*, "Aspects of antenna array configuration for UWB breast imaging," Proc. European Conf. on Antennas and Propagation, EuCAP 2012, Prague, pp. 1-4, 26-30 March 2012.

[106] M. Helbig, M. A. Hein, R. Herrmann *et al.*, "Experimental active antenna measurement setup for UWB breast cancer detection," *IEEE International Conference on Ultra-Wideband,* ICUWB 2012, Syracuse, USA, 17-20 Sept. 2012

[107] M. Lazebnik, D. Popovic, L. McCartney *et al.*, "A large-scale study of the ultrawideband microwave dielectric properties of normal, benign and malignant breast tissues obtained from cancer surgeries," *Physics in Medicine and Biology,* vol. 52, no. 20, pp. 6093-6115, Oct 21, 2007.

Chapter 12

ISOPerm: Non-Contacting Measurement of Dielectric Properties of Irregular Shaped Objects

Henning Mextorf, Frank Daschner, Mike Kent and Reinhard Knöchel

Additional information is available at the end of the chapter

http://dx.doi.org/10.5772/55079

1. Introduction

A mere glance at the contents of any of the conferences organised by ISEMA (International Society for Electromagnetic Aquametry) [1] shows that the measurement and control of water (its quantity and states) in materials is a very wide and active field. That water lends itself to study in this way is because of its very dominant dispersive dielectric properties and an unusually large dipole moment for such a small molecule (1.84 Debye units). At room temperature the dispersion is centred on $\sim$ 12.5GHz and the real part of the relative permittivity at its upper and lower frequency extremes is $\sim$ 4.3 and 80 respectively [2]. The complex dielectric properties characteristic of this dispersion are the properties that are measured and correlated with whatever aspect of the water content is of interest. Historically much of the work was carried out at one or two frequencies, mostly in X-band where the dielectric loss is at a maximum, but the advent of time domain reflectometry (TDR) [3] for broadband dielectric measurements in the microwave region eventually led to such measurements being made using open ended coaxial sensors [4], although a greater potential of such measurements was only realised later by the authors. The use of such sensors freed the experimenter from the difficult task of defining the sample shape by means of a sample cell; the measurements still required however that the sensor be in contact with the sample. As a true frequency domain UWB application, dielectric measurements of foodstuffs over a wide range of frequencies (100MHz to 20GHz) were made using network analysers and such coaxial sensors [5, 6]. Drawing on the experience in other chemometric applications such as NIR (near infra-red spectroscopy) [7], the dielectric spectra obtained were subjected to various multivariate analyses (PCR (principal component regression), PLSR (partial least squares regression), and ANNs (artificial neural networks both linear and non-linear)). Such analyses both compress the data into orthogonal factors and extract from those factors the best information to predict the composition of the foodstuffs. In such analysis the important variables are, not so much the dielectric properties at each sampled frequency, but rather the shape of the spectrum. In effect the data reduction provides suitable shape descriptors, which are in the case of foods, very dependent on the water content and its interaction with other constituents such as proteins and carbohydrates. Any that are dependent on other factors are eliminated in the regression analysis having no significant correlation with the material properties. Rather than using measurements in the frequency domain, of course it is possible to transform time domain measurements to the frequency domain using Fourier or other forms of transformation. For time varying data acquired by TDR the inverse Fourier transform in its most general form can be written as in equation 1.

$$h(t) = \frac{1}{2\pi} \int_{-\infty}^{+\infty} g(f) e^{2\pi i f t} \mathrm{d}f, \tag{1}$$

where h(t) is a time dependent function the Fourier transform of which is a frequency dependent spectrum g(f). Examination of equation (1) shows that at any instant t every component part of the spectrum contributes to the value of h. Because the subsequent multivariate analysis can be thought of as dealing with shape and is concerned only with the variations in g then transformation of h(t) to the frequency domain is not required, since related variations are present in h(t) and the shape of the time domain function is equally useful. A further justification for eliminating the transformation step concerns the difficulties with which it is fraught. Firstly, the truncation of the pulse after a finite measurement interval can introduce undesirable distortions in the integral from convolution of the pulse with the rectangular time window (windowing). Secondly, the act of sampling the pulse at regular intervals means that frequencies present with periods shorter than the sampling interval are incorporated as lower frequency information (aliasing), and thirdly, to accurately reconstruct the reflection coefficients in the frequency domain exact time referencing of the pulse is required, else phase errors cause large inaccuracies at the high frequency end of the spectrum. By carrying out the analysis on the raw, sampled TDR pulse, all the errors above can be avoided. This was tried and demonstrably gave the same results as the spectral data, with less computer effort and fewer error generating problems. The first problem to which this was applied was something less tangible than water content: it was in fact the quality of various seafoods as defined by more subjective methods [8]. The success of this approach naturally led to attempts to measure dielectric objects in a non-contacting fashion, using firstly transmitted UWB quasi-Gaussian pulses of 400ps width [9]. The transmitting and receiving antennas in this initial work were double ridged horns and the sample was arranged in a wide layer of uniform thickness. In such a situation, additional interfering variables can be the position of the antennae, polarization of the transmitted wave, multiple path effects and a host of others, all of which may be eliminated by multivariate analysis. Effects due to dielectric properties can be separated from those due to geometry and other exterior factors. Thereafter, further work followed with samples of increasingly complex shapes and different orientations, beginning with simple rectangles and progressing to other shapes such as triangles and circles [10, 11], albeit still of constant thickness. At the same time, various forms of UWB antennas were investigated but currently the choice is an array of simple dipoles for the receiving antennas with a horn antenna transmitting the pulse. The multivariate data analysis still uses PCA as a first step to reduce the data and provide shape descriptors, but because PCR is not entirely suitable for non-linear processes (being a linear regression of variables) the PCs are used as input to a non-linear ANN. A great deal of work has now been done, gradually broadening the application parameters until now it is possible to measure UWB dielectric properties of objects with any shape, thickness, orientation and without contact [12–20]. This has been the subject of the project 'ISOPerm' (irregularly shaped objects-permittivity) the methods and results of which will now be described.

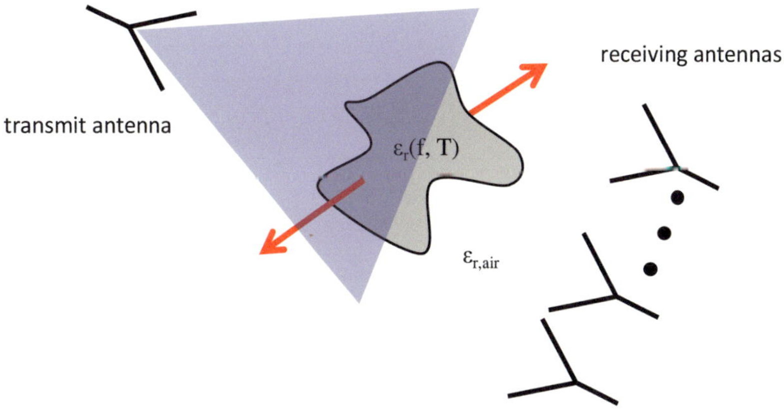

Figure 1. Irregularly shaped object illuminated by an electromagnetic wave.

2. Problem and approach

There is no appropriate method for the determination of the dielectric or related properties of irregular shaped objects in free-space. The investigated objects are considered to be small compared to the range of wavelengths and the footprints of the antennas used. A visualization of the problem is depicted in Figure 1. An electromagnetic wave illuminates an irregularly shaped object with its frequency and temperature dependent dielectric properties. It is surrounded by air. Portions of the scattered field are collected by multiple field probes, i.e. a line array of receiving antennas. The scattered signals contain information about geometry as well as the dielectric properties. Because of the complexity of the problem it is assumed that the development of a physical model (as used in conventional free-space methods having a plane parallel plate) would be too complex. Furthermore, an on-line method suitable for characterizing many objects in a short time is anticipated. Therefore, multivariate calibration methods are applied in order to separate dielectric from geometric influences. The objects are considered to be homogeneous and non-magnetic ($\mu_r = 1$). However, it is assumed that it would also be possible to predict both average permittivity and permeability. In research and industrial applications other properties, for example the water content, moisture, freshness or quality of foodstuffs, are of great interest. They are often strongly correlated to dielectric properties and can be determined directly without knowing the permittivity.

3. Measurement system and dedicated hardware

A measurement system has been built in order to verify the approach. Particular components, like the UWB-antennas used and the working principle of the whole set-up, are discussed in this section. A dedicated time domain transmission oscilloscope is presented. It is capable of transmitting and receiving UWB pulses with several gigahertz of bandwidth. It is specially tailored to the measurement problem and therefore functions with less hardware and software, is far more compact, and is cheaper compared to a universal laboratory instrument.

3.1. Compact ultra-wideband antennas and arrays

The sensor system should be able to transmit and receive ultra-wideband (UWB) signals in two orthogonal polarizations. This is of great importance if the orientation of the object

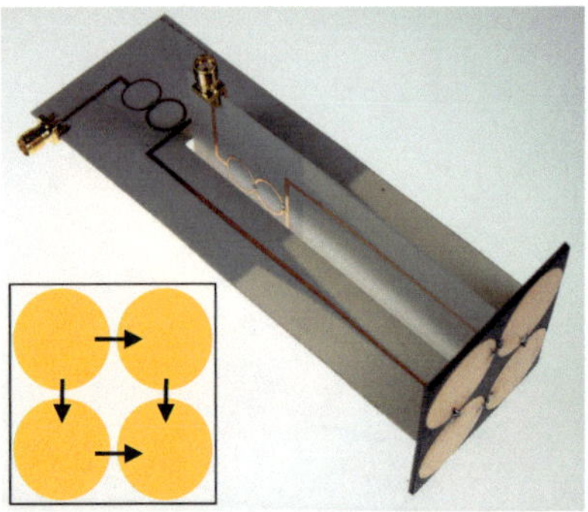

(a) Dual polarised dipole.

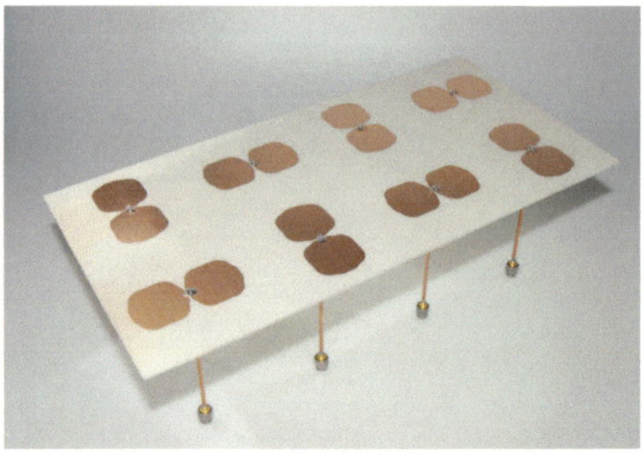

(b) Dual polarised array.

Figure 2. Ultra-wideband antenna structures.

under test (e.g. prolate ellipsoid) is not known. There are several polarization set-ups for receiving co- and cross-polarized parts of the scattered wave. In the set-up used latterly the polarization of the transmitting antenna is tilted at 45° with respect to the receiving array so that ideally, without an object under test, the receiving antennas receive the transmitted pulse in equal proportions for each polarization. Therefore, dual orthogonal UWB antennas have been developed. Furthermore the single antennas should be as compact as possible while operating in the lower gigahertz range. The arrays presented later are not group antennas as commonly used. The received signal of every antenna is sampled separately in order not to confound the information at different locations with respect to the object under test. Operating at frequencies in the lower GHz range is anticipated because the electromagnetic waves have a higher penetration depth (especially for lossy dielectrics with a high permittivity) and therefore will be more affected.

There are many UWB antennas suitable for building dual polarized antennas [21–25]. Dual polarized dipole (see [26–31]) and horn antennas (see [32–35]) can already be found in the literature. Most of these antennas have the disadvantage that they have crossed feed points. It is possible to overcome this problem with the configuration depicted in Figure 2. Both the four radiation elements and the whole configuration are two-fold symmetric. Two dipoles are excited in even mode in horizontal as well as in vertical polarization. Ideally, due to the symmetry, both planes are decoupled. The diameter of the radiation elements is 24mm; the whole PCB containing the four radiating elements measures 50mm × 50mm. The antenna is equipped with two orthogonal feed networks; each is composed of a two stage Wilkinson divider and two tapered baluns. The horizontal and vertical polarization feed networks have dimensions of 50mm × 142mm and 50mm × 103mm, respectively. In order to avoid backward radiation, absorber material is mounted appropriately (the absorber material was removed for the photograph shown in Figure 2).

The matching at both ports is better than 7dB in a frequency range from 2GHz to 5.7GHz; the isolation is better than 30dB. At 5GHz the radiated crosspolar portion is suppressed by more than 25dB (boresight); the gain is 5.2dB. Measurements with a line containing four of these antennas lead to excellent results but the whole antenna structure is rather complex. Since a scenario with an object moving orthogonally to the line array is anticipated (e.g. a conveyor belt), a simpler arrangement can be used. The antenna array shown in Figure 2 consists of eight single linear broadband dipoles. The geometry is optimized regarding the antenna matching. The radiating elements have a maximum diameter of 30mm. The feed is provided through a coaxial-slotline transition, so that no other components are necessary. The distance between two dipoles having an equal polarization is 80mm in each direction; between dipoles with the same polarization it is $\sqrt{2} \times 80$mm. The dimensions are overall 160mm $\times$ 320mm $\times$ 100mm. The matching is better than 10dB between 1.6GHz and 4.2GHz. The crosstalk between the single elements is maximum -20dB at 10GHz.

3.2. Signal generation and sampling

The core of the measurement system is the signal generation and sampling. The requirements for the system are that the step signals generated have a rise time in the range of 100ps, and a large amplitude. These signals are transmitted through the measurement path after which, the step response has to be sampled. The system should have a high bandwidth, low noise and low jitter. Furthermore it should be as compact and affordable as possible. Therefore, an impulse technique using equivalent time sampling is the method of choice. Classical swept sine wave techniques, as used in network analysers or real time digital sampling oscilloscopes, are too expensive, complex, and bulky.[1]

Equivalent time sampling is based on the repetitive stimulation by a measurement signal with a cycle duration of T_0. At every repetition of the measurement signal the moment of sampling is shifted by ΔT. The cycle of the sampling clock is then $T_1 = T_0 + \Delta T$. The sampled signal is therefore stretched to $T_2 = \frac{T_0}{\Delta T} T_1$. The effort regarding analog-to-digital conversion and the data transport and storage is greatly reduced compared to a real time oscilloscope. This technique has been used for decades; with recently available MMICs, both high performance and cheap hardware can be achieved. It is the method of choice for the sampling of signals with huge bandwidth in the GHz range with a high resolution. Laboratory instruments employing this technique have been used for preliminary investigations. In order to demonstrate the system performance and accuracy under practical conditions, dedicated hardware was developed. Two specially tailored systems were investigated. They differed

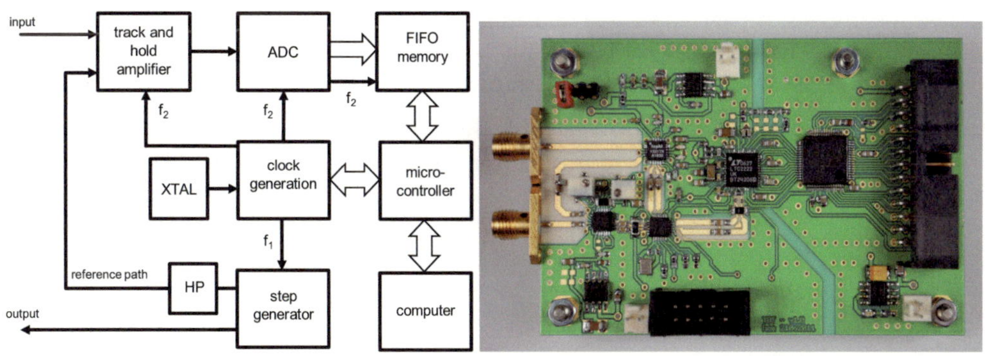

(a) Block diagram.

(b) Prototype: integrated on one PCB except microcontroller and computer.

Figure 3. Time domain transmission measurement system with a microcontroller.

[1] A rather exotic concept employing M-sequences can be found in this book (see chapter HALOS). There are also measurements carried out with this method which lead to excellent results [20].

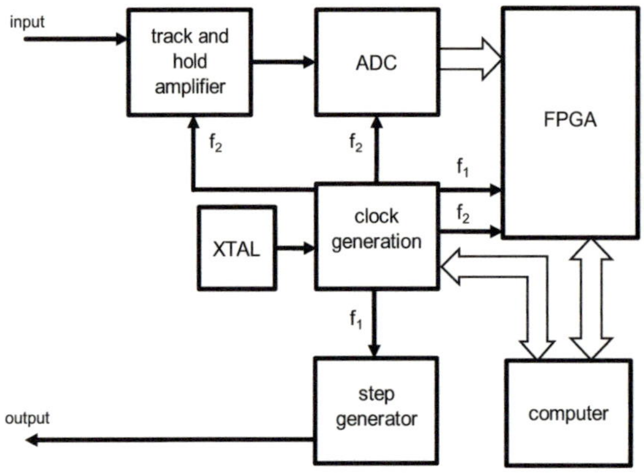

(a) Block diagram.

(b) Prototype: modular system except computer.

Figure 4. Time domain transmission measurement system with a FPGA.

mostly in the digital part behind the analog-digital converter (ADC); the front end was very similar.

A block diagram and a photograph of the first system using a microcontroller are shown in Figure 3. Two slightly detuned clocks are synthesized from a crystal. One clock triggers a step generator, which transmits a step signal with about 30ps rise time (20% to 80%) and up to 3V amplitude. The other triggers the track and hold amplifier (having an input bandwidth of 13GHz), the 12bit ADC, and the asynchronous FIFO (first in first out) memory with a width of 18bit and a depth of 32k. A microcontroller and a computer control the system. The FIFO memory can be read out, and the clock generation can be programmed via the I2C-bus. Clock signals up to hundreds of megahertz can be chosen with an accuracy of 1ppm. The dimensions of the RF-PCB are 90mm $\times$ 60mm. A reference path is necessary to measure time differences, because the phase of the two clock signals is not captured. A FIFO has to be used because the microcontroller is not able to read in 12bit words at tens of megahertz. Therefore, one waveform of up to 32k is captured at a time and is then transferred to the computer. Averaging, in order to improve the SNR, is carried out on the computer.

An improvement is the application of a field programmable gate array (FPGA). A modular system is shown in Figure 4. There is no more need for a fast external FIFO because data can be read into the FPGA directly up to about 66MHz. Furthermore, no reference path is needed, because the FPGA is able to discriminate the phase between both clock signals. Up to $2^{12} = 4096$ sampling points are possible while an averaging of up to $2^{13} = 8192$ waveforms can be carried out on the FPGA (this is restricted due to the internal resources of the FPGA used[2]).

Since both front ends are similar they offer a comparable performance. The RMS noise level of the receiver is < 0.9mV, the RMS jitter is < 0.7ps (no averaging). Figure 5 shows a comparison of sampled test signals between the proposed hardware and a Tektronix TDS8000. The frequencies are chosen as $f_1 = 50.01$MHz and $f_2 = 50$MHz, which leads to a resolution

[2] A Spartan 6 from Xilinx is used.

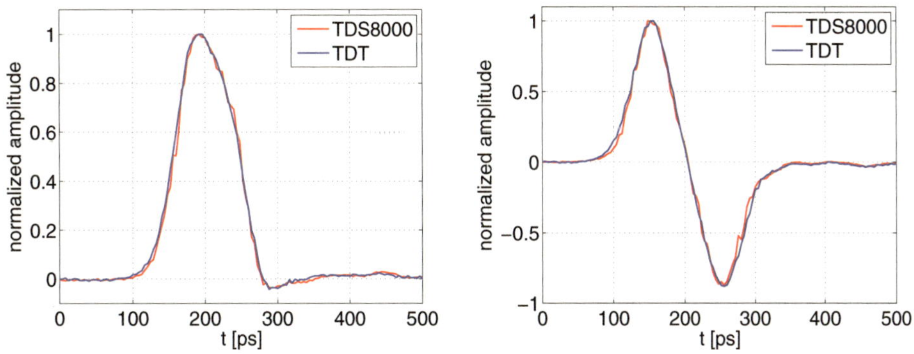

(a) Pulse.

(b) Monocycle.

Figure 5. Comparison between the proposed dedicated hardware (TDT) and a Textronix TDS8000 (no averaging).

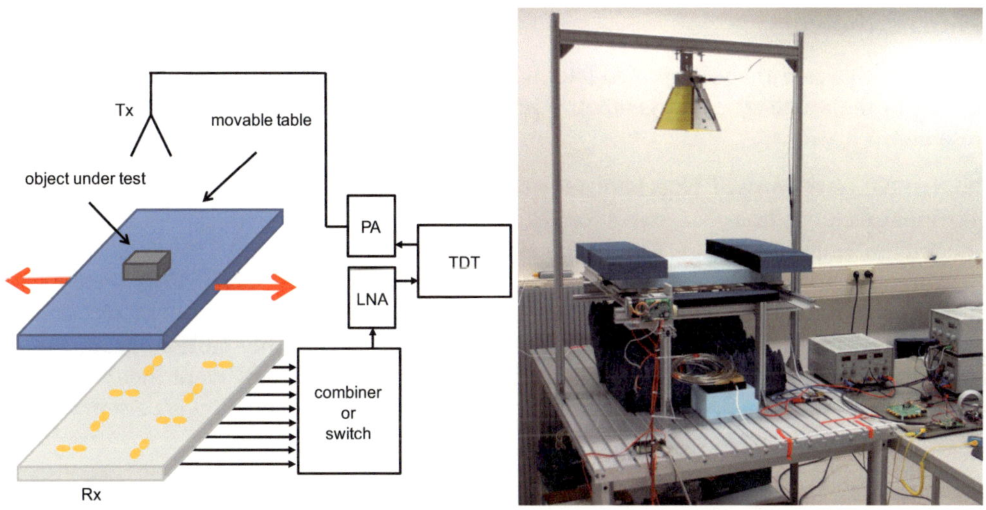

(a) Block diagram.

(b) Photograph.

Figure 6. Entire measurement set-up.

of $\Delta T = 4$ps. For the measurements presented later a much lower resolution is sufficient, e.g. $\Delta T = 40$ps.

3.3. Entire measurement setup

The entire measurement system is shown is Figure 6. The step signal of the TDT is amplified by a power amplifier (PA) before being transmitted through a horn antenna. The object under test is illuminated by the transmitted signal and portions of the scattered signal are received by an array of eight receiving antennas, which are arranged in two orthogonal polarizations. In order to emulate a 2-dimensional array, the movable table is moved orthogonally to the orientation of the receiving array. It is crucial to the multivariate calibration, which is applied later, that the received signals of each antenna placed at its unique position are sampled separately. Therefore, the antennas are switched sequentially onto the input of the TDT and are sampled there, after being amplified by a low noise amplifier (LNA). The MOSFET switch used requires relatively high hard- and software effort and it has to be synchronised with the other components included in the system. Furthermore, these switches have a latency of some nanoseconds. The switch has an insertion loss of 4dB at 2GHz and 8.6dB at 8GHz.

A simpler solution is to use a broadband combiner. The signals received by the individual antennas are combined using a broadband eight-way Wilkinson divider. Prior to combining there is a delay of $\tau = 2$ns between two adjacent inputs. It is possible to separate the individual pulses in time. Compared to using a receiver having more channels, or using a switch, the hardware effort is greatly reduced and the instantaneous sampling of all pulses is possible in about 20ns. One of the disadvantages that has to be considered is that there is an increased insertion loss of ideally 9dB; at 1GHz and 5GHz the measured insertion losses are 9.3dB and 10.7dB, respectively. There are also losses and temperature dependencies due to the delay lines, which have to be taken into account.

4. Measurements

For the multivariate analysis applied later it is necessary to measure a large number of objects varying in their dielectric and geometric properties. Here, two series of measurements are presented.

The first series comprised objects of moist clay granules. Five irregularly shaped moulds were manufactured from polystyrol foam[3]. They were then filled with clay granules having different amounts of moisture. One of the moulds is shown in Figure 7. The moisture content was varied in a range from about 4.5% to 24%[4] and reference measurements were carried out with a gravimetric method[5]. Furthermore, three rotation angles of 0°, 22.5° and 45° were applied to the test objects when putting them on the movable table in order to have more variation in the received signals. Overall 90 different objects were measured. The measurements were carried out in random order to avoid correlations with environmental effects, e.g. temperature variations.

Since there are eight receiving antennas and the movable table is moved to four positions, 32 pulses are received for every object under test when using a switch. With the Wilkison divider and delay lines, the pulses of the eight antennas are received instantaneously as a series of pulses. Two of the received series of pulses for two different objects having moisture contents of 4.88% and 20.93% are shown as examples in Figure 8. Although the variation in moisture content is high, the variation in the pulse shape is rather low.

The second series comprised plastic bottles filled with ethanol-water mixtures[6]. The water content was varied in a range from 2% to 20% in steps around 2%; reference measurements were carried out using a precision balance[7]. Ten bottles were filled with a total of around 190g

---

[3] In the following: *test series 1*

[4] All given moisture contents are on a wet basis.

[5] A Sartorius MA100 is used, accuracy of the weighing function: 0.1% for samples > 1g and 0.02% for samples > 5g. The weight of the samples was about $4 - 8$g.

[6] In the following: *test series 2*

[7] The gravimetric water content was determined using a Kern EW 4200-2NM. The repeatability is 0.01g.

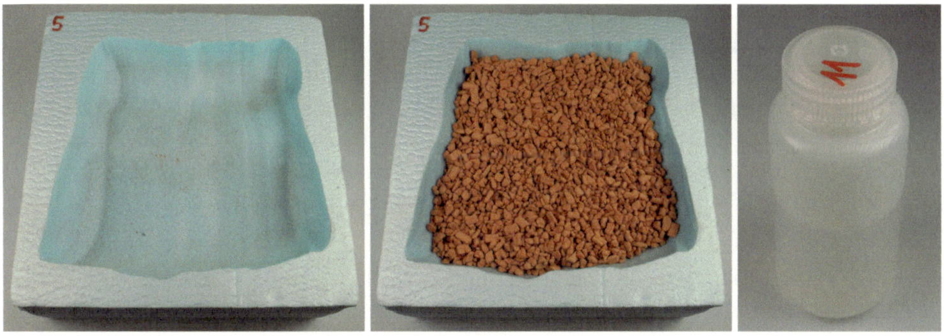

(a) One of the moulds manufactured from Styrofoam.
(b) Mould filled with moist clay granules.
(c) A bottle filled with an ethanol-water mixture.

Figure 7. Objects under test.

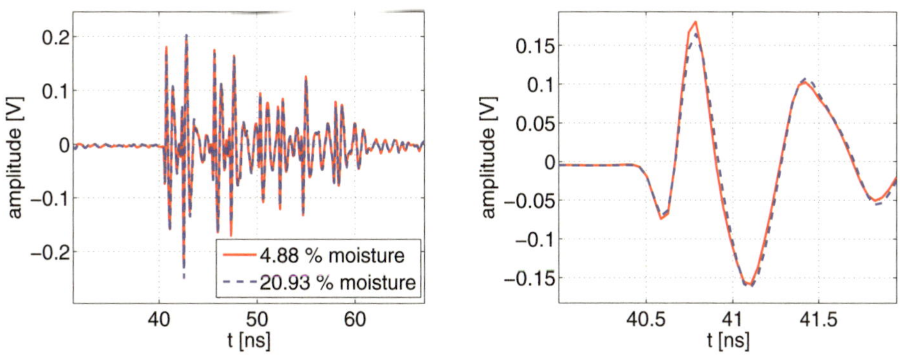

(a) Instantaneous sampled pulses received by the eight antennas using a broadband Wilkinson divider and delay lines.
(b) Closer look at the first received pulse.

Figure 8. Examples of received pulses from two objects having different moisture content.

of liquid, and another ten with about 140g. Therefore, overall 20 objects were measured. A photograph of one of the measured objects containing 113.15g ethanol and 27.94g of water is shown in Figure 7.

The time domain signals are gated in order to extract the time interval of interest: about 10 to 30 equidistant amplitude values per pulse are empirically chosen and subjected to the multivariate calibration methods which will be explained in the next section 5.

5. Multivariate data processing

The sequence of the eight sampled pulses consists of 217 sampling points. There are four positions of the movable table which then leads to $4 \times 217 = 868$ data points per object. These pulses are modified by many factors: the shape, the position, the rotation, and the intrinsic variables of the material under test. However, as can be seen in Figure 8 the value to be measured has only a relatively low influence on the apparent shape of the pulses, and practically all data points are modified when one or more of the factors mentioned above changes. Indeed each factor modifies the curve shape in a different manner. Therefore all measured points of the curve contribute in part to the variable(s) to be measured. Often these values consist only of one variable (for instance water content), a small set of variables (complex permittivity, quality) or an abstract class (shape of the object). Hence the challenge of data processing for the application discussed here is to extract the (hidden) relevant information from a huge data array. Due to the complexity physical modelling is impracticable.

Multivariate calibrations are established techniques for the extraction of relevant information from observed (measured) data without physical modelling. In the following, principal components analysis and regression (PCA/PCR), artificial neural networks (ANN) and partial least squares regression (PLSR) are applied to the data measured during the experiments described in section 4. Multivariate calibration methods have the disadvantage that they require a calibration procedure i.e. training. This means a portion of the measurements carried out on known samples need to be used to determine parameters or coefficients that enable a determination of the variable to be measured for unknown samples.

For this reason the measurements are divided randomly into a calibration and validation group. In general these two groups have an equal size. For test series 1 the number of data sets of the calibration and validation groups is $n_c = n_v = 45$ and for test series 2 they are $n_c = n_v = 10$. The more samples are available the more robust is the calibration and the meaningfulness of the validation.

In order to reduce the amount of data, in a pre-processing step the points having a lower variance may be removed from the input variables. For the experiments described here a majority of the 868 time points can be neglected when the threshold value of the standard deviation is set to 20% of the maximum standard deviation[8]. Thus for test series 1 and 2 the number of points used is $m_1 = 305$ and $m_2 = 187$, respectively. The raw matrix of the calibration data consists of the measured and pre-processed pulse sequence in each line. Hence the columns contain the data of the selected time points of a measurement:

$$Y_c = \begin{bmatrix} y_{11} & \cdots & y_{1m} \\ \vdots & \ddots & \vdots \\ y_{n_c 1} & \cdots & y_{n_c m} \end{bmatrix} . \tag{2}$$

Due to numerical reasons it is advantageous to standardize the raw data. Firstly the means of each column are calculated with

$$\overline{Y}_c = \left[\frac{1}{n_c} \sum_{i=1}^{n_c} y_{i1} \cdots \frac{1}{n_c} \sum_{i=1}^{n_c} y_{im} \right] . \tag{3}$$

The matrix of the normalized, standardized calibration data is calculated by subtracting the means of each column from each value of the columns, then dividing each by the standard deviations of the columns $\sigma_{c1} \cdots \sigma_{cm}$:

$$X_c = \left[Y_c - \begin{bmatrix} 1 \\ \vdots \\ 1 \end{bmatrix} \cdot \overline{Y}_c \right] \cdot \begin{bmatrix} \frac{1}{\sigma_{c1}} & \cdots & 0 \\ \vdots & \ddots & \vdots \\ 0 & \cdots & \frac{1}{\sigma_{cm}} \end{bmatrix} . \tag{4}$$

[8] which is 0.0037 and 0.0017 for test series 1 and 2, respectively.

The matrix of the pre-processed validation data is calculated similarly. However the data is normalized and standardized using the means and the standard deviations of the calibration set.

5.1. Principal component analysis and regression

As can be seen in Figure 8 neighbouring data points are highly correlated. Therefore it is not possible to use the selected data points directly in a linear regression to estimate the variable of interest[9]. For the two test series described in section 4 such a linear calibration equation would have $m_1 = 305$ and $m_2 = 187$ coefficients. Furthermore, and this aspect is more relevant, the calculation of the coefficients is numerically unstable because a matrix with correlated data needs to be inverted in the regression algorithm.

A solution to this problem is found using principal component analysis (PCA). The original data is linear transformed into a new set of variables

$$H_c = X_c \cdot P. \tag{5}$$

H_c comprises the so called principal components (the scores) and P is a matrix of the so called loadings. The scores have the advantage that they are uncorrelated and are arranged in such a way that the first principal component has the highest variance and the others are arranged in decreasing order of variance.

The matrix of the loadings P is composed of the eigenvectors. They are also orthogonal and of unit length. This transformation can be interpreted as a transformation into a new orthogonal coordinate system. The basis vectors of the new coordinate system are the eigenvectors and their direction is along the variances in decreasing order.

The properties of the principal components, their orthogonality and their arrangement regarding the variance, enable data reduction because the relevant information of the matrix is already described by the first few principal components. In order to obtain the properties of the transformed data as described above, the matrix P needs to be calculated by an eigenvalue decomposition [36–38] but this is not described in detail here. For the results calculated here the statistical toolbox of MATLAB is used. The eigenvalue decomposition of the PCA is processed without any consideration of the variable(s) of interest. This will be done in the next step of the data processing.

As mentioned above a multiple linear regression of the untransformed (therefore correlated) data is numerically unstable, but after the transformation (see eq. (5)) the data is uncorrelated and the value to be determined can be estimated by a linear combination of the principal components (principal component regression, PCR):

$$\hat{z}_c = \tilde{H}_c \cdot \beta, \tag{6}$$

where $\hat{z}_c$ is the estimated variable of interest (objective variable), $\tilde{H}_c$ is the matrix of the selected principal components and the vector β contains the coefficients of the linear equation.

[9] The variable of interest or objective variable is the parameter to be determined later, e.g. the moisture content.

The entries in the first column of $\tilde{H_c}$ are all unity in order to describe the mean of the value of interest in the linear equation.

In $\tilde{H_c}$ only the first k principal components are included. This selection leads to the desired data reduction. The value k need to be determined heuristically. Here for test series 1 the number of principal components used is $k_1 = 12$ and for test series 2 it is $k_2 = 2$.

The coefficients of the linear equation can be calculated by the following equation:

$$\beta = \left(\tilde{H_c}^T \cdot \tilde{H_c}\right)^{-1} \cdot \tilde{H_c} \cdot z_c, \tag{7}$$

where z_c consists of the variable of interest determined by a reference method, e.g. oven drying for moisture content.

After the calibration data is processed the system is essentially calibrated and is ready to handle unknown samples. However prior to that, the performance of the calibration still needs to be evaluated using the validation data. The target variables of interest are also determined for the pre-processed validation data using a reference method. The validation data (or later in use, the data of a measurement of an unknown sample) is processed in the following manner:

1. the scores are estimated using the loadings determined during the calibration procedure: $\hat{H_v} = X_v \cdot P$,
2. the unused principal components are removed and the unit column is added: $\hat{H_v} \Rightarrow \tilde{\hat{H_v}}$,
3. the value of interest is estimated by the linear equation $\hat{z_v} = \tilde{\hat{H_v}} \cdot \beta$.

For the evaluation of the quality of the calibration the root mean square error of calibration group $RMSE_c$ and the validation group $RMSE_v$ are calculated.

The results obtained with PCA/PCR for both test series are shown in Figure 9. The predicted moisture or water content is plotted vs. its true values. With perfect prediction all points of

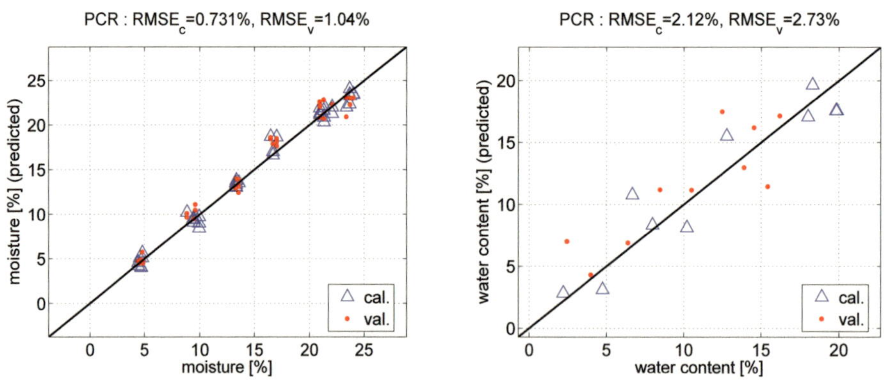

(a) Test series 1: clay granules. (b) Test series 2: ethanol-water mixtures.

Figure 9. Results obtained with PCA/PCR for both test series.

RER	Classification	Application
Up to 6	Very poor	Not recommended
7-12	Poor	Very rough screening
13-20	Fair	Screening
21-30	Good	Quality control
31-40	Very good	Process control
41+	Excellent	Any application

Table 1. *Classification using RER-values according to [39].*

the calibration and validation group would be on the so called quality line. For test series 1 a $RMSE_c = 0.731\%$ and $RMSE_v = 1.04\%$ is achieved. For test series 2 the errors are $RMSE_c = 2.12\%$ and $RMSE_v = 2.73\%$. The meaningfulness of $RMSE$ depends on the range of the variable to be predicted. Therefore the range error ratio RER is a better choice to evaluate the calibration. It is the ratio between the variable range $\Delta z = \max z - \min z$ and the RMSE[10]:

$$RER = \frac{\Delta z}{RMSE}. \tag{8}$$

The quality of the performance can be assessed using the ranges suggested in Table 1. For test series 1 the $RER_c = 26.7$ and $RER_v = 18.75$, hence the performance is *good*. But for test series 2 the accuracy obtained with PCA/PCR is only *poor* because $RER_c = 8.5$ and $RER_v = 6.6$.

5.2. Artificial neural networks

Although PCA/PCR is a linear operation it is more or less capable of processing non-linear data. However, when the unknown function describing the relationship between the pulse sequence and the value of interest is non-linear a purely linear method may not be the best choice. Artificial neural networks (ANN) can approximate unknown non-linear functions. For this application multilayer-feed-forward (MLFF) networks have a suitable architecture [40].

Such a network is shown in Figure 10. The input variables are weighted and processed by the neurons of the hidden layer. The activation functions of the neurons are non-linear[11]. This enables the non-linear function approximation. The output variable of the hidden layer is weighted again and processed by the neuron(s) of the output layer. The output of this layer needs to be post-processed (scaling and mean value) and the estimated variable of interest is available.

Due to their architecture ANN have several degrees of freedom: the number of hidden layers, the kind of activation function in each layer, and the number of neurons in the hidden layer. For the application discussed here one hidden layer and $n_{HL} = 10$ neurons in this hidden layer are sufficient. The problem is that the number of weighting factors between the layers increases with the number of neurons and for an optimal determination of the weighting factors a relatively large number of samples for the calibration (training) is necessary.

[10] In [39] the standard error is used instead of the root mean square error; for large numbers of samples there is practically no difference.

[11] tansig-function

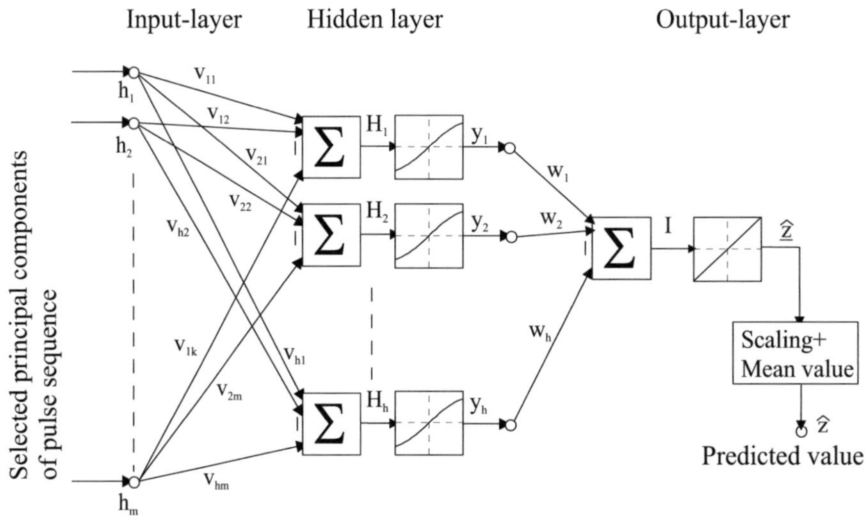

Figure 10. Architecture of the used multilayer-feed-forward-ANN.

For this reason a pre-processing of the data is recommended. If all selected time points were to be fed into the input layer $m_1 n_{HL} = 3050$ and $m_2 n_{HL} = 1870$ weighting factors would need to be found, with only $n_c = 45$ (test series 1) $n_c = 10$ training data sets. Therefore it is useful to feed the ANN with the selected principal components because they include the relevant information. This means the linear principal components regression is replaced by the non-linear ANN.

The training of the ANN has a relatively high calculation effort. Furthermore the starting values for the weighting factors are set randomly at the beginning of the training. This means the method is not strongly deterministic and it is not known for example, whether the optimal weighting factors were found because the training stopped in a local minimum of the error function. The training of the ANN was effected using the artificial neural network toolbox of MATLAB.

The results of the ANN are plotted in Figure 11. In comparison to the results of PCA/PCR there is an improvement observable:

- for test series 1 the RER_c increases to 37.8 (rating: *very good*) and the $RER_v = 22.4$ (*good*). This means there is a slight overfitting,
- for test series 2 the following ratings are obtained: $RER_c = 25.5$ (*good*) and $RER_v = 18.1$ (*fair*).

Despite the much higher calculation effort of ANN the improvements are not very satisfactory.

5.3. Partial least squares regression

PCA decorrelates the data by eigenvalue decomposition. Therefore the variable(s) of interest are not considered in this procedure. Only at the stage of PCR are they taken into account and a selection of relevant principal components is necessary. With the partial-least-squares

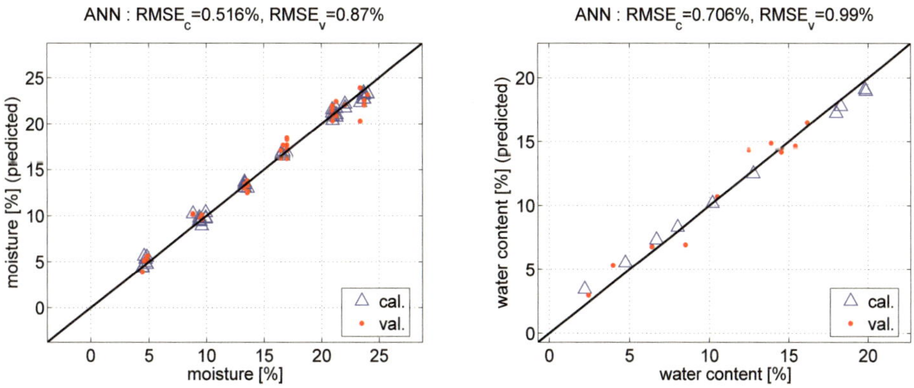

(a) Test series 1: clay granules.

(b) Test series 2: ethanol-water mixtures.

Figure 11. Results obtained with ANN for both test series.

regression (PLSR) the data is decorrelated regarding the variables(s) of interest. Several PLSR algorithms exist and sometimes the data is pre-processed non-linearly. Although PLSR was developed, more or less intuitively, in order to analyze economic data, in the meantime this method has also been used for several applications in other fields.

The algorithm used here for the processing of the measured data is described in [7] in detail and is only summarized in the following.

- Firstly, the input values are weighted in such a way that the covariance between them and the variable of interest is maximal.
- Secondly, the projection of the input values on the vector of the weighting values is called a *factor* or a *hidden path variable*. In the following, two regression analyses are considered:
  1. between the input variables and the factor, and
  2. between the variable of interest and the factor.

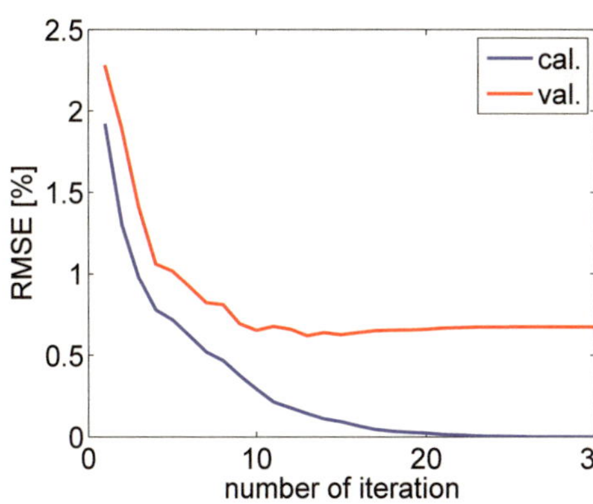

Figure 12. Influence of the number of factors H on the performance of the PLSR.

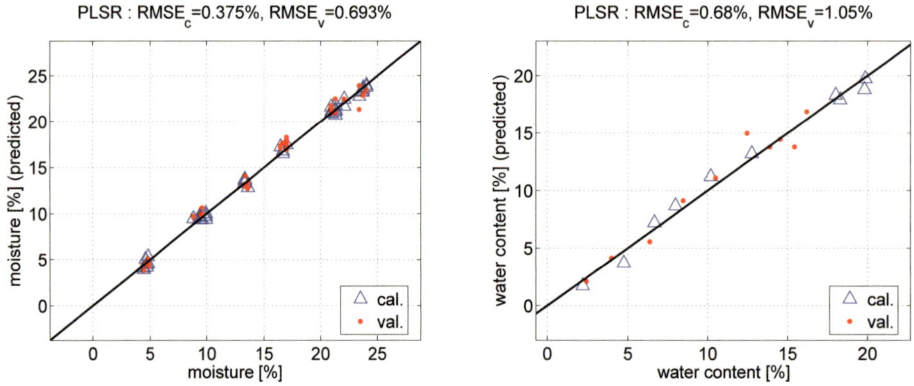

(a) Test series 1: clay granules.

(b) Test series 2: ethanol-water mixtures.

Figure 13. Results obtained with PLSR for both test series.

- Thirdly, the parts described by these linear models are subtracted from the measured values and the variables of interest and the algorithm is restarted using this new data in order to calculate the next factor,
- and finally, the procedure is repeated iteratively until a specified number H of hidden path variables is calculated.

All determined regression coefficients and weighting factors are used finally for the calculation of the regression equation. This means for the validation (and later application) that only a linear combination of the input values need to be calculated. Hence the calculation effort is much smaller in comparison to the ANN. The only degree of freedom is H, the number of factors to be used (number of iterations). When H is too high the $RMSE_c$ is significantly smaller than the $RMSE_v$. This means that overfitting occurs. However, as shown in Figure 12 H should be selected where $RMSE_v$ has a minimum. Furthermore $RMSE_c$ should not be much smaller (factor 1/2) than $RMSE_v$, otherwise the PLSR calibration could not handle unknown samples.

The performance of the PLSR is shown in Figure 13. For test series 1 $RER_c = 52$ (*excellent*) and $RER_v = 28.1$ (*good*). This is a further improvement in comparison to the ANN. For test series 2 the results stay similar to those of ANN: $RER_c = 26.5$ (*good*) and $RER_v = 17.1$ (*fair*).

5.4. Best calibration method

Principal component analysis and regression lead to acceptable results but the best calibrations were obtained with ANN and PLSR. However the computation effort is much higher with ANN and in general more samples are necessary for a successful training. PLSR is a linear operation and can be performed fast in real time. For this reason PLSR is the best choice for calibration of the application discussed here. In Table 2 the results of both test series are compared with similar experiments presented in several other publications. As can be seen, the performance obtained here is in the upper range. However one has to take into account the further advantages of the system discussed here: it is non-contacting, the objects can be rotated, and can have irregular shapes and sizes.

Publication	range [%]	$RMSE_c$ [%]	$RMSE_v$ [%]	RER_c	RER_v
[41]: Tobacco, PLSR	10-50	-	2	-	20
[42]: Scots pine, PLSR	0-15	0.46	0.74	32.6	20.3
[42]: Scots pine, PLSR	0-175	15.92	12.52	11	14
[9]: Clay granules, ANN	6.3-34.2	1.6	2.1	17.4	13.3
ISOperm:					
Clay granules, PLSR	4.5-24	0.38	0.69	52	28.1
Ethanol water mix. in bottle, ANN	2-20	0.71	1	25.5	18.1
[14], PLSR	70-100	1.28	2.55	23.4	11.8
[16], ANN	5-29.2	1.29	1.88	18.8	12.9
[17], PLSR	4.6-24.1	0.35	0.69	55.3	28.2
[19], PLSR	1.8-20.2	0.39	0.61	47.3	30.2
[20], PLSR	4.3-23.4	0.31	0.55	61.7	34.7
Other technologies:					
[43]: Wheat, admittance, PCR	9-20	-	0.39	-	28.2
[44]: Salmon, NIR, PLSR	61-70.8	-	0.98	-	10
[45]: Paper, NIR, PLSR	0-2.4	-	0.056	-	43.1
[46]: Theophyllin, NIR, ANN	1-22	0.45	0.83	46.7	25.3

Table 2. Comparison to other publications regarding the accuracy of the determination of moisture or water content. Except the method investigated in ISOPerm, all others are contacting and/or require a defined shape of the object under test.

6. Conclusions

Many industrial and scientific applications require extensive on-line process monitoring and quality control. Often the composition of goods (e.g. moisture content) is of great interest but also abstract parameters, for example quality or freshness, play an important role. The microwave sensor described is able to penetrate the investigated materials and by using UWB-techniques it is possible to gain information at various frequencies. The applied time domain techniques operate with low hardware effort and fast measurement speed while having a high accuracy. Using commercial MMICs signals exceeding a bandwidth of 10GHz can be generated and sampled with cheap and compact dedicated hardware. Today it is possible to employ multivariate calibration methods like artificial neural networks, which have a high computational effort, in real time. These methods are well established in, for example, NIR or image processing and are successfully adopted. The feasibility of the method has been successfully proven with accuracy even greater than in many previous publications using contacting methods. It has a great potential for many kinds of future applications in microwave sensing.

Author details

Henning Mextorf, Frank Daschner, Mike Kent and Reinhard Knöchel
University of Kiel, Germany

7. References

[1] Vainikainen, P. & Laitinen, T. [2009]. *Proceedings of the 8th International Conference on Electromagnetic Wave Interaction with Water and Moist Substances (ISEMA 2009)*, Espoo, Finland.

[2] Grant, E. H., Sheppard, R. J. & South, G. P. [1978]. *Dielectric behaviour of biological molecules in solution*, Clarendon Press, Oxford, UK.

[3] Fellner-Feldegg, H. [1969]. The measurement of dielectrics in the time domain, *J. Phys. Chem.* 73: 616 – 623.

[4] Burdette, E., Cain, F. & Seals, J. [1980]. In vivo probe measurement technique for determining dielectric properties at VHF through microwave frequencies, *IEEE Trans. Microw. Theory Tech.* 28(4).

[5] Kent, M. & Anderson, D. [1996]. Dielectric studies of added water in poultry meat and scallops, *Journal of food engineering* 28(3-4): 239–259.

[6] Kent, M., Knöchel, R., Daschner, F. & Berger, U.-K. [2000]. Composition of foods using microwave dielectric spectra, *Eur. Food Res. Technol.* (210): 359–366.

[7] Martens, H. & Naes, T. [1989]. *Multivariate Calibration*, John Wiley and Sons, Chichester.

[8] Kent, M., Knöchel, R., Daschner, F., Schimmer, O., Oelenschläger, J., Mierke-Klemeyer, S., Barr, U.-K., Floberg, P., Huidobro, M., Nunes, L., Batista, I. & Martins, A. [2004]. Time domain reflectometry as a tool for the estimation of quality in foods, *Agrophysics* 18(3).

[9] Schimmer, O., Gülck, A., Daschner, F., Piotrowski, J. & Knöchel, R. [2005]. Non-contacting determination of moisture content in bulk materials using sub-nanosecond UWB-pulses, *IEEE Trans. Microw. Theory Tech.* 53(6): 2107–2113.

[10] Mextorf, H., Martens, R., Daschner, F. & Knöchel, R. [2010]. Dual polarized UWB antenna for free-space characterization of dielectric objects, *Proc. German Microwave Conference 2010* pp. 162 – 165.

[11] Mextorf, H., Daschner, F., Kent, M. & Knöchel, R. [2010a]. Free-space determination of permittivity, size and orientation of rectangular shaped objects using multivariate analysis, *Proc. European Microwave Conference 2010* pp. 152 – 155.

[12] Mextorf, H., Daschner, F., Kent, M. & Knöchel, R. [2010b]. Non-contacting UWB-characterization of dielectric objects using multivariate calibration, *Proc. Aquametry 2010* pp. 136 – 144.

[13] Mextorf, H., Daschner, F., Kent, M. & Knöchel, R. [2011d]. UWB free-space characterization and shape recognition of dielectric objects using statistical methods, *IEEE Trans. Instrum. Meas.* 60(4): 1389 – 1396.

[14] Mextorf, H., Daschner, F., Kent, M. & Knöchel, R. [2011a]. Free-space prediction of the water content of irregularly shaped bodies filled with water-ethanol mixtures, *Proc. ISEMA 2011* pp. 162 – 169.

[15] Mextorf, H., Daschner, F., Kent, M. & Knöchel, R. [2011b]. New UWB free-space method for the classification and characterization of dielectric objects, *Proc. ICUWB 2011* pp. 410–414.

[16] Mextorf, H., Daschner, F., Kent, M. & Knöchel, R. [2011c]. Performance of multivariate calibration methods for the UWB characterization of dielectric objects, *Proc. CMM-Tagung 2011* pp. 105–112.

[17] Mextorf, H., Daschner, F., Kent, M. & Knöchel, R. [2012a]. Non-contacting moisture sensing using a dedicated UWB time domain instrument, *Proc. German Microwave Conference 2012* pp. 1–4.

[18] Mextorf, H., Daschner, F., Kent, M. & Knöchel, R. [2012b]. Signal quality considerations for free-space UWB moisture measurements, *Proc. Mikon 2012* pp. 627–630.
[19] Mextorf, H., Daschner, F., Kent, M. & Knöchel, R. [2012c]. UWB time domain transmission sensor for free-space moisture measurements, *IEEE MTT-S Int. Microw. Symp. Dig. 2012* pp. 1–3.
[20] Mextorf, H., Sachs, J., Daschner, F., Kent, M. & Knöchel, R. [2012]. Free-space moisture prediction of small objects using M-sequences, *Proc. ICUWB 2012* pp. 260–264.
[21] Schantz, H. G. [2003a]. Introduction to ultra-wideband antennas, *IEEE Conference on Ultra Wideband Systems and Technologies* pp. 1–9.
[22] Schantz, H. G. [2003b]. UWB magnetic antennas, *IEEE Antennas and Propagation Society International Symposium* 3: 604–607.
[23] Schantz, H. G. [2004]. A brief history of UWB antennas, *IEEE Aerospace and Electronic Systems Magazine* 19(4): 22–26.
[24] Schantz, H. G. [2005]. *The Art and Science of Ultrawideband Antennas*, Artech House, Norwood, MA.
[25] Wiesbeck, W., Adamiuk, G. & Sturm, C. [2009]. Basic properties and design principles of UWB antennas , *IProceedings of the IEEE* 97(2): 372–385.
[26] Perruisseau-Carrier, J., Hee, T. W. & Hall, P. S. [2003]. Dual-polarized broadband dipole, *IEEE Antennas Wireless Propag. Lett.* 2(1): 310–312.
[27] Woten, D. A. & El-Shenawee, M. [2008]. Broadband dual linear polarized antenna for statistical detection of breast cancer, *IEEE Trans. Antennas Propag.* 56(11): 3576–3590.
[28] Teo, P.-T., Lee, K.-S. & Lee, C.-K. [2003]. Maltese-cross coaxial balun-fed antenna for GPS and DCS1800 mobile communication, *IEEE Trans. Veh. Technol.* 52(4): 779–783.
[29] Suh, S.-Y., Stutzman, W., Davis, W., Walthot, A. & Schiffer, J. [2004]. A generalized crossed dipole antenna, the fourtear antenna, *IEEE APS 2004* 3: 2915–2918.
[30] Mak, K.-M., Wong, H. & Luk, K.-M. [2007]. A shorted bowtie patch antenna with a cross dipole for dual polarization, *IEEE Antennas Wireless Propag. Lett.* 6: 126–129.
[31] Adamiuk, G., Beer, S., Wiesbeck, W. & Zwick, T. [2009]. Dual-orthogonal polarized antenna for UWB-IR technology, *IEEE Antennas Wireless Propag. Lett.* 8: 981–984.
[32] Soroka, S. [1986]. A physically compact quad ridge horn design, *IEEE APS 1986* pp. 903–906.
[33] Shen, Z. & Feng, C. [2005]. A new dual-polarized broadband horn antenna, *IEEE Antennas and Wireless Propagation Letters* 4: 270–273.
[34] Schaubert, D., Elsallal, W., Kasturi, S., Boryssenko, A., Vouvakis, M. N. & Paraschos, G. [2008]. Wide bandwidth arrays of vivaldi antennas, *Institution of Engineering and Technology Seminar on Wideband, Multiband Antennas and Arrays for Defence or Civil Applications* pp. 1–20.
[35] Adamiuk, G., Zwick, T. & Wiesbeck, W. [2007]. Dual-orthogonal polarized vivaldi antenna for ultra wideband applications, *MIKON 2008* pp. 1–4.
[36] Götze, J. [1989]. *Orthogonale Matrizentransformationen*, R. Oldenbourg Verlag, München.
[37] G. Engeln-Muellges, F. R. [1996]. *Numerik Algorithmen*, VDI-Verlag, Düsseldorf.
[38] G.H. Golub, v. L. [1996]. *Matrix Computations*, The John Hopkins University Press, Baltimore.
[39] Williams, P. C. [2001]. *Near-Infrared Technology in the Agriculture and Food industries*, 2nd edn, American Association of Cereal Chemists, Inc., chapter Implementation of near-infrared technology, p. 165.
[40] Patterson, D. [1997]. *Künstliche neuronale Netze*, Prentice Hall Verlag, Haar.

[41] Dane, A. D., Rea, G. J., Walmsley, A. D. & Haswell, S. J. [2001]. The determination of moisture in tobacco by guided microwave spectroscopy and multivariate calibration, *Analytica Chimica Acta* 429(2): 185–194.

[42] Johansson, J., Hagmana, O. & Oja, J. [2003]. Predicting moisture content and density of Scots pine by microwave scanning of sawn timber, *Computers and Electronics in Agriculture* 41(1-3): 85–90.

[43] Lawrence, K. C., Windham, W. R. & Nelson, S. O. [1998]. Wheat moisture determination by 1- to 110-MHz swept-frequency admittance measurements, *Transactions of the ASAE* 41(1): 135–142.

[44] Wold, J. P. & Isaksson, T. [1997]. Non-destructive determination of fat and moisture in whole atlantic salmon by near-infrared diffuse spectroscopy, *Journal of Food Science* 62(4): 734–736.

[45] Neimanis, R., Lennholm, H. & Eriksson, R. [1999]. Determination of moisture content in impregnated paper using near infrared spectroscopy, *1999 Annual Report Conference on Electrical Insulation and Dielectric Phenomena* 1: 162–165.

[46] Rantanen, J., Rasanen, E., Antikainen, O., Mannermaa, J.-P. & Yliruusi, J. [2001]. In-line moisture measurement during granulation with a four-wavelength near-infrared sensor: an evaluation of process-related variables and a development of non-linear calibration model, *Chemometrics and Intelligent Laboratory Systems* 56(1): 51–58.

Chapter 13

Concepts and Components for Pulsed Angle Modulated Ultra Wideband Communication and Radar Systems

Alexander Esswein, Robert Weigel, Christian Carlowitz and Martin Vossiek

Additional information is available at the end of the chapter

http://dx.doi.org/10.5772/55082

1. Introduction

Ultra Wideband (UWB) systems have been utilized and commercialized since the beginning of the 1970s and have been successfully used in ground-, wall- and foliage-penetration, collision warning and avoidance, fluid level detection, intruder detection and vehicle radar and also for the topics of the intended research project, communication and position-location [1]. Up to now, the latter two fields have been treated separately in most developments.

UWB has the potential to yield solutions for the challenging problem of time dispersion caused by multipath propagation in indoor channels. For a local positioning system, multipath propagation determines the physical limit of the maximal accuracy that can be obtained at a given signal bandwidth [14].

There exist several techniques which are used to generate ultra wideband signals. Traditionally, UWB was defined as pulse based radio. Especially for radar and localization applications, the use of very narrow pulses is still the most dominant technique. In addition to that, there are UWB systems that use more complex modulation techniques, like multiband orthogonal frequency-division multiplexing (MB-OFDM) or direct sequence code-division multiple access (DS-CDMA) to spread the transmitted information over a large bandwidth. They are applied in communication systems whereas radar systems that use such techniques can hardly be found.

Recently there can be recognized an increasing interest for UWB technologies applied in mm-wave frequency bands. This interest is stimulated by novel regulation for future vehicular UWB systems in the 79 GHz band (77 - 81 GHz) [12], novel international allocation of unlicensed bands ranging from 57 - 66 GHz [9] and the attractive ISM bands at 122.5 GHz with 1 GHz bandwidth and at 244 GHz with 2 GHz bandwidth. Also, the 61.5 GHz ISM band with 500 MHz available bandwidth is often considered as a "de-facto" UWB band even though the bandwidth is just less than the bandwidth of 500 MHz usually demanded as the minimal bandwidth for UWB. The great advantage of mm-wave UWB bands is that they do not suffer from the severe power regulations known from standard UWB. At the above mentioned mm-wave UWB bands, the permitted maximum mean power density is at least 38 dB higher than in the UWB bands below 30 GHz.

Most of the mm-wave UWB communication and ranging systems published so far use a simple pulse generator as signal source. In the simplest case, a mm-wave CW carrier is modulated with an ASK (s. e.g [17]) or BPSK (s. e.g. [18])) sequence. A very interesting low-power approach that is somewhat related to the approach in this work is shown in [6] and [7]. Here, a 60 GHz oscillator itself is switched on and off. To guarantee a stable startup phase and to improve the phase noise, the oscillator is injection locked to a spurious harmonic of the switching signal. The benefit of the pulsed injection locking approach with respect to power consumption was impressively shown in this work. The general approach to obtain a stable pulse to pulse phase condition by injecting a spurious harmonic of the switching pulse into the oscillator is well known for a long time from low-power and low-cost microwave primary pulse radar systems. This basic principle can be extended in a way that frequency modulated signals can be generated based on a switched injection locked oscillator [19]. In this work, it is generalized for synthesizing arbitrarily phase modulated signals for integrated local positioning and communication. The fusion of positioning and communication capability is especially needed for future wireless devices applied in the "internet of things" or for advanced multimedia / augmented reality applications, for robot control and for vehicle2X / car2X applications.

Most existing UWB communication and ranging systems - especially those dedicated to low power consumption and mm-wave frequencies - employ simple impulse radios (IR). Popular IR-UWB modulation techniques include on-off keying (OOK), pulse-position modulation (PPM), pulse-amplitude modulation (PAM) and binary phase shift keying (BPSK) [5, 17, 18]. Their waveform can be synthesized using low complexity impulse generators and control circuitry, which comes at the cost of low spectral efficiency and severely limited control over spectral properties of the synthesized signals. Consequently, these transmitter cannot exhaust regulatory boundaries in all operation modes. High data rate synthesizers are often average power limited whereas low data rate implementations may be peak power limited [20].

2. Proposed concepts and components

In order to overcome these issues, pulsed angle modulated UWB signals are proposed to provide greater flexibility and better control over the spectral properties of the synthesized signals. Additionally, this signal type is well suited for both ranging and communication, since it allows synthesizing pulsed frequency modulated chirps that are attractive for ranging as well as digital phase modulation schemes for data transmission with the same hardware.

Since classic architectures containing VCOs, PLLs, mixer, linear amplifiers and switches are not suited for low complexity, low power systems, the switched injection-locked oscillator is suggested for signal synthesis. It regenerates and amplifies a weak phase-modulated signal. Consequently, the high frequency RF signal can be generated from a high power but efficiently synthesized low frequency phase modulated baseband signal in two simple stages - a lossy passive or low power frequency multiplier (harmonic generator) and a switched injection-locked oscillator as single stage pulsed high gain (> 50 dB) amplifier.

In this work, it is demonstrated that this approach allows synthesizing pulsed, arbitrarily phase modulated signals using the switched injection-locked harmonic sampling principle. The theory of this concept was investigated thoroughly and verified experimentally for the synthesis of phase shift keying (PSK) modulated communication signals and pulsed frequency modulated (PFM) radar signals with the same hardware. Regarding the switched injection-locked oscillator, implementations in planar surface mounted technology (6-7, 7-8 GHz) and integrated circuits (6-8 GHz, 63 GHz) were developed. Measurements with the first designs confirm the feasibility of the proposed concepts and already show promising results regarding transmitter signal to spur ratio and achievable ranging resolution and ranging uncertainty.

This work shows the half-term results of the ongoing project "Components and concepts for low-power mm-wave pulsed angle modulated ultra wideband communication and ranging (PAMUCOR)" within the DFG priority programme "Ultra-Wideband Radio Technologies for Communications, Localization and Sensor Applications"; for comparison, some results from the previous project "Concepts and components for pulsed frequency modulated ultra wideband secondary radar systems (PFM-USR)" are summarized.

3. Pulsed angle modulated UWB signals

3.1. Signal definition

Fig. 1 depicts a pulsed angle modulated UWB signal consisting of a sequence of short pulses (width T_d, period T_s), in which each pulse is an oscillation with the frequency ω_{osc} and the modulated initial phase φ_i:

$$s(t) = \sum_{i=0}^{N} \cos\left(\omega_{osc}\left(t - i \cdot T_s + \frac{T_d}{2}\right) + \varphi_i\right) \cdot \text{rect}\left(\frac{t - i \cdot T_s}{T_d}\right) \qquad (1)$$

with

$$\text{rect}(x) = \begin{cases} 1 & \text{for } -0.5 < x < 0.5 \\ 0 & \text{else} \end{cases}.$$

For flexible signal synthesis, initial phase modulation, pulse period, pulse width and oscillation frequency can be tuned.

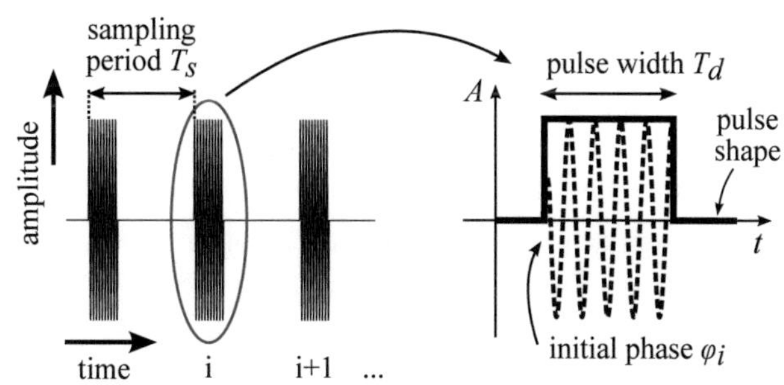

Figure 1. Pulsed angle modulated UWB signal - the modulated parameter is the initial phase φ_i of each pulse

3.2. SILO operation principle

The switched injection-locked oscillator (SILO) is basically a normal oscillator which is turned on and off while a weak reference signal is injected into its feedback loop (see Fig. 2). During startup of the oscillator, the injection signal provides an initial condition in the oscillator's resonator instead of noise like in oscillators without injection signal. This way, the instantaneous phase of the injection signal is adopted though the oscillator runs with its own natural frequency, which may differ from the injection signal's frequency. Since the power level of the injection signal is far too low to influence the oscillation as soon as the oscillator has reached its final amplitude, it performs only phase, but no frequency locking.

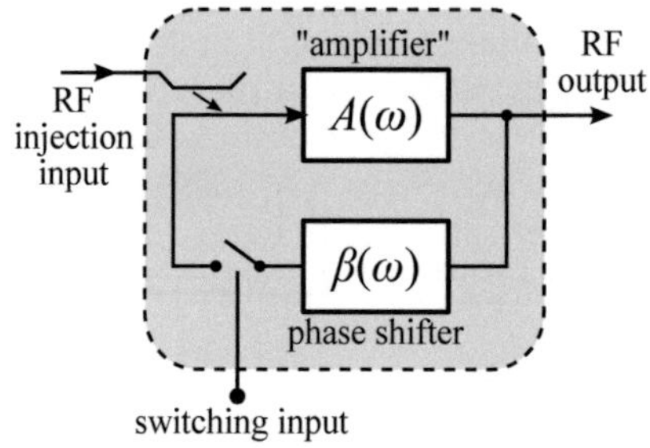

Figure 2. SILO principle

This behavior can be described theoretically by:

$$s(t) = \sum_{i=0}^{N} \cos\left(\omega_{osc}\left(t - i \cdot T_s + \frac{T_d}{2}\right) + \arg\left\{s_{inj}(t)\right\}\right) \cdot \mathrm{rect}\left(\frac{t - i \cdot T_s}{T_d}\right), \tag{2}$$

with the injection signal (center/reference frequency ω_{inj}, phase modulation $\varphi(t)$)

$$s_{inj}(t) = \cos\left(\omega_{inj}t + \varphi_{inj}(t)\right), \quad \arg\left\{s_{inj}(t)\right\} = \omega_{inj}t + \varphi_{inj}(t). \tag{3}$$

In spite of the fact that this model only describes the fundamental principle, the physical behavior of the oscillator is very similar in most operation modes. The most important disregarded physical effects observed in real implementations are:

- Due to balancing imperfections e.g. in differential oscillators, high order harmonics of the startup pulse turning on the circuit cause self-locking effects that degrade the SILO's performance at low injection levels. Hence, the rise time of the oscillator should not be too short in order to reduce the harmonic power level. Obviously, this leads to a trade-off with spectral bandwidth, minimum pulse width and maximum achievable pulse repetition rate.
- The phase sampling process is affected by the amplitude of the injection signal. In consequence, amplitude variations of the injection signal are converted into phase distortions. Therefore, constant amplitude injection signals should be used to mitigate these effects. Then there is only a constant phase offset between injection and regenerated signal.
- If the rise time of the oscillator is configured to be relatively long compared to the pulse width, there will be a noticeable dependence between the injection signal's power level and pulse width. With a large amplitude injection signal, the oscillator settles much faster than when starting from noise level. Again, constant amplitude injection signals are the preferred countermeasure to avoid pulse width jitter.

Thus, the simplifications of the proposed ideal model mainly affect time and frequency domain amplitude shape, which makes this model suitable for the analysis of the phase sampling process.

3.3. Phase sampling theory

In [3, 4, 19], the SILO's phase sampling principle and its applications have been investigated thoroughly. The most important results will be summarized and discussed in the following.

Starting from equations (2) and (3), the SILO's output signal can be expressed by (disregarding negative frequencies and finite time domain waveform length for sake of simplicity):

$$s(t) = \sum_{i=-\infty}^{+\infty} \left[a \cdot e^{j\left(\omega_{osc}t + (\omega_{inj}-\omega_{osc})\cdot\left(i\cdot T_s - \frac{T_d}{2}\right)\right)} \cdot e^{j\varphi_{inj}\left(i\cdot T_s - \frac{T_d}{2}\right)} \cdot \mathrm{rect}\left(\frac{t - i\cdot T_s}{T_d}\right)\right]. \quad (4)$$

This expression still suggests an oscillation with ω_{osc} - the presence of the injection signal regeneration feature that includes the frequency is not obvious. According to [4], the Fourier transform $F\{\cdot\}$ of (4) leads to:

$$S(\omega) = A \cdot \left[\mathrm{sinc}\left(\frac{(\omega-\omega_{osc})\cdot T_d}{2}\right) \cdot \left(e^{j(\omega_{inj}-\omega_{osc})\frac{T_d}{2}} \cdot F\{e^{j\varphi_{inj}\left(t-\frac{T_d}{2}\right)}\}(\omega) * \delta\left(\omega-\omega_{inj}\right) * \mathrm{III}_{\frac{1}{T_s}}\left(\frac{\omega}{2\pi}\right)\right)\right]. \quad (5)$$

The SILO output spectrum according to (5) consists of a convolution of the user-defined phase modulation spectrum with its center / carrier frequency signal and the sampling process' aliasing signal (Dirac comb, III), see Fig. 3. It is weighted with a sinc envelope centered at the oscillator's natural frequency ω_{osc}. Since this frequency only affects the envelope and a constant phase offset, the SILO can be regarded as a highly effective aliased regenerative amplifier. In consequence, an injected user-defined constant envelope phase modulated signal is reproduced correctly even with a free running oscillator with (in certain bounds) unknown natural frequency as long as Nyquist's sampling theorem is fulfilled (modulation bandwidth less than half pulse repetition frequency).

In general, this signal synthesis principle is not limited to phase modulated / constant envelope signal synthesis. For amplitude modulation, e.g. an electronically tuned attenuator at the SILO's output can be employed to manipulate the amplitude of each pulse synchronously to the pulse rate, which leads to a polar modulator. Since efficient pulse

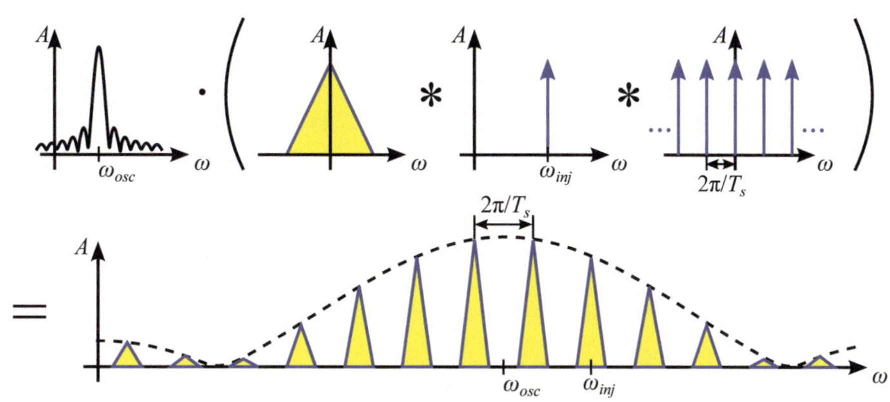

Figure 3. SILO output spectrum according to (5)

amplitude modulation is feasible for a long time in contrast to complex phase modulation and can be added independently, this work is concentrated on the latter aspect.

3.4. Phase modulated UWB communication signals

For the synthesis of communication signals [4], any phase modulated constant envelope signal that is bandwidth limited to half pulse repetition frequency can be chosen. The maximum possible symbol rate leads to one symbol per pulse.

Demodulation can be achieved similar to existing approaches that allow quadrature pulse demodulation (e.g. [11]). Basically, the phase of each pulse has to be sampled synchronously to the pulse sequence (i.e. during pulse duration), which can be realized e.g. by quadrature baseband down-conversion and synchronized sample acquisition. In this case, the sequence of received samples is given by

$$s_{recv}(k) = s\left(k \cdot T_s + \Delta t_{sync}\right) \cdot e^{-j\omega_{inj}\left(k \cdot T_s + \Delta t_{sync}\right)}, \quad k \in \mathbb{N}, \tag{6}$$

where Δt_{sync} denotes a modestly (uncertainty less than half pulse width) unknown synchronization error that has to be taken into account in practice. Inserting (4) in (6) leads to:

$$s_{recv}(k) = A_r \cdot e^{j\left(\varphi_{inj}\left(i \cdot T_s - \frac{T_d}{2}\right) + \left(\omega_{osc} - \omega_{inj}\right) \cdot \left(\Delta t_{sync} + \frac{T_d}{2}\right)\right)}. \tag{7}$$

Accordingly, the original phase modulation φ_{inj} is reconstructed correctly aside from a constant phase offset. Its constancy is guaranteed as long as the natural frequency of the unstabilized oscillator does not drift too fast, which is mostly given due to relatively slow changes in environmental parameters like temperature. For compensation, e.g. differential modulation schemes or short frames can be applied.

3.5. Frequency modulated UWB radar signals

Since the SILO based synthesizer is capable of generating any constant envelope phase modulated signals (within the bandwidth limit), even a frequency modulated radar signal with the bandwidth B, sweep duration T and phase

$$\varphi_{inj,FM}(t) = 2\pi \frac{B}{2T} t^2 \tag{8}$$

can be transmitted. At the receiver, the time delayed transmit signal $s(t)$ is mixed with a FMCW signal:

$$s_{recv,FM}(t) = s(t - t_d) \cdot e^{-j\left(\omega_{inj} t + \pi \frac{B}{T} t^2\right)}. \tag{9}$$

According to [3], the approximate resulting beat frequency spectrum (disregarding envelope)

$$S_{recv,FM}(\omega) = \underline{A} \cdot \delta\left(\omega + 2\pi \frac{B}{T}\left(t_d + \frac{T_d}{2}\right)\right) * \text{III}_{\frac{1}{T_s}}(\omega) \tag{10}$$

is equivalent to the conventional FMCW spectrum except for the aliases resulting from switched operation and a constant phase offset $\underline{A}$. The (one way) distance can be calculated from

$$f_b = \frac{B}{T}\left(t_d + \frac{T_d}{2}\right) \tag{11}$$

given that transmitter and receiver were precisely synchronized, which can be achieved through two-way synchronization like in [16].

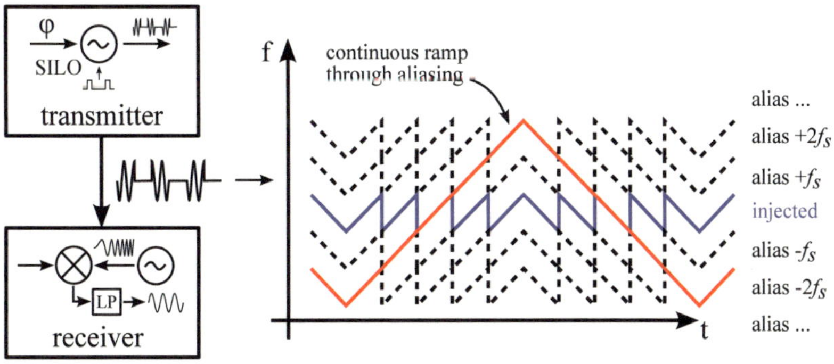

Figure 4. Exploiting sampling aliases to synthesize frequency modulated UWB radar signals with minimal effort

Strictly speaking, the sampling theorem is not met for a sweep bandwidth larger than the pulse repetition frequency. Though, aliasing can be exploited to minimize the ramp synthesis effort (see Fig. 4). The injected and regenerated signal is configured to represent a short chirp within the sampling bandwidth that is repeated continuously. Considering aliasing, the resulting signal appears to be continuous at the receiver when sweeping through all aliases.

The required effort can even be further reduced: Since the SILO only samples certain phase values, it is not necessary to actually generate continuous sweeps as intermediate signal. Instead, a CW injection signal with stepped phase modulation is sufficient as long as its phase (modulus 2π) equals (8) at sampling time. This approach results according to [3] in a short periodic sequence of samples (period $p \in \mathbb{N}^+$) under the condition that the term

$$\frac{pBT_s^2}{T} \tag{12}$$

is whole-number and p even. The sequence features a minimum period of

$$p_{min} = \frac{T}{BT_s^2}, \quad p_{min} \in \mathbb{N}^+. \tag{13}$$

The only restriction that results from exploiting aliases is a limitation in unambiguous range, i.e. maximum distance (phase velocity c_p):

$$d_{max} = \frac{c_p T}{BT_s}. \tag{14}$$

Considering a sampling period of 100 ns (T_s = 10 MHz), which is convenient for low power implementations, a sufficient maximum range of over 1 km can be achieved even at a high bandwidth of 2 GHz in 1 ms.

4. System concepts

In the following, concepts and implementations for the pulsed angle modulated signal synthesis principle are presented. Firstly, the harmonic sampling approach is presented, which is used to take advantage of all benefits of the switched injection-locked oscillator concept by generating a high power, high frequency signal efficiently from a low frequency intermediate signal (4.1). Secondly, a frequency modulated direct digital synthesis (DDS) based upconversion approach for radar applications from the preceding project (PFM-USR) is presented as starting point for the subsequent development (4.2). Thirdly, the recent hardware concept and implementation for phase stepped modulation is described, which allows for synthesizing both frequency modulated radar signals and phase modulated communication signals with the same simple communication signal generator hardware for integrated communication and ranging.

4.1. Harmonic sampling approach

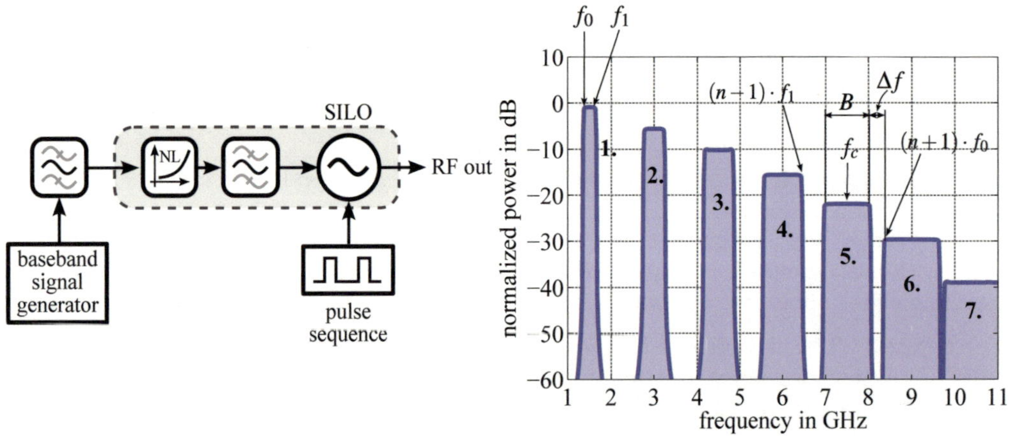

Figure 5. SILO based harmonic sampling; left: concept, right: spectrum of bandwidth limited signal after harmonic generator (here: FMCW sweep from f_0 to f_1) [2]

When synthesizing a high frequency pulsed angle modulated signal, classic approaches based on VCO, PLL, linear amplifier and pulsed switch are not suitable to meet goals like low complexity and low power hardware. Instead, a baseband modulator is proposed for signal generation that generates much lower frequencies than at the system's RF output, e.g. 5.8 GHz instead of 63.8 GHz. At lower frequency ranges, analog RF circuits are usually more efficient than their high frequency counterparts. The baseband signal is then applied to the input of a passive or low power non-linear element that generates harmonics, e.g. a diode or transistor (see Fig. 5). Finally, a SILO is used to amplify the upconverted signal by typically more than 50 dB (within pulse duration). Considering an instantaneous output power level of 0 to 5 dBm, an injection level of less than −45 dBm is sufficient, which allows for high losses and low power consumption in the preceding frequency multiplier stage.

In order to avoid strong intermodulation products caused by the baseband modulation, it should be "slow" compared to the center frequency of the baseband signal so that the non-linear element's instantaneous input and filtered output signal can be considered approximately single tone. This requirement is needed for the SILO, which can itself only correctly regenerate constant envelope signals (apart from the fact that intermodulation products are undesirable) that are stable during the startup phase of the oscillator, e.g. FMCW signals with low ramp slope or rectangular shaped PSK with symbol rate / pulse repetition frequency much smaller than RF frequency.

Regarding maximum baseband modulation bandwidth, there exists a limit for the frequency multiplication factor n in order to guarantee spectral separation, since the bandwidth increases with the harmonic order whereas the spacing of the harmonics' center frequencies is equidistant. According to [2] (see also Fig. 5 right), the upper boundary for the multiplication factor is (harmonic center frequency f_c, harmonic modulation bandwidth B):

$$n < \frac{f_c}{B} - \frac{1}{2}. \tag{15}$$

4.2. Frequency modulated baseband upconversion

The "classic" approach towards synthesizing linear frequency modulated signals (see Fig. 6) consists of a DDS generating a low frequency reference chirp, a PLL and VCO loop and a linear power amplifier. By adding a pulsed switch at the output, pulsed frequency modulation can be realized similar to section 4.2 as long as the pulse width is short enough (the latter signal has constant phase during the pulse, the first one features slight frequency modulation). Obviously, this classic approach has several disadvantages at high frequencies, especially power consuming linear amplifiers and a switch that dissipates more than 90% of the RF power at common pulse sequence duty cycles of less than 1:10.

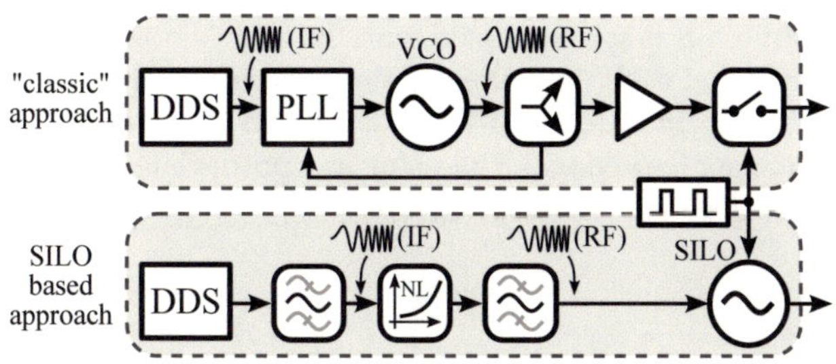

Figure 6. Comparison of classic and SILO based pulsed frequency modulated signal synthesis [2]

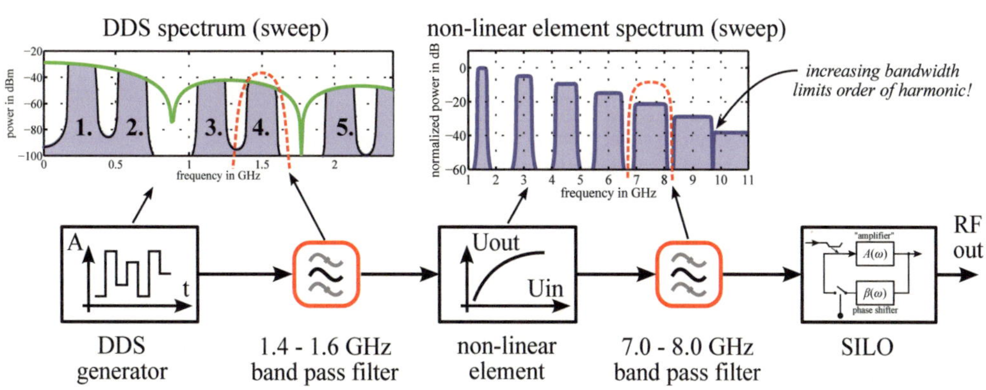

Figure 7. Harmonic sampling concept for FMCW baseband upconversion

Therefore, a harmonic sampling approach was proposed to directly synthesize the ramp from a DDS signal while avoiding PLLs and linear amplifiers at high frequencies [2]. Due to the bandwidth restrictions with harmonic sampling (see section 4.1), a single non-linear stage is not sufficient to generate a 7-8 GHz ramp with a commercially available 1 GS/s DDS circuit. Hence, a Nyquist image from the DDS is used to shift the baseband output frequency range to 1.4-1.6 GHz (see Fig. 7).

The main advantage of this concept is that the generated pulsed frequency modulated signal features a very good linearity in comparison to simple PLL control loops and that the only active component at output frequency is a simple, efficient oscillator (SILO). Despite the simplicity of this concept, its hardware design is quite challenging, since the amplitude of a wideband sweep is subject to many inherent sources of frequency dependent amplitude behavior like DDS spectral envelope, insufficient filter flatness and the non-linear element, which increases existing amplitude variations notably.

4.3. Phase stepped modulation with CW baseband for integrating radar and communication

For integrated communication and ranging, it is desirable to construct a hardware that can synthesize signals for both domains. In the past, they have mostly been developed separately with different hardware concepts. The previously proposed concept (4.2) is well suited for radar systems, but very specific to frequency modulated signals. In fact, angle modulated communication signals can be synthesized with further reduced effort (see Fig. 8) from a CW source with a phase shifter. It is synchronized with the SILO's pulse sequence and its offset is configured to guarantee that each new phase state is stable when the oscillator is turned on. This kind of modulation technique can also be employed to generate frequency modulated signals efficiently according to section 3.5 by using an appropriate sequence of phase samples that represent a frequency chirp.

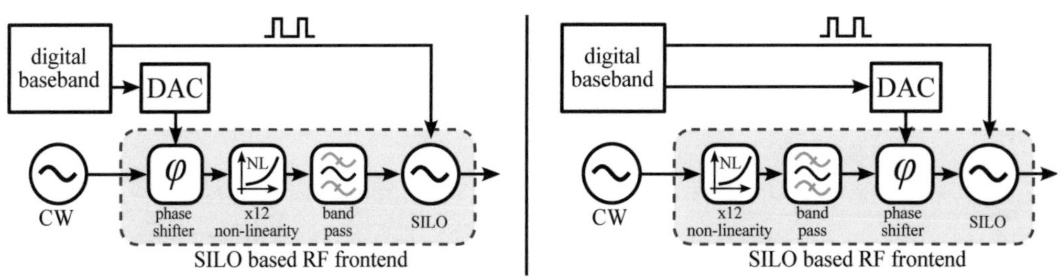

Figure 8. Concept for synthesizing pulsed angle modulated signals using a phase shifter; left: baseband phase modulation with 30 deg shifter, right: RF phase modulation with 360 deg shifter

Regarding hardware implementations, there are two major alternatives concerning the location of the phase shifter in the signal path. An attractive option is to add phase modulation before the frequency multiplication stage; this leads to a minimum amount of RF components and the phase shifter only needs to cover a shifting range of 30 degree, which is easy to design with good linearity. However, baseband modulation limits the multiplication factor (see section 4.1) and the phase shifter causes amplitude fluctuations that are increased in the subsequent non-linear stage. Alternatively, the phase shifter can be placed between RF filter and SILO, which allows for fast modulation, high multiplication factors with less effort (only a constant frequency single tone signal is applied), but requires a more sophisticated 360 degree phase shifter at RF frequency.

5. SILO concept and implementation

Consider the signal displayed in Fig. 1 and the basic SILO model depicted in Fig. 2. As a pulse width T_d of 1 ns and shorter was to be accomplished, the large parasitic capacitances associated with discrete components made it clear that only an integrated solution would be suitable for implementation of the SILO.

As a benchmark for the novel circuit concept of the SILO, some key components of a more conservative concept of generating pulsed frequency modulated signals were developed in an integrated circuit.

All integrated circuits were designed in Cadence Virtuoso and simulated using the Cadence Virtuoso Spectre Circuit Simulator (Cadence, Spectre and Virtuoso are registered trademarks of Cadence Design Systems, Inc). The transmission lines and passive baluns used in the 63.8 GHz-IC were simulated in the Sonnet Professional 2.5D field simulator.

5.1. The benchmark circuit: VCO with integrated switch

To evaluate the efficiency of the SILO approach, a conventional circuit using a VCO with wide tuning range and an output switch was designed. The system with the manufactured IC is shown in Fig. 9.

The schematic of the VCO can be seen in Fig. 10, together with the half-circuit of the designed output switch.

Parameter	$C^*_{var,min}$	$C^*_{var,max}$	$C_{var,min}$	$C_{var,max}$	L_B	R_E	R_{CC}
Value	65 fF	200 fF	145 fF	455 fF	0.41 nH	200 Ω	200 Ω
Parameter	C_1	C_2	C_3	C_4	V_{Bias}	V_{CC}	V_{tune}
Value	700 fF	200 fF	300 fF	300 fF	1.8 V	3.3 V	0 to 4 V

Table 1. VCO component parameters

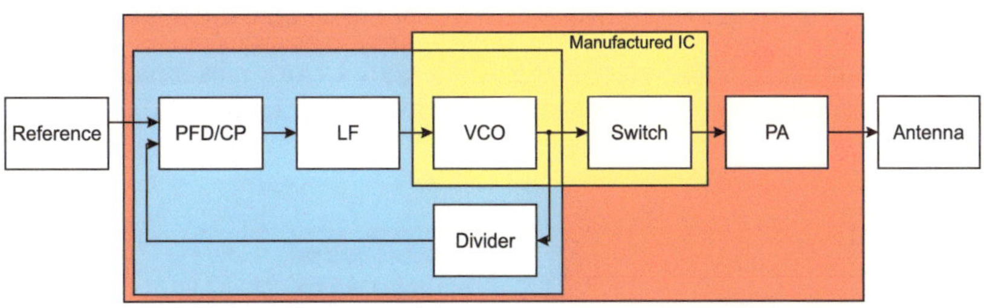

Figure 9. Pulsed frequency modulated continuous wave synthesizer system concept using an output switch

The VCO is based on a common collector Colpitts oscillator design, including a second varactor diode pair at the transistor base. It is described in detail in [8]. A short overview is given in the following.

A bipolar current mirror is used to drive the oscillator core. The emitter follower output buffer from [8] was replaced by a differential pair to increase common-mode rejection. The VCO frequency defining series resonant circuit consists of L_B and C_{in}:

$$f_{res} = \frac{1}{2\pi\sqrt{L_b C_{in}}} \tag{16}$$

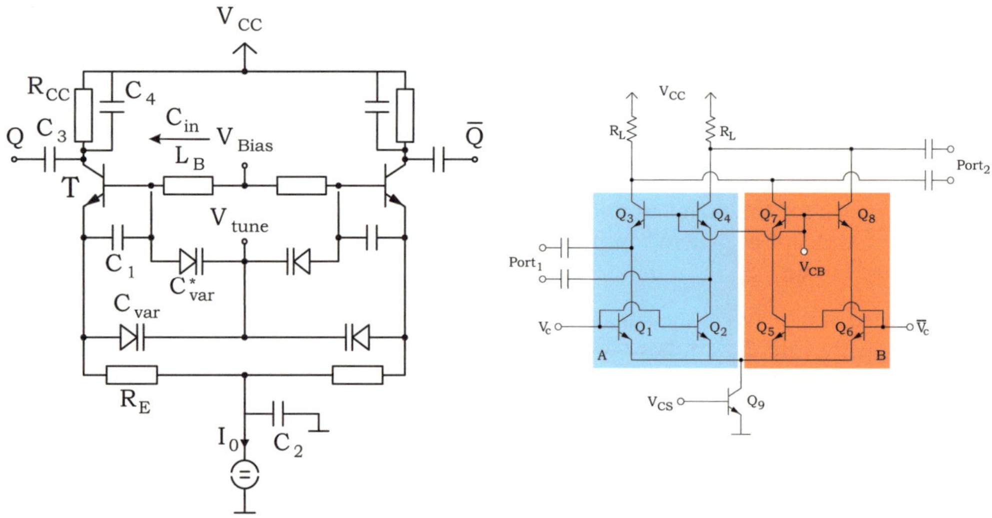

Figure 10. Synthesizer key components; left: VCO, right: half-circuit of single-pole double-throw switch

L_B is realized as a spiral inductor without tuning capability. Tuning is available by varying C_{in}, which has to be tuned over a wide tuning range using variable MOS-capacitance circuits.

For a minimum influence on the tuning range, C_P has to be minimized. It consists mainly of the collector base capacitance C_{CB} of transistor T and thus is given by size and bias conditions. C_S, which is determined mainly by C_{BE}, has to be maximized. Additionally, both varactor capacitance ranges have to be maximized. For a more detailed discussion, refer to [15]

The proposed pulsed ultra-wideband signal generation requires a switch after the frequency synthesizing PLL. The switch should have a minimum switching time in both on and off direction to enable the usage of very short pulses (in the $1-10\,\text{ns}$ range). Additionally, a constant input port impedance is important in order not to change the loading of the oscillator.

A switch circuit was designed based on [10]. The original work was aimed at a $22-29\,\text{GHz}$ UWB radar for automotive applications. Fig. 10, right, shows the half-circuit.

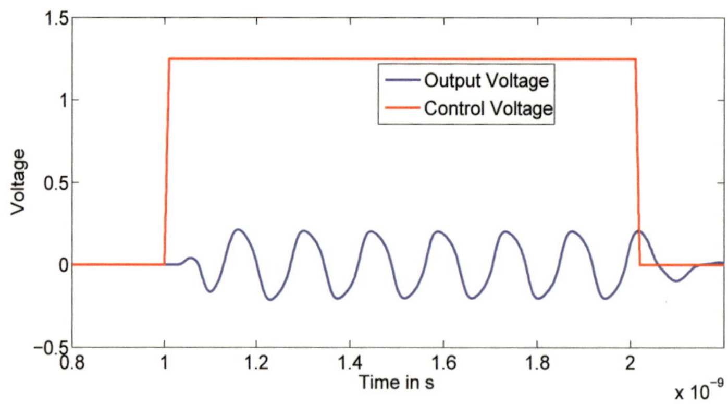

Figure 11. Switch transient simulation: Output voltage signal (blue) in reaction to control voltage (red) change.

The circuit works by switching the bias currents through branches A and B, implemented by transistors Q_1 to Q_4 and Q_5 to Q_8, respectively. This is done by alternating the control voltages applied to switching stages Q_1/Q_2 and Q_5/Q_6. The differential common base stages (Q_3/Q_4 and Q_7/Q_8) provide amplification and isolation, depending on the bias current. Transistor Q_9 provides the bias current, which is switched between the branches.

Fig. 11 shows the transient simulation of the output signal for a single rising V_C edge with a rise time of 5 ps. The delay between the control edge and a 90% of the output is below 250 ps. The addition of a matching network would improve insertion loss, but at the cost of worse area efficiency. The simulated input-referred noise was between 2.83 nV/$\sqrt{\text{Hz}}$ and 3.67 nV/$\sqrt{\text{Hz}}$.

A combination of VCO and output switch was simulated and then manufactured.

5.2. SILO oscillator concepts

As the injection locking property is universally stemming from oscillator theory, any oscillator can in theory be employed for switched-injection locking. There is an interesting trade-off to be made when considering an oscillator configuration for SILO building: The oscillator Q-factor should be high and excess loop gain should be low for better phase noise performance on the one hand, but a high-Q oscillator with low excess loop gain takes longer to begin oscillation, which is critical for pulsed angle modulated signal generation. A careful balance between the two qualities has to be found.

Another consideration has to be put into the point in the oscillator loop where the signal is injected into. In a cross-coupled oscillator, the resonator and gain stages are directly connected to the output. This means that there has to be a buffering circuit for the injected signal which provides backward isolation, in order to ensure the oscillation frequency of the oscillator is not influenced by the circuitry connected to the tank.

For the design of the SILO circuits, we concentrated on resonator-based oscillators, as they typically show better phase noise performance than inverter-based ring oscillators. A demonstrator implementation in discrete components was used for initial experimentation and verification of the viability of our approach. This circuit was aimed at a frequency range of 6 to 8 GHz. Subsequently, a SILO IC based on a pulse generator and a cross-coupled LC-oscillator was designed and manufactured. In a final step, a harmonics generator was combined with a Colpitts oscillator to sample a 5.8 GHz-signal and emit a 63.8 GHz-signal.

5.3. 6 and 8 GHz SMT SILO

For reference and for first experiments, SILO implementations based on surface mounted planar technology were realized. They are based on an ordinary common-collector Colpitts oscillator and designed for a natural frequency of 6 GHz respectively 7.5 GHz. In order to implement injection-locking, a directional coupler was added to apply the injection signal to the oscillator's output (see Fig. 12). The maximum achievable (10 dB) bandwidth is about 600 MHz at 7.5 ns pulse width.

Apart from parasitic technological limitations of lumped planar implementations, the single-ended design features an inherent source of self-locking to a harmonic of the switched power supply. Therefore, the pulse width is limited to about 10 to 20 ns in order to achieve a good compromise between bandwidth and minimum injection level. In consequence,

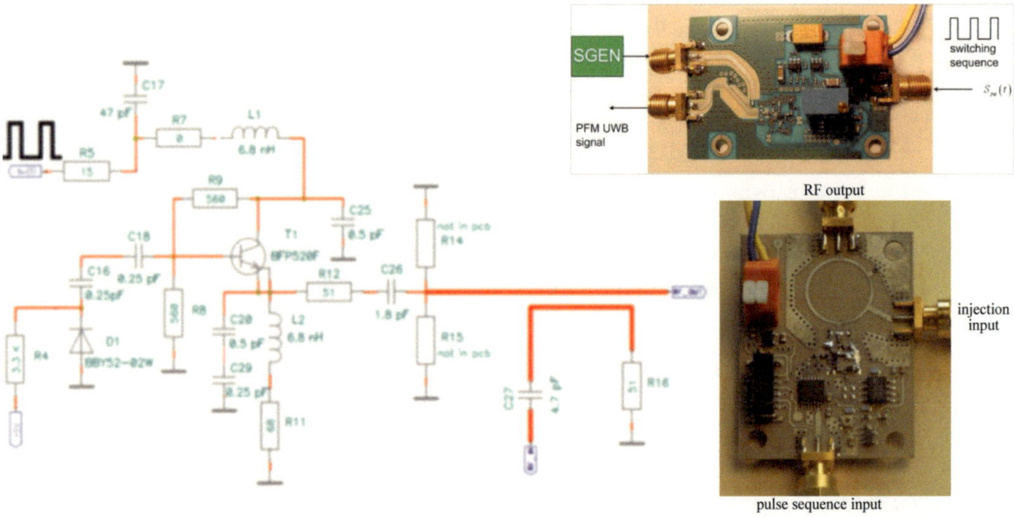

Figure 12. SILO SMT implementation; left: schematic of 7.5 GHz version; upper right: 7.5 GHz implementation; lower right: 6 GHz implementation

differential integrated circuit implementations are expected to deliver a significantly better self-locking suppression allowing much shorter pulsed in the order of 1 ns with comparable performance.

5.4. 7 GHz integrated circuit

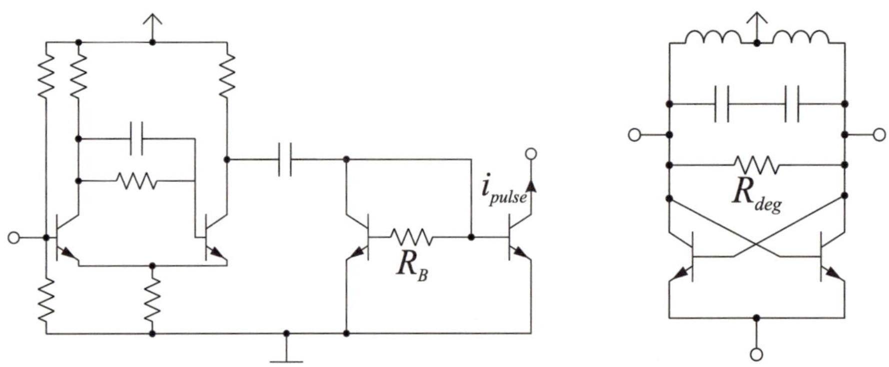

Figure 13. 7 GHz SILO circuits; left: Schmitt-trigger with peak generator, right: VCO with Q-degeneration resistor

The circuit consists of two active baluns for single-ended to differential and differential to single-ended conversion, a Schmitt-trigger with modified current mirror load for current peak generation and a simple cross-coupled oscillator for signal generation. It has an externally controllable pulse repetition rate and a pulse duration of approx. 1 ns. During operation it consumes 33 mA at 3.3 V supply voltage, while generating a $>$ 330 mV_{pp}signal. The generated signal has a 10 dB-bandwidth of over 2 GHz at 7.5 GHz center frequency.

Both Schmitt trigger with current peak generator and VCO with Q-degeneration circuits are shown in Fig. 13.

As efficient integrated circuits are built in a differential configuration but external circuitry and measurement equipment usually are only available in single-ended configuration, single-ended to differential (S2D) and differential to single-ended (D2S) conversion circuits are needed in the IC. We designed a simple active balun circuit that can act as both S2D-and D2S-converter. When employed as a S2D-converter, both outputs and one input are connected, when used as a D2S-converter, one output and both inputs are connected.

In order to control the pulse repetition rate externally, a Schmitt-trigger circuit with current peak generator was designed based on [13]. The circuit enables a wide variety of pulse repetition rates (1 − 80 MHz could be achieved with the measurement equipment at hand). The resistor R_B together with base-emitter capacitance C_{BE3} controls the time constant $\tau_{current}$ of the charging circuit:

$$\tau_{current} = R_B C_{BE3}. \tag{17}$$

The peak generator was designed for a pulse duration of 1 ns by selecting the size of the resistor $R_B = 5\,\text{k}\Omega$.

For the oscillator, a simple cross-coupled topology was chosen. As the oscillator has to lock to the injected phase, a low Q is preferable. In order to degenerate the Q, a resistor was connected in parallel to the LC-tank circuit. The current is provided by the peak generator. Fig. 13 shows the implementation.

A simple common-collector circuit is used as an output buffer to drive the 50 Ω load.

5.5. 63 GHz integrated circuit

The system developed for pulsed angle modulated signal generation at mm-wave frequency is shown in Fig. 14. The input signal of 5.8 GHz is coupled into the harmonics generator, which consists of a bipolar transistor with a resonant load. The load consisting of a transmission line of inductance L_1 and capacitors C_1 and C_2 is designed to couple the wanted 11th harmonic into the transformer. Fig. 17 shows the output power for the 1st, 10th, 11th and 12th harmonic depending on the input power. For an input power > -3 dBm, the 11th harmonic is the strongest. The now differential signal is used to lock the VCO shown in Fig. 15.

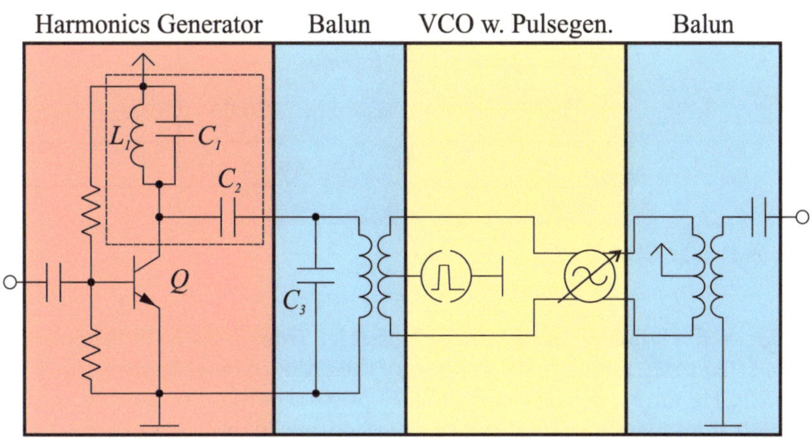

Figure 14. 63 GHz-system consisting of harmonics generator, baluns and VCO with pulse generator

The signal is coupled to the collector load transmission lines of the Colpitts oscillator using a transformer with a center tap. The center tap is connected to the pulsed current source of the oscillator.

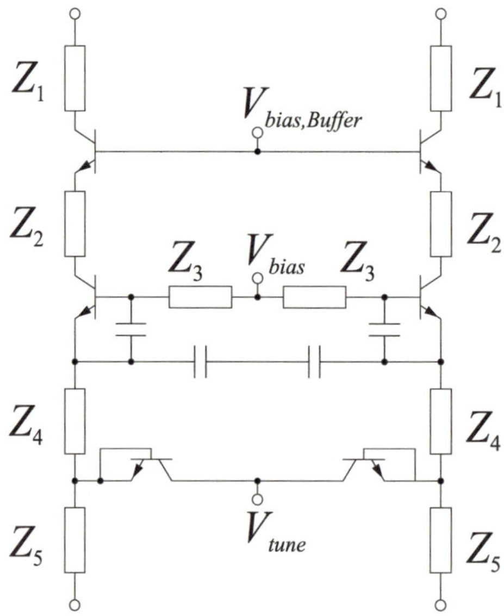

Figure 15. 63.8 GHz Colpitts voltage controlled oscillator schematic. Z_1 to Z_5 denote transmission lines

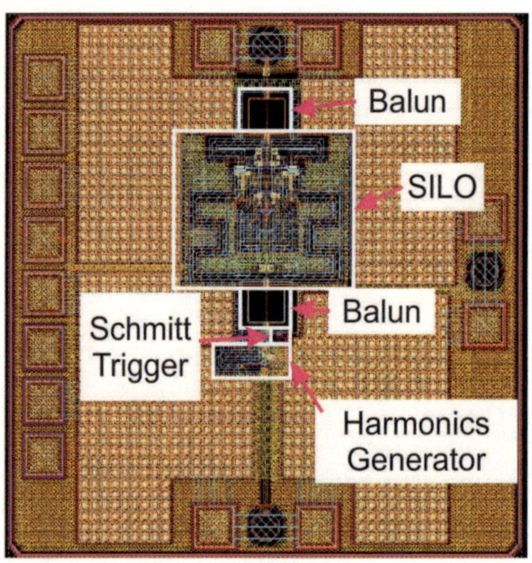

Figure 16. Layout of 63.8 GHz SILO

The simulation of the whole system was not possible. This is due to the fact that the system works in three frequency ranges, which differ by the order of magnitudes: The 5.8 GHz input signal, the 63.8 GHz output signal and the SILO pulse repetition frequency (10 – 100 MHz). Combined with the unknown modeling of switched injection-locking in the EDA software made it more viable to design each component (harmonics generator, VCO, pulse generator) separately. 16 shows the layout of the SILO circuit with its sub-components.

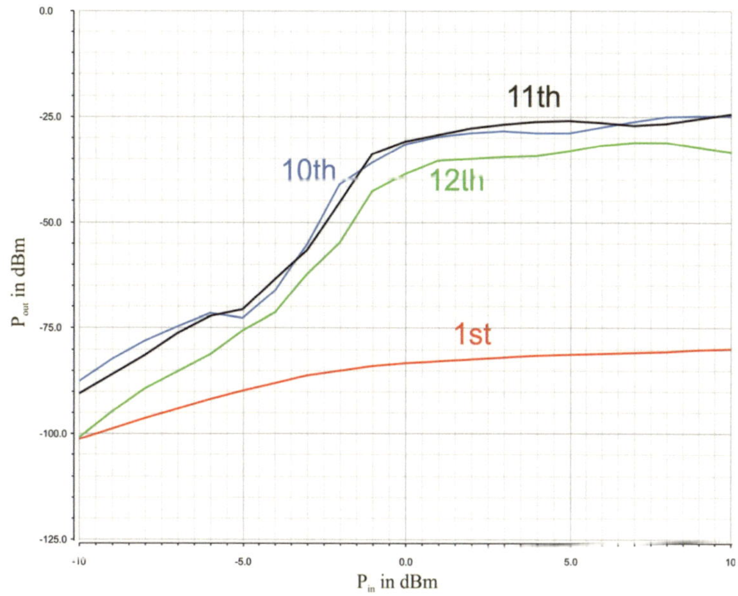

Figure 17. Power of the generated harmonics over the input power of the harmonics generator

6. Measurement setup and results

6.1. Verification of sampling theory

In order to verify the theoretical predictions concerning the switched injection locked harmonic sampling approach according to section 3.3, a demonstrator based on lumped planar components was built (see Fig. 18 and 19). It consists of a 480 MHz, 0 dBm signal source, a 10 MHz DAC modulated phase shifter, a single biased bipolar transistor frequency multiplier, a band pass filter (200 MHz @ 5.8 GHz) and the 5.8 GHz switched injection locked oscillator, which is turned on and off by the digital baseband synchronously to DAC modulation. Fig. 20 depicts the spectrum at the SILO's output. It features the typical sinc shaped peak comb in pulsed mode, which is aligned to and follows the injection frequency of 5.76 GHz when changed. When tuning the oscillators natural frequency (which is according to Fig. (20) different from the injection frequency) using a varactor diode, the sinc shape of the spectrum moves on the frequency axis while the peak positions do not change. These results prove most of the main claims of the generalized sampling theory according to (5) [4].

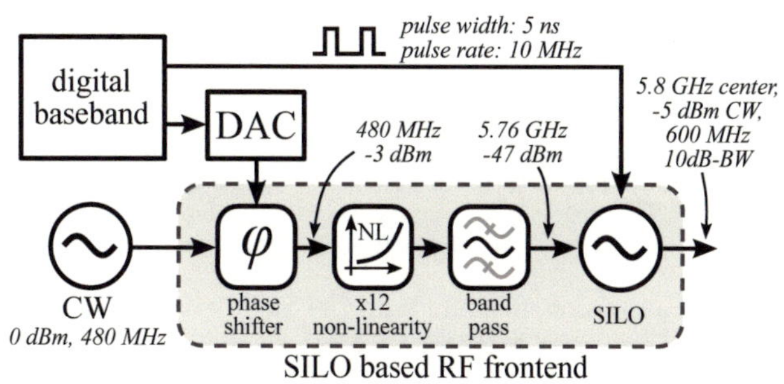

Figure 18. Implementation of communication and radar signal generator [4]

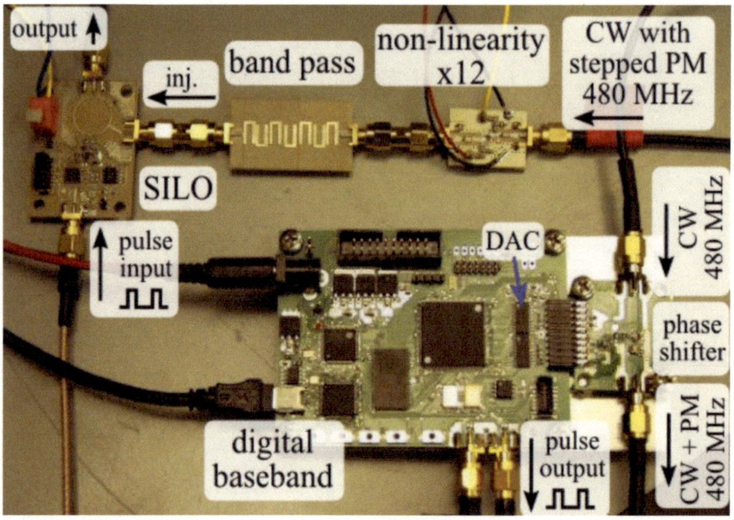

Figure 19. Hardware components for the 6 GHz transmitter system demonstrator (using lumped planar components SILO implementation)

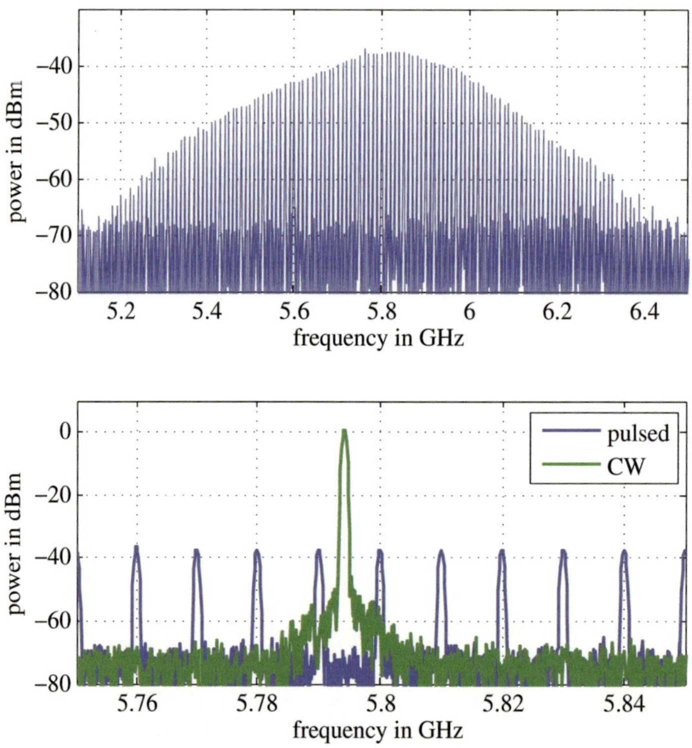

Figure 20. Spectrum of SILO based demonstrator with CW modulation; large peak: oscillator permanently on, comb: pulsed oscillator, background: comb zoomed out to show envelope, span 1.5 GHz [4]

6.2. Synthesis of communication signals

The synthesis of time domain communication signals was demonstrated using an 8 PSK modulation with cyclic transmission of all symbol values and maximum symbol rate, i.e. one symbol per pulse. The output signal of the demonstrator (Fig. 18, 19) was mixed to baseband using a quadrature mixer and displayed using an oscilloscope. Its waveform (Fig. 21) clearly shows the phase states and their repeatability in the IQ diagram. These results prove for the first time that it is feasible to generate UWB signals with more complex phase modulation than BPSK while at the same time keeping complexity and power consumption low.

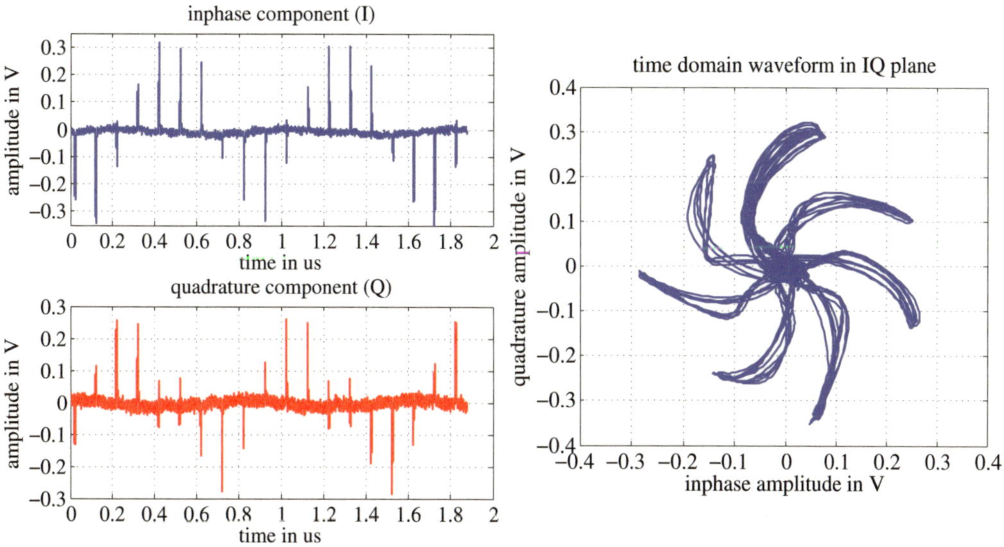

Figure 21. Generator output with 8 PSK modulation mixed to DC; left: inphase and quadrature component, right: IQ diagram [4]

6.3. Synthesis of radar signals

According to sections 3.5 and 4.3, the same simple hardware implementation used for communication signal synthesis (Fig. (18), (19)) can be employed to generate pulsed frequency modulated radar signals by repeatedly transmitting a limited list of phase samples. For a pulse rate of 10 MHz and a ramp slope of 20 MHz/μs, only 50 phase samples (one per pulse) are sufficient.

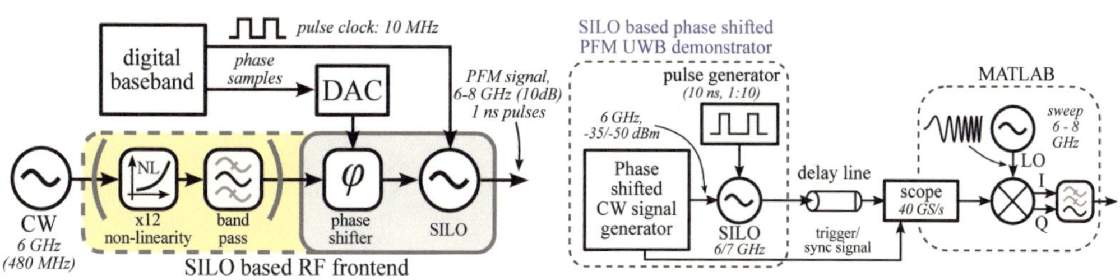

Figure 22. Measurement setup (on-waver) for the synthesis of radar signals using an integrated circuit SILO [3]

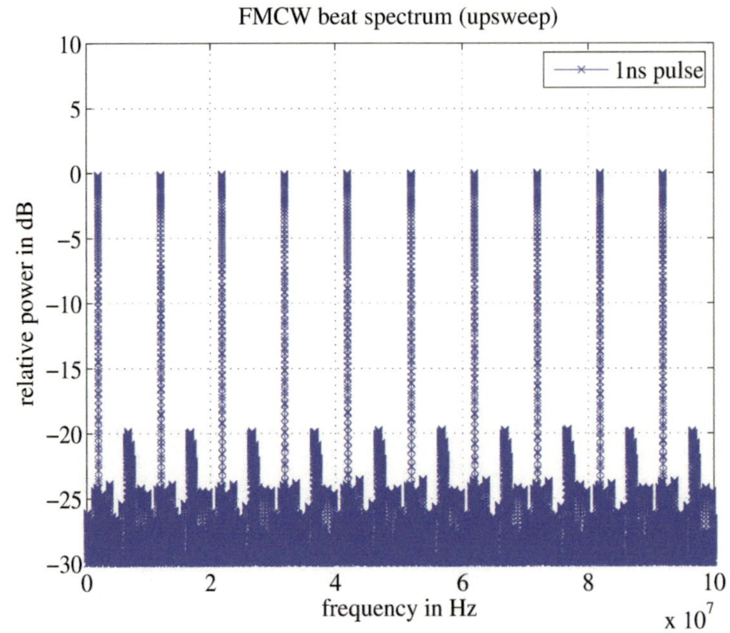

Figure 23. Beat spectrum (6-8 GHz SILO chip) of measured radar signal after mixing with linear sweep and before low pass filtering [3]

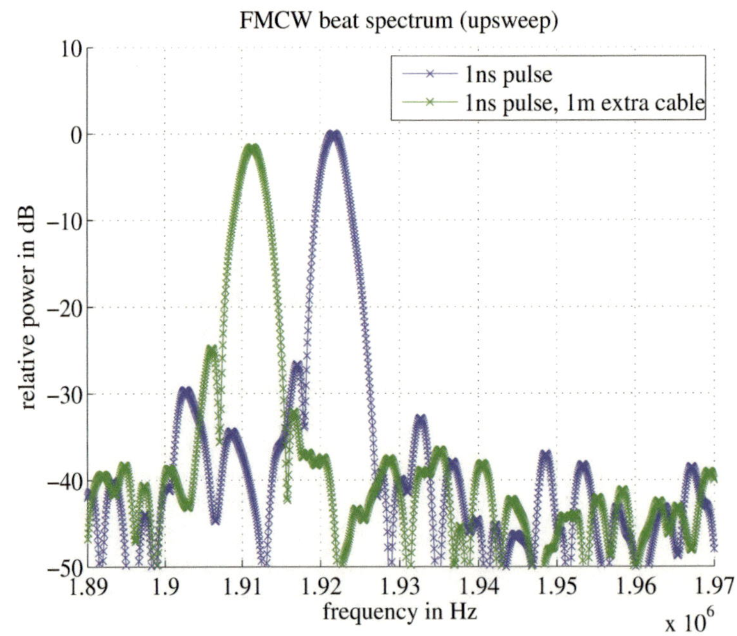

Figure 24. Zoomed beat spectrum (6-8 GHz SILO chip), comparison of two waveforms with different transmission delays [3]

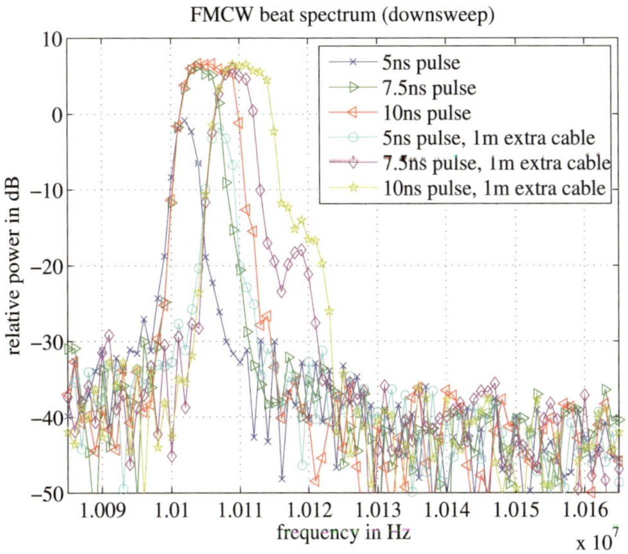

Figure 25. Comparison of 6-8 GHz chip (2 GHz bandwidth) with lumped implementation (wider pulses, smaller bandwidth) [3]

For verification, this approach was realized both using the previously employed lumped components SILO (6 GHz, 600 MHz bandwidth) and the first large bandwidth integrated circuit implementations (7 GHz, >2 GHz bandwidth) in order to demonstrate the resolution benefit for ranging. The setup for both experiments is depicted in Fig. 22; the generated and delayed signal is acquired using an oscilloscope and evaluated on a PC using a numerical computation software where it is mixed with a linear FMCW signal and analyzed in frequency domain (FFT).

Fig. 23 and 24 show the resulting beat frequency spectrum for the integrated circuit implementation using 1 ns pulses and 10 MHz pulse repetition rate. It corresponds to equation (10) except the small peaks that result from imperfections in the oscillator design leading to a slight turn-on pulse self-locking effect. Future designs are expected to fix this issue.

Comparing the results of the lumped and integrated circuit implementations (see Fig. 25), the benefit of much higher bandwidths regarding resolution becomes obvious. If the oscillator's spectral bandwidth is too small in relation to the sweep bandwidth, the beat frequency peak is broadened because of additional windowing through the narrowband SILO spectrum. Therefore, the oscillator bandwidth / pulse width should be adjusted to the desired sweep bandwidth in order to maximize spectral efficiency [3].

6.4. VCO with switch IC

The manufactured circuit is depicted in Fig.26. It measures $710 \times 1455\ \mu m^2$. For reasons of nonavailability of differential equipment, all measurements were done single-ended with the unused output terminated to ground with a 50 Ω resistor.

Fig. 27 shows the output power over the tuning range. The 10 dB decrease of output power compared to the previously published [8] VCO is attributed to the different VCO output buffer and the insertion loss of the switch.

Figure 26. VCO with switch circuit IC photograph

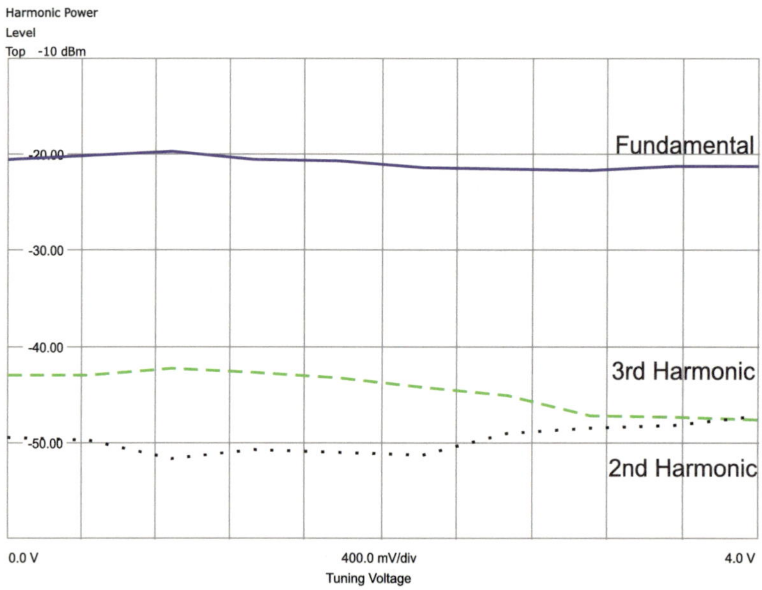

Figure 27. Measured output power of 1st, 2nd and 3rd harmonic of VCO with switch

The phase noise performance of the VCO with switch has deteriorated significantly from the previous [8] stand-alone VCO. This is mainly attributed to the new buffer structure which performed worse than anticipated.

6.5. 7 GHz SILO IC

The IHP Technologies SGB25V 250 nm SiGe:C BiCMOS process was chosen for manufacturing. It provides a cheap and flexible platform including one or two thick top metal layers consisting of aluminum. The advantage of using a BiCMOS process for a transmitter circuit is the possibility to build a system-on-a-chip (SoC) solution that integrates

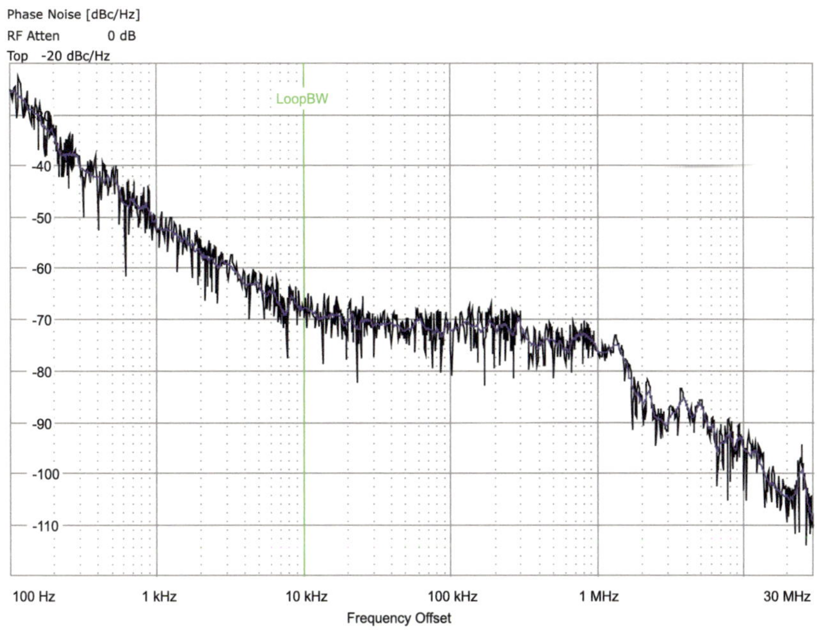

Figure 28. Measured phase noise of VCO with switch circuit

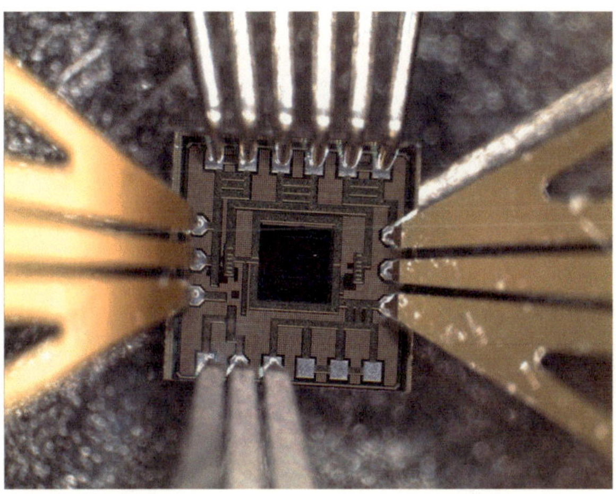

Figure 29. Manufactured 7 GHz SILO IC

digital baseband and analog RF circuits. Fig. 29 shows a chip photograph with connected measurement probes.

Fig. 30 shows the output power spectrum of the manufactured SILO.

The 10 dB-bandwidth stretches from 5to 8 GHz. A single pulse is shown in Fig. 31. A single cycle of oscillator start up, oscillation and decay has a duration of 1.5 ns.

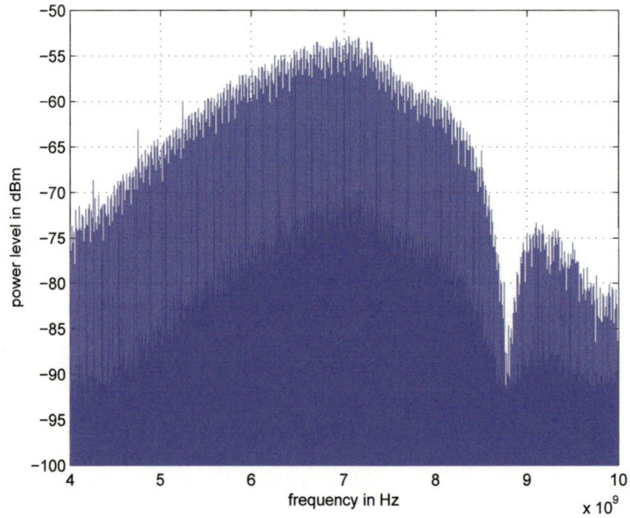

Figure 30. Measured output spectrum of 7 GHz SILO IC

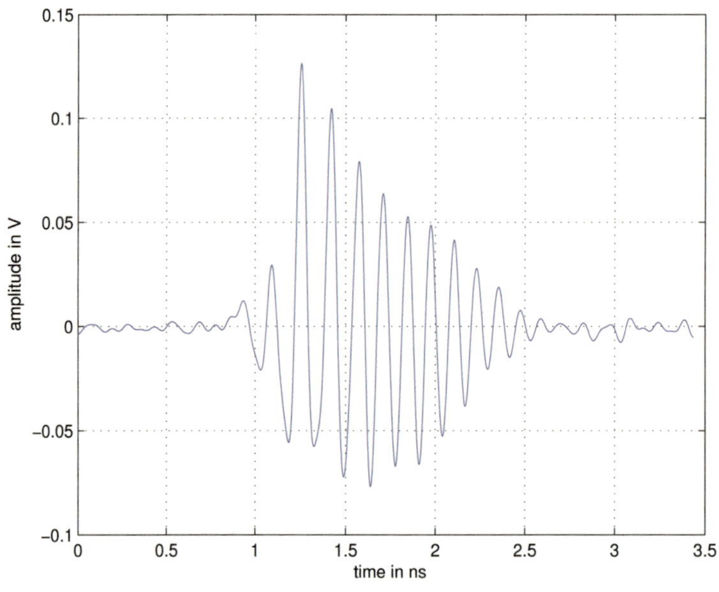

Figure 31. Transient output of 7 GHz SILO IC

7. Future work

Since this project is still ongoing, future work will cover further aspects that enhance theory and hardware implementation. Regarding pulsed angle modulated signals, more complex modulation schemes will be developed in conjunction with a more comprehensive study of error sources and their compensation. Furthermore, the first designs of the SILO circuit will be refined for an even better performance and higher integration level. Last but not least, hardware concepts for receiver technology are being developed.

Acknowledgement

This work was supported by the German Research Foundation (DFG - priority program SPP1202, grant VO 1453/3-2) within the project "Components and concepts for low-power mm-wave pulsed angle modulated ultra wideband communication and ranging".

Author details

Alexander Esswein and Robert Weigel
Institute for Electronics Engineering, University of Erlangen-Nuremberg, Germany

Christian Carlowitz and Martin Vossiek
Institute of Microwaves and Photonics, University of Erlangen-Nuremberg, Germany

8. References

[1] Barrett, T. W. [2000]. History of UltraWideBand (UWB) Radar & Communications: Pioneers and Innovators.

[2] Carlowitz, C., Esswein, A., Weigel, R. & Vossiek, M. [2012a]. A low power Pulse Frequency Modulated UWB radar transmitter concept based on switched injection locked harmonic sampling, *Microwave Conference (GeMiC), 2012 The 7th German*, pp. 1 –4.

[3] Carlowitz, C., Esswein, A., Weigel, R. & Vossiek, M. [2012b]. Synthesis of Pulsed Frequency Modulated Ultra Wideband Radar Signals Based on Stepped Phase Shifting, *IEEE International Conference on Ultra-Wideband (ICUWB)* .

[4] Carlowitz, C. & Vossiek, M. [2012]. Synthesis of Angle Modulated Ultra Wideband Signals Based on Regenerative Sampling, *IEEE International Microwave Symposium 2012.*

[5] Chandrakasan, A., Lee, F., Wentzloff, D., Sze, V., Ginsburg, B., Mercier, P., Daly, D. & Blazquez, R. [2009]. Low-Power Impulse UWB Architectures and Circuits, *Proceedings of the IEEE* 97(2): 332–352.

[6] Deparis, N., Loyez, C., Rolland, N. & Rolland, P.-A. [2008]. UWB in Millimeter Wave Band With Pulsed ILO, *Circuits and Systems II: Express Briefs, IEEE Transactions on* 55(4): 339 –343.

[7] Deparis, N., Siligarisy, A., Vincent, P. & Rolland, N. [2009]. A 2 pJ/bit pulsed ILO UWB transmitter at 60 GHz in 65-nm CMOS-SOI, *Ultra-Wideband, 2009. ICUWB 2009. IEEE International Conference on*, pp. 113 –117.

[8] Esswein, A., Dehm-Andone, G., Weigel, R., Aleksieieva, A. & Vossiek, M. [2010]. A low phase-noise SiGe Colpitts VCO with wide tuning range for UWB applications, *Wireless Technology Conference (EuWIT), 2010 European*, pp. 229 –232.

[9] Grass, E., Siaud, I., Glisic, S., Ehrig, M., Sun, Y., Lehmann, J., Hamon, M., Ulmer-Moll, A., Pagani, P., Kraemer, R. & Scheytt, C. [2008]. Asymmetric dual-band UWB / 60 GHz demonstrator, *Personal, Indoor and Mobile Radio Communications, 2008. PIMRC 2008. IEEE 19th International Symposium on*, pp. 1 –6.

[10] Hancock, T. & Rebeiz, G. [2005]. Design and Analysis of a 70-ps SiGe Differential RF Switch, *Microwave Theory and Techniques, IEEE Transactions on* 53(7): 2403 – 2410.

[11] J. Ryckaert et al [2006]. A 16mA UWB 3-to-5GHz 20Mpulses/s Quadrature Analog Correlation Receiver in 0.18/spl mu/m CMOS, *Solid-State Circuits Conference, 2006. ISSCC 2006. Digest of Technical Papers. IEEE International*, pp. 368 –377.

[12] Kohno, R. [2008]. Latest regulation and R&D for UWB inter-vehicle radar in millimeter wave band, *8th International Conference on ITS Telecommunications, ITST* .

[13] Lin, D., Schleicher, B., Trasser, A. & Schumacher, H. [2011]. Si/SiGe HBT UWB impulse generator tunable to FCC, ECC and Japanese spectral masks, *Radio and Wireless Symposium (RWS), 2011 IEEE*, pp. 66 –69.

[14] Miesen, R., Ebelt, R., Kirsch, F., Schaefer, T., Li, G., Wang, H. & Vossiek, M. [2011]. Where is the Tag? History, Modern Concepts, and Applications of Locatable RFIDs, *IEEE Microwave Magazine* . to be published.

[15] Pohl, N., Rein, H.-M., Musch, T., Aufinger, K. & Hausner, J. [2009]. SiGe Bipolar VCO With Ultra-Wide Tuning Range at 80 GHz Center Frequency, 44(10): 2655–2662.

[16] Roehr, S., Gulden, P. & Vossiek, M. [2008]. Precise Distance and Velocity Measurement for Real Time Locating Using a Frequency Modulated Continuous Wave Secondary Radar Approach, *IEEE Transactions on Microwave Theory and Techniques* 56(10): 2329–2339.

[17] Sewiolo, B., Hartmann, M., Guenther, O. & Weigel, R. [2006]. System Simulation of a 79 GHz UWB-Pulse Radar Transceiver Front-End for Automotive Applications, *VDE / ITG Diskussionssitzung Antennen und Messverfahren fuer Ultra-Wide-Band(UWB)-Systeme (UWB 2006) Kamp-Lintfort, Germany* .

[18] Trotta, S., Dehlink, B., Knapp, H., Aufinger, K., Meister, T., Bock, J., Simburger, W. & Scholtz, A. [2007]. SiGe Circuits for Spread Spectrum Automotive Radar, *Ultra-Wideband, 2007. ICUWB 2007. IEEE International Conference on*, pp. 523 –528.

[19] Vossiek, M. & Gulden, P. [2008]. The Switched Injection-Locked Oscillator: A Novel Versatile Concept for Wireless Transponder and Localization Systems, *Microwave Theory and Techniques, IEEE Transactions on* 56(4): 859 –866.

[20] Wentzloff, D. & Chandrakasan, A. [2006]. Gaussian pulse Generators for subbanded ultra-wideband transmitters, *Microwave Theory and Techniques, IEEE Transactions on* 54(4): 1647 – 1655.

Chapter 14

HaLoS – Integrated RF-Hardware Components for Ultra-Wideband Localization and Sensing

Stefan Heinen, Ralf Wunderlich, Markus Robens, Jürgen Sachs, Martin Kmec, Robert Weigel, Thomas Ußmüller, Benjamin Sewiolo, Mohamed Hamouda, Rolf Kraemer, Johann-Christoph Scheytt and Yevgen Borokhovych

Additional information is available at the end of the chapter

http://dx.doi.org/10.5772/54987

1. Introduction

Ultra-Wideband (UWB) sensors exploit very weak electromagnetic waves within the lower microwave range for sounding the objects or processes of interest. The interaction of electromagnetic waves with matter provides interesting options to gain information from a great deal of different scenarios. To mention only a few, it enables the assessment of the state of building materials and constructions, the investigation of biological tissue, the detection and localization of persons buried by rubble after an earthquake or unauthorized people hidden behind walls, and much more [1]. The advantage of such methods consists in their non-destructive and continuously running measurement procedure which may work at high speed and in contactless fashion.

Sensors applying electromagnetic interactions with the test object have been in use for a long time. However, most of such sensors are restricted to a relatively narrow bandwidth and, consequently, they can provide only a small amount of information about the test object. Sophisticated data processing supposed, UWB sensors may be able to provide more information and, therefore, to reduce ambiguities which are inherently part of indirect measurement methods such as electromagnetic sensing.

Depending on the actual tasks, the requirements on the sensing system may be quite different, such as the optimum operational frequency band, measurement speed, sensitivity, system costs, reliability, power consumption etc. There are several UWB sensing principles known, each having specific advantages and disadvantages. Generally, one can state that the usability of UWB-sensors will be largely improved with increasing degree of system integration regardless of the sensor principle. The HaLoS-project addresses this topic by investigating general purpose UWB sub-modules like amplifiers, ADCs, fast processing units etc. as well as an integration-friendly sensor concept based on ultra-wideband pseudo-noise codes.

The chapter is organized as follows. First, the most important performance figures of UWB sensors are introduced. Second, we give an overview of various UWB-sensor principles recently in use and explain the UWB pseudo-noise concept. Then, we address some specific topics like wideband receiver circuits, transmitter circuits and high-speed data capture. Finally, some aspects of monolithically integrated UWB-sensors are discussed.

2. Properties and basic concepts of UWB-sensors

2.1. Key figures of UWB-sensors

The UWB sensor configuration may be determined by demands which are guided by two different and partially conflicting aspects. On the one hand, these are the UWB radiation rules, and on the other one, we have to respect the physical constraints of the sensing problem. The radiation rules, which are not unique within different regions of the world, mainly limit spectral power emission, restrict the operation frequency band and require sounding signals of large instantaneous bandwidth. Seen from a physical point of view, we need an adequate operational frequency band which provides reasonable interaction between the sounding signal and the object of interest. This may lead to conflicting situations with the radiation rule for sensing tasks requiring wave penetration like through-wall radar or medical imaging. Thus, one has to search a proper compromise in the case of frequency mask violation. Though UWB sensors are banned from long-range applications due to low-radiation power, they promote biological and medical sensing since the target exposition is harmless. Furthermore, the interaction between sounding wave and target is based on linear phenomena. Hence, the sounding bandwidth may be provided instantaneously (complying with FCC or ECC radiation rules) or sequentially (violating these radiation rules) without affecting the measurement results as long as the scenario under test behaves stationary during the measurement. This paper is focused on techniques for information capture by exploiting electromagnetic interactions. Hence, we do not exclude sensor principles or frequency bands violating UWB radiation rules from our further discussions.

Spectral band and related parameters: As frequency diversity is a key issue of unambiguous information gathering by electric sensors, the widths and the occupation density of the spectral sounding band is of major interest. For the sake of brevity, we will deal here only with baseband signals (see [2] for deeper discussions) which we characterize by their two-sided bandwidth $\ddot{B}$ that can be linked to typical time domain parameters:

$$\ddot{B} \approx \begin{cases} t_w^{-1} & \text{for pulse shaped signal} \\ \tau_{coh}^{-1} & \text{for CW signal} \end{cases} \tag{1}$$

Here, t_w represents the width of a pulse, and τ_{coh} is the coherence time of a random or pseudo-random signal (i.e. the width of the auto-correlation function). The occupation density of the frequency band is given by the line spacing Δf which is either determined by the repetition rate f_0 of a periodic sounding signal (t_P - period duration) or via the Fourier Transform by the observation interval T of non-periodic signals:

$$\Delta f = \begin{cases} f_0 = t_P^{-1} & \text{periodic signal} \\ T^{-1} & \text{non-periodic signal} \end{cases} \tag{2}$$

As non-periodic signals are quite unusual in UWB sensing, we will avoid discussing them. The line spacing Δf gives the frequency resolution of the sensor or it determines the maximum observable length $T_W = \Delta f^{-1}$ of the impulse response $g(t)$ of a scenario under test. If $g(t)$ does not settle down within $T_w = t_P$, we have to anticipate time aliasing.

In the case of UWB radar sensing, we can convert (1) and (2) into corresponding spatial parameters. One of them assigns the range resolution δ_r, i.e. the capability of the radar to separate two close point targets of identical reflectivity. We will refer to the usual relation (c - wave velocity):

$$\delta_r \approx \frac{c}{2\,\underset{\cdot\cdot}{B}} \approx \begin{cases} \frac{1}{2}\tau_{coh}\,c & \text{for time stretched signal} \\ \frac{1}{2}t_w\,c & \text{for pulse shaped signal} \end{cases} \tag{3}$$

even if it should be considered with care. The relation originates from narrowband radar whose sounding signal suffers not from signal deformation neither by reflection at small bodies nor by antenna transmission. In contrast to that, a UWB signal bouncing a point scatterer will sustain a twofold differentiation and further deformations due to the antennas. The unambiguous rage r_{ua} of the UWB radar relates to the signal repetition by:

$$r_{ua} = \frac{1}{2}t_P\,c \tag{4}$$

Recording time: UWB sensors provide, depending on their principle of work, either the impulse response function (IRF) or the frequency response function (FRF) of the scenario under test. The time needed to collect all data for one IRF or FRF (including synchronous averaging of repetitive measurements) we call recording time T_R. Non-stationary test scenarios limit the recording time either to

$$T_R\,\underset{\cdot}{B}_{SC} \leq \frac{1}{2} \tag{5}$$

$\underset{\cdot}{B}_{SC}$ - physical (single-sided) bandwidth of the scenario variation or to

$$T_R \underset{\cdot\cdot}{B} \leq \frac{c}{2|v|} \tag{6}$$

$|v|$ - radial speed of a target

Equation (5) simply indicates the Nyquist theorem telling us that the refresh rate of the measurement $R = T_R^{-1}$ must be twice the bandwidth of the process to be observed. Relation (6) refers to the Doppler-effect. It is evoked from moving targets causing an expansion or compression of the scattered signal. If such signals are accumulated (by correlation or/and synchronous averaging) over a too long duration, they de-correlate resulting in an amplitude degradation of the receiving signal and finally in the loss of the target. Equation (6) should not be confused with Doppler ambiguity which is not relevant for UWB sensing.

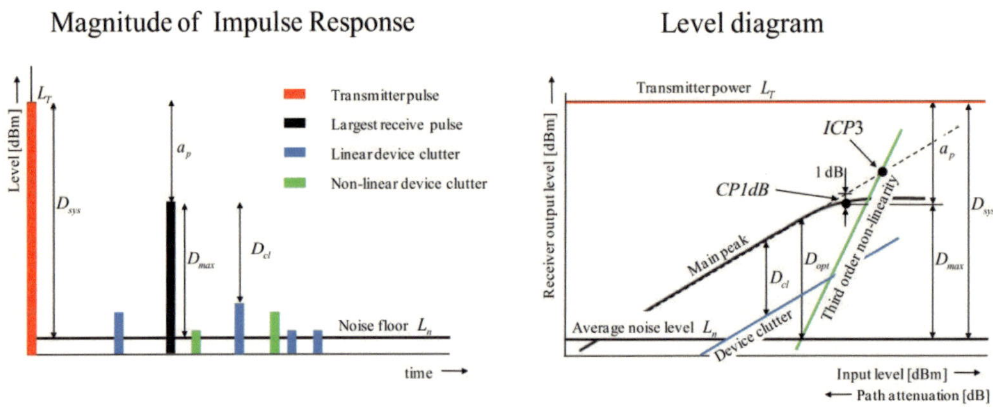

Figure 1. Stylized impulse response (left) and level diagram (right) of an UWB sensor

Dynamic range: Another group of important features relates to the sensitivity of weak signal detection. For illustration, we consider Fig. 1. The illustration on the left-hand side symbolizes the response of a single target. The red line represents the transmitter pulse (or also the auto-correlation function of a wideband CW-signal), and the black line is the target return which should be the only signal visible on the receiver screen. Obviously, we may detect many signal components hampering the detection of weak targets if there are some. These perturbing signals are random noise (electronic and quantization noise) and device internal clutter. It depends on the receiving signal and may be caused by the linear (internal mismatch, cross-coupling, frequency-dependent transmission behavior of electronic compounds) and non-linear effects (e.g. device saturation). Fig. 1 (right) depicts typical dependences of the perturbations from the level of the receiving signal. Based on this, we can derive various dynamic ranges:

- *Clutter-free dynamic range* D_{cl}: It refers to the level difference between receiving signal and the strongest internal clutter peak. D_{cl} determines the sensitivity to detect weak targets in the presence of a strong one, and also the strength of artefacts in radar images.
- *Optimum dynamic range* D_{opt}: Internal clutter caused by linear effects can be removed by sensor calibration as used in network analyzer measurements. A perfect calibration supposed, the erroneous signals are curtailed now by the noise floor and non-linear distortions. Hence, we get optimum conditions for a large dynamic range at the interception of noise and third-order distortion lines.
- *Maximum dynamic range* D_{max}: It is defined by the difference between 1 dB compression point and noise level. Its value gives a hint on the sensitivity to detect moving targets of weak reflectivity. In many cases, the strongest backscatter signals are caused by static objects. As long as the UWB sensor is not moved, these signals and their clutter contributions are stationary so that they may be simply removed by high-pass filtering in observation time. Hence, the detectability of moving targets is only limited by the noise level. The maximum dynamic range can be roughly estimated using the following relation [2]:

$$D_{\max} \approx \frac{2\,\eta_r\,T_R\,V_{1\mathrm{dB}}^2}{k\,T_0\,CF^2\,F\,R_0} = \frac{3\,\eta_r\,T_R\,\underset{..}{B}\,2^{2\,ENOB}}{CF^2} \tag{7}$$

η_r - receiver efficiency; T_R - recording time; $V_{1\mathrm{dB}}$ - input voltage at 1 dB compression point (before correlation); k - Boltzmann constant; T_0 - temperature; CF - crest factor; F - noise factor; R_0 - receiver input impedance; $\underset{..}{B}$ - receiver bandwidth; $ENOB$ - effective number of bits (before correlation).

The left part of eq. (7) applies performance parameters of analog receivers while the right part deals with the global effective number of bits merging the performance of analog and digital receiver components.

- *System performance* D_{sys}: It relates the transmitter level to the noise level. Hence, it is given by the maximum dynamic range and the attenuation of the strongest transmission path.

Time and frequency errors: While above mentioned device characteristics refer to ordinate quantities of a signal representation, the following features quantify the quality of the abscissa representation, i.e. the time or frequency axis. Related to this, we can observe systematic deviations like non-linear frequency or time axis representations resulting in non-equidistant sampling and distortions of frequency-time conversions. Random errors of the time or frequency axis representation, we call jitter or phase noise in the case of short time variation and drift for long term variations. Jitter (respectively phase noise) causes signal-dependent noise which is elevated at signal edges and disappears at flat signal parts.

Jitter limits the performance of super resolution techniques and reduces the sensor sensitivity to detect weak scattering targets in the vicinity of strong reflectors.

Efficiency: The term efficiency can be seen under different aspects. We will consider three of them here.

Receiver efficiency η_r (see also (7)): The receiver efficiency describes the capability of the receiver to exploit the incident signal energy. As the receiving signals are usually quite weak due to the restrictions of transmission power, one has to attach great importance to the receiver efficiency. It is determined by losses in the receiver front end, e.g. the insertion loss of filters or conversion loss of mixers or sampling gates. However, dead times for energy accumulation due to filter settling, incomplete data capture by reason of sub-sampling or incomplete exploitation of captured date due to serial instead of parallel data processing are much more important. Thus, the efficiency of recent UWB receivers is often reduced to values below 1 ‰ or even less which provides some potential for further improvements.

Figure of Merit FoM : In general terms, the Figure of Merit expresses the expense of energy which is required to achieve a certain effect. Two examples shall illustrate the approach. The first one deals with a Nyquist analog-to-digital converter which is aimed to digitize data with a certain rate f_s . An obvious definition of the Figure of Merit can be:

$$FoM_{flashADC} = \frac{P}{2^{ENOB} f_s} \left[\text{J/conversion}\right] \tag{8}$$

P - power dissipation of the ADC; $ENOB$ - effective number of bits of the ADC; f_s - sampling rate

Typical FoM-values for high speed ADCs are to be found at about 10 pJ/conversion. Hence, the power requirement of a 6 bit ADC @ 10 GHz is in the order of 6 W.

The second example relates to an amplifier whose FoM-value is expressed by:

$$FoM_{ampl} = \frac{P}{g\, \underset{\cdot}{B}\, CP_{1dB}} \quad \left[\text{Hz}^{-1}\right] \tag{9}$$

P - power dissipation of the amplifier; g - power gain in linear units; $\underset{\cdot}{B}$ - bandwidth; CP_{1dB} - 1 dB compression point in linear units.

The FoM-approach can be extended to further electronic components and numerical algorithms as well. We can conclude two things from FoM-philosophy. Firstly, the designer of an electronic sub-system or algorithm has to achieve a reasonable small FoM-value with his design. Secondly, the designer of the whole system gets some hints on the feasibility of his system conception and the scope of its features if the corresponding FoM-values are known.

Data throughput: UWB sensors provide lots of data particularly if they are assigned for MIMO-systems and high measurement rate. In order to conserve energy, memory space and data transmission capacity, the sensors should not provide unnecessary data. We have six basic options to reduce the data throughput:

- The data should be captured close to the Nyquist rate.
- The length of the measured impulse response should not be much longer than the settling time of the scenario under test.
- Synchronous averaging (if appropriate) should be performed immediately after data capture.
- A short word length of digitized data should be kept by avoiding high crest factor signals.
- Stationary data should be removed by feedback sampling or digital filtering immediately after data capture (see also chapter 3.3.4 in [2]), and
- Sparse or compressive sampling [3] should be performed. However, this point will not be considered here as it would go beyond the scope of this chapter.

Without going into detail, we would like to mention at least some further aspects that influence the performance of sensor operation, too. They concern interference issues like robustness against jamming and low probability of intercept (LPIR- low probability of intercept radar).

The performance figures summarized above are the basis for deciding on a certain sensor configuration for a specific application. In what follows, the most popular UWB sensor principles will be tabulated and assessed with respect to the introduced performance figures.

2.2. Principles of UWB-sensors

We divide the UWB sensor principles into two groups. While the sensors of the first group generate sounding signals of large instantaneous bandwidth, the devices belonging to the second group deal with narrowband signals swept over a large bandwidth. A thorough analysis of the different sensor concepts of both groups including a reference list can be found in [2]. Here, we will only give a short summary to get an impression of the most common sub-components of UWB sensors and to understand the advantages and disadvantages of the various principles.

2.2.1. Sensors of large instantaneous bandwidth

There are several UWB approaches known exploiting signals of large instantaneous bandwidth. Usually, they are denoted according the sounding signal applied by the sensor. Typical representatives of this signal class are:

- sub-nanosecond pulses
- very wideband pseudo-noise codes

- multi-carrier signals (also assigned as multi-sine), and
- white random noise.

By assumption, these signals have a bandwidth in the GHz range requiring often Nyquist rates of the measurement receivers above 10 GHz. Disregarding the device costs, this is hardly to achieve with the limited power budget and the restricted means of data handling (see section 2.1 – *Figure of Merit* and *Data throughput*) which a sensor usually has at its disposal. Hence, all these devices must reduce their data rates at the expense of receiver efficiency, which is reflected by a reduced dynamic range $D_{\max}$ (see(7)). The data rate reduction is either achieved by sub-sampling or by serializing the data recording.

Fig. 2 refers to three possible device conceptions for illustration. The two upper approaches require periodic sounding signals. Here, the signal shaper may be a pulse generator, a binary PN-generator or an arbitrary waveform generator. The most often found device implementations apply sub-nanosecond pulse generators. Indeed, the concept allows the implementation of very cost-effective and power saving sensors. However, their system performance often suffers from reduced dynamic range due to the large crest factor of the sounding signal (compare (7)); they do not provide jitter suppression (see also sub-chapter 2.3) and they are not robust against jamming. Wideband PN-generators are an interesting alternative to pulse generators since they provide powerful signals of low magnitude (i.e. of low crest factor). Arbitrary waveform generators are able to provide signals which can flexibly be adapted to the measurement problem. However, they are quite expensive, power hungry and limited with respect to the bandwidth. Hence, they have not been found in practically applicable sensor concepts recently.

Sub-sampling receiver: It is the most often applied UWB concept. It supposes periodic sounding signals (t_P - signal period). Typically, the measurement signals are captured by sequential sampling, providing one data sample per period whose time position is stepwise shifted over the whole signal. The actual sampling interval is $t_P + \Delta t$, while the equivalent sampling interval which has to meet the Nyquist criteria is Δt. Newer concepts apply interleaved sampling permitting higher sampling rates since more than one point per period is taken. The classical concept of time shift control uses the fast ramp-slow ramp approach which, however, tends to non-linear time axis representation, sampling jitter and time drift. A second method deals with two stable sine wave generators (e.g. Direct Digital Synthesizers of slightly different frequency ($f_1 = t_P^{-1}$; $f_2 = (t_P + \Delta t)^{-1}$). This reduces time drift and avoids time axis non-linearity. However, it still keeps the sampling jitter quite high since the trigger events launching the sounding pulse and activating the sampling gates are based on relative flat edges of the two sine waves of (comparatively low) frequency f_1 and f_2. Timing control based on digital counters for coarse timing exploits steep trigger edges improving the jitter performance. Then, the fine tuning is typically done by programmable delay chips which consist of hundreds of delay gates. As these gates are not absolutely identical, the delay line cannot ensure equidistant sampling.

Furthermore, the delay time depends on temperature, and the huge number of gates consumes plenty of energy.

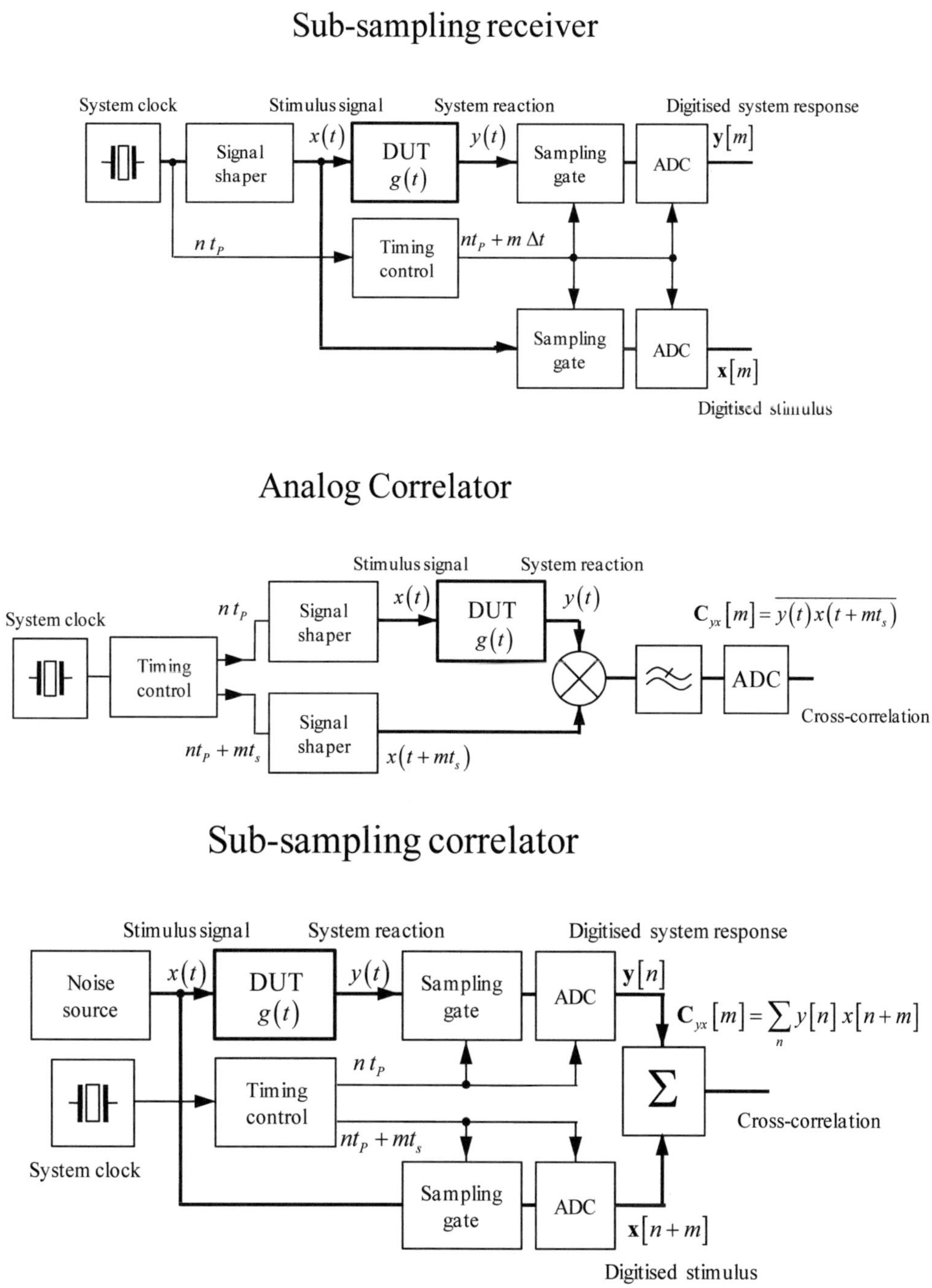

Figure 2. Three possible sensor structures exploiting signals of large instantaneous bandwidth.

Analog correlator: Due to the lag of programmable analog wideband delay lines, one applies two wideband sources (pulse or PN-sequence) providing two identical signals which are shifted in time. The time shift may be controlled by the same approaches as mentioned above. One of these signals stimulates the DUT, and the other one acts as reference in a correlator. Even if the mixer and the integrator do not waste signal energy, the correlator has about the same efficiency as a sequential sampling receiver as long as one does not deal with parallel correlation stages. We can find from eq. (7) that the correlation principle will provide the best dynamic range due to the large time-bandwidth product. But this benefit will be gambled away if sounding signals of large crest factors are applied.

Sub-sampling correlator: Here, we can use also random noise as stimulus. The time lag between measurement and reference signal is performed by shifting the sampling time as explained before. The correlation is done in the numerical domain. The approach is quite time consuming since the averaging time must be high in order to achieve a stable estimation.

2.2.2. *Sensors of narrow instantaneous bandwidth*

Strictly spoken, such sensors do not belong to UWB systems but they are doing the same job as real UWB devices if they are applied for sensing. Hence, they are worth being considered.

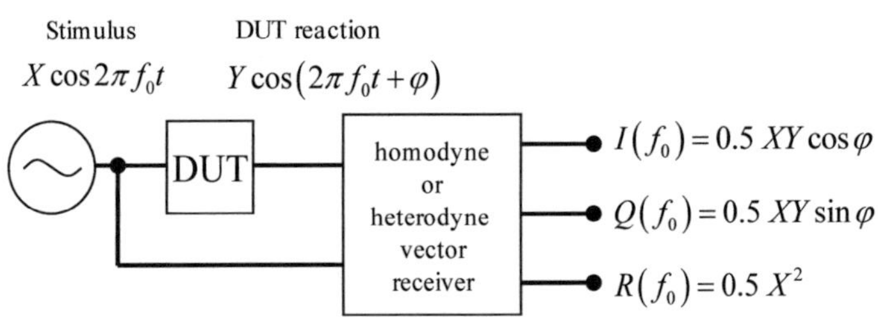

Figure 3. Principle of sine wave measurement.

Fig. 3 depicts the very basic principle. The signal source is a sine wave generator which steps or sweeps the signal frequency over the wanted bandwidth. Depending on the requirements (signal purity, frequency stability, frequency axis linearity, settling time etc.), free-running VCO's, synthesizers or DDS-circuits are in use. The receivers are based on homodyne or heterodyne down-conversion providing the complex valued frequency response function of the DUT:

$$\underline{G}(f) = \frac{I(f) + jQ(f)}{R(f)} \xrightarrow{IFFT} g(t) \tag{10}$$

which can be transformed via IFFT into the impulse response function. Simple implementations (e.g. many FMCW-radars) abstain from vector receivers. They only deal with the in-phase component.

Measurement principles applying sine waves provide the best suppression of noise and harmonic distortions due to narrowband filtering before signal capture. Their receiver efficiency tends to one as long as the settlement of resolution filters and signal source are negligible against the recording time. Hence, such devices often suffer from long measurement duration which leads to a strong range-Doppler coupling. The recording time can be reduced either by simultaneous measurements at different frequencies [7] (requiring complex parallel receiver and synthesizer) or by renouncing the narrowband filters (giving up the sensitivity benefits compared to the wideband approaches).

2.3. UWB pseudo-Noise Concept

Under the assumption of Pseudo-Noise (PN)-codes for sounding, Nyquist sampling for data capture and embedded pre-processing for data reduction, the principle depicted on the top of Fig. 2 seems to be the most promising if one trades the pros and cons of the various UWB principles with respect to monolithic integration, system performance, MIMO-capability and power consumption. Fig. 4 represents the modified structure adapted to the conditions mentioned above. The use of two receiver channels yields the best performance with respect to different application aspects like synchronous measurement of stimulus and reaction signal, opportunity of device calibration, difference or interferometric measurements as well as long term sensor stability.

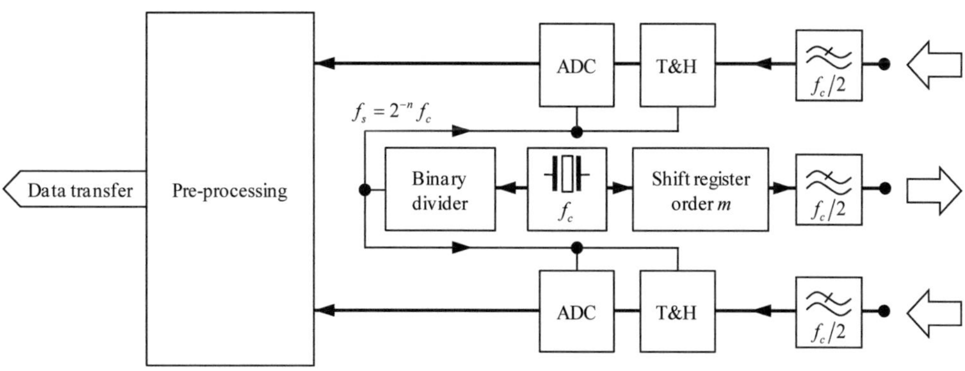

Figure 4. Basic structure of UWB PN-device.

A stable microwave oscillator controls the whole system. It has to provide only a single frequency f_c which allows the use of simple and stable generator concepts. The oscillator pushes a high-speed shift register. Depending on its feedback structure, it provides any binary sequence. Preferentially, M-sequences are used due to their favorable auto-correlation function. Other options could be Golay-codes [8] or Gold-codes if cross-correlation properties are in the foreground of interest.

For the sake of brevity, we restrict ourselves here solely to M-sequences. These codes have a sinc²-spectrum with the first zero located at f_c and they have concentrated about 80% of their energy in the spectral band from DC to $f_c/2$. Hence, we will not provoke a dramatic performance loss if we limit the upper frequency to $f_c/2$. On the contrary, the band limitation avoids disproportionate growth of noise against ever decreasing signal power.

The band limitation to half the clock rate touches several performance-relevant issues:

1. We need a sampling rate of at least f_c in order to meet the Nyquist theorem. In other words, it is sufficient to capture one sample per chip of the M-sequence. As the M-sequence is periodic, we can do this by sub-sampling. It is easy to show that a binary divider is sufficient for timing control since the number N of chips in the sequence is always one less than a power of two ($N = 2^m - 1$). If the order m of the shift register and the order n of the binary divider are identical, we have sequential sampling. For $n < m$, one speaks about interleaved sampling which takes more than one sample within one period.
2. Both, the signal edges of the microwave clock f_c as well as of the sampling clock f_s are quite steep. Hence, the trigger events activated by them are robust against jitter and drift.
3. The time axis of the receiver is defined by the sampling clock f_s. This clock originates from a stable RF-generator and a digital frequency divider which has to run trough all its states before it can launch a new impulse. Hence, any internal deviations between the involved flip-flops have no effect on the divided signal. Therefore, apart from the remaining jitter, we can expect exact equidistant sampling i.e. an absolutely linear time axis representation.
4. The principle of interleaved sampling allows the sampling rate to be varied by keeping the sensor concept. Thus, one can reduce the sampling rate in favor of reduced power consumption and device costs or it can also be increased to improve the receiver efficiency η_r depending on the development state of high-speed electronics.
5. Nyquist sampling provides the lowest possible data throughput[1] without violation of sampling theorem.

The embedded pre-processing is mainly aimed at data reduction by synchronous averaging (often the measurement rate is much higher than required by the time variance of the test scenarios), static background removal or signal transformations. It should, however, be respected that impulse compression (in order to get the impulse response) performed at this point will increase the data throughput toward the main processor since the word length of the data samples increases.

[1] Here, we refer to general measurement conditions. We disregard sparse sampling which is largely dependent on the measurement objects.

The sensor principle depicted in Fig. 4 is basically also able to deal with short sub-nanosecond pulses. However, this would greatly degrade the performance of the system which is largely determined by the amount of signal energy accumulated in the receiver. In the case of pulse signals, this requires amplifiers of high compression points and high resolution ADCs since the whole signal energy is concentrated in a short moment. Furthermore, the measurement object may be exposed to strong fields in the case of near-field measurements. The application of PN-codes avoids all these flaws since it carries enough energy even with small signal magnitudes. As the impulse compression (leading to high crest factor signals) is performed in the digital domain, the analog sensor components and test objects are spared from high voltage peaks.

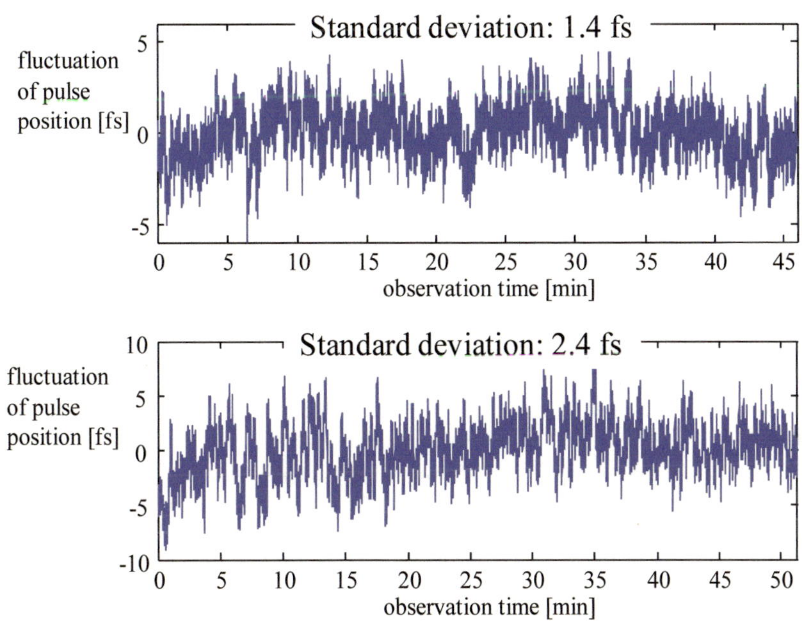

Figure 5. Stability of time position measurement. The delay time of a 50 cm long RF cable was measured at constant temperature. The RF clock was provided either by a sophisticated sine wave generator (SMP04 from Rhode & Schwarz) (top) or by a free running DRO (bottom).

It is well known that the impulse compression of time-extended wideband signals largely improves the dynamic range. As shown in [2] (chapter 4.7.3), it also reduces the jitter susceptibility. The impulse compression distributes the jitter power evenly over the whole signal like additive noise. However, a noise increase above the "natural" level of electronic and quantization noise cannot be observed since the jitter-induced perturbations remain quite low due to the measures described above. Hence, the edges of the impulse response of a DUT measured by the PN-principle are not affected by jitter as usually in pulse measurements. This favors the PN-sensor concept for applications dealing with super resolution techniques or micro-Doppler problems, particularly if weak scattering targets are overwhelmed by strong ones. Chapter11 *ultraMedis* gives some examples of related problems, and Fig. 5 illustrates the achieved short-time stability of an M-sequence sensor having a bandwidth of about 8 GHz. The short-time variance of the pulse position measurement was in the lower fs range corresponding to a distance variation below 1 μm.

The simple timing concept of the PN-sensors enables the implementation of large MIMO-arrays at which the number of cascaded measurement units is basically not limited. The principle is shown in Fig. 6. However, the data handling will be increasingly demanding with a rising number of channels. In a typical operation mode, the transmitters are sequentially activated while the receivers of all channels work in parallel. Some details of implemented MIMO-systems can be found in chapter 11 *ultraMedis*.

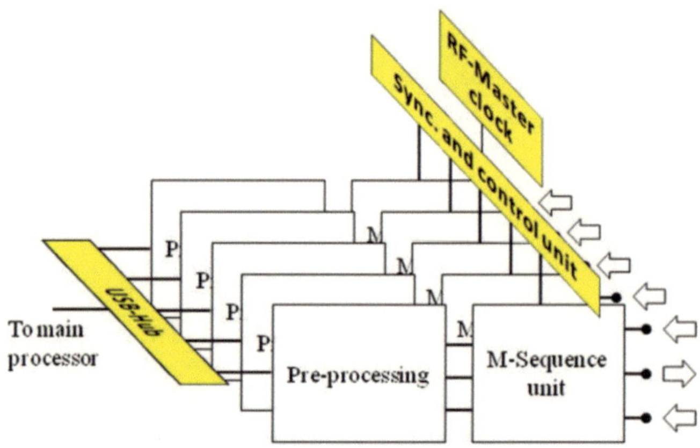

Figure 6. Creation of a MIMO-system by cascading M-sequence sensors.

The receiver of the UWB PN-sensor applies sub-sampling for data capture. Hence, its efficiency gives some potential of further improvements. This would, however, be connected with a considerable increase of the sampling rate f_s. As we can see from (8), the elevation of the sampling rate has to be done at the expense of the ADC resolution since the FoM-value is primarily fixed by the semi-conductor technology, while the maximum power is limited by the achievable heat transport. However, simply increasing the sampling rate based on low bit ADCs will not bring any profit with respect to the sensor performance, i.e. the opposite will happen.

As, however, the update rate of UWB PN-sensors is much higher than required by the time variance of the test object, the difference between two consecutive measurements is very low so that low resolution ADCs are sufficient for capturing these deviations. Anyway, this supposes a fast control loop and a (less power hungry) DAC of sufficient resolution which provides the captured signals from previous measurements for reference. Some basic considerations related to this type of feedback sampling can be found in [2]. Details of the layout and implementation of related sub-components are discussed in sections 3.4, 5 and 6.2.

3. Analog wideband receiver circuits

3.1. Introduction

In order to support system design in its individual stages, different amplifier versions have been developed. Classical ultra-wideband low-noise amplifiers (UWB-LNAs) have been implemented first to ensure early availability and to assess the SiGe BiCMOS technology applied. Then, new receiving components have been considered to address the requirements discovered in system design. This way, a new subtraction amplifier has been made available which allows for practical evaluation of the feedback sampling approach. **Fig. 7** illustrates the way in which LNAs and subtraction amplifiers are used as part of the system.

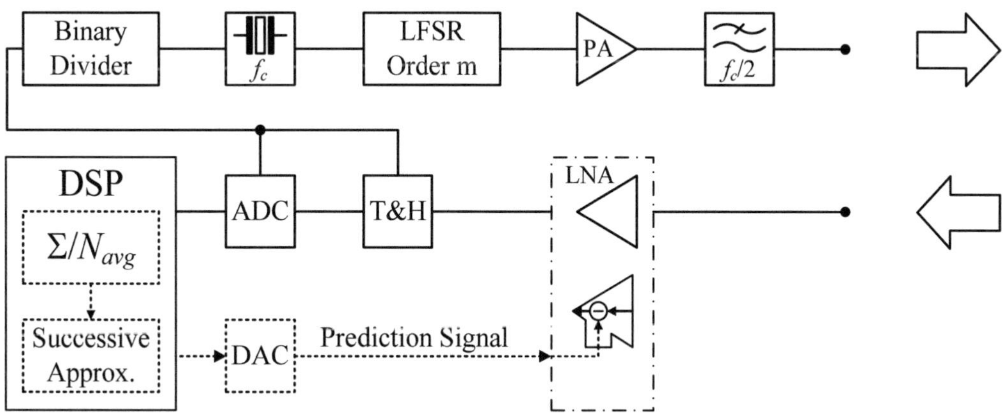

Figure 7. Transmit and receive path of the basic UWB PN-device from Fig. 4 (solid lines) and proposed extension to allow for feedback sampling (dashed lines – see subsection 6.2.3). Either the LNA component or the amplifier with integrated signal subtraction capability are present (dash-dotted box) according to the system state at which it is focused.

In this figure, the basic M-Sequence system is indicated by solid lines. Within this system, classical UWB-LNAs (the LNA component in the dash-dotted box) are used to support the way of operation presented in subsection 2.3. Possible implementations are covered in subsections 3.2.1 to 3.2.3. In addition, a system extension is indicated by the dashed lines, which tries to overcome the limitations imposed by the analog-to-digital converter. For this reason, a digitally calculated prediction signal is provided by a digital-to-analog converter and subtracted from the receive signal to generate a difference signal of highly reduced signal swing close to sampling time instances (see sections 5.1 and 6.2.3). Subtraction is performed by an amplifier with integrated signal subtraction capability, practical realizations of which are presented in sections 3.4.1 and 3.4.2.

The latest subtractor also accounts for low feeding point impedance, which is imperative for the conceptual design of dielectrically scaled antennas. The use of such a device is intended by the collaborative project *ultraMedis* (see chapter 11). For establishing common interface definitions, the performance of individual components has to be characterized by appropriate metrics. While those are well established for single-ended arrangements, this is not the case with the noise characterization of multiport or differential structures. Hence, a new de-embedding scheme for the noise figure of a differential device has been developed and will be presented.

3.2. LNAs for the basic M-Sequence system

Within the basic M-sequence system, low noise amplifiers (LNAs) perform the classical task of adapting the input signal swing to the dynamic range of the analog-to-digital converter (ADC) while adding only a minimum amount of excess noise and providing reasonable power-match conditions. If a high-gain LNA is used, the system also is less sensitive to noise added by succeeding components. Gain, in turn, is limited by the required linearity, and an appropriate compromise with respect to all counteracting requirements has to be found. While trading one parameter against the other, the conditions set by the technology have to be considered for the individual LNA. Resonant tuning and resistive feedback topologies are predominantly used in literature for mapping specifications to circuit designs. Though a resonant solution is favored in [11], the authors do admit that the parasitic base resistance of bipolar transistors causes a large contribution to output noise. Thus, the advantage of extraordinary low noise figures enabled by narrow-band resonant designs as compared to designs matched by resistive feedback is relativized. High magnetic field gradients potentially encountered in some of the applications, and the limited ability to use shielding as identified in [12] also make the use of inductors questionable. Therefore, resistive feedback solutions have been preferred as their use is additionally accompanied by notable die size advantages. While the design of individual amplifiers will be covered in the following subsections, general guidelines can be taken from standard textbooks. In [13], for example, the impact of feedback on noise and impedance match is analyzed in detail. For the design, too, a simplified version of the bipolar transistor small signal equivalent circuit model with additional noise sources as presented in [13] has been used.

3.2.1. Multiple resistive feedback LNA

One of the implemented amplifier versions which have been inspired by classical UWB-LNAs is depicted in Fig. 8. According to [10], this is a popular wideband amplifier topology often referred to as *Kukielka* amplifier. Due to numerous results reported in literature for this kind of amplifier, the impact of technology on circuit performance can be assessed. For comparison, especially SiGe implementations as presented in [9] are valuable. The main characteristics of this amplifier are set by the core circuit which comprises transistors Q_1 to Q_3. For analysis, the Darlington pair $Q_2 - Q_3$, which is used for gain-bandwidth extension of the second stage, is treated as single compound-transistor Q_{23}. Within the simplified circuit thus obtained, four feedback loops can be identified. For proper biasing, series-series feedback is applied to Q_1 and Q_{23}. By this measure, the bandwidth of both stages is improved. In turn, input and output impedances of the amplifier are increased rather than decreased as required for input and output power matching.

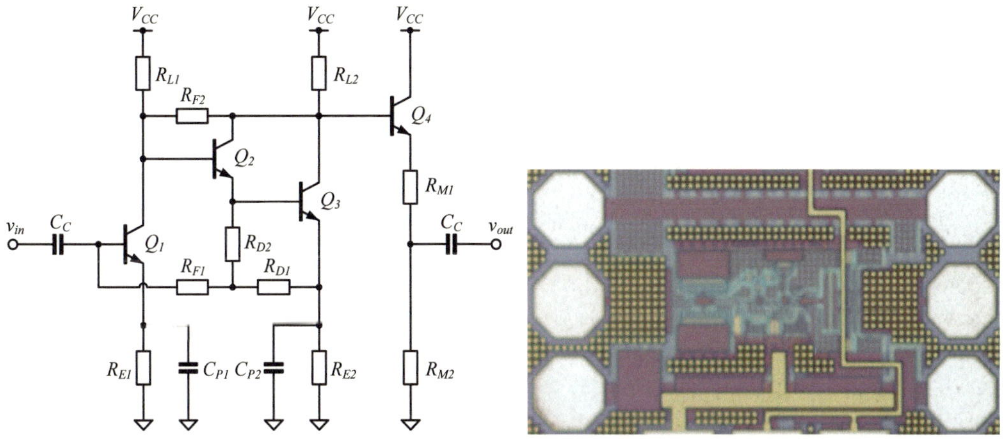

Figure 8. Schematic diagram (left) and chip photo (right) of the multiple resistive feedback LNA - the die area is only 530 μm × 280 μm.

Hence, local shunt-shunt feedback is applied to Q_{23} in order to reduce the output impedance. Finally, to enable input power matching, global shunt-series feedback from the emitter node of Q_{23} to the base of Q_1 is applied. According to [10], this configuration tends to present an overdamped response. For this reason, peaking capacitors C_{P1} and C_{P2} are inserted to improve the frequency behavior of the amplifier. The addition of peaking capacitors might, however, impair stability which has to be diligently observed during design for this reason as stated in [13]. In the same publication, an approximate calculation of the noise figure (NF) for this topology reveals that the latter is dominated by the noise properties of Q_1 as long as $R_{F1} \gg R_S$ and $R_{F1} \gg R_{eq}$. R_{eq} is related to the equivalent input noise voltage source v_{Q1} of transistor Q_1 by $R_{eq} = v_{Q1}^2 / (4kT)$. For this reason, the selection of transistor Q_1's bias current to yield optimal current density with respect to its noise properties should be a first step in design. After this initial step, one of the directed design procedures given in [10] or [9], respectively, can be used for further development. Those are derived from input and output power match conditions as well as from pole positions. In order to account for the characteristics of the Darlington pair transistors Q_2 and Q_3, substitutions $g_{m,23} \rightarrow g_{m,3}$ and $\omega_{T,23} \rightarrow 2 \cdot \omega_{T,3}$ can be applied according to [10]. In Fig. 8 (left), an emitter follower has been attached for further improving the output power matching in the technology used. Fig. 8 (right) shows the chip photograph of the implemented LNA without pads used for biasing. The dimensions of the displayed die area are $530\mu m \times 280\mu m$ only, which confirms the advantage resistive feedback amplifiers provide in view of die area as compared to resonant solutions. For the accurate characterization of the fabricated amplifier, on-waver measurements have been performed using a PM 8 probe station of Süss MicroTec (now acquired by Cascade Microtech). Due to the measurement arrangement, losses preceding and succeeding the device under test (DUT), i.e. the amplifier, cannot be avoided. However, their impact on the scattering parameters of the DUT can be eliminated by proper calibration of the network analyzer applied. Also, the spectrum analyzer with noise figure measurement personality in use allows for the specification of losses preceding and succeeding the DUT which are compensated for during measurement in this case[2]. Measurement results obtained in this way are shown in Fig. 9 together with results from post-layout simulation. Initially, they have been presented in [14].

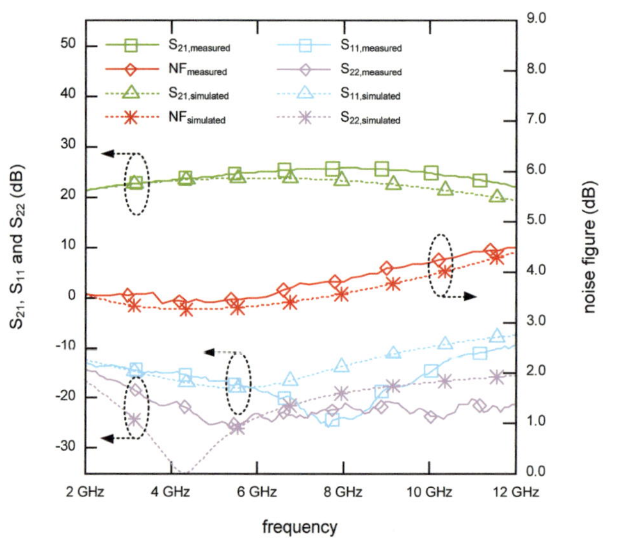

Figure 9. Measurement results of the multiple resistive feedback LNA. (The arrows point to the related axis scaling.)

From Fig. 9, peaking in the gain curve progression of the measurement data can be observed as compared to the results from post layout simulation. The maximum difference appears at about 8 GHz, which is the frequency at which a notch in measured S_{11} values also appears. In [14] it is thus suspected that this deviation arises due to the interaction of the test set-up with the DUT. In short summary, the results presented in Fig. 9 for the low-cost technology applied map pretty well the state-of-the-art performances reported in literature at that time. For a more detailed analysis, the reader may consult [14].

3.2.2. Active Feedback LNA

This amplifier has been inspired by the work presented in [15]. Due to the characteristics of the applied technology, certain adaptations have been required, though. Fig. 10 (left) shows the schematic diagram of the final design.

[2] Rhode & Schwarz ZVA-24 and FSQ-40 have been used for the measurements.

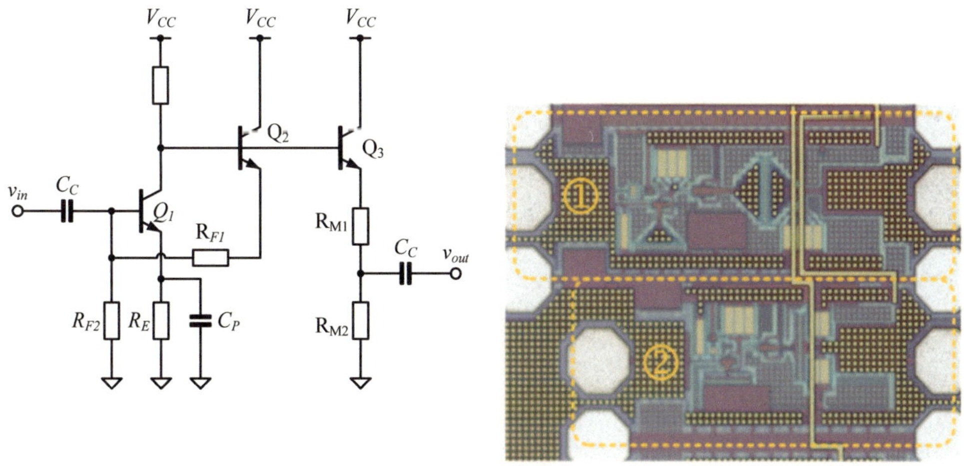

Figure 10. Schematic diagram (left) and chip photo of both active feedback LNA versions (right). The die areas are (1) $630\mu m \times 280\mu m$ and (2) $530\mu m \times 280\mu m$, respectively.

It is a one-stage amplifier with resistive emitter degeneration to ensure a stable DC operating point and to improve bandwidth. Due to the presence of the peaking capacitor C_P, degeneration is continuously shifted to higher frequencies. As in the case of the multiple resistive feedback amplifier, this technique has to be used with care to ensure that this measure does not impair amplifier stability. Input matching of Q_1 is achieved by feedback via transistor Q_2 as well as resistors R_{F1} and R_{F2}. The advantage of Q_2 is twofold: It improves the isolation between the input node and the output node in forward direction and, according to [15], it helps to enlarge the collector-emitter voltage of Q_1. Thus, the maximum oscillation frequency f_{max} is expected to be increased, and the large-signal behavior is said to be improved. The amplifier according to Fig. 10 was presented in [14] for the first time. Compared to the amplifier in [15], the inductor used for improving the frequency behavior has been abscised from the design while the peaking capacitor has been added. Also, for better output matching, emitter follower Q_3 has been attached. To avoid a lengthy discussion of circuit characteristics, Fig. 10 (left) uses an alternative way to depict the circuit as compared to [14] or [15]. This representation points out the large similarity of feedback paths in both amplifiers shown in Fig. 8 and Fig. 10. Though the actual implementations of the passive feedback networks differ, shunt-series feedback is applied for input matching in both cases, and many design steps can be executed by analogy. Fig. 10 (right) shows the chip photograph of two variants, which have been implemented to assess the impact of layout on circuit performance. In the first version, 90° lead corners are avoided and consecutive 45° lead corners are used instead. By this measure, the average lead length is increased. By contrast, a compact design has been targeted in the second layout version. As discussed in [14], results do not differ significantly as long as the length of the lead connecting the RF input with the first amplifying transistor is kept comparably long. Thus, only results for the second layout version are shown in Fig. 11.

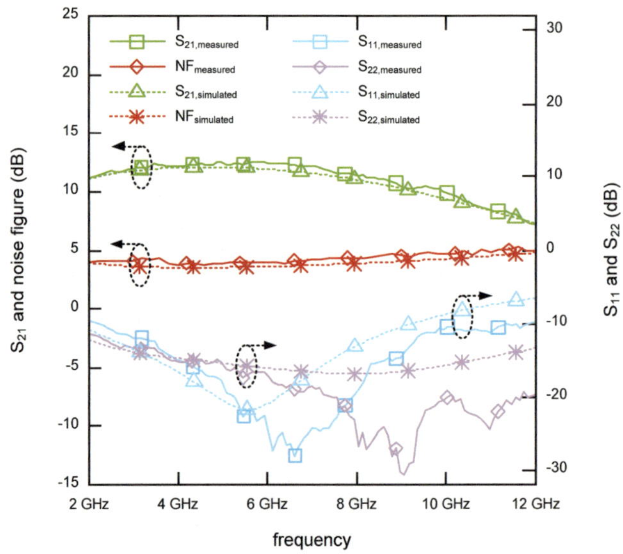

Figure 11. Measurement results of the active feedback LNA. (The arrows point to the related axis scaling.)

The die area occupied by the first layout version is $630\mu m \times 280\mu m$ while the second layout version only occupies $530\mu m \times 280\mu m$ (both excluding DC-pads). Measurements have been performed the same way as explained for the multiple resistive feedback LNA. Compared to Fig. 9, gain is much lower, which is expected due to the single stage nature of the active feedback amplifier. At the same time, the input referred $1\,dB$ compression point is improved notably. A more complete discussion of amplifier characteristics is presented in [14].

3.2.3. Pseudo-differential LNA

Core of the half-circuit shown in Fig. 12 (left) is the cascode amplifier with reactive shunt feedback on the left-hand side. As for the single-ended amplifiers, bias current of this arrangement should be selected due to noise and linearity considerations. While a more detailed analysis of this topology, as presented for the inductively degenerated cascade amplifier with capacitive shunt feedback in [16], might be desirable at this point, limited space for this section does not permit a lengthy discussion. Instead, we allude to the amplifier of [17] from which the topology of Fig. 12 (left) has been derived.

A pseudo-differential amplifier has been implemented to support the development of the basic M-sequence system by adding the capability to use differential circuitry, especially differential antennas. Many of its characteristics are inherited from the topology of [17]. However, emphasis with respect to certain design parameters has been shifted. The most peculiar aspect is the fact that the input matching network used in [17] could be abscised from the design. This modification was enabled by improved input-output isolation due to altered feedback tapping points. Some results of the manufactured chip, a photo of which is contained in Fig. 12 (right), are summarized in Fig. 13. Together with additional topological aspects, they are discussed in [18] in more detail. Measurements to gather those results have been performed on-waver at the PM 8 probe station using ground-signal-signal-ground (GSSG) probes. Similar to the single-ended case, scattering parameters could be determined directly by a calibrated network analyzer. By contrast, only single-ended equipment has been available for noise figure measurement, and it is left to the next subsection to discuss a method applicable to (pseudo-)differential amplifiers.

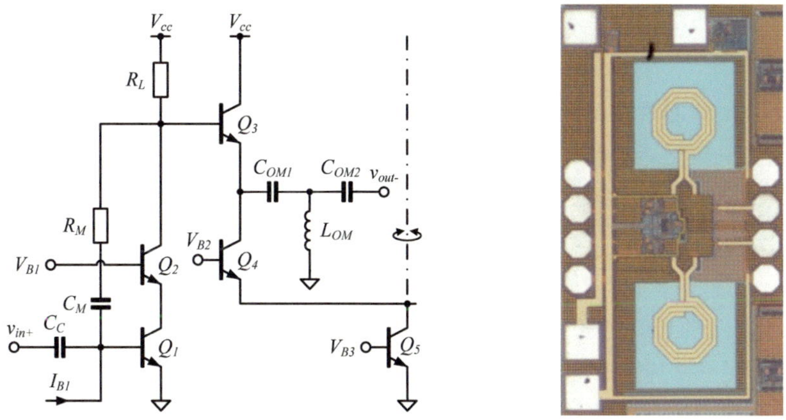

Figure 12. Half-circuit schematic diagram (left) and chip photo (right) of the pseudo-differential LNA.

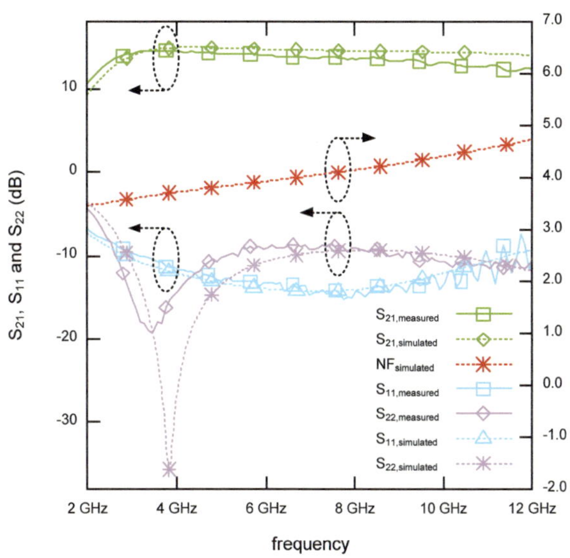

Figure 13. Measurement results of the pseudo-differential LNA. (The arrows point to the related axis scaling.)

3.3. Differential noise figure measurement

Due to space limitations, only a short introduction to this topic will be presented here. While there are alternative methods as presented in [23] and [24], for example, focus will be on a new powerful method proposed in [20]. A convenient definition for the noise figure of a differential (or multiport) device with respect to one of its ports has been given by Randa [22] and is reprinted in (11) with slight modifications:

$$F_i(\mathbf{\Gamma}_k,\mathbf{C}_k)=\frac{\left\{\left[\mathbf{I}-\mathbf{S}_k\mathbf{\Gamma}_k\right]^{-1}\left(\mathbf{S}_k\mathbf{C}_k\mathbf{S}_k^\dagger+\mathbf{C}_{S,a}\right)\left[\left[\mathbf{I}-\mathbf{S}_k\mathbf{\Gamma}_k\right]^{-1}\right]^\dagger\right\}_{ii}}{\left\{\left[\mathbf{I}-\mathbf{S}_k\mathbf{\Gamma}_k\right]^{-1}\mathbf{S}_k\mathbf{C}_k\mathbf{S}_k^\dagger\left[\left[\mathbf{I}-\mathbf{S}_k\mathbf{\Gamma}_k\right]^{-1}\right]^\dagger\right\}_{ii}} \tag{11}$$

The noise figure in (11) is parameterized by the matrix of reflection coefficients seen by the DUT into the ports of connected components $\mathbf{\Gamma}_k$ and the noise correlation matrix of incident noise waves injected by an external source $\mathbf{C}_k$. $\mathbf{I}$ is the identity matrix in (11), the dagger indicates the Hermitian conjugate, and $\mathbf{S}_k$ as well as $\mathbf{C}_{S,a}$ are the scattering matrix and the noise correlation matrix of emergent waves contributed by the DUT, respectively. To apply this definition, $\mathbf{C}_{S,a}$ has to be determined first. Therefore, the differential device has to be embedded into a network of passive components which provide the differential excitation as only single-ended measurement equipment is currently available. This is demonstrated in Fig. 14. The noise correlation matrix of the DUT then has to be de-embedded from the results measured for the component chain.

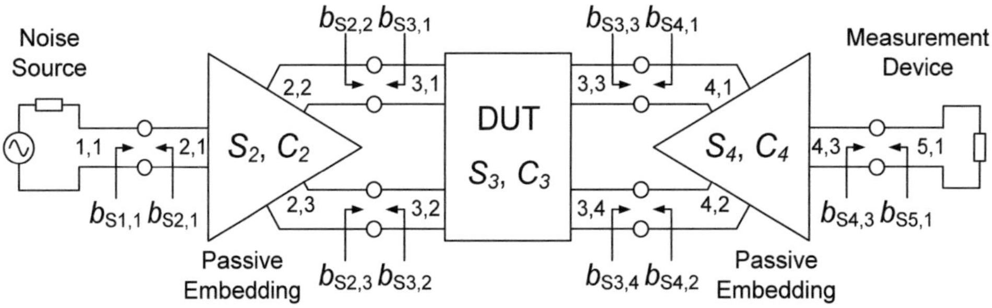

Figure 14. DUT embedded into passive components. $\mathbf{b}_{Sx,y}$ are equivalent generator noise waves caused by component x and ejected from the respective port y.

For this purpose, the noise distribution matrix defined in [21] is a convenient starting point. Multiplied by Boltzmann's constant k and the physical temperature T, it is the correlation matrix of emergent noise waves caused by a passive component which account for all the noise generated within the device. It can be related to the scattering matrix as shown by (12). A short and intuitive proof of this relation is contained in [25]:

$$\hat{\mathbf{N}} = \left(\overline{\mathbf{b}_S \mathbf{b}_S^\dagger}\right) / (kT) = \mathbf{I} - \mathbf{S}\mathbf{S}^\dagger \tag{12}$$

Making use of single-ended noise measurement equipment, the characteristic noise *equation reference goes here*temperature $\hat{T}_{chain}$, which characterizes noise from all the elements of the whole component chain, can also be determined. Thus, we are only left with the problem to describe signal transfer via the component chain to accomplish the goal of de-embedding $\mathbf{C}_{S,a}$. In [20], an approach based on the connection scattering matrix $\mathbf{W}$ has been presented for this purpose, which surpassed the method of [19], because it is applicable without simplifying assumptions. The connection scattering matrix was introduced long ago, and its use for computer-aided circuit analysis has been discussed in [26], for example. For all ports in a component network, it relates the incident to the impressed waves. To enable such matrix representation, incident, reflected, and impressed waves have to be composed to wave vectors $\mathbf{a}$, $\mathbf{b}$, and $\mathbf{b}_S$, respectively, which should be sorted in a component-wise way. For convenience, a component index is assigned to the vector entries, and it is assumed that elements corresponding to the DUT (index k) are placed at the bottom of each wave vector. An element is thus identified by two indices i,j representing the component and the respective port number. With this convention, all complex wave amplitudes can be related by the set of linear equations

$$\mathbf{b} = \mathbf{S} \cdot \mathbf{a} + \mathbf{b}_S \tag{13}$$

where $\mathbf{S}$ is a block-diagonal matrix assembled from the individual component S-parameter matrices. The connections between the single components impose additional constraints on the wave amplitudes. To account for them, the connection matrix $\mathbf{\Gamma}$ is used as in (14):

$$\mathbf{b} = \mathbf{\Gamma} \cdot \mathbf{a} \tag{14}$$

Most often, a common real reference impedance is applied for all components. Then, all entries of $\mathbf{\Gamma}$ are zero except those which refer to connected ports and, thus, are one. In this case, $\mathbf{\Gamma}$ is a permutation matrix with $\mathbf{\Gamma}^{-1} = \mathbf{\Gamma}^{\mathrm{T}}$. From (13) and (14), the incident wave vector $\mathbf{a}$ can be eliminated to get

$$\mathbf{W}\mathbf{\Gamma}^{-1}\mathbf{b} = \mathbf{b}_S, \textit{ with } \mathbf{W} = \mathbf{\Gamma} - \mathbf{S} \;\; \mathbf{W} = \mathbf{\Gamma} - \mathbf{S} \tag{15}$$

Noise generated by the source does not contribute to the characteristic noise temperature of the component chain. The respective entries of $\mathbf{b}_S$ can thus be set to zero. Furthermore, in the case of $\mathbf{S}_1$ containing zero entries[3], the corresponding rows of $\mathbf{W}\mathbf{\Gamma}^{-1}$ should be deleted to avoid rank deficient matrix problems in some of the computations. For convenience, we will refer to the matrix obtained from this operation by $\mathbf{V}$. After additional matrix partitioning, which is required later, (15) then becomes

---

[3] $\mathbf{S}_1$ is a submatrix of $\mathbf{S}$.

$$\begin{array}{c} \\ \\ \\ \mathbf{b}_{S,p} \\ \\ \\ \mathbf{b}_{S,a}\end{array}\begin{bmatrix}\mathbf{b}_{S,2}\\ \vdots \\ \mathbf{b}_{S,k-1}\\ \mathbf{b}_{S,k+1}\\ \vdots\\ \mathbf{b}_{S,n}\\ \hline \mathbf{b}_{S,k}\end{bmatrix} = \begin{bmatrix}\mathbf{V}_{pm} & \mathbf{V}_{pp} & \mathbf{V}_{pa}\\ \mathbf{V}_{am} & \mathbf{V}_{ap} & \mathbf{V}_{aa}\end{bmatrix}\begin{bmatrix}\mathbf{b}_1\\ \hline \mathbf{b}_2 \\ \vdots \\ \mathbf{b}_{k-1}\\ \mathbf{b}_{k+1}\\ \vdots\\ \mathbf{b}_n\\ \hline \mathbf{b}_k\end{bmatrix}\begin{array}{l}\mathbf{b}_m\\ \\ \\ \\ \mathbf{b}_p \\ \\ \\ \\ \mathbf{b}_a\end{array} \tag{16}$$

In (16), submatrix $\mathbf{V}_{am}$ will always be zero, because there is no direct connection from the input to the DUT. Hence, we are left with

$$\mathbf{b}_{S,p} = \mathbf{V}_{pm}\mathbf{b}_m + \left[\mathbf{V}_{pp} \mid \mathbf{V}_{pa}\right]\begin{bmatrix}\mathbf{b}_p \\ \mathbf{b}_a\end{bmatrix}, \tag{17}$$

and

$$\mathbf{b}_{S,a} = \left[\mathbf{V}_{ap} \mid \mathbf{V}_{aa}\right]\begin{bmatrix}\mathbf{b}_p \\ \mathbf{b}_a\end{bmatrix}. \tag{18}$$

Setting $\mathbf{Q} = \left[\mathbf{V}_{ap} \mid \mathbf{V}_{aa}\right]\left[\mathbf{V}_{pp} \mid \mathbf{V}_{pa}\right]^{-1}$ for abbreviation, (19) is obtained from (17) and (18) after some algebra[4]:

$$\mathbf{Q}\mathbf{V}_{pm}\mathbf{b}_m = \mathbf{Q}\mathbf{b}_{S,p} - \mathbf{b}_{S,a}. \tag{19}$$

Finally, since $\mathbf{b}_{S,p}$ and $\mathbf{b}_{S,a}$ arise from different sources and are thus uncorrelated, $\mathbf{C}_{S,a}$ can be determined from (20)

$$\mathbf{Q}\mathbf{V}_{pm}\mathbf{C}_m\mathbf{V}_{pm}^{\dagger}\mathbf{Q}^{\dagger} = \mathbf{Q}\mathbf{C}_{S,p}\mathbf{Q}^{\dagger} + \mathbf{C}_{S,a}$$
$$\text{with } \mathbf{C}_m = \overline{\mathbf{b}_m\mathbf{b}_m^{\dagger}},\ \mathbf{C}_{S,p} = \overline{\mathbf{b}_{S,p}\mathbf{b}_{S,p}^{\dagger}},\ \text{and } \mathbf{C}_{S,a} = \overline{\mathbf{b}_{S,a}\mathbf{b}_{S,a}^{\dagger}}. \tag{20}$$

In (20), noise correlation matrices of single passive components given by the product of kT and (12) are composed to the block diagonal matrix[5] $\mathbf{C}_{S,p}$. $\mathbf{C}_m$ accounts for noise from all

[4] As the right division function of a math program can be used to compute $\mathbf{Q}$, there is no need for an explicit inversion, and a minimum norm solution is obtained for a non-square system.

[5] Note that noise contributions of the output loads, according to [22], should not be considered for NF computation. With respect to the set-up of Fig. 14 this implies $\overline{b_{S,5,1}\, b_{S,5,1}^{*}} \overset{!}{=} 0$, which can be achieved by using the second stage correction of the NF meter.

components of the chain related back to the input. For the set-up of Fig. 14, $\mathbf{C}_m$ is a 1×1 matrix associated with the characteristic noise temperature as shown by (21):

$$\mathbf{C}_m \approx \left[k\hat{T}_{chain} \right] \tag{21}$$

For using (11) in noise figure computations, its parameter matrices $\mathbf{\Gamma}_k$ and $\boldsymbol{C}_k$ still need to be determined. $\mathbf{\Gamma}_k$ contains reflection coefficients, which relate waves injected from the DUT $\boldsymbol{b}_{S,a}$ to those reflected back from the embedding network $\boldsymbol{a}_a$. It follows from inspection that $\mathbf{\Gamma}_k$ is a submatrix of $\mathbf{W}^{-1}$. Focusing on $\boldsymbol{C}_k$, a simple argument leads to some confidence that (22) is a reasonable choice: Assume that the power splitter at the input of Fig. 14 only excites the differential mode. In this case, noise at the input of the DUT is completely correlated. Also, there should be a noise power of kT available at the input of the DUT in differential mode to stay comparable to the standard noise figure definition. Then, due to the properties of mixed-mode transformation, (22) is the evident solution. This is discussed in [20] in more detail.

$$\mathbf{C}_k = \begin{bmatrix} 0.5kT & -0.5kT & 0 & 0 \\ -0.5kT & 0.5kT & 0 & 0 \\ 0 & 0 & 0 & 0 \\ 0 & 0 & 0 & 0 \end{bmatrix} \tag{22}$$

Thus, all inputs required for (11) are determined, and the noise figure with respect to a certain output port can be calculated. Instead of a physical port, also a logical port can be considered. For this purpose, matrices in the numerator and the denominator of (11) have to be transformed by an appropriate transformation. In view of Fig. 14, mixed-mode transformation provides noise power spectral densities in the differential mode at the selected output of the DUT for both cases: Noise generated by the DUT and the input sources, as well as noise generated by the input sources alone. Their ratio finally determines the differential noise figure of the device. This approach has been applied to the pseudo-differential amplifier shown in Fig. 15. The result is contained in Fig. 16 together with the noise figure measured from one signal branch, which is given for comparison. In the measurements, the losses of the probe heads have been appropriately taken into account.

3.4. Solutions for the feedback-sampling approach

The introduction of the feedback-sampling concept by system design spawned the requirement of signal subtraction at the input of the receive path. Hence, the demand for new components equipped with two inputs arose - one for the RF signal taken from the antenna, and another one for a digital prediction signal provided by the signal processing via a digital-to-analog converter (DAC). In theory, subtraction results in an output signal of highly reduced voltage swing to which the analog-to-digital converter (ADC) used for signal acquisition will be exposed. To confirm this theory in practice, two versions of an input subtractor have been implemented and will be presented in the following subsections.

3.4.1. *Pseudo-differential feedback-sampling amplifier*

Similar to the design of amplifiers for the basic system, the development of the feedback-sampling amplifier has been guided by the assumption that the receive signal in the RF path is rather week and sufficient amplification has to be provided, while the least amount of excess noise should be added. Hence, the amplifier of Fig. 12 has been reviewed and was deemed to be suited as RF input stage of the new topology. In Fig. 15, it can be identified in the dashed box on the left-hand side. Some adjustments - especially with respect to the values of components in the feedback network - had to be made, though, as the application required a shift in the covered frequency range. At that time, the prediction signal had to be provided by a current steering DAC and can be assumed to be of rather large signal swing. Hence, no amplification is provided for this signal. Promoted by the nature of the prediction signal, the required signal subtraction is performed in the current domain. In Fig. 15, two current mirrors inject their output signals into a common output node for this purpose. While this implies signal addition instead of signal subtraction at first glance, signal addition can be turned into signal subtraction by simple sign inversion, which is enabled either by exploiting the properties of the pseudo-differential amplifier structure itself or by sign selection of the prediction signal in the digital domain. The special current-mirror arrangement of Fig. 15 has been chosen for balancing the maximum output powers which is important to ensure that the signal from the DAC input can cancel the signal from the RF input. Linearity of both signal paths, transconductance from the DAC input to the common output and maximum output currents of the DAC have to be harmonized to account for this requirement.

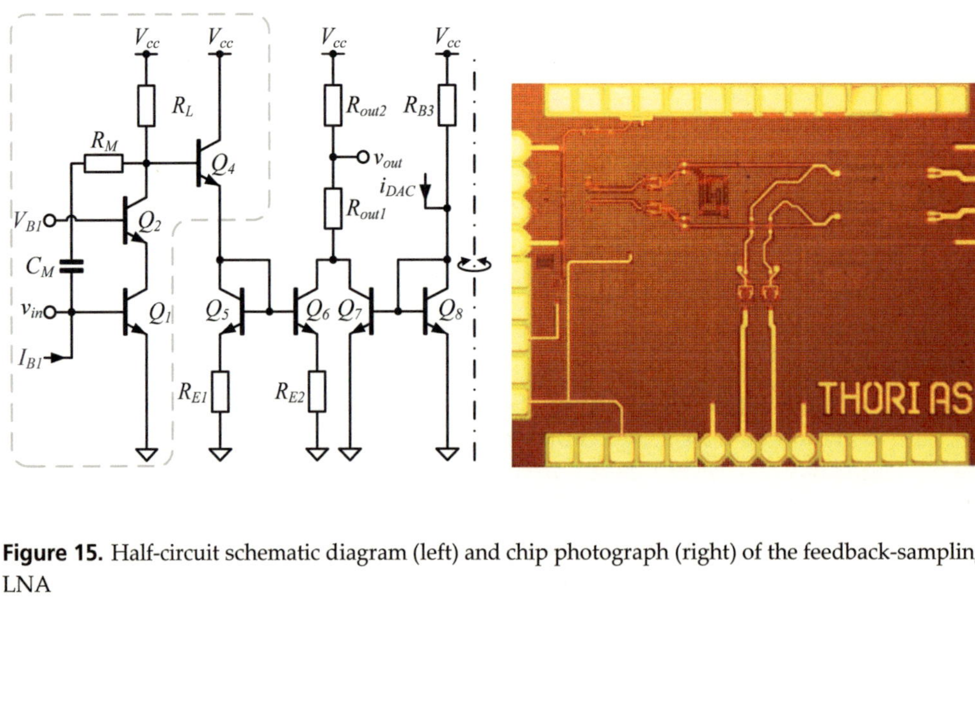

Figure 15. Half-circuit schematic diagram (left) and chip photograph (right) of the feedback-sampling LNA

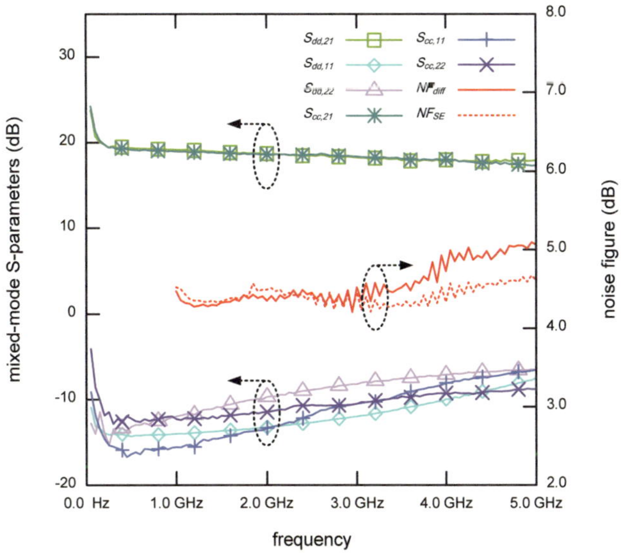

Figure 16. Measurement results for the RF path of the feedback-sampling amplifier. (The arrows point to the related axis scaling.)

This topology has been characterized in detail. Results for the RF signal path are presented in Fig. 16.

As explained for the pseudo-differential amplifier in section 3.2.3, the PM 8 probe station equipped with GSSG-probes can be used to examine an individual differential signal path. In order to detect the mixed-mode parameters given in Fig. 16, the device has been exposed to true mode excitation provided by the network analyzer, while calibration data have been applied to compensate for losses caused by the test set-up. Noise characterization can be performed by the method presented in section 3.3 and has been discussed in [19][20]. A remarkable feature of the results of this measurement is the fact that the differential noise figure NF_{diff} does not coincide with the noise figure measured from one signal path[6] NF_{SE}. Especially at higher frequencies, a large deviation occurs. In [20], this is explained by cross-talk caused by parasitics. So, the use of differential de-embedding schemes is recommended instead of a single-ended noise figure measurement from one signal branch. However, in view of the aim to assess the capabilities of the feedback-sampling approach, the signal subtraction itself is most interesting. In [27], we presented different test set-ups for verification purposes. Fig. 17 shows four representative results which confirm the ability to cancel the RF signal by an appropriate prediction signal.

[6] In Fig. 16, NF curve progressions start at 1 GHz because this is the lower corner frequency of the hybrid couplers used for measurement.

Those measurements were obtained from a test set-up incorporating the PM 8 probe station with GSSG-probes, in which two signal generators[7] synchronized by a frequency standard provided the input signals to both inputs at appropriate power levels via two hybrid couplers. The differential output signal of the DUT was recombined by a third hybrid coupler and displayed by a signal analyzer[8]. In Fig. 17, no loss compensation is applied and results are clipped to 100 kHz span. Two cases can be distinguished: First, the digital prediction signal has been switched off (DAC_{off}) and only the RF signal has been present at the inputs. Then, also the prediction signal has been applied (DAC_{on}) and a notable reduction in output signal power can be observed for all frequencies.

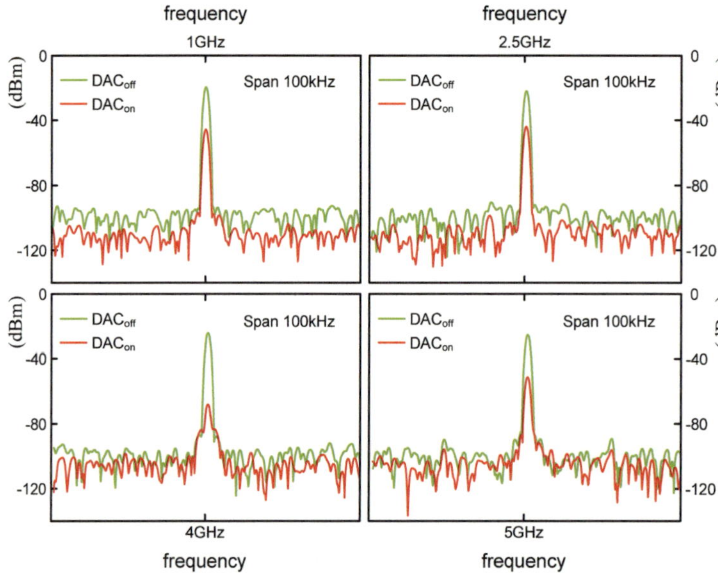

Figure 17. Signal subtraction enabled by the feedback-sampling amplifier.

3.4.2. Subtractor with Low Impedance antenna interface

The feedback-sampling amplifier of the preceding subsection is expected to perform well as long as the assumption of (reasonable) small input signals is justified. A key requirement in the feedback-sampling concept is linearity preceding the signal subtraction in order not to distort the zero crossings which are sampled by the analog-to-digital converter. However, as soon as array operation is considered, antenna cross-talk is likely to violate this assumption. In addition, a dense antenna array requires the antennas to have small outer dimensions. This can be achieved by dielectrically scaling the antennas, which - in turn - leads to a low (7 Ω) feeding point impedance. The latter has to be interfaced by the subtraction circuit. The topology shown in Fig. 18 is a first approach towards an analog subtractor which provides appropriate single-ended inputs to interface with both, a dielectrically scaled antenna and

[7] Rhode & Schwarz SMJ100A

[8] Rhode & Schwarz FSV

the DAC. In this implementation, noise figure is traded against linearity, as input signals close to 0 dBm might occur. For its implementation, component parameters have been determined by a semi-automated procedure, in which the input stage - a common base configuration - was optimized with respect to input power matching, while an upper bound for NF_{min} was respected and noise matching was clearly observed. As in the case of the feedback-sampling amplifier according to Fig. 15, signal subtraction is performed in the current domain using a common output node.

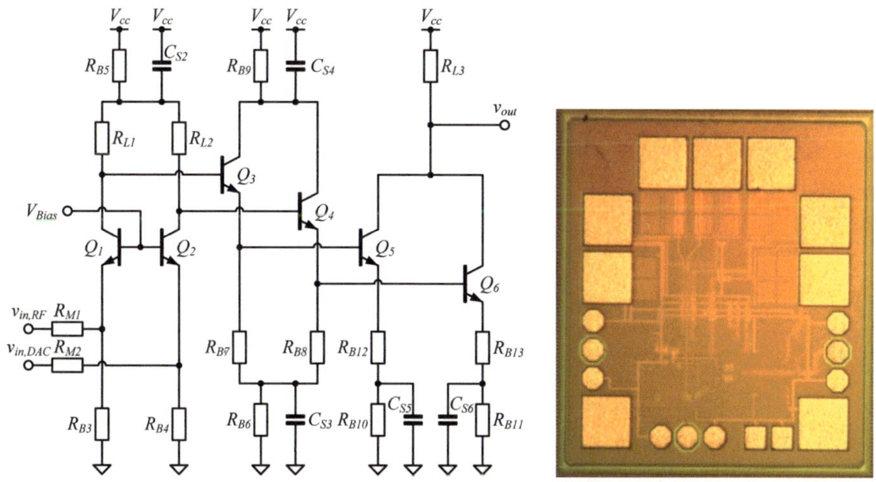

Figure 18. Schematic diagram (left) and chip photo (right) of the subtractor.

A printed circuit board has been designed to enable joint performance evaluation of this amplifier and the 7 Ω antenna. Due to the low feeding point impedance, separate characterization is less useful. To avoid problems involved in interconnecting devices with 7 Ω reference impedance, the amplifier should directly be attached to the antenna, which is supported by the board. Thus, both evaluation and refinement of this circuit will have to be performed in close collaboration with our partners from the *ultraMedis* project.

4. Transmitter circuits

4.1. Introduction

The circuits introduced in this section serve for the M-sequence topology. They have been implemented in a cost-efficient 0.25 μm Silicon Germanium BiCMOS technology, which opens up new fields of ultra-wideband radar applications. In the following sub-chapters, the design of different hardware blocks for the ultra-wideband radar front-end is presented. The design of a multi-purpose M-Sequence generator is presented which acts as a pulse compression modulator and exhibits an up-conversion mixer. A highly efficient power-distributed amplifier has been implemented utilizing a novel cascode power matching approach to achieve superior output power performance. Additionally, a fully differential broadband amplifier using cascaded emitter followers has been designed that exhibits a variable gain control and excellent broadband performance.

4.2. M-sequence generator

The well-known very broadband spectrum of M-sequences is widely used for testing the correct functionality of broadband integrated circuits, such as amplifiers, multiplexers, and transceivers. The run for higher data rates and amplifiers with broader bandwidth often outperforms commercially available test equipment and necessitates some sources to test these circuits. The measurement equipment vendors cannot supply data sources as fast as the technology evolves. The application which is targeted in this chapter is that M-sequences are used for pulse compression in ultra-wideband radar systems. For this application, it is important that the generator consumes little energy only, and it should generate a sequence of appropriate length (see (2)). Early high-speed PRBS generators for high data rates have been employed in III/V HBT technologies [28]. Moreover, a 110 Gb/s PRBS generator has been published in [29] using InP HBT technology with a transit frequency (f_T) more than 300 GHz. Recently, several PRBS generator circuits have been published in SiGe bipolar technology for test purposes in fiber-optic communications. In [30] a 100 Gb/s 2^7-1 PRBS generator has been implemented in a 200 GHz f_T SiGe bipolar technology. As in the 80 Gb/s $2^{31}-1$ pseudo random binary sequence generator introduced in [31], the output of the shift register has been multiplexed to achieve a higher maximum data rate. However, these circuits have a power consumption of 1.9-9.8W and utilize cost-intensive high-end processes. A 4×23 Gb/s 2^7-1 PRBS generator with a power consumption of 60 mW per lane has been publicized in [32] utilizing a 150 GHz f_T SiGe BiCMOS technology. A 2^7-1 multiplexed PRBS generator in 0.13 µm bulk CMOS exhibits 24 Gb/s output data rate [33]. In the following section, the circuit implementation with measurement results of the M-Sequence generator is presented.

4.2.1. *Upconverted M-sequence generator*

A simple way to generate M-sequences is to utilize a digital linear feedback shift register (LFSR), as depicted in Fig. 19. This device generates a binary pseudo-random code of length 2^n-1, where n is the number of stages in the shift register. Feedback is provided by adding the output of the shift register, modulo two, to the output of one of the previous stages. The actual sequence obtained depends on both the feedback connections and the initial loading of the register.

The proposed architecture depicted in Fig. 20 consists of serially connected shift registers with the characteristic polynomial

$$f(x)=x^9\oplus x^5\oplus 1 \tag{23}$$

an additional XOR gate acting as a modulo-2 adder to yield the delayed sequence, and one multiplexer. The selected feedback in the proposed architecture enables to generate two M-

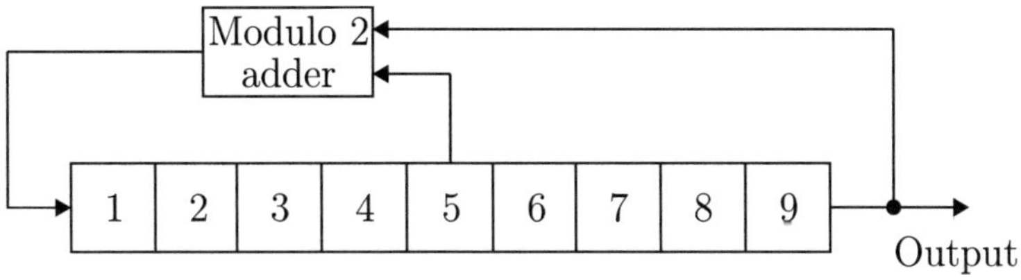

Figure 19. Nine bit linear feedback shift register for 9th order M-Sequence.

sequences with a mutual shift of half the word length. Those are multiplexed to yield the same M-sequence at twice the data rate. The input for the multiplexer is set between the two latches of the fifth flip-flop highlighted Fig. 20. This leads to a phase shift of half the pulse width in order to achieve the maximum voltage swing at the input of the multiplexer. Thus, the proposed architecture makes it possible to boost the circuit performance at the cost of an additional adder and a multiplexer. The architecture is extended to provide the possibility of upconversion for the generated M-sequence. This has been facilitated by implementing a mixer core at the output of the multiplexed LFSR. The mixer performs a BPSK modulation of the 9th order M-sequence signal generated by the multiplexed shift register. The circuit was implemented as an XOR gate instead of a conventional Gilbert cell as opposed to [34]. The actual circuit is nearly identical but the XOR operates in the limiting region compared to the small-signal operation of the Gilbert cell. The limiting behavior simplifies the design and requirements of the mixer, and results in lower power consumption. No emitter degeneration has to be implemented to increase linearity for large signal inputs. The XOR gate is driven by a LO buffer that can be digitally controlled to allow the generation of baseband M-sequence signals without the need for up-conversion. An additional output buffer with a resistively matched output has been included in order to control the output voltage swing in a wide range.

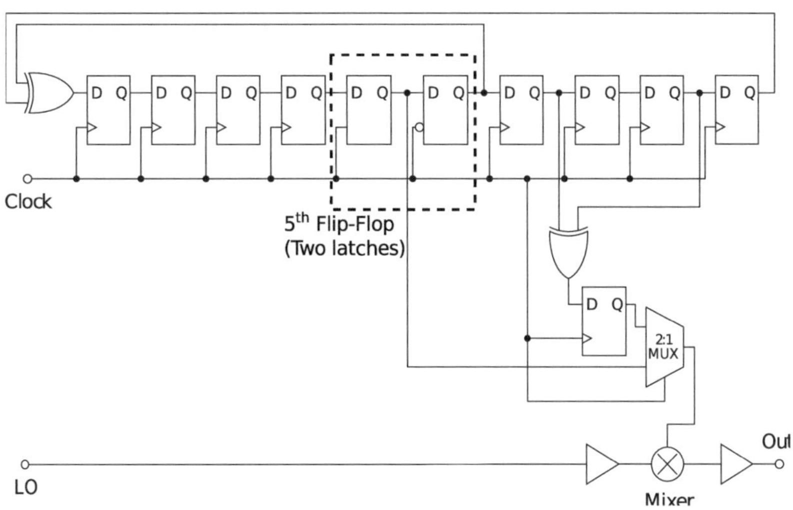

Figure 20. Circuit architecture of the proposed $2^9 - 1$ M-sequence generator providing the possibility of up-conversion for the radar signal.

Once the functional simulations have been completed, each individual block is designed on the transistor level. As the multiplexing architecture has been chosen, the flip-flops only have to work at half the operation speed. Standard CML flip-flops have been designed consisting of two latches, which inhibit two differential pairs. A schematic diagram of a CML flip-flop is depicted in Fig. 21. The flip-flops used in the LFSR are designed to offer a differential output voltage of 2×300 mV. According to [31], the tail current I_T and the emitter area A_e are related by

$$A_e = l_e \times w_e = \frac{I_T}{1.5 J_{peak\, f_T}} \tag{24}$$

for fastest switching time such that $J_{peak\, f_T}$ is the current density for maximum f_T. The lowest tail current is set by the minimum allowed transistor size, which is $l_e \times w_e = 0.84\ \mu m \times 0.42\ \mu m$. Thus, the tail current is chosen to be 0.85 mA whereby the collector current for the maximum transit frequency f_T is 1.25 mA for a $84\ \mu m \times 0.42\ \mu m$ transistor. The output voltage swing of each flip-flop was set to 2×250 mV. Simulations indicate that the latches work up to 12.5 Gb/s, which is sufficient for a 25 Gb/s multiplexed M-sequence.

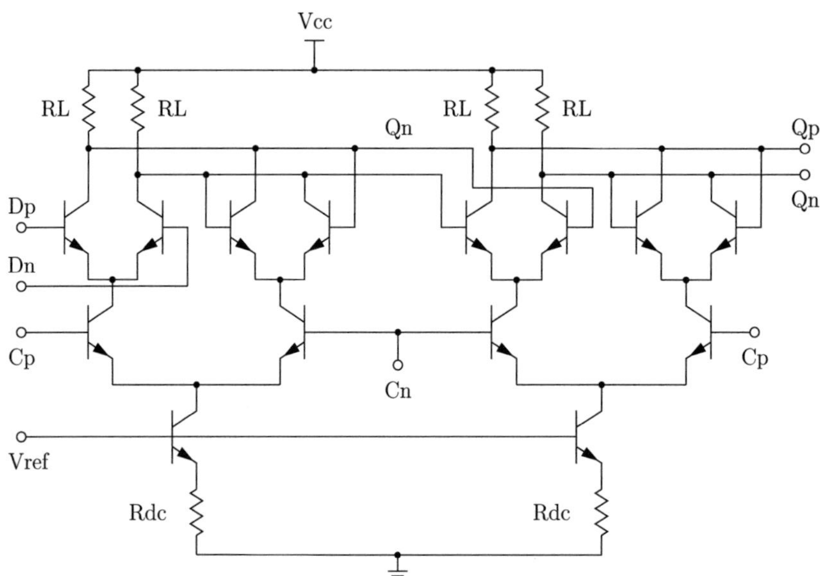

Figure 21. Schematic diagram of a CML flip-flop.

4.2.2. *Simulated and measured results*

The M-sequence generator has been simulated in time domain to find out the maximum data rate and to verify the correct function of the register. At clock frequencies higher than the maximum allowable clock frequency, the PRBS register does not work as expected and the output is not an M-sequence. A simulated waveform of a 25 Gb/s M-sequence is shown in Fig. 22 using the techniques described above. It can be seen that every bit can be distinguished from each other. A single bit has a slightly lower output voltage than a bit sequence with the same value, which is caused by the limited output bandwidth. As mentioned before, the mixer and the output buffer both have a limiting character which attenuates the flip-flop glitches. Thus, the waveform exhibits low ripple, actually at sequences with a series of equal bits, which indicates that the clock feed through is very low. However, the output waveform exhibits some deviations, which is quite common for circuits at this high data rate. This behavior may result from the slightly inductive behavior of an emitter follower in the signal chain. As long as the circuit is stable, this does not cause problems.

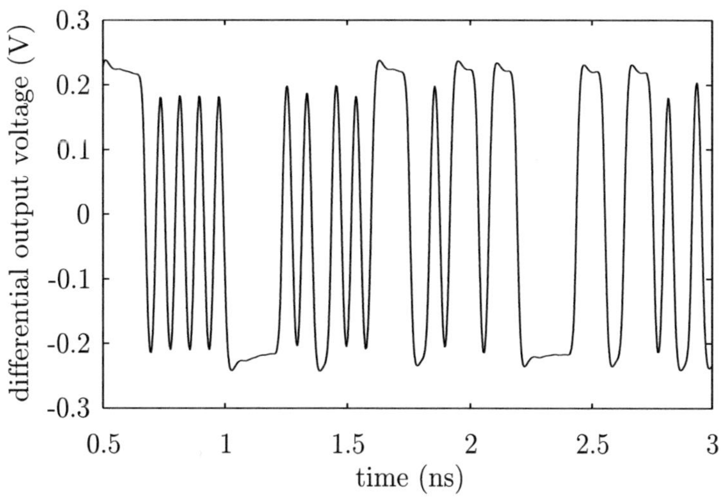

Figure 22. Simulated M-sequence waveform at 25 Gb/s data rate.

The M-sequence generator chip is placed on a Rogers TMM10i ceramic substrate for wire bonding. In order to protect the circuit mechanically and keep the bond wires as short as possible, it is placed in a topside cavity and fixed utilizing an electrical and thermal conductive epoxy glue, as shown in Fig. 23. The thickness of the ceramic substrate has been chosen to be 381 μm, which is almost equal to the chip height of 370 μm and the glue. Thus, the distance between the substrate edge and the bond pad can be reduced. The continuous ground plane below ensures a good thermal conduction, and 1.2 mm thick *FR*4 stabilize the brittle ceramic substrate.

The correct function of the generator can be checked through calculation of the normalized autocorrelation function of one complete M-sequence. The calculation of the ACF has been implemented in Matlab. As the correlation properties are of substantial interest for radar applications utilizing pulse compressed waveforms, the PRBS signal is measured with an Agilent DSO 91204A oscilloscope and compared with the simulation results. The simulated and measured 10 Gb/s waveforms together with the normalized cross correlation function of the measured signal are presented in Fig. 24.

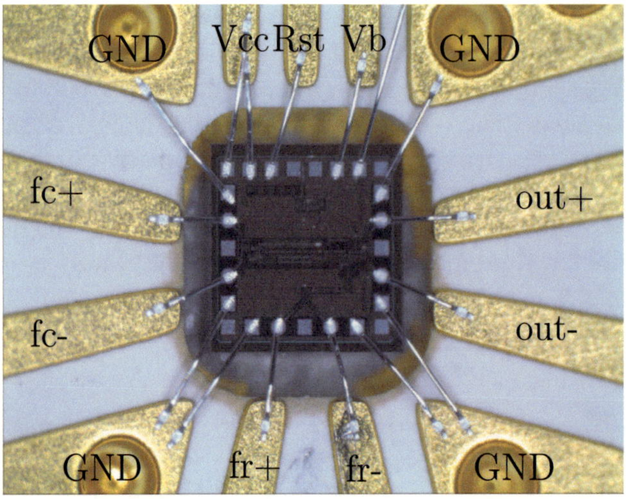

Figure 23. Photograph of the bonded M-sequence generator chip placed in a topside cavity on a ceramic substrate.

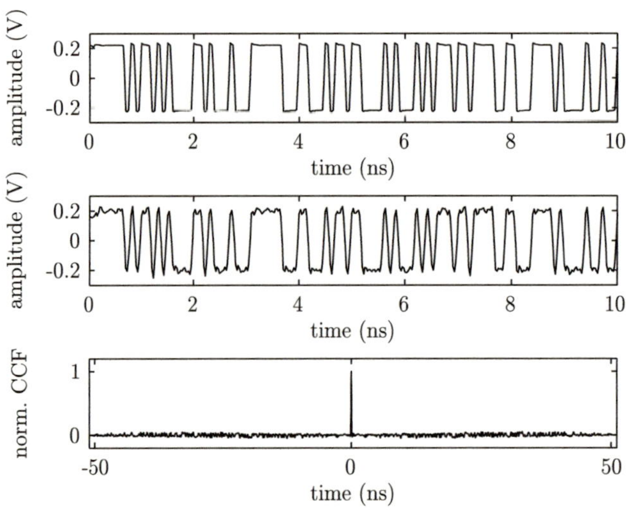

Figure 24. Simulated (top) and measured (middle) time domain representation of the proposed 10 Gb/s M-Sequence and their normalized cross correlation function.

4.3. Distributed power amplifier

The transmitted random sequence is subjected to high losses especially when transmitted through human body cells. Thus, a power amplifier is required to be placed directly before the antenna to increase the signal power. Distributed amplifiers (DAs) are appealing aspirants for UWB systems due to their inherently large bandwidth. The two major challenges in designing distributed power amplifiers are maintaining high linearity over the entire bandwidth, since narrowband linearization techniques cannot be utilized, and achieving high output power and efficiency. In order to increase the HBT distributed power amplifier performance, which is limited by the characteristics of the active cells used [35], alternative structures are investigated. The cascode cell is an appealing circuit due to its higher output impedance, higher breakdown voltage, and reduced Miller effect. Moreover, loading the two transistors by the required impedance for optimum power leads to an output power twice as high as compared to a single transistor. However, the conventional cascode configuration does not meet these conditions since the common base transistor's (T_{cb}) low-input impedance restricts the output voltage excursion of the common emitter transistor (T_{ce}). Therefore, it does not see its optimum power load impedance. In addition, the power performance of the cascode cell becomes one of the most important challenges to obtain maximum output power over the required bandwidth. To be power optimized, another series capacitor C_a is inserted on the base of T_{cb} to avoid its early power saturation compared to T_{ce}. A small signal model of the modified cascode gain cell is depicted in Fig. 25. The input impedance of the common base transistor can be calculated as follows:

$$z_{in,cb} = r_{be,2} + R_{b,2} + \frac{1}{j\omega C_a} \tag{25}$$

The capacitor C_a and input impedance $z_{in,cb}$ act as a voltage divider between the optimum values for $v_{ce,1}$ and $v_{be,2}$.

In order to achieve higher gain and greater bandwidth, an additional inductor is added between the collector of T_{ce} and the emitter of T_{cb}. The influence of a 1.5 nH inductor and various capacitances on the voltage gain of the cascode cell is demonstrated in Fig. 26. The small-signal schematic diagram of the modified cascode cell with inductive peaking is presented in Fig. 27a. The output resistance of the modified cascode circuit can be written as

$$z_{out} \approx z_{ce,2} + \left[\left(j\omega L_a + z_{ce,1} \right) || \left(z_{be,2} + R_{b,2} + \frac{1}{j\omega C_a} \right) \right] \tag{26}$$

neglecting the influence of miller capacitance c_{cb}. This leads to a resonance effect which is dominated by L_a, C_a and c_{be}. The self-resonant frequency f_r of the LC low pass can be calculated as

$$f_r = \frac{1}{\pi \sqrt{\frac{L_a C_a c_{be,.2}}{C_a + c_{be,2}}}} \tag{27}$$

The resonant frequency shows good agreement with the theoretical considerations set out in (27).

Another effect is that the output impedance of the cascode cell increases significantly from 2 to 15 GHz under the influence of the 1.5 nH inductor. The initial values for L_a and C_a are then optimized under large-signal conditions using nonlinear simulations of the inductively

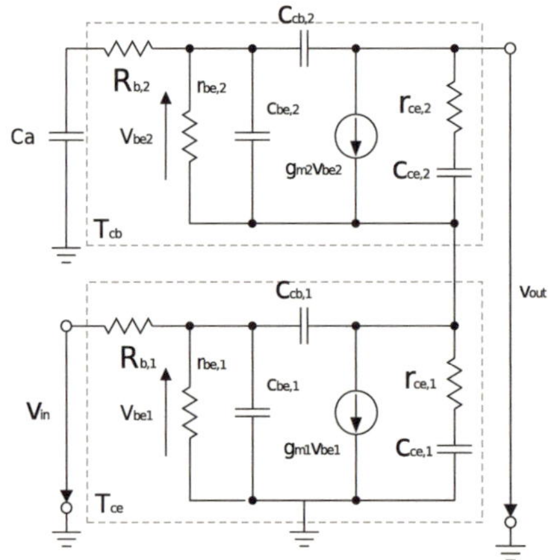

Figure 25. Small signal equivalent circuit of the modified cascode gain cell.

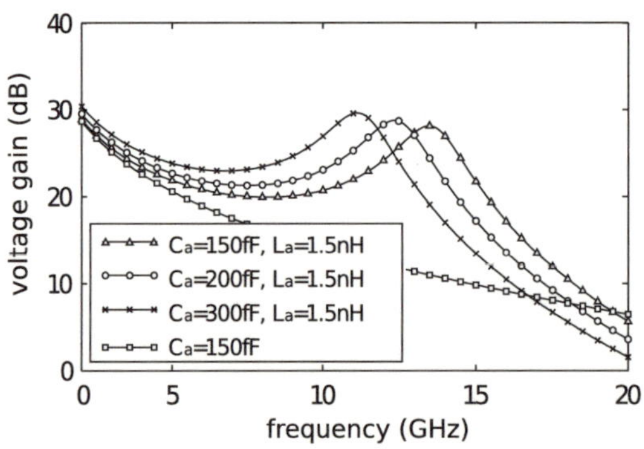

Figure 26. Voltage gain of different cascode cells with and without additional inductor $L_a = 1.5\ nH$ and various capacitances C_a.

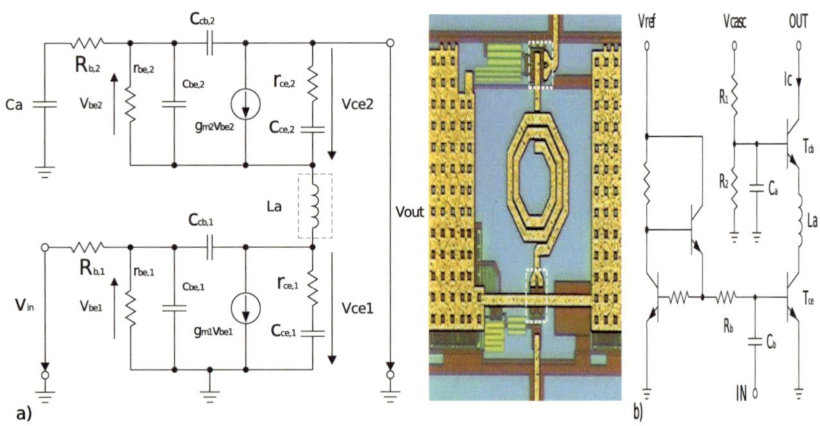

Figure 27. a) Small signal model of the modified cascode gain cell with inductive peaking and b) the schematic view.

peaked cascode circuit close to 1 dB compression point in order to obtain the maximum output power and efficiency. Accordingly, one single cell of a simple common emitter stage is used to synthesize the required ratio between the values of the inductor and capacitor. The test circuit is terminated with a 200 Ω resistor. The goal is to have equal deflections of the load lines both for the common emitter and the common base transistor. These results demonstrate that the proposed cascode configuration can obtain twice the output voltage swing compared to a single common emitter transistor at the same collector current so that twice the output power can be achieved.

A demonstrator chip has been implemented utilizing the methodology described previously. Fig. 28 depicts a schematic diagram of the four-stage tapered collector-line traveling wave amplifier with capacitive coupling and power-matched cascode gain cells. Each gain cell consists of two 26.4 μm² standard purpose transistors with BV_{ceo} of 4 V and a peak f_t value of 45 GHz, which are connected by a 1 nH inductor L_a . A single gain cell is depicted in Fig. 27b. The octagonal spiral inductor exhibits a Q-factor of 20 at 12 GHz.

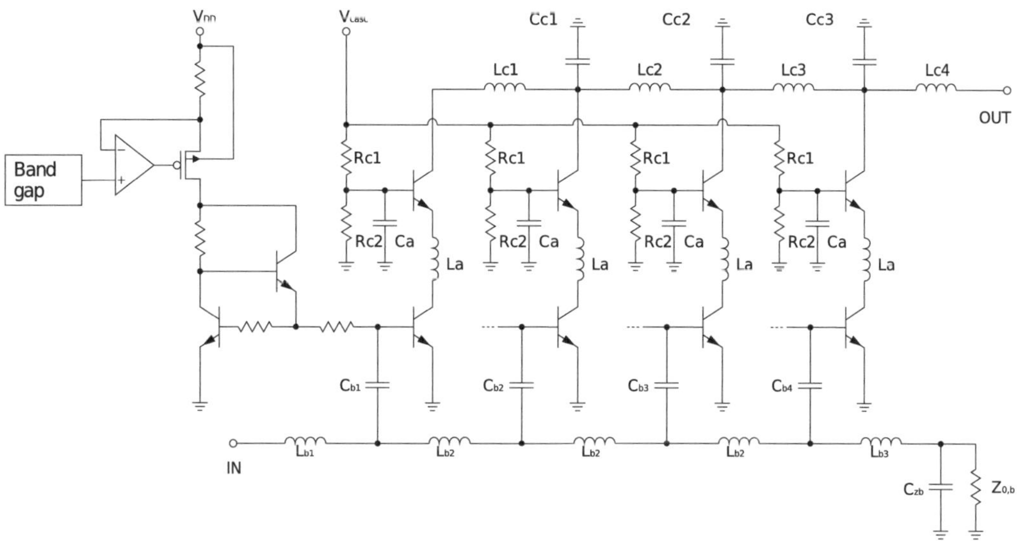

Figure 28. Complete schematic representation of the implemented distributed amplifier structure.

The tapered collector line has been realized using staggered inductors $L_{C1} > L_{C2} > L_{C3} > L_{C4}$ together with shunt capacitances $C_{ci} \parallel C_{ce}\, ; (1 \le i \le 3)$ in order to achieve a coherent addition of the collector currents and a flat gain over the entire bandwidth. Biasing is implemented using three transistor current mirrors with ratio of 32:1 and a low dropout (LDO) voltage reference driven by a band-gap voltage source. A chip microphotograph of the complete circuit is shown in Fig. 29. The transistors are biased through the collector line by means of an external bias-tee. The bias point was selected at $VCC = 5\,V$ and $VDD = 2.6\,V$. A power and ground grid facilitates a low impedance connection and - due to the low distance between the congruent metal grids - a large capacitor is shaped. The chip size of the amplifier circuit is 2.1mm².

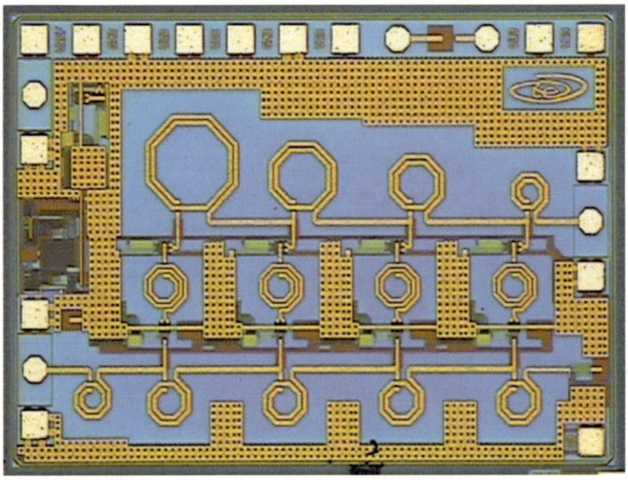

Figure 29. Microphotograph of the manufactured 2.1 mm² traveling wave power amplifier.

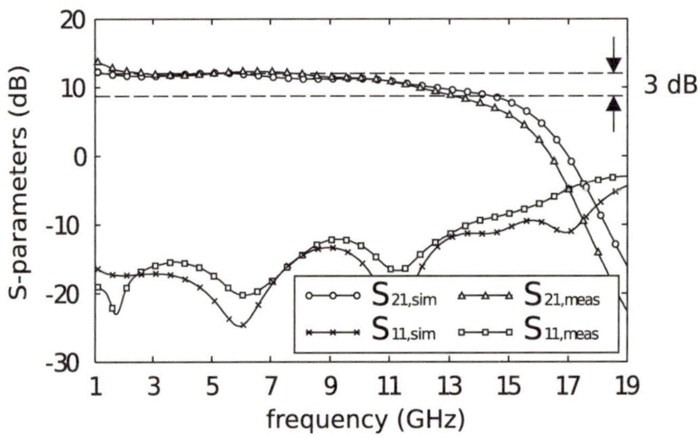

Figure 30. Simulated and measured small signal gain S_{21} and input reflection coefficient S_{11}.

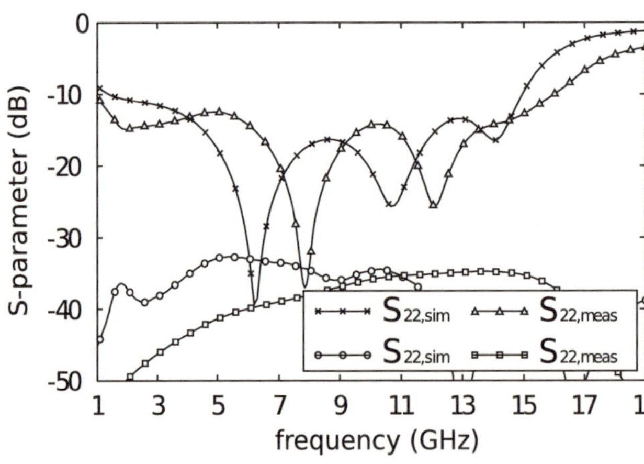

Figure 31. Simulated and measured output reflection coefficient S_{22} and isolation S_{12}.

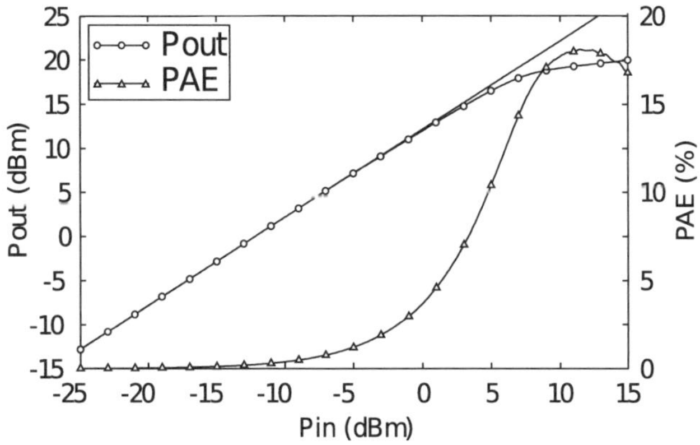

Figure 32. Output power P_{out} and power added efficiency PAE at a center frequency of 7 GHz.

The distributed power amplifier chip was tested via on-wafer probing. The measurements of the circuit were carried out using an Agilent N5242A PNA-X vector network analyzer. Fig. 30 shows the simulated and measured small signal gain S_{21} and input return loss S_{11}. The traveling wave power amplifier exhibits a measured gain of 11 dB with a gain ripple of ±1 dB up to 12 GHz and a 3 dB bandwidth of 13 GHz. The simulated and measured output return loss S_{22} and the reverse isolation S_{12} are illustrated in Fig. 31. Both the input and output return loss are below -12 dB over the entire frequency range. The measured reverse isolation S_{12} remains below -35 dB. The circuit is unconditionally stable, also verified for large RF input signals. Fig. 32 shows the output power Pout and power-added efficiency PAE at a center frequency of 7 GHz. The 1 dB compression point is at 17.45 dBm with an associated power-added efficiency of 13.9%. The saturated output power P_{sat} is 20 dBm and the maximum power-added efficiency is 22.1%.

4.4. Differential broad band amplifier

Broadband variable gain amplifiers are key components for ultra-wideband radar applications and important building blocks to increase the dynamic range. Especially for M-sequence based radar systems without upconversion, the lower frequency range, which contains most of the signal energy [36], has to be considered. Biomedical and ground penetrating radar systems necessitate a lower frequency boundary of less than 1 GHz [37], [38].

Moreover, the broadband variable gain amplifier (VGA) should be fully differential. Great care has to be taken to avoid the distortion of the signal shape through gain ripple and group delay variation. In this section, the analysis, design and measurement results of a fully differential broadband VGA are presented. After some considerations about mismatching in broadband amplifiers have been made, the frequency behavior of cascaded emitter followers is investigated, and the implementation of a variable gain control is explained. Finally, the implementation of the broadband amplifier is presented, introducing the circuit architecture and presenting measurement results. The amplifier is

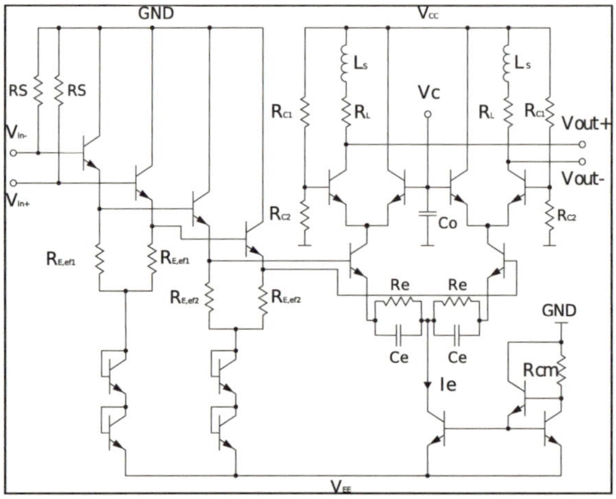

Figure 33. Simplified schematic diagram of the proposed broadband variable gain amplifier.

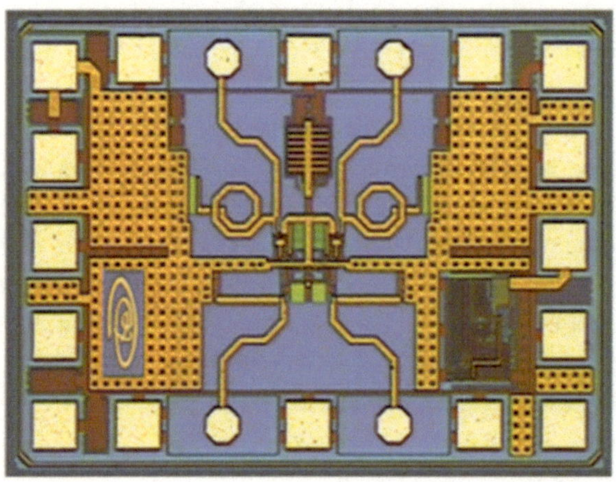

Figure 34. Microphotograph of the manufactured 0.75 mm² variable gain amplifier.

fully differential and based on a cascode configuration as depicted in Fig. 33. This is useful for high frequency circuit design, because this multi device configuration has small high-frequency feedback, achieved by the negligible Miller effect, and a large bandwidth. Driving the cascode stage with cascaded emitter-followers leads to an enhancement of bandwidth and provides dc level shifting [39].

The voltage gain of the cascaded emitter-followers has a frequency dependence that is similar to the frequency dependence of the transfer function of an RLC series resonance circuit [40]. This can be used to provide gain peaking at the desired frequency. The transfer function depends on the transistor parameters, the biasing current, the resistors, and the load. The main problem using emitter-followers to drive cascode stages is that the circuit might become unstable. This will be the case if the negative input resistance of the second emitter-follower stage becomes larger than the positive output resistance of the first stage at a certain frequency, which is shown in Fig. 35. The frequency is determined by the point where the imaginary parts, which are the reactances of the series resonator, cancel each other out:

$$Re\{Z_{11}(\omega_o)\} = -Re\{Z_{22}(\omega_o)\} \tag{28}$$

$$Im\{Z_{11}(\omega_o)\} = Im\{Z_{22}(\omega_o)\} \tag{29}$$

The core of the broadband amplifier is a signal summing VGA as illustrated in Fig. 33, where the gain is controlled by applying an analog dc voltage at V_C. The amplifier gain can be set from 0 to maximum gain whereby it behaves like a cascode differential stage when the control voltage is set to 0 and all current flows through the load resistors R_L. Furthermore, capacitive emitter degeneration is used to attain additional gain at high frequencies for a higher cut-off frequency. Tuning the values of R_e and $C_{e'}$, introduces trade-off between high gain, bandwidth and stability, because C_e influences the capacitive load of the cascaded emitter followers. Inductive peaking is carefully applied using small inductors L_C in the collector branches in order to avoid high group delay [41]. In order to achieve a high output swing, high currents in the differential amplifier are necessary.

The circuit is implemented in the 0.25 µm IHP SGB25V value technology. A chip photograph of the broadband variable gain amplifier is depicted in Fig. 34. The circuit elements composing the amplifier core have been arranged symmetrically to maximize the even mode suppression. The mixed-mode S-Parameters are measured on-wafer using 150 µm GSGSG probes.

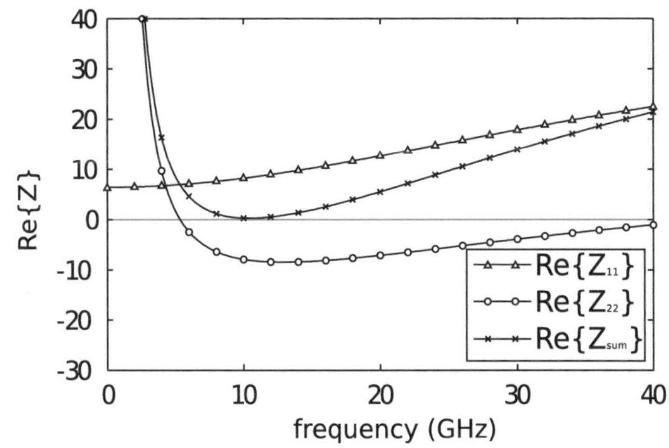

Figure 35. Simulated impedances Z_{11} and Z_{22} illustrating the stability relations of the emitter followers.

Fig. 36a illustrates the differential simulated and measured gain S_{dd21} as well as the input and output return loss at 100 Ω differential source and load impedance. The measured differential gain is 11.5 dB with a gain flatness of ±1.5 dB. The 3 dB cut-off frequency is 30 GHz, which results in a gain-bandwidth product (GBP) of 113 GHz which is 1.5 times the

f_t of the transistor. The corresponding measured and simulated group delay is shown in Fig. 36b. The measured group delay variation is 35 ps, which is higher than that in the simulation and also induced by the stronger resonance behavior.

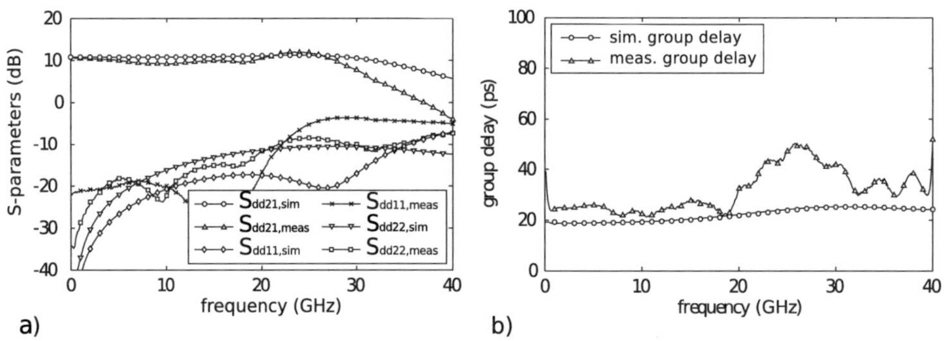

Figure 36. a) Simulated and measured mixed mode S-parameters. b) Simulated and measured group delay.

As depicted in Fig. 37, the amplifier gain can be adjusted between 0 and 11.5 dB. The large signal behavior is measured on-wafer. An output 1 dB compression point of 12 dBm has been measured up to 20 GHz.

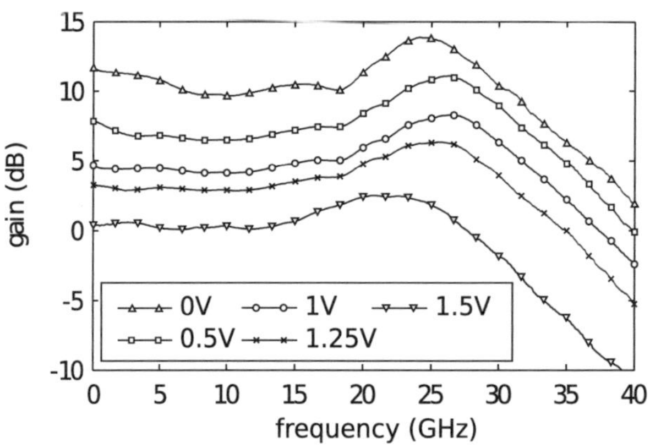

Figure 37. Measured gain by sweeping the control voltage V_c from 0 to 1.25 V

5. High-speed data capturing

5.1. Introduction of data capturing device using feedback principle

A straightforward data capture and digitizing can be directly performed by a conventional analog-to-digital converter (ADC). There are a number of limitations which arise using this method. The first and most crucial one is an inverse relationship between accuracy and speed of the conversion. In terms of the ADC, it is the inverse relationship between resolution and bandwidth. It is impossible to realize a high-speed ADC with the resolution which fulfils the sensor specification in modern technologies.

To overcome this limitation, a more complicated method of data capture based on "stroboscopic feedback loop" can be used. This method utilizes a feedback loop to relax accuracy requirements of the ADC (see [2] and chapter 6.2.3). The digital output of the data capturing device is represented by two summands: the value of the first summand is measured by the ADC; the value of the second summand is calculated based on its previous state and on the first one. The ratio between predicted and measured summands, i.e. between the resolution of the ADC and DAC can be calculated from the conversion efficiency of the both converters [42].

The block diagram of the data capturing device with feedback is depicted in Fig. 38. It consists of 3 logical parts, highlighted in colors in Fig. 38: Signal Processing, ADC and DAC, and LNA with subtraction amplifier. Although the subtraction amplifier belongs to the data capturing block, it has been integrated into LNA and moved to the receiver part of the sensor.

The data capturing device works as following:

- A capturing block digitizes a difference (residue) between the received and predicted values. This function is performed by a high-speed low-resolution analog-to-digital converter.
- A digital predictor evaluates the data from the ADC and makes a prognosis about the value to be expected next.
- The predicted value is converted into an analog form with a high-speed DAC.
- In analog domain, the predicted value is subtracted from the received signal with a subtraction amplifier.

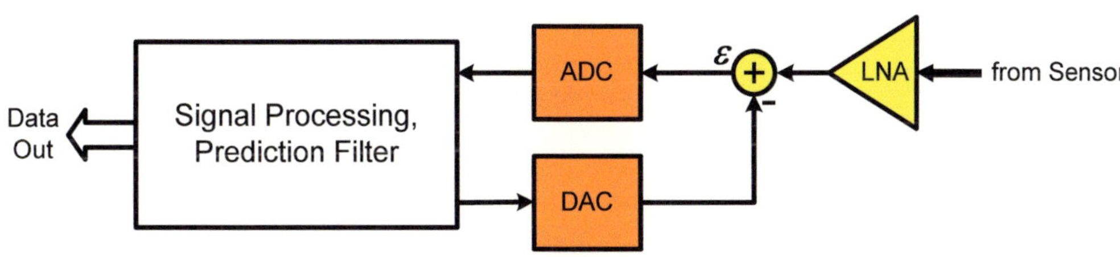

Figure 38. Block diagram of the data capturing device

5.2. High-speed 4-bit analog-to-digital converter

5.2.1. High-speed analog-to-digital converter

The fastest type of the A/D converters is a full flash ADC. A block diagram of a typical full flash converter is shown in Fig. 39. It consists of a reference network, a bank of comparators, correction and encoding logic and test buffers. The challenges of the implementation of the ADC are usually related to the analog part of the converter, namely to the reference network and to the bank of the comparators. It is possible to implement the high-speed comparator in the selected technology which will satisfy all requirements, but the reference network is a bottle-neck of the converter.

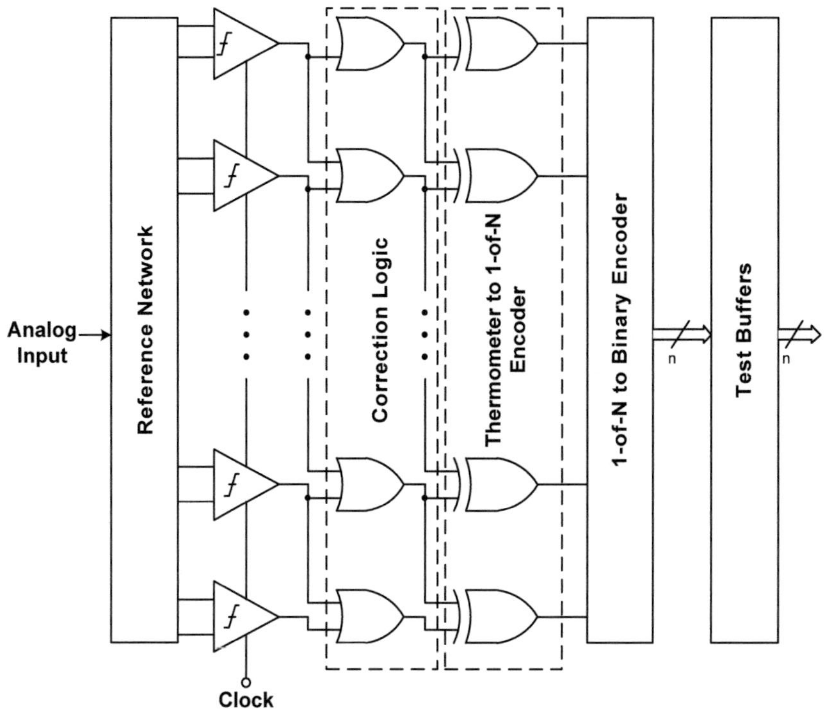

Figure 39. Block diagram of the full flash ADC.

5.2.2. *Reference network*

The task of the reference network is to provide equidistant reference voltages which will be further processed by the comparators.

There are two conventional implementations of the reference network. First, the simplest way is a Kelvin divider or resistor ladder. It suffers from several drawbacks, such as DC-bowing, clock and input feed-through [43]. Furthermore, is not well suited for the high-speed ADCs.

A second configuration is a differential one. It consists of two branches; each has a driver loaded with a chain of serially connected tap resistors. Both branches are equal, only outputs of the second branch are "inverted" or mirrored with respect to the middle point [43]. The main problem related to the differential network is its bandwidth, which often becomes a bottleneck of the system. The reference network has to drive a big parasitic capacitive load caused by the bank of comparators. In the full flash ADC, it is one of the main limitations, because the number of comparators is doubled when increasing the resolution by 1 bit.

The second problem of such network is the non-equal transfer characteristic of the output nodes [42].

5.2.3. Proposed bandwidth enhancement technique

Drawbacks of the conventional differential reference network are mainly due to its serial configuration; a change in one component will affect the others. This inherent property of the serial connection makes individual adjustments and compensations impossible. To overcome this limitation, a new configuration of the reference network is proposed. An idea is to build the resistor network in a segmented serial-parallel configuration and substitute one driver (emitter follower) with several drivers, connected in parallel. A full overview of the possible configurations is described in [44]. Among this variety one configuration should be highlighted, namely the configuration illustrated in Fig. 40 where each segment contains one tap resistor and one current source. The reference network is fully parallel, thus allowing the maximum speed to be achieved.

The main feature of the parallel network is the flexibility to choose component values. This freedom gives the possibility of equalizing the bandwidth of an individual segment that leads to the optimal speed at given power dissipation.

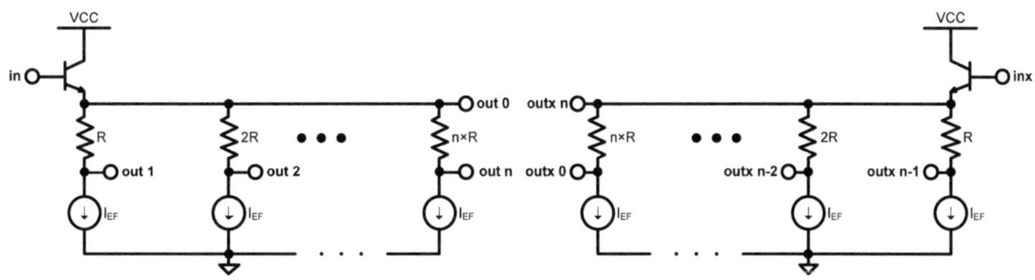

Figure 40. Parallel configuration of the network with one resistor per segment.

Fig. 40 shows the case when all driver currents are equal. In practice, it is more useful not to keep the currents in all segments equal, but to equalize the bandwidths in each segment instead. The network, however, does not only present good advantages, it also has some drawbacks. Flexibility of adjusting different parameters leads to different geometries of the resistors. In the case of the conventional network, all resistors have the same value and the same geometry. Proper layout minimizes the mismatch between them. The proposed network cannot benefit from this feature.

5.2.4. Design of comparator

Signals from the reference network are led to a bank of *n* comparators. Comparators decide if the input is above or below the reference. For decreasing the probability of errors, a master-slave comparator with a preamplifier is used. An overall schematic diagram is shown in Fig. 41. The role of the preamplifier for the comparator is twofold: It works as a limiting amplifier, and it provides an additional amplification of the input signal. Another important function of the preamplifier is isolating the reference network from kick-back noise, produced by the master latch. In this particular example, the Cherry-Hooper amplifier with emitter follower feedback is used as preamplifier.

The master latch has an auxiliary current source I_{aux}. This current source prevents the cross-coupled differential pair from being completely switched off, thus keeping the base-emitter capacitance charged. The time to charge this capacitance is decreased, and as a result the overall speed of the latch is increased. The I_{aux} has to be sufficiently small because it adds hysteresis which decreases the sensitivity of the comparator. Setting the value of I_{aux} equal to 10 % of I_{EE2} is a good compromise between speed and sensitivity. In the slave latch, there is no auxiliary current source because the input signal of the slave latch is relatively large, and an auxiliary current source does not have a strong influence as in the case of the master latch.

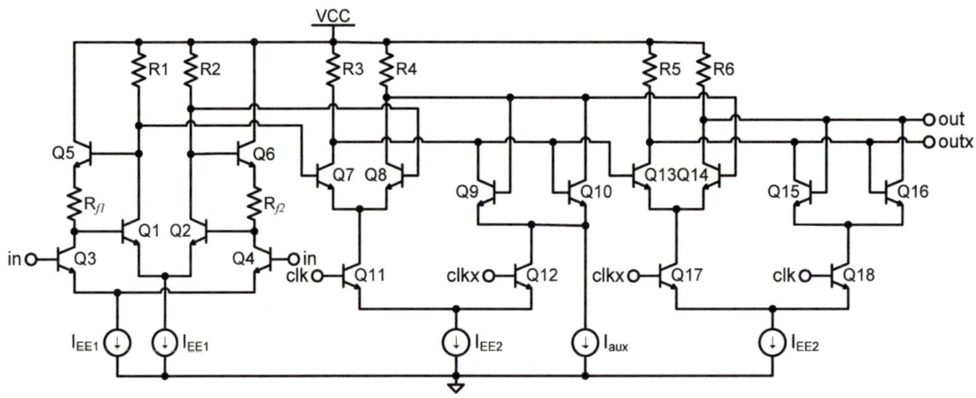

Figure 41. Master-slave comparator, full schematic diagram.

5.2.5. Experimental results

The ADC with the proposed parallel reference network was implemented in 0.25 μm SiGe BiCMOS technology. The Chip micrograph of the ADC is depicted in Fig. 42.

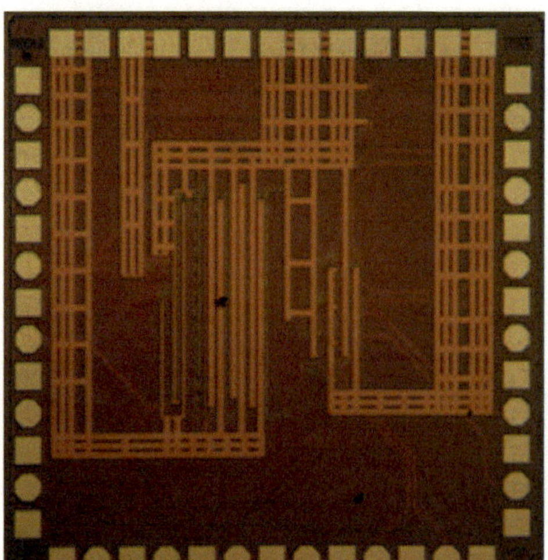

Figure 42. ADC's chip micrograph.

Static measurements: For measuring static errors of the ADCs, a low frequency 50 MHz sine signal was applied to the input of the converter at 5 GS/s sample rate. A deviation of a transition from the mean value, the differential nonlinearity (DNL), was calculated for each step. A cumulative sum of differential errors represents the integral nonlinearity (INL). The results are graphically presented in Fig. 43.

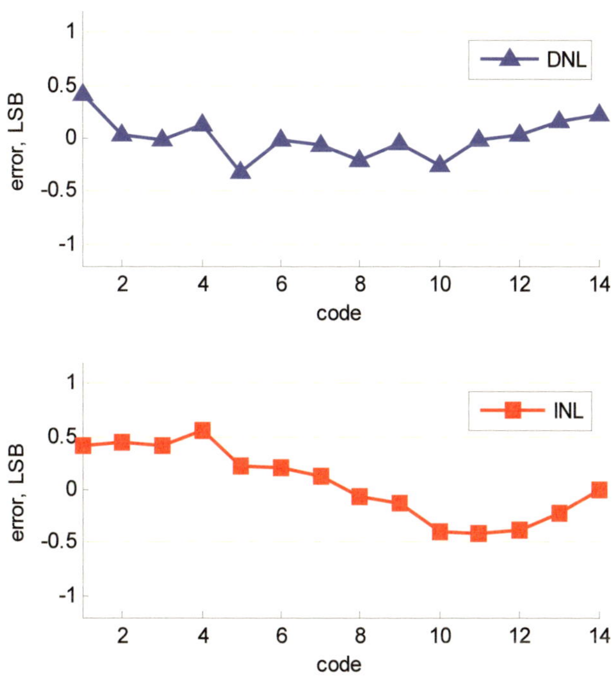

Figure 43. DNL and INL from histogram testing of ADC.

Dynamic measurements: Signal-to-noise and distortion ratio (SINAD) of the test circuit was measured over the frequency range up to 6 GHz at a constant sample rate of 15.01 GS/s. The small frequency offset of 10 MHz was made to accumulate quantization errors over the whole dynamic range. The measurement results are presented in Fig. 44, which shows SINAD of the converter up to the input frequency of 6 GHz. The dashed line shows a level where SINAD drops 3 dB below its value at low frequency. The frequency where SINAD crosses the 3 dB line indicates the effective resolution bandwidth of the converter, which in this case is greater than 6 GHz.

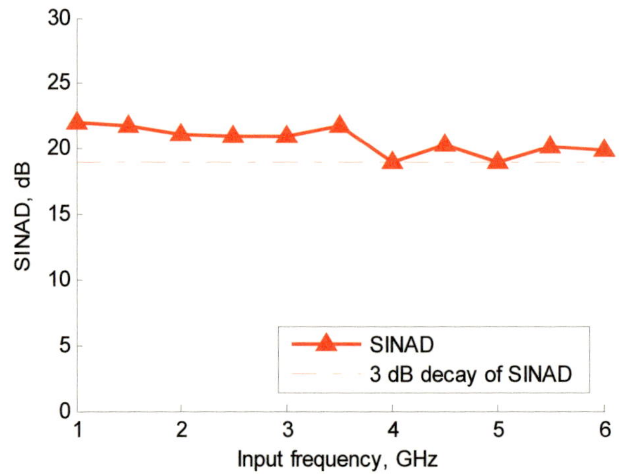

Figure 44. Measured SINAD over a frequency range up to 6 GHz.

5.3. High-speed predictor

The main function of the predictor of predicting the part of the received value was described above. The predictor also carries out two additional functions:

- Making subsequent averaging of the digitized values, increasing signal-to-noise ratio of the measured signal.
- Decreasing the data throughput for further data processing.

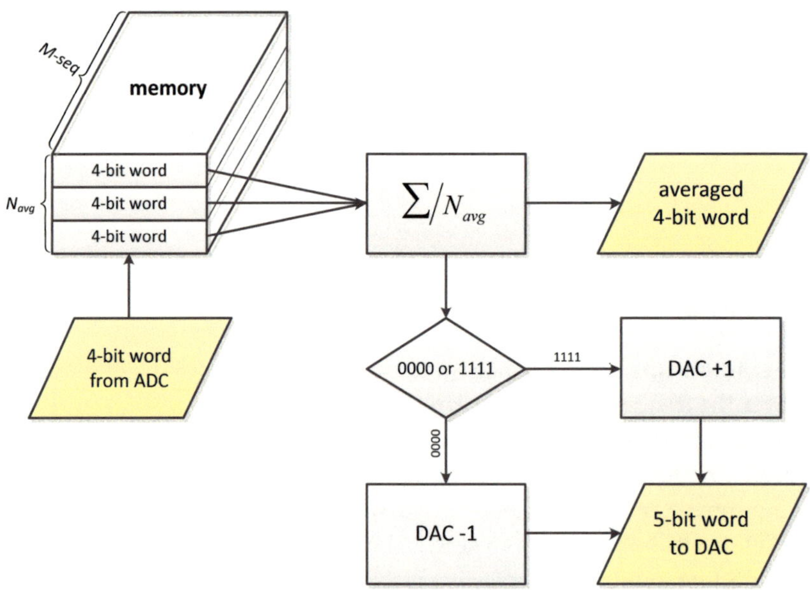

Figure 45. Functional diagram of the predictor.

The functional diagram of the predictor is depicted in Fig. 45. The predictor consists of the memory where the data from the ADC are accumulated; an averaging block, which makes averaging of the accumulated data; and a block where the output DAC value is calculated. An algorithm to calculate the DAC value is a modified version of the successive approximation algorithm with a constant +LSB step.

The described functionality is coded using VHDL language and implemented using ECL library available in IHP BiCMOS Technology. For speed purposes, the predictor was divided into several sub-blocks which were implemented separately. This method decreases the complexity of the separate sub-block, and achieves a higher operational speed. The block diagram of the predictor is depicted in Fig. 46. The predictor consists of a demultiplexer, a bank of predicting blocks and a multiplexer. A predicting block carries out three functions: accumulation, averaging, and prediction. The demultiplexer deserializes the M-Sequence and commutates M-Sequence parts (chips) to the individual predicting blocks so that each has to work with only one defined chip. The multiplexer reverses the parallel processing and serializes the predicted values which finally fed the DAC.

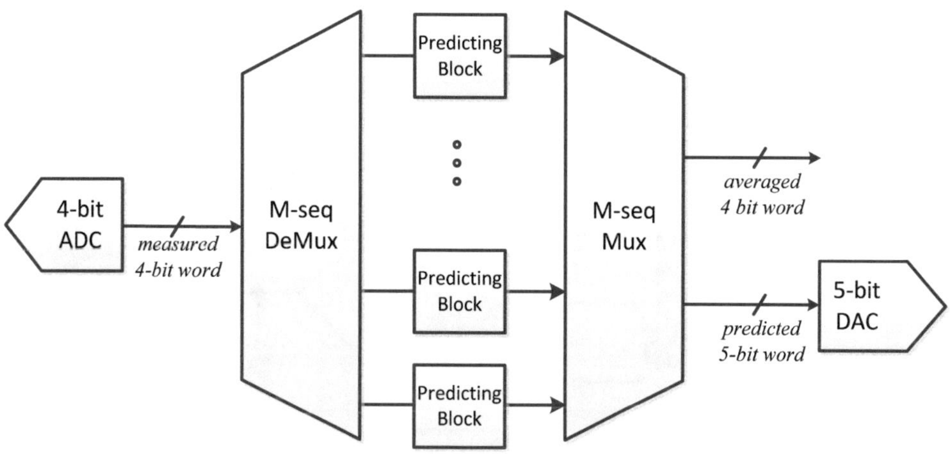

Figure 46. Structure diagram of the predictor.

5.4. High-speed digital-to-analog converter with off-chip calibration

The digital-to-analog converter transforms a digitally predicted value into the analog domain. To prevent information loss, the accuracy of the DAC should correspond to the accuracy of the whole capturing device. Simultaneously, the DAC should work at 10 GS/s. To satisfy both requirements, the converter is implemented using a segmented current steering architecture. The block diagram of the converter is depicted in Fig. 47. It consists of the two segments: A unary sub-converter and an R-2R sub-converter. The current sources of the both sub-converters are connected to the summing node. As will be seen later from measurements relying only on technology, component matching would give insufficient accuracy, which in this particular case is 10 times lower than required. Therefore, an additional calibration of the current sources is implemented. The current sources are realized as voltage controlled current sources. The controlling voltages are produced by auxiliary low-power µDACs which are externally controlled via SPI interface.

The calibration algorithm could be characterized as successive approximation of the DAC output to the reference value. The detailed calibration flow of the each current source is as follows:

1. The current source under calibration (CSUC) is disconnected from the summing node. For this purpose, the corresponding digital input is applied to the DAC.
2. The analog output of the DAC is measured and stored in memory as "zero-value". The measurement is performed with a 14-bit ADC on an FPGA board.
3. The CSUC is connected to the summing node.
4. According to a binary search algorithm, the MSB of µDAC is set to "1".
5. The output of ADC is measured again, and the difference between the stored "zero-value" and the measured value is calculated.
6. Depending on this difference, the decision concerning the value of the MSB of the µDAC is made.
7. Steps 4-6 are repeated for the remaining 9 bits of µDAC.
8. Steps 1-7 are repeated for each current source.

A set-up to implement the proposed calibration scheme is depicted in Fig. 47. The calibration algorithm is implemented on a Spartan-3AN Starter Kit board.

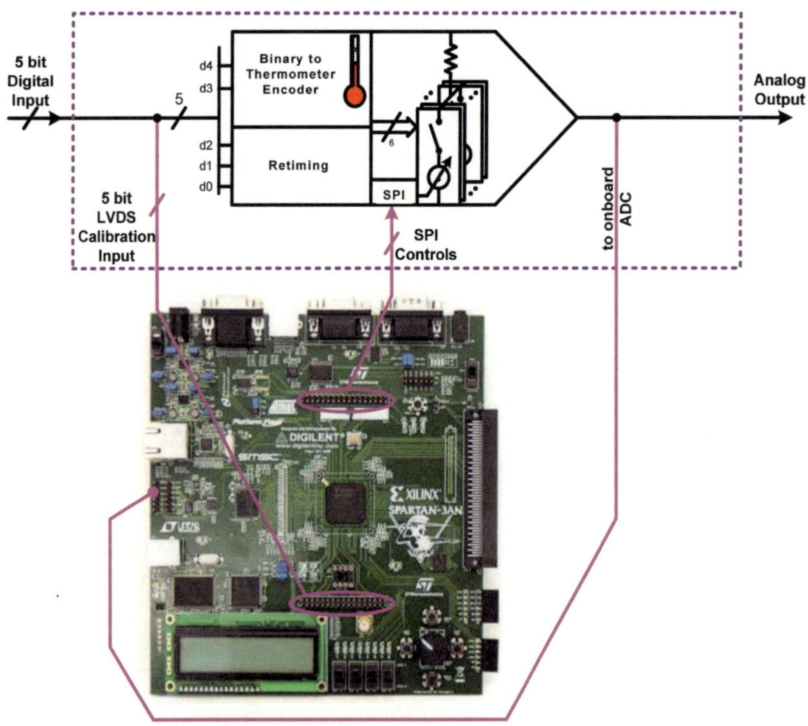

Figure 47. Off-chip calibration of the DAC.

5.4.1. Experimental results

Static measurements: The DAC test chip was implemented in SiGe 0.25 μm BiCMOS IHP technology. The chip was mounted on a test board and connected to the FPGA board. The results of static tests are given in Fig. 48 where both DNL and INL values before and after calibration are given. INL errors were also recalculated in percent of the input range. To

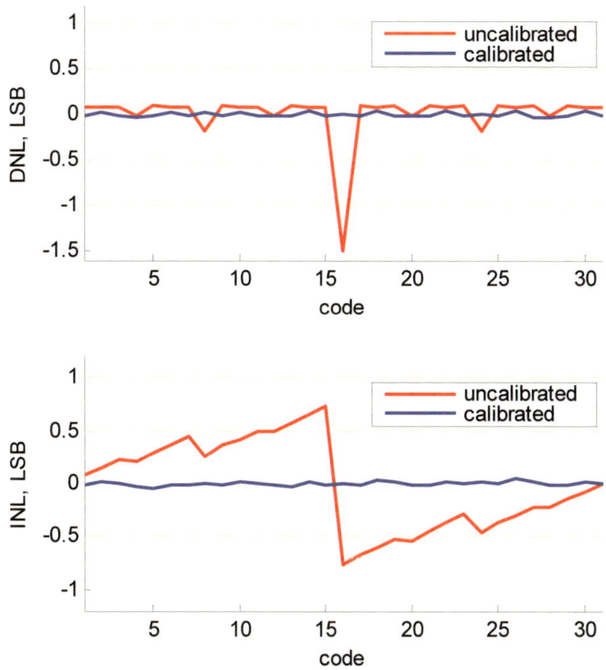

Figure 48. DNL and INL of uncalibrated and calibrated DAC output.

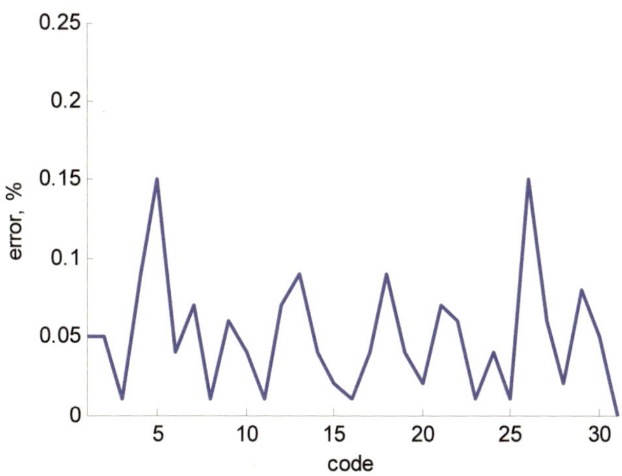

Figure 49. Static accuracy of the calibrated DAC.

achieve 9-bit overall resolution, the DAC should have an error below 0.2 %. The measured error after calibration is depicted in Fig. 49. The error is below 0.15 %.

Dynamic measurements: Dynamic characteristics of the DAC were measured together with the 4-bit ADC under the assumption that LSB usually works faster than MSB. The direct measurement of the spurious free dynamic range (SFDR) has no practical sense since the ADC limits the overall performance. For estimating the performance, an envelope test was applied [45]. Proper work of the converters assumes the presence of the all transition steps at the frequency of interest. Fig. 50 shows DAC outputs at 5 GHz and 5.5 GHz. Both converters (ADC and DAC) have all 16 transition levels up to 5.5 GHz input. Only the amplitude at 5.5 GHz starts to decay.

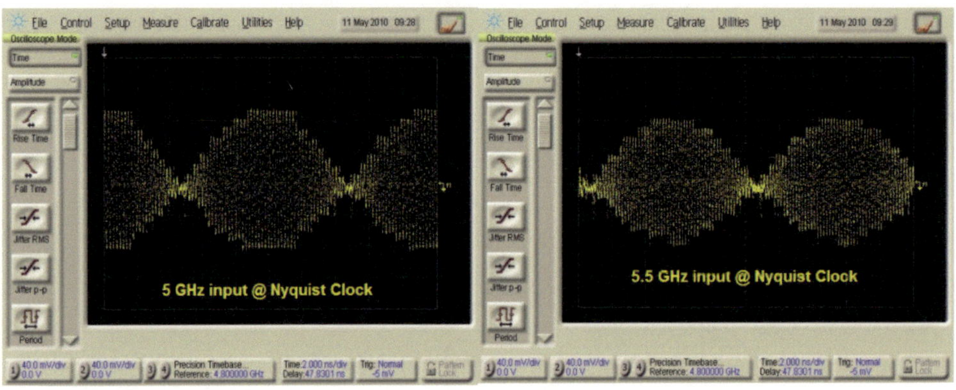

a) b)

Figure 50. Envelope test of ADC-DAC at **a)** 5 GHz and **b)** 5.5 GHz.

5.5. Conclusion

The design and measurements of the high-speed data capturing device for the M-sequence sensor are described in this chapter. The data capturing device utilizes the "stroboscopic feedback loop" for achieving high dynamic range together with high sampling rate.

A number of different techniques are used to achieve the desired performance of the separate components.

To achieve a high effective resolution bandwidth of the analog-to-digital converter, the new segmented reference network was proposed. The new network, implemented in the ADC [46] allows increasing the effective resolution bandwidth several times compared to the similar conventional one [47], while the power dissipation is only slightly increased.

The high-speed predictor was described in VHDL and implemented using a high-speed ECL library. Despite the disadvantage of the power dissipation, the ECL implementation allows speeds of up to 10 GS/s to be achieved. Furthermore, it is simple to modify the predictor to comply with different system parameters, such as the M-sequence length or averaging factor.

An off-chip calibration was implemented for the high-speed digital-to-analog converter. The calibration is implemented on an FPGA-board. After having been modified slightly, it could be integrated into the DAC. The static errors of the DAC after calibration are lower than 0.15 % which allows the use of a converter in the data capturing device with a target resolution of 9 bits.

6. M-sequence devices

6.1. Introduction

While previous sections were aimed to discuss specific sub-components such as individual semi-conductor chips of an UWB-sensor, we would like to consider some aspects of the whole sensor electronics here. For that purpose, several M-Sequence devices were implemented at different integration levels, and some Ukolos-partners (*ultraMedis, CoLoR*) were provided with demonstrator devices for their own use. In order to have a running sensor system, the device implementation has to cover the whole manufacturing cycle from chip-design and manufacture, chip housing, RF-PCB-design and assembly, design and implementation of the digital components (ADC, FPGA, interfaces etc) up to the programming of sensor internal pre-processing, the data transfer to the host PC and application-specific software for data evaluation and visualization. Furthermore, device specific test and evaluation methods and routines had to be developed and implemented in order to perform high-resolution device characterization (e.g. [48])

In what follows, we will first introduce an experimental device which is aimed to evaluate new concepts or modifications under real conditions. Secondly, we refer to a device configuration which implements the principle depicted in Fig. 4 for the practical use by other Ukolos-projects and finally, there will be some discussions toward single chip solutions.

6.2. Experimental demonstrator device

6.2.1. Device concept and aim

The aim of an experimental demonstrator device is to investigate the impact of individual sub-components on the performance of the whole device, as well as to have the opportunity to flexibly perform device modifications without the need of redesigning complex RF-PCBs. The device is organized in a modular concept as symbolically depicted in Fig. 51. Fig. 52 shows an example of a demonstrator implementation of such kind.

The individual sub-components as e.g. shift register for stimulus generation, T&H-circuits, RF-power distribution, RF-synchronization etc. are organized as plug-ins. Hence, one can simply replace a device component by a new one if improved circuits, better IC-housing or

RF-PCBs are available. Furthermore, the various modules may be interconnected to different device structures as shown in Fig. 4 or Fig. 55.

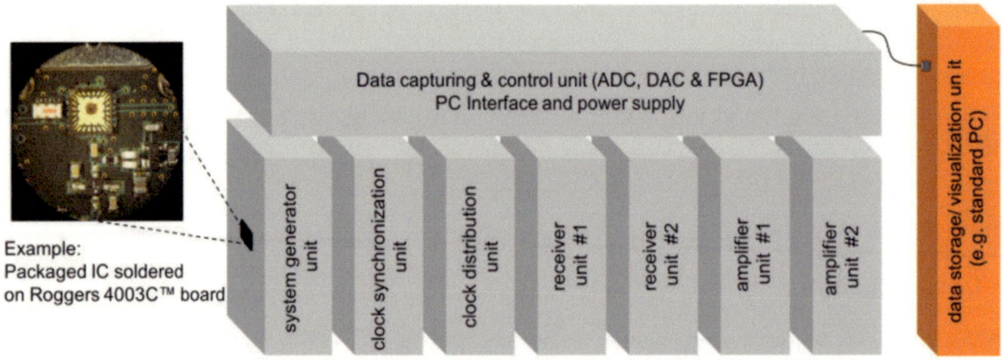

Figure 51. Modular concept of the experimental system.

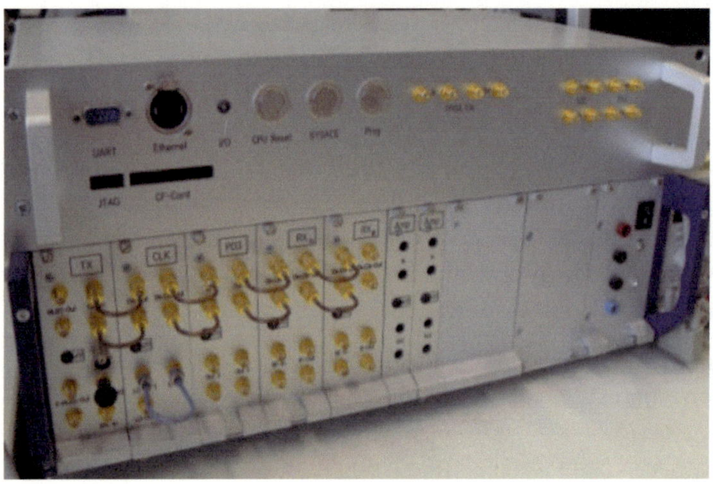

Figure 52. Photograph of the demonstrator implementation example.

6.2.2. Demonstrator performance

The particular RF plug-ins of the demonstrator are designed to operate with signals of large fractional bandwidth at the lower end of microwave frequencies or with toggle frequencies up to about 20 GHz. The generator unit provides periodic M-sequences of length 2^m-1, where m represents the order of the sequence. The demonstrator has optionally implemented 9th or 12th order generators, which accordingly produce signals with periods 511 or 4095 chips. The generator plug-in operates with toggle rates between 500 MHz and 20 GHz for the 9th order M-sequence, and the 12th order device may be operated between 500 MHz and 16 GHz. In the case of radar applications, the unambiguity range (4) of the measurement may cover values from 3.8 m (related to 9th order M-sequence and 20 GHz clock) up to 1.2 km (12th order M-sequence and 500 MHz clock).

The clock synchronization unit which precisely defines the receiver sampling points is a 9th order binary divider with a maximal toggle rate of 24 GHz. Random fluctuations of the sampling point (jitter) could be reduced down to some tens of femtoseconds [48] due to the balanced circuit topology and the optimized architecture of the timing system (see [2], [49] detailed discussions). Note, that the time position uncertainty of the measured impulse response (compare Fig. 5) is father decreased as consequence of the impulse compression (i.e. correlation; see [2] for discussion).

The clock distribution plug-in is an active device which recovers and distributes the sampling clock among the receivers and the analog-to-digital converters. The unit can handle clock pulses with 20 ps falling/rising edges and features wideband reverse signal rejection better than 40 dB per branch.

The receivers are ultra wideband sampling gates with an 18 GHz analog input bandwidth, better than -40 dB signal feed-through over the full bandwidth, -15 dBm input compression points and a decay rate of about 20 % per ms relative to full scale (i.e. 5…200 ppm per sampling cycle depending on the clock rate (0.5 – 20 GHz) of the system). Other potential components of the experimental demonstrator are discussed in sub-chapters 3 to 5.

The transmitter-receiver and receiver-receiver cross-talk is better than 130 dB over the full operational band. In order to achieve this value, attention was paid to RF-housing, clock signal distribution and power supply decoupling (see also Fig. 57). The recent configuration of the demonstrator RF electronics is able to handle (internally) up to about 70 000 IRFs per second (9th order M-sequence at 18 GHz system clock). The data transfer to a host PC (based on commercial standard interfaces like USB and LAN) reduces, however, the actually achievable update rate to about 300 IRFs/s. The corresponding gap is filled by synchronous averaging in order to use the available data amount for noise suppression. The achievable receiver dynamic is about 114 dB @ 1 IRF/s. It has to be noted that device non-linearity is classically qualified by the intercept point which is based on a Taylor-series model of the device under test and sine wave stimulation. In order to keep this established philosophy, the approach was extended to wideband signals [48].

This is illustrated by Fig. 53. In the example at the top, the Tx- and RX-port of an M-sequence device were connected via a variable attenuator and the impulse response was recorded for attenuator values between 0 and 120 dB. In the case of very weak input signals (large attenuation), we can only observe noise and device internal cross-talk. If we reduce the attenuation, the wanted signal peak (it is called "main pulse" in Fig. 53) appears and increases linearly with the signal level while the cross-talk level remains constant. By reducing the attenuation further, other signal parts become to protrude from noise. They also increase linearity at the beginning. These signals are caused from device internal reflections, deviations from the ideal time shape of an M-sequence and misalignments of the ADC timing (refer to Fig. 54). We call them device internal clutter. For very high signal levels, the receiver will tend to saturate which leads to the compression of the main peak and the internal clutter signals. Furthermore, the appearing non-linear distortions create new signal peaks which leave an apparently chaotic mark (see [2] and Fig. 1 for details). While the cross-talk and the internal clutter may be removed by device calibration [50] since they are caused by linear effects, the non-linear distortions should be avoided by respecting corresponding input levels of the measurement signal.

The level diagram at the bottom of Fig. 53 refers to the non-compressed receiver signal. It shows the strength of the linear, quadratic and cubic signal parts in dependency from the signal power (see [48] for details).

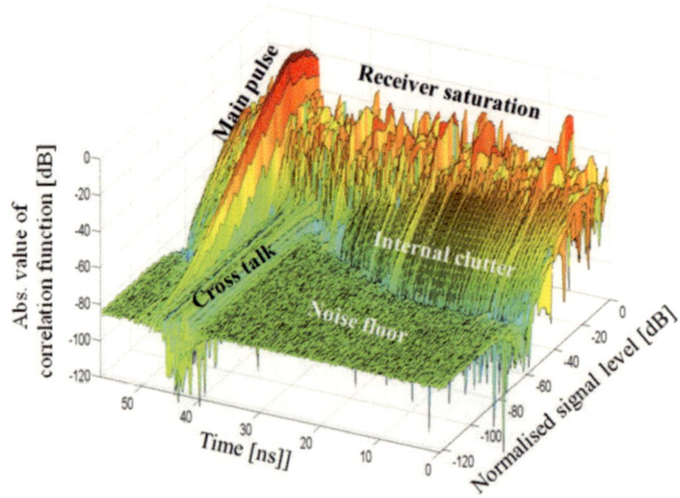

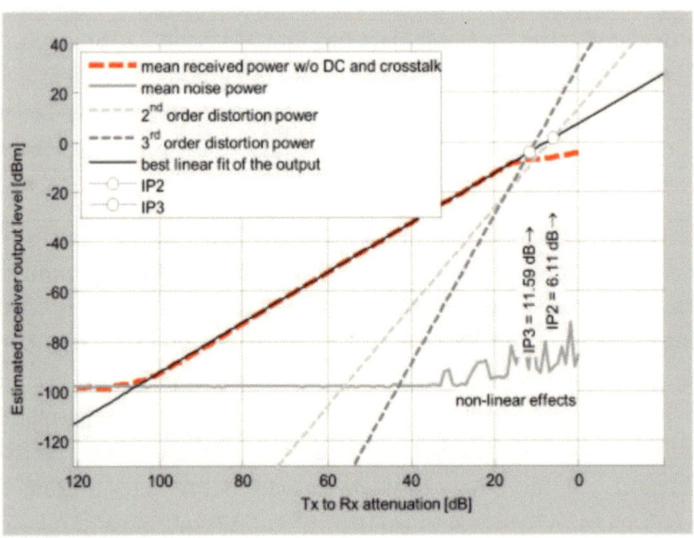

Figure 53. Top: Sensor pulse response of system as a function of the input level. Bottom: Level diagram (see also Fig. 1) of raw data (i.e. without impulse compression). In the shown case, the input related 1dB compression point is -14 dB below the transmitter power.

The effect of ADC timing misalignment is illustrated in Fig. 54. Theoretically, the ADC could capture the voltage sample at any time point within the hold phase of the T&H-circuit since by definition the signal level should keep a constant value during the hold interval. Unfortunately this is not case as demonstrated by Fig. 54. Here, the impulse response (i.e. the correlation function) of the M-sequence device was recorded by insertion a variable delay between the start of the hold phase and the trigger of the ADC. Ideally, we should see only a single pulse as long as the ADC is triggered within the hold phase and noise within the track phase (which is however out of interest here). But actually, some spurious signals appear whose strength and time position depend on the delay between "hold-start" and ADC-trigger. Hence, by selecting a reasonable delay between T&H and ADC, we can minimize these spurious signals.

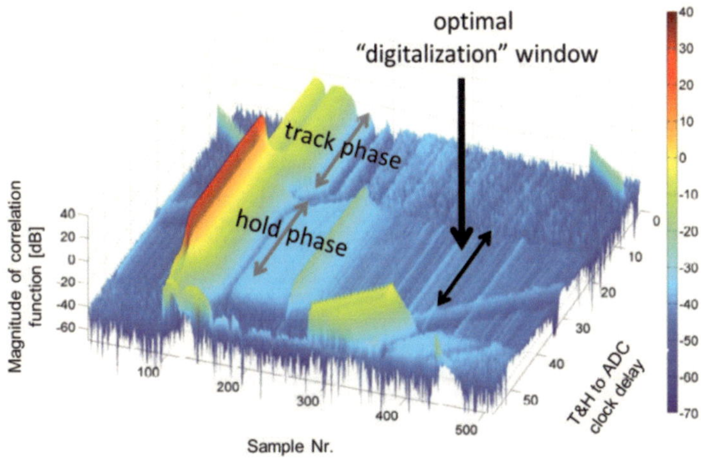

Figure 54. Sensor pulse response as a function of delay between T&H and ADC.

6.2.3. M-sequence feedback-sampling

This sub-chapter gives an example of the usefulness of the modular experimental device. It deals with feedback sampling. Feedback loops have been used for a long time in sampling circuits. However, they were usually restricted to sequential sampling having very large Nyquist rates so that only minor signal variations between consecutive samples appear. Only these variations are captured by that approach (see [2] for details).

In our case, this simple method cannot be applied since the voltage steps between two consecutive samples may cover the full receiver input range as we firstly apply Nyquist sampling and secondly, the natural order of the data samples may be disrupted due to interleaved sampling. Hence, we need some modifications of the principle which pose some challenges to the practical implementation.

For the purpose of feedback sampling, the data capturing & control unit was additionally equipped with a digital-to-analog converter which has to provide the feedback signal. The principle and the device structure are depicted in Fig. 55. The idea behind the digital feedback sampling implementation is to deal with high-speed signals (analog and digital) of low dynamic range (i.e. low amplitude) and to exploit the fact that the temporal variations of the scenarios under test are of the orders smaller than the measurement speed. This implies for the radargram (see Fig. 55, on the left) that adjacent samples at a horizontal line undergo only minor variations (instead of consecutive samples in sequential sampling). Thus, it will be possible to predict the measurement values along the observation time axis. This is the reason to insert a DAC into the feedback loop which converts the predicted digital values into analog ones. If the predicted signal levels are subtracted from the received signal, only the prediction error has to be captured by the ADC and processed by a digital high-speed system.

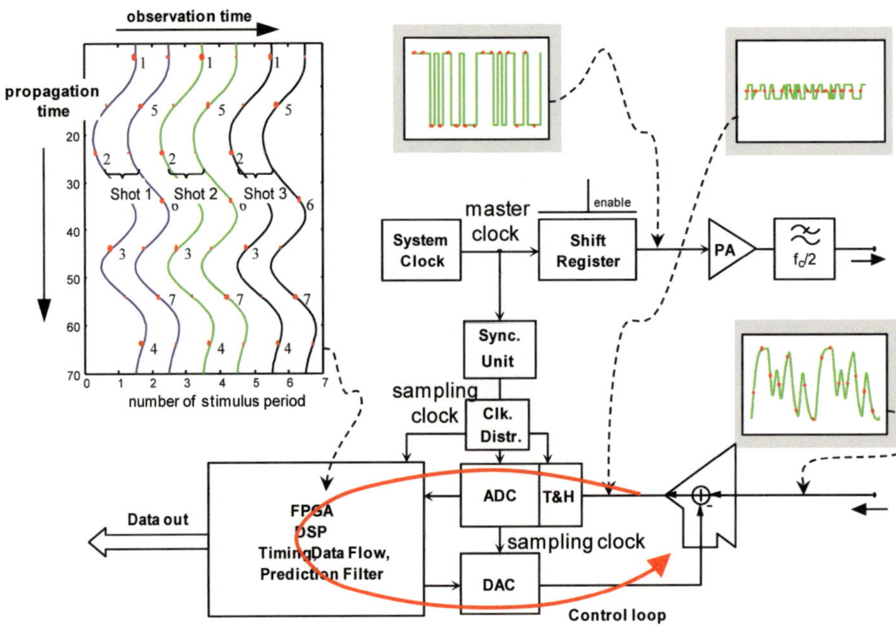

Figure 55. Basic structure of an M-Sequence feedback approach.

Fig. 56 gives an example of the output signals of the T&H-circuit. The constant voltage during the hold phase must be captured by the ADC. In the open loop example (above), we can observe that the hold voltage jumps from sample to sample. Hence, the ADC must be able to convert voltages within a large range. The second example shows the closed loop operation. Now, the predicted value is subtracted before AD-conversion, and we actually get a voltage during the hold phase which is always at about the same level. Under optimum conditions, the magnitude of the prediction error is determined by the strength of random noise which is usually quit weak. Therefore the requirements onto the dynamic range of the receive electronics can be relaxed.

Under optimum conditions, the magnitude of the prediction error is in the same order as random noise. Therefore, the demands made on the dynamic range of the receive electronics can be relaxed.

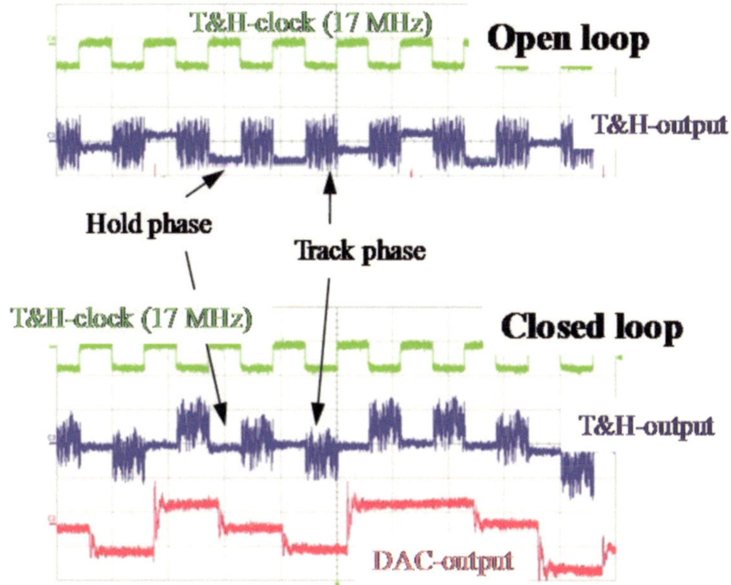

Figure 56. Digital feedback sampling. Above: open loop; below: closed loop.

6.3. Prototype devices

Fig. 57 shows a photograph of a primary (1Tx 2Rx) M-sequence RF board and corresponding ADC PCB with PC Interface (USB). The RF board is designed for assembly with the *HaLoS*-project originating ICs. Each of the board layouts corresponds to the architecture shown in Fig. 4, so that both boards connected together represent the basic M-sequence working unit. This unit is considered as main integral part of the UWB devices provided for partner projects within the UKoLoS- and other scientific projects.

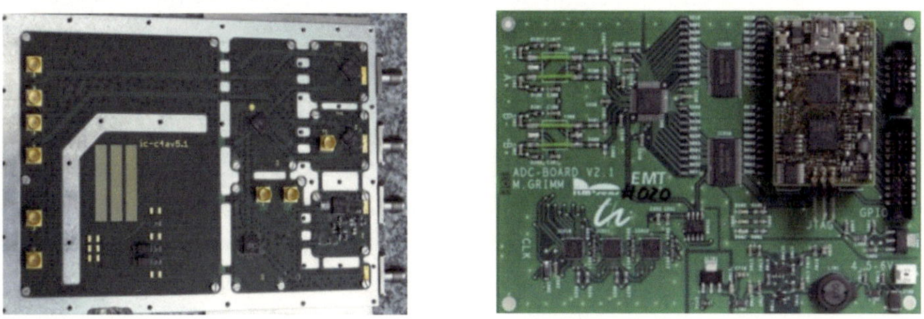

Figure 57. Photograph of the primary 1Tx 2Rx RF board (left-hand side) and corresponding ADC board (right-hand side).

While in the early project phase the sensor devices were finalized in cooperation with Meodat GmbH (Ilmenau, Germany), the final device assembly was performed by ILMSENS (TU Ilmenau service GmbH, Ilmenau, Germany) later. To give the reader an impression, some device examples are depicted in Fig. 58 as well in the chapter 11.

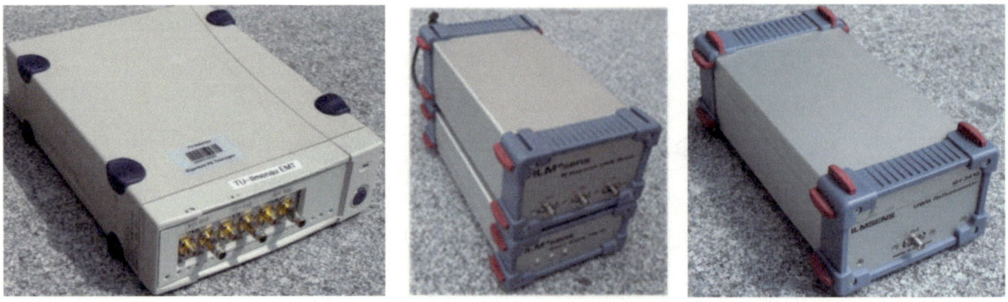

Figure 58. Left-hand side: 1Tx 2 Rx UWB sensor unit (MEODAT). Middle and right-hand side: 1Tx 2Rx UWB sensor unit with portable power supply and UWB reflectometer (ILMSENS).

6.4. Single-chip sensor head

The ability to create an optimized multi chip sensor is apparent, but the manufacturability of such system is much more difficult with a longer parts list and more complex assembly as for instance in the case of the construction of complex MIMO sensing systems (see 8Tx 16Rx system in the chapter 11). One promising way is to realize all active high-frequency system components (i.e. components on the primary RF board – see Fig. 57 left-hand side) onto one chip. This will enhance the overall system performance, reliability, robustness and assembling yields. By contrast, however, increased complexity on the single die means more second-order effects that have not been studied so far. For example, undesired on-chip coupling interactions between the different constituent system components become more pronounced and are more challenging to manage especially because of dealing with ultra-wideband signals. Such unwanted signal coupling or cross-talk can degrade the performance of the sensitive receive circuitry and, consequently, of the whole system. The aim to study such interactions which have not been considered so far, the expected advantages but also the knowledge gained from multi-chip approach analyses, have motivated the first monolithic integration of the complete RF-part of the M-sequence UWB radar electronics into one silicon die.

6.4.1. UWB single-chip head architecture

Fig. 59 shows the simplified block topology of the realized M-Sequence based single-chip transceiver head (alias System-on-Chip, SoC head). In correspondence with the system topology depicted in Fig. 4, the M-sequence transceiver SoC contains one transmitter and two receiver circuits (commonly assigned as 1Tx 2Rx configuration).

According to our experience, the 1Tx-2Rx topology of the primary sensor cell represents the optimum regarding achievable performance and circuit complexity. Moreover, the implementation of 1Tx 2Rx structure on one die has the advantage of permitting both cross-talk investigations between active and passive circuit parts (i.e. transmitter and receiver) as well as between two passive parts (i.e. receiver 1 and 2). From a practical point of view, the stand-alone 1Tx 2Tx devices are suitable for implementations where two receive channels are needed a priori, e.g. for simple localization tasks or in material testing (see chapter 11) in which the second (slave) receive channel can be used for device online calibration purposes. The desired MIMO usability as for instance in novel UWB-arrays for high-resolution near-field imaging (*ultraMedis*) or localizations (*CoLoR*) with 1Tx 2Rx constellation of primary sensing cells is also given.

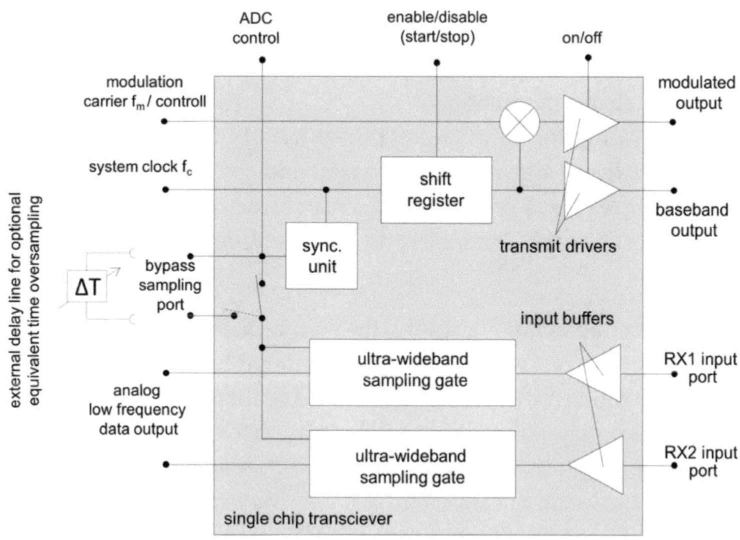

Figure 59. Simplified architecture of a fully monolithically integrated UWB M-Sequence based transceiver.

6.4.2. Design philosophy

It is apparent that the presented single chip architecture envisages ultra-fast switching cells (i.e. stimulus generator and synchronization unit) with their relatively high signal swing output buffers as well as very sensitive analog input blocks integrated on the same chip substrate. So, the undesired signal coupling or cross-talk can degrade the performance of the sensitive receive circuitry and, consequently, of the whole system. Especially in the case of analog devices, which handle the ultra-wideband signals, the on-chip interferences can be catastrophic. For example, intermodulation/interaction of noise components with the measured signal within the frequency band of interest may cause device saturation. Therefore, special emphasis is put on the isolation of the SoC channels during the design phase ,as discussed in [2] or [49].

6.4.3. Individual functional block peculiarities

Particular functional circuit cores of the SoC transceiver components are designed to fulfill at minimum the parameters of the demonstrator plug-in blocks discussed above. Additionally, the SoC transceiver includes additional build-in options to open further functionalities as e.g. (equivalent time) oversampling [51] or frequency conversion in order to meet the UWB radiation rules [52]-[55]. In particular, the SoC concept is intended for very wideband material investigations and MIMO-applications like in medical microwave imaging [56]-[61].

In summary, the goals of the single-chip integration are:

- to improve the synchronization between transmitter and receiver due to shorter interconnections with steeper signal edges,
- to provide means of a flexible adaptation of the operational frequency band by introducing a wideband modulator,
- to save power consumption by avoiding power hungry PCB-interconnection lines,
- to investigate broadband signal leakage on chip and cross-talk due to the housing,
- to avoid temperature effects on calibrated sensor systems due to temperature difference between the measurement channels and temperature expansion of device internal cables.

Thus, to achieve the desired MIMO usability, the shift register may be enabled and disabled, and transmitter buffers can be switched off (power down) by simple TTL-signals so that no external RF-switch is required to operate in a MIMO system. The transceivers are designed in such a way that they may either work while being driven individually or they may be cascaded with respect to the master system clock so that all units of a MIMO array work synchronously. Once the array is calibrated, a power down feature will be used for active transmitter selection. Thus, all receivers of MIMO array work in parallel and capture permanently data in order to get maximum measurement speed. As shown in Fig. 59, the transceiver IC is equipped with a wideband multiplier which optionally allows the sensor stimulus frequency band to be shifted and doubled ([2] or [49], [55]) or the operational band to be adapted to a specific application [50] in conformity with regulation requirements [52] - [54]. The channel is designed for operation up to 18 GHz.

Moreover, the multiplier can invert the stimulus M-sequence by implementing simple ECL signals on the control port. This feature may be useful to provide uncorrelated transmit signals in MIMO arrays. In addition, the sampling timing control chain is equipped with optional switchable shunt path. This add-on allows direct clock supplying from chip periphery. Thereby, user-selectable sampling rates or enhanced signal capturing approaches (e. g. equivalent time oversampling approach [50]) are possible without IC redesign. The analog receivers are designed to operate with wideband signals up to 18 GHz. The maximum linear operation input signal peak-peak swing is 60 mV.

Fig. 60 shows the chip die micrograph of the discussed transceiver with marked particular functional blocks and well visible top metal of a decoupling guard between the transmitter and receiver (line in the middle). The transmitter and receiver cores as well as their particular I/O pads are placed on the opposite die sides to minimize mutual on-chip coupling as well as inductive coupling between the bond wires after packaging. As extensively discussed in [62] or [63], [64], the decoupling guard is a guard well in a trench between the noisy transmitter and sensitive receivers. In the final assembly, the guard is connected to the quiet potential in order to fix the voltage of the substrate between the Tx and Rx die part by absorbing potential substrate fluctuations. The transceiver die occupies an area of about $2000\mu m \times 1200\mu m$ and the build-in circuits consume in total about 300 mA from 3 V supply.

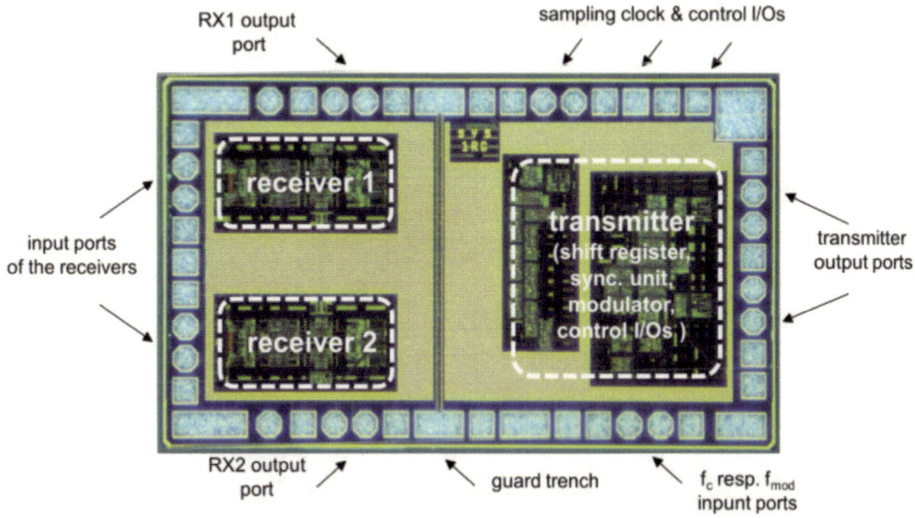

Figure 60. Microphotograph of the SiGe M-Sequence based transceiver head die with depicted particular functional blocks. The die size is 2×1.2 mm².

6.4.4. *Single chip transceiver head evaluation*

For the sensor head prototype evaluation, the transceiver chip has been measured on wafer as well as packaged with well-established chip-on-board technology using an optimized composite 4-layer carrier made from Rogers 4003C™, FR4 laminate and ultrasonic bonding procedure with 1-mil aluminum wires. The bond wire landing areas for RF ports on the board are designed to match as closely as possible (i.e. realizable) the pitch on the IC to avoid long wire connections. The cavity approach has not been implemented because of challenging technological realization on the selected carrier board. Fig. 61 shows the test board whereat the wired die is zoomed out for better visualization. The die is located in the center of the photo and top glue is used to protect the bond wires. It is mounted on a metal patch which is connected with VEE. This allows a direct connection between the substrate and the board's lowest potential. The top layer is mainly used for RF signal routing whereas the bottom layer is used for control lines. The inner plane layer below the top layer is the common board GND, and it also provides the GND reference for the RF signal lines. The other inner plane is the supply layer. The supply is bypassed to the GND with a 0.1 μF ceramic capacitor placed as close as possible to the die.

A basic test set-up for the sensor head prototype parameter evaluation is symbolically depicted in Fig. 62. The photograph on the right side shows an example of such test assembly. The evaluation board is connected to a 10-bit data "digitizer" (ADC), FPGA control and pre-processing unit which is equipped also with the PC interface. Moreover, for the parameter characterization, stable sinusoidal reference has to be connected to the system clock port (not shown in the photograph). This signal comes on board through an SMP connector and toggles the analyzed assembly. The toggle rate for the packaged prototype can be chosen quite flexibly between 0.5 and about 19 GHz, which implies a good compliance with the actual and intended applications needs.

Figure 61. Evaluation board for Single-Chip sensor head. The wired die (close-up photograph) is protected by top glue.

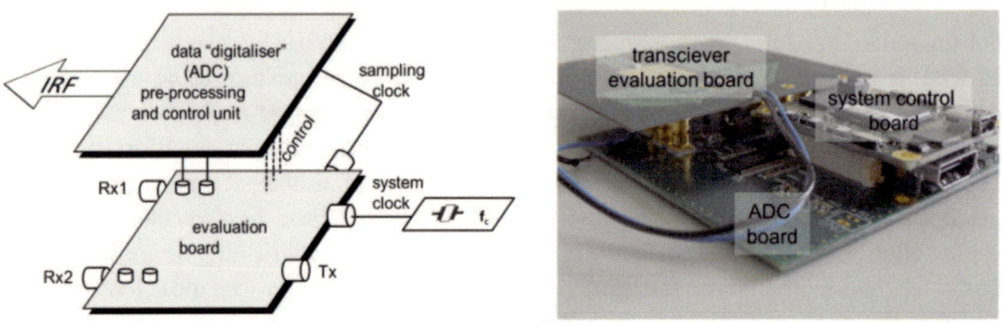

Figure 62. Generic transceiver test configuration (left) and an example of an experimental M-sequence based sensing unit assembly (right).

Interchannel cross-talk plays an important role in many applications. As Tx-Rx-decoupling up to 130 dB could be reached if the individual components are properly shielded (see Fig. 57, left), an interesting question is how the single chip devices behave with respect to that problem even though decoupling design techniques are implemented [2]. Fig. 63 shows the results for on-wafer measurements and the housed chip. Obviously, the chip design outperforms the quality of the chip wiring with respect to the cross-talk performance. The impulse response function of the housed chip is also shown in Fig. 63 (left). It was gained using the configuration as depicted in Fig. 62 (left). The cross-talk pulse can clearly be identified. However, it should be noted that it can largely be suppressed by post-processing via system calibration. Thus, achieved spurious free system dynamic range is comparable to that of the demonstrator device.

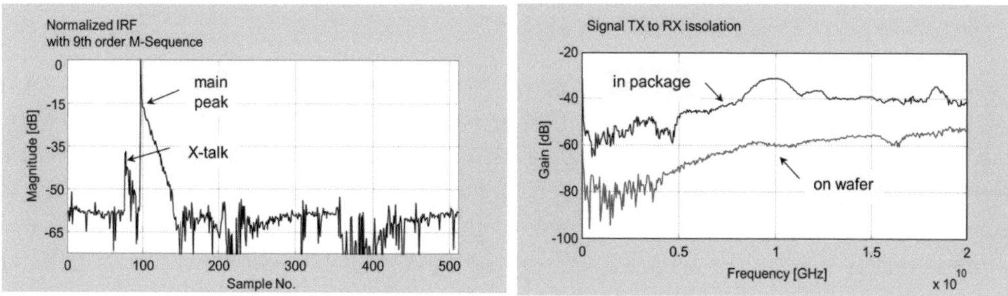

Figure 63. Example of normalized IRF captured with a 9th order M-Sequence based experimental single chip assembly and broadband measured TX to RX isolation.

As a result, it can be concluded that we have successfully realized a novel functional hardware platform, both multi-chip and single-chip based, for current (e.g. [56], [60], [61]) as well as future scientific investigations in the complex field of an ultra-wideband MIMO sensing and localizations.

7. Summary

Electromagnetic sounding for non-destructive and remote sensing, respectively, has been exploited for a long time. However, its practical application was mostly restricted to narrow-band sensors or it was banned to the laboratory in the case of wideband examinations. The reason for this limitation has been the lack of reasonable wideband measurement equipment.

The first field deployable ultra wideband devices were used in ground-penetrating radar (GPR). They mostly exploited powerful nanosecond or sub-nanosecond pulses to feed the transmission antenna. Meanwhile, several other UWB-sensor techniques have been introduced. Section 2 summarizes the most popular of them. The challenges of corresponding research and development are mainly to be seen in the performance improvement of the sensor electronics and its monolithic integration aimed at cost and power reduction.

The main part of the chapter deals with a pseudo-noise UWB approach and its main components. The pseudo-noise concept is an interesting alternative to other wideband sensing principles promoting both high device performance and monolithic integration. Due to its simple and rigid synchronization, it provides exact and time-stable signal generation and signal capture which promotes:

- simple adaptation of bandwidth, signal duration (period duration) and recording time to the needs of the actual application,
- the implementation of large MIMO-arrays,

- data processing in the time and frequency domain,
- device calibrations as usually with network analyzers,
- high-range precision and super-resolution capabilities, and
- excellent micro Doppler performance.

The most relevant RF components of a pseudo-noise sensor cover the test signal generation (i.e. pseudo noise code), the analog handling of the receive signals, and the high-speed conversion of the analog signals to the digital domain. Device concepts suited for these tasks are discussed in sections 3 to 5. Due to special requirements set by the application and the applied semiconductor technology, innovative solutions are presented. Among those are a distributed power amplifier with a novel cascode gain cell, new subtraction amplifiers, an analog-to-digital converter with a new reference network, and a high-speed predictor. Also, appropriate verification schemes are presented. A final section referring to implemented devices as they were applied in other UoKoLoS-projects suggests some first steps toward a fully integrated pseudo-noise sensor device.

Author details

Stefan Heinen, Ralf Wunderlich and Markus Robens
RWTH Aachen University, Germany

Jürgen Sachs and Martin Kmec
Ilmenau University of Technology, Germany

Robert Weigel, Thomas Ußmüller, Benjamin Sewiolo and Mohamed Hamouda
Friedrich-Alexander Universität Erlangen, Germany

Rolf Kraemer, Johann-Christoph Scheytt and Yevgen Borokhovych
Brandenburgische Technische Universität Cottbus, Germany

Acknowledgement

This work has been funded by the German Research Foundation (DFG) in the framework of the priority program UKoLoS (SPP 1202), project acronym HaLoS. The authors greatly appreciate the discussions and valuable support in electronic design given by M. Grimm, R. Herrmann, P. Rauschenbach and K. Schilling, as well as the support given by K. Borkowski in the manufacture of the electronic and mechanical components.

8. References

[1] J. Sachs, "Ultra-Wideband Sensing: The Road to new Radar Applications," in *IRS 2010 International Radar Symposium*, Vilnius (Lithuania), 2010.

[2] J. Sachs, *Handbook of Ultra-Wideband Short-Range Sensing - Theory, Sensors, Applications.* Berlin: Wiley-VCH, 2012.

[3] E. J. Candes, and M. B. Wakin, "An Introduction To Compressive Sampling," *Signal Processing Magazine, IEEE,* vol. 25, no. 2, pp. 21-30, 2008.

[4] E. J. Candes, "Compressive sampling," in *International Congress of Mathematicians,* Madrid, Spain, 2006.

[5] J. Ender, "Do we still need Nyquist and Kotelnikov? - Compressive Sensing applied to RADAR," in *8th European Conference on Synthetic Aperture Radar EUSAR 2010,* Aachen, Germany, 2010.

[6] H. Qiong, Q. Lele, W. Bingheng *et al.,* "UWB Through-Wall Imaging Based on Compressive Sensing," *Geoscience and Remote Sensing, IEEE Transactions on,* vol. 48, no. 3, pp. 1408-1415, 2010.

[7] P. v. Genderen, "Multi-Waveform SFCW radar," in *33rd European Microwave Conference,* 2003, pp. 849-852.

[8] S. Foster, "Impulse response measurement using Golay codes," in *IEEE International Conference on Acoustics, Speech, and Signal Processing,* ICASSP '86., pp. 929-932.

[9] J. Lee and J. D. Cressler, "Analysis and Design of an Ultra-Wideband Low-Noise Amplifier Using Resistive Feedback in SiGe HBT Technology," in *IEEE Trans. Microw. Theory Tech.,* Vol. 54, No. 3, pp. 1262-1268, March 2006.

[10] M.-C. Chiang et al., "Analysis, Design, and Optimization of InGaP-GaAs HBT Matched-Impedance Wide-Band Amplifiers with Multiple Feedback Loops," in *IEEE J. Solid-State Circuits,* Vol. 37, No. 6, pp. 694-701, June 2002.

[11] A. Ismail and A. A. Abidi, "A 3-10-GHz Low-Noise Amplifier With Wideband LC-Ladder Matching Network," in *IEEE J. Solid-State Circuits,* Vol. 39, No. 12, pp. 2269-2277, December 2004.

[12] F. Thiel, O. Kosch, and F. Seifert, "Ultra-Wideband Sensors for Improved Magnetic Resonance Imaging, Cardiovascular Monitoring and Tumour Diagnostics," in *Sensors 10,* pp. 10778-10802, December 2010.

[13] P. R. Gray et al., "Analysis and Design of Analog Integrated Circuits," 5th edit., John Wiley & Sons, 2010.

[14] M. Robens, R. Wunderlich, and S. Heinen, "UWB LNAs for Ground Penetrating Radar," in *IEEE Int. Symp. on Circuits and Systems,* Taipei, Taiwan, pp. 229-232, May 2009.

[15] H. Knapp et al., "15 GHz Wideband Amplifier with 2.8 dB Noise Figure in SiGe Bipolar Technology," in *IEEE Radio Freq. Integr. Circuits Symp.,* pp. 287-290, 2001.

[16] B. Kang et al., "Design and Analysis of a Cascode Bipolar Low-Noise Amplifier With Capacitive Shunt Feedback Under Power-Constraint," in *IEEE Trans. Microw. Theory Tech.,* Vol. 59, No. 6, pp. 1539-1551, June 2011.

[17] P. K. Datta and G. Fischer, "An Ultra-Wideband Low Power Consumption Differential Low Noise Amplifier in SiGe:C BiCMOS Technology," in *IEEE Radio and Wireless Symp.,* pp. 107-110, January 2006.

[18] M. Robens et al., "Differential UWB-LNA for M-Sequence Radar," in *Semicond. Conf. Dresden,* April 2009.

[19] M. Robens, R. Wunderlich, and S. Heinen, "Differential Noise Figure Measurement: A Matrix Based Approach," in *IEEE MTT-S Int. Microw. Symp. Dig.*, Anaheim, CA, pp. 385-388, May 2010.

[20] M. Robens, R. Wunderlich, and S. Heinen, "Differential Noise Figure De-Embedding: A Comparison of Available Approaches," in *IEEE Trans. Microw. Theory Tech.*, Vol. 59, No. 5, pp. 1397-1407, May 2011.

[21] H. Bosma, "On the Theory of Linear Noisy Systems," in Ph. D. dissertation, Dept. Elect. Eng., Eindhoven Univ. Technology, 1967.

[22] J. Randa, "Noise Characterization of Multiport Amplifiers," in *IEEE Trans. Microw. Theory Tech*, Vol. 49, No. 10, pp. 1757-1763, October 2001.

[23] A. A. Abidi and J. C. Leete, "De-Embedding the Noise Figure of Differential Amplifiers," in *IEEE J. Solid-State Circuits*, Vol. 34, No. 6, pp. 882-885, June 1999.

[24] J. Dunsmore and S. Wood, "Vector Corrected Noise Figure and Noise Parameter Measurements of Differential Amplifiers," in *European Microwave Conf.*, Rome, Italy, pp. 707-710, September 2009.

[25] S. W. Wedge and D. B. Rutledge, "Noise Waves and Passive Linear Multiports," in *IEEE Microw. and Guided Wave Letters*, Vol. 1, No. 5, pp. 117-119, May 1991.

[26] V. A. Monaco and P. Tiberio, "Computer-Aided Analysis of Microwave Circuits," in *IEEE Trans. Microw. Theory Tech.*, Vol. 22, No. 3, pp. 249-263, March 1974.

[27] M. Robens, R. Wunderlich, and S. Heinen, "Input Amplifier for Sensitivity Improvement in an M-Sequence Radar Front-End," in *Int. Conf. Indoor Positioning and Indoor Navigation*, pp. 450-451, September 2010.

[28] M. G. Chen and J. K. Notthoff, "A 3.3–V 21– Gb/s PRBS Generator in Al-GaAs/GaAs HBT Technology," *IEEE J. solid-State Circuits*, vol. 35, no. 09, pp. 1266–1270, Sept. 2000.

[29] T. Kjellberg, J. Hallin, and T. Swahn, "104 Gb/s 211-1 and 110 Gb/s 29-1 PRBS Generator in InP HBT Technology," in *IEEE International Solid-State Circuits Conference. Digest of Technical Papers*. 2004, Feb. 2006, pp. 2160–2169.

[30] H. Knapp, M. Wurzer, W. Perndl, K. Aufinger, J. Bock, and T. F. Meister, "100–Gb/s 27-1 and 54– Gb/s 211-1 PRBS Generators in SiGe Bipolar Technology," *IEEE J. Solid-State Circuits*, vol. 40, no. 10, pp. 2118–2125, Oct. 2005.

[31] T. O. Dickson, E. Laskin, I. Khalid, R. Beerkens, X. Jingqiong, B. Karajica, and S. Voinigescu, "An 80 Gb/s 231-1 Pseudorandom Binary Sequence Generator in SiGe BiCMOS Technology," *IEEE J. Solid-State Circuits*, vol. 40, no. 12, pp. 2735–2745, Dec. 2005.

[32] E. Laskin and S. P. Voinigescu, "A 60mW per Lane, 4×, 23– Gb/s 27-1 PRBS Generator," *IEEE J. Solid-State Circuits*, vol. 41, no. 10, pp. 2198–2208, Oct. 2006.

[33] F. Weiss, H.-D. Wohlmuth, D. Kehrer, and A. L. Scholtz, "A 24- Gb/s 27-1 Pseudo Random Bit Sequence Generator IC in 0.13 μm Bulk CMOS." Proceedings of the *32nd European Solid-State Circuits Conference*, ESSCIRC 2006, p. 468-471

[34] H. Veenstra, E. van der Hejden, and D. van Goor, "A 15–27GHz Pseuo-Noise UWB Transmitter for Short-Range Automotive Radar in a Production SiGe Technology," in Proceeding of the *31st European Solid-State Circuits Conference*, 2005., Sept. 2005, pp. 275–278.

[35] J. P. Fraysse, J. P. Viaud, M. Campovecchio, P. Auxemery, and R. Qur, "A 2W, High Efficiency, 2-8GHz, Cascode HBT MMIC Power Distributed Amplifier," in *2000 IEEE MTT-S Int. Microwave Symp. Dig.*, vol. 1, June 2000, pp. 529–532.

[36] R. Zetik, J. Sachs, and R. S. Thomae, "UWB Short Range Radar Sensing," *Instrumentation and Measurement Magazine*, IEEE, vol. 10, pp. 39–45, Apr. 2007.

[37] J. Sachs, R. Herrmann, M. Kmec, M. Helbig, and K. Schilling, "Recent Advances and Applications of M-Sequence Based Ultra-Wideband Sensors." *IEEE International Conference on Ultra-Wideband*, ICUWB 2007.

[38] U. Schwarz, M. Helbig, J. Sachs, R. Stephan, and M. A. Hein, "Design and Application of Dielectrically Scaled Double-Ridged Horn Antennas for Biomedical UWB Radar Applications," in *IEEE International Conference on Ultra-Wideband,* ICUWB 2009., Sept. 2009, pp. 150–154.

[39] B. Sewiolo, J. Rascher, G. Fischer, and R. Weigel, "A 30 GHz bandwidth driver amplifier with high output voltage swing for ultra-wideband localization- and sensor systems," in *Proc. Asia-Pacific Conf.* (APMC'08), Dec. 2008, pp. 1–4.

[40] S. Trotta, H. Knapp, K. Aufinger, T. F.Meister, J. Bock, B. Dehlink, W. Simburger, and A. L. Scholtz, "An 84 GHz bandwidth and 20 dB gain broadband amplifier in SiGe bipolar technology," *IEEE J. Solid-State Circuits,* vol. 42, no. 10, pp. 2099–2106, Oct. 2007.

[41] C. Schick, T. Feger, E. Sonmez, K.-B. Schad, A. Trasser, and H. Schumacher, "Broadband SiGe HBT amplifier concepts for 40 Gbit/s fibreoptic communication systems," in *Proc. 35th Eur. Microw. Conf.*, Oct. 2005, vol. 1, pp. 113–116.

[42] Borokhovych Y. High-Speed Data Capturing Components for Super Resolution Maximum Length Binary Sequence UWB Radar. Brandenburgischen Technischen Universität; 2011. Available from: http://systems.ihp-microelectronics.com/web/-index.php5?id=16.

[43] Razavi B. Principles of Data Conversion System Design. IEEE Press; 1995.

[44] Analog-Digital-Umsetzer mit breitbandigem Eingangsnetzwerk. DE102009002062A1 07.10.2010; 2010.

[45] Kester W. Analog-Digital Conversion. ADI Central Applications Department; 2004.

[46] Borokhovych Y, Gustat H, Scheytt C. 4-bit, 16 GS/s ADC with new Parallel Reference Network. In: Proc. *IEEE Int. Conf. Microwaves, Communications, Antennas and Electronics Systems* COMCAS; 2009. p. 1–4.

[47] Borokhovych Y, H Gustat. 4-bit, 15 GS/s ADC in SiGe. In: *Proc. NORCHIP*; 2008. p. 268–271.

[48] Streng, B.: "Entwicklung und Implementierung eines Messplatzes zur Charakterisierung des HF-Teils von M-Sequenzmessköpfen in MATLAB," TU Ilmenau, Fak. EI, FG EMT, diploma thesis, 2009

[49] M. Kmec, M. Helbig, R. Herrmann, P. Rauschenbach, J. Sachs, K. Schilling, "M-Sequence Based Single Chip UWB-Radar Sensor," *ANTEM/AMEREM 2010 Conference*, July 5 – 9, 2010, Ottawa, ON, Canada.

[50] R. Herrmann, "Development of a 12 GHz bandwidth M-sequence based ultra-wideband radar and its application to crack detection in salt mines", TU Ilmenau, dissertation, 2011

[51] R. Herrmann, J. Sachs, K. Schilling, and F. Bonitz, "New extended M-sequence ultra wideband radar and its application to the disaggregation zone in salt rock," Proc. *12th International Conference on Ground Penetrating Radar*. Birmingham, UK, Jun. 2008.
[52] FCC News Release, "New Public Safety Applications and Broadband Internet Access Among Uses Envisioned by FCC Authorization of Ultra-Wideband Technology", 14 Feb 2002
[53] FCC 02-48, "Revision of Part 15 of the Commission's Rules Regarding Ultra-Wideband Transmission Systems", First Report & Order, Washington DC, Adopted 14 Feb 2002, Released 22 April 2002
[54] Electronic Communication Committee, "ECC Decision of 24 March 2006 amended 6 July 2007 at Constanta on the harmonized conditions for devices using Ultra ◎Wideband (UWB) technology in bands below 10.6 GHz, " July, 6, 2007
[55] M. Kmec, J. Sachs, P. Peyerl, P. Rauschenbach, R. Thomä, R. Zetik, "A Novel Ultra-Wideband Real-Time MIMO Channel Sounder Architecture," *XXVIIIth URSI General Assembly 2005,* Oct. 23-29, New Delhi, October 2005.
[56] M. Helbig, I. Hilger, M. Kmec, G. Rimkus, J. Sachs, "Experimental phantom trials for UWB breast cancer detection," *German Microwave Conference*, GeMiC 2012, Ilmenau
[57] M. Lazebnik, E.L. Madsen, G.R. Frank et al.: „Tissue-mimicking phantom materials for narrowband and ultrawideband microwave applications" *Physics in medicine and biology* , vol. 50, 2005, 4245-4258
[58] E. C. Fear, S. C. Hagness, P. M. Meaney, M. Okoniewski, and M. A. Stuchly, "Enhancing breast tumor detection with near-field imaging," *IEEE Microwave Magazine*, vol. 3, no. 1, pp. 48–56, Mar. 2002
[59] I.J. Klemm J.A. Craddock, A. Leendertz et al.: "Radar-based breast cancer detection using a hemispherical antenna array – experimental results," *IEEE Trans on Antennas and Propagation*, vol. 57, 2009, 1692-1704
[60] Kosch O., Thiel F., Ittermann B., and Seifert F., "Non-contact cardiac gating with ultra-wideband radar sensors for high field MRI", *Proc. Intl. Soc. Mag. Reson. Med. 19*, (ISMRM), Montreal, Canada, ISSN 1545-4428, p.1804, (2011)
[61] M. Helbig, M. Kmec, J. Sachs, C. Geyer, I. Hilger, G. Rimkus, "Aspects of antenna array configuration for UWB breast imaging Brust", *6th European Conference on Antennas and Propagation,* Prague, March 2012
[62] Afzali-Kusha, A.; Nagata, M.; Verghese, N.K.; Allstot, D.J.; , "Substrate Noise Coupling in SoC Design: Modeling, Avoidance, and Validation," *Proceedings of the IEEE* , vol.94, no.12, pp.2109-2138, Dec. 2006
[63] S.M. Sze, *Physics of Semiconductor Devices 2nd edn*, New York, Wiley, 1981
[64] J.A. Olmstead and S. Vulih, "Noise problems in mixed analog-digital integrated circuits," in Proc. *IEEE Custom Integrated Circuits Conf.*, 1987, pp. 659-662

UWB in Medicine – High Performance UWB Systems for Biomedical Diagnostics and Short Range Communications

Dayang Lin, Michael Mirbach, Thanawat Thiasiriphet, Jürgen Lindner, Wolfgang Menzel, Hermann Schumacher, Mario Leib and Bernd Schleicher

Additional information is available at the end of the chapter

http://dx.doi.org/10.5772/55078

1. Introduction

This chapter presents scientific achievements in the field of UWB radar and communication systems for biomedical applications. These contributions focus on low-power MMIC designs, novel antenna structures and competitive approaches for communication and imaging.

The first section describes components for UWB radar sensors and communication systems, namely antennas and integrated circuits. Novel broadband antenna concepts for UWB radar and communication applications are presented. Symmetrical UWB antenna structures for free space propagation with improved performance compared to existing antennas regarding radiation pattern stability over frequency are designed, realized and successfully characterized. Novel differential feeding concepts are applied, suppressing parasitic radiation by cable currents on feed lines. For applications such as communication with implants and catheter localization, a miniaturized antenna optimized for radiation in human tissue is designed. The radiation characteristics of the antenna are measured using an automated setup embedded in a liquid consisting of sugar and water, mimicking the dielectric properties of biological tissue. For UWB radar transmitters, a differential and low-power impulse generator IC is realized addressing the FCC spectral mask based on a quenched cross-coupled LC oscillator. The total power consumption is only 6 mW at an impulse repetition rate of 100 MHz. By adding a simple phase control circuit setting the start-up phase condition of the LC oscillator, an impulse generator with a bi-phase modulation scheme is achieved. A further modification introduces a variable width of the pulse envelope as well as a variable oscillation frequency. The corresponding spectra have controllable 10 dB bandwidths and center frequencies fitting the different spectral allocations in the USA, Europe and Japan. On the receiver side, both a fully differential correlation-based and an energy detection receiver for the 3.1-10.6 GHz band are designed. Monostatic UWB radar systems require transmit/receive turn-around times in the nanosecond regime. Integrated front-ends which successfully address this issue are presented here for the first time.

The second section deals with signal processing. As, due to the large RF bandwidth, direct analog to digital conversion and digital signal processing are not feasible (at least not at reasonable power consumption), analog signal processing is one focus. For communication, detection methods based on analog correlation require channel estimation, storing of impulse responses and also precise time synchronization. Therefore methods based on energy detection are developed which require no or little channel knowledge, having low complexity, robustness to multipath propagation and high resistance to synchronization and symbol clock errors. New modulation techniques are described, which can cope with interchip and intersymbol interference. Also a novel support by a comb filter resulting in significant SNR improvements in interference and multiuser scenarios is presented. The methods developed for communication applications can also be used in the radar context. For detection and tracking of moving targets (e.g. heart in the body) new algorithms based on particle filtering are developed for the digital signal processing part. It is shown that the accuracy, the resolution and robustness can be improved compared to conventional methods. For the objective of catheter localization, the knowledge of the shape and position of the human body surface is inevitable. A UWB imaging algorithm for the detection and estimation of this surface has been developed based on trilateration and is also described in this second section. Furthermore, building on this surface estimation algorithm, a new method for the localization of transmitters in dielectric media is presented. Taking into account the refraction effects on the boundary surface, the algorithm uses the impulse time of arrival to determine the transmitter position inside of the dielectric medium.

The third section finally describes the design of bistatic UWB radar systems using the components presented in the first section. Single-ended and differential radar demonstrators are developed, with which the potential of impulse-radio UWB sensing is evaluated. Measurements aimed at applications of the developed hardware such as vital sign monitoring and communication with implants are presented. Further measurements are performed to prove the functionality of the imaging algorithms derived in the second section. For surface estimation, a single radar sensor is moved around a highly reflective target in order to emulate a whole sensor array. For the verification of subsurface transmitter localization, a transmitter is placed inside of a container filled with tissue mimicking liquid, and its position is visualized with respect to the estimated container surface.

2. Circuit and component design

2.1. UWB antennas

Concepts for antennas with an ultra-wideband behavior are well-known and established [30]. However, advancements focusing on specific applications and specific performance parameters are inevitable to keep pace with the requirements of modern communication, radar, and localization systems. In particular in the medical environment, application-oriented antennas are mandatory to cope with the need for reliable systems (e. g. health monitoring systems) or with extreme environmental conditions (e. g. implanted systems). In the following, three novel UWB antennas are introduced targeting different tasks in the medical field and showing outstanding characteristics with respect to certain key antenna parameters.

2.1.1. Circular slot antenna excited with a dipole element

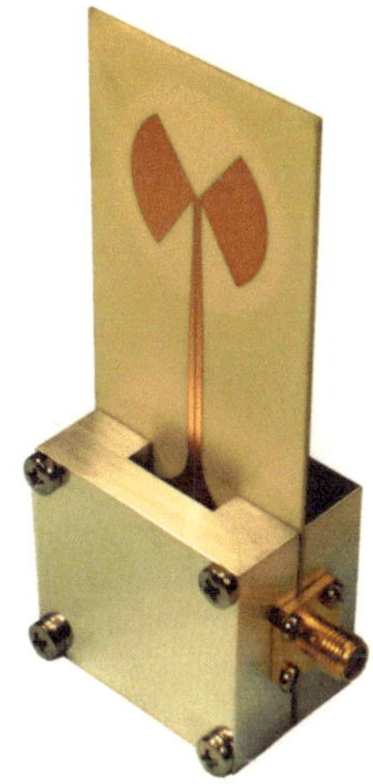

Figure 1. Dipole slot antenna.

Planar broad monpoles or dipoles are favored UWB antennas for communication systems with high data rates, e. g. potentially used in base stations for patient monitoring. However, broad monopoles fed single-endedly are prone to cable currents on the feeding line disturbing the radiation characteristic in the lower frequency range [15], while dipoles behave like a λ-radiator with a zero in main beam direction in the upper frequency range. Both effects lead to an undesired change of the radiation pattern in the operational frequency range reducing the effective bandwidth. For the widely-used impulse based UWB systems, this leads to a broadening of the impulse and consequently to a degraded system performance.

A practical solution to overcome the described parasitic radiation pattern performance is the combination of a circular slot antenna with a dipole feeding element as depicted in Fig. 1. The circular slot behaves like a broad monopole according to Babinet's principle. A broad dipole located in the center of the circular slot and consisting of two circular segments excites the slot antenna. The inherent symmetrical feeding of the dipole avoids the propagation of cable currents due to the virtual ground plane in between the transmission lines and results in an uniform radiation characteristic over the UWB frequency range.

The length of the exciting dipole is designed to be $\lambda/2$ at the center of the FCC UWB frequency range. Therefore, the dipole is smaller than pure UWB dipole antennas with a typical length of $\lambda/2$ at the lower edge of the FCC UWB frequency range. The perimeter of the circular slot is about λ at the lower edge of the FCC UWB frequency range leading to a resonance at 4.3 GHz (see simulation result for $|S_{11}|$ in Fig. 2(a)), and hence, to a return loss better than 10 dB at 3.1 GHz. Additional resonances with a low qualtiy factor are arising if the perimeter of the slot is a multiple of the wavelength (see at 6.9 GHz and 9.8 GHz in Fig. 2(a)). Therefore, a UWB behavior regarding return loss is achieved.

In order to characterize the dipole slot antenna with a single-ended coaxial line, a common UWB planar transition from coplanar stripline to a microstrip line based on [32] is used. A metallic shielding around this balun suppresses any parasitic radiation (see Fig. 1). The

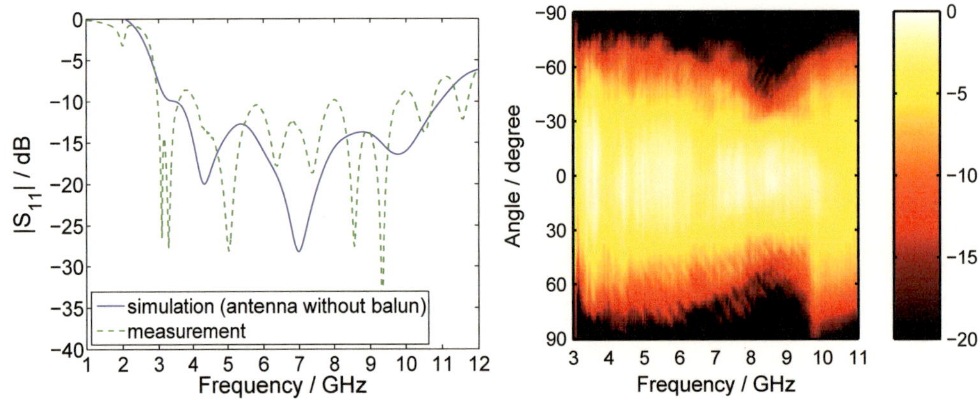

(a) Reflection coefficient

(b) Measured radiation pattern in dB (E-plane)

Figure 2. Reflection and radiation performance of the dipole slot antenna.

measured return loss behavior of the antenna including balun is presented in Fig. 2(a) and shows the predicted UWB performance.

The measured radiation pattern demonstrates the desired uniform characteristic over the FCC frequency range, as can be seen in Fig. 2(b) for the E-plane. In the H-plane a similar pattern is obtained having a slightly narrower beam with a mean 3 dB beamwidth of 60° in contrast to 65° for the E-plane. The uniform radiation behavior in the frequency domain results in a short antenna impulse response making the antenna a suitable candidate for impulse-based communication systems.

An upgrade of this single polarized antenna for horizontal and vertical polarizations is possible, if two narrow feeding dipoles are placed in orthogonal position to each other inside the circular slot. The expense of the additional feature is a deterioration in the return loss and radiation characteristic (see [20]).

2.1.2. Dielectric rod antenna fed by a planar circular slot

Emerging applications in medicine are vital sign monitoring [33], breast cancer detection [11], and tracking of inner organs for improved magnetic resonance tomography [38]. For these radar based sensing systems, directive antennas with a small beamwidth in both planes are compulsory. A promising approach to achieve this attribute is to exploit dielectric rod antennas either fed by a tapered slot antenna [14] or by a biconical dipole [4]. Both ideas show very good electrical characteristics, but suffer from necessary sophisticated fabrication and assembly.

A new antenna proposal is shown in Fig. 3(a). In this concept, the planar circular slot antenna presented in Sec. 2.1.1 acts as feed for a circular dielectric waveguide. The electrical field distribution inside the circular slot is similar to the fundamental mode H_{11} of the dielectric waveguide. Hence, this mode is predominantly excited. Due to that and since the H_{11}-mode possesses no cutoff frequency, in general an ultra-wideband performance is obtained. The diameter of the dielectric rod is chosen to 43 mm, which is a compromise between a good return loss behavior and single-mode operation of the dielectric waveguide. The conical shape at the end of the rod acts as smooth transition of the waveguide impedance to the free space impedance.

Due to the rod permittivity of 2.8, the electromagnetic field is mainly concentrated within the rod. Therefore, the antenna primarily radiates in forward direction along the rod. The unidirectionality and, hence, the gain is slightly improved – especially for lower frequencies – placing a metallic reflector at the backside of the antenna (see Fig. 3(a)). The reflector distance for the shown structure is $\lambda/4$ at 5.35 GHz leading to an optimized mean gain.

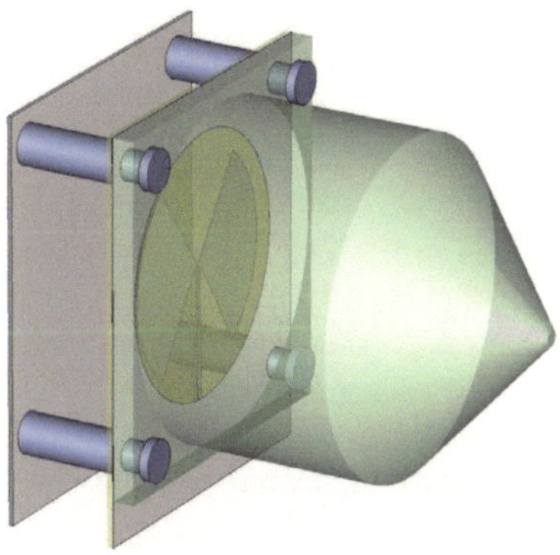

(a) Sketch with transparent rod

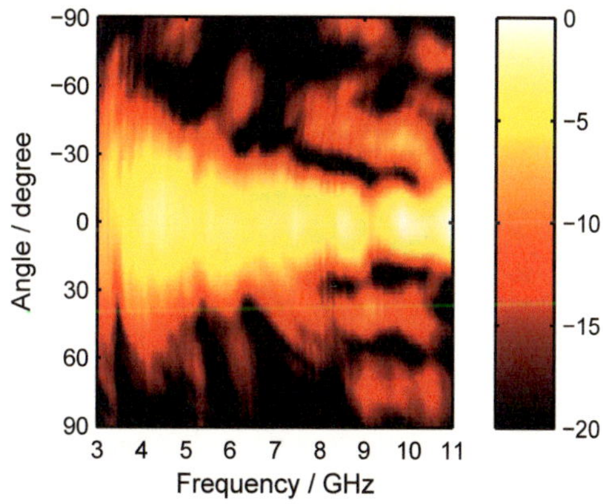

(b) Measured radiation pattern in dB (H-plane)

Figure 3. Sketch and radiation performance of the dielectric rod antenna fed by the dipole slot antenna.

The radiation pattern in Fig. 3(b) shows the obtained small beamwidth of the antenna for the H-plane. The increased side lobe level at frequencies above 8 GHz is caused by a parasitic leaky wave and is hardly avoidable for rod antennas. The given radiation characteristics in Fig. 3(b) are typical for both planes and results in a high mean gain of 8.7 dBi including the return loss. Due to the fact that the return loss is better than 10 dB from 3.5 GHz to 11.8 GHz, the influence of the return loss on the realized gain is negligible.

Besides this good electrical performance, the major benefits of this antenna in contrast to [14] and [4] are the compactness, the low weight, and the ease of fabrication. These attributes make the antenna also interesting for all kinds of industrial applications.

2.1.3. UWB slot antenna optimized for radiation in human tissue

High data rate communication for implanted devices [5] or precise catheter localization inside the human body are futuristic topics in medicine. There, impulse-based UWB technology is advantageous compared to narrow band systems due to the low power consumption caused by the simple system architecture. However, UWB antennas optimized for radiation in human tissue are hardly investigated.

In Fig. 4(a) a UWB antenna is proposed for radiation in human tissue, which is based on a similar concept as the dipole slot antenna in Sec. 2.1.1. An elliptical slot antenna is fed by a broad monopole located in the center of the slot. A monopole is chosen to obtain a small structure and to avoid a bulky balun for characterization purposes. Instead of a single layer structure, two substrate layers with slots in the top and bottom layer metalizations are used. The monopole is arranged in the center metalization layer and is fed by a triplate line. In this way, the buried feeding is insulated from the adjacent highly lossy human tissue. The antenna dimensions need to be optimized according to the surrounding medium of the antenna. The width of the antenna is 11 mm assuming skin tissue around the antenna with a typical permittivity of 28 at 6.85 GHz. The size reduction factor compared to an antenna designed for air instead of skin tissue is 5.4 leading automatically to a miniaturized UWB antenna.

(a) Sketch (b) Photograph

Figure 4. Sketch of the tissue optimized antenna and photograph including microstrip-to-coaxial transition.

In order to connect the antenna to a coaxial cable, a broadband transition from the triplate line to a coaxial line is applied [19]. The realized antenna including this transition is depicted in Fig. 4(b). The characterization of the tissue optimized antenna is performed in a tank filled with tissue-mimicking liquid approximating the permittivity and loss behavior of skin tissue. The chosen liquid is a 50% sugar solution in water [20]. Fig. 5(a) shows the return loss of the antenna being inside the sugar-water solution and compares its performance with a measurement, where the antenna is placed on both sides on human skin. Both measurement results agree very well and show a return loss of more than 10 dB above 3.8 GHz.

The radiation pattern is also measured in the tissue-mimicking liquid using two identical antennas and applying the two-antenna method. The obtained radiation pattern for the H-plane is presented in Fig. 5(b). There, the losses of the tissue-mimicking medium are compensated. Since the losses are increasing significantly with increasing frequency, measurements only up to 9 GHz are possible limited by the dynamic range of the measurement setup. Within this frequency range, a desired uniform and broad characteristic is achieved. Hence, UWB performance for a sufficiently small antenna for implants is demonstrated. For catheter localization additional miniaturization is required. A possible approach as well as more details about all presented antennas in Sec. 2.1 can be found in [20].

2.2. Transmitter MMICs

All integrated circuits reported in this section were realized in an inexpensive Si/SiGe HBT technology offered by Telefunken Semiconductors GmbH. Two kinds of transistors, with high f_T ($f_T = 80\,\text{GHz}$, $BV_{CEO} = 2.4\,\text{V}$) and with high breakdown voltage ($f_T = 50\,\text{GHz}$, $BV_{CEO} = 4.3\,\text{V}$) are available simultaneously. The process incorporates 4 types of resistors, MIM capacitors, as well as 3 metalization layers. All the devices were fabricated on a low resistivity $20\Omega\,\text{cm}$

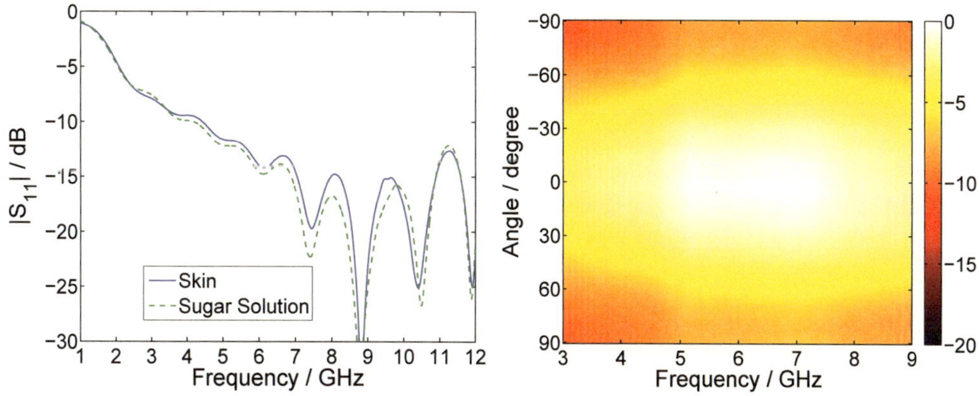

(a) Reflection coefficient

(b) Measured radiation pattern in dB (H-plane)

Figure 5. Reflection and radiation performance of the tissue optimized antenna.

substrate. The technology is fully adequate for impulse-radio-ultra-wideband (IR-UWB) applications.

Generating short time-domain impulses making efficient use of the spectral mask is the key challenge in IR-UWB systems. Approaches include the up-conversion of base band pulses to the allocated UWB frequency band using an oscillator and mixer [39] and direct generation based on damped relaxation oscillator [8]. Here impulse generators based on a quenched-oscillator concept with great circuit simplicity are presented. A cross-coupled LC oscillator is chosen as the core to introduce tunability of the waveform and the inherent convenience of achieving biphase modulation.

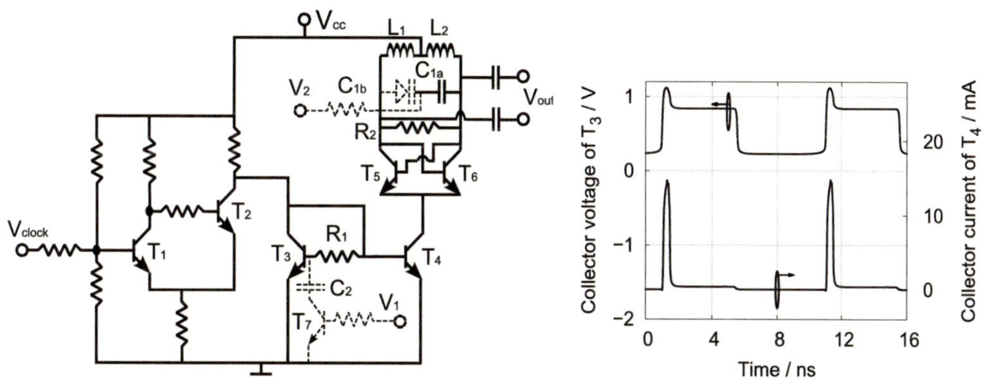

(a) Circuit schematic

(b) Simulation

Figure 6. Complete circuit schematic of the UWB impulse generator. The dashed components (C_{1b}, C_2 and T_7) show the extension for tunability of the impulse shape for different spectral masks and transistor level simulation: the collector potential of T_3 and the collector current of T_4.

Fig. 6(a) shows the impulse generator circuit. First, disregard the components with a dashed line, which are the extention for tunability of the impulse shape. T_1 and T_2 form a Schmitt trigger, creating a fast rising edge at the collector of T_2 when a positive input signal triggers T_1 to be on. This reduces the effect of the time-domain influence of the input clock signal on the output impulses. After an initial sharp increase, the collector potential of T_2 is pulled down again to a lower constant value by the current mirror formed by T_3 and T_4. So the collector potential of T_3 has a spike performance before it becomes stable, as shown in Fig. 6(b), which correspondingly generates a collector current spike at T_4, creating the envelope of the output impulse. During the rest of the period, the collector current of T_4 is too low to turn the oscillator on since T_3 is chosen much larger than T_4. The width of the current spike is determined predominately by the time constant τ of the charging circuit including the resistor R_1 and the base-emitter capacitor C_{be3} of T_3. τ can be written as

$$\tau = R_1 C_{be3}. \tag{1}$$

The amplitude of the current spike can be easily adjusted by changing the size of T_4. The repetition rate of the current spike train is equal to the input signal frequency and limited by the spike width.

The LC oscillator is activated by the current spikes once the collector currents of the cross-coupled pair (T_5 and T_6) are high enough to create a negative real part of the impedance. A slight asymmetry in the cross-coupled pair ensures that the oscillation always starts with the same phase and shortens the start-up time which in turn reduces power consumption because the necessary current spike width for a given output impulse envelope is shortened. R_2 is placed to quench the oscillator off more quickly immediately after the current spikes have disappeared. Thus, short-time domain impulses with a repetition rate equal to the input signal frequency are generated. The center frequency ω_0 of the oscillation is mainly determined by L_1, L_2, C_{1a} and the parasitic capacitance from the cross-coupled pair C_{para}. ω_0 can be expressed as

$$\omega_0 = \frac{1}{\sqrt{(C_{1a} + C_{para})(L_1 + L_2)}}. \tag{2}$$

It is designed to be around 6 GHz to fit the FCC spectral mask. The microphotograph of this realized impulse generator is shown in Fig. 7(a). It is a quite compact design with an area of 0.50 x 0.60 mm^2 due to a simple circuit topology.

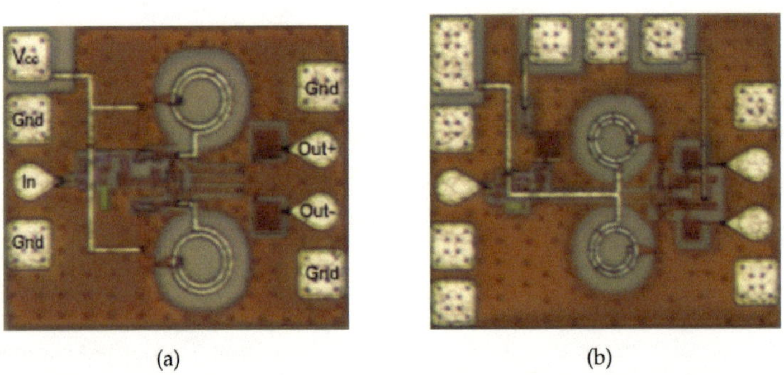

Figure 7. Microphotographs of (a) the realized UWB impulse generator shown in Fig. 6(a) excluding the components with a dashed outline and (b) the impulse generator tunable to FCC, ECC and Japanese spectral masks.

The impulses measured on-chip in time domain are shown in Fig. 8(a). The impulse generator is fed with a 100 MHz and 1.3 GHz sinusoidal signal separately. The differential signal is

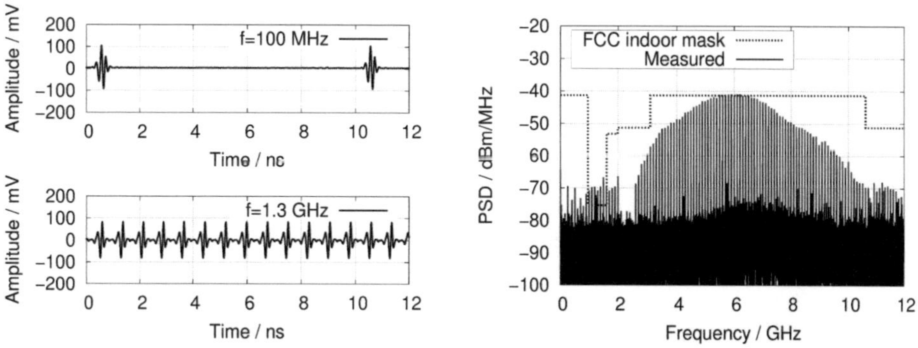

(a) Time domain (b) Spectrum

Figure 8. Measured results of time-domain output impulse waveforms at 100 MHz and 1.3 GHz repetition rate and the spectrum of the 100 MHz impulse train, demonstrating compliance with the FCC indoor spectral mask. The impulse generator is shown in Fig. 7(a)

displayed using the subtraction feature of a real time oscilloscope. The measured results show a peak-peak amplitude of 200 mV, and a full width at half maximum (FWHM) of the envelope of 0.3 ns. This circuit has a very low power consumption: 6 mW at 100 MHz and 10 mW at 1.3 GHz. The spectrum of the measured 100 MHz impulse train can be seen in Fig. 8(b). The maximum power spectral density (PSD) is -41.3 dBm per spectral line, and it has a 10 dB bandwith of 4.9 GHz from 3.5-8.4 GHz. It shows that the output spectrum complies well with the FCC mask for indoor UWB applications.

The two parameters (τ and ω_0) which determine the impulse parameters (envelope and osillation frequency) are easily modified. This is shown by the dashed components in Fig. 6(a). C_2 and T_7 are introduced to modify the envelope by changing the capacitance between the base of T_3 and ground, switching C_2 in and out. When $V_1 = 0$, T_7 is off, τ is the same as before, resulting the emitted impulses to conform to the FCC mask. When $V_1 = 1$ V, T_7 is on, the charging circuit will include R_1 and C_{be3} in parallel with C_2. In this case, the time constant τ_1 can be written as

$$\tau_1 = R_1(C_{be3} + C_2). \tag{3}$$

Since C_2 is chosen much larger than C_{be3}, the envelope width of the impulses is larger in this situation, suiting for ECC or Japanese masks depending on the center frequency adjustment. The tank circuit capacitance is now formed by C_{1a} in series with a varactor C_{1b}. The oscillation frequency can be expressed as

$$\omega_0 = \frac{1}{\sqrt{(\frac{C_{1a}C_{1b}}{C_{1a}+C_{1b}} + C_{para})(L_1 + L_2)}}. \tag{4}$$

Through changing the varactor capacitance C_{1b} with V_2, the center frequency ω_0 is adjustable. Depending on the applied voltages V_1 and V_2, the generated impulses conform to the FCC, ECC mask, or Japanese mask. The microphotogragh of this tunable impulse generator can be seen in Fig. 7(b). It is quite compact with an area of 0.53 x 0.61 mm^2.

By setting $V_1 = 0$ and $V_2 = 2$ V, the LC oscillator is triggered with a shorter current spike. So the generated waveform is similar as shown in Fig. 8(a), targetting the FCC mask. This impulse generator is suitable for the ECC mask when $V_1 = 1$ V and $V_2 = 2.3$ V. Under these conditions, the LC oscillator is triggered by a longer current spike with a FWHM of 2 ns. The measured output impulse train can be seen in Fig. 9(a). The impulses have a peak-peak amplitude of

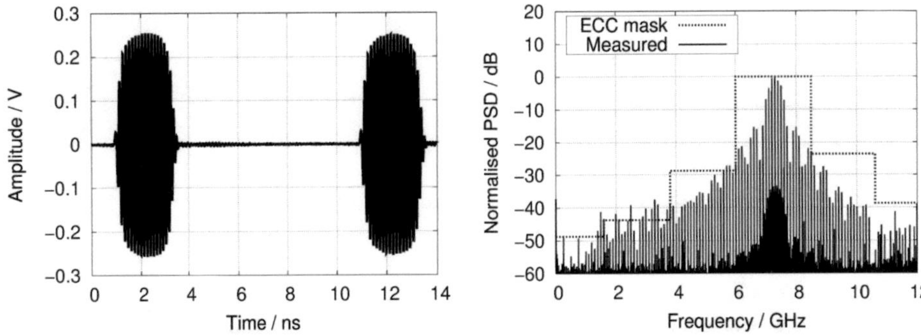

(a) Time domain (b) Spectrum

Figure 9. Measured results of impulse waveform in time domain and normalized PSD of the tunable impulse generator output signal setting for compliance with the ECC UWB mask.

0.5 V. The circuit has a total power consumption of 10 mW and a maximum output impulse repetition rate exceeding 300 MHz in this case. The normalized PSD of the impulse train has a center frequency around 7 GHz with a 10 dB bandwidth of 600 MHz, shown in Fig. 9(b). It fits well into the ECC mask. By changing the value of V_2, the center frequency of the impulses will be shifted, this makes the circuit usable for the Japanese mask. The measured output signal

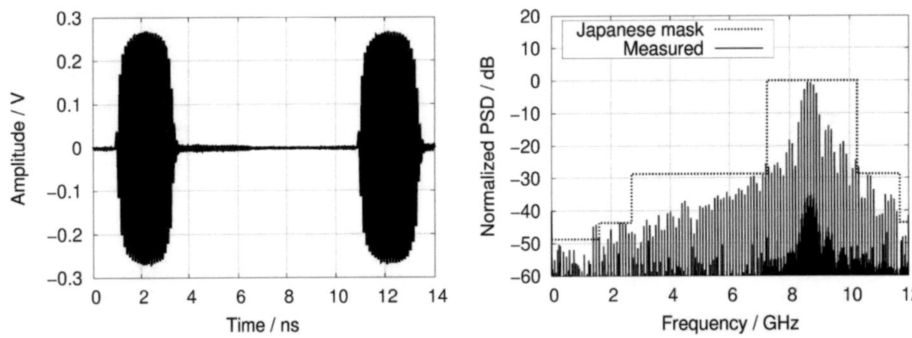

(a) Time domain (b) Spectrum

Figure 10. Measured results of time domain waveform and normalized PSD of the output signal targeting the Japanese UWB mask.

in the time domain with $V_{tune2} = 6$ V is shown in Fig. 10(a). The measured impulse train has a similar envelope as the mode targeting the ECC mask because the triggered current spike has the same width. The peak-peak amplitude of the impulses whose envelope has a FWHM of 2 ns is 0.5 V. The complete power consumption in this mode is 10 mW. The spectrum is presented in Fig. 10(b). It shows that the center frequency is shifted to 8.7 GHz for a good fit to the Japanese mask.

The performance under these three modes is summarized in Tab. 1. This impulse generator can be used for on-off keying (OOK) and pulse-position modulation (PPM) in all these three modes.

Mode (setting)	10dB bandwidth (GHz)	V_{PP} (V)	power cons. (mW)
FCC V_1=0, V_2=2.0 V	4.2	0.36	6
ECC V_1=1 V, V_2=2.3 V	0.6	0.5	10
Japanese mask V_1=1 V, V_2=6 V	0.6	0.5	10

Table 1. Performance summary of the tunable impulse generator.

Biphase modulation capability can be introduced by modifying the DC currents flowing in the individual branches of the differential LC oscillator. As shown in Fig. 11(a), additional branch currents are set through current mirrors T_7, T_9 and T_8, T_{10}. When the input data signal is

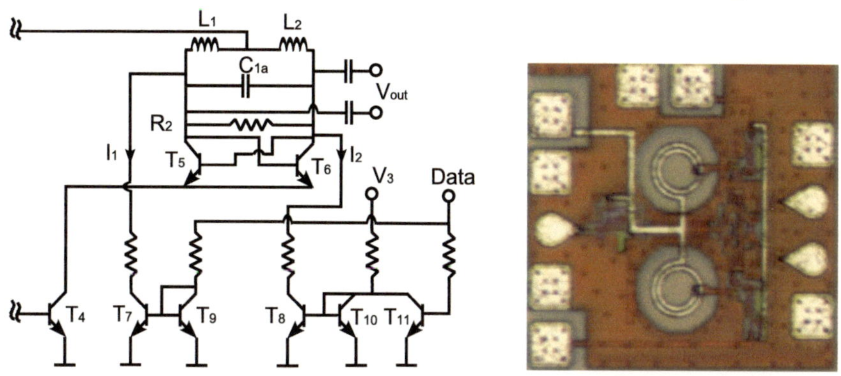

(a) Circuit schematic (b) Microphotograph

Figure 11. Adding biphase capability to the impulse generator in Fig. 6 and the microphotograph of the impulse generator with biphase modulation function.

low, transistor T_9 is off, and the collector current I_1 of T_7 is zero because of the current mirror configuration to T_9. Meanwhile, the applied voltage V_3 will generate a collector current I_2 in T_8 through the current mirror configuration of T_8 and T_{10}, because T_{11} is off. When the data signal is high, T_{10} switches into saturation and T_8 blocks, which causes I_2 to be zero. At the same time, the high potential at the base of T_9 introduces a collector current I_1 in T_7. Thus, oscillation will start in one of these two phase states once the tail transistor T_4 is turned on, constituting the biphase modulation. Additionally, this asymmetry shortens the start-up time, which in turn reduces the power consumption. The fabricated IC is shown in Fig. 11(b). It has an area of 0.56 X 0.53 mm^2.

The measured time domain waveforms with different voltage potentials applied to the data port can be seen in Fig. 12(a). The results show a peak-peak amplitude of 260 mV and an envelope width of 0.3 ns FWHM. The orientation of the impulses is clearly reversed, showing a perfect biphase modulation. Fig. 12(b) shows the spectrum of a 200 MHz impulse train with the data port connected to ground. It is centered around 6 GHz with a 10 dB bandwidth of

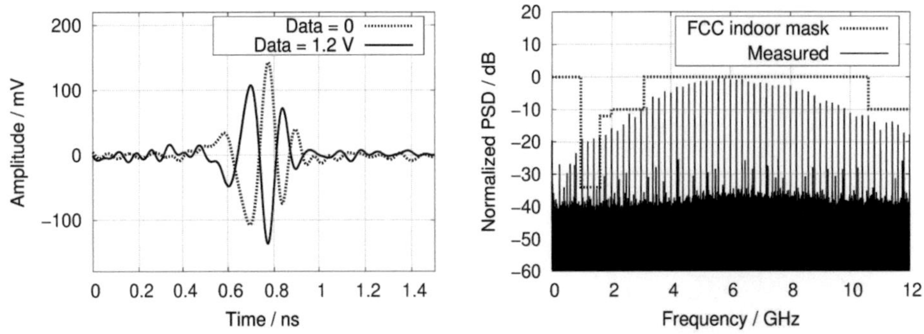

(a) Time domain (b) Spectrum

Figure 12. Measured time-domain results of biphase modulated impulses with different applied voltages at the data port and the spectrum information of a 200 MHz impulse train with the data port connected to ground.

6.7 GHz from 3.1 - 9.8 GHz, which complies well with the FCC spectral mask for indoor UWB systems.

2.3. Receiver MMICs

Two different types of receivers will be described in this section. Energy detection receivers are implemented for on-off keying (OOK) communications and localization applications, they are conceptually simple and do not require syncronization, but are also sensitive to interference. Correlation detection receivers are introduced to solve this problem, they are more robust to interference, but require accurate timing synchronization with the transmitter. This problem is eliminated in radar applications, because transmitter and receiver are co-located and synchronized with a common reference.

A fully differential UWB low-noise amplifier is a key element for both receivers. The LNA should provide a low noise figure, a high gain, a flat frequency response, and a small group delay variation within the complete frequency range. Another key component is a four quadrant analog multiplier, which performs the squaring operation in the energy detection receiver and the multiplication operation in the correlation receiver. Detailed explanation of these components will be described below.

2.3.1. Fully differential UWB low-noise amplifier

Fig. 13(a) shows the fully differential UWB low-noise amplifier schematic. It consists of a differential cascode, followed by two emitter follower stages as buffers. Input and output are differential as the LNA will be connected to a symmetrical antenna, and shall feed a Gilbert cell type analog multiplier directly, without an unbal circuit. The symmetry of the emitter-coupled pair is achieved by placing identical transistors and passive components in the two branches.

T_1 through T_4 form the differential cascode which is biased by the stacked current mirror. The primary reason of the cascode configuration is to reduce the Miller effect at the input port, increasing the bandwidth. The shunt-shunt feedback (R_1, C_1 and R_2, C_2) further broadens

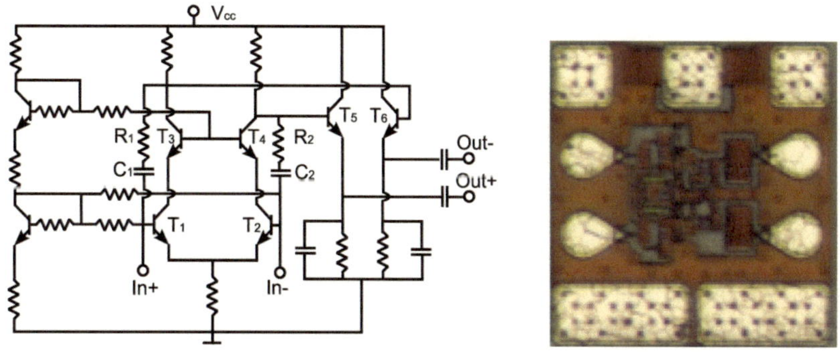

(a) Circuit schematic (b) Microphotograph

Figure 13. Complete circuit schematic and the chip microphotograph of the fully differential low-noise amplifier.

the bandwidth and improves the input matching simultaneously. Careful selection of input transistor size and adjusting the bias point was done as a compromise between optimum current density for minimum noise figure, noise-matched input impedance and achievable bandwidth. The emitter size of T_1 is chosen to be 0.5 μm x 24.7 μm and the emitter current is 5 mA. The wide band noise and input power match were accomplished by the selection of input transistor with suitable biasing and shunt-shunt resistive feedback. A negligible penalty, with a maximum value of 0.2 dB, is achieved within the entire band for not achieving noise match exactly. T_5, T_6 form a differential emitter follower buffer. The emitter degeneration capacitors are used to improve the buffer bandwidth.

The microphotograph of this differential LNA is shown in Fig. 13(b). Because this design is completely inductor-less, the IC has an extremely small size of 0.37 X 0.38 mm^2 including all bound pads. The lowest available metal layer was placed below the large-sized bonding pads to provide a ground shield, as otherwise the noise figure may be deteriorated by the substrate noise pick-up.

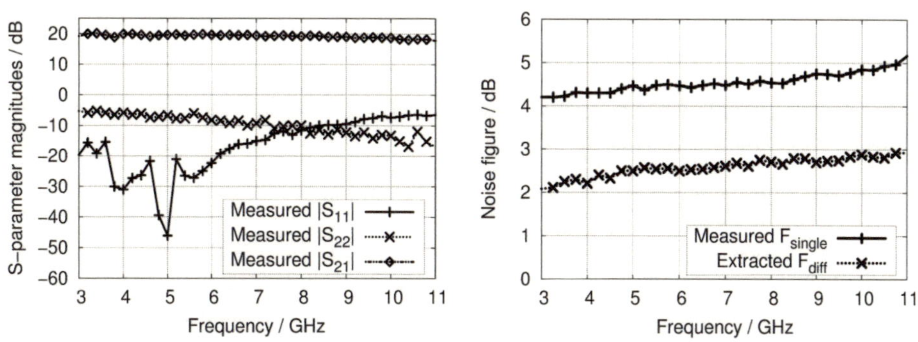

(a) S-parameters (b) Noise figures

Figure 14. Measurement results of the S-parameter magnitudes and single-ended and extracted differential noise figures.

One drawback of the differential configuration is a complex measurement setup. Two identical passive microstrip line UWB baluns are used for differential S-parameter measurement. The influence of the baluns is removed during the calibration process. The measurement is restricted to 3 - 11 GHz due to the operating range of the UWB baluns. Fig. 14(a) shows the measured S-parameters. The non-ideal performance of the UWB baluns introduces ripples in the measured curves. The measurement results show a differential gain of 19.9 dB with a 1.8 dB variation, the input matching has a value of smaller than -7 dB and the output one is smaller than -6 dB in the complete FCC allocated frequency range. The method in [2] is adopted to extract the differential noise figure. First a single-ended noise figure F_{single} is measured from port In+ to Out- with the other ports terminated by 50 Ω resistors. Then, by measuring the transducer gain from port In+ to Out- (G_{31}) and Out+ to In+ (G_{32}), the differential noise figure can be extracted as

$$F_{diff} = 1 + \frac{1}{G_{31} + G_{32}}(F_{single}G_{31} - G_{31} - G_{32}). \quad (5)$$

Fig. 14(b) shows the information of the noise figures. The differential noise figure varies from 2 dB at 3 GHz to 2.9 dB at 10.6 GHz. Small group delay variation within the entire band is required for single-band IR-UWB systems. As depicted in Fig. 15(a), the group delay variation is smaller than 15 ps within the complete band. Fig. 15(b) shows the measured large signal behavior at 7 GHz of this differential amplifier. The input 1 dB compression point is -17.5 dBm. The complete power consumption of this differential LNA is 77 mW.

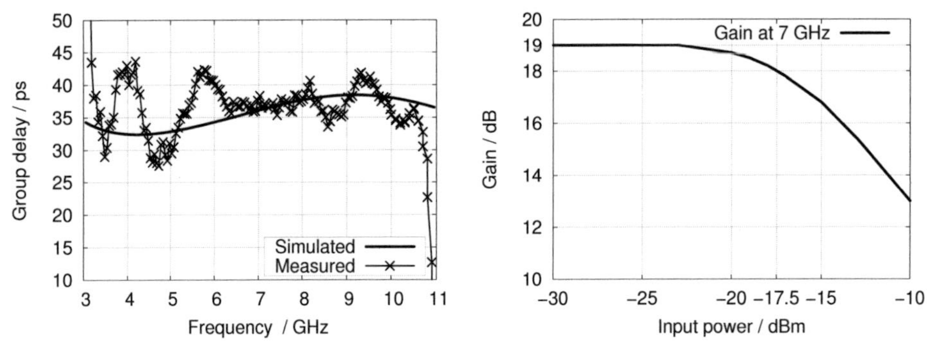

(a) Group delay (b) Gain

Figure 15. Measured results of group delay versus frequency and gain depending on the input power.

2.3.2. *Energy detection receiver*

The core of energy detection receivers is a squaring circuit. Fig. 16(a) displays the squaring circuit based on a Gilbert cell four quadrant multiplier comprising two differential stages in parallel with cross-coupled output, complemented by a low-pass filter and a differential output buffer. The squaring operation is realized by connecting the same signal to both inputs of the Gilbert cell. The signal fed to the lower pair of the Gilbert cell is taken directly from the LNA output transistors, while the signal fed to the top quad is passed first through the emitter follower buffer. Both paths introduce almost the same group delay. Thus, the two branches of the input signal arrive simultaneously at the multiplier, ensuring an exact squaring operation. The load resistors (R_1, R_2) of the Gilbert cell, together with shunt capacitors (C_1, C_2) of the

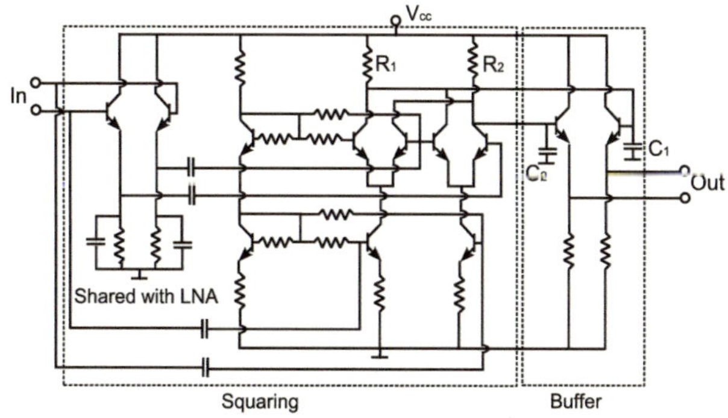

(a) Circuit schematic

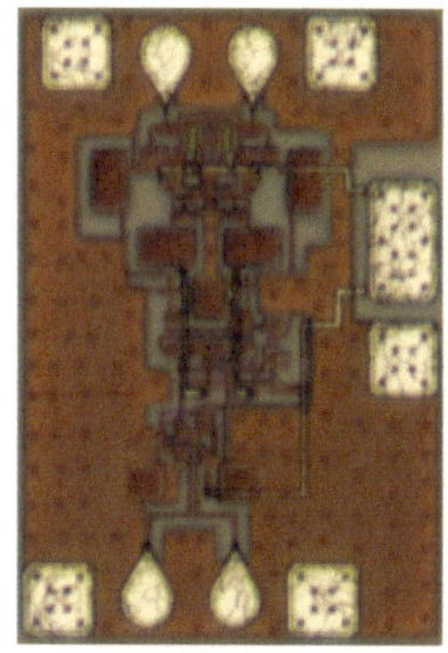

(b) Microphotograph

Figure 16. Squaring circuit, low-pass filter and buffer of the energy detection receiver.

output buffer form low-pass filters with 1 GHz 3 dB bandwidth, which are needed for the envelope detection. The LNA from Fig. 13(a) is added to complete the energy detection receiver, which totally consumes 108 mW. Fig. 16(b) shows the microphotograph of the fabricated receiver IC, it measures 0.43 mm x 0.61 mm, including bond pads.

For testing the energy detection receiver, a 700 Mbit/s return-to-zero (RZ) impulse train was generated by the impulse generator shown in Fig. 6(a), which has a power comsumption of 7.5 mW at this rate. The transmitter and receiver ICs are seperately mounted on Rogers RO4003C substrates which also carry the dipole-fed circular slot antennas discussed in 2.1.1, and are wire-bonded to microstrip transmission lines feeding the antennas. The two antennas are placed at a distance of 30 cm. Fig. 17(a) shows the input data sequence from a pattern

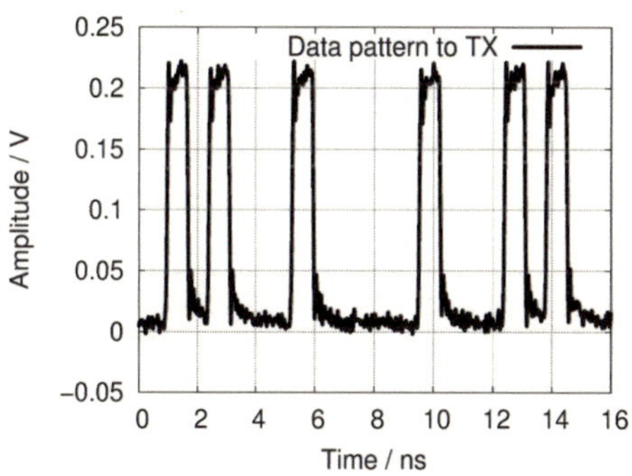

(a) Data sequence

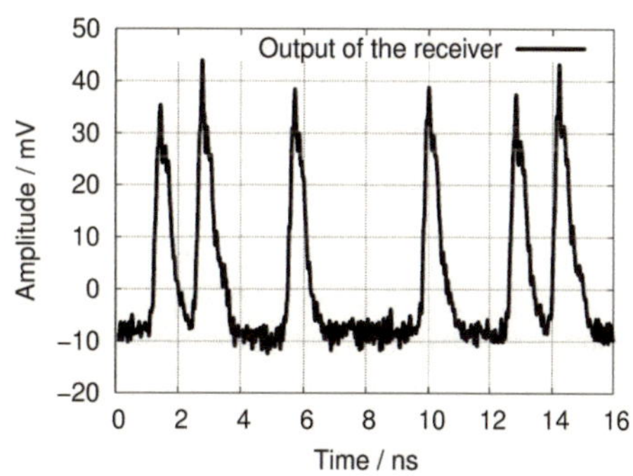

(b) Output

Figure 17. OOK transmission experiment at 700 Mbit/s over 30 cm, data sequence (700 Mbit/s data rate) applied to the transmitter and detected signal at the receiver output.

generator. The corresponding detected impulse envelopes with a peak amplitude of 40 mV at the output of the receiver IC can be seen in Fig. 17(b). This experiment clearly demonstrates that the simple transmitter/receiver combination can be used to transmit significant bit rates over short distances. Detailed measured results of the receiver are shown in [23].

2.3.3. Correlation detection receiver

Coherent detection receivers are based on the cross-correlation realized by feeding the received signal and the on-chip generated template impulse into a wideband analog four-quadrant multiplier and subsequent low-pass filtering. Fig. 18(a) shows the block diagram of the correlation receiver. The multiplier-based correlation is done in the RF domain, which leads to an energy efficient solution by omitting power-hungry wideband ADCs. In a radar setup, the transmit and receive clocks need to be phase adjusted, which in practice is done by a DDS board. The complete schematic of the UWB analog correlator circuit

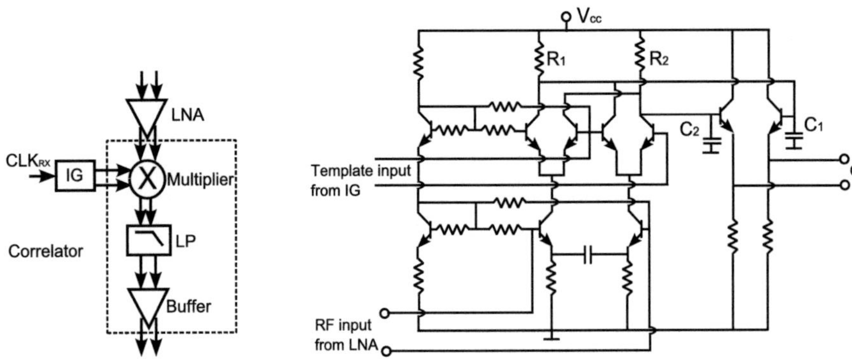

(a) Diagram (b) Circuit Schematic

Figure 18. Architecture of the correlation receiver system and the schematic of the correlator with a true multiplier, a low-pass filter and a buffer.

can be seen in Fig. 18(b). The core of the correlator is again a Gilbert cell which acts as a wide-band multiplier with proper template impulse amplitude applied to the switching quad. Capacitively shunted resistive emitter degeneration results in the necessary gain flatness over the whole UWB frequency band. The low-pass filters are formed by the load resistors (R_1, R_2) of the Gilbert-cell with the shunt capacitors (C_1, C_2) of the buffer, the same as shown for the case of the energy detection above. The correlator consumes 35 mW.

The complete receiver IC, including the differential LNA, the correlator and the template impulse generator, is shown in Fig. 19(a). It measures $0.43 \times 0.61\,\text{mm}^2$ and consumes a total DC power of 130 mW. To demonstrate the correlation performance of the receiver, the receiver IC was connected to a dipole-fed slot antenna, and placed at a distance of 20 cm from the transmitter. A small offset frequency of 100 Hz was introduced between the transmitter and receiver clocks, making the template impulses continuously sweep through the received signal. The measured cross-correlation can be seen in Fig. 19(b). More details of the correlation receiver can be found in [24].

2.4. Monostatic radar MMICs

All UWB radar sets reported so far use a bistatic antenna configuration. A monostatic UWB radar would significantly reduce the size of IR-UWB sensors because of the elimination of one antenna. However, implementation of rapid switching between the transmit and the receive path is difficult to realize in either this low-cost bipolar-only or CMOS technology. Here, a

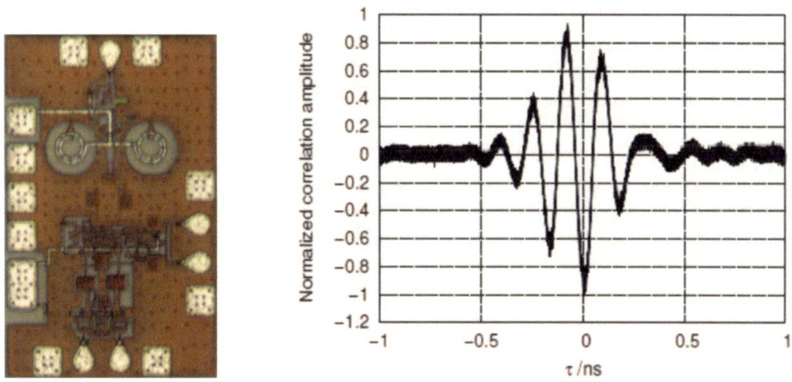

(a) Microphotograph (b) Correlation

Figure 19. Chip micrograph of the fully integrated correlation receiver and the measured normalized cross-correlation of the received impulse with the template impulse.

novel front-end concept based on a merged impulse generator/low noise amplifier, shown in Fig. 20(a) is proposed. In this design, the input of the differential low-noise amplifier is tied

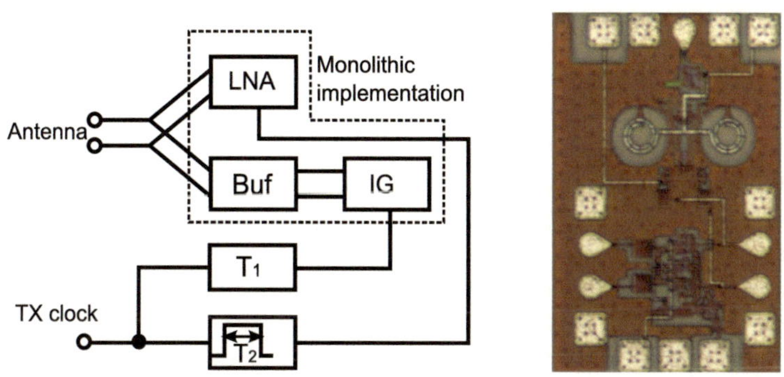

(a) Diagram (b) Microphotogragh

Figure 20. Block diagram and chip photo of the proposed monostatic UWB radar front-end.

together with the output of a buffer amplifier following the impulse generator. An external monoflop and a bandgap reference circuit ensure that the LNA is disabled during the impulse emission. The LNA bias is recovered after the impulse has been transmitted, and it returns to full gain within 1.5 ns. The added parasitics of the buffer are included in the design of LNA.

Fig. 20(b) shows the fabricated IC. In the experimental test, the antenna terminal is connected to two short coaxial cables, each of which feeds into a 10 dB attenuator shorted at the far end. An approximately 1 ns delay is generated by the coaxial cable and attenuator, corresponding to a distance of 30 cm in air. The measured time domain trace at the output of the differential LNA can be seen in Fig. 21. The significant common-mode transients due to the bias switching are completely invisible owing to the balanced setup. The result clearly shows the received impulse echo. Due to a high isolation of the 'cold' low-noise amplifier, the crosstalk from the transmitted pulse is barely visible and will not influence the further processing.

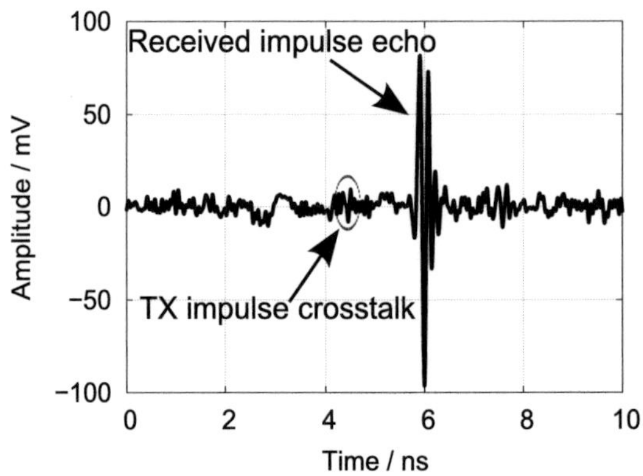

Figure 21. Real time oscilloscope trace showing the functional test of the monostatic radar frontend, displaying transmitted pulse crosstalk and received pulse echo.

3. Signal processing

3.1. Energy detection for UWB communications

For UWB communications, the detection of transmit symbols can be done with a coherent or a non-coherent method. Typically, with perfect synchronization, the coherent detection based on correlation gives a better performance than non-coherent detection. Coherent detection is known to be the optimum method with respect to the bit error rate (BER) for AWGN channels. In case of multipath propagation channels, the transmit impulse and the channel impulse response together (convolved) must be used for correlation detection, and in general more complex signal processing is required. Analog signal processing is considered here because analog to digital conversion for UWB signals is hardly available and requires high power consumption. Unfortunately, the coherent approaches are not well suited for our large bandwidth analog signal processing within the receiving frontend. Energy detection is usually preferred because, if applied in a proper way, no channel impulse response is needed in the receiver. Moreover, energy detection is also robust with respect to synchronization accuracy. In its basic form the energy detector consists of a squaring device, an integrator, a sampler and a decision device.

Pulse position modulation (PPM) and On-off keying (OOK) are modulation techniques that are typically used in combination with energy detection. The data modulation is performed by changing the position or amplitude for PPM and OOK respectively. The detection of PPM is easier to perform since comparing the signal energy at two different intervals is enough, while OOK requires threshold estimation. We look at the BER performance of PPM with respect to synchronization and multipath propagation. Different levels of synchronization in AWGN channels (perfect, ± 20 ps, and ± 40 ps) and a perfect synchronization for a multipath channel are considered. The errors in the synchronization accuracy are uniformly distributed. The transmitted signal is an impulse train of fifth derivative Gaussian functions ($\sigma = 51$ ps) with PPM modulation. The tested channels AWGN channel and a multipath channel with delay spread are assumed to be shorter than 4 ns, which is half of the impulse repetition interval. The correlation receiver uses a template impulse (fifth derivative Gaussian) for correlation, and the correlation is centered at the first strongest path of the channel. The BER for all settings is shown in Fig. 22.

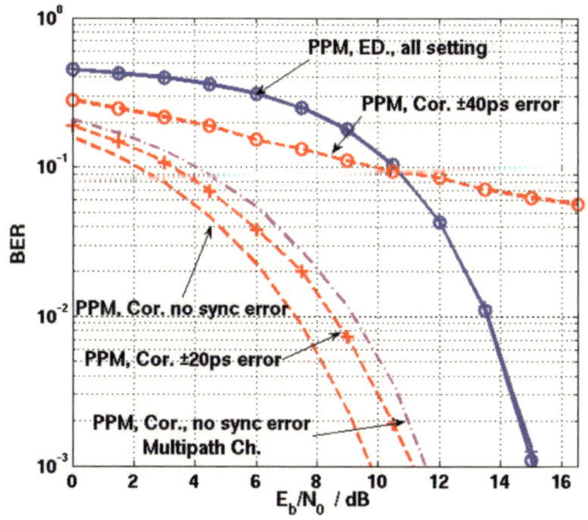

Figure 22. BER performance of PPM modulation for correlator and energy detection with AWGN channel and multipath channel.

The performance of the correlation receiver suffers from synchronization uncertainty and multipath propagation. This is due to the fact that the impulse used in UWB systems is very narrow and the impulse correlation receiver cannot capture all of the signal energy. On the other hand, the energy detector shows good performance also for non-perfect synchronization and multipath channel. We can conclude from the results that, with a performance trade-off, the energy detection is much more robust. Other challenges for implementing energy detection are multiuser capability and interference cancellation. These problems can be solved by the comb filter receiver presented in the following part.

3.2. Comb filter

The received signal power for medical applications are expected to be very small due to high attenuation in human tissue. We propose a receiver based on a comb filter to improve Signal-to-Noise ratio (SNR) before further processing. The comb filter is a feedback loop with an analogue delay and a constant loop gain of one for all frequencies. It is used to perform a coherent combination of the incoming UWB impulses. The feedback loop sums up the number of impulses used for the transmission of a data symbol/measurement and is reset after this. The coherent combination results in SNR improvement, interference suppression which come from different transmitters in a multiuser environment or narrowband interfering signals. Several UWB impulses are transmitted for one data symbol/measurement. One important feature of the concept is that the individual UWB impulses are weighted by +1 or -1 according to a spreading sequence. The UWB transmit signal $s(t)$ can be written as

$$s(t) = \sum_{k=-\infty}^{\infty} \sum_{n=0}^{N-1} c_n p(t - nT_c - kT_s), \qquad (6)$$

where $p(t)$ is a UWB impulse and c_n is the spreading sequence. T_c is the period between two UWB impulses or 'chip period' and $T_s = NT_c$ is symbol period for communications or measurement period for radar/localization application. For communication, assuming a binary transmission, the impulse train of each data symbol can be modulated in different ways. We consider three modulations using the direct sequence spread spectrum technique based on OOK, PPM, Code shift keying (CSK) ,i.e. DS-OOK, DS-PPM and DS-CSK. For DS-OOK, transmitting a train of impulses represents the data '1', while no impulse signals means '0'. For DS-PPM, the two basic waveforms for a binary transmission are different by time shift. For DS-CSK, the two waveforms result from two different spreading sequences. A decision threshold is not required for DS-PPM and DS-CSK which is a big advantage compared to DS-OOK. The basic waveforms for different modulation techniques are illustrated in Fig. 23.

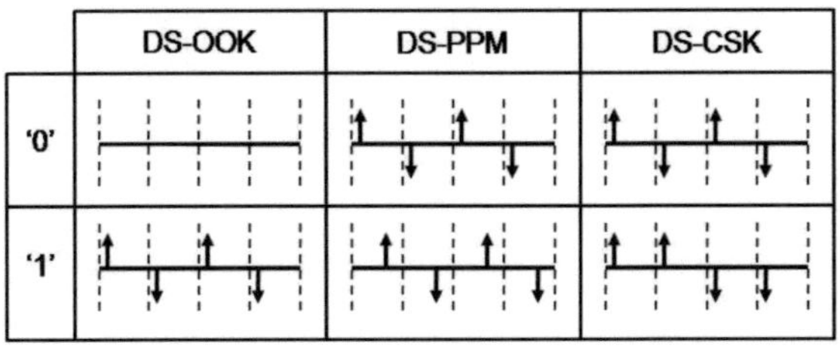

Figure 23. Comparison of UWB transmit signals for different modulation techniques.

Transmitted through the channel, the impulses are affected by the channel impulse response which is expected to be a multipath environment with or without a line-of-sight path. The chip period is chosen to be larger than the multipath spread, with the result that no interchip interference (i.e. no overlapping of channel impulse responses) occurs. We assume the channel to be time invariant within the symbol period. This means that the signal or basic waveform which represents one data symbol/measurement in the received signal consists of a corresponding number of channel impulse responses.

3.2.1. Basic comb filter receiver

The receiver based on the comb filter with remodulation is shown in Fig. 24. The receiver consists of an antenna, a LNA, a multiplier, the comb filter, the energy detector for communications and a correlator for radar/localization applications. This concept allows energy detection in a multipath and multi-sensor environment and amplification of the impulse response in the comb filter delay loop.

In the receiver, the received signal is multiplied with the spreading waveform (i.e. the sequence with rectangular "chips"). If the spreading sequence matches, the result is a periodically repeated channel impulse response with period T_c. After multiplication, the received impulses are delayed and summed up by the comb filter. At the output of the comb filter, only the last T_c period of every T_S is used for further processing which is expected to be an amplified and improved SNR channel impulse response. Interference from different UWB transmitters and other systems are eliminated at the comb filter. The process of remodulation and coherent combination at the comb filter is illustrated in Fig. 25.

Spreading is used not only for the multiuser or multi-sensor purpose, but also for shaping the power spectral density of the transmitted signal. The impulse train may have high spikes in the power density spectrum, because it has a periodic behavior over a longer time interval,

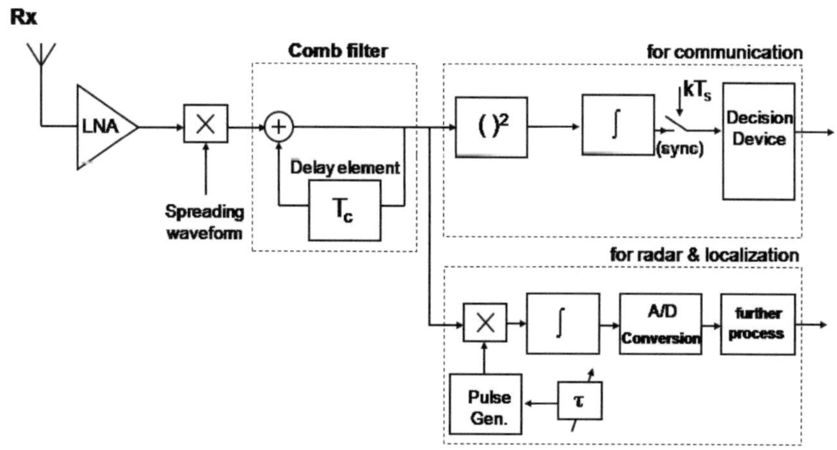

Figure 24. Block diagram of the basic comb filter based receiver.

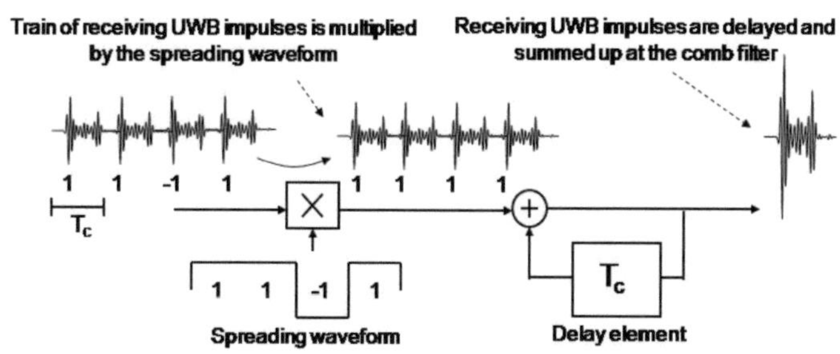

Figure 25. Comb filter principle in the receiver.

and hence it could violate the spectral mask requirement. The longer and more random the spreading sequence, the better. Of course, a direct restriction is given by the desired data rate.

3.2.2. Multiuser interference

In general we will have many transmitters or many sensors and different corresponding spreading sequences. The sequences are selected such that the mutual cross-correlation values are as small as possible. This guarantees that many transmitters can be used at the same time in the same area, e.g. several implants for digital transmission. The signals from all sensors are assumed to be transmitted in parallel with synchronization in a certain time window. The remodulation and the comb filter at the receiver is the crucial part in suppressing the multiuser interference (MUI). The weighting factors of the channel impulse response train after the multiplication are a scrambled version between transmit and receive sequence, if they are not the same. The accumulation at the comb filter would eliminate the MUI and the result of the weight is the cross-correlation values. The illustration of multiplication with the spreading sequence and comb filter accumulation of a signal with two different sequences is shown in Fig. 26.

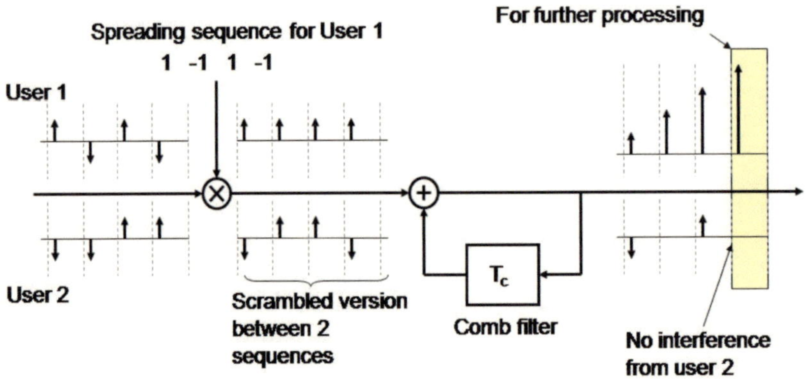

Figure 26. Illustration of multiuser interference suppression at the comb filter.

We assume synchronization in a certain time window between different users. With this assumption, we mainly consider m-sequences because they have low crosscorrelation and are simple to generate. Alternatively, binary zero-correlation-zone (ZCZ) sequence can also be considered. Unlike m-sequences, these sequences have a crosscorrelation of zero within a small window but the drawback is a smaller number of sequence in one set [40]. Binary ZCZ sequence sets were investigated in [37] for communication because of their potential to completely eliminate multi-sensor and interchip interference (ICI). A near-far problem was taken into account and we can see an advantage of this sequence set.

3.2.3. Signal-to-noise ratio improvement

The SNR of received signals can be significantly improved after the coherent combination in the comb filter, since the signal power grows quadratically, while the noise power grows only linearly. We first consider ideal components where there is no distortion in delay line and the comb filter has no loop gain. For the investigation of the SNR improvement, an AWGN channel is sufficient. A train of modulated UWB impulses and additive noise is the input signal to the multiplier. The SNR at the input of the multiplier is compared to the SNR after the comb filter to calculate the comb filter processing gain G_p. Fig. 27(a) shows an example of an input signal with a SNR = -15 dB (upper) compared to the output signal (lower). The SNR improvement can be seen clearly as the receiving impulse becomes visible after a few iterations. This comb filter signal processing results in a SNR improvement of $10\log(N)$ dB, where N is the number of iterations. The UWB signal is algebraically added, therefore the signal energy is increased by a factor of N^2. On the other hand, the noise contributions in each chip are added in power, and therefore the noise energy within the symbol interval is increased only by a factor of N. For communications, the BER of PPM with correlation and energy detection also show the same improvement. The BER performance for DS-PPM with $N = 1$ and $N = 63$ using M sequence is shown in Fig. 27(b) . Using several chips per symbol with the comb filter approach gives a performance gain with respect to SNR but not to E_b/N_0 due to the difference in the data rate. For UWB transmission, the SNR cannot be improved by increasing the transmitted energy because of the spectral mask limitation. A trade-off between data rate and the SNR improvement at the receiving side has to be made.

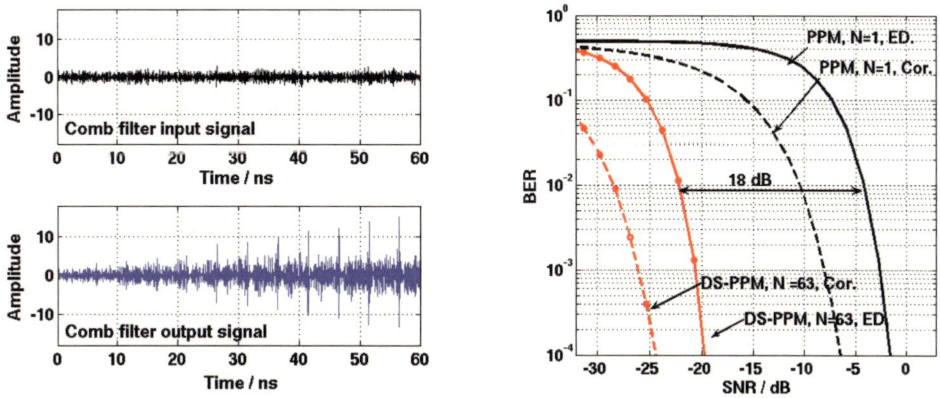

(a) (b)

Figure 27. (a) Input and output of the comb filter with remodulation. (b) BER performance of DS-PPM with N = 1, 63.

For physical systems this calculation is only valid if no instability is introduced. Therefore the values of the gain G_c of the comb filter can only be in the range of 0-1, because otherwise an oscillation occurs. G_c is designed to be one but considering real components, a loss in the loop is possible. The comb filter processing gain G_p is a function of the gain G_c of the comb filter and the number of iterations N. It can be calculated as

$$G_p = 10 \cdot \log_{10} \left(\frac{\left(\sum_{n=0}^{N-1} G_c^n \right)^2}{\sum_{n=0}^{N-1} G_c^{2n}} \right) \tag{7}$$

The maximum processing gain G_p of the comb filter is 10 log(N) dB, when $G_c = 1$. The processing gain is reduced if G_c is less than one. The relations between G_p, G_c and N are shown in Fig. 28. We can see from Fig. 28(a) that the processing gain G_p saturated in the case where G_c is less than one because the impulse energy vanishes after some iterations. With a higher number of impulses per symbol, the comb filter loop gain has to be controlled more precisely as shown in Fig. 28(b).

3.2.4. Narrowband interference

Since UWB covers a very large bandwidth, strong interference within the band is possible and can cause problems at the receiver. Comb filter in combination with multiplication with the spreading waveform can suppress narrowband interference very well. Only a periodic signal that has a period which equals one or multiples of the comb filter delay can go through the comb filter. The comb filter transfer function $H(f)$ is given as follows:

$$H(f) = \sum_{k=-\infty}^{\infty} T_s \cdot \mathrm{sinc}(T_s(f - k/T_c)) \exp(-j\pi f T_s) \tag{8}$$

We can see that the transfer function of the comb filter consists of several peaks. The peaks could be seen as tunnels that allow only signals with specific frequencies to pass. After the

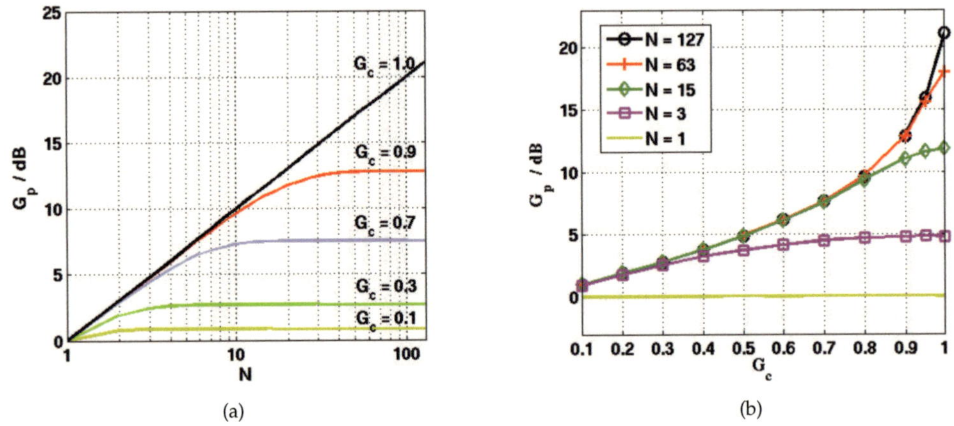

Figure 28. Relationship between processing gain G_p, comb filter loop gain G_c and impulses per symbol N.

multiplication with the spreading waveform in the receiver, the spectrum of the UWB signal has a shape matched to the transfer function of the comb filter (signal with period T_c). On the other hand, the narrowband interferer is spread by this multiplication and only some frequency components could go through the comb filter. The width of each peak depends on the number of chips per symbol. It gets smaller as the number of summation steps in the comb filter increases and as a result more interference is suppressed.

The improvement of the signal-to-interference ratio (SIR) is demonstrated in Fig. 29(a). The signals at the input of the comb filter are the UWB signal and a narrowband interference consisting of an IEEE 802.11a OFDM WLAN signal with a bandwidth of 16.66 MHz and a center frequency of 5.2 GHz. In this example, the input SIR of -15 dB is improved by 10 dB for $N = 63$. In addition, the BER performance of DS-OOK and DS-PPM in an AWGN channel with the same interference is shown in Fig. 29(b). The performance for both methods is improved with increasing number of chips per symbol. The degradation of the performance due to the narrowband interference for DS-PPM is much less than that for DS-OOK. The narrowband interferer gives a contribution to both integrator outputs for DS-PPM, and because the outputs are compared, the influence is reduced. For the DS-OOK the influence remains.

3.2.5. Shortened delay comb filter

The main challenge for implementing the comb filter based receiver is to realize the true wideband analog delay element. Shortening the delay means that overlapping channel impulse responses can occur at the receiver, the channel impulse response becomes longer than the chip interval T_c. As a result, after the comb filtering we get an amplified window-cut-out of the true impulse response with the window width T_c. It is shown in [35] that we can control the position of this window by adjusting the spreading sequence at the remodulation. The property of being able to extract different parts of the channel impulse response gives an opportunity to construct a receiver by using a rake concept. This means that for each part of the impulse response we have one rake branch where we calculate the energy. The modulation technique that is well-suited for this structure is DS-CSK. The block diagram

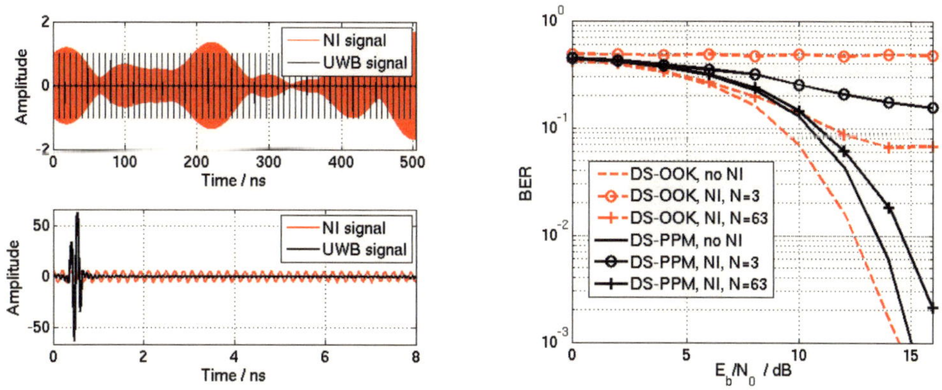

(a) (b)

Figure 29. (a) Input and output of the comb filter with remodulation. (b) BER performance with SIR = -15 dB.

of the receiver for this concept is shown in Fig. 30. Note that the spreading sequences for the rake branches are cyclic-shifted versions of one single sequence. For DS-CSK, two parallel branches with different corresponding spreading sequence sets are needed. The results from two branches are then compared.

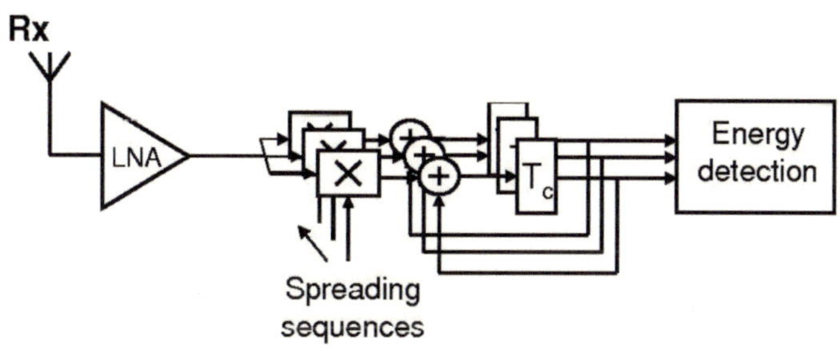

Figure 30. Block diagram of a rake-like comb filter receiver with energy detection.

In comparison to our basic concept described before, the delay is shorter. If the delay is shortened, for example, from 12 ns to 2 ns, this is much more realistic with regard to the realization. The performance of the new concept was verified by simulations. We used UWB impulses within a band from 4 GHz to 6 GHz. Note that this band is due to the Wireless Body Area Networks (WBANs) channel model we used [12] but it is not important for our concept, it could be any. The channel model consists of two groups of paths with a lognormal fading statistic for each group. We used 4000 channel realizations. The total propagation time of the channels is 12 ns. Eight users/sensors with the same average receive energy are considered. The spreading sequences were m-sequences with $N = 127$. The receive signal at the input of the comb filter is shown in the upper part of Fig. 31(a). Additive noise is not considered here to focus only on the ICI and MUI suppression. We can see that both, ICI and MUI are very strong here. The SIR in this example is -17 dB. In the lower part of Fig. 31(a) the original channel impulse response (black) is compared to the results taken from different rake branches (red). From the results we can see that the interference is suppressed very well.

We also look at the BER performance of this new concept in Fig. 31. A WBAN channel is used. For our original basic comb filter concept, $T_c = 12$ ns and 6 ns are considered. We label them as 'Basic 1' and 'Basic 2', respectively. For the rake-like receiver, we consider $T_c = 2$ ns with 6 rake branches which yield equivalent integration time of 12 ns. We can see that the performance of our original concept is getting worse if the delay is shortened. There is a loss of about 4 dB at the BER for 10^{-4}. The result for 'Rake-like' is comparable to 'Basic 1'. The results show that we can achieve a very similar performance to the original concept with much shorter analog delay elements.

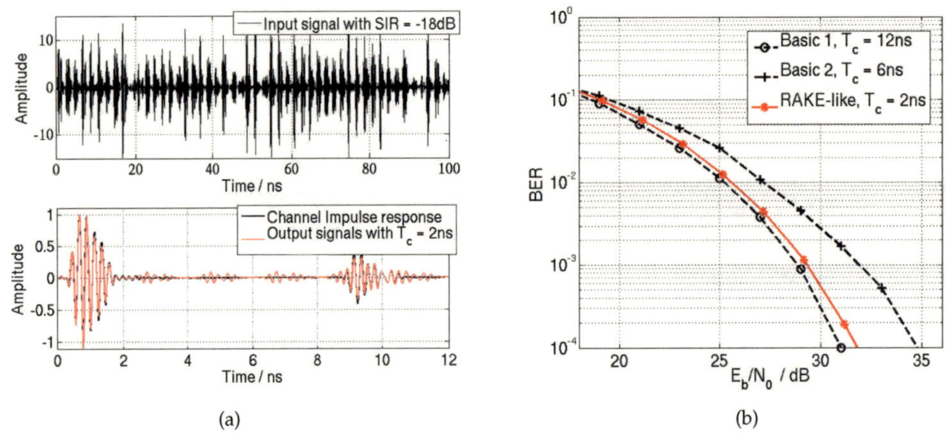

Figure 31. (a) Input signal of the comb filter (upper). Comparison between the actual channel impulse response and the output signal of the rake branches (lower). (b) Comparison of the bit error rate from different receivers.

3.3. Particle filtering

In this part we want to address an algorithm for movement tracking which can be used for radar and imaging application. The background is the application of IR-UWB for biomedical diagnostics, e.g. vital sign detection (like breathing/heart rate) and also tracking of body movements to compensate for errors which degrade the quality of inner body imaging. Simple methods such as tracking the maximum/minimum of the receive signal in a specific area perform reasonably well. A problem occurs if the peak of the signal cannot be easily identified. In cases with high attenuation, the estimation can become very bad for those simple methods. Another problem is caused by stationary echoes (clutter). Particle filtering can help in this situation. It is a technique which implements a sequential Bayesian estimation by using Monte-Carlo methods. Particle filtering is commonly known to be used in localization applications and tracking in dynamic scenarios [18]. The estimation uses a movement model to incorporate the temporal correlation of the change of unknown parameters.

The Bayesian estimator finds unknown parameters $\boldsymbol{x}_k$ (signal delays) from a set of measurement signals $\boldsymbol{z}_{1:k}$ using a posterior density function $p(\boldsymbol{x}_k|\boldsymbol{z}_{1:k})$. It is very common that the algorithm is processed in a recursive manner. In summary, the algorithm consists of a 'prediction' phase and an 'update' phase, where the prior density $p(\boldsymbol{x}_k|\boldsymbol{z}_{1:k-1})$ and the posterior density $p(\boldsymbol{x}_k|\boldsymbol{z}_{1:k})$ are estimated every time step k.

Prediction:

$$p(\boldsymbol{x}_k|\boldsymbol{z}_{1:k-1}) = \int p(\boldsymbol{x}_k|\boldsymbol{x}_{k-1})p(\boldsymbol{x}_{k-1}|\boldsymbol{z}_{k-1})d\boldsymbol{x}_{k-1} \tag{9}$$

Update:

$$p(\boldsymbol{x}_k|\boldsymbol{z}_{1:k}) = \frac{p(\boldsymbol{z}_k|\boldsymbol{x}_k)p(\boldsymbol{x}_k|\boldsymbol{z}_{1:k-1})}{p(\boldsymbol{z}_{1:k}|\boldsymbol{z}_{1:k-1})} \tag{10}$$

We can see that the likelihood function $p(\boldsymbol{z}_k|\boldsymbol{x}_k)$ and the movement model $p(\boldsymbol{x}_k|\boldsymbol{x}_{k-1})$ are important items in Bayesian estimation. Unfortunately the posterior density is usually intractable but there are several ways to implement the algorithm. Particle filtering deals with this problem by using samples (particles) with associated weights to represent the posterior density. If the number of particles is sufficiently large, the estimates reach optimal Bayesian estimation.

A setup consisting of a bistatic UWB transceiver and a moving metal plate is used for demonstrating the use of particle filtering in IR-UWB tracking. A 5th derivative Gaussian impulse fitting to the FCC mask is used and the repetition rate is 200 MHz (i.e. $T_s = 5$ ns). The corresponding distance is 75 cm. The metal plate moves periodically within 5 mm range. A schematic block diagram of the setup is shown in Fig. 32(a). The receive signal consists of two main contributions which can be seen in each period, corresponding to two-path propagation on the channel. The first path is the direct path between the transmit and the receive antennas (direct coupling signal). This signal has very strong visible ringing due to signal reflection from the impedance mismatch. The second signal is a signal reflected from the moving metal plate. An example of a receiving signal with a target 30 cm away is shown in Fig. 32(b). More details about the setup can be found in Sec. 4.1.

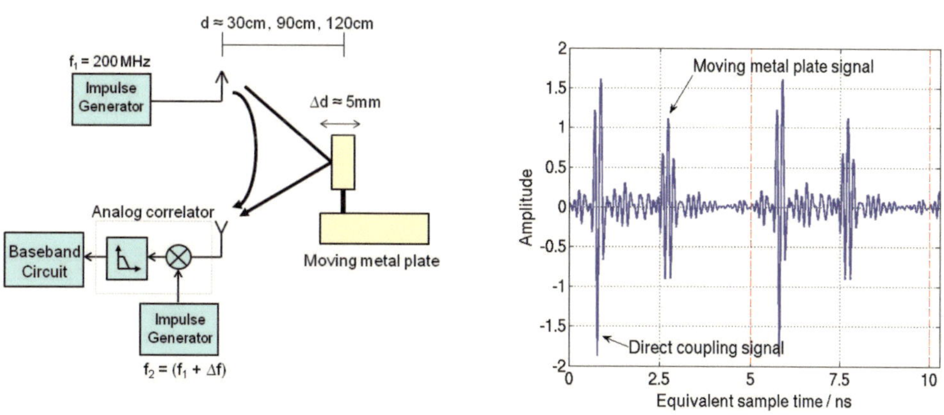

(a) Measurement setup (b) Receive signal

Figure 32. Measurement setup for target movement tracking and example of a periodic receive signal after the correlator.

Particle filtering with a 2-path model is used for tracking the transmission delay of the two paths. The signal after the impulse correlation can be represented in discrete-time in each period and is considered as measurement signal z_k. The delay of each paths form the state vector x_k. More details on the setup can be found in [36].

In the following, tracking results for a moving metal plate with distances of approximately 90 cm and 120 cm are discussed. Since the distance of ambiguity for the measurement setup is 75 cm, we use the knowledge that the target is in the 75-150 cm range. This is sufficiently large for our target application. The reflected signal appears one period after the original impulse was transmitted. Particle filtering with 1000 particles is considered and the results are compared with a conventional maximum tracking method. The movement of the first path (direct coupling) and the second path (metal plate) are tracked simultaneously.

We first consider a setup with the moving target at approximately 120 cm. An example for one period of the receive signal is shown in Fig. 33(a). The reflected signal is located at around 8.8 ns. In this example, the reflected signal can be easily recognized. Fig. 33(b). shows the tracking results of the moving metal plate from both methods (particle filtering and maximum tracking for comparison). The tracking results fit well with each other and the small movement of 5 mm was estimated correctly. We can see that the particle filtering is more robust. This improvement comes from the fact that the movement model incorporates the temporal correlation of the change of the channel delays in different time steps. The conventional method does not use this information and the results can change rapidly. The particle filter needs some iterations to converge to the correct estimate.

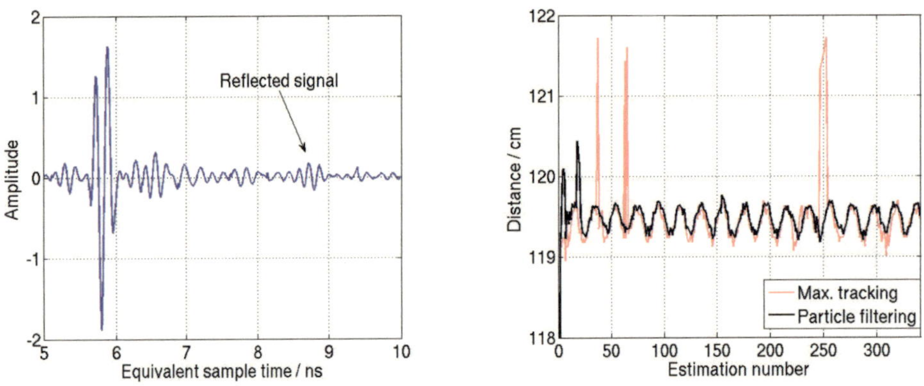

(a) Receive signal

(b) Tracking results

Figure 33. Receive signal with target distance $\approx$ 120 cm and tracking result for target distance $\approx$ 120 cm from conventional maximum tracking and particle filtering.

In this part, we consider the case where the noise is strong and the peak of the signal cannot be easily identified. To reduce the SNR of the signal, Gaussian noise was added to the measurement data of the previous part, so that the SNR was approximately 0dB. An example for one period of the receive signal z_k is shown in Fig. 34(a). We can see that the target signal is not clearly visible anymore and the peak value is disturbed strongly by noise. The tracking results are shown in Fig. 34(b). The conventional method does not work. The particle filtering still gives good estimates, because it does not consider only the maximum point in the target signal but the waveform as a whole. The movement model also plays a role in this improvement.

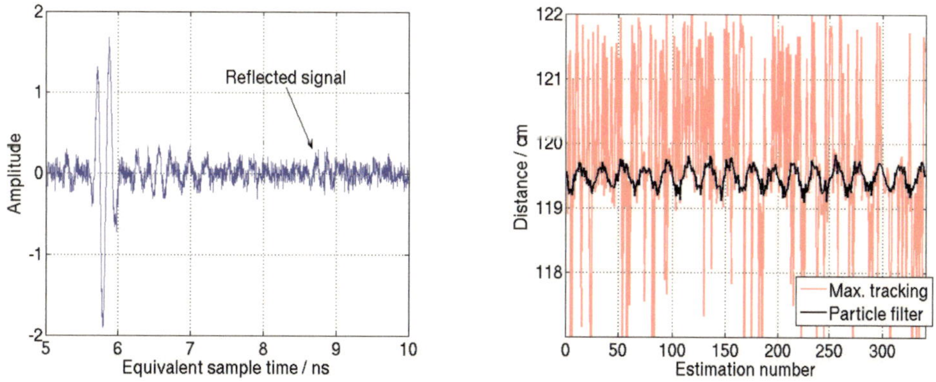

(a) Receive signal

(b) Tracking results

Figure 34. Receive signal with target distance ≈ 120 cm with additional noise and tracking result for target signal ≈ 120 cm with additional noise from conventional maximum tracking and particle filtering.

Now we consider the tracking of the reflected signal for the moving metal plate at a distance of 90 cm. An example of a receive signal is shown in Fig. 35(a). We can see that the target signal is in the same interval as the ringing of the direct coupling signal. The amplitudes of the target signal and the ringing are comparable. It is not easy to distinguish between these two signals anymore. This is the same situation as in radar where we have cluttering. It can cause a bias to the estimation. A comparison of the tracking results from maximum tracking and particle filtering is shown in Fig. 35(b). We can see that the maximum tracking performs very badly because of the bias. The tracking results from particle filtering are very good. The use of the multipath propagation model eliminates the error bias given by the clutter (direct coupling).

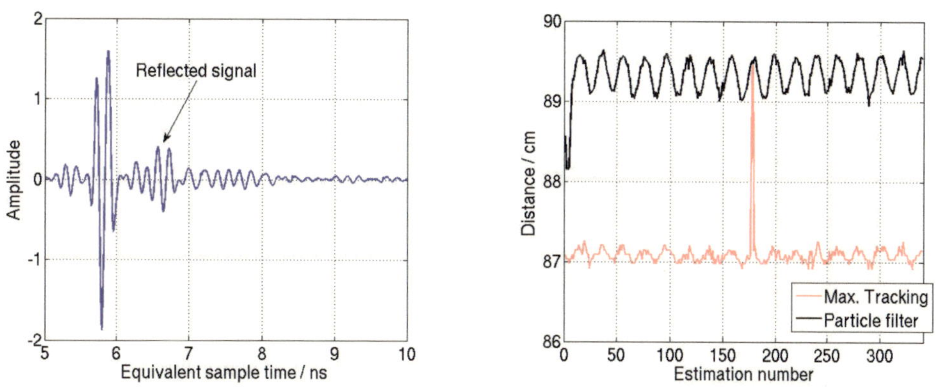

(a) Receive signal

(b) Tracking results

Figure 35. Receive signal with target distance ≈ 90 cm and tracking result for target distance ≈ 90 cm from conventional maximum tracking and particle filtering.

Usually, the drawback of the particle filter is its complexity. In the applications considered here, this drawback is not so serious, because our system model is relatively simple. Using particle filtering in parallel with conventional methods and exchange information between both methods can also help to reduce complexity.

3.4. Surface estimation and subsurface localization algorithms

In this section we present algorithms which can be applied for the localization of actively transmitting beacons inside of the human body. The targeted application is the tracking of catheters equiped with UWB transmitters. In this context, the use of active transmitters would mitigate the challenges related to the high attenuation of electromagnetic waves in human tissue, which makes purely passive localization extremely difficult [3].

A similar approach has been investigated in the field of ultrasonics, where catheter-mounted ultrasound transducers in combination with external arrays of imaging transducers are used to track the catheter position [25]. The advantage of UWB catheter localization is a contactless measurement setup with receivers placed in air around the patient while ultrasound transducers have to be placed directly on the body.

In subsurface imaging and localization problems, where sensors are not directly in contact with the medium, the permittivity contrast between air and the medium cannot be neglected as it leads to a different wave propagation velocity inside of the material. In case of medical applications, a relative permittivity of human tissue between 30 and 50 in the FCC UWB frequency range has to be considered [13]. UWB signals are therefore strongly reflected at the air-to-body interface. These reflections are beneficial to surface estimation applications using UWB pulse radars. For in-body localization, however, they constrain a signal emission from inside of the body. In order to overcome the strong reflection losses, we therefore propose a system that combines an active transmitter inside of the body with an array of radar transceivers outside of the body. The sensor array acts as both a surface scanner and a receiver recording the time of arrival (ToA) of a signal transmitted from inside of the body. A determination of the exact body shape prior to localization is necessary to cope with refraction effects at its surface and the change of wave propagation speed from air to tissue. This distiguishes the proposed in-body localization method from most through-dielectric localization problems like through-the-wall imaging where a plane boundary between air and the target medium is assumed [1].

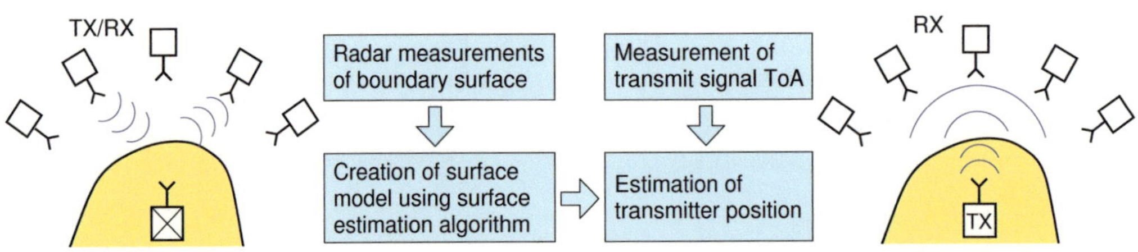

Figure 36. Proposed localization procedure of a transmitter buried in a dielectric medium. In the first step the sensor array is used to scan the surface (left), in the second step it only receives the signal transmitted from inside of the medium (right).

Fig. 36 summarizes the proposed localization approach of a transmitter placed in a dielectric medium. In a first step, the array of radar sensors is used to measure the distance to the boundary surface. These measured distances together with the known antenna positions are the input of the surface estimation algorithm which returns a model of the boundary surface.

In the second step, the transmitter inside of the dielectric is switched on and the radar sensors operate in receive mode recording the time of flight of the transmitted signal. Finally, we analyze all the acquired data and determine the position of the transmitter.

Several methods to estimate the surface of a highly reflective medium using UWB pulse radar sensors have been investigated in recent years [17, 29]. Some of these imaging algorithms, however, need extensive preprocessing of the measurement data or suffer from high complexity and computation time. In the first part of this section, we derive a simple and easy to implement 3D surface estimation algorithm based on trilateration. In the second part, building on this surface estimation method, we present an approach for the localization of transmitters inside an arbitrarily shaped dielectric medium taking into account its surface profile.

3.4.1. Surface estimation algorithm based on trilateration

Radar measurements with quasi-omnidirectional antennas only provide information about the target distance, but not about its direction. This makes surface imaging an inverse problem which can only be solved by combining measurement results from different antenna positions. In this context, target ranging using trilateration means determining the intersections of spheres, the radii of which correspond to measured target distances. The underlying assumption for using trilateration as a surface estimation method is that two neighboring antennas are "seeing" the same scattering center. As with other imaging algorithms this assumption can lead to inaccuracies of estimated target points.

The imaging principle shall first be explained using a two-dimensional example. Fig. 37(a) shows the measurement scenario of a linear array of monostatic radar transceivers arranged along the x-axis scanning the surface of a target in z-direction. Each array element measures the distance to the closest point on the target. Two exemplary measurement points X_n and X_{n+1} are picked out, and semi circles with radii corresponding to the measured target distances are plotted around the antennas. The estimated surface point is the intersection of the two circles.

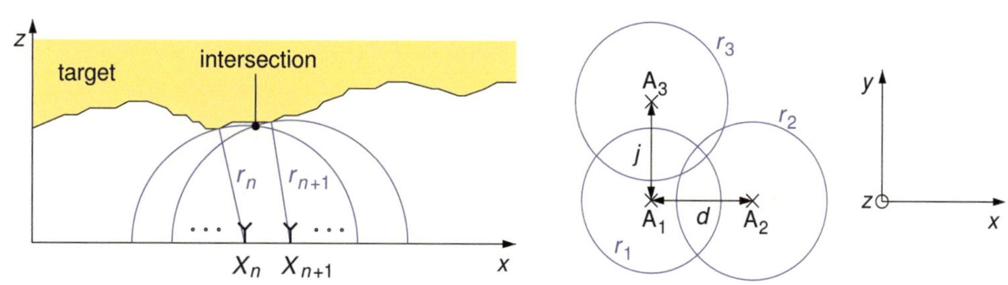

(a) 2D imaging principle

(b) Top view of 3D imaging setup

Figure 37. Cross section of a 2D imaging problem using a linear array of monostatic radar transceivers along the x-axis (a) and top view of a 3D imaging setup showing three measuring points of an antenna array in the x-y-plane (b).

Three-dimensional imaging demands for a third antenna position located in a different dimension. This setup is shown as a top view in Fig. 37(b). At each of the three positions a target distance r_i is measured which leads to a set of spheres with radius r_i around the respective antenna position A_i, as defined by the system of equations

$$r_1^2 = x^2 + y^2 + z^2, \tag{11}$$

$$r_2^2 = (x-d)^2 + y^2 + z^2, \tag{12}$$

$$r_3^2 = x^2 + (y-j)^2 + z^2, \tag{13}$$

where d and j are the distances between two antennas in x- and y-direction, respectively. For simplicity, the first antenna position A_1 shall be at the center of the coordinate system. The above equation system is valid for a planar antenna array. In case of a curved antenna array an offset z-value has to be inserted.

The initial assumption that all three antennas are "seeing" the same target has to be assured by comparing the measured target distances $r_{1\text{-}3}$. If the difference between these distances is small enough, the assumption can be considered valid. For the above equations this precondition can be formulated as follows:

$$|r_1 - r_2| \leq T_{s,x} \quad \text{and} \quad |r_1 - r_3| \leq T_{s,y} \tag{14}$$

A threshold T_s in the range of about half the antenna distance has shown good results.

If the conditions in (14) are fulfilled, the target surface point of interest can be calculated by intersecting the three spheres. The coordinates (x, y, z) of the intersection are

$$x = \frac{r_1^2 - r_2^2 + d^2}{2d}, \tag{15}$$

$$y = \frac{r_1^2 - r_3^2 + j^2}{2j}, \tag{16}$$

$$z = \pm\sqrt{r_1^2 - x^2 - y^2}. \tag{17}$$

These coordinates are offsets refering to the position of the first antenna A_1. The sign in (17) depends on the arrangement of the radar transceivers. Here, we assume that the antennas are oriented towards positive z-values.

The necessary steps of the presented trilateration algorithm can be summarized as follows:

1. Pick three neighboring measurement points in two different dimensions (here: along the x- and y-axis).
2. Extract the target distance from the recorded radar measurement data at each antenna position. Multiple target responses per measurement are possible.
3. Check if the differences between the measured distances satisfy the trilateration condition in eq. (14).
4. If the previous condition is fulfilled, calculate the target coordinates using eq. (15)-(17).
5. Repeat the two previous steps if higher order reflections exist, or otherwise start over with the next three measurement positions.

3.4.2. Subsurface localization algorithm

The given parameters of the localization problem are the shape of the dielectric medium containing the transmitter, its distance to the antenna array and the ToA of the localization signal at each array element. Since all the wave propagation effects are reciprocal, our problem can also be regarded in a reverse way: At each receiver position the transmission of a short pulse with a delay corresponding to the respective previously measured ToA is assumed. In order to get the original beacon position we have to find the spot where all these virtual pulses would superimpose, i.e. the intersection of the impulse wavefronts inside of the medium [27].

According to the Huygens-Fresnel principle a refracted wavefront can be represented by an infinite number of spherical waves which originate from points on the boundary surface reached by the incoming wave. This is shown in Fig. 38(a) for a pulse transmitted from an antenna at position (3,0) towards a dielectric half space. In this 2D example, 20 source points of radial waves on the dielectric surface are considered. The radii $r_{i,n}$ of the secondary waves are calculated from the measured ToA at the receiving antenna and the length of the ray $p_{i,n}$ between the respective antenna position n and surface point i:

$$r_{i,n} = \frac{1}{\sqrt{\varepsilon_r}} \left(c_0 \cdot \mathrm{ToA}_n - p_{i,n} \right), \tag{18}$$

where c_0 is the speed of light. The division by $\sqrt{\varepsilon_r}$ accounts for the different wave propagation speed in the dielectric medium. The envelope of all radial waves corresponds to the wavefront we are looking for. By repeating the procedure for every antenna element of the receiver array we get a set of wavefronts as illustrated in Fig. 38(b). Finally, we determine the intersection of these wavefronts to obtain the transmitter position inside of the medium. A 3D localization problem is solved in an analog way with envelopes of spheres leading to intersecting 3D wavefronts. Here, however, the derivation of the localization procedure shall be limited to the 2D case because of simpler graphical representations.

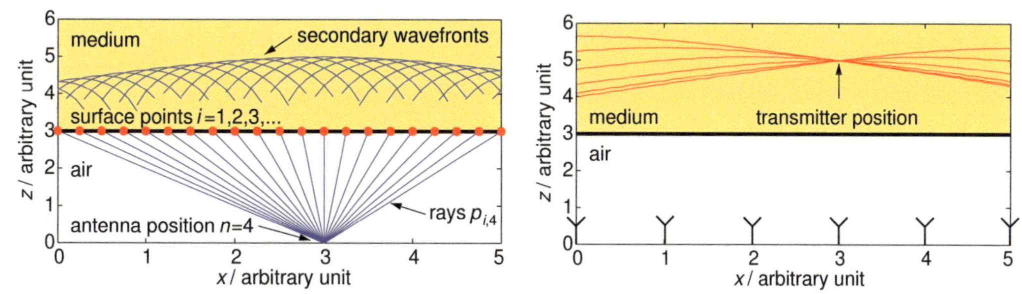

(a) Determination of single wavefront

(b) Localization by intersecting wavefronts

Figure 38. Determination of the 2D wavefront shape inside a medium with $\varepsilon_r = 30$ for a signal transmitted in air at position (3,0) by evaluating the envelope of radial secondary wavefronts (a). Intersections of wavefronts corresponding to six different antenna positions indicate the transmitter location at (3,5) (b).

The wavefront shapes in Fig. 38(b) agree with the hyperbolic approximation of wavefronts in dielectric half spaces [28]. With more complex boundaries, however, the analytical calculation of refracted wavefronts is no longer practical, while the approach presented here is independent of the surface shape. An example of a transmitter placed behind a more complex surface is presented in Fig. 39. The times of flight between the transmitter at (5 cm,5 cm) and the individual elements of the receiver array at $z = -20$ cm are calculated using electromagnetic field simulation software [6].

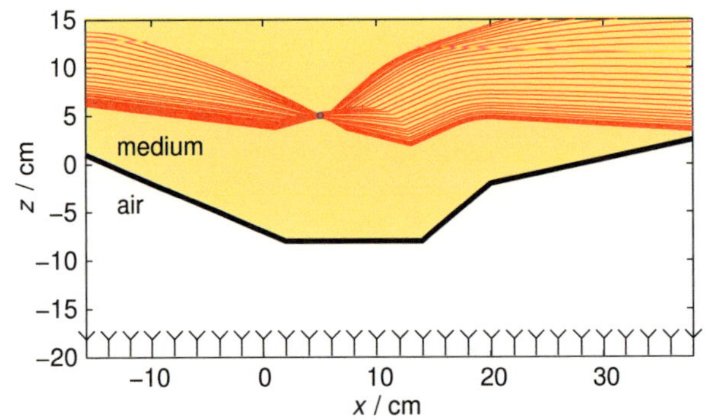

Figure 39. Calculated wavefronts in a medium with ε_r=10 based on EM field simulations. The receiving antennas are placed along the x-axis at z=-20, the transmitter is positioned at (5 cm,5 cm).

The above mentioned 2D localization procedure leads to a belt of refracted wavefronts with a focussing point where the receiver has been placed in the simulation. To decrease calculation time it is also possible to search for the narrowest point in the wavefront belt instead of calculating the intersections. This leads to an estimated transmitter position of (4.60 cm,4.94 cm), having an error of about 4 mm.

It is obvious that in the example of Fig. 39 a smaller number of wavefronts and thus less receiving antennas would be sufficient for a successful localization of the transmitter. But in practical applications this high number of antennas might still be required as a dense sensor array is rather needed for surface estimation than for solving the localization problem.

4. Systems design and measurement results

4.1. Single-ended bistatic radar system

At first a single-ended radar demonstrator was developed in the project, with which the potential of impulse-radio UWB sensing is evaluated. The demonstrator is built combining commercially available components for the low frequency operation control and components tailored for UWB operation dealing with the signals in the 3.1-10.6 GHz band. The UWB components are developed and fabricated in the aforementioned SiGe HBT production technology. A block diagram of the single-ended bistatic radar system is depicted in Fig. 40. The sensor uses separate antennas for transmitter and receiver to avoid losses due to power divider structures on the feedline of a single transceiver antenna. Besides a heavy crosstalk into the low-noise amplifier (LNA) of the receiver is avoided by using two antennas.

An ultra-broadband Vivaldi antenna is chosen, which consists of an exponentially tapered slot on a microstrip substrate. The transition from microstrip to slot line is done by a Marchand balun, as discussed in [22]. On the feeding line of the transmit antenna an impulse generator IC is mounted, which emits an impulse with a shape very similar to the fifth derivative of a Gaussian bell shape with a standard deviation $\sigma = 51$ ps. This impulse shape fits into

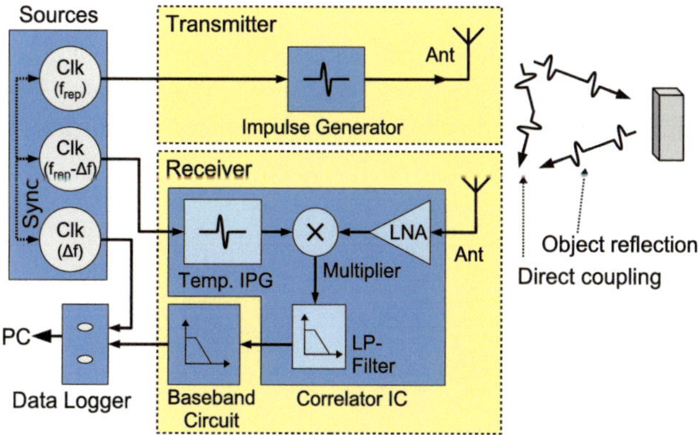

Figure 40. Block diagram of the single-ended bistatic radar system.

the 3.1-10.6 GHz UWB spectral mask allocated by the FCC in the United States and has a voltage amplitude of 600 mV peak-to-peak [8]. The impulse generator radiates an impulse at every rising slope of an input trigger signal. Here a sinusoidal signal is used to trigger the impulse generator, so at every rising edge of the sinusoidal signal an impulse is emitted which results in a continuous impulse train. The sinusoidal signal is supplied from one source of the direct-digital-synthesizer (DDS) AD9959. All four clock sources of the AD9959 (three are used) are synchronized among each other to allow a phase and frequency stable operation between the signals. The transmitter is adjusted to generate an impulse train with a repetition rate of $f_{rep} = 200$ MHz. To reduce impulse-to-impulse jitter, spurious emissions of the 200 MHz sinusoidal trigger signal are filtered by a narrowband helix filter. The generated impulse train is continuously radiated by the antennas, is reflected at the desired object and enters the receiver. The reflection at the object causes a phase inversion to the impulses, therefore it is received with inverted amplitude. Additionally the impulse train is fed to the receiver by a direct and non-inversed coupling between the antennas.

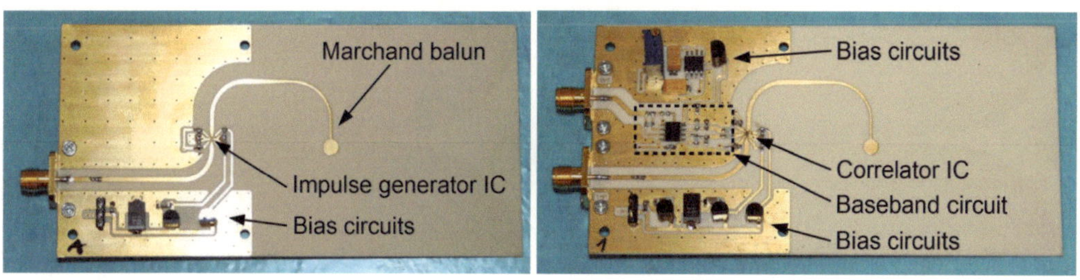

(a) (b)

Figure 41. Picture of (a) Vivaldi transmit antenna with mounted impulse generator IC and (b) Vivaldi receive antenna with correlator IC and baseband circuit.

These signals are processed in the receiver by a monolithic correlator IC, which consist of a UWB LNA, a four-quadrant multiplier, a template impulse generator generating a fifth Gaussian derivative impulse corresponding to the transmit impulse, and a first integrating low-pass filter with a cut-off frequency of 800 MHz [9]. The template impulse generator is driven by a second clock source of the AD9959 at a repetition rate of

$f_{rep} - \Delta f = 200\,\text{MHz} - 25\,\text{Hz}$. This sinusoidal clock signal is filtered as well with a helix filter to improve jitter performance. The output signal of the correlator IC is processed with a baseband circuit, where it is amplified and further integrated by a low-pass filter with a cut-off frequency of 25 kHz. After this it is fed to a data-logger, which samples the generated signals with a sampling period of 30 μs and transfers them to a PC for further processing. Additionally a synchronizing signal of $\Delta f = 25\,\text{Hz}$ from the third DDS clock signal is sampled

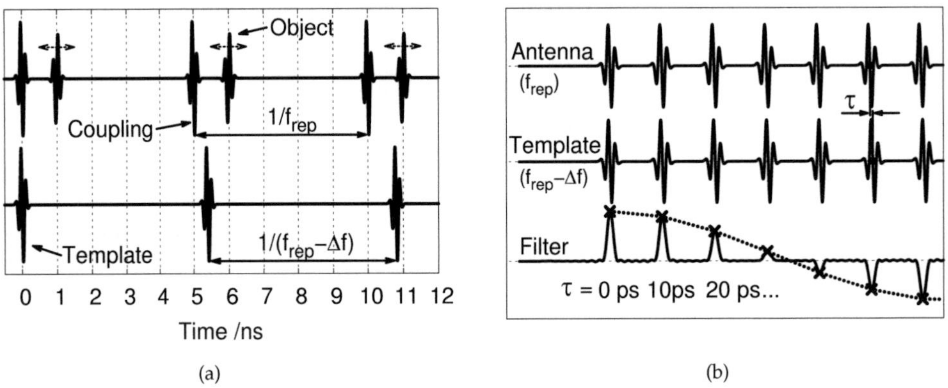

Figure 42. (a) Illustration of the template impulse sweeping through the antenna receive signal. (b) Illustration of the evolving correlation signal from receive and template signal in the region where the impulses overlap.

and transferred to the PC as well. In the PC a post-processing of the continuous data stream is done using a custom software written in Lab Windows/CVI. Pictures of transmit and receive antenna with mounted components are shown in Fig. 41. A detailed description and supporting measurements of the hardware can be found in [7, 31].

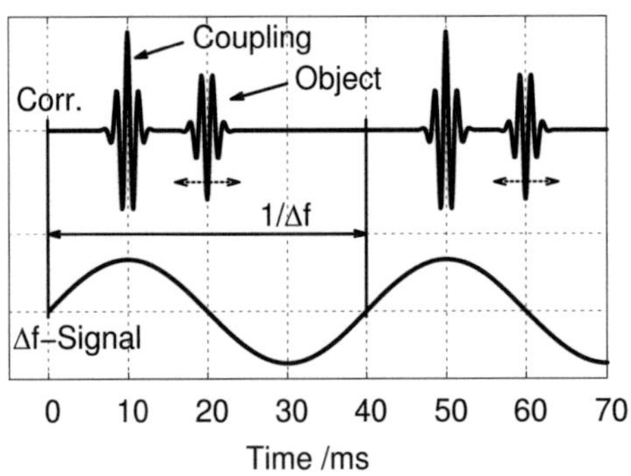

Figure 43. Illustration of the correlation signal at the receiver antenna output port, together with the Δf-signal for separation of the repetition.

To determine the distance between sensor and desired object and the variation of the distance between them, a time-of-flight measurement is applied using a sweeping impulse correlation as explained in [10]. The sweeping impulse correlation is very similar to an undersampling technique and avoids high-speed sampling of the gigahertz-range UWB impulses. This method is illustrated in Fig. 42 by showing the signals appearing in the receiver. In the upper trace of Fig. 42(a) the received signal at the output of the LNA in the correlator IC is shown. An impulse from both, the non-inverted direct coupling and the inverted reflection at the object is shown. The impulses are repeated at the repetition rate f_{rep}. The template impulse in the correlator IC, shown in the lower part of Fig. 42(a), is operating at a repetition rate $f_{rep} - \Delta f$. Therefore it is continuously changing its time alignment to the received signals and appears sweeping through the impulse sequence.

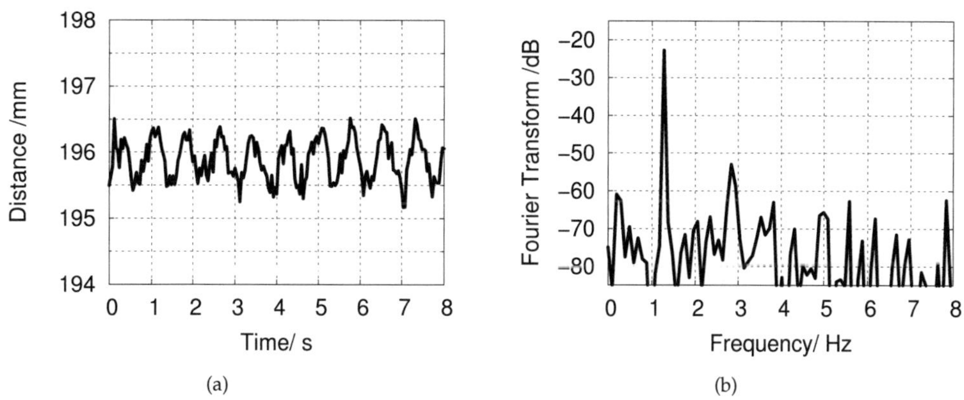

Figure 44. Measurement of sinusoidally moving metal plate placed in front of the antennas at a distance of 19.6 cm and a deviation of approx. 1 mm in (a) time and (b) frequency domain.

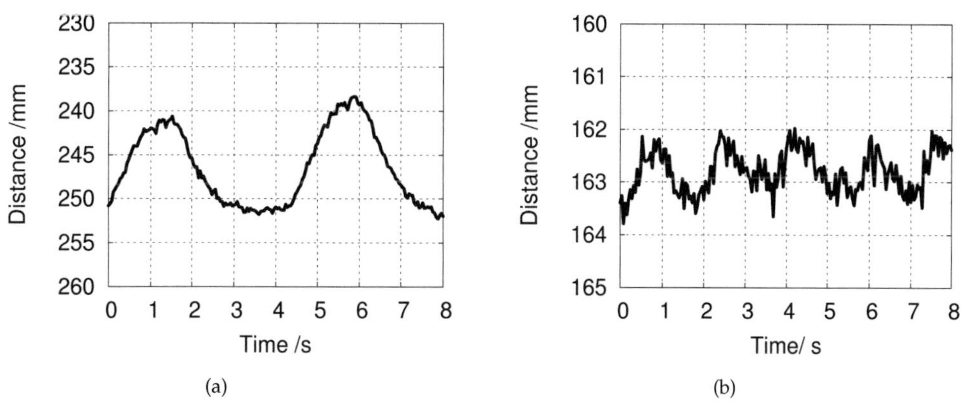

Figure 45. Time domain breathing measurements of (a) a male test person lying on the back and (b) a seven-week old child sleeping in bed.

At each position of the impulses the cross-correlation is computed by multiplying the signals and integrating them. This operation can be seen in Fig. 42(b) in the region where one received impulse and the template impulse are overlapping each other. The lower part of the figure illustrates the output of both filters. The solid line represents the output signal of the 800 MHz filter which integrates each single impulse. The dashed line shows the output signal of the baseband filter with a cut-off frequency of 25 kHz, which integrates a multitude of single impulses. By this procedure, a cross-correlation curve of two fifth Gaussian derivatives is developing. The correlated curve of the above illustration example is depicted in Fig. 43. The correlation impulses from the direct coupling and the reflection at the object can be distinguished. As discussed, the reflection at the object is mirrored, because the impulse is inverted by the reflection. This signal is present at the output of the receive antenna and is sampled by the data-logger. The correlation signal is continuously repeated with a repetition rate of Δf. For a separation of the correlation sweeps, the Δf-signal is sampled as well by the data-logger. When the object under investigation is now moving, the part of the correlation signal coming from the object reflection is correspondingly changing its alignment to the Δf-signal and the movement can be measured. The movement determination of the object is continuously done by software in the PC. First a separation of the correlation sweeps by the rising slope of the Δf-signal is performed. Then both slopes of the correlation curves, the slope from the object and the slope from the direct coupling, are tracked and their positions continuously monitored. Tracking both slopes yielded best precision, compared to tracking the minimum of the correlation curve or only the slope of the correlation signal from the object [31].

To measure the precision of the sensor a metal plate is placed in front, which is mounted on a sledge driven by an eccentric disk, moving the metal plate forward and backward with an approximately sinusoidal deviation. Fig. 44(a) shows a time domain record of a movement measurement with the metal plate placed at a distance of 19.6 cm, a deviation amplitude of approximately 1 mm and a repetition rate of around 1.35 cycles/s. The movement is clearly resolved by the measurement. In Fig. 44(b) the calculated spectrum of the measurement can be seen. The frequency maximum is very clearly visible and verifies a precision of the demonstrator in the millimeter to sub-millimeter range.

In a further measurement the sensor is pointed to the abdomen of a male test person lying on the back at a distance of approximately 25 cm[1]. At the abdomen the largest breathing amplitude occurs. Fig. 45(a) shows a breathing measurement in case the person is breathing normally. The breathing amplitude exceeds 10 mm and the repetition rate is around 2.5 cycles/s. For a further measurement the demonstrator is placed towards a sleeping seven-week old child lying on the back at a distance of approximately 16.3 cm. Fig. 45(b) shows a rhythmic breathing period with a movement of around 1 mm in the direction of the sensor and a repetition rate of 1.5 cycles/s. These measurements show, that the sensor can be used to monitor the breathing of adults and infants lying on the back and that breathing patterns can clearly be detected using the single-ended bistatic impulse-radio UWB radar demonstrator.

4.2. Differential bistatic radar system

A differential bistatic radar system for detecting vital signs was also developed. It follows the approach described in Sec. 4.1, but with fully differential ICs as described in Sec. 2.2 and Sec. 2.3.3 and with significantly reduced power consumption. Here two dipole-fed circular slot antennas are chosen instead of the the Vivaldi antennas. Both the differential impulse

[1] Measurements on humans using the single-ended bistatic radar sensor have been approved by the ethic commission of Ulm University.

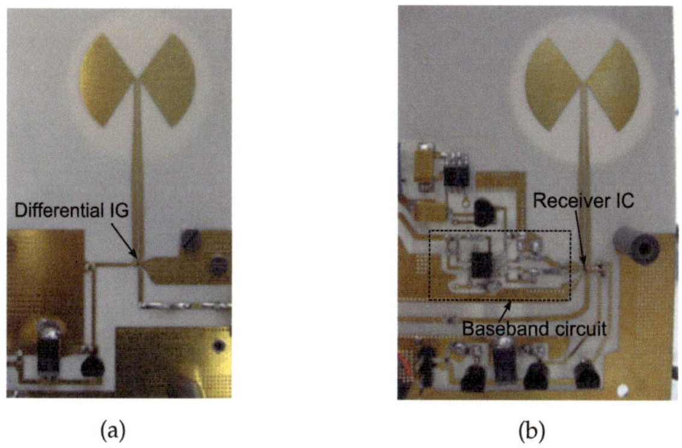

Figure 46. Photographs of transmit antenna with mounted differential impulse generator IC and the complete receiver with RF frontend IC and baseband circuit.

generator and the correlation receiver front-end are mounted chip-on-board at the feeding points of the dipole antennas. Fig. 46 shows the pictures of transmit and receive antennas with mounted differential ICs.

Using the same DDS clock generator and post-processing software as described in Sec. 4.1, the ability of the realized differential bistatic radar system for tracking a metal plate which moves back and forth with a sinusoidal deviation is demonstrated. The measured result in the time domain can be seen in Fig. 47(a). A deviation amplitude of around 1.5 mm can be

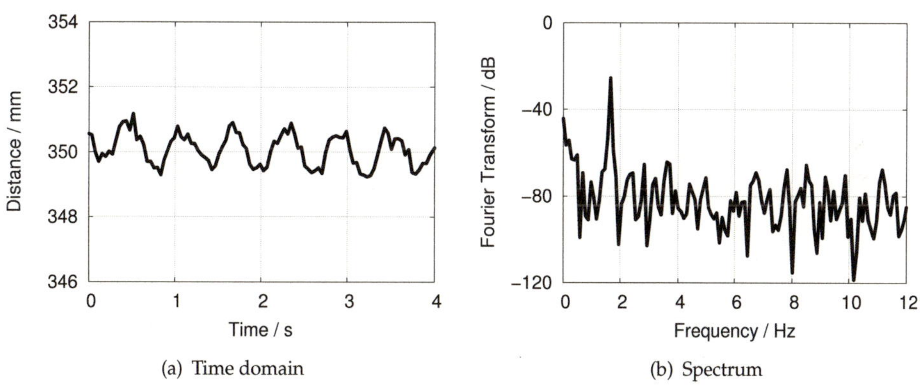

Figure 47. Measurement results of a moving metal plate in front of the bistatic radar with a distance of 35 cm in time and frequency domains.

clearly seen. Fig. 47(b) shows the calculated spectrum information of the measurement. The maximum point is clearly visible and indicates the movement frequency of the metal plate.

A common application for this radar system is the detection of vital signs. Here, an adult male with pronounced tachypnea is seated 5 cm from the radar. Fig. 48(a) shows the recorded time domain data. The breathing pattern is clearly visible and its amplitude is around 5 mm. The time domain data is Fourier transformed to frequency domain as shown in Fig. 48(b). It clearly indicates that the respiration rate is around 35/min.

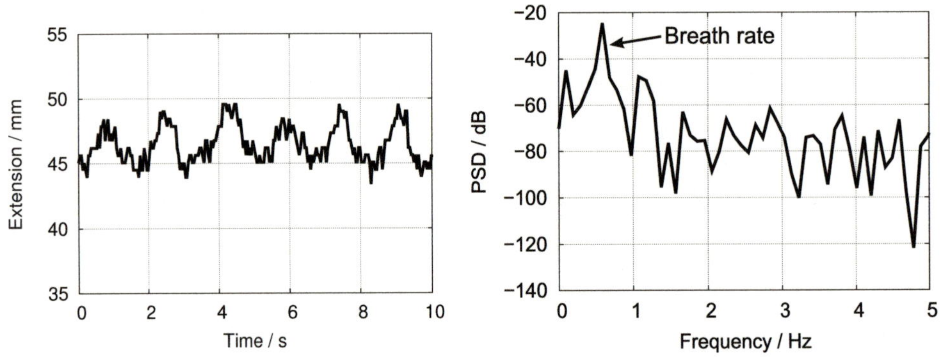

(a) Time domain (b) Spectrum

Figure 48. Time domain and frequency domain measurements of vital signs of a male test person standing in front of the bistatic radar.

In a further measurement, surface estimation of a container filled with a sugar solution whose properties are similar to those of human tissue was performed by moving the radar up and down in 2 cm steps along both the x- and y-axis. For demonstrating that the radar system is capable for this application, only the lower part of the container is scanned. The trilateration-based imaging algorithm derived in Sec. 3.4 is applied. The photograph of the container is shown in Fig. 49(a), and a cloud of estimated surface points representing the front of the target can be seen in in Fig. 49(b). The result clearly indicates the distance from the radar sensor to the container and the planar surface structure.

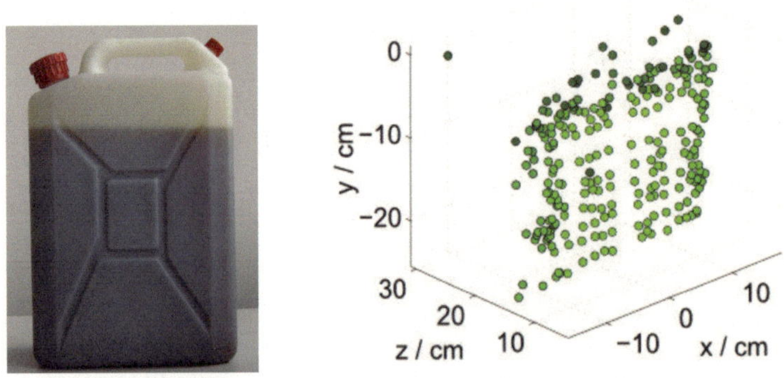

(a) Photograph (b) Measurement

Figure 49. Photograph of the liquid container and measured cloud of estimated surface points.

4.3. Communcation with implants

In this section we address another application of UWB technology in medicine, the communication with implants. Impulse-based UWB technology is a promising solution for future implanted medical devices demanding data rates in the Mbit/s range and a low power consumption. Here, we present a demonstration system for uni-directional data transmission between a transmitter inside tissue-mimicking liquid and a receiver placed inside or outside the medium. These two measurement scenarios are illustrated in Fig. 50.

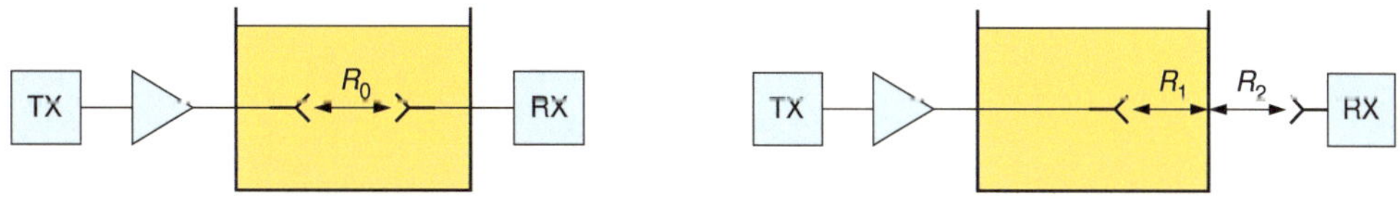

(a) Tissue to tissue (b) Tissue to air

Figure 50. Measurement scenarios for the demonstration of impulse-based data transmission for implants using an energy detector. An amplifier is applied on the transmit side to cope with the high attenuation of the phantom liquid.

As discussed in Sec. 3.1, many different approaches for IR-UWB receivers are feasible. Non-coherent detection is chosen here because of the challenges regarding synchronization in coherent receiver setups and because of the dispersive properties of human tissue, which lead to unpredictable shapes of the received signals. Even though non-coherent detectors are suboptimal, they are insusceptible to dispersive effects of the channel, and therefore well-suited for communication with implants.

The simplest non-coherent receiver concept is an energy detector, which basically consists of a squaring device and an integrator (see Sec. 2.3). Besides its simplicity, the advantage of an energy detector is that possible narrow-band interferers within the operating frequency band could be suppressed by using the comb filter approach described in Sec. 3.2. This would improve the SNR additionally. Hence, the energy detector is chosen as the receiver in the demonstration setup for communication with implants. In this demonstration, on-off keying is used as the modulation scheme and a unidirectional transmission with a data rate of 100 Mbit/s is selected. This data rate and a unidirectional transmission are sufficient for most medical applications.

The transmitter in Fig. 50 consists of a UWB impulse generator (IG) and an antenna. The transmit data are generated externally by a bit pattern generator, clocked by a signal generator. The receiver topology is the same as shown in Sec. 2.3.2, but in a single-ended configuration. The output signal is amplified and displayed on an oscilloscope. For further signal processing in the digital domain, a comparator circuit with a properly adjusted threshold voltage can be applied. A more detailed description of the transmitter and receiver structures used can be found in [21].

To demonstrate the transmission in an environment similar to human tissue with a highly dispersive and lossy behavior, two antennas are immersed in a tissue-mimicking liquid. This liquid consists mainly of sugar and water. The properties of the liquid are similar to skin tissue with a relative dielectric constant ε_r of 28 at 7 GHz and an attenuation of about 22 $\frac{\text{dB}}{\text{cm}}$. The influence of the dispersive behavior on the impulse shape is illustrated in Fig. 51 in time and frequency domain. There, the output signal after the transmission through the phantom liquid is shown in comparison to the input signal. Due to increased losses for higher frequencies, the impulse shape is significantly broadened and the amplitude is decreased by approximately 60 dB for the used path distance of 23.5 mm between the two immersed antennas.

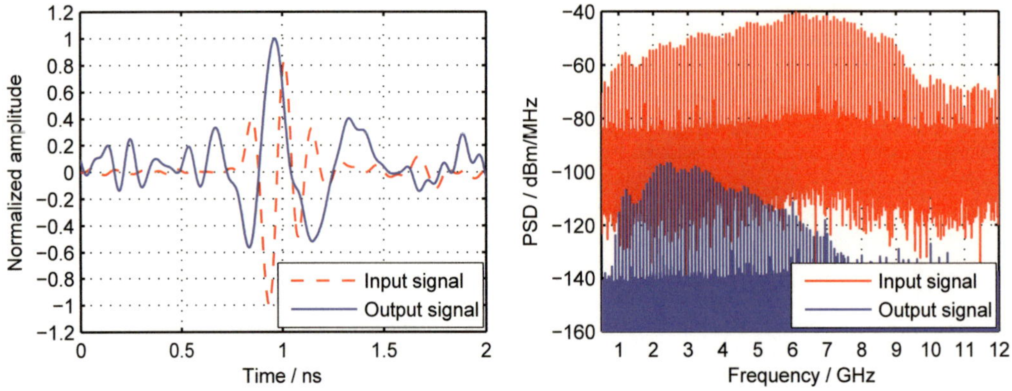

(a) Time domain

(b) Frequency domain

Figure 51. Typical received signal after the transmission of an impulse through tissue-mimicking liquid. The distance between transmitter and receiver is $R_0 = 23.5$ mm.

A data transmission is still achievable in these cases as illustrated by the reception of a typical bit pattern in Fig. 52(a). To this end, an additional amplifier with 12 dB gain is inserted after the impulse generator in order to compensate for the high losses. This measurement setup demonstrates the transmission from a deeply implanted device to a reading device placed directly on the human body, e.g. applicable for capsular endoscopy. In a final scenario, the communication of a less deeply implanted device with a base station outside the human body is considered. There, one antenna is immersed in the phantom liquid and a second one is located outside in free space (see Fig 50(b)). In this setup, the distance of the immersed antenna to free space is fixed to 15 mm, and the location of the outer antenna is varied. Similar measurement results as before are obtained here. In Fig. 52(b) the output voltage of the energy detector is plotted against the distance R_2 between the medium surface and receiver in air. The observed maximum distance of the base station for a reception is 15 cm, then the attenuation limit is reached.

4.4. Surface estimation and subsurface localization measurements

4.4.1. Measurement setup

For the evaluation of the surface estimation and subsurface localization algorithms presented in Sec. 3.4, measurements are performed using a similar bistatic UWB radar sensor as described in Sec. 4.1 [22] and the miniaturized antenna optimized for radiation in human tissue, presented in Sec. 2.1.3. The measurement setup is illustrated in Fig. 53. In this setup we use one single radar sensor which is emulating a whole sensor array by being moved along the x- and y-axis in front of the target. The target object is a container filled with the same tissue-mimicking liquid already used in the measurements of Sec. 4.3. A control unit providing clock signals for the radar sensor is connected to a computer where the signal processing and visualization is performed. In the measurements for subsurface transmitter localization, the upper elements of the setup in Fig. 53 are switched on. The miniaturized antenna placed inside of the tissue-mimicking liquid is now transmitting a 5th derivative of a Gaussian pulse generated by the impulse generator (IG). Because of the high loss in the

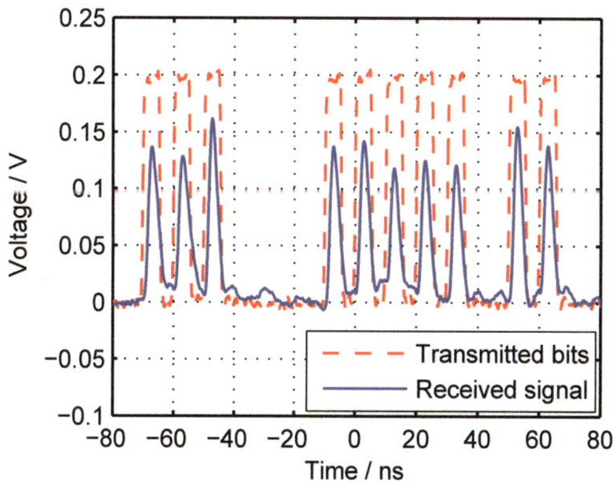

(a) Tissue to tissue

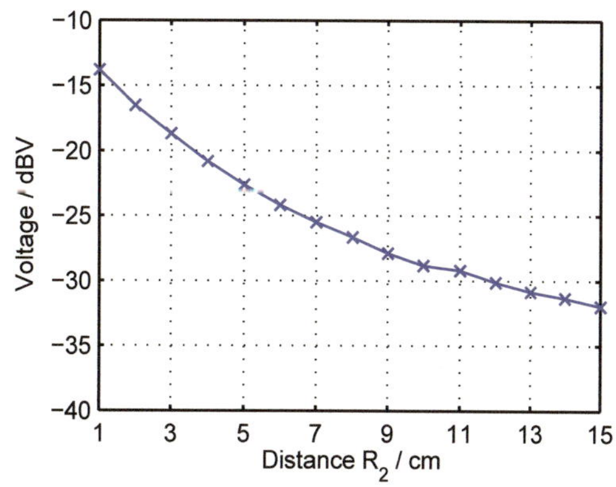

(b) Tissue to air

Figure 52. Measured output voltage of the energy detector according to the measurement scenarios in Fig. 50 using a data rate of 100 Mbit/s.

medium of about 22 $\frac{\text{dB}}{\text{cm}}$, an amplifier is used. For localization, the radar sensor outside of the container is operating in receive mode measuring the ToA of the transmitted impulse.

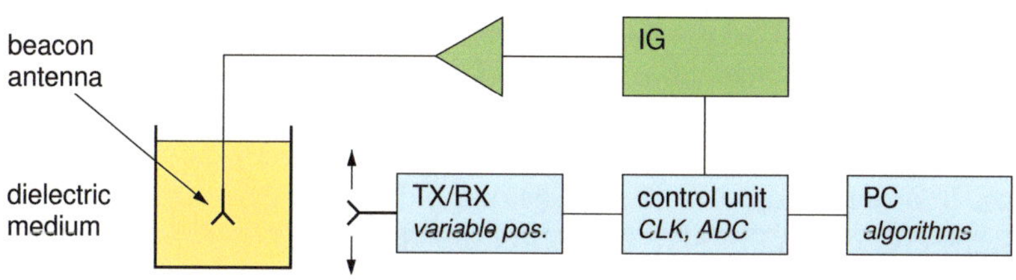

Figure 53. Measurement setup for surface estimation and subsurface transmitter localization. The radar sensor is moved in front of the liquid container to scan its surface. For localization, the upper part consisting of an impulse generator and an amplifier is switched on.

4.4.2. Surface estimation results

In 3D surface measurements the performance of the trilateration-based imaging algorithm derived in Sec. 3.4.1 is verified. As a target object a plastic dummy of a female torso of about 60 cm height as pictured in Fig. 54(a) has been chosen. In order to increase the target's reflectivity the surface of the dummy has been treated with highly conductive copper laquer. Radar scans of the 3D surface are performed by moving the radar sensor up and down in 1 cm-steps along the y-axis and by rotating the target object in 5°-steps. As a result, a cylindrical array of radar sensors is emulated.

Fig. 54(b) shows the output of the imaging algorithm, a cloud of estimated surface points representing the front side of the target. An interpolation of these points is necessary to obtain a surface model needed for the subsurface localization application. The interpolated surface is illustrated in Fig. 54(c) showing a good agreement with the original target object.

The performance of the proposed algorithm regarding its accuracy is compared to the well-established imaging algorithms "Seabed" [29] and "Envelope of Spheres" [16] by evaluating the surface measurement of a metal sphere with a known diameter of 35 cm. A

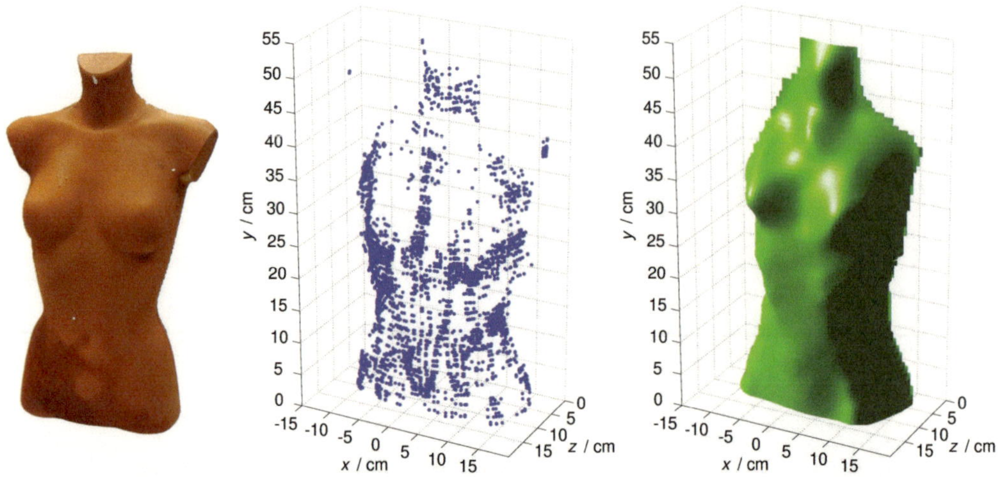

(a) Photograph (b) Scattered surface points (c) Interpolated surface

Figure 54. Surface estimation result of the proposed trilateration-based imaging algorithm for a radar measurement of a human torso dummy using a circular antenna arrangement.

spherical target was chosen here since a mathematical surface of the torso dummy model is unavailable. The error distance between estimated points and the ideal surface of the target sphere is calculated and plotted in Fig. 55 for different densities of measurement points. The graphs show the percentage of estimated points having a certain deviation in cm from the ideal surface. In order to compare the results of the different algorithms at exactly the same coordinates, the estimated points have been interpolated on an identical coordinate grid.

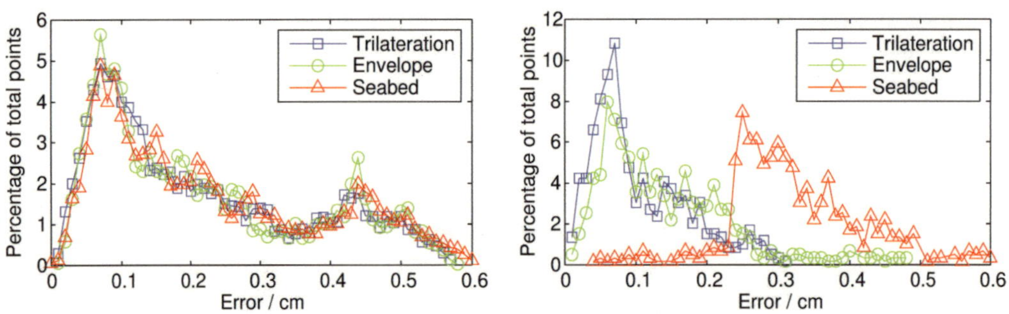

(a) Antenna distance $d = 1.5$ cm (b) Antenna distance $d = 6$ cm

Figure 55. Percentaged distribution of the error between estimated surface points and the actual surface of a trapezoidal target using a monostatic radar setup with two different distances d between measuring points. The performance of the proposed algorithm is compared to state-of-the-art imaging algorithms.

While in the first measurement in Fig. 55(a) a relatively small step width of $d = 1.5$ cm between two measuring positions has been used, the measurement results in Fig. 55(b) show the errors obtained with a quadrupled step width of 6 cm. It can be seen that with a high density of measurement points there is no significant difference of estimation errors between the three compared algorithms. The deviations from the ideal surface points are in a low millimeter range. However, with an increased distance between measurement points the performance of "Seabed" is degrading significantly, while the errors obtained with trilateration and the "Envelope of Spheres" algorithm remain in the same range.

It can be seen that the trilateration-based imaging algorithm achieves similar results as state-of-the-art surface estimation algorithms while the compexity is reduced since here no preprocessing of measurement data is needed. A more detailed description of the trilateration-based imaging algorithm and further measurement results can be found in [26].

4.4.3. Transmitter localization results

In the measurements for the evaluation of 3D transmitter localization, a liquid container whose concave surface roughly approximates the surface of a human body is chosen. Fig. 56 shows a top and side view of the measuring setup with the transmitter placed in a glass fish bowl of 21 cm diameter. The radar sensor is scanning in 1 cm steps in x- and y-direction emulating an antenna array with 14×11 elements in total.

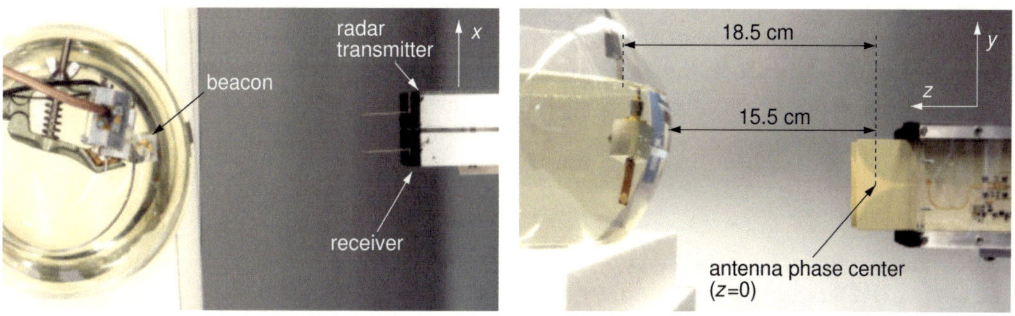

(a) Top view (b) Side view

Figure 56. Photographs of the 3D measurement setup with a transmitter placed in a concave liquid container.

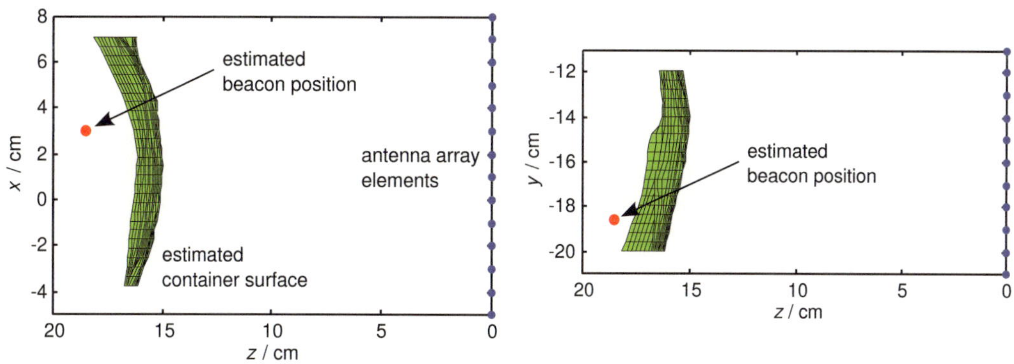

(a) Top view (b) Side view

Figure 57. Measurement results in a top and side view corresponding to the photographs in Fig. 56 showing the estimated location of the beacon behind the container surface. The dots at $z = 0$ represent the sensor positions.

The surface points found by the trilateration-based imaging algorithm are interpolated to create the boundary shape depicted in Fig. 57. These graphs follow the perspectives of the photographs in Fig. 56 allowing a rough evaluation of the localization result. By applying the localization method of Sec. 3.4.2 based on the evaluation of 3D wavefronts inside of the medium, an estimated position of the transmitter of $(x,y,z) = (3.0\,\text{cm},-18.4\,\text{cm},18.5\,\text{cm})$ is calculated. Even though an exact verification of the antenna position inside of the liquid is difficult, the manually measured z-distance of 18.5 cm between the transmitter and the receiver agrees with the estimated beacon position. In the photographs of Fig. 56 the receiver antenna is positioned at the x- and y-position closest to the beacon antenna in the container. These known sensor coordinates of (3 cm,-19 cm) also coincide well with the estimated beacon position. In case of a convex surface like the fish bowl an even better localization can be expected when using a circular or spherical arrangement of the receivers around the medium, instead of a planar arrangement.

Due to the high attenuation of signals transmitted through tissue-mimicking liquid we can only localize positions close to the container surface. In our measurements the signal-to-noise ratio of impulses running through more than 3 cm of tissue-mimicking liquid became too low to be detected. One possible way to increase the maximum transmitter distance from the surface would be the use of a cascade of multiple amplifiers to realize a higher transmitter power output. Regarding the optimization of signal processing an approach based on compressed sensing is investigated to cope with lower signal-to-noise ratios at the receiver [34].

5. Conclusion

In this chapter, novel hardware components, algorithms, systems and possible approaches in the field of UWB radar and communications for biomedical applications have been presented.

Three novel UWB antenna concepts have been introduced targeting different tasks in the medical field. For communication purposes, a circular slot antenna excited with a dipole element has been presented. Its novel differential feeding concept suppresses parasitic radiation by cable currents on the feed lines. In a new concept for applications requiring directive antennas with small beamwidths, a dielectric rod has been added to the circular slot antenna, resulting in a compact and easy-to-fabricate antenna with a high mean gain of 8.7 dBi. Furthermore, a miniaturized UWB slot antenna, optimized for the radiation in human tissue has been designed.

A flexible, differential chipset using Si/SiGe HBT technology for IR-UWB applications has been presented. On the transmit side, a low power impulse generator based on cross-coupled LC oscillator is successfully realized. It generates ns-duration and stable impulses with a spectrum well fitting the FCC mask. This impulse generator has been successfully extended to include tunability to the FCC, ECC and Japanese UWB masks as well as a biphase modulation function. On the receive side, an energy detection receiver optimized for simple on-off keying communications and a correlation detection receiver for short range radar applications have been presented. Both receivers are based on a fully differential UWB low-noise amplifier and a four-quadrant RF multiplier which performs either squaring or multiplication operations.

Using the aforementioned components, two bistatic radar systems, a single-ended and a differential configuration, have been built. Their performance has been demonstrated in a setup for vital sign detection.

Regarding UWB communications, a proposal for a transmission scheme has been discussed, using a special spread spectrum method and energy detection combined with a comb filter, which improves the SNR and rejects narrowband interference. The robustness of this concept has been demonstrated for multipath propagation channels as well as for narrowband interference, noise, and synchronization errors. The approach fits well to medical applications, because small multipath delay spreading promises an easier realization of the analog time delay needed for the comb filter. A trade-off between the number of UWB impulses per symbol (bit) and the data rate requirement has to be made for different applications.

In addition, a new concept for an impulse-radio transmission based on code shift keying with a comb filter receiver has been introduced. In this concept, the delay of the analog delay element could be shorter than the channel impulse response. It has been shown and verified by simulation that the performance and the resistance against multipath propagation, noise, narrow band and multisensor/multiuser interference are the same as in the original approach with longer delays in the comb filter loop.

Simulation results have shown that particle filtering can improve the ranging and tracking performance of an impulse UWB radar substantially in scenarios with low signal-to-noise ratio and cluttering in comparison with more conventional methods. A trade-off between realization complexity and performance can be adjusted thanks to the flexibility of the proposed algorithm.

It has been shown practically that a non-coherent energy detector is a suitable receiver concept for UWB communication with implants. The energy detector operates without synchronization and is insusceptible to dispersive effects of the channel. Demonstrational measurements in tissue-mimicking liquid have been performed with a data rate of 100 Mbit/s meeting the requirements for modern medical devices.

A 3D surface estimation algorithm based on trilateration for ultra-wideband pulse radars has been presented and derived mathematically. Since this method needs no preprocessing of measurement data its implementation is very simple. In 3D surface measurements the performance of the proposed algorithm has been verified, and comparisons with established algorithms have shown a similar performance regarding estimation errors.

As a next step towards the targeted application of catheter tracking, a method for the localization of UWB transmitters buried in homogeneous dielectric media has been presented. With the aid of surface estimation algorithms a localization behind an arbitrarily shaped medium boundary is possible. For this purpose we have proposed a system consisting of an array of UWB radar sensors outside the medium and a beacon inside the medium transmitting a short UWB pulse. The external sensors serve for surface scanning and for measuring the time of arrival of the transmitted signal. The performance of the proposed localization algorithm has been verified using electromagnetic field simulations and measurements, in which a transmitter has been placed in tissue-mimicking liquid.

Author details

Dayang Lin, Michael Mirbach, Thanawat Thiasiriphet, Jürgen Lindner, Wolfgang Menzel and Hermann Schumacher
Ulm University, Germany

Mario Leib
EADS Deutschland GmbH, Division Cassidian, Ulm, Germany

Bernd Schleicher
TriQuint Semiconductor GmbH, Munich, Germany

6. References

[1] Ahmad, F. & Amin, M. G. [2006]. Noncoherent approach to through-the-wall radar localization, *IEEE Transactions on Aerospace and Electronic Systems* 42(4): 1405–1419.

[2] Belostotski, L. & Haslett, J. [2009]. A technique for differential noise figure measurement with a noise figure analyzer, *IEEE Microwave Magazine* pp. 158–161.

[3] Bilich, C. G. [2006]. UWB radars for Bio-Medical Sensing: Attenuation Model for Wave Propagation in the body at 4GHz, *University of Trento Technical report DIT-06-051* .

[4] Blech, M. D. & Eibert, T. F. [2007]. A Dipole Excited Ultrawideband Dielectric Rod Antenna With Reflector, *IEEE Transactions on Antennas and Propagation* 55(7): 1948–1954.

[5] Buchegger, T., Ossberger, G., Reisenzahn, A., Hochmaier, E., Stelzer, A. & Springer, A. [2005]. Ultra-Wideband Transceivers for Cochlear Implants, *EURASIP Journal on Applied Signal Processing* 18: 3069–3075.

[6] CST [2011]. *CST MICROWAVE STUDIO®, User Manual.* Darmstadt, Germany.

[7] Dederer, J. [2009]. *Si/SiGe HBT ICs for Impulse Ultra-Wideband (I-UWB) Communications and Sensing*, Cuvillier Verlag, Göttingen, Germany.

[8] Dederer, J., Schleicher, B., Santos, F., Trasser, A. & Schumacher, H. [2007]. *FCC compliant 3.1-10.6 GHz UWB Pulse Radar using Correlation Detection*, Proc. IEEE MTT-S Int. Microw. Symp., Honolulu, HI, pp. 1471–1474.

[9] Dederer, J., Schleicher, B., Trasser, A., Feger, T. & Schumacher, H. [2008]. A Fully Monolithic 3.1-10.6 GHz UWB Si/SiGe HBT Impulse-UWB Correlation Receiver, *IEEE International Conference on Ultra-Wideband (ICUWB)*, Vol. 1, pp. 33–36.

[10] Devine, P. [2000]. *Radar Level Measurement: The User's Guide*, VEGA Controls.

[11] Fear, E. C., Hagness, S. C., Meaney, P. M., Okoniewski, M. & Stuchly, M. A. [2002]. Enhancing Breast Tumor Detection With Near-field Imaging, *IEEE Microwave Magazine* 3(1): 48–56.

[12] Fort, A., Desset, C., Ryckaert, J., Doncker, P. D., Biesen, L. V. & Wambacq, P. [2005]. Characterization of the ultra wideband body area propagation channel, *Ultra-Wideband, 2005. ICU 2005. 2005 IEEE International Conference on*, p. 6 pp.

[13] Gabriel, S., Lau, R. W. & Gabriel, C. [1996]. The dielectric properties of biological tissues: II. Measurements in the frequency range 10 Hz to 20 GHz, *Phys. Med. Biol.* 41: 2251.

[14] Hees, A., Hasch, J. & Detlefsen, J. [2008]. Characteristics of a Corrugated Tapered Slot Antenna with Dielectric Rod and Metallic Reflector, *IEEE International Conference on Ultra-Wideband*, Vol. 1.

[15] Hertel, T. W. [2005]. Cable-current Effects of Miniature UWB Antennas, *IEEE International Symposium on Antennas and Propagation*, Vol. 3A, pp. 524–527.

[16] Kidera, S., Sakamoto, T. & Sato, T. [2009]. High-Resolution 3-D Imaging Algorithm With an Envelope of Modified Spheres for UWB Through-the-Wall Radars, *Antennas and Propagation, IEEE Transactions on* 57(11): 3520–3529.

[17] Kidera, S., Sakamoto, T. & Sato, T. [2010]. Accurate UWB Radar Three-Dimensional Imaging Algorithm for a Complex Boundary Without Range Point Connections, *Geoscience and Remote Sensing, IEEE Transactions on* 48(4): 1993–2004.

[18] Krach, B., Lentmaier, M. & Robertson, P. [2008]. Joint bayesian positioning and multipath mitigation in gnss, *Acoustics, Speech and Signal Processing, 2008. ICASSP 2008. IEEE International Conference on*, pp. 3437–3440.

[19] Leberer, R. [2005]. *Untersuchung von quasi-planaren Antennen mit sektorförmiger und omnidirektionaler Strahlungscharakteristik im Millimeterwellenbereich*, Dissertation, Universität Ulm, Institut für Mikrowellentechnik.

[20] Leib, M. [2011]. *Ultrabreitbandige Antennen für Kommunikation und Sensorik in der Medizintechnik*, Dissertation, Universität Ulm, Institut für Mikrowellentechnik.

[21] Leib, M., Mach, T., Schleicher, B., Ulusoy, C. A., Menzel, W. & Schumacher, H. [2010]. Demonstration of UWB communication for implants using an energy detector, *German Microwave Conference, 2010*, pp. 158–161.

[22] Leib, M., Schmitt, E., Gronau, A., Dederer, J., Schleicher, B., Schumacher, H. & Menzel, W. [2009]. A compact ultra-wideband radar for medical applications, *Frequenz* 63(1-2): 2–8.

[23] Lin, D., Trasser, A. & Schumacher, H. [2011a]. *A Fully Differential IR-UWB Front-end for Noncoherent Communication and Localization*, Proc. IEEE ICUWB 2011, Bologna, Italy, pp. 116–120.

[24] Lin, D., Trasser, A. & Schumacher, H. [2011b]. *Low Power, Fully Differential SiGe IR-UWB Transmitter and Correlation Receiver ICs*, Proc. IEEE RFIC, Baltimore, MD, pp. 101–104.

[25] Merdes, C. L. & Wolf, P. D. [2001]. Locating a catheter transducer in a three-dimensional ultrasound imaging field, *Biomedical Engineering, IEEE Transactions on* 48(12): 1444–1452.

[26] Mirbach, M. & Menzel, W. [2011]. A simple surface estimation algorithm for UWB pulse radars based on trilateration, *ICUWB 2011, 2011 IEEE International Conference on Ultra-Wideband*, pp. 273–277.

[27] Mirbach, M. & Menzel, W. [2012]. Time of Arrival Based Localization of UWB Transmitters Buried in Lossy Dielectric Media, *ICUWB 2012, 2012 IEEE International Conference on Ultra-Wideband*, pp. 328–332.

[28] Rappaport, C. [2004]. A simple approximation of transmitted wavefront shape from point sources above lossy half spaces, *Geoscience and Remote Sensing Symposium, 2004. IGARSS '04. Proceedings. 2004 IEEE International*, Vol. 1, pp. 421–424.

[29] Sakamoto, T. [2007]. A fast algorithm for 3-dimensional imaging with UWB pulse radar systems, *IEICE Transactions on Communications* E90-B(3): 636–644.

[30] Schantz, H. [2005]. *The Art and Science of Ultra-Wideband Antennas*, Artech House, Norwood, USA.

[31] Schleicher, B. [2012]. *Impulse-Radio Ultra-Wideband Systems for Vital-Sign Monitoring and Short-Range Communications*, Verlag Dr. Hut, München, Germany.

[32] Schueppert, B. [1988]. Microstrip/Slotline Transitions: Modeling and Experimental Investigation, *IEEE Transactions on Microwave Theory and Techniques* 36(8): 1272–1282.

[33] Staderini, E. M. [2002]. UWB Radars in Medicine, *IEEE Aerospace and Electronic Systems Magazine* 17(1): 13–18.

[34] Thiasiriphet, T., Ibrahim, M. & Lindner, J. [2012]. *Compressed Sensing for UWB Medical Radar Applications*.

[35] Thiasiriphet, T. & Lindner, J. [2010]. An uwb receiver based on a comb filter with shortened delay, *Ultra-Wideband (ICUWB), 2010 IEEE International Conference on*, Vol. 1, pp. 1–4.

[36] Thiasiriphet, T. & Lindner, J. [2011]. Particle filtering for uwb radar applications, *Ultra-Wideband (ICUWB), 2011 IEEE International Conference on*, pp. 248–252.

[37] Thiasiriphet, T., Zhan, R. & Lindner, J. [2009]. Interference suppression in ds-ppm uwb systems, *Ultra-Wideband. ICUWB 2009. 2009 IEEE International Conference on*, pp. 718–722.

[38] Thiel, F., Hein, M., Schwarz, U., Sachs, J. & Seifert, F. [2008]. Fusion of Magnetic Resonance Imaging and Ultra-Wideband-Radar for Biomedical Applications, *IEEE International Conference on Ultra-Wideband*, Vol. 1, pp. 97–100.

[39] Zheng, Y., Wong, K., Asaru, M., Shen, D., Zhao, W., The, Y., Andrew, P., Lin, F., Yeoh, W. & Singh, R. [2007]. *A 0.18um CMOS Dual-Band UWB Transceiver*, IEEE ISSCC Dig. Tech. Papers, San Francisco, CA, pp. 114–115.

[40] Zhou, Z., Pan, Z. & Tang, X. [2007]. A new family of optimal zero correlation zone sequences from perfect sequences based on interleaved technique, *Signal Design and Its Applications in Communications, 2007. IWSDA 2007. 3rd International Workshop on*, pp. 195–199.